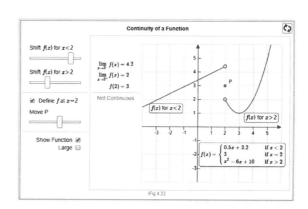

Calculus with Applications
Brief Version

ELEVENTH EDITION

Calculus with Applications
Brief Version

ELEVENTH EDITION

Margaret L. Lial
American River College

Raymond N. Greenwell
Hofstra University

Nathan P. Ritchey
Edinboro University

Boston Columbus Indianapolis New York San Francisco
Amsterdam Cape Town Dubai London Madrid Milan Munich Paris Montréal Toronto
Delhi Mexico City São Paulo Sydney Hong Kong Seoul Singapore Taipei Tokyo

Editorial Director: Chris Hoag
Editor in Chief: Deirdre Lynch
Acquisitions Editor: Jeff Weidenaar
Editorial Assistant: Alison Oehmen
Program Manager: Tatiana Anacki
Project Manager: Christine O'Brien
Program Management Team Lead: Karen Wernholm
Project Management Team Lead: Peter Silvia
Media Producer: Stephanie Green
TestGen Content Manager: John Flanagan
MathXL Content Manager: Kristina Evans
Marketing Manager: Claire Kozar
Marketing Assistant: Fiona Murray
Senior Author Support/Technology Specialist: Joe Vetere
Rights and Permissions Project Manager: Gina Cheselka
Procurement Specialist: Carol Melville
Associate Director of Design: Andrea Nix
Program Design Lead: Heather Scott
Text Design: Cenveo Publisher Services
Composition: Cenveo Publisher Services
Illustrations: Cenveo Publisher Services
Cover Design: Tamara Newnam
Cover Image: Hans Huber, Getty Images

Library of Congress Cataloging-in-Publication Data
Lial, Margaret L.
  Calculus with applications, brief version / Margaret L. Lial, American River College, Raymond N. Greenwell, Hofstra
University, Nathan P. Ritchey, Edinboro University. — Eleventh edition.
    pages cm
Includes index.
  ISBN 978-0-321-97941-4 (hardcover) — ISBN 0-321-97941-9 (hardcover)
1. Calculus—Textbooks. I. Greenwell, Raymond N. II. Ritchey, Nathan P. III. Title.
QA303.2.L535 2016
515–dc23
                                                            2014038787

6

www.pearsonhighered.com

ISBN 13: 978-0-321-97941-4
ISBN 10: 0-321-97941-9

Contents

Preface

Calculus with Applications, Brief Version is a thorough, applications-oriented text for students majoring in business, management, economics, or the life or social sciences. In addition to its clear exposition, this text consistently connects the mathematics to career and everyday-life situations. A prerequisite of two years of high school algebra is assumed. A greatly enhanced MyMathLab course, new applications and exercises, and other new learning tools make this 11th edition a rich learning resource for students.

Our Approach

Our main goal is to present applied calculus in a concise and meaningful way so that students can understand the full picture of the concepts they are learning and apply them to real-life situations. This is done through a variety of means.

Focus on Applications Making this course meaningful to students is critical to their success. Applications of the mathematics are integrated throughout the text in the exposition, the examples, the exercise sets, and the supplementary resources. We are constantly on the lookout for novel applications, and the text reflects our efforts to infuse it with relevance. Our research is showcased in the Index of Applications at the back of the book and the extended list of sources of real-world data on www.pearsonhighered.com/mathstatsresources. *Calculus with Applications, Brief Version* presents students with myriad opportunities to relate what they're learning to career situations through the *Apply It* question at the beginning of sections, the applied examples and exercises, and the *Extended Application* at the end of each chapter.

Pedagogy to Support Students Students need careful explanations of the mathematics along with examples presented in a clear and consistent manner. Additionally, students and instructors should have a means to assess the basic prerequisite skills needed for the course content. This can be done with the *Prerequisite Skills Diagnostic Test,* located just prior to Chapter R. If the diagnostic test reveals gaps in basic skills, students can find help right within the text. Further, *Warm-Up Exercises* are now included at the beginning of many exercise sets. Within MyMathLab are additional diagnostic tests (one per chapter), and remediation is automatically personalized to meet student needs. Students will appreciate the many annotated examples within the text, the *Your Turn* exercises that follow examples, the *For Review* references, and the wealth of learning resources within MyMathLab.

Beyond the Textbook Students want resources at their fingertips and, for them, that means digital access. So Pearson has developed a robust MyMathLab course for *Calculus with Applications, Brief Version.* MyMathLab has a well-established and well-documented track record of helping students succeed in mathematics. The MyMathLab online course for this text contains over 1800 exercises to challenge students and provides help when they need it. Students who learn best through video can view (and review) section- and example-level videos within MyMathLab. These and other resources are available to students as a unified and reliable tool for their success.

New to the Eleventh Edition

Based on our experience in the classroom along with feedback from many instructors across the country, the focus of this revision is to improve the clarity of the presentation and provide students with more opportunities to learn, practice, and apply what they've learned on their own. We do this both in the presentation of the content and in the new features added to the text.

New Features

- *Warm-Up Exercises* were added to many exercise sets to provide an opportunity for students to refresh key prerequisite skills at "point of use."

- Graphing calculator screens have been updated to reflect the TI-84 Plus C, which features color and a higher screen resolution. Additionally, the graphing calculator notes have been updated throughout.

- We added more "help text" annotations to examples. These notes, set in small blue type, appear next to the steps within worked-out examples and provide an additional aid for students with weaker algebra skills.

- For many years this text has featured enormous amounts of real data used in examples and exercises. The 11th edition will not disappoint in this area. We have added or updated 137 (18.3%) of the application exercises throughout the text.

- We updated exercises and examples based on user feedback and other factors. Of the 2720 exercises within the sections, 304 (11.2%) are new or updated. Of the 320 examples in the text, 48 (15%) are new or updated.

- MyMathLab contains a wealth of new resources to help students learn and to help you as you teach. Some resources were added or revised based on student usage of the *previous* edition of the MyMathLab course. For example, more exercises were added to those chapters and sections that are more widely assigned.

  - Hundreds of new exercises were added to the course to provide you with more options for assignments, including:
    - More application exercises throughout the text
    - *Setup & Solve* exercises that require students to specify how to set up a problem as well as solve it
    - Exercises that take advantage of the enhanced graphing tool

  - An Integrated Review version of the course contains preassigned diagnostic and remediation resources for key prerequisite skills. Skills Check Quizzes help diagnose gaps in skills prior to each chapter. MyMathLab then provides personalized help on only those skills that a student has not mastered.

  - The videos for the course have increased in number, type, and quality:
    - New videos feature more applications and more challenging examples.
    - In addition to full-length lecture videos, MyMathLab now includes assignable, shorter video clips that focus on a specific concept or example.
    - MathTalk Videos help motivate students by pointing out relevant connections to their majors—especially business. The videos feature Andrea Young from Ripon College (WI), a dynamic math professor (and actor!). The videos can be used as lecture starters or as part of homework assignments (in regular or flipped classes). Assignable exercises that accompany the videos help make these videos a part of homework assignments.
    - A Guide to Video-Based Instruction shows which exercises correspond to each video, making it easy to assess students after they watch an instructional video. This is perfect for flipped-classroom situations.

  - Learning Catalytics is a "bring your own device" student engagement, assessment, and classroom intelligence system. Students can use any web-enabled device—laptop, smartphone, or tablet—that they already have. Those with access to MyMathLab have instant access to Learning Catalytics and can log in using their MyMathLab username and password. With Learning Catalytics, you assess students in real time, using open-ended tasks to probe student understanding. It allows you to engage students by creating open-ended questions that ask for numerical, algebraic, textual, or graphical responses—or just simple multiple-choice. Learning Catalytics contains Pearson-created content for calculus so you can take advantage of this exciting technology immediately.

New and Revised Content

The chapters and sections in the text are in the same order as the previous edition, making it easy for users to transition to the new edition. In addition to revising exercises and examples throughout, updating and adding real-world data, we made the following changes:

Chapter R
- Added new *Your Turn* exercises to ensure that there is a student assessment for each major concept.
- Added more detail to R.2 on factoring perfect squares.

Chapter 1
- Rewrote the part of 1.1 involving graphing lines, emphasizing different methods for graphing.
- Rewrote 1.2 on supply, demand, break-even analysis, and equilibrium; giving formal definitions that match what students would see in business and economics courses. All of the business applications were revised, according to recommendations from reviewers, to be more in line with business texts. Also added a new Example 6 on finding a cost function.
- Added color for pedagogical reasons to make content easier to follow.

Chapter 2
- Updated the introduction to 2.1, rewriting it as an example to make it easier for students to reference the necessary skills to identify nonlinear functions, determine the domain and range, and estimate values from a graph.
- In 2.2, added another approach to graphing parabolas by splitting former Example 4 into two separate examples. The new Example 5 illustrates how to graph a parabola by first finding its characteristics (including orientation, intercepts, vertex, and axis of symmetry). The characteristics are highlighted in a box for easy reference.
- Added quadratic regression to 2.2. Example 9 includes a by-hand method and a method using technology.
- Rewrote Example 10 in 2.2, which illustrates translations and reflections of a graph, by breaking it into three parts. The first part is a basic transformation, and the ensuing parts build in complexity.
- Added the definition of a real root to 2.3 and added a Technology Note to illustrate how to use a graphing calculator to approximate the roots of higher degree polynomials.
- Added cubic regression to 2.3 (Example 5).

Chapter 3
- Added Caution note to 3.1 and added a new solution method to Example 9.
- Added new Example 2 to 3.3, using recent data.
- Updated Example 4 in 3.3 to use clearer wording.

Chapter 4
- Clarified the rules for differentiation in 4.1, 4.2, and 4.3 and added a new Example 8.
- Expanded Example 9 in 4.1 to include a new graph.
- Updated Example 10 in 4.1 and Example 4 in 4.5.

Chapter 5
- Added new examples to 5.2 (Example 3(c)) and 5.3 (Example 6(b)).
- Expanded Example 6(a) in 5.4 to show the inflection point.

Chapter 6
- Updated Example 3 in 6.1 to show an application of the concept.
- Modified examples in 6.2 (Example 3), 6.4 (Example 2), and 6.6 (Example 1).

Chapter 7
- Added annotations and comments to Example 10 in 7.1.
- Simplified Examples 1, 2, 3, and 6 in 7.2 and added annotations and comments.
- Added a "For Review" box to 7.3.
- Enlarged all small integral signs throughout the chapter for clarity.
- Updated Example 7 in 7.4 and Example 5 in 7.5.
- Added more explanation of the consumer surplus to 7.5.

Chapter 8
- Added annotations to several examples in 8.1 to denote steps in integration by parts.
- Revised the solutions to Examples 4 and 5 in 8.3, giving more detail and adding annotation to denote the steps in determining the accumulated amount of money flow.

Chapter 9
- Rewrote and expanded Exercise 8 in 9.1, on the Cobb-Douglas Production Function, emphasizing the interpretation of the solutions.
- Added three new exercises to 9.1 on exponential and logarithmic functions of several variables.
- Revised the solution to Example 4 in 9.3, giving more detail.
- Rewrote the solution to Example 3 in 9.4, illustrating how to find the extrema of a constrained function of one or more variables using a spreadsheet.

Features of *Calculus with Applications*

Chapter Opener

Each chapter opens with a quick introduction that relates to an application presented in the chapter.

Apply It

An Apply It question, typically at the start of a section, motivates the math content of the section by posing a real-world question that is then answered within the examples or exercises.

Teaching Tips

Teaching Tips are provided in the margins of the Annotated Instructor's Edition for those who are new to teaching this course.

For Review

For Review boxes are provided in the margin as appropriate, giving students just-in-time help with skills they should already know but may have forgotten. For Review comments sometimes include an explanation, while others refer students back to earlier parts of the book for a more thorough review.

> **FOR REVIEW**
>
> Recall that $e^x > 0$ for all x, so there can never be a solution to $e^{g(x)} = 0$ for any function $g(x)$.

Caution

Caution notes provide students with a quick "heads up" to common difficulties and errors.

CAUTION Notice from Example 5(c) that $g(x + h)$ is *not* the same as $g(x) + h$, which equals $-x^2 + 4x - 5 + h$. There is a significant difference between applying a function to the quantity $x + h$ and applying a function to x and adding h afterward.

Your Turn Exercises

These exercises follow selected examples and provide students with an easy way to quickly stop and check their understanding. Answers are provided at the end of the section's exercises.

Technology Notes

Material on graphing calculators or Microsoft Excel is clearly labeled to make it easier for instructors to use this material (or not).

• New The figures depicting calculator screens now reflect the TI-84 Plus C, which features color and higher pixel counts.

FIGURE 56

Exercise Sets

Basic exercises are followed by an Applications section, which is grouped by subheads such as "Business and Economics." Other types of exercises include the following:

• New **Warm-Up** exercises at the beginning of most sections provide a chance for students to refresh the key prerequisite skills needed for the section's exercises.

• **Connections** exercises integrate topics presented in different sections or chapters and are indicated with ⬭.

• **Writing** exercises, labeled with ✎, provide students with an opportunity to explain important mathematical ideas.

• **Technology** exercises are labeled ⬭ for graphing calculator and ⬭ for spreadsheets.

• Exercises that are particularly challenging are denoted with a + in the Annotated Instructor's Edition only.

• The Annotated Instructor's Edition contains most answers right on the page. Overflow answers are at the back of the book.

Chapter Summary and Review

• The end-of-chapter **Summary** provides students with a quick summary of the key ideas of the chapter followed by a list of key definitions, terms, and examples.

• Chapter **Review Exercises** include Concept Check exercises and an ample set of Practice and Exploration exercises. This arrangement provides students with a comprehensive set of exercises to prepare for chapter exams.

Extended Applications

• Extended Applications are provided at the end of every chapter as in-depth applied exercises to help stimulate student interest. These activities can be completed individually or as a group project. Additional Extended Applications for the text can be found online at www.pearsonhighered.com/mathstatsresources.

Supplements

FOR STUDENTS	FOR INSTRUCTORS

FOR STUDENTS

Student's Solutions Manual (in print and electronically within MyMathLab)

- Provides detailed solutions to all odd-numbered text exercises and sample chapter tests with answers.
- ISBN 0133864537 / 9780133864533

Graphing Calculator Manual for Applied Mathematics (downloadable)

- Contains detailed instruction for using the TI-83/ TI-83+/ TI-84+C
- Instructions are organized by topic.
- Downloadable from within MyMathLab

Excel Spreadsheet Manual for Applied Mathematics (downloadable)

- Contains detailed instruction for using Excel 2013
- Instructions are organized by topic.
- Downloadable from within MyMathLab

FOR INSTRUCTORS

Annotated Instructor's Edition

- Includes all the answers, usually on the same page as the exercises, for quick reference
- Challenge problems labeled with a + (plus sign)
- Numerous teaching tips
- ISBN 0321998774 / 9780321998774

Instructor's Solutions and Resource Manual (Download Only)

- Provides complete solutions to all exercises, two versions of a pre-test and final exam, and teaching tips.
- Available to qualified instructors within MyMathLab or through the Pearson Instructor Resource Center (www.pearsonhighered.com/irc).
- ISBN 0133864650 / 9780133864656

PowerPoint Presentations (Download Only)

- Includes lecture content and key graphics from the book.
- Available to qualified instructors within MyMathLab or through the Pearson Instructor Resource Center (www.pearsonhighered.com/irc).
- ISBN 0133983862 / 9780133983869

TestGen Computerized Test Bank

- TestGen® (www.pearsoned.com/testgen) enables instructors to build, edit, print, and administer tests using a computerized bank of questions developed to cover all the objectives of the text.
- TestGen is algorithmically based, allowing instructors to create multiple but equivalent versions of the same question or test with the click of a button. Instructors can also modify test bank questions or add new questions.
- The software and testbank are available to qualified instructors within MyMathLab or through the Pearson Instructor Resource Center (www.pearsonhighered.com/irc).
- ISBN 0133864626 / 9780133864625

MyMathLab®Online Course (access code required)

MyMathLab delivers **proven results** in helping individual students succeed.

- MyMathLab has a consistently positive impact on the quality of learning in higher education math instruction. MyMathLab can be successfully implemented in any environment—lab-based, hybrid, fully online, traditional—and demonstrates the quantifiable difference that integrated usage has on student retention, subsequent success, and overall achievement.
- MyMathLab's comprehensive online gradebook automatically tracks your students' results on tests, quizzes, homework, and in the study plan. You can use the gradebook to quickly

intervene if your students have trouble, or to provide positive feedback on a job well done. The data within MyMathLab's gradebook are easily exported to a variety of spreadsheet programs.

MyMathLab provides **engaging experiences** that personalize, stimulate, and measure learning for each student.

- **Personalized Learning:** MyMathLab offers several features that support adaptive learning: personalized homework and the adaptive study plan. These features allow your students to work on just what they need to learn when it makes the most sense, maximizing their potential for understanding and success.

- **Exercises:** The homework and practice exercises in MyMathLab are correlated to the exercises in the textbook, and they regenerate algorithmically to give students unlimited opportunity for practice and mastery. The software provides helpful feedback when students enter incorrect answers and includes optional learning aids including guided solutions, sample problems, animations, videos, and eText.

- **Learning and Teaching Tools** include:
  - **Learning Catalytics**—a "bring your own device" student engagement, assessment, and classroom intelligence system, included within MyMathLab. Includes questions written specifically for this course.
  - **Instructional videos**—full-length lecture videos as well as shorter example-based videos.
  - **MathTalk videos**—connect the math to the real world (particularly business). Also include assignable exercises to gauge student understanding of video content.
  - **Help for Gaps in Prerequisite Skills**—diagnostic quizzes tied to personalized assignments help address gaps in algebra skills that might otherwise impede success.
  - **Excel Spreadsheet Manual**—specifically written for this course.
  - **Graphing Calculator Manual**—specifically written for this course.
  - **Interactive Figures**—illustrate key concepts and allow manipulation for use as teaching and learning tools. Includes assignable exercises that require use of the figures.

- **Complete eText** is available to students through MyMathLab courses for the lifetime of the edition, giving students unlimited access to the eBook within any course using that edition of the textbook.

- **MyMathLab Accessibility:** MyMathLab is compatible with the JAWS screen reader, and enables multiple-choice and free-response problem types to be read and interacted with via keyboard controls and math notation input. MyMathLab also works with screen enlargers, including ZoomText, MAGic, and SuperNova. And all MyMathLab videos have closed captioning. More information on this functionality is available at http://mymathlab.com/accessibility.

MyMathLab comes from an **experienced partner** with educational expertise and an eye on the future.

- Knowing that you are using a Pearson product means knowing that you are using quality content. This means that our eTexts are accurate and our assessment tools work. It means we are committed to making MyMathLab as accessible as possible.

- Whether you are just getting started with MyMathLab, or have a question along the way, we're here to help you learn about our technologies and how to incorporate them into your course.

- To learn more about how MyMathLab combines proven learning applications with powerful assessment and continuously adaptive capabilities, visit www.mymathlab.com or contact your Pearson representative.

Additional Online Courseware Options (access code required)

- **MathXL®** is the homework and assessment engine that runs MyMathLab. (MyMathLab is MathXL plus a learning management system.)

- **MyMathLab Integrated Review Course** contains preassigned diagnostic and remediation resources for key prerequisite skills. Skills Check Quizzes help diagnose gaps in skills prior to each chapter. MyMathLab then provides personalized help on just those skills that a student has not mastered.
- **MyMathLab®Plus** contains all the features of MyMathLab with convenient management tools and a dedicated services team. It includes batch enrollment so everyone can be ready to start class on the first day, login directly from your campus portal, advanced reporting features, and 24/7 support by email or online chat.

Acknowledgments

We wish to thank the following professors for their contributions in reviewing portions of this text:

John Alford, *Sam Houston State University*
Robert David Borgersen, *University of Manitoba*
Jeffrey Breeding, *Fordham University*
C. T. Bruns, *University of Colorado, Boulder*
Nurit Budinsky, *University of Massachusetts, Dartmouth*
Martha Morrow Chalhoub, *Collin College, Preston Ridge Campus*
Scott E. Clark, *University of Arizona*
Karabi Datta, *Northern Illinois University*
James "Rob" Ely, *Blinn College—Bryan Campus*
Sam Evers, *The University of Alabama*
Kevin Farrell, *Lyndon State College*
Chris Ferbrache, *Fresno City College*
Peter Gomez, *Houston Community College, Northwest*
Sharda K. Gudehithlu, *Wilbur Wright College*
Mary Beth Headlee, *State College of Florida*
David L. Jones, *University of Kansas*
Karla Karstens, *University of Vermont*
Monika Keindl, *Northern Arizona University*
Lynette J. King, *Gadsden State Community College*
Jason Knapp, *University of Virginia*
Mark C. Lammers, *University of North Carolina, Wilmington*
Rebecca E. Lynn, *Colorado State University*
Rodolfo Maglio, *Northeastern Illinois University*
Cyrus Malek, *Collin College*
Lawrence Marx, *University of California, Davis*
Javad Namazi, *Fairleigh Dickinson University*
Dana Nimic, *Southeast Community College, Lincoln*
Leonard Nissim, *Fordham University*
Lisa Nix, *Shelton State Community College*
Sam Northshield, *SUNY, Plattsburgh*
Susan Ojala, *University of Vermont*
Jigarkumar Patel, *University of Texas, Dallas*
Brooke Quinlan, *Hillsborough Community College*
Candace Rainer, *Meridian Community College*
Brian S. Rickard, *University of Arkansas*
Arthur J. Rosenthal, *Salem State College*
Theresa Rushing, *The University of Tennessee at Martin*
Katherine E. Schultz, *Pensacola Junior College*
Barbara Dinneen Sehr, *Indiana University, Kokomo*
Gordon H. Shumard, *Kennesaw State University*
Walter Sizer, *Minnesota State University, Moorhead*

Jennifer Strehler, *Oakton Community College*
Antonis P. Stylianou, *University of Missouri—Kansas City*
Darren Tapp, *Hesser College*
Jason Terry, *Central New Mexico Community College*
Yan Tian, *Palomar College*
Sara Van Asten, *North Hennepin Community College*
Charles K. Walsh, *College of Southern Maryland*
Amanda Wheeler, *Amarillo College*
Douglas Williams, *Arizona State University*
Roger Zarnowski, *Angelo State University*

The following faculty members provided direction on the development of the MyMathLab course for this edition:

Frederick Adkins, *Indiana University of Pennsylvania*
Rachelle Bouchat, *Indiana University of Pennsylvania*
Pete Bouzar, *Golden West College*
Raghu Gompa, *Jackson State University*
Brian Hagelstrom, *North Dakota State College of Science*
Thomas Hartfield, *University of North Georgia – Gainesville*
Weihu Hong, *Clayton State University*
Cheryl Kane, *University of New England*
Karla Karstens, *University of Vermont*
Lidiya Klinger, *Fullerton College*
Carrie Lahnovych, *Rochester Institute of Technology*
Fred Mohanespour, *Indiana University – Purdue University Fort Wayne*
Gina Monks, *Pennsylvania State University – Hazleton*
Duc Phan, *Collin College*
Michael Puente, *Richland College*
John Racquet, *University at Albany*
Christian Roettger, *Iowa State University*
Amit Saini, *University of Nevada – Reno*
Jamal Salahat, *Owens State Community College*
Jack Saraceno, *Shelton State Community College*
Sulakshana Sen, *Bethune Cookman University*
Olga Tsukernik, *Rochester Institute of Technology*
Dennis Ward, *St. Petersburg College*
Martin Wesche, *Clayton State University*
Greg Wisloski, *Indiana University of Pennsylvania*
Dennis Wolf, *Indiana University – South Bend*
Dinesh Yadav, *Dallas County Community College*

We also thank Elka Block and Frank Purcell for doing an excellent job updating the Student's Solutions Manual and Instructor's Solutions and Resource Manual. Further thanks go to our accuracy checkers Lisa Collette, Damon Demas, Paul Lorczak, and Rhea Meyerholtz. We are grateful to Karla Harby and Mary Ann Ritchey for their editorial assistance. We especially appreciate the staff at Pearson, whose contributions have been very important in bringing this project to a successful conclusion.

Raymond N. Greenwell
Nathan P. Ritchey

Prerequisite Skills Diagnostic Test

Below is a very brief test to help you recognize which, if any, prerequisite skills you may need to remediate in order to be successful in this course. After completing the test, check your answers in the back of the book. In addition to the answers, we have also provided the solutions to these problems in Appendix A. These solutions should help remind you how to solve the problems. For problems 5-26, the answers are followed by references to sections within Chapter R where you can find guidance on how to solve the problem and/ or additional instruction. Addressing any weak prerequisite skills now will make a positive impact on your success as you progress through this course.

1. What percent of 50 is 10?

2. Simplify $\dfrac{13}{7} - \dfrac{2}{5}$.

3. Let x be the number of apples and y be the number of oranges. Write the following statement as an algebraic equation: "The total number of apples and oranges is 75."

4. Let s be the number of students and p be the number of professors. Write the following statement as an algebraic equation: "There are at least four times as many students as professors."

5. Solve for k: $7k + 8 = -4(3 - k)$.

6. Solve for x: $\dfrac{5}{8}x + \dfrac{1}{16}x = \dfrac{11}{16} + x$.

7. Write in interval notation: $-2 < x \le 5$.

8. Using the variable x, write the following interval as an inequality: $(-\infty, -3]$.

9. Solve for y: $5(y - 2) + 1 \le 7y + 8$.

10. Solve for p: $\dfrac{2}{3}(5p - 3) > \dfrac{3}{4}(2p + 1)$.

11. Carry out the operations and simplify: $(5y^2 - 6y - 4) - 2(3y^2 - 5y + 1)$.

12. Multiply out and simplify $(x^2 - 2x + 3)(x + 1)$.

13. Multiply out and simplify $(a - 2b)^2$.

14. Factor $3pq + 6p^2q + 9pq^2$.

15. Factor $3x^2 - x - 10$.

16. Perform the operation and simplify: $\dfrac{a^2 - 6a}{a^2 - 4} \cdot \dfrac{a - 2}{a}$.

17. Perform the operation and simplify: $\dfrac{x + 3}{x^2 - 1} + \dfrac{2}{x^2 + x}$.

18. Solve for x: $3x^2 + 4x = 1$.

19. Solve for z: $\dfrac{8z}{z + 3} \le 2$.

20. Simplify $\dfrac{4^{-1}(x^2y^3)^2}{x^{-2}y^5}$.

21. Simplify $\dfrac{4^{1/4}(p^{2/3}q^{-1/3})^{-1}}{4^{-1/4}\,p^{4/3}q^{4/3}}$.

22. Simplify as a single term without negative exponents: $k^{-1} - m^{-1}$.

23. Factor $(x^2 + 1)^{-1/2}(x + 2) + 3(x^2 + 1)^{1/2}$.

24. Simplify $\sqrt[3]{64b^6}$.

25. Rationalize the denominator: $\dfrac{2}{4 - \sqrt{10}}$.

26. Simplify $\sqrt{y^2 - 10y + 25}$.

R Algebra Reference

In this chapter, we will review the most important topics in algebra. Knowing algebra is a fundamental prerequisite to success in higher mathematics. This algebra reference is designed for self-study; study it all at once or refer to it when needed throughout the course. Since this is a review, answers to all exercises are given in the answer section at the back of the book.

R.1 | Polynomials

An expression such as $9p^4$ is a **term**; the number 9 is the **coefficient**, p is the **variable**, and 4 is the **exponent**. The expression p^4 means $p \cdot p \cdot p \cdot p$, while p^2 means $p \cdot p$, and so on. Terms having the same variable and the same exponent, such as $9x^4$ and $-3x^4$, are **like terms**. Terms that do not have both the same variable and the same exponent, such as m^2 and m^4, are **unlike terms**.

A **polynomial** is a term or a finite sum of terms in which all variables have whole number exponents, and no variables appear in denominators. Examples of polynomials include

$$5x^4 + 2x^3 + 6x, \qquad 8m^3 + 9m^2n - 6mn^2 + 3n^3, \qquad 10p, \qquad \text{and} \qquad -9.$$

Order of Operations
Algebra is a language, and you must be familiar with its rules to correctly interpret algebraic statements. The following order of operations has been agreed upon through centuries of usage.

- Expressions in **parentheses** (or other grouping symbols) are calculated first, working from the inside out. The numerator and denominator of a fraction are treated as expressions in parentheses.
- **Powers** are performed next, going from left to right.
- **Multiplication** and **division** are performed next, going from left to right.
- **Addition** and **subtraction** are performed last, going from left to right.

For example, in the expression $[6(x + 1)^2 + 3x - 22]^2$, suppose x has the value of 2. We would evaluate this as follows:

$$
\begin{aligned}
[6(2 + 1)^2 + 3(2) - 22]^2 &= [6(3)^2 + 3(2) - 22]^2 &&\text{Evaluate the expression in the innermost parentheses.} \\
&= [6(9) + 3(2) - 22]^2 &&\text{Evaluate 3 raised to a power.} \\
&= (54 + 6 - 22)^2 &&\text{Perform the multiplications.} \\
&= (38)^2 &&\text{Perform the addition and subtraction from left to right.} \\
&= 1444 &&\text{Evaluate the power.}
\end{aligned}
$$

In the expression $\dfrac{x^2 + 3x + 6}{x + 6}$, suppose x has the value of 2. We would evaluate this as follows:

$$
\begin{aligned}
\frac{2^2 + 3(2) + 6}{2 + 6} &= \frac{16}{8} &&\text{Evaluate the numerator and the denominator.} \\
&= 2 &&\text{Simplify the fraction.}
\end{aligned}
$$

Adding and Subtracting Polynomials
The following properties of real numbers are useful for performing operations on polynomials.

Properties of Real Numbers

For all real numbers a, b, and c:

1. $a + b = b + a$;   **Commutative properties**
   $ab = ba$;

2. $(a + b) + c = a + (b + c)$;   **Associative properties**
   $(ab)c = a(bc)$;

3. $a(b + c) = ab + ac$.   **Distributive property**

EXAMPLE 1   Properties of Real Numbers

(a) $2 + x = x + 2$ — Commutative property of addition

(b) $x \cdot 3 = 3x$ — Commutative property of multiplication

(c) $(7x)x = 7(x \cdot x) = 7x^2$ — Associative property of multiplication

(d) $3(x + 4) = 3x + 12$ — Distributive property

One use of the distributive property is to add or subtract polynomials. Only like terms may be added or subtracted. For example,

$$12y^4 + 6y^4 = (12 + 6)y^4 = 18y^4,$$

and

$$-2m^2 + 8m^2 = (-2 + 8)m^2 = 6m^2,$$

but the polynomial $8y^4 + 2y^5$ cannot be further simplified. To subtract polynomials, we use the facts that $-(a + b) = -a - b$ and $-(a - b) = -a + b$. In the next example, we show how to add and subtract polynomials.

EXAMPLE 2   Adding and Subtracting Polynomials

Add or subtract as indicated.

(a) $(8x^3 - 4x^2 + 6x) + (3x^3 + 5x^2 - 9x + 8)$

SOLUTION   Combine like terms.

$$(8x^3 - 4x^2 + 6x) + (3x^3 + 5x^2 - 9x + 8)$$
$$= (8x^3 + 3x^3) + (-4x^2 + 5x^2) + (6x - 9x) + 8$$
$$= 11x^3 + x^2 - 3x + 8$$

(b) $2(-4x^4 + 6x^3 - 9x^2 - 12) + 3(-3x^3 + 8x^2 - 11x + 7)$

SOLUTION   Multiply each polynomial by the factor in front of the polynomial, and then combine terms as before.

$$2(-4x^4 + 6x^3 - 9x^2 - 12) + 3(-3x^3 + 8x^2 - 11x + 7)$$
$$= -8x^4 + 12x^3 - 18x^2 - 24 - 9x^3 + 24x^2 - 33x + 21$$
$$= -8x^4 + 3x^3 + 6x^2 - 33x - 3$$

(c) $(2x^2 - 11x + 8) - (7x^2 - 6x + 2)$

YOUR TURN 1 Perform the operation $3(x^2 - 4x - 5) - 4(3x^2 - 5x - 7)$.

SOLUTION   Distributing the minus sign and combining like terms yields

$$(2x^2 - 11x + 8) + (-7x^2 + 6x - 2)$$
$$= -5x^2 - 5x + 6.$$ TRY YOUR TURN 1

Multiplying Polynomials
The distributive property is also used to multiply polynomials, along with the fact that $a^m \cdot a^n = a^{m+n}$. For example,

$$x \cdot x = x^1 \cdot x^1 = x^{1+1} = x^2 \quad \text{and} \quad x^2 \cdot x^5 = x^{2+5} = x^7.$$

EXAMPLE 3   Multiplying Polynomials

Multiply.

(a) $8x(6x - 4)$

SOLUTION   Using the distributive property yields

$$8x(6x - 4) = 8x(6x) - 8x(4)$$
$$= 48x^2 - 32x.$$

(b) $(3p - 2)(p^2 + 5p - 1)$

SOLUTION   Using the distributive property yields

$$
\begin{aligned}
(3p - 2&)(p^2 + 5p - 1) \\
&= 3p(p^2 + 5p - 1) - 2(p^2 + 5p - 1) \\
&= 3p(p^2) + 3p(5p) + 3p(-1) - 2(p^2) - 2(5p) - 2(-1) \\
&= 3p^3 + 15p^2 - 3p - 2p^2 - 10p + 2 \\
&= 3p^3 + 13p^2 - 13p + 2.
\end{aligned}
$$

(c) $(x + 2)(x + 3)(x - 4)$

SOLUTION   Multiplying the first two polynomials and then multiplying their product by the third polynomial yields

$$
\begin{aligned}
(x + 2&)(x + 3)(x - 4) \\
&= [(x + 2)(x + 3)](x - 4) \\
&= (x^2 + 2x + 3x + 6)(x - 4) \\
&= (x^2 + 5x + 6)(x - 4) \\
&= x^3 - 4x^2 + 5x^2 - 20x + 6x - 24 \\
&= x^3 + x^2 - 14x - 24.
\end{aligned}
$$

YOUR TURN 2   Perform the operation $(3y + 2)(4y^2 - 2y - 5)$.

TRY YOUR TURN 2

A **binomial** is a polynomial with exactly two terms, such as $2x + 1$ or $m + n$. When two binomials are multiplied, the FOIL method (First, Outer, Inner, Last) is used as a memory aid.

EXAMPLE 4   Multiplying Polynomials

Find $(2m - 5)(m + 4)$ using the FOIL method.

SOLUTION

$$
\begin{aligned}
(2m - 5)(m + 4) &= \overset{F}{(2m)(m)} + \overset{O}{(2m)(4)} + \overset{I}{(-5)(m)} + \overset{L}{(-5)(4)} \\
&= 2m^2 + 8m - 5m - 20 \\
&= 2m^2 + 3m - 20
\end{aligned}
$$

YOUR TURN 3   Find $(2x + 7)(3x - 1)$ using the FOIL method.

TRY YOUR TURN 3

EXAMPLE 5   Multiplying Polynomials

Find $(2k - 5m)^3$.

SOLUTION   Write $(2k - 5m)^3$ as $(2k - 5m)(2k - 5m)(2k - 5m)$. Then multiply the first two factors using FOIL.

$$
\begin{aligned}
(2k - 5m)(2k - 5m) &= 4k^2 - 10km - 10km + 25m^2 \\
&= 4k^2 - 20km + 25m^2
\end{aligned}
$$

Now multiply this last result by $(2k - 5m)$ using the distributive property, as in Example 3(c).

$$
\begin{aligned}
(4k^2 &- 20km + 25m^2)(2k - 5m) \\
&= 4k^2(2k - 5m) - 20km(2k - 5m) + 25m^2(2k - 5m) \\
&= 8k^3 - 20k^2m - 40k^2m + 100km^2 + 50km^2 - 125m^3 \\
&= 8k^3 - 60k^2m + 150km^2 - 125m^3 \quad \text{Combine like terms.}
\end{aligned}
$$

YOUR TURN 4   Find $(3x + 2y)^3$.

TRY YOUR TURN 4

Notice in the first part of Example 5, when we multiplied $(2k - 5m)$ by itself, that the product of the square of a binomial is the square of the first term, $(2k)^2$, plus twice the product of the two terms, $(2)(2k)(-5m)$, plus the square of the last term, $(-5k)^2$.

CAUTION | Avoid the common error of writing $(x + y)^2 = x^2 + y^2$. As the first step of Example 5 shows, the square of a binomial has three terms, so

$$(x + y)^2 = x^2 + 2xy + y^2.$$

Furthermore, higher powers of a binomial also result in more than two terms. For example, verify by multiplication that

$$(x + y)^3 = x^3 + 3x^2y + 3xy^2 + y^3.$$

Remember, for any value of $n \neq 1$,

$$(x + y)^n \neq x^n + y^n.$$

R.1 EXERCISES

Perform the indicated operations.

1. $(2x^2 - 6x + 11) + (-3x^2 + 7x - 2)$

2. $(-4y^2 - 3y + 8) - (2y^2 - 6y - 2)$

3. $-6(2q^2 + 4q - 3) + 4(-q^2 + 7q - 3)$

4. $2(3r^2 + 4r + 2) - 3(-r^2 + 4r - 5)$

5. $(0.613x^2 - 4.215x + 0.892) - 0.47(2x^2 - 3x + 5)$

6. $0.5(5r^2 + 3.2r - 6) - (1.7r^2 - 2r - 1.5)$

7. $-9m(2m^2 + 3m - 1)$

8. $6x(-2x^3 + 5x + 6)$

9. $(3t - 2y)(3t + 5y)$

10. $(9k + q)(2k - q)$

11. $(2 - 3x)(2 + 3x)$

12. $(6m + 5)(6m - 5)$

13. $\left(\dfrac{2}{5}y + \dfrac{1}{8}z\right)\left(\dfrac{3}{5}y + \dfrac{1}{2}z\right)$

14. $\left(\dfrac{3}{4}r - \dfrac{2}{3}s\right)\left(\dfrac{5}{4}r + \dfrac{1}{3}s\right)$

15. $(3p - 1)(9p^2 + 3p + 1)$

16. $(3p + 2)(5p^2 + p - 4)$

17. $(2m + 1)(4m^2 - 2m + 1)$

18. $(k + 2)(12k^3 - 3k^2 + k + 1)$

19. $(x + y + z)(3x - 2y - z)$

20. $(r + 2s - 3t)(2r - 2s + t)$

21. $(x + 1)(x + 2)(x + 3)$

22. $(x - 1)(x + 2)(x - 3)$

23. $(x + 2)^2$

24. $(2a - 4b)^2$

25. $(x - 2y)^3$

26. $(3x + y)^3$

1. $-9x^2 + 8x + 13$ 2. $12y^3 + 2y^2 - 19y - 10$

3. $6x^2 + 19x - 7$

4. $27x^3 + 54x^2y + 36xy^2 + 8y^3$

R.2 Factoring

Multiplication of polynomials relies on the distributive property. The reverse process, where a polynomial is written as a product of other polynomials, is called **factoring**. For example, one way to factor the number 18 is to write it as the product $9 \cdot 2$; both 9 and 2 are **factors** of 18. Usually, only integers are used as factors of integers. The number 18 can also be written with three integer factors as $2 \cdot 3 \cdot 3$.

The Greatest Common Factor
To factor the algebraic expression $15m + 45$, first note that both $15m$ and 45 are divisible by 15; $15m = 15 \cdot m$ and $45 = 15 \cdot 3$. By the distributive property,

$$15m + 45 = 15 \cdot m + 15 \cdot 3 = 15(m + 3).$$

Both 15 and $m + 3$ are factors of $15m + 45$. Since 15 divides into both terms of $15m + 45$ (and is the largest number that will do so), 15 is the **greatest common factor** for the polynomial $15m + 45$. The process of writing $15m + 45$ as $15(m + 3)$ is often called **factoring out** the greatest common factor.

EXAMPLE 1 Factoring

Factor out the greatest common factor.

(a) $12p - 18q$

SOLUTION Both $12p$ and $18q$ are divisible by 6. Therefore,

$$12p - 18q = 6 \cdot 2p - 6 \cdot 3q = 6(2p - 3q).$$

(b) $8x^3 - 9x^2 + 15x$

SOLUTION Each of these terms is divisible by x.

$$8x^3 - 9x^2 + 15x = (8x^2) \cdot x - (9x) \cdot x + 15 \cdot x$$
$$= x(8x^2 - 9x + 15) \quad \text{or} \quad (8x^2 - 9x + 15)x$$

YOUR TURN 1 Factor $4z^4 + 4z^3 + 18z^2$.

TRY YOUR TURN 1

One can always check factorization by finding the product of the factors and comparing it to the original expression.

| CAUTION | When factoring out the greatest common factor in an expression like $2x^2 + x$, be careful to remember the 1 in the second term. |

$$2x^2 + x = 2x^2 + 1x = x(2x + 1), \quad \text{not } x(2x).$$

Factoring Trinomials

A polynomial that has no greatest common factor (other than 1) may still be factorable. For example, the polynomial $x^2 + 5x + 6$ can be factored as $(x + 2)(x + 3)$. To see that this is correct, find the product $(x + 2)(x + 3)$; you should get $x^2 + 5x + 6$. A polynomial such as this with three terms is called a **trinomial**. To factor a trinomial of the form $x^2 + bx + c$, where the coefficient of x^2 is 1, use FOIL backwards. We look for two factors of c whose sum is b. When c is positive, its factors must have the same sign. Since b is the sum of these two factors, the factors will have the same sign as b. When c is negative, its factors have opposite signs. Again, since b is the sum of these two factors, the factor with the greater absolute value will have the same sign as b.

EXAMPLE 2 Factoring a Trinomial

Factor $y^2 + 8y + 15$.

SOLUTION Since the coefficient of y^2 is 1, factor by finding two numbers whose *product* is 15 and whose *sum* is 8. Because the constant and the middle term are positive, the numbers must both be positive. Begin by listing all pairs of positive integers having a product of 15. As you do this, also form the sum of each pair of numbers.

Products	Sums
$15 \cdot 1 = 15$	$15 + 1 = 16$
$5 \cdot 3 = 15$	$5 + 3 = 8$

The numbers 5 and 3 have a product of 15 and a sum of 8. Thus, $y^2 + 8y + 15$ factors as

$$y^2 + 8y + 15 = (y + 5)(y + 3).$$

YOUR TURN 2 Factor $x^2 - 3x - 10$.

The answer can also be written as $(y + 3)(y + 5)$.

TRY YOUR TURN 2

If the coefficient of the squared term is *not* 1, work as shown on the next page.

EXAMPLE 3   Factoring a Trinomial

Factor $4x^2 + 8xy - 5y^2$.

SOLUTION   The possible factors of $4x^2$ are $4x$ and x or $2x$ and $2x$; the possible factors of $-5y^2$ are $-5y$ and y or $5y$ and $-y$. Try various combinations of these factors until one works (if, indeed, any work). For example, try the product $(x + 5y)(4x - y)$.

$$(x + 5y)(4x - y) = 4x^2 - xy + 20xy - 5y^2$$
$$= 4x^2 + 19xy - 5y^2$$

This product is not correct, so try another combination.

$$(2x - y)(2x + 5y) = 4x^2 + 10xy - 2xy - 5y^2$$
$$= 4x^2 + 8xy - 5y^2$$

YOUR TURN 3 Factor $6a^2 + 5ab - 4b^2$.

Since this combination gives the correct polynomial,

$$4x^2 + 8xy - 5y^2 = (2x - y)(2x + 5y). \qquad \text{TRY YOUR TURN 3}$$

Special Factorizations   Four special factorizations occur so often that they are listed here for future reference.

Special Factorizations

$x^2 - y^2 = (x + y)(x - y)$	**Difference of two squares**
$x^2 + 2xy + y^2 = (x + y)^2$	**Perfect square**
$x^3 - y^3 = (x - y)(x^2 + xy + y^2)$	**Difference of two cubes**
$x^3 + y^3 = (x + y)(x^2 - xy + y^2)$	**Sum of two cubes**

A polynomial that cannot be factored is called a **prime polynomial**.

EXAMPLE 4   Factoring Polynomials

Factor each polynomial, if possible.

(a) $64p^2 - 49q^2 = (8p)^2 - (7q)^2 = (8p + 7q)(8p - 7q)$ Difference of two squares

(b) $x^2 + 36$ is a prime polynomial.

(c) $x^2 + 12x + 36 = x^2 + 2(x)(6) + 6^2 = (x + 6)^2$ Perfect square

(d) $9y^2 - 24yz + 16z^2 = (3y)^2 + 2(3y)(-4z) + (-4z)^2$ Perfect square
$$= [3y + (-4z)]^2 = (3y - 4z)^2$$

(e) $y^3 - 8 = y^3 - 2^3 = (y - 2)(y^2 + 2y + 4)$ Difference of two cubes

(f) $m^3 + 125 = m^3 + 5^3 = (m + 5)(m^2 - 5m + 25)$ Sum of two cubes

(g) $8k^3 - 27z^3 = (2k)^3 - (3z)^3 = (2k - 3z)(4k^2 + 6kz + 9z^2)$ Difference of two cubes

(h) $p^4 - 1 = (p^2 + 1)(p^2 - 1) = (p^2 + 1)(p + 1)(p - 1)$ Difference of two squares

> **CAUTION**   In factoring, always look for a common factor first. Since $36x^2 - 4y^2$ has a common factor of 4,
>
> $$36x^2 - 4y^2 = 4(9x^2 - y^2) = 4(3x + y)(3x - y).$$
>
> It would be incomplete to factor it as
>
> $$36x^2 - 4y^2 = (6x + 2y)(6x - 2y),$$
>
> since each factor can be factored still further. To *factor* means to factor completely, so that each polynomial factor is prime.

R.2 EXERCISES

Factor each polynomial. If a polynomial cannot be factored, write *prime*. Factor out the greatest common factor as necessary.

1. $7a^3 + 14a^2$

2. $3y^3 + 24y^2 + 9y$

3. $13p^4q^2 - 39p^3q + 26p^2q^2$

4. $60m^4 - 120m^3n + 50m^2n^2$

5. $m^2 - 5m - 14$

6. $x^2 + 4x - 5$

7. $z^2 + 9z + 20$

8. $b^2 - 8b + 7$

9. $a^2 - 6ab + 5b^2$

10. $s^2 + 2st - 35t^2$

11. $y^2 - 4yz - 21z^2$

12. $3x^2 + 4x - 7$

13. $3a^2 + 10a + 7$

14. $15y^2 + y - 2$

15. $21m^2 + 13mn + 2n^2$

16. $6a^2 - 48a - 120$

17. $3m^3 + 12m^2 + 9m$

18. $4a^2 + 10a + 6$

19. $24a^4 + 10a^3b - 4a^2b^2$

20. $24x^4 + 36x^3y - 60x^2y^2$

21. $x^2 - 64$ **22.** $9m^2 - 25$

23. $10x^2 - 160$ **24.** $9x^2 + 64$

25. $z^2 + 14zy + 49y^2$ **26.** $s^2 - 10st + 25t^2$

27. $9p^2 - 24p + 16$ **28.** $a^3 - 216$

29. $27r^3 - 64s^3$ **30.** $3m^3 + 375$

31. $x^4 - y^4$ **32.** $16a^4 - 81b^4$

YOUR TURN ANSWERS

1. $2z^2(2z^2 + 2z + 9)$ **2.** $(x + 2)(x - 5)$

3. $(2a - b)(3a + 4b)$

R.3 Rational Expressions

Many algebraic fractions are **rational expressions**, which are quotients of polynomials with nonzero denominators. Examples include

$$\frac{8}{x - 1}, \qquad \frac{3x^2 + 4x}{5x - 6}, \qquad \text{and} \qquad \frac{2y + 1}{y^2}.$$

Next, we summarize properties for working with rational expressions.

Properties of Rational Expressions

For all mathematical expressions P, Q, R, and S, with $Q \neq 0$ and $S \neq 0$:

$$\frac{P}{Q} = \frac{PS}{QS} \qquad \textbf{Fundamental property}$$

$$\frac{P}{Q} + \frac{R}{Q} = \frac{P + R}{Q} \qquad \textbf{Addition}$$

$$\frac{P}{Q} - \frac{R}{Q} = \frac{P - R}{Q} \qquad \textbf{Subtraction}$$

$$\frac{P}{Q} \cdot \frac{R}{S} = \frac{PR}{QS} \qquad \textbf{Multiplication}$$

$$\frac{P}{Q} \div \frac{R}{S} = \frac{P}{Q} \cdot \frac{S}{R} \quad (R \neq 0) \qquad \textbf{Division}$$

When writing a rational expression in lowest terms, we may need to use the fact that $\frac{a^m}{a^n} = a^{m-n}$. For example,

$$\frac{x^4}{3x} = \frac{1x^4}{3x} = \frac{1}{3} \cdot \frac{x^4}{x} = \frac{1}{3} \cdot x^{4-1} = \frac{1}{3}x^3 = \frac{x^3}{3}.$$

EXAMPLE 1 Reducing Rational Expressions

Write each rational expression in lowest terms, that is, reduce the expression as much as possible.

(a) $\dfrac{8x + 16}{4} = \dfrac{8(x + 2)}{4} = \dfrac{4 \cdot 2(x + 2)}{4} = 2(x + 2)$

Factor both the numerator and denominator in order to identify any common factors, which have a quotient of 1. The answer could also be written as $2x + 4$.

YOUR TURN 1 Write in lowest terms $\dfrac{z^2 + 5z + 6}{2z^2 + 7z + 3}$.

(b) $\dfrac{k^2 + 7k + 12}{k^2 + 2k - 3} = \dfrac{(k + 4)(k + 3)}{(k - 1)(k + 3)} = \dfrac{k + 4}{k - 1}$

The answer cannot be further reduced. TRY YOUR TURN 1

CAUTION One of the most common errors in algebra involves incorrect use of the fundamental property of rational expressions. Only common *factors* may be divided or "canceled." It is essential to factor rational expressions before writing them in lowest terms. In Example 1(b), for instance, it is not correct to "cancel" k^2 (or cancel k, or divide 12 by -3) because the additions and subtraction must be performed first. Here they cannot be performed, so it is not possible to divide. After factoring, however, the fundamental property can be used to write the expression in lowest terms.

EXAMPLE 2 Combining Rational Expressions

Perform each operation.

(a) $\dfrac{3y + 9}{6} \cdot \dfrac{18}{5y + 15}$

SOLUTION Factor where possible, then multiply numerators and denominators and reduce to lowest terms.

$$\dfrac{3y + 9}{6} \cdot \dfrac{18}{5y + 15} = \dfrac{3(y + 3)}{6} \cdot \dfrac{18}{5(y + 3)} \qquad \text{Factor.}$$

$$= \dfrac{3 \cdot 18(y + 3)}{6 \cdot 5(y + 3)} \qquad \text{Multiply.}$$

$$= \dfrac{3 \cdot \cancel{6} \cdot 3\cancel{(y + 3)}}{\cancel{6} \cdot 5\cancel{(y + 3)}} = \dfrac{3 \cdot 3}{5} = \dfrac{9}{5} \qquad \text{Reduce to lowest terms.}$$

(b) $\dfrac{m^2 + 5m + 6}{m + 3} \cdot \dfrac{m}{m^2 + 3m + 2}$

SOLUTION Factor where possible.

$$\dfrac{(m + 2)(m + 3)}{m + 3} \cdot \dfrac{m}{(m + 2)(m + 1)} \qquad \text{Factor.}$$

$$= \dfrac{m\cancel{(m + 2)}\cancel{(m + 3)}}{\cancel{(m + 3)}\cancel{(m + 2)}(m + 1)} = \dfrac{m}{m + 1} \qquad \text{Reduce to lowest terms.}$$

(c) $\dfrac{9p - 36}{12} \div \dfrac{5(p - 4)}{18}$

SOLUTION Use the division property of rational expressions.

$$\dfrac{9p - 36}{12} \div \dfrac{5(p - 4)}{18} = \dfrac{9p - 36}{12} \cdot \dfrac{18}{5(p - 4)} \qquad \text{Invert and multiply.}$$

$$= \dfrac{9\cancel{(p - 4)}}{\cancel{6} \cdot 2} \cdot \dfrac{\cancel{6} \cdot 3}{5\cancel{(p - 4)}} = \dfrac{27}{10} \qquad \text{Factor and reduce to lowest terms.}$$

(d) $\dfrac{4}{5k} - \dfrac{11}{5k}$

SOLUTION   As shown in the list of properties, to subtract two rational expressions that have the same denominators, subtract the numerators while keeping the same denominator.

$$\frac{4}{5k} - \frac{11}{5k} = \frac{4 - 11}{5k} = -\frac{7}{5k}$$

(e) $\dfrac{7}{p} + \dfrac{9}{2p} + \dfrac{1}{3p}$

SOLUTION   These three fractions cannot be added until their denominators are the same. A **common denominator** into which p, $2p$, and $3p$ all divide is $6p$. Note that $12p$ is also a common denominator, but $6p$ is the **least common denominator**. Use the fundamental property to rewrite each rational expression with a denominator of $6p$.

$$\frac{7}{p} + \frac{9}{2p} + \frac{1}{3p} = \frac{6 \cdot 7}{6 \cdot p} + \frac{3 \cdot 9}{3 \cdot 2p} + \frac{2 \cdot 1}{2 \cdot 3p} \qquad \text{\small Rewrite with common denominator } 6p.$$

$$= \frac{42}{6p} + \frac{27}{6p} + \frac{2}{6p}$$

$$= \frac{42 + 27 + 2}{6p}$$

$$= \frac{71}{6p}$$

(f) $\dfrac{x + 1}{x^2 + 5x + 6} - \dfrac{5x - 1}{x^2 - x - 12}$

SOLUTION   To find the least common denominator, we first factor each denominator. Then we change each fraction so they all have the same denominator, being careful to multiply only by quotients that equal 1.

$$\frac{x + 1}{x^2 + 5x + 6} - \frac{5x - 1}{x^2 - x - 12}$$

$$= \frac{x + 1}{(x + 2)(x + 3)} - \frac{5x - 1}{(x + 3)(x - 4)} \qquad \text{\small Factor denominators.}$$

$$- \frac{x + 1}{(x + 2)(x + 3)} \cdot \frac{(x - 4)}{(x - 4)} \quad \frac{5x - 1}{(x + 3)(x - 4)} \cdot \frac{(x + 2)}{(x + 2)} \qquad \text{\small Rewrite with common denominators.}$$

$$= \frac{(x^2 - 3x - 4) - (5x^2 + 9x - 2)}{(x + 2)(x + 3)(x - 4)} \qquad \text{\small Multiply numerators.}$$

$$= \frac{-4x^2 - 12x - 2}{(x + 2)(x + 3)(x - 4)} \qquad \text{\small Subtract.}$$

$$= \frac{-2(2x^2 + 6x + 1)}{(x + 2)(x + 3)(x - 4)} \qquad \text{\small Factor numerator.}$$

YOUR TURN 2 Perform each of the following operations.

(a) $\dfrac{z^2 + 5z + 6}{2z^2 - 5z - 3} \cdot \dfrac{2z^2 - z - 1}{z^2 + 2z - 3}.$

(b) $\dfrac{a - 3}{a^2 + 3a + 2} + \dfrac{5a}{a^2 - 4}.$

Because the numerator cannot be factored further, we leave our answer in this form. We could also multiply out the denominator, but factored form is usually more useful.

TRY YOUR TURN 2

R.3   EXERCISES

Write each rational expression in lowest terms.

1. $\dfrac{5v^2}{35v}$

2. $\dfrac{25p^3}{10p^2}$

3. $\dfrac{8k + 16}{9k + 18}$

4. $\dfrac{2(t - 15)}{(t - 15)(t + 2)}$

5. $\dfrac{4x^3 - 8x^2}{4x^2}$

6. $\dfrac{36y^2 + 72y}{9y}$

7. $\dfrac{m^2 - 4m + 4}{m^2 + m - 6}$

8. $\dfrac{r^2 - r - 6}{r^2 + r - 12}$

9. $\dfrac{3x^2 + 3x - 6}{x^2 - 4}$

10. $\dfrac{z^2 - 5z + 6}{z^2 - 4}$

11. $\dfrac{m^4 - 16}{4m^2 - 16}$

12. $\dfrac{6y^2 + 11y + 4}{3y^2 + 7y + 4}$

Perform the indicated operations.

13. $\dfrac{9k^2}{25} \cdot \dfrac{5}{3k}$

14. $\dfrac{15p^3}{9p^2} \div \dfrac{6p}{10p^2}$

15. $\dfrac{3a + 3b}{4c} \cdot \dfrac{12}{5(a + b)}$

16. $\dfrac{a - 3}{16} \div \dfrac{a - 3}{32}$

17. $\dfrac{2k - 16}{6} \div \dfrac{4k - 32}{3}$

18. $\dfrac{9y - 18}{6y + 12} \cdot \dfrac{3y + 6}{15y - 30}$

19. $\dfrac{4a + 12}{2a - 10} \div \dfrac{a^2 - 9}{a^2 - a - 20}$

20. $\dfrac{6r - 18}{9r^2 + 6r - 24} \cdot \dfrac{12r - 16}{4r - 12}$

21. $\dfrac{k^2 + 4k - 12}{k^2 + 10k + 24} \cdot \dfrac{k^2 + k - 12}{k^2 - 9}$

22. $\dfrac{m^2 + 3m + 2}{m^2 + 5m + 4} \div \dfrac{m^2 + 5m + 6}{m^2 + 10m + 24}$

23. $\dfrac{2m^2 - 5m - 12}{m^2 - 10m + 24} \div \dfrac{4m^2 - 9}{m^2 - 9m + 18}$

24. $\dfrac{4n^2 + 4n - 3}{6n^2 - n - 15} \cdot \dfrac{8n^2 + 32n + 30}{4n^2 + 16n + 15}$

25. $\dfrac{a + 1}{2} - \dfrac{a - 1}{2}$

26. $\dfrac{3}{p} + \dfrac{1}{2}$

27. $\dfrac{6}{5y} - \dfrac{3}{2}$

28. $\dfrac{1}{6m} + \dfrac{2}{5m} + \dfrac{4}{m}$

29. $\dfrac{1}{m - 1} + \dfrac{2}{m}$

30. $\dfrac{5}{2r + 3} - \dfrac{2}{r}$

31. $\dfrac{8}{3(a - 1)} + \dfrac{2}{a - 1}$

32. $\dfrac{2}{5(k - 2)} + \dfrac{3}{4(k - 2)}$

33. $\dfrac{4}{x^2 + 4x + 3} + \dfrac{3}{x^2 - x - 2}$

34. $\dfrac{y}{y^2 + 2y - 3} - \dfrac{1}{y^2 + 4y + 3}$

35. $\dfrac{3k}{2k^2 + 3k - 2} - \dfrac{2k}{2k^2 - 7k + 3}$

36. $\dfrac{4m}{3m^2 + 7m - 6} - \dfrac{m}{3m^2 - 14m + 8}$

37. $\dfrac{2}{a + 2} + \dfrac{1}{a} + \dfrac{a - 1}{a^2 + 2a}$

38. $\dfrac{5x + 2}{x^2 - 1} + \dfrac{3}{x^2 + x} - \dfrac{1}{x^2 - x}$

YOUR TURN ANSWERS

1. $(z + 2)/(2z + 1)$

2. (a) $(z + 2)/(z - 3)$

  (b) $6(a^2 + 1)/[(a - 2)(a + 2)(a + 1)]$

R.4 Equations

Linear Equations

Equations that can be written in the form $ax + b = 0$, where a and b are real numbers, with $a \neq 0$, are **linear equations**. Examples of linear equations include $5y + 9 = 16$, $8x = 4$, and $-3p + 5 = -8$. Equations that are *not* linear include absolute value equations such as $|x| = 4$. The following properties are used to solve linear equations.

Properties of Equality

For all real numbers a, b, and c:

1. If $a = b$, then $a + c = b + c$.   **Addition property of equality**
  (The same number may be added
  to both sides of an equation.)

2. If $a = b$, then $ac = bc$.   **Multiplication property of equality**
  (Both sides of an equation may be
  multiplied by the same number.)

EXAMPLE 1   Solving Linear Equations

Solve the following equations.

(a) $x - 2 = 3$

SOLUTION   The goal is to isolate the variable. Using the addition property of equality yields

$$x - 2 + 2 = 3 + 2, \quad \text{or} \quad x = 5.$$

(b) $\dfrac{x}{2} = 3$

SOLUTION   Using the multiplication property of equality yields

$$2 \cdot \frac{x}{2} = 2 \cdot 3, \quad \text{or} \quad x = 6.$$

The following example shows how these properties are used to solve linear equations. The goal is to isolate the variable. The solutions should always be checked by substitution into the original equation.

EXAMPLE 2   Solving a Linear Equation

Solve $2x - 5 + 8 = 3x + 2(2 - 3x)$.

SOLUTION

$$
\begin{aligned}
2x - 5 + 8 &= 3x + 4 - 6x & \text{Distributive property} \\
2x + 3 &= -3x + 4 & \text{Combine like terms.} \\
5x + 3 &= 4 & \text{Add } 3x \text{ to both sides.} \\
5x &= 1 & \text{Add } -3 \text{ to both sides.} \\
x &= \frac{1}{5} & \text{Multiply both sides by } \tfrac{1}{5}.
\end{aligned}
$$

YOUR TURN 1 Solve $3x - 7 = 4(5x + 2) - 7x$.

Check by substituting into the original equation. The left side becomes $2(1/5) - 5 + 8$ and the right side becomes $3(1/5) + 2[2 - 3(1/5)]$. Verify that both of these expressions simplify to 17/5.   **TRY YOUR TURN 1**

Quadratic Equations
An equation with 2 as the greatest exponent of the variable is a *quadratic equation*. A **quadratic equation** has the form $ax^2 + bx + c = 0$, where a, b, and c are real numbers and $a \neq 0$. A quadratic equation written in the form $ax^2 + bx + c = 0$ is said to be in **standard form**.

The simplest way to solve a quadratic equation, but one that is not always applicable, is by factoring. This method depends on the **zero-factor property**.

Zero-Factor Property
If a and b are real numbers, with $ab = 0$, then either

$$a = 0 \quad \text{or} \quad b = 0 \quad \text{(or both)}.$$

EXAMPLE 3   Solving a Quadratic Equation

Solve $6r^2 + 7r = 3$.

SOLUTION   First write the equation in standard form.

$$6r^2 + 7r - 3 = 0$$

Now factor $6r^2 + 7r - 3$ to get

$$(3r - 1)(2r + 3) = 0.$$

By the zero-factor property, the product $(3r - 1)(2r + 3)$ can equal 0 if and only if

$$3r - 1 = 0 \quad \text{or} \quad 2r + 3 = 0.$$

YOUR TURN 2 Solve $2m^2 + 7m = 15$.

Solve each of these equations separately to find that the solutions are $1/3$ and $-3/2$. Check these solutions by substituting them into the original equation.   TRY YOUR TURN 2

CAUTION Remember, the zero-factor property requires that the product of two (or more) factors be equal to *zero*, not some other quantity. It would be incorrect to use the zero-factor property with an equation in the form $(x + 3)(x - 1) = 4$, for example.

If a quadratic equation cannot be solved easily by factoring, use the *quadratic formula*. (The derivation of the quadratic formula is given in most algebra books.)

Quadratic Formula

The solutions of the quadratic equation $ax^2 + bx + c = 0$, where $a \neq 0$, are given by

$$x = \frac{-b \pm \sqrt{b^2 - 4ac}}{2a}.$$

EXAMPLE 4   Quadratic Formula

Solve $x^2 - 4x - 5 = 0$ by the quadratic formula.

SOLUTION   The equation is already in standard form (it has 0 alone on one side of the equal sign), so the values of a, b, and c from the quadratic formula are easily identified. The coefficient of the squared term gives the value of a; here, $a = 1$. Also, $b = -4$ and $c = -5$, where b is the coefficient of the linear term and c is the constant coefficient. (Be careful to use the correct signs.) Substitute these values into the quadratic formula.

$$x = \frac{-(-4) \pm \sqrt{(-4)^2 - 4(1)(-5)}}{2(1)} \qquad a = 1, b = -4, c = -5$$

$$x = \frac{4 \pm \sqrt{16 + 20}}{2} \qquad (-4)^2 = (-4)(-4) = 16$$

$$x = \frac{4 \pm 6}{2} \qquad \sqrt{16 + 20} = \sqrt{36} = 6$$

The $\pm$ sign represents the two solutions of the equation. To find both of the solutions, first use $+$ and then use $-$.

$$x = \frac{4 + 6}{2} = \frac{10}{2} = 5 \quad \text{or} \quad x = \frac{4 - 6}{2} = \frac{-2}{2} = -1$$

The two solutions are 5 and -1.

CAUTION Notice in the quadratic formula that the square root is added to or subtracted from the value of $-b$ *before* dividing by $2a$.

EXAMPLE 5   **Quadratic Formula**

Solve $x^2 + 1 = 4x$.

SOLUTION   First, add $-4x$ on both sides of the equal sign in order to get the equation in standard form.

$$x^2 - 4x + 1 = 0$$

Now identify the values of a, b, and c. Here $a = 1$, $b = -4$, and $c = 1$. Substitute these numbers into the quadratic formula.

$$x = \frac{-(-4) \pm \sqrt{(-4)^2 - 4(1)(1)}}{2(1)} \qquad a = 1, b = -4, c = 1$$

$$= \frac{4 \pm \sqrt{16 - 4}}{2}$$

$$= \frac{4 \pm \sqrt{12}}{2}$$

Simplify the solutions by writing $\sqrt{12}$ as $\sqrt{4 \cdot 3} = \sqrt{4} \cdot \sqrt{3} = 2\sqrt{3}$. Substituting $2\sqrt{3}$ for $\sqrt{12}$ gives

$$x = \frac{4 \pm 2\sqrt{3}}{2}$$

$$= \frac{2(2 \pm \sqrt{3})}{2} \qquad \text{Factor } 4 \pm 2\sqrt{3}.$$

$$= 2 \pm \sqrt{3}. \qquad \text{Reduce to lowest terms.}$$

The two solutions are $2 + \sqrt{3}$ and $2 - \sqrt{3}$.

The exact values of the solutions are $2 + \sqrt{3}$ and $2 - \sqrt{3}$. The $\sqrt{}$ key on a calculator gives decimal approximations of these solutions (to the nearest thousandth):

$$2 + \sqrt{3} \approx 2 + 1.732 = 3.732*$$

$$2 - \sqrt{3} \approx 2 - 1.732 = 0.268 \qquad \text{TRY YOUR TURN 3}$$

YOUR TURN 3 Solve $z^2 + 6 = 8z$.

NOTE Sometimes the quadratic formula will give a result with a negative number under the radical sign, such as $3 \pm \sqrt{-5}$. A solution of this type is a complex number. Since this text deals only with real numbers, such solutions cannot be used.

Equations with Fractions

When an equation includes fractions, first eliminate all denominators by multiplying both sides of the equation by a common denominator, a number that can be divided (with no remainder) by each denominator in the equation. When an equation involves fractions with variable denominators, it is *necessary* to check all solutions in the original equation to be sure that no solution will lead to a zero denominator.

EXAMPLE 6   **Solving Rational Equations**

Solve each equation.

(a) $\dfrac{r}{10} - \dfrac{2}{15} = \dfrac{3r}{20} - \dfrac{1}{5}$

SOLUTION   The denominators are 10, 15, 20, and 5. Each of these numbers can be divided into 60, so 60 is a common denominator. Multiply both sides of the equation by

*The symbol $\approx$ means "is approximately equal to."

60 and use the distributive property. (If a common denominator cannot be found easily, all the denominators in the problem can be multiplied together to produce one.)

$$\frac{r}{10} - \frac{2}{15} = \frac{3r}{20} - \frac{1}{5}$$

$$60\left(\frac{r}{10} - \frac{2}{15}\right) = 60\left(\frac{3r}{20} - \frac{1}{5}\right) \qquad \text{Multiply by the common denominator.}$$

$$60\left(\frac{r}{10}\right) - 60\left(\frac{2}{15}\right) = 60\left(\frac{3r}{20}\right) - 60\left(\frac{1}{5}\right) \qquad \text{Distributive property}$$

$$6r - 8 = 9r - 12$$

Add $-9r$ and 8 to both sides.

$$6r - 8 + (-9r) + 8 = 9r - 12 + (-9r) + 8$$

$$-3r = -4$$

$$r = \frac{4}{3} \qquad \text{Multiply each side by } -\tfrac{1}{3}.$$

Check by substituting into the original equation.

(b) $\dfrac{3}{x^2} - 12 = 0$

SOLUTION   Begin by multiplying both sides of the equation by x^2 to get $3 - 12x^2 = 0$. This equation could be solved by using the quadratic formula with $a = -12$, $b = 0$, and $c = 3$. Another method that works well for the type of quadratic equation in which $b = 0$ is shown below.

$$3 - 12x^2 = 0$$

$$3 = 12x^2 \qquad \text{Add } 12x^2.$$

$$\frac{1}{4} = x^2 \qquad \text{Multiply by } \tfrac{1}{12}.$$

$$\pm\frac{1}{2} = x \qquad \text{Take square roots.}$$

Verify that there are two solutions, $-1/2$ and $1/2$.

(c) $\dfrac{2}{k} - \dfrac{3k}{k + 2} = \dfrac{k}{k^2 + 2k}$

SOLUTION   Factor $k^2 + 2k$ as $k(k + 2)$. The least common denominator for all the fractions is $k(k + 2)$. Multiplying both sides by $k(k + 2)$ gives the following:

$$k(k + 2) \cdot \left(\frac{2}{k} - \frac{3k}{k + 2}\right) = k(k + 2) \cdot \frac{k}{k^2 + 2k}$$

$$2(k + 2) - 3k(k) = k$$

$$2k + 4 - 3k^2 = k \qquad \text{Distributive property}$$

$$-3k^2 + k + 4 = 0 \qquad \text{Add } -k; \text{ rearrange terms.}$$

$$3k^2 - k - 4 = 0 \qquad \text{Multiply by } -1.$$

$$(3k - 4)(k + 1) = 0 \qquad \text{Factor.}$$

$$3k - 4 = 0 \qquad \text{or} \qquad k + 1 = 0$$

$$k = \frac{4}{3} \qquad k = -1$$

YOUR TURN 4 Solve $\dfrac{1}{x^2 - 4} + \dfrac{2}{x - 2} = \dfrac{1}{x}$.

Verify that the solutions are $4/3$ and -1.   TRY YOUR TURN 4

| CAUTION | It is possible to get, as a solution of a rational equation, a number that makes one or more of the denominators in the original equation equal to zero. That number is not a solution, so it is *necessary* to check all potential solutions of rational equations. These introduced solutions are called **extraneous solutions**. |

EXAMPLE 7 **Solving a Rational Equation**

Solve $\dfrac{2}{x-3} + \dfrac{1}{x} = \dfrac{6}{x(x-3)}$.

SOLUTION The common denominator is $x(x-3)$. Multiply both sides by $x(x-3)$ and solve the resulting equation.

$$x(x-3) \cdot \left(\frac{2}{x-3} + \frac{1}{x} \right) = x(x-3) \cdot \left[\frac{6}{x(x-3)} \right]$$
$$2x + x - 3 = 6$$
$$3x = 9$$
$$x = 3$$

Checking this potential solution by substitution into the original equation shows that 3 makes two denominators 0. Thus, 3 cannot be a solution, so there is no solution for this equation.

R.4 EXERCISES

Solve each equation.

1. $2x + 8 = x - 4$

2. $5x + 2 = 8 - 3x$

3. $0.2m - 0.5 = 0.1m + 0.7$

4. $\dfrac{2}{3}k - k + \dfrac{3}{8} = \dfrac{1}{2}$

5. $3r + 2 - 5(r + 1) = 6r + 4$

6. $5(a + 3) + 4a - 5 = -(2a - 4)$

7. $2[3m - 2(3 - m) - 4] = 6m - 4$

8. $4[2p - (3 - p) + 5] = -7p - 2$

Solve each equation by factoring or by using the quadratic formula. If the solutions involve square roots, give both the exact solutions and the approximate solutions to three decimal places.

9. $x^2 + 5x + 6 = 0$

10. $x^2 = 3 + 2x$

11. $m^2 = 14m - 49$

12. $2k^2 - k = 10$

13. $12x^2 - 5x = 2$

14. $m(m - 7) = -10$

15. $4x^2 - 36 = 0$

16. $z(2z + 7) = 4$

17. $12y^2 - 48y = 0$

18. $3x^2 - 5x + 1 = 0$

19. $2m^2 - 4m = 3$

20. $p^2 + p - 1 = 0$

21. $k^2 - 10k = -20$

22. $5x^2 - 8x + 2 = 0$

23. $2r^2 - 7r + 5 = 0$

24. $2x^2 - 7x + 30 = 0$

25. $3k^2 + k = 6$

26. $5m^2 + 5m = 0$

Solve each equation.

27. $\dfrac{3x - 2}{7} = \dfrac{x + 2}{5}$

28. $\dfrac{x}{3} - 7 = 6 - \dfrac{3x}{4}$

29. $\dfrac{4}{x - 3} - \dfrac{8}{2x + 5} + \dfrac{3}{x - 3} = 0$

30. $\dfrac{5}{p - 2} - \dfrac{7}{p + 2} = \dfrac{12}{p^2 - 4}$

31. $\dfrac{2m}{m - 2} - \dfrac{6}{m} = \dfrac{12}{m^2 - 2m}$

32. $\dfrac{2y}{y - 1} = \dfrac{5}{y} + \dfrac{10 - 8y}{y^2 - y}$

33. $\dfrac{1}{x - 2} - \dfrac{3x}{x - 1} = \dfrac{2x + 1}{x^2 - 3x + 2}$

34. $\dfrac{5}{a} + \dfrac{-7}{a + 1} = \dfrac{a^2 - 2a + 4}{a^2 + a}$

35. $\dfrac{5}{b + 5} - \dfrac{4}{b^2 + 2b} = \dfrac{6}{b^2 + 7b + 10}$

36. $\dfrac{2}{x^2 - 2x - 3} + \dfrac{5}{x^2 - x - 6} = \dfrac{1}{x^2 + 3x + 2}$

37. $\dfrac{4}{2x^2 + 3x - 9} + \dfrac{2}{2x^2 - x - 3} = \dfrac{3}{x^2 + 4x + 3}$

YOUR TURN ANSWERS

1. $-3/2$

2. $3/2, -5$

3. $4 \pm \sqrt{10}$

4. $-1, -4$

R.5 Inequalities

To write that one number is greater than or less than another number, we use the following symbols.

Inequality Symbols

$<$ means *is less than* $\qquad$ $\leq$ means *is less than or equal to*
$>$ means *is greater than* $\qquad$ $\geq$ means *is greater than or equal to*

Linear Inequalities

An equation states that two expressions are equal; an **inequality** states that they are unequal. A **linear inequality** is an inequality that can be simplified to the form $ax < b$. (Properties introduced in this section are given only for $<$, but they are equally valid for $>$, $\leq$, or $\geq$.) Linear inequalities are solved with the following properties.

Properties of Inequality

For all real numbers a, b, and c:

1. If $a < b$, then $a + c < b + c$.
2. If $a < b$ and if $c > 0$, then $ac < bc$.
3. If $a < b$ and if $c < 0$, then $ac < bc$.

Pay careful attention to property 3; it says that if both sides of an inequality are multiplied by a negative number, the direction of the inequality symbol must be reversed.

EXAMPLE 1 Solving a Linear Inequality

Solve $4 - 3y \leq 7 + 2y$.

SOLUTION Use the properties of inequality.

$$4 - 3y + (-4) \leq 7 + 2y + (-4) \qquad \text{Add } -4 \text{ to both sides.}$$
$$-3y \leq 3 + 2y$$

Remember that *adding* the same number to both sides never changes the direction of the inequality symbol.

$$-3y + (-2y) \leq 3 + 2y + (-2y) \qquad \text{Add } -2y \text{ to both sides.}$$
$$-5y \leq 3$$

Multiply both sides by $-1/5$. Since $-1/5$ is negative, change the direction of the inequality symbol.

$$-\frac{1}{5}(-5y) \geq -\frac{1}{5}(3)$$

$$y \geq -\frac{3}{5}$$

TRY YOUR TURN 1

YOUR TURN 1 Solve $3z - 2 > 5z + 7$.

CAUTION It is a common error to forget to reverse the direction of the inequality sign when multiplying or dividing by a negative number. For example, to solve $-4x \leq 12$, we must multiply by $-1/4$ on both sides *and* reverse the inequality symbol to get $x \geq -3$.

The solution $y \geq -3/5$ in Example 1 represents an interval on the number line. **Interval notation** often is used for writing intervals. With interval notation, $y \geq -3/5$ is written as $[-3/5, \infty)$. This is an example of a **half-open interval**, since one endpoint, $-3/5$, is included. The **open interval** $(2, 5)$ corresponds to $2 < x < 5$, with neither endpoint included. The **closed interval** $[2, 5]$ includes both endpoints and corresponds to $2 \leq x \leq 5$.

The **graph** of an interval shows all points on a number line that correspond to the numbers in the interval. To graph the interval $[-3/5, \infty)$, for example, use a solid circle at $-3/5$, since $-3/5$ is part of the solution. To show that the solution includes all real numbers greater than or equal to $-3/5$, draw a heavy arrow pointing to the right (the positive direction). See Figure 1.

FIGURE 1

EXAMPLE 2   Graphing a Linear Inequality

Solve $-2 < 5 + 3m < 20$. Graph the solution.

SOLUTION   The inequality $-2 < 5 + 3m < 20$ says that $5 + 3m$ is *between* -2 and 20. Solve this inequality with an extension of the properties given above. Work as follows, first adding -5 to each part.

$$-2 + (-5) < 5 + 3m + (-5) < 20 + (-5)$$
$$-7 < 3m < 15$$

Now multiply each part by $1/3$.

$$-\frac{7}{3} < m < 5$$

A graph of the solution is given in Figure 2; here open circles are used to show that $-7/3$ and 5 are *not* part of the graph.*

FIGURE 2

Quadratic Inequalities

A **quadratic inequality** has the form $ax^2 + bx + c > 0$ (or $<$, or $\leq$, or $\geq$). The greatest exponent is 2. The next few examples show how to solve quadratic inequalities.

EXAMPLE 3   Solving a Quadratic Inequality

Solve the quadratic inequality $x^2 - x < 12$.

SOLUTION   Write the inequality with 0 on one side, as $x^2 - x - 12 < 0$. This inequality is solved with values of x that make $x^2 - x - 12$ negative (< 0). The quantity $x^2 - x - 12$ changes from positive to negative or from negative to positive at the points where it equals 0. For this reason, first solve the *equation* $x^2 - x - 12 = 0$.

$$x^2 - x - 12 = 0$$
$$(x - 4)(x + 3) = 0$$
$$x = 4 \quad \text{or} \quad x = -3$$

FIGURE 3

Locating -3 and 4 on a number line, as shown in Figure 3, determines three intervals A, B, and C. Decide which intervals include numbers that make $x^2 - x - 12$ negative by substituting any number from each interval into the polynomial. For example,

choose -4 from interval A: $(-4)^2 - (-4) - 12 = 8 > 0$;

choose 0 from interval B: $0^2 - 0 - 12 = -12 < 0$;

choose 5 from interval C: $5^2 - 5 - 12 = 8 > 0$.

FIGURE 4

YOUR TURN 2 Solve $3y^2 \leq 16y + 12$.

Only numbers in interval B satisfy the given inequality, so the solution is $(-3, 4)$. A graph of this solution is shown in Figure 4.   **TRY YOUR TURN 2**

*Some textbooks use brackets in place of solid circles for the graph of a closed interval, and parentheses in place of open circles for the graph of an open interval.

EXAMPLE 4   Solving a Polynomial Inequality

Solve the inequality $x^3 + 2x^2 - 3x \geq 0$.

SOLUTION   This is not a quadratic inequality because of the x^3 term, but we solve it in a similar way by first factoring the polynomial.

$$x^3 + 2x^2 - 3x = x(x^2 + 2x - 3) \qquad \text{Factor out the common factor.}$$
$$= x(x - 1)(x + 3) \qquad \text{Factor the quadratic.}$$

Now solve the corresponding equation.

$$x(x - 1)(x + 3) = 0$$

$$x = 0 \qquad \text{or} \qquad x - 1 = 0 \qquad \text{or} \qquad x + 3 = 0$$
$$x = 1 \qquad\qquad\qquad x = -3$$

FIGURE 5

These three solutions determine four intervals on the number line: $(-\infty, -3)$, $(-3, 0)$, $(0, 1)$, and $(1, \infty)$. Substitute a number from each interval into the original inequality to determine that the solution consists of the numbers between -3 and 0 (including the endpoints) and all numbers that are greater than or equal to 1. See Figure 5. In interval notation, the solution is

$$[-3, 0] \cup [1, \infty).*$$

Inequalities with Fractions   Inequalities with fractions are solved in a similar manner as quadratic inequalities.

EXAMPLE 5   Solving a Rational Inequality

Solve $\dfrac{2x - 3}{x} \geq 1$.

SOLUTION   First solve the corresponding equation.

$$\frac{2x - 3}{x} = 1$$

$$2x - 3 = x \qquad \text{Multiply both sides by } x.$$
$$x = 3 \qquad \text{Solve for } x.$$

The solution, $x = 3$, determines the intervals on the number line where the fraction may change from greater than 1 to less than 1. This change also may occur on either side of a number that makes the denominator equal 0. Here, the x-value that makes the denominator 0 is $x = 0$. Test each of the three intervals determined by the numbers 0 and 3.

$$\text{For } (-\infty, 0), \text{ choose } -1: \frac{2(-1) - 3}{-1} = 5 \geq 1.$$

$$\text{For } (0, 3), \quad \text{choose} \quad 1: \frac{2(1) - 3}{1} = -1 \ngeq 1.$$

$$\text{For } (3, \infty), \quad \text{choose} \quad 4: \frac{2(4) - 3}{4} = \frac{5}{4} \geq 1.$$

FIGURE 6

The symbol $\ngeq$ means "is *not* greater than or equal to." Testing the endpoints 0 and 3 shows that the solution is $(-\infty, 0) \cup [3, \infty)$, as shown in Figure 6.

> **CAUTION**   A common error is to try to solve the inequality in Example 5 by multiplying both sides by x. The reason this is wrong is that we don't know in the beginning whether x is positive, negative, or 0. If x is negative, the $\geq$ would change to $\leq$ according to the third property of inequality listed at the beginning of this section.

*The symbol $\cup$ indicates the *union* of two sets, which includes all elements in either set.

EXAMPLE 6   Solving a Rational Inequality

Solve $\dfrac{(x-1)(x+1)}{x} \leq 0$.

SOLUTION   We first solve the corresponding equation.

$$\frac{(x-1)(x+1)}{x} = 0$$

$$(x-1)(x+1) = 0 \qquad \text{Multiply both sides by } x.$$

$$x = 1 \quad \text{or} \quad x = -1 \qquad \text{Use the zero-factor property.}$$

Setting the denominator equal to 0 gives $x = 0$, so the intervals of interest are $(-\infty, -1)$, $(-1, 0)$, $(0, 1)$, and $(1, \infty)$. Testing a number from each region in the original inequality and checking the endpoints, we find the solution is

$$(-\infty, -1] \cup (0, 1],$$

FIGURE 7

as shown in Figure 7.

CAUTION   Remember to solve the equation formed by setting the *denominator* equal to zero. Any number that makes the denominator zero always creates two intervals on the number line. For instance, in Example 6, substituting $x = 0$ makes the denominator of the rational inequality equal to 0, so we know that there may be a sign change from one side of 0 to the other (as was indeed the case).

EXAMPLE 7   Solving a Rational Inequality

Solve $\dfrac{x^2 - 3x}{x^2 - 9} < 4$.

SOLUTION   Solve the corresponding equation.

$$\frac{x^2 - 3x}{x^2 - 9} = 4$$

$$x^2 - 3x = 4x^2 - 36 \qquad \text{Multiply by } x^2 - 9.$$

$$0 = 3x^2 + 3x - 36 \qquad \text{Get 0 on one side.}$$

$$0 = x^2 + x - 12 \qquad \text{Multiply by } \tfrac{1}{3}.$$

$$0 = (x + 4)(x - 3) \qquad \text{Factor.}$$

$$x = -4 \quad \text{or} \quad x = 3$$

Now set the denominator equal to 0 and solve that equation.

$$x^2 - 9 = 0$$

$$(x - 3)(x + 3) = 0$$

$$x = 3 \quad \text{or} \quad x = -3$$

FIGURE 8

The intervals determined by the three (different) solutions are $(-\infty, -4)$, $(-4, -3)$, $(-3, 3)$, and $(3, \infty)$. Testing a number from each interval in the given inequality shows that the solution is

$$(-\infty, -4) \cup (-3, 3) \cup (3, \infty),$$

YOUR TURN 3 Solve
$\dfrac{k^2 - 35}{k} \geq 2$.

as shown in Figure 8. For this example, none of the endpoints are part of the solution because $x = 3$ and $x = -3$ make the denominator zero and $x = -4$ produces an equality.

TRY YOUR TURN 3

R.5 EXERCISES

Write each expression in interval notation. Graph each interval.

1. $x < 4$

2. $x \geq -3$

3. $1 \leq x < 2$

4. $-2 \leq x \leq 3$

5. $-9 > x$

6. $6 \leq x$

Using the variable x, write each interval as an inequality.

7. $[-7, -3]$

8. $[4, 10)$

9. $(-\infty, -1]$

10. $(3, \infty)$

11.

12.

13.

14.

Solve each inequality and graph the solution.

15. $6p + 7 \leq 19$

16. $6k - 4 < 3k - 1$

17. $m - (3m - 2) + 6 < 7m - 19$

18. $-2(3y - 8) \geq 5(4y - 2)$

19. $3p - 1 < 6p + 2(p - 1)$

20. $x + 5(x + 1) > 4(2 - x) + x$

21. $-11 < y - 7 < -1$

22. $8 \leq 3r + 1 \leq 13$

23. $-2 < \dfrac{1 - 3k}{4} \leq 4$

24. $-1 \leq \dfrac{5y + 2}{3} \leq 4$

25. $\dfrac{3}{5}(2p + 3) \geq \dfrac{1}{10}(5p + 1)$

26. $\dfrac{8}{3}(z - 4) \leq \dfrac{2}{9}(3z + 2)$

Solve each inequality. Graph each solution.

27. $(m - 3)(m + 5) < 0$

28. $(t + 6)(t - 1) \geq 0$

29. $y^2 - 3y + 2 < 0$

30. $2k^2 + 7k - 4 > 0$

31. $x^2 - 16 > 0$

32. $2k^2 - 7k - 15 \leq 0$

33. $x^2 - 4x \geq 5$

34. $10r^2 + r \leq 2$

35. $3x^2 + 2x > 1$

36. $3a^2 + a > 10$

37. $9 - x^2 \leq 0$

38. $p^2 - 16p > 0$

39. $x^3 - 4x \geq 0$

40. $x^3 + 7x^2 + 12x \leq 0$

41. $2x^3 - 14x^2 + 12x < 0$

42. $3x^3 - 9x^2 - 12x > 0$

Solve each inequality.

43. $\dfrac{m - 3}{m + 5} \leq 0$

44. $\dfrac{r + 1}{r - 1} > 0$

45. $\dfrac{k - 1}{k + 2} > 1$

46. $\dfrac{a - 5}{a + 2} < -1$

47. $\dfrac{2y + 3}{y - 5} \leq 1$

48. $\dfrac{a + 2}{3 + 2a} \leq 5$

49. $\dfrac{2k}{k - 3} \leq \dfrac{4}{k - 3}$

50. $\dfrac{5}{p + 1} > \dfrac{12}{p + 1}$

51. $\dfrac{2x}{x^2 - x - 6} \geq 0$

52. $\dfrac{8}{p^2 + 2p} > 1$

53. $\dfrac{z^2 + z}{z^2 - 1} \geq 3$

54. $\dfrac{a^2 + 2a}{a^2 - 4} \leq 2$

YOUR TURN ANSWERS

1. $z < -9/2$ **2.** $[-2/3, 6]$ **3.** $[-5, 0) \cup [7, \infty)$

R.6 Exponents

Integer Exponents

Recall that $a^2 = a \cdot a$, while $a^3 = a \cdot a \cdot a$, and so on. In this section, a more general meaning is given to the symbol a^n.

Definition of Exponent

If n is a natural number, then

$$a^n = a \cdot a \cdot a \cdot \cdots \cdot a,$$

where a appears as a factor n times.

In the expression a^n, the power n is the **exponent** and a is the **base**. This definition can be extended by defining a^n for zero and negative integer values of n.

Zero and Negative Exponents

If a is any nonzero real number, and if n is a positive integer, then

$$a^0 = 1 \quad \text{and} \quad a^{-n} = \frac{1}{a^n}.$$

(The symbol 0^0 is meaningless.)

YOUR TURN 1 Find $\left(\frac{2}{3}\right)^{-3}$.

EXAMPLE 1 Exponents

(a) $6^0 = 1$

(b) $(-9)^0 = 1$

(c) $3^{-2} = \frac{1}{3^2} = \frac{1}{9}$

(d) $9^{-1} = \frac{1}{9^1} = \frac{1}{9}$

(e) $\left(\frac{3}{4}\right)^{-1} = \frac{1}{(3/4)^1} = \frac{1}{3/4} = \frac{4}{3}$

TRY YOUR TURN 1

The following properties follow from the definitions of exponents given above.

Properties of Exponents

For any integers m and n, and any real numbers a and b for which the following exist:

1. $a^m \cdot a^n = a^{m+n}$

4. $(ab)^m = a^m \cdot b^m$

2. $\dfrac{a^m}{a^n} = a^{m-n}$

5. $\left(\dfrac{a}{b}\right)^m = \dfrac{a^m}{b^m}$

3. $(a^m)^n = a^{mn}$

Note that $(-a)^n = a^n$ if n is an even integer, but $(-a)^n = -a^n$ if n is an odd integer.

EXAMPLE 2 Simplifying Exponential Expressions

Use the properties of exponents to simplify each expression. Leave answers with positive exponents. Assume that all variables represent positive real numbers.

(a) $7^4 \cdot 7^6 = 7^{4+6} = 7^{10}$ (or 282,475,249)   Property 1

(b) $\dfrac{9^{14}}{9^6} = 9^{14-6} = 9^8$ (or 43,046,721)   Property 2

(c) $\dfrac{r^9}{r^{17}} = r^{9-17} = r^{-8} = \dfrac{1}{r^8}$ Property 2

(d) $(2m^3)^4 = 2^4 \cdot (m^3)^4 = 16m^{12}$ Properties 3 and 4

(e) $(3x)^4 = 3^4 \cdot x^4 = 81x^4$ Property 4

(f) $\left(\dfrac{x^2}{y^3}\right)^6 = \dfrac{(x^2)^6}{(y^3)^6} = \dfrac{x^{2\cdot6}}{y^{3\cdot6}} = \dfrac{x^{12}}{y^{18}}$ Properties 3 and 5

(g) $\dfrac{a^{-3}b^5}{a^4b^{-7}} = \dfrac{b^{5-(-7)}}{a^{4-(-3)}} = \dfrac{b^{5+7}}{a^{4+3}} = \dfrac{b^{12}}{a^7}$ Property 2

(h) $p^{-1} + q^{-1} = \dfrac{1}{p} + \dfrac{1}{q}$ *Definition of a^{-n}.*

$= \dfrac{1}{p} \cdot \dfrac{q}{q} + \dfrac{1}{q} \cdot \dfrac{p}{p}$ *Get common denominator.*

$= \dfrac{q}{pq} + \dfrac{p}{pq} = \dfrac{p + q}{pq}$ *Add.*

(i) $\dfrac{x^{-2} - y^{-2}}{x^{-1} - y^{-1}} = \dfrac{\dfrac{1}{x^2} - \dfrac{1}{y^2}}{\dfrac{1}{x} - \dfrac{1}{y}}$ *Definition of a^{-n}*

$= \dfrac{\dfrac{y^2 - x^2}{x^2 y^2}}{\dfrac{y - x}{xy}}$ *Get common denominators and combine terms.*

$= \dfrac{y^2 - x^2}{x^2 y^2} \cdot \dfrac{xy}{y - x}$ *Invert and multiply.*

$= \dfrac{(y - x)(y + x)}{x^2 y^2} \cdot \dfrac{xy}{y - x}$ *Factor.*

$= \dfrac{x + y}{xy}$ *Simplify.*   **TRY YOUR TURN 2**

YOUR TURN 2 Simplify $\left(\dfrac{y^2 z^{-4}}{y^{-3} z^4}\right)^{-2}$.

| **CAUTION** | If Example 2(e) were written $3x^4$, the properties of exponents would not apply. When no parentheses are used, the exponent refers only to the factor closest to it. Also notice in Examples 2(c), 2(g), 2(h), and 2(i) that a negative exponent does *not* indicate a negative number. |

Roots For *even* values of n and nonnegative values of a, the expression $a^{1/n}$ is defined to be the **positive nth root** of a or the **principal nth root** of a. For example, $a^{1/2}$ denotes the positive second root, or **square root**, of a, while $a^{1/4}$ is the positive fourth root of a. When n is *odd*, there is only one nth root, which has the same sign as a. For example, $a^{1/3}$, the **cube root** of a, has the same sign as a. By definition, if $b = a^{1/n}$, then $b^n = a$. On a calculator, a number is raised to a power using a key labeled x^y, y^x, or $\wedge$. For example, to take the fourth root of 6 on a TI-84 Plus C calculator, enter $6 \wedge (1/4)$ to get the result 1.56508458.

EXAMPLE 3   **Calculations with Exponents**

(a) $121^{1/2} = 11$, since 11 is positive and $11^2 = 121$.

(b) $625^{1/4} = 5$, since $5^4 = 625$.

(c) $256^{1/4} = 4$ **(d)** $64^{1/6} = 2$

(e) $27^{1/3} = 3$ **(f)** $(-32)^{1/5} = -2$

(g) $128^{1/7} = 2$ **(h)** $(-49)^{1/2}$ is not a real number.

                                  TRY YOUR TURN 3

YOUR TURN 3 Find $125^{1/3}$.

Rational Exponents In the following definition, the domain of an exponent is extended to include all rational numbers.

> **Definition of** $a^{m/n}$
>
> For all real numbers a for which the indicated roots exist, and for any rational number m/n,
>
> $$a^{m/n} = (a^{1/n})^m.$$

YOUR TURN 4 Find $16^{-3/4}$.

EXAMPLE 4   Calculations with Exponents

(a) $27^{2/3} = (27^{1/3})^2 = 3^2 = 9$ **(b)** $32^{2/5} = (32^{1/5})^2 = 2^2 = 4$

(c) $64^{4/3} = (64^{1/3})^4 = 4^4 = 256$ **(d)** $25^{3/2} = (25^{1/2})^3 = 5^3 = 125$

TRY YOUR TURN 4

NOTE $27^{2/3}$ could also be evaluated as $(27^2)^{1/3}$, but this is more difficult to perform without a calculator because it involves squaring 27 and then taking the cube root of this large number. On the other hand, when we evaluate it as $(27^{1/3})^2$, we know that the cube root of 27 is 3 without using a calculator, and squaring 3 is easy.

All the properties for integer exponents given in this section also apply to any rational exponent on a nonnegative real-number base.

EXAMPLE 5   Simplifying Exponential Expressions

(a) $\dfrac{y^{1/3}y^{5/3}}{y^3} = \dfrac{y^{1/3+5/3}}{y^3} = \dfrac{y^2}{y^3} = y^{2-3} = y^{-1} = \dfrac{1}{y}$

YOUR TURN 5 Simplify $\left(\dfrac{x^{1/2}x^4}{x^{3/2}}\right)^{1/3}$.

(b) $m^{2/3}(m^{7/3} + 2m^{1/3}) = m^{2/3+7/3} + 2m^{2/3+1/3} = m^3 + 2m$

(c) $\left(\dfrac{m^7 n^{-2}}{m^{-5} n^2}\right)^{1/4} = \left(\dfrac{m^{7-(-5)}}{n^{2-(-2)}}\right)^{1/4} = \left(\dfrac{m^{12}}{n^4}\right)^{1/4} = \dfrac{(m^{12})^{1/4}}{(n^4)^{1/4}} = \dfrac{m^{12/4}}{n^{4/4}} = \dfrac{m^3}{n}$

TRY YOUR TURN 5

In calculus, it is often necessary to factor expressions involving fractional exponents.

EXAMPLE 6   Simplifying Exponential Expressions

Factor out the smallest power of the variable, assuming all variables represent positive real numbers.

(a) $4m^{1/2} + 3m^{3/2}$

SOLUTION   The smallest exponent is 1/2. Factoring out $m^{1/2}$ yields

$$4m^{1/2} + 3m^{3/2} = m^{1/2}(4m^{1/2-1/2} + 3m^{3/2-1/2})$$
$$= m^{1/2}(4 + 3m).$$

Check this result by multiplying $m^{1/2}$ by $4 + 3m$.

(b) $9x^{-2} - 6x^{-3}$

SOLUTION   The smallest exponent here is -3. Since 3 is a common numerical factor, factor out $3x^{-3}$.

$$9x^{-2} - 6x^{-3} = 3x^{-3}(3x^{-2-(-3)} - 2x^{-3-(-3)}) = 3x^{-3}(3x - 2)$$

Check by multiplying. The factored form can be written without negative exponents as

$$\frac{3(3x - 2)}{x^3}.$$

(c) $(x^2 + 5)(3x - 1)^{-1/2}(2) + (3x - 1)^{1/2}(2x)$

SOLUTION   There is a common factor of 2. Also, $(3x - 1)^{-1/2}$ and $(3x - 1)^{1/2}$ have a common factor. Always factor out the quantity to the *smallest* exponent. Here $-1/2 < 1/2$, so the common factor is $2(3x - 1)^{-1/2}$ and the factored form is

$$2(3x - 1)^{-1/2}[(x^2 + 5) + (3x - 1)x] = 2(3x - 1)^{-1/2}(4x^2 - x + 5).$$

YOUR TURN 6 Factor $5z^{1/3} + 4z^{-2/3}$.

TRY YOUR TURN 6

R.6 EXERCISES

Evaluate each expression. Write all answers without exponents.

1. 8^{-2}

2. 3^{-4}

3. 5^0

4. $\left(-\dfrac{3}{4}\right)^0$

5. $-(-3)^{-2}$

6. $-(-3^{-2})$

7. $\left(\dfrac{1}{6}\right)^{-2}$

8. $\left(\dfrac{4}{3}\right)^{-3}$

Simplify each expression. Assume that all variables represent positive real numbers. Write answers with only positive exponents.

9. $\dfrac{4^{-2}}{4}$

10. $\dfrac{8^9 \cdot 8^{-7}}{8^{-3}}$

11. $\dfrac{10^8 \cdot 10^{-10}}{10^4 \cdot 10^2}$

12. $\left(\dfrac{7^{-12} \cdot 7^3}{7^{-8}}\right)^{-1}$

13. $\dfrac{x^4 \cdot x^3}{x^5}$

14. $\dfrac{y^{10} \cdot y^{-4}}{y^6}$

15. $\dfrac{(4k^{-1})^2}{2k^{-5}}$

16. $\dfrac{(3z^2)^{-1}}{z^5}$

17. $\dfrac{3^{-1} \cdot x \cdot y^2}{x^{-4} \cdot y^5}$

18. $\dfrac{5^{-2}m^2y^{-2}}{5^2m^{-1}y^{-2}}$

19. $\left(\dfrac{a^{-1}}{b^2}\right)^{-3}$

20. $\left(\dfrac{c^3}{7d^{-2}}\right)^{-2}$

Simplify each expression, writing the answer as a single term without negative exponents.

21. $a^{-1} + b^{-1}$

22. $b^{-2} - a$

23. $\dfrac{2n^{-1} - 2m^{-1}}{m + n^2}$

24. $\left(\dfrac{m}{3}\right)^{-1} + \left(\dfrac{n}{2}\right)^{-2}$

25. $(x^{-1} - y^{-1})^{-1}$

26. $(x \cdot y^{-1} - y^{-2})^{-2}$

Write each number without exponents.

27. $121^{1/2}$

28. $27^{1/3}$

29. $32^{2/5}$

30. $-125^{2/3}$

31. $\left(\dfrac{36}{144}\right)^{1/2}$

32. $\left(\dfrac{64}{27}\right)^{1/3}$

33. $8^{-4/3}$

34. $625^{-1/4}$

35. $\left(\dfrac{27}{64}\right)^{-1/3}$

36. $\left(\dfrac{121}{100}\right)^{-3/2}$

Simplify each expression. Write all answers with only positive exponents. Assume that all variables represent positive real numbers.

37. $3^{2/3} \cdot 3^{4/3}$

38. $27^{2/3} \cdot 27^{-1/3}$

39. $\dfrac{4^{9/4} \cdot 4^{-7/4}}{4^{-10/4}}$

40. $\dfrac{3^{-5/2} \cdot 3^{3/2}}{3^{7/2} \cdot 3^{-9/2}}$

41. $\left(\dfrac{x^6 y^{-3}}{x^{-2} y^5}\right)^{1/2}$

42. $\left(\dfrac{a^{-7} b^{-1}}{b^{-4} a^2}\right)^{1/3}$

43. $\dfrac{7^{-1/3} \cdot 7r^{-3}}{7^{2/3} \cdot (r^{-2})^2}$

44. $\dfrac{12^{3/4} \cdot 12^{5/4} \cdot y^{-2}}{12^{-1} \cdot (y^{-3})^{-2}}$

45. $\dfrac{3k^2 \cdot (4k^{-3})^{-1}}{4^{1/2} \cdot k^{7/2}}$

46. $\dfrac{8p^{-3} \cdot (4p^2)^{-2}}{p^{-5}}$

47. $\dfrac{a^{4/3} \cdot b^{1/2}}{a^{2/3} \cdot b^{-3/2}}$

48. $\dfrac{x^{3/2} \cdot y^{4/5} \cdot z^{-3/4}}{x^{5/3} \cdot y^{-6/5} \cdot z^{1/2}}$

49. $\dfrac{k^{-3/5} \cdot h^{-1/3} \cdot t^{2/5}}{k^{-1/5} \cdot h^{-2/3} \cdot t^{1/5}}$

50. $\dfrac{m^{7/3} \cdot n^{-2/5} \cdot p^{3/8}}{m^{-2/3} \cdot n^{3/5} \cdot p^{-5/8}}$

Factor each expression.

51. $3x^3(x^2 + 3x)^2 - 15x(x^2 + 3x)^2$

52. $6x(x^3 + 7)^2 - 6x^2(3x^2 + 5)(x^3 + 7)$

53. $10x^3(x^2 - 1)^{-1/2} - 5x(x^2 - 1)^{1/2}$

54. $9(6x + 2)^{1/2} + 3(9x - 1)(6x + 2)^{-1/2}$

55. $x(2x + 5)^2(x^2 - 4)^{-1/2} + 2(x^2 - 4)^{1/2}(2x + 5)$

56. $(4x^2 + 1)^2(2x - 1)^{-1/2} + 16x(4x^2 + 1)(2x - 1)^{1/2}$

YOUR TURN ANSWERS

1. $27/8$

2. z^{16}/y^{10}

3. 5

4. $1/8$

5. x

6. $z^{-2/3}(5z + 4)$

R.7 Radicals

We have defined $a^{1/n}$ as the positive or principal nth root of a for appropriate values of a and n. An alternative notation for $a^{1/n}$ uses radicals.

Radicals

If n is an even natural number and $a > 0$, or n is an odd natural number, then

$$a^{1/n} = \sqrt[n]{a}.$$

The symbol $\sqrt[n]{}$ is a **radical sign**, the number a is the **radicand**, and n is the **index** of the radical. The familiar symbol $\sqrt{a}$ is used instead of $\sqrt[2]{a}$.

EXAMPLE 1   Radical Calculations

(a) $\sqrt[4]{16} = 16^{1/4} = 2$

(b) $\sqrt[5]{-32} = -2$

(c) $\sqrt[3]{1000} = 10$

(d) $\sqrt[6]{\dfrac{64}{729}} = \dfrac{2}{3}$

With $a^{1/n}$ written as $\sqrt[n]{a}$, the expression $a^{m/n}$ also can be written using radicals.

$$a^{m/n} = (\sqrt[n]{a})^m \qquad \text{or} \qquad a^{m/n} = \sqrt[n]{a^m}$$

The following properties of radicals depend on the definitions and properties of exponents.

Properties of Radicals

For all real numbers a and b and natural numbers m and n such that $\sqrt[n]{a}$ and $\sqrt[n]{b}$ are real numbers:

1. $(\sqrt[n]{a})^n = a$

2. $\sqrt[n]{a^n} = \begin{cases} |a| & \text{if } n \text{ is even} \\ a & \text{if } n \text{ is odd} \end{cases}$

3. $\sqrt[n]{a} \cdot \sqrt[n]{b} = \sqrt[n]{ab}$

4. $\dfrac{\sqrt[n]{a}}{\sqrt[n]{b}} = \sqrt[n]{\dfrac{a}{b}} \quad (b \neq 0)$

5. $\sqrt[m]{\sqrt[n]{a}} = \sqrt[mn]{a}$

Property 3 can be used to simplify certain radicals. For example, since $48 = 16 \cdot 3$,

$$\sqrt{48} = \sqrt{16 \cdot 3} = \sqrt{16} \cdot \sqrt{3} = 4\sqrt{3}.$$

To some extent, simplification is in the eye of the beholder, and $\sqrt{48}$ might be considered as simple as $4\sqrt{3}$. In this textbook, we will consider an expression to be simpler when we have removed as many factors as possible from under the radical.

EXAMPLE 2 Radical Calculations

(a) $\sqrt{1000} = \sqrt{100 \cdot 10} = \sqrt{100} \cdot \sqrt{10} = 10\sqrt{10}$

(b) $\sqrt{128} = \sqrt{64 \cdot 2} = 8\sqrt{2}$

(c) $\sqrt{2} \cdot \sqrt{18} = \sqrt{2 \cdot 18} = \sqrt{36} = 6$

(d) $\sqrt[3]{54} = \sqrt[3]{27 \cdot 2} = \sqrt[3]{27} \cdot \sqrt[3]{2} = 3\sqrt[3]{2}$

(e) $\sqrt{288m^5} = \sqrt{144 \cdot m^4 \cdot 2m} = 12m^2\sqrt{2m}$

(f) $2\sqrt{18} - 5\sqrt{32} = 2\sqrt{9 \cdot 2} - 5\sqrt{16 \cdot 2}$

$$= 2\sqrt{9} \cdot \sqrt{2} - 5\sqrt{16} \cdot \sqrt{2}$$

$$= 2(3)\sqrt{2} - 5(4)\sqrt{2} = -14\sqrt{2}$$

YOUR TURN 1 Simplify $\sqrt{28x^9y^5}$.

(g) $\sqrt{x^5} \cdot \sqrt[3]{x^5} = x^{5/2} \cdot x^{5/3} = x^{5/2+5/3} = x^{25/6} = \sqrt[6]{x^{25}} = x^4\sqrt[6]{x}$

TRY YOUR TURN 1

When simplifying a square root, keep in mind that $\sqrt{x}$ is nonnegative by definition. Also, $\sqrt{x^2}$ is not x, but $|x|$, the **absolute value of x**, defined as

$$|x| = \begin{cases} x & \text{if } x \geq 0 \\ -x & \text{if } x < 0. \end{cases}$$

For example, $\sqrt{(-5)^2} = |-5| = 5$. It is correct, however, to simplify $\sqrt{x^4} = x^2$. We need not write $|x^2|$ because x^2 is always nonnegative.

EXAMPLE 3 Simplifying by Factoring

Simplify $\sqrt{m^2 - 4m + 4}$.

SOLUTION Factor the polynomial as $m^2 - 4m + 4 = (m - 2)^2$. Then by property 2 of radicals and the definition of absolute value,

$$\sqrt{(m-2)^2} = |m - 2| = \begin{cases} m - 2 & \text{if } m - 2 \geq 0 \\ -(m-2) = 2 - m & \text{if } m - 2 < 0. \end{cases}$$

CAUTION Avoid the common error of writing $\sqrt{a^2 + b^2}$ as $\sqrt{a^2} + \sqrt{b^2}$. We must add a^2 and b^2 *before* taking the square root. For example, $\sqrt{16 + 9} = \sqrt{25} = 5$, *not* $\sqrt{16} + \sqrt{9} = 4 + 3 = 7$. This idea applies as well to higher roots. For example, in general,

$$\sqrt[3]{a^3 + b^3} \neq \sqrt[3]{a^3} + \sqrt[3]{b^3},$$

$$\sqrt[4]{a^4 + b^4} \neq \sqrt[4]{a^4} + \sqrt[4]{b^4}.$$

Also, $\qquad \sqrt{a + b} \neq \sqrt{a} + \sqrt{b}.$

Rationalizing Denominators

The next example shows how to *rationalize* (remove all radicals from) the denominator in an expression containing radicals.

EXAMPLE 4   Rationalizing Denominators

Simplify each expression by rationalizing the denominator.

(a) $\dfrac{4}{\sqrt{3}}$

SOLUTION   To rationalize the denominator, multiply by $\sqrt{3}/\sqrt{3}$ (or 1) so the denominator of the product is a rational number.

$$\frac{4}{\sqrt{3}} \cdot \frac{\sqrt{3}}{\sqrt{3}} = \frac{4\sqrt{3}}{3} \qquad \sqrt{3} \cdot \sqrt{3} = \sqrt{9} = 3$$

(b) $\dfrac{2}{\sqrt[3]{x}}$

SOLUTION   Here, we need a perfect cube under the radical sign to rationalize the denominator. Multiplying by $\sqrt[3]{x^2}/\sqrt[3]{x^2}$ gives

$$\frac{2}{\sqrt[3]{x}} \cdot \frac{\sqrt[3]{x^2}}{\sqrt[3]{x^2}} = \frac{2\sqrt[3]{x^2}}{\sqrt[3]{x^3}} = \frac{2\sqrt[3]{x^2}}{x}.$$

(c) $\dfrac{1}{1 - \sqrt{2}}$

SOLUTION   The best approach here is to multiply both numerator and denominator by the number $1 + \sqrt{2}$. The expressions $1 + \sqrt{2}$ and $1 - \sqrt{2}$ are conjugates,* and their product is $1^2 - (\sqrt{2})^2 = 1 - 2 = -1$. Thus,

$$\frac{1}{1 - \sqrt{2}} = \frac{1(1 + \sqrt{2})}{(1 - \sqrt{2})(1 + \sqrt{2})} = \frac{1 + \sqrt{2}}{1 - 2} = -1 - \sqrt{2}.$$

TRY YOUR TURN 2

YOUR TURN 2 Rationalize the denominator in
$$\frac{5}{\sqrt{x} - \sqrt{y}}.$$

Sometimes it is advantageous to rationalize the *numerator* of a rational expression. The following example arises in calculus when evaluating a *limit*.

EXAMPLE 5   Rationalizing Numerators

Rationalize each numerator.

(a) $\dfrac{\sqrt{x} - 3}{x - 9}.$

SOLUTION   Multiply the numerator and denominator by the conjugate of the numerator, $\sqrt{x} + 3$.

$$\frac{\sqrt{x} - 3}{x - 9} \cdot \frac{\sqrt{x} + 3}{\sqrt{x} + 3} = \frac{(\sqrt{x})^2 - 3^2}{(x - 9)(\sqrt{x} + 3)} \qquad (a - b)(a + b) = a^2 - b^2$$

$$= \frac{x - 9}{(x - 9)(\sqrt{x} + 3)}$$

$$= \frac{1}{\sqrt{x} + 3}$$

*If a and b are real numbers, the *conjugate* of $a + b$ is $a - b$.

(b) $\dfrac{\sqrt{3} + \sqrt{x + 3}}{\sqrt{3} - \sqrt{x + 3}}$

SOLUTION   Multiply the numerator and denominator by the conjugate of the numerator, $\sqrt{3} - \sqrt{x + 3}$.

$$\frac{\sqrt{3} + \sqrt{x + 3}}{\sqrt{3} - \sqrt{x + 3}} \cdot \frac{\sqrt{3} - \sqrt{x + 3}}{\sqrt{3} - \sqrt{x + 3}} = \frac{3 - (x + 3)}{3 - 2\sqrt{3}\sqrt{x + 3} + (x + 3)}$$

$$= \frac{-x}{6 + x - 2\sqrt{3(x + 3)}}$$

YOUR TURN 3 Rationalize the numerator in $\dfrac{4 + \sqrt{x}}{16 - x}$.

TRY YOUR TURN 3

R.7   EXERCISES

Simplify each expression by removing as many factors as possible from under the radical. Assume that all variables represent positive real numbers.

1. $\sqrt[3]{125}$

2. $\sqrt[4]{1296}$

3. $\sqrt[5]{-3125}$

4. $\sqrt{50}$

5. $\sqrt{2000}$

6. $\sqrt{32y^5}$

7. $\sqrt{27} \cdot \sqrt{3}$

8. $\sqrt{2} \cdot \sqrt{32}$

9. $7\sqrt{2} - 8\sqrt{18} + 4\sqrt{72}$

10. $4\sqrt{3} - 5\sqrt{12} + 3\sqrt{75}$

11. $4\sqrt{7} - \sqrt{28} + \sqrt{343}$

12. $3\sqrt{28} - 4\sqrt{63} + \sqrt{112}$

13. $\sqrt[3]{2} - \sqrt[3]{16} + 2\sqrt[3]{54}$

14. $2\sqrt[3]{5} - 4\sqrt[3]{40} + 3\sqrt[3]{135}$

15. $\sqrt{2x^3y^2z^4}$

16. $\sqrt{160r^7s^9t^{12}}$

17. $\sqrt[3]{128x^3y^8z^9}$

18. $\sqrt[4]{x^8y^7z^{11}}$

19. $\sqrt{a^3b^5} - 2\sqrt{a^7b^3} + \sqrt{a^3b^9}$

20. $\sqrt{p^7q^3} - \sqrt{p^5q^9} + \sqrt{p^9q}$

21. $\sqrt{a} \cdot \sqrt[3]{a}$

22. $\sqrt{b^3} \cdot \sqrt[4]{b^3}$

Simplify each root, if possible.

23. $\sqrt{16 - 8x + x^2}$

24. $\sqrt{9y^2 + 30y + 25}$

25. $\sqrt{4 - 25z^2}$

26. $\sqrt{9k^2 + h^2}$

Rationalize each denominator. Assume that all radicands represent positive real numbers.

27. $\dfrac{5}{\sqrt{7}}$

28. $\dfrac{5}{\sqrt{10}}$

29. $\dfrac{-3}{\sqrt{12}}$

30. $\dfrac{4}{\sqrt{8}}$

31. $\dfrac{3}{1 - \sqrt{2}}$

32. $\dfrac{5}{2 - \sqrt{6}}$

33. $\dfrac{6}{2 + \sqrt{2}}$

34. $\dfrac{\sqrt{5}}{\sqrt{5} + \sqrt{2}}$

35. $\dfrac{1}{\sqrt{r} - \sqrt{3}}$

36. $\dfrac{5}{\sqrt{m} - \sqrt{5}}$

37. $\dfrac{y - 5}{\sqrt{y} - \sqrt{5}}$

38. $\dfrac{\sqrt{z} - 1}{\sqrt{z} - \sqrt{5}}$

39. $\dfrac{\sqrt{x} + \sqrt{x + 1}}{\sqrt{x} - \sqrt{x + 1}}$

40. $\dfrac{\sqrt{p} + \sqrt{p^2 - 1}}{\sqrt{p} - \sqrt{p^2 - 1}}$

Rationalize each numerator. Assume that all radicands represent positive real numbers.

41. $\dfrac{1 + \sqrt{2}}{2}$

42. $\dfrac{3 - \sqrt{3}}{6}$

43. $\dfrac{\sqrt{x} + \sqrt{x + 1}}{\sqrt{x} - \sqrt{x + 1}}$

44. $\dfrac{\sqrt{p} - \sqrt{p - 2}}{\sqrt{p}}$

YOUR TURN ANSWERS

1. $2x^4y^2\sqrt{7xy}$ **2.** $5(\sqrt{x} + \sqrt{y})/(x - y)$ **3.** $1/(4 - \sqrt{x})$

Linear Functions

Over short time intervals, many changes in the economy are well modeled by linear functions. In an exercise in the first section of this chapter, we will examine a linear model that predicts the number of cellular telephone users in the United States. Such predictions are important tools for cellular telephone company executives and planners.

Before using mathematics to solve a real-world problem, we must usually set up a **mathematical model**, a mathematical description of the situation. In this chapter we look at some mathematics of *linear* models, which are used for data whose graphs can be approximated by straight lines. Linear models have an immense number of applications, because even when the underlying phenomenon is not linear, a linear model often provides an approximation that is sufficiently accurate and much simpler to use.

1.1 Slopes and Equations of Lines

APPLY IT How fast has tuition at public colleges been increasing in recent years, and how well can we predict tuition in the future?

In Example 13 of this section, we will answer these questions using the equation of a line.

There are many everyday situations in which two quantities are related. For example, if a bank account pays 6% simple interest per year, then the interest I that a deposit of P dollars would earn in one year is given by

$$I = 0.06 \cdot P, \quad \text{or} \quad I = 0.06P.$$

The formula $I = 0.06P$ describes the relationship between interest and the amount of money deposited.

Using this formula, we see, for example, that if $P = \$100$, then $I = \$6$, and if $P = \$200$, then $I = \$12$. These corresponding pairs of numbers can be written as **ordered pairs**, $(100, 6)$ and $(200, 12)$, whose order is important. The first number denotes the value of P and the second number the value of I.

Ordered pairs are graphed with the perpendicular number lines of a **Cartesian coordinate system**, shown in Figure 1.* The horizontal number line, or ***x*-axis**, represents the first components of the ordered pairs, while the vertical or ***y*-axis** represents the second components. The point where the number lines cross is the zero point on both lines; this point is called the **origin**.

Each point on the *xy*-plane corresponds to an ordered pair of numbers, where the *x*-value is written first. From now on, we will refer to the point corresponding to the ordered pair (x, y) as "the point (x, y)."

Locate the point $(-2, 4)$ on the coordinate system by starting at the origin and counting 2 units to the left on the horizontal axis and 4 units upward, parallel to the vertical axis. This point is shown in Figure 1, along with several other sample points. The number -2 is the ***x*-coordinate** and the number 4 is the ***y*-coordinate** of the point $(-2, 4)$.

The *x*-axis and *y*-axis divide the plane into four parts, or **quadrants**. For example, quadrant I includes all those points whose *x*- and *y*-coordinates are both positive. The quadrants are numbered as shown in Figure 1. The points on the axes themselves belong to no quadrant. The set of points corresponding to the ordered pairs of an equation is the **graph** of the equation.

The *x*- and *y*-values of the points where the graph of an equation crosses the axes are called the ***x*-intercept** and ***y*-intercept**, respectively.** See Figure 2.

---

*The name "Cartesian" honors René Descartes (1596–1650), one of the greatest mathematicians of the 17th century. According to legend, Descartes was lying in bed when he noticed an insect crawling on the ceiling and realized that if he could determine the distance from the bug to each of two perpendicular walls, he could describe its position at any given moment. The same idea can be used to locate a point in a plane.
**Some people prefer to define the intercepts as ordered pairs, rather than as numbers.

FIGURE 1

FIGURE 2

FIGURE 3

Slope of a Line

An important characteristic of a straight line is its *slope*, a number that represents the "steepness" of the line. To see how slope is defined, look at the line in Figure 3. The line passes through the points $(x_1, y_1) = (-3, 5)$ and $(x_2, y_2) = (2, -4)$. The difference in the two x-values,

$$x_2 - x_1 = 2 - (-3) = 5$$

in this example, is called the **change in x**. The symbol Δx (read "delta x") is used to represent the change in x. In the same way, Δy represents the **change in y**. In our example,

$$\Delta y = y_2 - y_1$$
$$= -4 - 5$$
$$= -9.$$

These symbols, Δx and Δy, are used in the following definition of slope.

Slope of a Nonvertical Line

The **slope** of a nonvertical line is defined as the vertical change (the "rise") over the horizontal change (the "run") as one travels along the line. In symbols, taking two different points (x_1, y_1) and (x_2, y_2) on the line, the slope is

$$m = \frac{\textbf{Change in } y}{\textbf{Change in } x} = \frac{\Delta y}{\Delta x} = \frac{y_2 - y_1}{x_2 - x_1},$$

where $x_1 \neq x_2$.

By this definition, the slope of the line in Figure 3 is

$$m = \frac{\Delta y}{\Delta x} = \frac{-4 - 5}{2 - (-3)} = -\frac{9}{5}.$$

The slope of a line tells how fast y changes for each unit of change in x.

NOTE Using similar triangles, it can be shown that the slope of a line is independent of the choice of points on the line. That is, the same slope will be obtained for *any* choice of two different points on the line.

EXAMPLE 1 Slope

Find the slope of the line through each pair of points.

(a) $(7, 6)$ and $(-4, 5)$

SOLUTION Let $(x_1, y_1) = (7, 6)$ and $(x_2, y_2) = (-4, 5)$. Use the definition of slope.

$$m = \frac{\Delta y}{\Delta x} = \frac{5 - 6}{-4 - 7} = \frac{-1}{-11} = \frac{1}{11}$$

(b) $(5, -3)$ and $(-2, -3)$

SOLUTION Let $(x_1, y_1) = (5, -3)$ and $(x_2, y_2) = (-2, -3)$. Then

$$m = \frac{-3 - (-3)}{-2 - 5} = \frac{0}{-7} = 0.$$

Lines with zero slope are horizontal (parallel to the x-axis).

(c) $(2, -4)$ and $(2, 3)$

SOLUTION Let $(x_1, y_1) = (2, -4)$ and $(x_2, y_2) = (2, 3)$. Then

$$m = \frac{3 - (-4)}{2 - 2} = \frac{7}{0},$$

YOUR TURN 1 Find the slope of the line through $(1, 5)$ and $(4, 6)$.

which is undefined. This happens when the line is vertical (parallel to the y-axis).

TRY YOUR TURN 1

CAUTION	The phrase "no slope" should be avoided; specify instead whether the slope is zero or undefined.

In finding the slope of the line in Example 1(a), we could have let $(x_1, y_1) = (-4, 5)$ and $(x_2, y_2) = (7, 6)$. In that case,

$$m = \frac{6 - 5}{7 - (-4)} = \frac{1}{11},$$

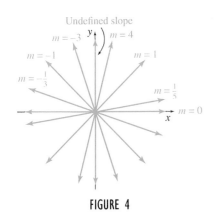

FIGURE 4

the same answer as before. The order in which coordinates are subtracted does not matter, as long as it is done consistently.

Figure 4 shows examples of lines with different slopes. Lines with positive slopes rise from left to right, while lines with negative slopes fall from left to right.

It might help you to compare slope with the percent grade of a hill. If a sign says a hill has a 10% grade uphill, this means the slope is 0.10, or 1/10, so the hill rises 1 foot for every 10 feet horizontally. A 15% grade downhill means the slope is -0.15.

Equations of a Line
An equation in two first-degree variables, such as $4x + 7y = 20$, has a line as its graph, so it is called a **linear equation**. In the rest of this section, we consider various forms of the equation of a line.

Suppose a line has a slope m and y-intercept b. This means that it passes through the point $(0, b)$. If (x, y) is any other point on the line, then the definition of slope tells us that

$$m = \frac{y - b}{x - 0}.$$

FOR REVIEW

For review on solving a linear equation, see Section R.4.

We can simplify this equation by multiplying both sides by x and adding b to both sides. The result is

$$y = mx + b,$$

which we call the *slope-intercept* form of a line. This is the most common form for writing the equation of a line.

Slope-Intercept Form

If a line has slope m and y-intercept b, then the equation of the line in **slope-intercept form** is

$$y = mx + b.$$

When $b = 0$, we say that y is **proportional** to x.

EXAMPLE 2   **Equation of a Line**

Find an equation in slope-intercept form for each line.

(a) Through $(0, -3)$ with slope $3/4$

SOLUTION We recognize $(0, -3)$ as the y-intercept because it's the point with 0 as its x-coordinate, so $b = -3$. The slope is $3/4$, so $m = 3/4$. Substituting these values into $y = mx + b$ gives

$$y = \frac{3}{4}x + (-3) = \frac{3}{4}x - 3.$$

(b) With x-intercept 7 and y-intercept 2

SOLUTION Notice that $b = 2$. To find m, use the definition of slope after writing the x-intercept as $(7, 0)$ (because the y-coordinate is 0 where the line crosses the x-axis) and the y-intercept as $(0, 2)$.

$$m = \frac{0 - 2}{7 - 0} = -\frac{2}{7}$$

YOUR TURN 2 Find the equation of the line with x-intercept -4 and y-intercept 6.

Substituting these values into $y = mx + b$, we have

$$y = -\frac{2}{7}x + 2.$$

TRY YOUR TURN 2

EXAMPLE 3   **Finding the Slope**

Find the slope of the line whose equation is $3x - 4y = 12$.

SOLUTION To find the slope, solve the equation for y.

$$3x - 4y = 12$$
$$-4y = -3x + 12 \qquad \text{Subtract } 3x \text{ from both sides.}$$
$$y = \frac{3}{4}x - 3 \qquad \text{Divide both sides by } -4.$$

YOUR TURN 3 Find the slope of the line whose equation is $8x + 3y = 5$.

The coefficient of x is $3/4$, which is the slope of the line. Notice that this is the same line as in Example 2(a).

TRY YOUR TURN 3

The slope-intercept form of the equation of a line involves the slope and the y-intercept. Sometimes, however, the slope of a line is known, together with one point (perhaps *not* the y-intercept) that the line passes through. The *point-slope form* of the equation of a line is used to find the equation in this case. Let (x_1, y_1) be any fixed point on the line, and let (x, y) represent any other point on the line. If m is the slope of the line, then by the definition of slope,

$$\frac{y - y_1}{x - x_1} = m,$$

or

$$y - y_1 = m(x - x_1). \qquad \text{Multiply both sides by } x - x_1.$$

Point-Slope Form

If a line has slope m and passes through the point (x_1, y_1), then an equation of the line is given by

$$y - y_1 = m(x - x_1),$$

the **point-slope form** of the equation of a line.

EXAMPLE 4 Point-Slope Form

Find an equation of the line that passes through the point $(3, -7)$ and has slope $m = 5/4$.

SOLUTION Use the point-slope form.

$$y - y_1 = m(x - x_1)$$

$$y - (-7) = \frac{5}{4}(x - 3) \qquad y_1 = -7, m = \tfrac{5}{4}, x_1 = 3$$

$$y + 7 = \frac{5}{4}(x - 3)$$

$$4y + 28 = 5(x - 3) \qquad \text{Multiply both sides by 4.}$$

$$4y + 28 = 5x - 15 \qquad \text{Distribute.}$$

$$4y = 5x - 43 \qquad \text{Subtract 28 from both sides.}$$

$$y = \frac{5}{4}x - \frac{43}{4} \qquad \text{Divide both sides by 4.} \qquad \blacksquare$$

FOR REVIEW

See Section R.4 for details on eliminating denominators in an equation.

The point-slope form also can be useful to find an equation of a line if we know two different points that the line goes through, as in the next example.

EXAMPLE 5 Point-Slope Form with Two Points

Find an equation of the line through $(5, 4)$ and $(-10, -2)$.

SOLUTION Begin by using the definition of slope to find the slope of the line that passes through the given points.

$$\text{Slope} = m = \frac{-2 - 4}{-10 - 5} = \frac{-6}{-15} = \frac{2}{5}$$

Either $(5, 4)$ or $(-10, -2)$ can be used in the point-slope form with $m = 2/5$. If $(x_1, y_1) = (5, 4)$, then

$$y - y_1 = m(x - x_1)$$

$$y - 4 = \frac{2}{5}(x - 5) \qquad y_1 = 4, m = \tfrac{2}{5}, x_1 = 5$$

$$y - 4 = \frac{2}{5}x - 2 \qquad \text{Distributive property}$$

$$y = \frac{2}{5}x + 2. \qquad \text{Add 4 to both sides.}$$

YOUR TURN 4 Find the equation of the line through $(2, 9)$ and $(5, 3)$. Put your answer in slope-intercept form. $\blacksquare$

Check that the same result is found if $(x_1, y_1) = (-10, -2)$.   **TRY YOUR TURN 4** ▬

EXAMPLE 6   Horizontal Line

Find an equation of the line through $(8, -4)$ and $(-2, -4)$.

SOLUTION   Find the slope.

$$m = \frac{-4 - (-4)}{-2 - 8} = \frac{0}{-10} = 0$$

Choose, say, $(8, -4)$ as (x_1, y_1).

$$y - y_1 = m(x - x_1)$$
$$y - (-4) = 0(x - 8) \qquad y_1 = -4, m = 0, x_1 = 8$$
$$y + 4 = 0 \qquad 0(x - 8) = 0$$
$$y = -4$$

Plotting the given ordered pairs and drawing a line through the points show that the equation $y = -4$ represents a horizontal line. See Figure 5(a). Every horizontal line has a slope of zero and an equation of the form $y = k$, where k is the y-value of all ordered pairs on the line.

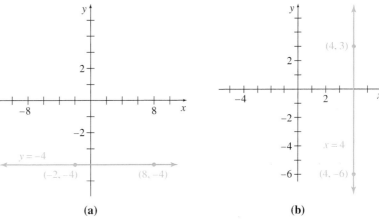

FIGURE 5

EXAMPLE 7   Vertical Line

Find an equation of the line through $(4, 3)$ and $(4, -6)$.

SOLUTION   The slope of the line is

$$m = \frac{-6 - 3}{4 - 4} = \frac{-9}{0},$$

which is undefined. Since both ordered pairs have x-coordinate 4, the equation is $x = 4$. Because the slope is undefined, the equation of this line cannot be written in the slope-intercept form.

Again, plotting the given ordered pairs and drawing a line through them show that the graph of $x = 4$ is a vertical line. See Figure 5(b).

Slope of Horizontal and Vertical Lines

The slope of a horizontal line is 0.

The slope of a vertical line is undefined.

The different forms of linear equations discussed in this section are summarized below. The slope-intercept and point-slope forms are equivalent ways to express the equation of a nonvertical line. The slope-intercept form is simpler for a final answer, but you may find the point-slope form easier to use when you know the slope of a line and a point through which the line passes. The slope-intercept form is often considered the standard form. Any line that is not vertical has a unique slope-intercept form but can have many point-slope forms for its equation.

Equations of Lines

Equation	Description
$y = mx + b$	**Slope-intercept form:** slope m, y-intercept b
$y - y_1 = m(x - x_1)$	**Point-slope form:** slope m, line passes through (x_1, y_1)
$x = k$	**Vertical line:** x-intercept k, no y-intercept (except when $k = 0$), undefined slope
$y = k$	**Horizontal line:** y-intercept k, no x-intercept (except when $k = 0$), slope 0

Parallel and Perpendicular Lines

One application of slope involves deciding whether two lines are parallel, which means that they never intersect. Since two parallel lines are equally "steep," they should have the same slope. Also, two lines with the same "steepness" are parallel.

Parallel Lines

Two lines are **parallel** if and only if they have the same slope, or if they are both vertical.

EXAMPLE 8 Parallel Line

Find the equation of the line that passes through the point $(3, 5)$ and is parallel to the line $2x + 5y = 4$.

SOLUTION The slope of $2x + 5y = 4$ can be found by writing the equation in slope-intercept form. To put the equation in this form, solve for y.

$$2x + 5y = 4$$

$$y = -\frac{2}{5}x + \frac{4}{5} \qquad \text{Subtract } 2x \text{ from both sides and divide both sides by 5.}$$

This result shows that the slope is $-2/5$, the coefficient of x. Since the lines are parallel, $-2/5$ is also the slope of the line whose equation we want. This line passes through $(3, 5)$. Substituting $m = -2/5$, $x_1 = 3$, and $y_1 = 5$ into the point-slope form gives

$$y - y_1 = m(x - x_1)$$

$$y - 5 = -\frac{2}{5}(x - 3)$$

$$y - 5 = -\frac{2}{5}x + \frac{6}{5} \qquad \text{Distributive property}$$

$$y = -\frac{2}{5}x + \frac{6}{5} + 5 \cdot \frac{5}{5} \qquad \text{Add 5 to both sides and get a common denominator.}$$

$$y = -\frac{2}{5}x + \frac{31}{5}.$$

YOUR TURN 5 Find the equation of the line that passes through the point $(4, 5)$ and is parallel to the line $3x - 6y = 7$. Put your answer in slope-intercept form.

TRY YOUR TURN 5

As already mentioned, two nonvertical lines are parallel if and only if they have the same slope. Two lines having slopes with a product of -1 are perpendicular. A proof of this fact, which depends on similar triangles from geometry, is given as Exercise 43 in this section.

Perpendicular Lines

Two lines are **perpendicular** if and only if the product of their slopes is -1, or if one is vertical and the other horizontal.

▌ EXAMPLE 9   Perpendicular Line

Find the equation of the line L passing through the point $(3, 7)$ and perpendicular to the line having the equation $5x - y = 4$.

SOLUTION   To find the slope, write $5x - y = 4$ in slope-intercept form:

$$y = 5x - 4.$$

The slope is 5. Since the lines are perpendicular, if line L has slope m, then

$$5m = -1$$

$$m = -\frac{1}{5}.$$

Now substitute $m = -1/5$, $x_1 = 3$, and $y_1 = 7$ into the point-slope form.

$$y - 7 = -\frac{1}{5}(x - 3)$$

YOUR TURN 6 Find the equation of the line passing through the point $(3, 2)$ and perpendicular to the line having the equation $2x + 3y = 4$.

$$y - 7 = -\frac{1}{5}x + \frac{3}{5} \qquad \text{Distribute.}$$

$$y = -\frac{1}{5}x + \frac{3}{5} + 7 \cdot \frac{5}{5} \qquad \begin{array}{l}\text{Add 7 to both sides and get a}\\ \text{common denominator.}\end{array}$$

$$y = -\frac{1}{5}x + \frac{38}{5} \qquad\qquad\qquad \text{TRY YOUR TURN 6} \ \blacksquare$$

The next example uses the equation of a line to analyze real-world data. In this example, we are looking at how one variable changes over time. To simplify the arithmetic, we will *rescale* the variable representing time, although computers and calculators have made rescaling less important than in the past. Here it allows us to work with smaller numbers, and, as you will see, find the y-intercept of the line more easily. We will use rescaling on many examples throughout this book. When we do, it is important to be consistent.

▌ EXAMPLE 10   Prevalence of Cigarette Smoking

In recent years, the percentage of the U.S. population age 18 and older who smoke has decreased at a roughly constant rate, from 20.9% in 2005 to 18.1% in 2012. *Source: Centers for Disease Control and Prevention*.

(a) Find the equation describing this linear relationship.

SOLUTION   Let t represent time in years, with $t = 0$ representing 2000. With this rescaling, the year 2005 corresponds to $t = 5$ and the year 2012 corresponds to $t = 12$. Let y represent the percentage of the population who smoke. The two ordered pairs representing the given information are then $(5, 20.9)$ and $(12, 18.1)$. The slope of the line through these points is

$$m = \frac{18.1 - 20.9}{12 - 5} = \frac{-2.8}{7} = -0.4.$$

This means that, on average, the percentage of the adult population who smoke is decreasing by about 0.4% per year.

Using $m = -0.4$ in the point-slope form, and choosing $(t_1, y_1) = (5, 20.9)$, gives the required equation.

$$y - 20.9 = -0.4(t - 5)$$
$$y - 20.9 = -0.4t + 2 \qquad \text{Distributive property}$$
$$y = -0.4t + 22.9 \qquad \text{Add 20.9 to both sides.}$$

We could have used the other point $(12, 18.1)$ and found the same answer. Instead, we'll use this to check our answer by observing that $-0.4(12) + 22.9 = 18.1$, which agrees with the y-value at $t = 12$.

(b) One objective of Healthy People 2020 (a campaign of the U.S. Department of Health and Human Services) is to reduce the percentage of U.S. adults who smoke to 12% or less by the year 2020. If this decline in smoking continued at the same rate, will they meet this objective?

SOLUTION   Using the same rescaling, $t = 20$ corresponds to the year 2020. Substituting this value into the above equation gives

$$y = -0.4(20) + 22.9 = 14.9.$$

Continuing at this rate, an estimated 14.9% of the adult population will still smoke in 2020, and the objective of Healthy People 2020 will not be met.   ▬▬

Notice that if the formula from part (a) of Example 10 is valid for all nonnegative t, then eventually y becomes 0:

$$-0.4t + 22.9 = 0$$
$$-0.4t = -22.9 \qquad \text{Subtract 22.9 from both sides.}$$
$$t = \frac{-22.9}{-0.4} = 57.25 \approx 57^*, \qquad \text{Divide both sides by } -0.4.$$

which indicates that 57 years from 2000 (in the year 2057), 0% of the U.S. adult population will smoke. Of course, it is still possible that in 2057 there will be adults who smoke; the trend of recent years may not continue. Most equations are valid for some specific set of numbers. It is highly speculative to extrapolate beyond those values.

On the other hand, people in business and government often need to make some prediction about what will happen in the future, so a tentative conclusion based on past trends may be better than no conclusion at all. There are also circumstances, particularly in the physical sciences, in which theoretical reasons imply that the trend will continue.

Graph of a Line

It can be shown that every equation of the form $ax + by = c$ has a straight line as its graph, assuming a and b are not both 0. Although just two points are needed to determine a line, it is a good idea to plot a third point as a check.

EXAMPLE 11   **Graph of a Line**

Graph $y = -\dfrac{3}{2}x + 6$.

SOLUTION   There are several ways to graph a line.

Method 1
Plot Points   Note that the y-intercept is 6, so the point $(0, 6)$ is on the line. Next, by substituting $x = 2$ and $x = 4$ into the equation, we find the points $(2, 3)$ and $(4, 0)$. (We could use any values for x, but we chose even numbers so the value of y would be an integer.) These three points are

*The symbol $\approx$ means "is approximately equal to."

plotted in Figure 6(a). A line is drawn through them in Figure 6(b). (We need to find only two points on the line. The third point gives a confirmation of our work.)

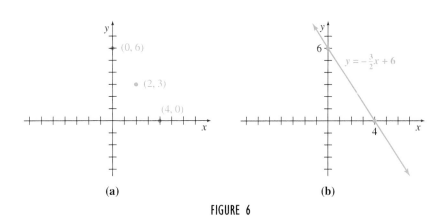

(a)   (b)

FIGURE 6

Method 2
Find Intercepts

We already found the y-intercept $(0, 6)$ by setting $x = 0$. To find the x-intercept, let $y = 0$.

$$0 = -\frac{3}{2}x + 6$$

$$\frac{3}{2}x = 6 \qquad \text{Add } \tfrac{3}{2}x \text{ to both sides.}$$

$$x = \frac{6}{3/2} = 6\left(\frac{2}{3}\right) = 4 \qquad \text{Divide both sides by } \tfrac{3}{2}.$$

This gives the point $(4, 0)$ that we found earlier. Once the x- and y-intercepts are found, we can draw a line through them.

Method 3
Use the Slope and y-Intercept

Observe that the slope of $-3/2$ means that every time x increases by 2, y decreases by 3. So start at the y-intercept $(0, 6)$ and go 2 across and 3 down to the point $(2, 3)$. Again, go 2 across and 3 down to the point $(4, 0)$. Draw the line through these points. ▬

Not every line has two distinct intercepts; the graph in the next example does not cross the x-axis, and so it has no x-intercept.

EXAMPLE 12   **Graph of a Horizontal Line**

Graph $y = -3$.

SOLUTION   The equation $y = -3$, or equivalently, $y = 0x - 3$, always gives the same y-value, -3, for any value of x. Therefore, no value of x will make $y = 0$, so the graph has no x-intercept. As we saw in Example 6, the graph of such an equation is a horizontal line parallel to the x-axis. In this case the y-intercept is -3, as shown in Figure 7. ▬

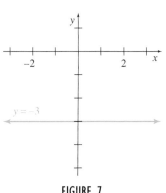

FIGURE 7

Linear equations allow us to set up simple mathematical models for real-life situations. In almost every case, linear (or any other reasonably simple) equations provide only approximations to real-world situations. Nevertheless, these are often remarkably useful approximations.

EXAMPLE 13 Tuition

APPLY IT

The table on the left lists the average annual cost (in dollars) of tuition and fees at public four-year colleges for selected years. *Source: The College Board*.

(a) Plot the cost of public colleges by letting $t = 0$ correspond to 2000. Are the data *exactly* linear? Could the data be *approximated* by a linear equation?

SOLUTION The data are plotted in Figure 8(a) in a figure known as a **scatterplot**. Although it is not exactly linear, it is approximately linear and could be approximated by a linear equation.

Cost of Public College

Year	Tuition and Fees
2000	3508
2002	4098
2004	5126
2006	5804
2008	6599
2010	7629
2012	8646
2013	8893

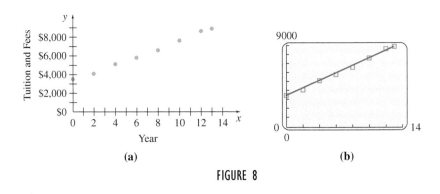

(a) (b)

FIGURE 8

(b) Use the points $(0, 3508)$ and $(13, 8893)$ to determine an equation that models the data.

SOLUTION We first find the slope of the line as follows:

$$m = \frac{8893 - 3508}{13 - 0} = \frac{5385}{13} \approx 414.2.$$

We have rounded to four digits, noting that we cannot expect more accuracy in our answer than in the data, which are accurate to four digits. The slope indicates that the average annual cost of tuition and fees at public four-year colleges is increasing by about $414 per year.

Using the slope-intercept form of the line, $y = mt + b$, with $m = 414.2$ and $b = 3508$, gives

$$y = 414.2t + 3508.$$

TECHNOLOGY NOTE

A graphing calculator plot of this line and the data points are shown in Figure 8(b). Notice that the points closely fit the line. More details on how to construct this graphing calculator plot are given at the end of this example.

(c) Discuss the accuracy of using this equation to estimate the cost of public colleges in the year 2030.

SOLUTION The year 2030 corresponds to the year $t = 30$, for which the equation predicts a cost of

$$y = 414.2(30) + 3508 = 15,934, \quad \text{or} \quad \$15,934.$$

The year 2030 is many years in the future, however. Many factors could affect the tuition, and the actual figure for 2030 could turn out to be very different from our prediction.

⊹ **TECHNOLOGY NOTE**

You can plot data with a TI-84 Plus C graphing calculator using the following steps.

1. Store the data in lists. Select the `EDIT` command in the `STAT` menu. Enter the x-values in `L1` and y-values in `L2`.

2. Define the stat plot. Go to `STAT PLOT` and select `Plot1`. Select `On`. Be sure that `xlist` is `L1` and `ylist` is `L2`.

3. Define the viewing window.

4. To graph the line, select `Y=` and enter the equation in `Y1`.

5. Display the graph by selecting `GRAPH`.

Consult the calculator's instruction booklet or the *Graphing Calculator and Excel Spreadsheet Manual*, available with this book, for specific instructions. See the calculator–generated graph in Figure 8(b), which includes the points and line from Example 13. Notice how the line closely approximates the data.

1.1 WARM-UP EXERCISES

W1. Evaluate $\dfrac{15 - (-3)}{-2 - 4}$. *(Sec. R.1)*

Solve each equation for *y*. *(Sec. R.4)*

W2. $y - (-3) = -2(x + 5)$

W3. $y - \dfrac{1}{2} = \dfrac{2}{5}\left(x + \dfrac{1}{3}\right)$

W4. $2x - 3y = 7$

1.1   EXERCISES

Find the slope of each line.

1. Through $(4, 5)$ and $(-1, 2)$

2. Through $(5, -4)$ and $(1, 3)$

3. Through $(8, 4)$ and $(8, -7)$

4. Through $(1, 5)$ and $(-2, 5)$

5. $y = x$

6. $y = 3x - 2$

7. $5x - 9y = 11$

8. $4x + 7y = 1$

9. $x = 5$

10. The x-axis

11. $y = 8$

12. $y = -6$

13. A line parallel to $6x - 3y = 12$

14. A line perpendicular to $8x = 2y - 5$

In Exercises 15–24, find an equation in slope-intercept form for each line.

15. Through $(1, 3)$, $m = -2$

16. Through $(2, 4)$, $m = -1$

17. Through $(-5, -7)$, $m = 0$

18. Through $(-8, 1)$, with undefined slope

19. Through $(4, 2)$ and $(1, 3)$

20. Through $(8, -1)$ and $(4, 3)$

21. Through $(2/3, 1/2)$ and $(1/4, -2)$

22. Through $(-2, 3/4)$ and $(2/3, 5/2)$

23. Through $(-8, 4)$ and $(-8, 6)$

24. Through $(-1, 3)$ and $(0, 3)$

In Exercises 25–34, find an equation for each line in the form $ax + by = c$, where $a, b,$ and c are integers with no factor common to all three and $a \geq 0$.

25. x-intercept -6, y-intercept -3

26. x-intercept -2, y-intercept 4

27. Vertical, through $(-6, 5)$

28. Horizontal, through $(8, 7)$

29. Through $(-4, 6)$, parallel to $3x + 2y = 13$

30. Through $(2, -5)$, parallel to $2x - y = -4$

31. Through $(3, -4)$, perpendicular to $x + y = 4$

32. Through $(-2, 6)$, perpendicular to $2x - 3y = 5$

33. The line with y-intercept 4 and perpendicular to $x + 5y = 7$

34. The line with x-intercept $-2/3$ and perpendicular to $2x - y = 4$

35. Do the points $(4, 3)$, $(2, 0)$, and $(-18, -12)$ lie on the same line? Explain why or why not. (*Hint:* Find the slopes between the points.)

36. Find k so that the line through $(4, -1)$ and $(k, 2)$ is

(a) parallel to $2x + 3y = 6$,

(b) perpendicular to $5x - 2y = -1$.

37. Use slopes to show that the quadrilateral with vertices at $(1, 3)$, $(-5/2, 2)$, $(-7/2, 4)$, and $(2, 1)$ is a parallelogram.

38. Use slopes to show that the square with vertices at $(-2, 5)$, $(4, 5)$, $(4, -1)$, and $(-2, -1)$ has diagonals that are perpendicular.

For the lines in Exercises 39 and 40, which of the following is closest to the slope of the line? (a) 1 (b) 2 (c) 3 (d) 21 (e) 22 (f) −3

39.

40.

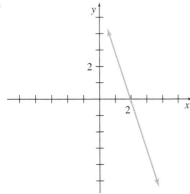

In Exercises 41 and 42, estimate the slope of the lines.

41.

42.

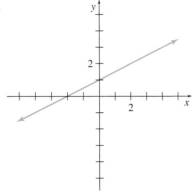

43. To show that two perpendicular lines, neither of which is vertical, have slopes with a product of −1, go through the following steps. Let line L_1 have equation $y = m_1x + b_1$, and let L_2 have equation $y = m_2x + b_2$, with $m_1 > 0$ and $m_2 < 0$. Assume that L_1 and L_2 are perpendicular, and use right triangle MPN shown in the figure. Prove each of the following statements.

(a) MQ has length m_1.

(b) QN has length $-m_2$.

(c) Triangles MPQ and PNQ are similar.

(d) $m_1/1 = 1/(-m_2)$ and $m_1m_2 = -1$

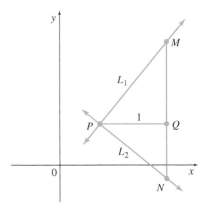

44. Consider the equation $\dfrac{x}{a} + \dfrac{y}{b} = 1$.

(a) Show that this equation represents a line by writing it in the form $y = mx + b$.

(b) Find the x- and y-intercepts of this line.

(c) Explain in your own words why the equation in this exercise is known as the intercept form of a line.

Graph each equation.

45. $y = x - 1$ **46.** $y = 4x + 5$

47. $y = -4x + 9$ **48.** $y = -6x + 12$

49. $2x - 3y = 12$ **50.** $3x - y = -9$

51. $3y - 7x = -21$ **52.** $5y + 6x = 11$

53. $y = -2$ **54.** $x = 4$

55. $x + 5 = 0$ **56.** $y + 8 = 0$

57. $y = 2x$ **58.** $y = -5x$

59. $x + 4y = 0$ **60.** $3x - 5y = 0$

APPLICATIONS

Business and Economics

61. Sales The sales of a small company were $27,000 in its second year of operation and $63,000 in its fifth year. Let y represent sales in the xth year of operation. Assume that the data can be approximated by a straight line.

(a) Find the slope of the sales line, and give an equation for the line in the form $y = mx + b$.

(b) Use your answer from part (a) to find out how many years must pass before the sales surpass $100,000.

62. Cost The total cost for a bakery to produce 100 gourmet cupcakes is $126, while the total cost to produce 120 gourmet cupcakes is $144. Let y represent the total cost for x gourmet cupcakes. Assume that a straight line can approximate the data.

  (a) Find and interpret the slope of the cost line for the gourmet cupcakes.

  (b) Determine an equation that models the data. Write the equation in the slope-intercept form.

  (c) Use your answer from part (b) to determine an approximation of the cost of 180 gourmet cupcakes.

63. Tuition The table lists the annual cost (in dollars) of tuition and fees at private four-year colleges for selected years. (See Example 13.) *Source: The College Board.*

Year	Tuition and Fees
2000	16,072
2002	18,060
2004	20,045
2006	22,308
2008	24,818
2010	26,766
2012	28,989
2013	30,094

  (a) Sketch a graph of the data. Do the data appear to lie roughly along a straight line?

  (b) Let $t = 0$ correspond to the year 2000. Use the points $(0, 16{,}072)$ and $(13, 30{,}094)$ to determine a linear equation that models the data. What does the slope of the graph of the equation indicate?

  (c) Discuss the accuracy of using this equation to estimate the cost of private college in 2025.

64. Use of Cellular Telephones The following table shows the subscribership of cellular telephones in the United States (in millions) for selected years between 2000 and 2012. *Source: The World Almanac and Book of Facts 2014.*

Year	2000	2004	2008	2012
Subscribers (in millions)	109.48	182.14	270.33	326.48

  (a) Plot the data by letting $t = 0$ correspond to 2000. Discuss how well the data fit a straight line.

  (b) Determine a linear equation that approximates the number of subscribers using the points $(0, 109.48)$ and $(12, 326.48)$.

  (c) Repeat part (b) using the points $(4, 182.14)$ and $(12, 326.48)$.

  (d) Discuss why your answers to parts (b) and (c) are similar but not identical.

  (e) Using your equations from parts (b) and (c), approximate the number of cellular phone subscribers in the year 2010. Compare your result with the actual value of 296.29 million.

65. Consumer Price Index The Consumer Price Index (CPI) is a measure of the change in the cost of goods over time. The index was 100 for the three-year period centered on 1983. For simplicity, we will assume that the CPI was exactly 100 in 1983. Then the CPI of 229.6 in 2012 indicates that an item that cost $1.00 in 1983 would cost $2.30 in 2012. The CPI has been increasing approximately linearly over the last few decades. *Source: The World Almanac and Book of Facts 2014.*

  (a) Use this information to determine an equation for the CPI in terms of t, which represents the years since 1980.

  (b) Based on the answer to part (a), what was the predicted value of the CPI in 2000? Compare this estimate with the actual CPI of 172.2.

  (c) Describe the rate at which the annual CPI is changing.

Life Sciences

66. HIV Infection The time interval between a person's initial infection with HIV and that person's eventual development of AIDS symptoms is an important issue. The method of infection with HIV affects the time interval before AIDS develops. One study of HIV patients who were infected by intravenous drug use found that 17% of the patients had AIDS after 4 years, and 33% had developed the disease after 7 years. The relationship between the time interval and the percentage of patients with AIDS can be modeled accurately with a linear equation. *Source: Epidemiologic Review.*

  (a) Write a linear equation $y = mt + b$ that models these data, using the ordered pairs $(4, 0.17)$ and $(7, 0.33)$.

  (b) Use your equation from part (a) to predict the number of years before half of these patients will have AIDS.

67. Exercise Heart Rate To achieve the maximum benefit for the heart when exercising, your heart rate (in beats per minute) should be in the target heart rate zone. The lower limit of this zone is found by taking 70% of the difference between 220 and your age. The upper limit is found by using 85%. *Source: Physical Fitness.*

  (a) Find formulas for the upper and lower limits (u and l) as linear equations involving the age x.

  (b) What is the target heart rate zone for a 20-year-old?

  (c) What is the target heart rate zone for a 40-year-old?

  (d) Two women in an aerobics class stop to take their pulse and are surprised to find that they have the same pulse. One woman is 36 years older than the other and is working at the upper limit of her target heart rate zone. The younger woman is working at the lower limit of her target heart rate zone. What are the ages of the two women, and what is their pulse?

  (e) Run for 10 minutes, take your pulse, and see if it is in your target heart rate zone. (After all, this is listed as an exercise!)

68. Ponies Trotting A study found that the peak vertical force on a trotting Shetland pony increased linearly with the pony's speed, and that when the force reached a critical level, the pony switched from a trot to a gallop. For one pony, the critical force was 1.16 times its body weight. It experienced a force of 0.75 times its body weight at a speed of 2 meters per second and a force of 0.93 times its body weight at 3 meters per second. At what speed did the pony switch from a trot to a gallop? *Source: Science*.

69. Life Expectancy Some scientists believe there is a limit to how long humans can live. One supporting argument is that during the past century, life expectancy from age 65 has increased more slowly than life expectancy from birth, so eventually these two will be equal, at which point, according to these scientists, life expectancy should increase no further. In 1900, life expectancy at birth was 46 yr, and life expectancy at age 65 was 76 yr. In 2010, these figures had risen to 78.7 and 84.1, respectively. In both cases, the increase in life expectancy has been linear. Using these assumptions and the data given, find the maximum life expectancy for humans. *Source: Science*.

Social Sciences

70. Child Mortality Rate The mortality rate for children under 5 years of age around the world has been declining in a roughly linear fashion in recent years. The rate per 1000 live births was 90 in 1990 and 48 in 2012. *Source: World Health Organization*.

(a) Determine a linear equation that approximates the mortality rate in terms of time t, where t represents the number of years since 1900.

(b) One of the Millennium Development Goals (MDG) of the World Health Organization is to reduce the mortality rate for children under 5 years of age to 30 by 2015. If this trend were to continue, in what year would this goal be reached?

71. Immigration In 1950, there were 249,187 immigrants admitted to the United States. In 2012, the number was 1,031,631. *Source: 2012 Yearbook of Immigration Statistics*.

(a) Assuming that the change in immigration is linear, write an equation expressing the number of immigrants, y, in terms of t, the number of years after 1900.

(b) Use your result in part (a) to predict the number of immigrants admitted to the United States in 2015.

(c) Considering the value of the y-intercept in your answer to part (a), discuss the validity of using this equation to model the number of immigrants throughout the entire 20th century.

72. Marriage The following table lists the U.S. median age at first marriage for men and women. The age at which both groups marry for the first time seems to be increasing at a roughly linear rate in recent decades. Let t correspond to the number of years since 1980. *Source: U.S. Census Bureau*.

Age at First Marriage			
Year	1980	1990	2000	2010
Men	24.7	26.1	26.8	28.2
Women	22.0	23.9	25.1	26.1

(a) Find a linear equation that approximates the data for men, using the data for the years 1980 and 2010.

(b) Repeat part (a) using the data for women.

(c) Which group seems to have the faster increase in median age at first marriage?

(d) According to the equation from part (a), in what year will the men's median age at first marriage reach 30?

(e) When the men's median age at first marriage is 30, what will the median age be for women?

Physical Sciences

73. Galactic Distance The table lists the distances (in megaparsecs where 1 megaparsec $\approx 3.1 \times 10^{19}$ km) and velocities (in kilometers per second) of four galaxies moving rapidly away from Earth. *Source: Astronomical Methods and Calculations, and Fundamental Astronomy*.

Galaxy	Distance	Velocity
Virga	15	1600
Ursa Minor	200	15,000
Corona Borealis	290	24,000
Bootes	520	40,000

(a) Plot the data points, letting x represent distance and y represent velocity. Do the points lie in an approximately linear pattern?

(b) Write a linear equation $y = mx$ to model these data, using the ordered pair (520, 40,000).

(c) The galaxy Hydra has a velocity of 60,000 km per sec. Use your equation to approximate how far away it is from Earth.

(d) The value of m in the equation is called the *Hubble constant*. The Hubble constant can be used to estimate the age of the universe A (in years) using the formula

$$A = \frac{9.5 \times 10^{11}}{m}.$$

Approximate A using your value of m.

74. Global Warming In 1990, the Intergovernmental Panel on Climate Change predicted that the average temperature on Earth would rise 0.3°C per decade in the absence of international controls on greenhouse emissions. Let t measure the time in years since 1970, when the average global temperature was 15°C. *Source: Science News*.

(a) Find a linear equation giving the average global temperature in degrees Celsius in terms of t, the number of years since 1970.

(b) Scientists have estimated that the sea level will rise by 65 cm if the average global temperature rises to 19°C. According to your answer to part (a), when would this occur?

General Interest

75. News Sources A survey asked respondents where they got news "yesterday" (the day before they participated in the survey). In 2006, about 40% of respondents got their news from the newspaper, while 23% got their news online. In 2012, about 29% got their news from the newspaper, while 39% got their news online. Both news sources changed at a roughly linear rate. *Source: State of the Media*.

(a) Find a linear equation expressing the percent of respondents, y_n, who got their news from the newspaper in terms of t, the years since 2000.

(b) Find a linear equation expressing the percent of respondents, y_o, who got their news online in terms of t, the years since 2000.

(c) Find the rate of change over time for the percentage of respondents for each source of news.

1.2 Linear Functions and Applications

APPLY IT How many units must be sold for a firm to break even?

In Example 7 in this section, this question will be answered using a linear function.

As we saw in the previous section, many situations involve two variables related by a linear equation. For such a relationship, when we express the variable y in terms of x, we say that y is a **linear function** of x. This means that for any allowed value of x (the **independent variable**), we can use the equation to find the corresponding value of y (the **dependent variable**). Examples of equations defining linear functions include $y = 2x + 3$, $y = -5$, and $2x - 3y = 7$, which can be written as $y = (2/3)x - (7/3)$. Equations in the form $x = k$, where k is a constant, do not define linear functions. All other linear equations define linear functions.

$f(x)$ Notation

Letters such as f, g, or h are often used to name functions. For example, f might be used to name the function defined by

$$y = 5 - 3x.$$

To show that this function is named f, it is common to replace y with $f(x)$ (read "f of x") to get

$$f(x) = 5 - 3x.$$

By choosing **2** as a value of x, $f(x)$ becomes

$$f(2) = 5 - 3 \cdot 2 = 5 - 6 = -1.$$

The corresponding ordered pair is $(2, -1)$. In a similar manner,

$$f(-4) = 5 - 3(-4) = 17, \qquad f(0) = 5, \qquad f(-6) = 23,$$

and so on.

EXAMPLE 1 Function Notation

Let $g(x) = -4x + 5$. Find $g(3)$, $g(0)$, $g(-2)$, and $g(b)$.

SOLUTION To find $g(3)$, substitute 3 for x.

$$g(3) = -4(3) + 5 = -12 + 5 = -7$$

Similarly,

$$g(0) = -4(0) + 5 = 0 + 5 = 5,$$
$$g(-2) = -4(-2) + 5 = 8 + 5 = 13,$$

YOUR TURN 1
Calculate $g(-5)$.

and

$$g(b) = -4b + 5.$$

TRY YOUR TURN 1

We summarize the discussion below.

> ### Linear Function
> A relationship f defined by
>
> $$y = f(x) = mx + b,$$
>
> for real numbers m and b, is a **linear function**.

Linear functions are used to model many applications in business and science, as shown in the next examples.

Supply and Demand

The supply and demand relationship is one of the fundamental concepts of economics. The **supply** is the quantity of a certain good that producers are willing to provide. The **demand** is the quantity of the good that consumers are willing to buy.

The **Law of Demand** states that, if all other factors remain equal, as the price of an item increases, consumers are less likely to buy an increasingly expensive item, and so the demand for the item decreases. On the other hand, the **Law of Supply** states that as the price of an item increases, producers are more likely to see a profit in selling the item, and so the supply of the item increases. It is important to note that suppliers cannot always react quickly to a change in demand or a change in price.

Economists study the interaction between price, supply, and demand. If the price increases, the increase in the quantity supplied and decrease in the quantity demanded can eventually result in a surplus, which causes the price to fall. These countervailing trends tend to move the price, as well as the quantity supplied and demanded, toward an equilibrium value.

For example, during the late 1980s and early 1990s, the consumer demand for cranberries (and all of their healthy benefits) soared. The quantity demanded surpassed the quantity supplied, causing a shortage, and cranberry prices rose dramatically. As prices increased, growers wanted to increase their profits, so they planted more acres of cranberries. Unfortunately, cranberries take 3 to 5 years from planting until they can first be harvested. As growers waited and prices increased, consumer demand decreased. When the cranberries were finally harvested, the supply overwhelmed the demand and a huge surplus occurred, causing the price of cranberries to drop in the late 1990s. Other factors were involved in this situation, but the relationship between price, supply, and demand was nonetheless typical. *Source: Agricultural Marketing Resource Center*.

Although economists consider price to be the independent variable, they have the unfortunate habit of plotting price, usually denoted by p, on the vertical axis, while everyone else graphs the independent variable on the horizontal axis. This custom was started by the English economist Alfred Marshall (1842–1924). To abide by this custom, we will write p, the price, as a function of q, the quantity produced, and plot p on the vertical axis. But remember, it is really *price* that determines how much consumers demand and producers supply, not the other way around.

Supply and demand functions are not necessarily linear, the simplest kind of function. Yet most functions are approximately linear if a small enough piece of the graph is taken, allowing applied mathematicians to often use linear functions for simplicity. That approach will be taken in this chapter.

EXAMPLE 2 Supply and Demand

Suppose that Greg Tobin, manager of a giant supermarket chain, has studied the supply and demand for watermelons. He has noticed that the demand increases as the price decreases. He has determined that the quantity (in thousands) demanded weekly, q, and the price (in dollars) per watermelon, p, are related by the linear function

$$p = D(q) = 9 - 0.75q. \quad \text{Demand function}$$

(a) Find the quantity demanded at a price of $5.25 per watermelon and at a price of $3.75 per watermelon.

SOLUTION To find the quantity demanded at a price of $5.25 per watermelon, replace p in the demand function with 5.25 and solve for q.

$$5.25 = 9 - 0.75q$$
$$-3.75 = -0.75q \quad \text{Subtract 9 from both sides.}$$
$$5 = q \quad \text{Divide both sides by } -0.75.$$

Thus, at a price of $5.25, the quantity demanded is 5000 watermelons.

Similarly, replace p with 3.75 to find the demand when the price is $3.75. Verify that this leads to $q = 7$. When the price is lowered from $5.25 to $3.75 per watermelon, the quantity demanded increases from 5000 to 7000 watermelons.

(b) Greg also noticed that the quantity of watermelons supplied decreased as the price decreased. Price p and supply q are related by the linear function

$$p = S(q) = 0.75q. \quad \text{Supply function}$$

Find the quantity supplied at a price of $5.25 per watermelon and at a price of $3.00 per watermelon.

SOLUTION Substitute 5.25 for p in the supply function, $p = 0.75q$, to find that $q = 7$, so the quantity supplied is 7000 watermelons. Similarly, replacing p with 3 in the supply equation gives a quantity supplied of 4000 watermelons. If the price decreases from $5.25 to $3.00 per watermelon, the quantity supplied also decreases, from 7000 to 4000 watermelons.

(c) Graph both functions on the same axes.

SOLUTION The results of part (a) are written as the ordered pairs $(5, 5.25)$ and $(7, 3.75)$. The line through those points is the graph of the demand function, $p = 9 - 0.75q$, shown in red in Figure 9(a). We used the ordered pairs $(7, 5.25)$ and $(4, 3)$ from the work in part (b) to graph the supply function, $p = 0.75q$, shown in blue in Figure 9(a).

TRY YOUR TURN 2

(a)

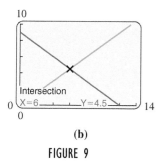

(b)

FIGURE 9

YOUR TURN 2 Find the quantity of watermelon demanded and supplied at a price of $3.30 per watermelon.

TECHNOLOGY NOTE

A calculator-generated graph of the lines representing the supply and demand functions in Example 2 is shown in Figure 9(b). To get this graph, the equation of each line, using x and y instead of q and p, was entered, along with an appropriate window. The coordinates of the intersection point, as shown at the bottom of the graph, are found by selecting `intersect` from the `CALC` menu.

NOTE Not all supply and demand problems will have the same scale on both axes. It helps to consider the intercepts of both the supply graph and the demand graph to decide what scale to use. For example, in Figure 9, the y-intercept of the demand function is 9, so the scale should allow values from 0 to at least 9 on the vertical axis. The x-intercept of the demand function is 12, so the values on the x-axis must go from 0 to 12.

FIGURE 10

As shown in the graphs of Figure 9, both the supply graph and the demand graph pass through the point $(6, 4.5)$. If the price of a watermelon is more than $4.50, the quantity supplied will exceed the quantity demanded and there will be a **surplus** of watermelons. At a price less than $4.50, the quantity demanded will exceed the quantity supplied and there will be a **shortage** of watermelons. Only at a price of $4.50 will quantity demanded and supplied be equal. For this reason, $4.50 is called the *equilibrium price*. When the price is $4.50, quantity demanded and supplied both equal 6000 watermelons, the *equilibrium quantity*. Figure 10 illustrates a general supply and demand situation.

Equilibrium

When supply and demand are equal, the economy is said to be at **equilibrium**. The **equilibrium price** of the commodity is the price found at the point where the supply and demand graphs for that commodity intersect. The **equilibrium quantity** is the quantity demanded and supplied at that same point.

Note that in the real marketplace, prices are constantly changing in reaction to fluctuations in the supply and demand, and so therefore, equilibrium can rarely be achieved except in theory.

EXAMPLE 3 Equilibrium Quantity

Use algebra to find the equilibrium quantity and price for the watermelons in Example 2.

SOLUTION The equilibrium quantity is found when the prices from both supply and demand are equal. Set the two expressions for p equal to each other and solve.

$$9 - 0.75q = 0.75q$$

$$9 = 1.5q \qquad \text{Add } 0.75q \text{ to both sides.}$$

$$6 = q \qquad \text{Divide both sides by 1.5.}$$

The equilibrium quantity is 6000 watermelons, the same answer found earlier.

The equilibrium price can be found by plugging the value of $q = 6$ into either the demand or the supply function. Using the demand function,

$$p = D(6) = 9 - 0.75(6) = 4.5.$$

The equilibrium price is $4.50, as we found earlier. Check your work by also plugging $q = 6$ into the supply function.   **TRY YOUR TURN 3**

YOUR TURN 3 Repeat Example 3 using the demand equation $D(q) = 10 - 0.85q$ and the supply equation $S(q) = 0.4q$.

TECHNOLOGY NOTE

You may prefer to find the equilibrium quantity by solving the equation with your calculator. Or, if your calculator has a TABLE feature, you can use it to find the value of q that makes the two expressions equal.

Another important issue is how, in practice, the equations of the supply and demand functions can be found. Data need to be collected, and if they lie perfectly along a line, then the equation can easily be found with any two points. What usually happens, however, is that the data are scattered, and there is no line that goes through all the points. In this case we must find a line that approximates the linear trend of the data as closely as possible (assuming the points lie approximately along a line) as in Example 13 in the previous section. This is usually done by the *method of least squares*, also referred to as *linear regression*. We will discuss this method in Section 1.3.

Cost Analysis The cost of manufacturing an item commonly consists of two parts. The first is a **fixed cost** for designing the product, setting up a factory, training workers, and so on. Within broad limits, the fixed cost is constant for a particular product and does not change as more items are made. The second part is a *cost per item* for labor, materials, packing, shipping, and so on. The total value of this second cost *does* depend on the number of items made.

EXAMPLE 4 Cost Analysis

A small company decides to produce video games. The owners find that the fixed cost for creating the game is $5000, after which they must spend $12 to produce each individual copy of the game. Find a formula $C(x)$ for the cost as a linear function of x, the number of games produced.

SOLUTION   Notice that $C(0) = 5000$, since \$5000 must be spent even if no games are produced. Also, $C(1) = 5000 + 12 = 5012$, and $C(2) = 5000 + 2 \cdot 12 = 5024$. In general,

$$C(x) = 5000 + 12x,$$

because every time x increases by 1, the cost should increase by \$12. The number 12 is also the slope of the graph of the cost function; the slope gives us the cost to produce one additional item.

In economics, **marginal cost** is the rate of change of cost $C(x)$ at a level of production x and is equal to the slope of the cost function at x. It approximates the cost of producing one additional item. In fact, some books define the marginal cost to be the cost of producing one additional item. With *linear functions*, these two definitions are equivalent, and the marginal cost, which is equal to the slope of the cost function, is *constant*. For instance, in the video game example, the marginal cost of each game is \$12. For other types of functions, these two definitions are only approximately equal. Marginal cost is important to management in making decisions in areas such as cost control, pricing, and production planning.

The work in Example 4 can be generalized. Suppose the total cost to make x items is given by the linear cost function $C(x) = mx + b$. The fixed cost is found by letting $x = 0$:

$$C(0) = m \cdot 0 + b = b;$$

thus, the fixed cost is b dollars. The additional cost of each additional item, the marginal cost, is m, the slope of the line $C(x) = mx + b$.

Linear Cost Function

In a cost function of the form $C(x) = mx + b$, the m represents the marginal cost and b the fixed cost. Conversely, if the fixed cost of producing an item is b and the marginal cost is m, then the **linear cost function** $C(x)$ for producing x items is $C(x) = mx + b$.

EXAMPLE 5   Cost Function

The marginal cost to make x batches of a prescription medication is \$10 per batch, while the cost to produce 100 batches is \$1500. Find the cost function $C(x)$, given that it is linear.

SOLUTION   Since the cost function is linear, it can be expressed in the form $C(x) = mx + b$. The marginal cost is \$10 per batch, which gives the value for m. Using $x = 100$ and $C(x) = 1500$ in the point-slope form of the line gives

$$
\begin{aligned}
C(x) - 1500 &= 10(x - 100) \\
C(x) - 1500 &= 10x - 1000 \qquad \text{\small Distributive property} \\
C(x) &= 10x + 500. \qquad \text{\small Add 1500 to both sides.}
\end{aligned}
$$

YOUR TURN 4   Repeat Example 5, using a marginal cost of \$15 per batch and a cost of \$1930 to produce 80 batches.

The cost function is given by $C(x) = 10x + 500$, where the fixed cost is \$500.

TRY YOUR TURN 4

EXAMPLE 6   Cost Function

A company has fixed costs of \$12,500. The total cost to produce 1000 widgets is \$13,260. (A widget is a term often used by economists to refer to an abstract unit of production.) Find the cost function $C(x)$, given that it is linear.

SOLUTION   The linear cost function is $C(x) = mx + b$. Here the fixed cost is given: $b = 12,500$. We need to find the marginal cost, m. One way to find the marginal cost is to use the value of b, along with $x = 1000$ and $C(x) = 13,260$, in the linear cost function form and solve for m.

$$C(x) = mx + b$$

$$13,260 = m \cdot 1000 + 12,500 \qquad \text{\small $C(x) = 13{,}260, x = 1000, \text{ and } b = 12{,}500$}$$

$$760 = 1000m \qquad \text{\small Subtract 12,500 from both sides.}$$

$$0.76 = m \qquad \text{\small Divide both sides by 1000.}$$

YOUR TURN 5 Repeat Example 6, using a fixed cost of $7145 and a cost of $7965 to produce 100 items.

The cost function is given by $C(x) = 0.76x + 12,500$, where the marginal cost is $0.76.

Note that the marginal cost, m, could have also been found by using the slope formula, $m = (y_2 - y_1)/(x_2 - x_1)$, with the points $(0, 12,500)$ and $(1000, 13,260)$.

TRY YOUR TURN 5 ▬

Break-Even Analysis

The goal of every company is to make a profit. A company can make a profit only if the revenue received from its customers exceeds the cost of producing and selling its goods and services. Companies use break-even analysis to analyze the revenue and cost of sales, and to determine the amount that revenues can fall while still making a profit.

Break-Even Analysis

Let p be the price per unit and let x be the number of units sold (demand).

The **revenue,** $R(x)$, is the amount of money that a company receives; it is the product of the price per unit p and the number of units sold x. That is,

$$R(x) = px.$$

The **profit,** $P(x)$, is the money a company makes after paying its costs; it is the difference between revenue $R(x)$ and cost $C(x)$. That is,

$$P(x) = R(x) - C(x).$$

The number of units, x, at which revenue just equals cost is the **break-even quantity**. It is found by setting revenue equal to cost. That is,

$$R(x) = C(x).$$

(The break-even quantity can also be found by setting profit equal to zero; $P(x) = 0$.)

The **break-even point** is the corresponding ordered pair, and indicates the point of zero loss or profit.

EXAMPLE 7 Break-Even Analysis

A firm producing poultry feed finds that the total cost $C(x)$ in dollars of producing and selling x units is given by

$$C(x) = 20x + 100.$$

Management plans to charge $24 per unit for the feed.

(a) Determine the revenue function.

SOLUTION Revenue is the product of the price per unit, $p = 24$, and the number of units sold, x. That is,

$$R(x) = 24x.$$

(b) How many units must be sold for the firm to break even?

SOLUTION The firm will break even (no profit and no loss) as long as revenue just equals cost. Set the revenue function equal to the cost function and solve.

$$R(x) = C(x)$$
$$24x = 20x + 100$$
$$4x = 100$$
$$x = 25$$

The firm breaks even by selling 25 units, which is the break-even quantity. The graphs of $R(x) = 24x$ and $C(x) = 20x + 100$ are shown in Figure 11. The break-even point (where $x = 25$) is shown on the graph. If the company sells more than 25 units (if $x > 25$), it makes a profit. If it sells fewer than 25 units, it loses money.

(c) What is the profit if 100 units of feed are sold?

SOLUTION Use the formula for profit $P(x)$.

$$P(x) = R(x) - C(x)$$
$$= 24x - (20x + 100)$$
$$= 4x - 100$$

Then $P(100) = 4(100) - 100 = 300$. The firm will make a profit of $300 from the sale of 100 units of feed.

(d) How many units must be sold to produce a profit of $900?

SOLUTION Let $P(x) = 900$ in the equation $P(x) = 4x - 100$ and solve for x.

$$900 = 4x - 100$$
$$1000 = 4x$$
$$x = 250$$

Sales of 250 units will produce $900 profit. **TRY YOUR TURN 6**

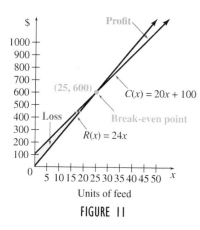

$C(x) = 20x + 100$
$R(x) = 24x$
Units of feed
FIGURE 11

YOUR TURN 6 Repeat Example 7(d), using a cost function $C(x) = 35x + 250$, a charge of $58 per unit, and a profit of $8030.

Temperature One of the most common linear relationships found in everyday situations deals with temperature. Recall that water freezes at 32° Fahrenheit and 0° Celsius, while it boils at 212° Fahrenheit and 100° Celsius.* The ordered pairs $(0, 32)$ and $(100, 212)$ are graphed in Figure 12 on axes showing Fahrenheit (F) as a function of Celsius (C). The line joining them is the graph of the function.

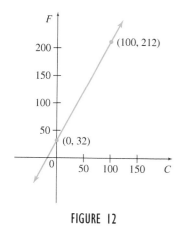

FIGURE 12

EXAMPLE 8 Temperature

Derive an equation relating F and C.

SOLUTION To derive the required linear equation, first find the slope using the given ordered pairs, $(0, 32)$ and $(100, 212)$.

$$m = \frac{212 - 32}{100 - 0} = \frac{180}{100} = \frac{9}{5}$$

*Gabriel Fahrenheit (1686–1736), a German physicist, invented his scale with 0° representing the temperature of an equal mixture of ice and ammonium chloride (a type of salt), and 96° as the temperature of the human body. (It is often said, erroneously, that Fahrenheit set 100° as the temperature of the human body. Fahrenheit's own words are quoted in *A History of the Thermometer and Its Use in Meteorology* by W. E. Knowles, Middleton: The Johns Hopkins Press, 1966, p. 75.) The Swedish astronomer Anders Celsius (1701–1744) set 0° and 100° as the freezing and boiling points of water.

The *F*-intercept of the graph is 32, so by the slope-intercept form, the equation of the line is

$$F = \frac{9}{5}C + 32.$$

With simple algebra this equation can be rewritten to give *C* in terms of *F*:

$$C = \frac{5}{9}(F - 32).$$

1.2 WARM-UP EXERCISES

W1. Evaluate $3(x - 2)^2 + 6(x + 4) - 5x + 4$ for $x = 5$.
(Sec. R.1)

W2. Graph the line $y = 7 - 2.5x$.
(Sec. 1.1)

1.2 EXERCISES

For Exercises 1–10, let $f(x) = 7 - 5x$ and $g(x) = 2x - 3$. Find the following.

1. $f(2)$　　　　　　**2.** $f(4)$

3. $f(-3)$　　　　　**4.** $f(-1)$

5. $g(1.5)$　　　　　**6.** $g(2.5)$

7. $g(-1/2)$　　　　**8.** $g(-3/4)$

9. $f(t)$　　　　　　**10.** $g(k^2)$

In Exercises 11–14, decide whether the statement is true or false.

11. To find the *x*-intercept of the graph of a linear function, we solve $y = f(x) = 0$, and to find the *y*-intercept, we evaluate $f(0)$.

12. The graph of $f(x) = -5$ is a vertical line.

13. The slope of the graph of a linear function cannot be undefined.

14. The graph of $f(x) = ax$ is a straight line that passes through the origin.

15. Describe what fixed costs and marginal costs mean to a company.

16. In a few sentences, explain why the price of a commodity not already at its equilibrium price should move in that direction.

17. Explain why a linear function may not be adequate for describing the supply and demand functions.

18. In your own words, describe the break-even quantity, how to find it, and what it indicates.

Write a linear cost function for each situation. Identify all variables used.

19. A Lake Tahoe resort charges a snowboard rental fee of $10 plus $2.25 per hour.

20. An Internet site for downloading music charges a $10 registration fee plus 99 cents per downloaded song.

21. A parking garage charges 2 dollars plus 75 cents per half-hour.

22. For a one-day rental, a car rental firm charges $44 plus 28 cents per mile.

Assume that each situation can be expressed as a linear cost function. Find the cost function in each case.

23. Fixed cost: $100; 50 items cost $1600 to produce.

24. Fixed cost: $35; 8 items cost $395 to produce.

25. Marginal cost: $75; 50 items cost $4300 to produce.

26. Marginal cost: $120; 700 items cost $96,500 to produce.

APPLICATIONS

Business and Economics

27. Supply and Demand　Suppose that the demand and price for a certain model of a youth wristwatch are related by

$$p = D(q) = 16 - 1.25q,$$

where *p* is the price (in dollars) and *q* is the quantity demanded (in hundreds). Find the price at each level of demand.

(a) 0 watches　　**(b)** 400 watches　　**(c)** 800 watches

Find the quantity demanded for the watch at each price.

(d) $8　　　　　　　　**(e)** $10

(f) $12　　　　　　　　**(g)** Graph $p = 16 - 1.25q$.

Suppose the price and supply of the watch are related by

$$p = S(q) = 0.75q,$$

where *p* is the price (in dollars) and *q* is the quantity supplied (in hundreds) of watches. Find the quantity supplied at each price.

(h) $0   **(i)** $10   **(j)** $20

(k) Graph $p = 0.75q$ on the same axis used for part (g).

(l) Find the equilibrium quantity and the equilibrium price.

28. Supply and Demand Suppose that the demand and price for strawberries are related by

$$p = D(q) = 5 - 0.25q,$$

where p is the price (in dollars) and q is the quantity demanded (in hundreds of quarts). Find the price at each level of demand.

(a) 0 quarts   **(b)** 400 quarts   **(c)** 840 quarts

Find the quantity demanded for the strawberries at each price.

(d) $4.50   **(e)** $3.25

(f) $2.40   **(g)** Graph $p = 5 - 0.25q$.

Suppose the price and supply of strawberries are related by

$$p = S(q) = 0.25q,$$

where p is the price (in dollars) and q is the quantity supplied (in hundreds of quarts) of strawberries. Find the quantity supplied at each price.

(h) $0   **(i)** $2   **(j)** $4.50

(k) Graph $p = 0.75q$ on the same axis used for part (g).

(l) Find the equilibrium quantity and the equilibrium price.

29. Supply and Demand Let the supply and demand functions for butter pecan ice cream be given by

$$p = S(q) = \frac{2}{5}q \quad \text{and} \quad p = D(q) = 100 - \frac{2}{5}q,$$

where p is the price in dollars and q is the number of 10-gallon tubs.

(a) Graph these on the same axes.

(b) Find the equilibrium quantity and the equilibrium price. (*Hint:* The way to divide by a fraction is to multiply by its reciprocal.)

30. Supply and Demand Let the supply and demand functions for sugar be given by

$$p = S(q) = 1.4q - 0.6 \quad \text{and}$$
$$p = D(q) = -2q + 3.2,$$

where p is the price per pound and q is the quantity in thousands of pounds.

(a) Graph these on the same axes.

(b) Find the equilibrium quantity and the equilibrium price.

31. Supply and Demand Suppose that the supply function for honey is $p = S(q) = 0.3q + 2.7$, where p is the price in dollars for an 8-oz container and q is the quantity in barrels. Suppose also that the equilibrium price is $4.50 and the demand is 2 barrels when the price is $6.10. Find an equation for the demand function, assuming it is linear.

32. Supply and Demand Suppose that the supply function for walnuts is $p = S(q) = 0.25q + 3.6$, where p is the price in

dollars per pound and q is the quantity in bushels. Suppose also that the equilibrium price is $5.85, and the demand is 4 bushels when the price is $7.60. Find an equation for the demand function, assuming it is linear.

33. Break-Even Analysis Producing x units of tacos costs $C(x) = 5x + 20$; revenue is $R(x) = 15x$, where $C(x)$ and $R(x)$ are in dollars.

(a) What is the break-even quantity?

(b) What is the profit from 100 units?

(c) How many units will produce a profit of $500?

34. Break-Even Analysis To produce x units of a religious medal costs $C(x) = 12x + 39$. The revenue is $R(x) = 25x$. Both $C(x)$ and $R(x)$ are in dollars.

(a) Find the break-even quantity.

(b) Find the profit from 250 units.

(c) Find the number of units that must be produced for a profit of $130.

35. T-Shirt Cost Joanne Wendelken sells silk-screened T-shirts at community festivals and crafts fairs. Her marginal cost to produce one T-shirt is $3.50. Her total cost to produce 60 T-shirts is $300, and she sells them for $9 each.

(a) Find the linear cost function for Joanne's T-shirt production.

(b) How many T-shirts must she produce and sell in order to break even?

(c) How many T-shirts must she produce and sell to make a profit of $500?

36. Publishing Costs Alfred Juarez owns a small publishing house specializing in Latin American poetry. His fixed cost to produce a typical poetry volume is $525, and his total cost to produce 1000 copies of the book is $2675. His books sell for $4.95 each.

(a) Find the linear cost function for Alfred's book production.

(b) How many poetry books must he produce and sell in order to break even?

(c) How many books must he produce and sell to make a profit of $1000?

37. Marginal Cost of Coffee The manager of a restaurant found that the cost to produce 100 cups of coffee is $11.02, while the cost to produce 400 cups is $40.12. Assume the cost $C(x)$ is a linear function of x, the number of cups produced.

(a) Find a formula for $C(x)$.

(b) What is the fixed cost?

(c) Find the total cost of producing 1000 cups.

(d) Find the total cost of producing 1001 cups.

(e) Find the marginal cost of the 1001st cup.

(f) What is the marginal cost of *any* cup and what does this mean to the manager?

38. Marginal Cost of a New Plant In deciding whether to set up a new manufacturing plant, company analysts have decided that a linear function is a reasonable estimation for the total cost $C(x)$ in dollars to produce x items. They estimate the cost to produce

10,000 items as $547,500, and the cost to produce 50,000 items as $737,500.

(a) Find a formula for $C(x)$.

(b) Find the fixed cost.

(c) Find the total cost to produce 100,000 items.

(d) Find the marginal cost of the items to be produced in this plant and what does this mean to the manager?

Break-Even Analysis **You are the manager of a firm. You are considering the manufacture of a new product, so you ask the accounting department for cost estimates and the sales department for sales estimates. After you receive the data, you must decide whether to go ahead with production of the new product. Analyze the data in Exercises 39–42 (find a break-even quantity) and then decide what you would do in each case. Also write the profit function.**

39. $C(x) = 85x + 900$; $R(x) = 105x$; no more than 38 units can be sold.

40. $C(x) = 105x + 6000$; $R(x) = 250x$; no more than 400 units can be sold.

41. $C(x) = 70x + 500$; $R(x) = 60x$ (*Hint*: What does a negative break-even quantity mean?)

42. $C(x) = 1000x + 5000$; $R(x) = 900x$

43. Break-Even Analysis Suppose that the fixed cost for a product is $400 and the break-even quantity is 80. Find the marginal profit (the slope of the linear profit function).

44. Break-Even Analysis Suppose that the fixed cost for a product is $650 and the break-even quantity is 25. Find the marginal profit (the slope of the linear profit function).

Life Sciences

45. Deer Ticks Deer ticks are of concern because they can carry the parasite that causes Lyme disease. One study found a relationship between the density of acorns produced in the fall and the density of deer tick larvae the following spring. The relationship can be approximated by the linear function

$$f(x) = 34x + 230,$$

where x is the number of acorns per square meter in the fall, and $f(x)$ is the number of deer tick larvae per 400 square meters the following spring. *Source: Science*.

(a) Approximately how many acorns per square meter would result in 1000 deer tick larvae per 400 square meters?

(b) Find and interpret the slope of the line.

46. Tobacco Deaths The U.S. Centers for Disease Control and Prevention projects that tobacco could soon be the leading cause of death in the world. In 1990, 35 million years of healthy life were lost globally due to tobacco. This quantity was rising linearly at a rate of about 28 million years each decade. In contrast, 100 million years of healthy life were lost due to

diarrhea, with the rate falling linearly 22 million years each decade. *Source: Science*.

(a) Write the years of healthy life in millions lost globally to tobacco as a linear function $f_t(t)$ of the years, t, since 1990.

(b) Write the years of healthy life in millions lost globally to diarrhea as a linear function $f_d(t)$ of the years, t, since 1990.

(c) Using your answers to parts (a) and (b), find in what year the amount of healthy life lost to tobacco was expected to first equal that lost to diarrhea.

Physical Sciences

47. Temperature Use the formula for conversion between Fahrenheit and Celsius derived in Example 8 to convert each temperature.

(a) 58°F to Celsius   (b) −20°F to Celsius

(c) 50°C to Fahrenheit

48. Body Temperature You may have heard that the average temperature of the human body is 98.6°. Recent experiments show that the actual figure is closer to 98.2°. The figure of 98.6 comes from experiments done by Carl Wunderlich in 1868. But Wunderlich measured the temperatures in degrees Celsius and rounded the average to the nearest degree, giving 37°C as the average temperature. *Source: Science News*.

(a) What is the Fahrenheit equivalent of 37°C?

(b) Given that Wunderlich rounded to the nearest degree Celsius, his experiments tell us that the actual average human body temperature is somewhere between 36.5°C and 37.5°C. Find what this range corresponds to in degrees Fahrenheit.

49. Temperature Find the temperature at which the Celsius and Fahrenheit temperatures are numerically equal.

General Interest

50. Education Cost A recent budget for the California State University system projected a fixed cost of $486,000 at each of five off-campus centers, plus a marginal cost of $1140 per student. *Source: California State University*.

(a) Find a formula for the cost at each center, $C(x)$, as a linear function of x, the number of students.

(b) The budget projected 500 students at each center. Calculate the total cost at each center.

(c) Suppose, due to budget cuts, that each center is limited to $1 million. What is the maximum number of students that each center can then support?

YOUR TURN ANSWERS ▰▰▰

1. 25

2. 7600 and 4400

3. 8000 watermelons and $3.20 per watermelon

4. $C(x) = 15x + 730$ **5.** $C(x) = 8.2x + 7145$ **6.** 360

1.3 The Least Squares Line

In Example 1 in this section, we show how to answer such questions using the method of least squares.

We use past data to find trends and to make tentative predictions about the future. The only assumption we make is that the data are related linearly—that is, if we plot pairs of data, the resulting points will lie close to some line. This method cannot give exact answers. The best we can expect is that, if we are careful, we will get a reasonable approximation.

The table lists the number of accidental deaths per 100,000 people in the United States through the past century. ***Source: National Center for Health Statistics.*** If you were a manager at an insurance company, these data could be very important. You might need to make some predictions about how much you will pay out next year in accidental death benefits, and even a very tentative prediction based on past trends is better than no prediction at all.

The first step is to draw a **scatterplot**, as we have done in Figure 13. Notice that the points lie approximately along a line, which means that a linear function may give a good approximation of the data. If we select two points and find the line that passes through them, as we did in Section 1.1, we will get a different line for each pair of points, and in some cases the lines will be very different. We want to draw one line that is simultaneously close to all the points on the graph, but many such lines are possible, depending upon how we define the phrase "simultaneously close to all the points." How do we decide on the best possible line? Before going on, you might want to try drawing the line you think is best on Figure 13 below.

The line used most often in applications is that in which the sum of the squares of the vertical distances from the data points to the line is as small as possible. Such a line is called the **least squares line**. The least squares line for the data in Figure 13 is drawn in Figure 14. How does the line compare with the one you drew on Figure 13? It may not be exactly the same, but should appear similar.

In Figure 14, the vertical distances from the points to the line are indicated by d_1, d_2, and so on, up through d_{10} (read "d-sub-one, d-sub-two, d-sub-three," and so on). For n points, corresponding to the n pairs of data, the least squares line is found by minimizing the sum $(d_1)^2 + (d_2)^2 + (d_3)^2 + \cdots + (d_n)^2$.

We often use **summation notation** to write the sum of a list of numbers. The Greek letter sigma, Σ, is used to indicate "the sum of." For example, we write the sum $x_1 + x_2 + \cdots + x_n$, where n is the number of data points, as

$$x_1 + x_2 + \cdots + x_n = \Sigma x.$$

Similarly, Σxy means $x_1 y_1 + x_2 y_2 + \cdots + x_n y_n$, and so on.

Accidental Death Rate	
Year	**Death Rate**
1920	71.2
1930	80.5
1940	73.4
1950	60.3
1960	52.1
1970	56.2
1980	46.5
1990	36.9
2000	34.0
2010	39.1

FIGURE 13

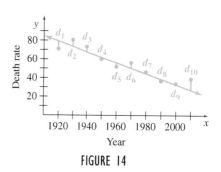

FIGURE 14

CAUTION | Note that Σx^2 means $x_1^2 + x_2^2 + \cdots + x_n^2$, which is *not* the same as squaring Σx. When we square Σx, we write it as $(\Sigma x)^2$.

For the least squares line, the sum of the distances we are to minimize, $d_1^2 + d_2^2 + \cdots + d_n^2$, is written as

$$d_1^2 + d_2^2 + \cdots + d_n^2 = \Sigma d^2.$$

To calculate the distances, we let $(x_1, y_1), (x_2, y_2), \cdots, (x_n, y_n)$ be the actual data points and we let the least squares line be $Y = mx + b$. We use Y in the equation instead of y to distinguish the predicted values (Y) from the y-value of the given data points. The predicted value of Y at x_1 is $Y_1 = mx_1 + b$, and the distance, d_1, between the actual y-value y_1 and the predicted value Y_1 is

$$d_1 = |Y_1 - y_1| = |mx_1 + b - y_1|.$$

Likewise,

$$d_2 = |Y_2 - y_2| = |mx_2 + b - y_2|,$$

and

$$d_n = |Y_n - y_n| = |mx_n + b - y_n|.$$

The sum to be minimized becomes

$$\Sigma d^2 = (mx_1 + b - y_1)^2 + (mx_2 + b - y_2)^2 + \cdots + (mx_n + b - y_n)^2$$
$$= \Sigma(mx + b - y)^2,$$

where $(x_1, y_1), (x_2, y_2), \cdots, (x_n, y_n)$ are known and m and b are to be found.

The method of minimizing this sum requires advanced techniques in multivariable calculus and is not given here. To obtain the equation for the least squares line, a system of equations must be solved, producing the following formulas for determining the slope m and y-intercept b.*

Least Squares Line

The **least squares line** $Y = mx + b$ that gives the best fit to the data points $(x_1, y_1), (x_2, y_2), \ldots, (x_n, y_n)$ has slope m and y-intercept b given by

$$m = \frac{n(\Sigma xy) - (\Sigma x)(\Sigma y)}{n(\Sigma x^2) - (\Sigma x)^2} \quad \text{and} \quad b = \frac{\Sigma y - m(\Sigma x)}{n}.$$

EXAMPLE 1   Least Squares Line

APPLY IT   Calculate the least squares line for the accidental death rate data.

SOLUTION

Method I
Calculating by Hand

To find the least squares line for the given data, we first find the required sums. To reduce the size of the numbers, we rescale the year data. Let x represent the years since 1900, so that, for example, $x = 20$ corresponds to the year 1920. Let y represent the death rate. We then calculate the values in the xy, x^2, and y^2 columns and find their totals. (The column headed y^2 will be used later.) Note that the number of data points is $n = 10$.

*See Exercise 9 at the end of this section.

		Least Squares Calculations		
x	y	xy	x^2	y^2
20	71.2	1424	400	5069.44
30	80.5	2415	900	6480.25
40	73.4	2936	1600	5387.56
50	60.3	3015	2500	3636.09
60	52.1	3126	3600	2714.41
70	56.2	3934	4900	3158.44
80	46.5	3720	6400	2162.25
90	36.9	3321	8100	1361.61
100	34.0	3400	10,000	1156.00
110	39.1	4301	12,100	1528.81
$\Sigma x = 650$	$\Sigma y = 550.2$	$\Sigma xy = 31{,}592$	$\Sigma x^2 = 50{,}500$	$\Sigma y^2 = 32{,}654.86$

Putting the column totals into the formula for the slope m, we get

$$m = \frac{n(\Sigma xy) - (\Sigma x)(\Sigma y)}{n(\Sigma x^2) - (\Sigma x)^2} \qquad \text{Formula for } m$$

$$= \frac{10(31{,}592) - (650)(550.2)}{10(50{,}500) - (650)^2} \qquad \text{Substitute from the table.}$$

$$= \frac{315{,}920 - 357{,}630}{505{,}000 - 422{,}500} \qquad \text{Multiply.}$$

$$= \frac{-41{,}710}{82{,}500} \qquad \text{Subtract.}$$

$$= -0.5055757576 \approx -0.506.$$

The significance of m is that the death rate per 100,000 people is tending to drop (because of the negative) at a rate of 0.506 per year.

Now substitute the value of m and the column totals in the formula for b:

$$b = \frac{\Sigma y - m(\Sigma x)}{n} \qquad \text{Formula for } b$$

$$= \frac{550.2 - (-0.5055757576)(650)}{10} \qquad \text{Substitute.}$$

$$= \frac{550.2 - (-328.6242424)}{10} \qquad \text{Multiply.}$$

$$= \frac{878.8242424}{10} = 87.88242424 \approx 87.9.$$

Substitute m and b into the least squares line, $Y = mx + b$; the least squares line that best fits the 10 data points has equation

$$Y = -0.506x + 87.9.$$

This gives a mathematical description of the relationship between the year and the number of accidental deaths per 100,000 people. The equation can be used to predict y from a given value of x, as we will show in Example 2. As we mentioned before, however, caution must be exercised when using the least squares equation to predict data points that are far from the range of points on which the equation was modeled.

> **CAUTION** In computing m and b, we rounded the final answer to three digits because the original data were known only to three digits. It is important, however, *not* to round any of the intermediate results (such as Σx^2) because round-off error may have a detrimental effect on the accuracy of the answer. Similarly, it is important not to use a rounded-off value of m when computing b.

Method 2
Graphing Calculator

FIGURE 15

The calculations for finding the least squares line are often tedious, even with the aid of a calculator. Fortunately, many calculators can calculate the least squares line with just a few keystrokes. For purposes of illustration, we will show how the least squares line in this example is found with a TI-84 Plus C graphing calculator.

We begin by entering the data into the calculator. We will be using the first two lists, called L_1 and L_2. Choosing the STAT menu, then choosing the fourth entry ClrList, we enter L_1, L_2, to indicate the lists to be cleared. Now we press STAT again and choose the first entry EDIT, which brings up the blank lists. As before, rescale the years (letting 20 correspond to 1920, and so on) and then enter the numbers in L_1. We put the death rate in L_2, giving the screen shown in Figure 15.

Quit the editor, press STAT again, and choose CALC instead of EDIT. Then choose item 4 LinReg $(ax + b)$ to get the values of a (the slope) and b (the y-intercept) for the least squares line, as shown in Figure 16. With a and b rounded to three decimal places, the least squares line is $Y = -0.506x + 87.9$. A graph of the data points and the line is shown in Figure 17.

FIGURE 16

FIGURE 17

For more details on finding the least squares line with a graphing calculator, see the *Graphing Calculator and Excel Spreadsheet Manual* available with this book.

Method 3
Spreadsheet

Many computer spreadsheet programs can also find the least squares line. Figure 18 shows the scatterplot and least squares line for the accidental death rate data using an Excel spreadsheet. The scatterplot was found using the Marked Scatter chart from the Gallery and the line was found using the Add Trendline command under the Chart menu. For details, see the *Graphing Calculator and Excel Spreadsheet Manual* available with this book.

YOUR TURN 1 Calculate the least squares line for the following data.

x	1	2	3	4	5	6
y	3	4	6	5	7	8

TRY YOUR TURN 1

FIGURE 18

EXAMPLE 2   Least Squares Line

What do we predict the accidental death rate to be in 2020?

SOLUTION   Use the least squares line equation given above with $x = 120$.

$$Y = -0.506x + 87.9$$
$$= -0.506(120) + 87.9$$
$$\approx 27.2$$

The accidental death rate in 2020 is predicted to be about 27.2 per 100,000 population. In this case, we will have to wait until the 2020 data become available to see how accurate our prediction is. We have observed, however, that the accidental death rate began to go up after 2000 and was 39.1 per 100,000 population in 2010. This illustrates the danger of extrapolating beyond the data.

EXAMPLE 3   Least Squares Line

In what year is the death rate predicted to drop below 26 per 100,000 population?

SOLUTION   Let $Y = 26$ in the equation above and solve for x.

$$26 = -0.506x + 87.9$$
$$-61.9 = -0.506x \qquad \text{Subtract 87.9 from both sides.}$$
$$x \approx 122.3 \qquad \text{Divide both sides by } -0.506.$$

This implies that sometime in the year 2022 (122 years after 1900) the death rate drops below 26 per 100,000 population.

Correlation   Although the least squares line can always be found, it may not be a good

model. For example, if the data points are widely scattered, no straight line will model the data accurately. One measure of how well the original data fits a straight line is the **correlation coefficient**, denoted by r, which can be calculated by the following formula.

Correlation Coefficient

$$r = \frac{n(\Sigma xy) - (\Sigma x)(\Sigma y)}{\sqrt{n(\Sigma x^2) - (\Sigma x)^2} \cdot \sqrt{n(\Sigma y^2) - (\Sigma y)^2}}$$

Although the expression for r looks daunting, remember that each of the summations, Σx, Σy, Σxy, and so on, are just the totals from a table like the one we prepared for the data on accidental deaths. Also, with a calculator, the arithmetic is no problem! Furthermore, statistics software and many calculators can calculate the correlation coefficient for you.

The correlation coefficient measures the strength of the linear relationship between two variables. It was developed by statistics pioneer Karl Pearson (1857–1936). The correlation coefficient r is between 1 and -1 or is equal to 1 or -1. Values of exactly 1 or -1 indicate that the data points lie *exactly* on the least squares line. If $r = 1$, the least squares line has a positive slope; $r = -1$ gives a negative slope. If $r = 0$, there is no linear correlation between the data points (but some *nonlinear* function might provide an excellent fit for the data). A correlation coefficient of zero may also indicate that the data fit a horizontal line. To investigate what is happening, it is always helpful to sketch a scatterplot of the data. Some scatterplots that correspond to these values of r are shown in Figure 19 on the next page.

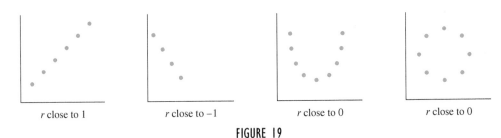

| r close to 1 | r close to -1 | r close to 0 | r close to 0 |

FIGURE 19

A value of r close to 1 or -1 indicates the presence of a linear relationship. The exact value of r necessary to conclude that there is a linear relationship depends upon n, the number of data points, as well as how confident we want to be of our conclusion. For details, consult a text on statistics.*

EXAMPLE 4 Correlation Coefficient

Find r for the data on accidental death rates in Example 1.

SOLUTION

Method 1
Calculating by Hand

From the table in Example 1,

$$\Sigma x = 650, \ \Sigma y = 550.2, \ \Sigma xy = 31{,}592, \ \Sigma x^2 = 50{,}500,$$
$$\Sigma y^2 = 32{,}654.86, \quad \text{and} \quad n = 10.$$

Substituting these values into the formula for r gives

$$r = \frac{n(\Sigma xy) - (\Sigma x)(\Sigma y)}{\sqrt{n(\Sigma x^2) - (\Sigma x)^2} \cdot \sqrt{n(\Sigma y^2) - (\Sigma y)^2}} \qquad \text{Formula for } r$$

$$= \frac{10(31{,}592) - (650)(550.2)}{\sqrt{10(50{,}500) - (650)^2} \cdot \sqrt{10(32{,}654.86) - (550.2)^2}} \qquad \text{Substitute.}$$

$$= \frac{315{,}920 - 357{,}630}{\sqrt{505{,}000 - 422{,}500} \cdot \sqrt{326{,}548.6 - 302{,}720.04}} \qquad \text{Multiply.}$$

$$= \frac{-41{,}710}{\sqrt{82{,}500} \cdot \sqrt{23{,}828.56}} \qquad \text{Subtract.}$$

$$= \frac{-41{,}710}{44{,}337.97695} \qquad \text{Take square roots and multiply.}$$

$$= -0.9407285327 \approx -0.941.$$

This is a high correlation, which agrees with our observation that the data fit a line quite well.

Method 2
Graphing Calculator

Most calculators that give the least squares line will also give the correlation coefficient. To do this on the TI-84 Plus C, press the second function CATALOG and go down the list to the entry DiagnosticOn. Press ENTER at that point, then press STAT, CALC, and choose item 4 to get the display in Figure 20. The result is the same as we got by hand. The command DiagnosticOn need be entered only once, and the correlation coefficient will always appear in the future.

```
LinReg
 y=ax+b
 a=-.5055757576
 b=87.88242424
 r²=.8849701723
 r=-.9407285327
 ■
```

FIGURE 20

*For example, see *Introductory Statistics*, 9th edition, by Neil A. Weiss, Boston, Mass.: Pearson, 2012.

Method 3
Spreadsheet

YOUR TURN 2 Find *r* for the following data.

x	1	2	3	4	5	6
y	3	4	6	5	7	8

Many computer spreadsheet programs have a built-in command to find the correlation coefficient. For example, in Excel, use the command "= CORREL(A1:A10,B1:B10)" to find the correlation of the 10 data points stored in columns A and B. For more details, see the *Graphing Calculator and Excel Spreadsheet Manual* available with this text.

TRY YOUR TURN 2

The square of the correlation coefficient gives the fraction of the variation in *y* that is explained by the linear relationship between *x* and *y*. Consider Example 4, where $r^2 = (-0.941)^2 = 0.885$. This means that 88.5% of the variation in *y* is explained by the linear relationship found earlier in Example 1. The remaining 11.5% comes from the scattering of the points about the line.

EXAMPLE 5 Average Expenditure per Pupil Versus Test Scores

Many states and school districts debate whether or not increasing the amount of money spent per student will guarantee academic success. The following scatterplot shows the average eighth grade reading score on the National Assessment of Education Progress (NAEP) for the 50 states and the District of Columbia in 2011 plotted against the average expenditure per pupil in 2011. Explore how the correlation coefficient is affected by the inclusion of the District of Columbia, which spent $18,475 per pupil and had a NAEP score of 242. *Source: U.S. Census Bureau and National Center for Education Statistics.*

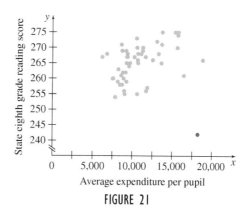

FIGURE 21

SOLUTION A spreadsheet was used to create a plot of the points shown in Figure 21. Washington D.C. corresponds to the red point in the lower right, which is noticeably separate from all the other points. Using the original data, the correlation coefficient when Washington D.C. is included is 0.1798, indicating that there is not a strong linear correlation. Excluding Washington D.C. raises the correlation coefficient to 0.4427, which is a somewhat stronger indication of a linear correlation. This illustrates that one extreme data point that is separate from the others, known as an **outlier**, can have a strong effect on the correlation coefficient.

Even if the correlation between average expenditure per pupil and reading score in Example 5 was high, this would not prove that spending more per pupil causes high reading scores. To prove this would require further research. It is a common statistical fallacy to assume that correlation implies causation. Perhaps the correlation is due to a third underlying variable. In Example 5, perhaps states with wealthier families spend more per pupil, and the students read better because wealthier families have greater access to reading material. Determining the truth requires careful research methods that are beyond the scope of this textbook.

1.3 EXERCISES

1. Suppose a positive linear correlation is found between two quantities. Does this mean that one of the quantities increasing causes the other to increase? If not, what does it mean?

2. Given a set of points, the least squares line formed by letting x be the independent variable will not necessarily be the same as the least squares line formed by letting y be the independent variable. Give an example to show why this is true.

3. For the following table of data,

x	1	2	3	4	5	6	7	8	9	10
y	0	0.5	1	2	2.5	3	3	4	4.5	5

   (a) draw a scatterplot.

   (b) calculate the correlation coefficient.

   (c) calculate the least squares line and graph it on the scatterplot.

   (d) predict the y-value when x is 11.

The following problem is reprinted from the November 1989 *Actuarial Examination on Applied Statistical Methods*. *Source: Society of Actuaries*.

4. You are given

X	6.8	7.0	7.1	7.2	7.4
Y	0.8	1.2	0.9	0.9	1.5

   Determine r^2, the coefficient of determination for the regression of Y on X. Choose one of the following. (*Note:* The coefficient of determination is defined as the square of the correlation coefficient.)

   (a) 0.3   (b) 0.4   (c) 0.5   (d) 0.6   (e) 0.7

5. Consider the following table of data.

x	1	1	2	2	9
y	1	2	1	2	9

   (a) Calculate the least squares line and the correlation coefficient.

   (b) Repeat part (a), but this time delete the last point.

   (c) Draw a graph of the data, and use it to explain the dramatic difference between the answers to parts (a) and (b).

6. Consider the following table of data.

x	1	2	3	4	9
y	1	2	3	4	-20

   (a) Calculate the least squares line and the correlation coefficient.

   (b) Repeat part (a), but this time delete the last point.

   (c) Draw a graph of the data, and use it to explain the dramatic difference between the answers to parts (a) and (b).

7. Consider the following table of data.

x	1	2	3	4
y	1	1	1	1.1

   (a) Calculate the correlation coefficient.

   (b) Sketch a graph of the data.

   (c) Based on how closely the data fits a straight line, is your answer to part (a) surprising? Discuss the extent to which the correlation coefficient describes how well the data fit a horizontal line.

8. Consider the following table of data.

x	0	1	2	3	4
y	4	1	0	1	4

   (a) Calculate the least squares line and the correlation coefficient.

   (b) Sketch a graph of the data.

   (c) Comparing your answers to parts (a) and (b), does a correlation coefficient of 0 mean that there is no relationship between the x- and y- values? Would some curve other than a line fit the data better? Explain.

9. The formulas for the least squares line were found by solving the system of equations

$$nb + (\Sigma x)m = \Sigma y$$
$$(\Sigma x)b + (\Sigma x^2)m = \Sigma xy.$$

   Solve the above system for b and m to show that

$$m = \frac{n(\Sigma xy) - (\Sigma x)(\Sigma y)}{n(\Sigma x^2) - (\Sigma x)^2} \quad \text{and}$$

$$b = \frac{\Sigma y - m(\Sigma x)}{n}.$$

APPLICATIONS

Business and Economics

10. **Consumer Durable Goods** The total value of consumer durable goods (goods purchased by households for their nonbusiness use with a life expectancy of at least three years) has grown at an approximately linear rate in recent years. The annual data for the years 2003 through 2012 can be summarized as follows, where x represents the years since 2000 and y the total value of consumer durable goods in trillions of dollars. *Source: Bureau of Economic Analysis*.

   $n = 10$ $\Sigma x = 75$ $\Sigma x^2 = 645$
   $\Sigma y = 43.792$ $\Sigma y^2 = 193.0496$ $\Sigma xy = 338.271$

   (a) Find an equation for the least squares line.

   (b) Use your result from part (a) to predict the total value of consumer durable goods in the year 2020.

(c) If this growth continues linearly, in what year will the total value of consumer durable goods first reach at least 6 trillion dollars?

(d) Find and interpret the correlation coefficient.

11. Decrease in Banks The number of banks in the United States has been dropping steadily since 1984, and the trend in recent years has been roughly linear. The annual data for the years 2003 through 2012 can be summarized as follows, where x represents the years since 2000 and y the number of banks, in thousands, in the United States. *Source: FDIC.*

$$n = 10 \qquad \Sigma x = 75 \qquad \Sigma x^2 = 645$$
$$\Sigma y = 70.457 \qquad \Sigma y^2 = 499.481335 \qquad \Sigma xy = 512.775$$

(a) Find an equation for the least squares line.

(b) Use your result from part (a) to predict the number of U.S. banks in the year 2020.

(c) If this trend continues linearly, in what year will the number of U.S. banks drop below 4000?

(d) Find and interpret the correlation coefficient.

12. Internet The percent of households with Internet use at home has been growing steadily, as shown by the following table. *Source: U.S. Census Bureau.*

Year	2000	2003	2007	2009	2012
Percent of Households	41.5	54.7	61.7	68.7	74.8

(a) Find an equation for the least squares line, letting x equal the number of years since 2000.

(b) Based on your answer to part (a), at approximately what rate is the percent of households with Internet use at home growing per year?

(c) Use your result from part (a) to predict the percent of households with Internet use at home in the year 2015.

(d) If this trend continues linearly, in what year will the percent of households with Internet use at home first exceed 90%?

(e) Find and interpret the correlation coefficient.

13. Landlines The percent of U.S. households with telephone landlines has decreased at a roughly linear rate, as shown by the following table. *Source: Centers for Disease Control and Prevention.*

Year	Percent of Households
2005	89.7
2006	84.1
2007	81.9
2008	77.9
2009	73.5
2010	68.2
2011	63.8
2012	59.6

(a) Find the equation of the least squares line, letting x equal the number of years since 2000.

(b) Based on your answer to part (a), at approximately what rate is the percent of households with landlines decreasing per year?

(c) Use your result from part (a) to predict the percent of households with landlines in the year 2015.

(d) If this trend continues linearly, in what year will the percent of households with landlines first dip below 40%?

(e) Find and interpret the correlation coefficient.

14. Consumer Credit The total amount of consumer credit has been increasing steadily in recent years. The following table gives the total U.S. outstanding consumer credit (in billions of dollars). *Source: Federal Reserve.*

Year	Consumer Credit	Year	Consumer Credit
2004	2219.5	2009	2553.5
2005	2319.8	2010	2648.1
2006	2415.0	2011	2757.0
2007	2551.9	2012	2924.3
2008	2592.1	2013	3099.2

(a) Find an equation for the least squares line, letting x equal the number of years since 2000.

(b) Based on your answer to part (a), at approximately what rate is consumer credit growing per year?

(c) Use your result from part (a) to predict the amount of consumer credit in 2015.

(d) If this trend continues linearly, in what year will the total debt first exceed $4000 billion?

(e) Find and interpret the correlation coefficient.

15. Mean Earnings The mean earnings (in dollars) of workers 18 years and over, by educational attainment, have increased steadily over the years and are given in the following table. Let x equal the number of years since 1900. *Source: U.S. Census.*

Year	High School Graduate	Bachelor's Degree	Year	High School Graduate	Bachelor's Degree
1976	8393	13,033	1996	22,154	38,112
1978	9834	15,291	1998	23,594	43,782
1980	11,314	18,075	2000	25,692	49,595
1982	12,560	20,272	2002	27,280	51,194
1984	13,893	23,072	2004	28,631	51,568
1986	15,120	26,511	2006	31,071	56,788
1988	16,750	28,344	2008	31,283	58,613
1990	17,820	31,112	2010	31,003	57,621
1992	18,737	32,629	2012	32,630	60,159
1994	20,248	37,224			

(a) Plot the data for high school graduates and for workers with a bachelor's degree. Do the data points lie in a linear pattern?

(b) Find an equation for the least squares line for the mean earnings of high school graduates. Calculate and interpret the correlation coefficient.

(c) Based on your answer from part (b), at approximately what rate are the mean earnings of high school graduates growing per year?

(d) Find an equation for the least squares line for the mean earnings of workers with a bachelor's degree. Calculate and interpret the correlation coefficient.

(e) Based on your answer from part (d), at approximately what rate are the mean earnings of workers with a bachelor's degree growing per year?

(f) If these trends continue linearly, in what year will the mean earnings for high school graduates exceed $75,000? In what year will workers with a bachelor's degree exceed $75,000?

16. Air Fares Using Expedia, a discount travel website, the American Airline prices for a one-way flight from New York City to various cities were recorded. The following table gives the distances from New York City to 14 selected cities, along with the airfare to each of these cities.

City	Distance, x (miles)	Price, y (dollars)
Chicago	802	191
Denver	1771	183
Kansas City	1198	193
Little Rock	1238	247
Los Angeles	2786	316
Minneapolis	1207	206
Nashville	892	151
Philadelphia	95	290
Phoenix	2411	223
Portland	2885	321
Reno	2705	288
St. Louis	948	152
San Diego	2762	267
Seattle	2815	341

(a) Plot the data. Do the data points lie in a linear pattern?

(b) Find the correlation coefficient. Combining this with your answer to part (a), does the cost of a ticket tend to go up with the distance flown?

(c) Find the equation for the least squares line, and use it to find the approximate marginal cost per mile to fly.

(d) Identify the outlier in the scatterplot. Discuss the reason why there would be a difference in price to this city.

Life Sciences

17. Bird Eggs The average length and width of various bird eggs are given in the following table. *Source: National Council of Teachers of Mathematics*.

Bird Name	Width (cm)	Length (cm)
Canada goose	5.8	8.6
Robin	1.5	1.9
Turtledove	2.3	3.1
Hummingbird	1.0	1.0
Raven	3.3	5.0

(a) Plot the points, putting the length on the y-axis and the width on the x-axis. Do the data appear to be linear?

(b) Find the least squares line, and plot it on the same graph as the data.

(c) Suppose there are birds with eggs even smaller than those of hummingbirds. Would the equation found in part (b) continue to make sense for all positive widths, no matter how small? Explain.

(d) Find the correlation coefficient.

18. Size of Hunting Parties In the 1960s, the famous researcher Jane Goodall observed that chimpanzees hunt and eat meat as part of their regular diet. Sometimes chimpanzees hunt alone, while other times they form hunting parties. The following table summarizes research on chimpanzee hunting parties, giving the size of the hunting party and the percentage of successful hunts. *Source: American Scientist and Mathematics Teacher*.

Number of Chimps in Hunting Party	Percentage of Successful Hunts
1	20
2	30
3	28
4	42
5	40
6	58
7	45
8	62
9	65
10	63
12	75
13	75
14	78
15	75
16	82

(a) Plot the data. Do the data points lie in a linear pattern?

(b) Find the correlation coefficient. Combining this with your answer to part (a), does the percentage of successful hunts tend to increase with the size of the hunting party?

(c) Find the equation of the least squares line, and graph it on your scatterplot.

19. Crickets Chirping Biologists have observed a linear relationship between the temperature and the frequency with which a cricket chirps. The following data were measured for the striped ground cricket. *Source: The Song of Insects*.

Temperature °F (x)	Chirps per Second (y)
88.6	20.0
71.6	16.0
93.3	19.8
84.3	18.4
80.6	17.1
75.2	15.5
69.7	14.7
82.0	17.1
69.4	15.4
83.3	16.2
79.6	15.0
82.6	17.2
80.6	16.0
83.5	17.0
76.3	14.4

(a) Find the equation for the least squares line for the data.

(b) Use the results of part (a) to determine how many chirps per second you would expect to hear from the striped ground cricket if the temperature were 73°F.

(c) Use the results of part (a) to determine what the temperature is when the striped ground crickets are chirping at a rate of 18 times per sec.

(d) Find the correlation coefficient.

Social Sciences

20. Pupil-Teacher Ratios The following table gives the national average pupil-teacher ratio in public schools over selected years. *Source: National Center for Education Statistics*.

Year	Ratio
1990	17.2
1994	17.3
1998	16.4
2002	15.9
2006	15.6
2010	15.3

(a) Find the equation for the least squares line. Let x correspond to the number of years since 1990 and let y correspond to the average number of pupils per teacher.

(b) Use your answer from part (a) to predict the pupil-teacher ratio in 2020. Does this seem realistic?

(c) Calculate and interpret the correlation coefficient.

21. Poverty Levels The following table lists how poverty level income cutoffs (in dollars) for a family of four have changed over time. *Source: U.S. Census Bureau*.

Year	Income
1980	8414
1985	10,989
1990	13,359
1995	15,569
2000	17,604
2005	19,971
2010	22,315

Let x represent the year, with $x = 0$ corresponding to 1980, and y represent the income in thousands of dollars.

(a) Plot the data. Do the data appear to lie along a straight line?

(b) Calculate the correlation coefficient. Does your result agree with your answer to part (a)?

(c) Find the equation of the least squares line.

(d) Use your answer from part (c) to predict the poverty level in the year 2018.

22. Ideal Partner Height In an introductory statistics course at Cornell University, 147 undergraduates were asked their own height and the ideal height for their ideal spouse or partner. For this exercise, we are including the data for only a representative sample of 10 of the students, as given in the following table. All heights are in inches. *Source: Chance*.

Height	Ideal Partner's Height
59	66
62	71
66	72
68	73
71	75
67	63
70	63
71	67
73	66
75	66

(a) Find the regression line and correlation coefficient for these data. What strange phenomenon do you observe?

(b) The first five data pairs are for female students and the second five for male students. Find the regression line and correlation coefficient for each set of data.

(c) Plot all the data on one graph, using different types of points to distinguish the data for the males and for the females. Using this plot and the results from part (b), explain the strange phenomenon that you observed in part (a).

23. SAT Scores At Hofstra University, all students take the math SAT before entrance, and most students take a mathematics placement test before registration. Recently, one professor collected the following data for 19 students in his Finite Mathematics class:

Math SAT	Placement Test	Math SAT	Placement Test	Math SAT	Placement Test
540	20	580	8	440	10
510	16	680	15	520	11
490	10	560	8	620	11
560	8	560	13	680	8
470	12	500	14	550	8
600	11	470	10	620	7
540	10				

(a) Find an equation for the least squares line. Let x be the math SAT and y be the placement test score.

(b) Use your answer from part (a) to predict the mathematics placement test score for a student with a math SAT score of 420.

(c) Use your answer from part (a) to predict the mathematics placement test score for a student with a math SAT score of 620.

(d) Calculate the correlation coefficient.

(e) Based on your answer to part (d), what can you conclude about the relationship between a student's math SAT and mathematics placement test score?

Physical Sciences

24. Length of a Pendulum Grandfather clocks use pendulums to keep accurate time. The relationship between the length of a pendulum L and the time T for one complete oscillation can be determined from the data in the table.* *Source: Gary Rockswold*.

L (ft)	T (sec)
1.0	1.11
1.5	1.36
2.0	1.57
2.5	1.76
3.0	1.92
3.5	2.08
4.0	2.22

(a) Plot the data from the table with L as the horizontal axis and T as the vertical axis.

(b) Find the least squares line equation and graph it simultaneously, if possible, with the data points. Does it seem to fit the data?

*The actual relationship is $L = 0.81T^2$, which is not a linear relationship. This illustrates that even if the relationship is not linear, a line can give a good approximation.

(c) Find the correlation coefficient and interpret it. Does it confirm your answer to part (b)?

25. Air Conditioning While shopping for an air conditioner, Adam Bryer consulted the following table, which gives a machine's BTUs and the square footage (ft^2) that it would cool.

ft² (x)	BTUs (y)
150	5000
175	5500
215	6000
250	6500
280	7000
310	7500
350	8000
370	8500
420	9000
450	9500

(a) Find the equation for the least squares line for the data.

(b) To check the fit of the data to the line, use the results from part (a) to find the BTUs required to cool a room of 150 ft², 280 ft², and 420 ft². How well do the actual data agree with the predicted values?

(c) Suppose Adam's room measures 230 ft². Use the results from part (a) to decide how many BTUs it requires. If air conditioners are available only with the BTU choices in the table, which would Adam choose?

(d) Why do you think the table gives ft² instead of ft³, which would give the volume of the room?

General Interest

26. Football The following data give the expected points for a football team with first down and 10 yards to go from various points on the field. (*Note:* $\Sigma x = 500$, $\Sigma x^2 = 33{,}250$, $\Sigma y = 20.668$, $\Sigma y^2 = 91.927042$, $\Sigma xy = 399.16$.) *Source: Operations Research*.

Yards from Goal (x)	Expected Points (y)
5	6.041
15	4.572
25	3.681
35	3.167
45	2.392
55	1.538
65	0.923
75	0.236
85	−0.637
95	−1.245

(a) Calculate the correlation coefficient. Does there appear to be a linear correlation?

(b) Find the equation of the least squares line.

(c) Use your answer from part (a) to predict the expected points when a team is at the 50-yd line.

27. Athletic Records The table shows the men's and women's outdoor world records (in seconds) in the 800-m run. *Source: Nature, Track and Field Athletics, Statistics in Sports, and The World Almanac and Book of Facts 2014.*

Year	Men's Record	Women's Record
1905	113.4	—
1915	111.9	—
1925	111.9	144
1935	109.7	135.6
1945	106.6	132
1955	105.7	125
1965	104.3	118
1975	103.7	117.48
1985	101.73	113.28
1995	101.11	113.28
2005	101.11	113.28
2013	100.91	113.28

Let x be the year, with $x = 0$ corresponding to 1900.

(a) Find the equation for the least squares line for the men's record (y) in terms of the year (x).

(b) Find the equation for the least squares line for the women's record.

(c) Suppose the men's and women's records continue to improve as predicted by the equations found in parts (a) and (b). In what year will the women's record catch up with the men's record? Do you believe that will happen? Why or why not?

(d) Calculate the correlation coefficient for both the men's and the women's record. What do these numbers tell you?

(e) Draw a plot of the data, and discuss to what extent a linear function describes the trend in the data.

28. Athletic Records An analysis similar to that of the previous exercise can be done for the 100-meter sprint, for which the winning times in the Summer Olympic Games are listed in the table in the next column. Let x be the number of years since 1900. *Source: The Olympics; Athens to Athens 1896–2004 and The World Almanac and Book of Facts 2014.*

(a) Find the regression line for the men's winning time.

(b) Find the regression line for the women's winning time.

Year	Men's Winning Time	Women's Winning Time
1900	11.00	
1904	11.00	
1908	10.80	
1912	10.80	
1920	10.80	
1924	10.60	
1928	10.80	12.20
1932	10.30	11.90
1936	10.30	11.50
1948	10.30	11.90
1952	10.40	11.50
1956	10.50	11.50
1960	10.20	11.00
1964	10.00	11.40
1968	9.95	11.08
1972	10.14	11.07
1976	10.06	11.08
1980	10.25	11.06
1984	9.99	10.97
1988	9.92	10.54
1992	9.96	10.82
1996	9.84	10.94
2000	9.87	10.75
2004	9.85	10.93
2008	9.69	10.78
2012	9.63	10.75

(c) Suppose the men's and women's winning times continue to improve as predicted by the equations found in parts (a) and (b). In what year will the women's winning time beat the men's winning time? (Remember that the Summer Olympic Games occur only every four years.)

(d) Calculate the correlation coefficient for both the men's and the women's winning time. What do these numbers tell you?

(e) Draw a plot of the data on the window $0 \leq x \leq 115$, $9 \leq Y \leq 12.5$, including the regression lines, and discuss to what extent a linear function describes the trend in the data.

29. Running If you think a marathon is a long race, consider the Hardrock 100, a 100.5 mile running race held in southwestern Colorado. The chart on the next page lists the times that the 2008 winner, Kyle Skaggs, arrived at various mileage points along the way. *Source: www.run100s.com.*

(a) What was Skagg's average speed?

(b) Graph the data, plotting time on the x-axis and distance on the y-axis. Do the data appear to lie approximately on a straight line?

(c) Find the equation for the least squares line, fitting distance as a linear function of time.

(d) Calculate the correlation coefficient. Does it indicate a good fit of the least squares line to the data?

(e) Based on your answer to part (d), what is a good value for Skagg's average speed? Compare this with your answer to part (a). Which answer do you think is better? Explain your reasoning.

Time (hours)	Miles
0	0
2.317	11.5
3.72	18.9
5.6	27.8
7.08	32.8
7.5	36.0
8.5	43.9
10.6	51.5
11.93	58.4
15.23	71.8
17.82	80.9
18.97	85.2
20.83	91.3
23.38	100.5

YOUR TURN ANSWERS ▬
1. $Y = 0.9429x + 2.2$

2. $r = 0.9429$

CHAPTER REVIEW

SUMMARY

In this chapter we studied linear functions, whose graphs are straight lines. We developed the slope-intercept and point-slope formulas, which can be used to find the equation of a line, given a point and the slope or given two points. We saw that lines have many applications in virtually every discipline. Lines are used through the rest of this book, so fluency in their use is important. We concluded the chapter by introducing the method of least squares, which is used to find an equation of the line that best fits a given set of data.

Slope of a Nonvertical Line   The slope of a nonvertical line is defined as the vertical change (the "rise") over the horizontal change (the "run") as one travels along the line. In symbols, taking two different points (x_1, y_1) and (x_2, y_2) on the line, the slope is

$$m = \frac{y_2 - y_1}{x_2 - x_1},$$

where $x_1 \neq x_2$.

Equations of Lines	**Equation**	**Description**
	$y = mx + b$	Slope-intercept form: slope m and y-intercept b.
	$y - y_1 = m(x - x_1)$	Point-slope form: slope m and line passes through (x_1, y_1).
	$x = k$	Vertical line: x-intercept k, no y-intercept (except when $k = 0$), undefined slope.
	$y = k$	Horizontal line: y-intercept k, no x-intercept (except when $k = 0$), slope 0.

Parallel Lines	Two lines are parallel if and only if they have the same slope, or if they are both vertical.

Perpendicular Lines	Two lines are perpendicular if and only if the product of their slopes is -1, or if one is vertical and the other horizontal.

Linear Function   A relationship f defined by

$$y = f(x) = mx + b,$$

for real numbers m and b, is a linear function.

Equilibrium   When supply and demand are equal, the economy is said to be at equilibrium. The equilibrium price of the commodity is the price found at the point where the supply and demand graphs for that commodity intersect. The equilibrium quantity is the quantity demanded and supplied at that same point.

Linear Cost Function   In a cost function of the form $C(x) = mx + b$, the m represents the marginal cost and b represents the fixed cost.

Revenue   The revenue, $R(x)$, is the amount of money that a company receives; it is the product of the price per unit p and the number of units sold x. That is,

$$R(x) = px.$$

Profit   The profit, $P(x)$, is the money a company makes after paying its costs; it is the difference between revenue $R(x)$ and cost $C(x)$. That is,

$$P(x) = R(x) - C(x).$$

Break-Even Quantity   The number of units, x, at which revenue just equals cost is the break-even quantity. It is found by setting revenue equal to cost. That is,

$$R(x) = C(x).$$

(The break-even quantity can also be found by setting profit equal to zero; $P(x) = 0$.)

Least Squares Line   The least squares line $Y = mx + b$ that gives the best fit to the data points (x_1, y_1), (x_2, y_2), . . . , (x_n, y_n) has slope m and y-intercept b given by the equations

$$m = \frac{n(\Sigma xy) - (\Sigma x)(\Sigma y)}{n(\Sigma x^2) - (\Sigma x)^2}$$

$$b = \frac{\Sigma y - m(\Sigma x)}{n}.$$

Correlation Coefficient

$$r = \frac{n(\Sigma xy) - (\Sigma x)(\Sigma y)}{\sqrt{n(\Sigma x^2) - (\Sigma x)^2} \; \sqrt{n(\Sigma y^2) - (\Sigma y)^2}}$$

KEY TERMS

To understand the concepts presented in this chapter, you should know the meaning and use of the following terms.
For easy reference, the section in the chapter where a word (or expression) was first used is provided.

1.1
mathematical model
ordered pair
Cartesian coordinate system
axes
origin
coordinates
quadrants
graph
intercepts
slope

linear equation
slope-intercept form
proportional
point-slope form
parallel
perpendicular
scatterplot

1.2
linear function
independent variable
dependent variable

supply
demand
Law of Supply
Law of Demand
surplus
shortage
equilibrium
equilibrium price
equilibrium quantity
fixed cost
marginal cost

linear cost function
revenue
profit
break-even quantity
break-even point

1.3
least squares line
summation notation
correlation coefficient
outlier

REVIEW EXERCISES

CONCEPT CHECK

Determine whether each statement is true or false, and explain why.

1. A given line can have more than one slope.

2. The equation $y = 3x + 4$ represents the equation of a line with slope 4.

3. The line $y = -2x + 5$ intersects the point $(3, -1)$.

4. The line that intersects the points $(2, 3)$ and $(2, 5)$ is a horizontal line.

5. The line that intersects the points $(4, 6)$ and $(5, 6)$ is a horizontal line.

6. The x-intercept of the line $y = 8x + 9$ is 9.

7. The function $f(x) = \pi x + 4$ represents a linear function.

8. The function $f(x) = 2x^2 + 3$ represents a linear function.

9. The lines $y = 3x + 17$ and $y = -3x + 8$ are perpendicular.

10. The lines $4x + 3y = 8$ and $4x + y = 5$ are parallel.

11. A correlation coefficient of zero indicates a perfect fit with the data.

12. It is not possible to get a correlation coefficient of -1.5 for a set of data.

PRACTICE AND EXPLORATIONS

13. What is marginal cost? Fixed cost?

14. What six quantities are needed to compute a correlation coefficient?

Find the slope for each line that has a slope.

15. Through $(-3, 7)$ and $(2, 12)$

16. Through $(4, -1)$ and $(3, -3)$

17. Through the origin and $(11, -2)$

18. Through the origin and $(0, 7)$

19. $4x + 3y = 6$ 20. $4x - y = 7$

21. $y + 4 = 9$ 22. $3y - 1 = 14$

23. $y = 5x + 4$ 24. $x = 5y$

Find an equation in the form $y = mx + b$ for each line.

25. Through $(5, -1)$; slope $= 2/3$

26. Through $(8, 0)$; slope $= -1/4$

27. Through $(-6, 3)$ and $(2, -5)$

28. Through $(2, -3)$ and $(-3, 4)$

29. Through $(2, -10)$, perpendicular to a line with undefined slope

30. Through $(-2, 5)$; slope $= 0$

Find an equation for each line in the form $ax + by = c$, where a, b, and c are integers with no factor common to all three and $a \geq 0$.

31. Through $(3, -4)$, parallel to $4x - 2y = 9$

32. Through $(0, 5)$, perpendicular to $8x + 5y = 3$

33. Through $(-1, 4)$; undefined slope

34. Through $(7, -6)$, parallel to a line with undefined slope

35. Through $(3, -5)$, parallel to $y = 4$

36. Through $(-3, 5)$, perpendicular to $y = -2$

Graph each linear equation defined as follows.

37. $y = 4x + 3$ 38. $y = 6 - 2x$

39. $3x - 5y = 15$ 40. $4x + 6y = 12$

41. $x - 3 = 0$ 42. $y = 1$

43. $y = 2x$ 44. $x + 3y = 0$

APPLICATIONS

Business and Economics

45. **U.S. Imports from China**   The United States is China's largest export market. Imports from China have grown from about $100 billion in 2000 to $440 billion in 2013. This growth has been approximately linear. *Source: U.S. Census Bureau.*

   (a) Determine a linear equation that approximates this growth in imports in terms of t, where t represents the number of years since 2000.

   (b) Describe the rate at which the imports from China are changing.

   (c) Based on the answer to part (a), what is the predicted amount of imports from China in 2015?

   (d) If this trend were to continue, in what year would imports from China be at least $600 billion?

46. **U.S. Exports to China**   U.S. exports to China have grown (although at a slower rate than imports) since 2000. In 2000, about $16 billion of goods were exported to China. By 2013, this amount had grown to $122 billion. This growth has been approximately linear. *Source: U.S. Census Bureau.*

   (a) Determine a linear equation that approximates this growth in exports in terms of t, where t represents the number of years since 2000.

   (b) Describe the rate at which the exports to China are changing.

   (c) Based on the answer to part (a), what is the predicted amount of exports in 2015?

   (d) If this trend were to continue, in what year would exports from China be at least $200 billion?

47. Median Income In recent years, the median income for all U.S. households has declined. In 2007, the median income (in 2012 dollars) was $55,627. By 2012, the median income had fallen to $51,017. This decline has been roughly linear. *Source: U.S. Census Bureau.*

(a) Determine a linear equation that approximates this decline in the median income in terms of t, where t represents the number of years since 2000.

(b) Describe the rate at which the median income for all U.S. households is changing.

(c) Based on the answer to part (a), what is the predicted median income in 2015?

(d) If this trend were to continue, in what year would the median income drop below $40,000?

48. Supply and Demand The supply and demand for fresh jumbo lump crabmeat in a local fish store are related by the equations

$$\text{Supply:} \quad p = S(q) = 4q + 10$$

and

$$\text{Demand:} \quad p = D(q) = 40 - 4q,$$

where p represents the price in dollars per pound and q represents the quantity of crabmeat in pounds per day. Find the quantity supplied and demanded at each of the following prices.

(a) $20

(b) $24

(c) $32

(d) Graph both the supply and demand functions on the same axes.

(e) Find the equilibrium price.

(f) Find the equilibrium quantity.

49. Supply and Demand A company is manufacturing a new dietary supplement pill that promotes body fat reduction.

(a) The company will supply 60 pills for a price of $40, and will supply 100 pills for a price of $60. Determine a linear supply function, $p = S(q)$, for this product.

(b) The demand for this new dietary supplement pill is 50 pills at a price of $47.50 and 80 pills at a price of $32.50. Determine a linear demand function, $p = D(q)$, for this product.

(c) Find the equilibrium price and quantity for the new dietary supplement pill.

Cost In Exercises 50–53, find a linear cost function.

50. Eight units cost $300; fixed cost is $60.

51. Fixed cost is $2000; 36 units cost $8480.

52. Twelve units cost $445; 50 units cost $1585.

53. Thirty units cost $1500; 120 units cost $5640.

54. Break-Even Analysis The cost of producing x cartons of CDs is $C(x)$ dollars, where $C(x) = 200x + 1000$. The CDs sell for $400 per carton.

(a) Find the break-even quantity.

(b) What revenue will the company receive if it sells just that number of cartons?

55. Break-Even Analysis The cost function for flavored coffee at an upscale coffeehouse is given in dollars by $C(x) = 3x + 160$, where x is in pounds. The coffee sells for $7 per pound.

(a) Find the break-even quantity.

(b) What will the revenue be at that point?

56. Profit To manufacture x thousand computer chips requires fixed expenditures of $352 plus $42 per thousand chips. Receipts from the sale of x thousand chips amount to $130 per thousand.

(a) Find the linear expenditure function, $E(x)$.

(b) Find the linear receipt function, $R(x)$.

(c) For profit to be made, receipts must be greater than expenditures. How many chips must be sold to produce a profit?

57. New Car Cost The average new car cost (in dollars) from 2002 to 2012 is given in the table. *Source: National Automobile Dealers Association.*

Year	Cost
2002	26,150
2003	27,550
2004	28,050
2005	28,400
2006	28,450
2007	28,200
2008	28,350
2009	28,966
2010	29,793
2011	30,659
2012	30,910

(a) Find a linear equation for the average new car cost in terms of t, the number of years since 2000, using the data for 2002 and 2012.

(b) Repeat part (a), using the data for 2004 and 2012.

(c) Find the equation of the least squares line using all the data.

(d) Use a graphing calculator to plot the data and the three lines from parts (a)–(c).

(e) Discuss which of the three lines found in parts (a)–(c) best describes the data, as well as to what extent a linear model accurately describes the data.

(f) Calculate the correlation coefficient.

58. Social Security   The average monthly Social Security payments (in 2012 dollars) to Old-Age, Survivors, and Disability Insurance (OASDI) recipients for selected years are given in the following table. *Source: Social Security Administration*.

Year	Average	Year	Average
1940	293	1980	819
1945	231	1985	907
1950	341	1990	954
1955	434	1995	980
1960	479	2000	1027
1965	518	2005	1084
1970	581	2010	1136
1975	746	2012	1153

(a) Find an equation for the least squares line, letting x equal the number of years since 1900.

(b) Use your result from part (a) to predict the average OASDI payment in 2015.

(c) Plot the data, along with the least squares line. Do the data points lie in a linear pattern?

(d) Find and interpret the correlation coefficient.

59. Meat Consumption   The U.S. per capita consumption (in pounds) of beef, pork, and chicken for 2000 and 2012 is given in the following table. Assume that the changes in consumption are approximately linear. *Source: U.S. Department of Agriculture*.

	2000	2012
Beef	64.5	54.5
Pork	47.8	42.6
Chicken	54.2	56.6

(a) Determine three linear functions, $b(t)$, $p(t)$, and $c(t)$, which approximate the pounds of beef, pork, and chicken, respectively, consumed for time t, where t is the number of years since 2000.

(b) Describe the rates at which the consumption of each type of meat is changing.

(c) In what year did the consumption of chicken surpass the consumption of beef?

(d) If these trends were to continue, approximate the consumption of each type of meat in 2015.

Life Sciences

60. World Health   In general, people tend to live longer in countries that have a greater supply of food. Listed in the table at the top of the next column is the 2009 daily calorie supply and 2009 life expectancy at birth for 10 randomly selected countries. *Source: Food and Agriculture Organization*.

Country	Calories (x)	Life Expectancy (y)
Belize	2680	75.6
Cambodia	2382	62.7
France	3531	81.2
India	2321	64.7
Mexico	3146	76.5
New Zealand	3172	80.4
Peru	2563	73.6
Sweden	3125	81.1
Tanzania	2137	56.6
United States	3668	78.2

(a) Find the correlation coefficient. Do the data seem to fit a straight line?

(b) Draw a scatterplot of the data. Combining this with your results from part (a), do the data seem to fit a straight line?

(c) Find the equation of the least squares line.

(d) Use your answer from part (c) to predict the life expectancy in Canada, which has a daily calorie supply of 3399. Compare your answer with the actual value of 80.8 years.

(e) Briefly explain why countries with a higher daily calorie supply might tend to have a longer life expectancy. Is this trend likely to continue to higher calorie levels? Do you think that an American who eats 5000 calories a day is likely to live longer than one who eats 3600 calories? Why or why not?

(f) (For the ambitious!) Find the correlation coefficient and least squares line using the data for a larger sample of countries, as found in an almanac or other reference. Is the result in general agreement with the previous results?

61. Blood Sugar and Cholesterol Levels   The following data show the connection between blood sugar levels and cholesterol levels for eight different patients.

Patient	Blood Sugar Level (x)	Cholesterol Level (y)
1	130	170
2	138	160
3	142	173
4	159	181
5	165	201
6	200	192
7	210	240
8	250	290

For the data given in the preceding table, $\Sigma x = 1394$, $\Sigma y = 1607$, $\Sigma xy = 291{,}990$, $\Sigma x^2 = 255{,}214$, and $\Sigma y^2 = 336{,}155$.

(a) Find the equation of the least squares line.

(b) Predict the cholesterol level for a person whose blood sugar level is 190.

(c) Find the correlation coefficient.

Social Sciences

62. **Marital Status**   More people are staying single longer in the United States. The following table gives the percent of never-married adults, age 15 and over, by gender, for the years 1995 and 2013. The growth has been approximately linear. *Source: U.S. Census.*

Year	Males	Females
1995	31.0	23.5
2013	34.4	28.6

(a) Find a linear function, $m(t)$, for the percent of never-married males in terms of t, the number of years since 1990.

(b) Determine the rate of change for the percent of never-married males.

(c) Find a linear function, $f(t)$, for the percent of never-married females in terms of t, the number of years since 1990.

(d) Determine the rate of change for the percent of never-married females.

(e) Use the results from parts (a) and (c) to estimate the percent of never-married males and females in 2015.

63. **Poverty**   The following table gives the number (in thousands) of families under the poverty level in the United States in recent years. *Source: U.S. Census Bureau.*

Year	Families Below Poverty Level (in thousands)
2000	6400
2001	6813
2002	7229
2003	7607
2004	7835
2005	7657
2006	7668
2007	7623
2008	8147
2009	8792
2010	9400
2011	9497
2012	9520

(a) Find a linear equation for the number of families below poverty level (in thousands) in terms of t, the number of years since 2000, using the data for 2000 and 2012.

(b) Repeat part (a), using the data for 2004 and 2012.

(c) Find the equation of the least squares line using all the data. Then plot the data and the three lines from parts (a)–(c) on a graphing calculator.

(d) Discuss which of the three lines found in parts (a)–(c) best describes the data, as well as to what extent a linear model accurately describes the data.

(e) Calculate the correlation coefficient.

64. **Governor's Salaries**   Listed in the table below are the estimated 2013 populations (in millions) and the salary of the governor (in thousands of dollars) for ten randomly selected states. *Source: U.S. Census Bureau and The World Almanac and Book of Facts 2014.*

State	Population, x	Governor's Salary, y
AZ	6.63	95
CT	3.60	150
IA	3.09	130
KS	2.89	100
LA	4.63	130
MA	6.69	140
NY	19.65	179
PA	12.77	187
TN	6.50	178
WY	0.58	105

(a) Find the correlation coefficient. Do the data seem to fit a straight line?

(b) Draw a scatterplot of the data. Compare this with your answer from part (a).

(c) Find the equation for the least squares line.

(d) Based on your answer to part (c), how much does a governor's salary increase, on average, for each additional million in population?

(e) Use your answer from part (c) to predict the governor's salary in your state. Based on your answers from parts (a) and (b), would this prediction be very accurate? Compare with the actual salary, as listed in an almanac or other reference.

(f) (For the ambitious!) Find the correlation coefficient and least squares line using the data for all 50 states, as found in an almanac or other reference. Is the result in general agreement with the previous results?

65. Movies A mathematician exploring the relationship between ratings of movies, their year of release, and their length discovered a paradox. Rather than list the data set of 100 movies in the original research, we have created a sample of size 10 that captures the properties of the original dataset. In the table to the right, the rating is a score from 1 to 10, and the length is in minutes. *Source: Journal of Statistics Education*.

(a) Find the correlation coefficient between the years since 2000 and the length.

(b) Find the correlation coefficient between the length and the rating.

(c) Given that you found a positive correlation between the year and the length in part (a), and a positive correlation between the length and the rating in part (b), what would you expect about the correlation between the year and the rating? Calculate this correlation. Are you surprised?

(d) Discuss the paradoxical result in part (c). Write out in words what each correlation tells you. Try to explain what

Year	Rating	Length
2001	10	120
2003	5	85
2004	3	100
2004	6	105
2005	4	110
2005	8	115
2006	6	135
2007	2	105
2007	5	125
2008	6	130

is happening. You may want to look at a scatterplot between the year and the rating, and consider which points on the scatterplot represent movies of length no more than 110 minutes, and which represent movies of length 115 minutes or more.

EXTENDED APPLICATION
USING EXTRAPOLATION TO PREDICT LIFE EXPECTANCY

One reason for developing a mathematical model is to make predictions. If your model is a least squares line, you can predict the y-value corresponding to some new x by substituting this x into an equation of the form $Y = mx + b$. (We use a capital Y to remind us that we're getting a predicted value rather than an actual data value.) Data analysts distinguish between two very different kinds of prediction, *interpolation*, and *extrapolation*. An interpolation uses a new x inside the x range of your original data. For example, if you have inflation data at 5-year intervals from 1950 to 2015, estimating the rate of inflation in 1957 is an interpolation problem. But if you use the same data to estimate what the inflation rate was in 1920, or what it will be in 2020, you are extrapolating.

In general, interpolation is much safer than extrapolation, because data that are approximately linear over a short interval may be nonlinear over a larger interval. One way to detect nonlinearity is to look at *residuals*, which are the differences between the actual data values and the values predicted by the line of best fit. Here is a simple example.

The regression equation for the linear fit in Figure 22 is $Y = 3.431 + 1.334x$. Since the r-value for this regression line is 0.93, our linear model fits the data very well. But we might notice

that the predictions are a bit low at the ends and high in the middle. We can get a better look at this pattern by plotting the residuals. To find them, we put each value of the independent variable into the regression equation, calculate the predicted value Y, and subtract it from the actual y-value. The residual plot is shown in Figure 23,

FIGURE 22

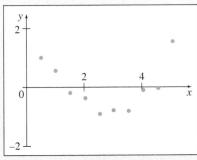

FIGURE 23

with the vertical axis rescaled to exaggerate the pattern. The residuals indicate that our data have a nonlinear, U-shaped component that is not captured by the linear fit. Extrapolating from this data set is probably not a good idea; our linear prediction for the value of y when x is 10 may be much too low.

EXERCISES

The following table gives the life expectancy at birth of females born in the United States in various years from 1970 to 2010. **Source: National Center for Health Statistics.**

Year of Birth	Life Expectancy (years)
1970	74.7
1975	76.6
1980	77.4
1985	78.2
1990	78.8
1995	78.9
2000	79.3
2005	80.1
2010	81.0

1. Find an equation for the least squares line for these data, using year of birth as the independent variable.

2. Use your regression equation to guess a value for the life expectancy of females born in 1900.

3. Compare your answer with the actual life expectancy for females born in 1900, which was 48.3 years. Are you surprised?

4. How accurate do you think it will be to predict the life expectancy at birth of females born in 2025?

5. Find the life expectancy predicted by your regression equation for each year in the table, and subtract it from the actual value in the second column. This gives you a table of residuals. Plot your residuals as points on a graph.

6. What will happen if you try linear regression on the *residuals*? If you're not sure, use your calculator or software to find the regression equation for the residuals. Why does this result make sense?

7. Now look at the residuals as a fresh data set, and see if you can sketch the graph of a smooth function that fits the residuals well. Use a graphing calculator or software to see if a quadratic or cubic regression equation would fit the residuals.

8. Since most of the females born in 1995 are still alive, how did the Public Health Service come up with a life expectancy of 78.9 years for these women?

9. Go to the website WolframAlpha.com and enter: "linear fit {1970, 74.7}, {1975, 76.6}, etc.," putting in all the data from the table. Discuss how the solution compares with the solutions provided by a graphing calculator and by Microsoft Excel.

DIRECTIONS FOR GROUP PROJECT

Assume that you and your group (3–5 students) are preparing a report for a local health agency that is interested in using linear regression to predict life expectancy. Using the questions above as a guide, write a report that addresses the spirit of each question and any issues related to that question. The report should be mathematically sound, grammatically correct, and professionally crafted. Provide recommendations as to whether the health agency should proceed with the linear equation or whether it should seek other means of making such predictions.

2

Nonlinear Functions

There are fourteen mountain peaks over 8000 meters on the earth's surface. At these altitudes climbers face the challenge of "thin air," since atmospheric pressure is about one-third of the pressure at sea level. An exercise in Section 4 of this chapter shows how the change in atmospheric pressure with altitude can be modeled with an exponential function.

Figure I below shows the average price of gold for each year in the past few decades. The graph is not a straight line and illustrates a function that, unlike those studied in Chapter I, is *nonlinear*. Linear functions are simple to study, and they can be used to approximate many functions over short intervals. But most functions exhibit behavior that, in the long run, does not follow a straight line. In this chapter we will study some of the most common nonlinear functions. ***Source: kitco.com.***

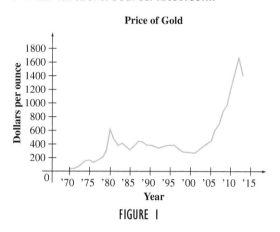

FIGURE I

2.1 Properties of Functions

APPLY IT   How are the energy consumptions of the United States, China, and India changing over time?

We will analyze this question in Exercise 72 in this section, after developing the concept of nonlinear functions.

As we saw in Chapter 1, the linear cost function $C(x) = 12x + 5000$ for video games is related to the number of items produced. The number of games produced is the independent variable and the total cost is the dependent variable because it depends on the number produced. When a specific number of games (say 1000) is substituted for x, the cost $C(x)$ has one specific value ($12 \cdot 1000 + 5000$). Because of this, the variable $C(x)$ is said to be a *function* of x.

Function

A **function** is a rule that assigns to each element from one set exactly one element from another set.

In most cases in this book, the "rule" mentioned in the box is expressed as an equation, such as $C(x) = 12x + 5000$, and each set mentioned in the definition will ordinarily be the real numbers or some subset of the reals. When an equation is given for a function, we say that the equation *defines* the function. Whenever x and y are used in this book to define a function, x represents the independent variable and y the dependent variable. Of course, letters other than x and y could be used and are often more meaningful. For example, if the independent variable represents the number of items sold for $4 each and the dependent variable represents revenue, we might write $R(s) = 4s$.

The independent variable in a function can take on any value within a specified set of values called the *domain*.

Domain and Range

The set of all possible values of the independent variable in a function is called the **domain** of the function, and the resulting set of possible values of the dependent variable is called the **range**.

EXAMPLE 1 Dow Jones Industrial Average

An important function to investors around the world is the Dow Jones industrial average, a performance measure of the stock market. Figure 2 shows how this average varied over the year 2013. Label this function $y = f(t)$, where y is the Dow Jones industrial average and t is the time in days from the beginning of 2013. *Source: Yahoo!Finance.*

FIGURE 2

(a) Is $f(t)$ a linear function?

SOLUTION Notice that the function increases and decreases during the year, so it is not linear, although a linear function could be used as a very rough approximation. Such a function, whose graph is not a straight line, is called a **nonlinear function**. The independent variable here is t, the time in days. The dependent variable is y, the Dow Jones industrial average at any time.

(b) Determine the domain and range of the function.

SOLUTION As with linear functions, the domain is mapped along the horizontal axis and the range along the vertical axis. The domain is $\{t \mid 0 \leq t \leq 365\}$, or $[0, 365]$; $t = 0$ corresponds to the beginning of the day on January 1, and $t = 365$ corresponds to the end of the day on December 31. By looking for the lowest and highest values of the function, we estimate the range to be approximately $\{y \mid 13{,}100 \leq y \leq 16{,}600\}$, or $[13{,}100, 16{,}600]$.

(c) Estimate $f(14)$.

SOLUTION We do not have a formula for $f(t)$. (If we had possessed such a formula at the beginning of 2013, we could have made a lot of money!) Instead, we can use the graph to estimate values of the function. To estimate $f(14)$, draw a vertical line from $t = 14$ (January 14), as shown in Figure 3(a) on the next page. The y-coordinate seems to be roughly 13,500, so we estimate $f(14) \approx 13{,}500$.

(d) Solve the equation $f(t) = 14{,}250$.

SOLUTION To solve the equation $f(t) = 14{,}250$, look for points on the graph that have a y-coordinate of 14,250. As Figure 3(b) shows, this occurs around March 5 (the 64th day of the year). Thus, $f(t) = 14{,}250$ when $t = 64$.

FIGURE 3

Dow Jones Industrial Average	
Day (t)	Close (y)
0	13,104.14
1	13,104.14
2	13,412.55
3	13,391.36
4	13,435.21
5	13,435.21
6	13,435.21
7	13,384.29

The function in Example 1 can also be given as a table. The table in the margin shows the value of the function for several values of t.

Notice from the table that $f(0) = f(1) = 13{,}104.14$. The stock market was closed for the first day of 2013, so the Dow Jones average did not change. This illustrates an important property of functions: Several different values of the independent variable can have the same value for the dependent variable. On the other hand, we cannot have several different y-values corresponding to the same value of t; if we did, this would not be a function.

What is $f(6.5)$? We do not know. When the stock market closed on January 6, the Dow Jones industrial average was 13,435.21. The closing value the following day was 13,384.29. We do not know what happened in between, although this information is recorded by the New York Stock Exchange.

Functions arise in numerous applications, and an understanding of them is critical for understanding calculus. The following example shows some of the ways functions can be represented and will help you in determining whether a relationship between two variables is a function or not.

EXAMPLE 2   **Functions**

Which of the following are functions?

(a)

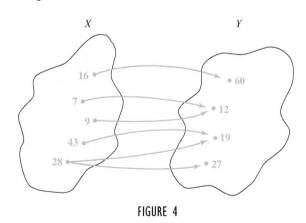

FIGURE 4

SOLUTION   Figure 4 shows that an x-value of 28 corresponds to *two* y-values, 19 and 27. In a function, each x must correspond to exactly one y, so this correspondence is not a function.

(b) The x^2 key on a calculator

SOLUTION This correspondence between input and output is a function because the calculator produces just one x^2 (one y-value) for each x-value entered. Notice also that two x-values, such as 3 and -3, produce the same y-value of 9, but this does not violate the definition of a function.

(c)

x	1	1	2	2	3	3
y	3	-3	5	-5	8	-8

SOLUTION Since at least one x-value corresponds to more than one y-value, this table does not define a function.

(d) The set of ordered pairs with first elements mothers and second elements their children

SOLUTION Here the mother is the independent variable and the child is the dependent variable. For a given mother, there may be several children, so this correspondence is not a function.

(e) The set of ordered pairs with first elements children and second elements their birth mothers

SOLUTION In this case the child is the independent variable and the mother is the dependent variable. Since each child has only one birth mother, this is a function.

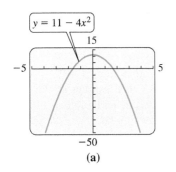
$y = 11 - 4x^2$
(a)

EXAMPLE 3 Functions

Decide whether each equation or graph represents a function. (Assume that x represents the independent variable here, an assumption we shall make throughout this book.) Give the domain and range of any functions.

(a) $y = 11 - 4x^2$

SOLUTION For a given value of x, calculating $11 - 4x^2$ produces exactly one value of y. (For example, if $x = -7$, then $y = 11 - 4(-7)^2 = -185$, so $f(-7) = -185$.) Since one value of the independent variable leads to exactly one value of the dependent variable, $y = 11 - 4x^2$ meets the definition of a function.

Because x can take on any real-number value, the domain of this function is the set of all real numbers. Finding the range is more difficult. One way to find it would be to ask what possible values of y could come out of this function. Notice that the value of y is 11 minus a quantity that is always 0 or positive, since $4x^2$ can never be negative. There is no limit to how large $4x^2$ can be, so the range is $(-\infty, 11]$.

Another way to find the range would be to examine the graph. Figure 5(a) shows a graphing calculator view of this function, and we can see that the function takes on y-values of 11 or less. The calculator cannot tell us, however, whether the function continues to go down past the viewing window, or turns back up. To find out, we need to study this type of function more carefully, as we will do in the next section.

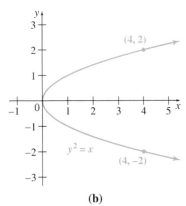
$y^2 = x$
(b)

(b) $y^2 = x$

SOLUTION Suppose $x = 4$. Then $y^2 = x$ becomes $y^2 = 4$, from which $y = 2$ or $y = -2$. Since one value of the independent variable can lead to two values of the dependent variable, $y^2 = x$ does not represent a function. A graph of $y^2 = x$ is shown in Figure 5(b).

(c) $y = 7$

SOLUTION No matter what the value of x, the value of y is always 7. This is indeed a function; it assigns exactly one element, 7, to each value of x. Such a function is known as a **constant function**. The domain is the set of all real numbers, and the range is the set $\{7\}$. Its graph is the horizontal line that intersects the y-axis at $y = 7$, as shown in Figure 5(c). Every constant function has a horizontal line for its graph.

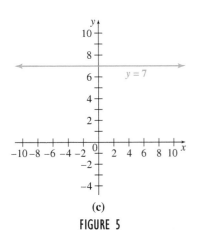
$y = 7$
(c)

FIGURE 5

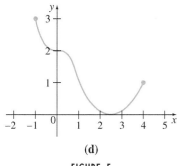

(d)

FIGURE 5

(d) The graph in Figure 5(d).

SOLUTION For each value of x, there is only one value of y. For example, the point $(-1, 3)$ on the graph shows that $f(-1) = 3$. Therefore, the graph represents a function. From the graph, we see that the values of x go from -1 to 4, so the domain is $[-1, 4]$. By looking at the values of y, we see that the range is $[0, 3]$.

The following agreement on domains is customary.

Agreement on Domains

Unless otherwise stated, assume that the domain of all functions defined by an equation is the greatest subset of real numbers that are meaningful replacements for the independent variable.

For example, suppose

$$y = \frac{-4x}{2x - 3}.$$

Any real number can be used for x except $x = 3/2$, which makes the denominator equal 0. By the agreement on domains, the domain of this function is the set of all real numbers except $3/2$, which we denote $\{x \mid x \neq 3/2\}$, $\{x \neq 3/2\}$, or $(-\infty, 3/2) \cup (3/2, \infty)$.*

| CAUTION | When finding the domain of a function, there are two operations to avoid: (1) dividing by zero; and (2) taking the square root (or any even root) of a negative number. Later sections will present other functions, such as logarithms, which require further restrictions on the domain. For now, just remember these two restrictions on the domain. |

EXAMPLE 4 **Domain and Range**

Find the domain and range for each function defined as follows.

(a) $f(x) = x^2$

SOLUTION Any number may be squared, so the domain is the set of all real numbers, written $(-\infty, \infty)$. Since $x^2 \geq 0$ for every value of x, the range is $[0, \infty)$.

(b) $y = x^2$, with the domain specified as $\{-2, -1, 0, 1, 2\}$.

SOLUTION With the domain specified, the range is the set of values found by applying the function to the domain. Since $f(0) = 0$, $f(-1) = f(1) = 1$, and $f(-2) = f(2) = 4$, the range is $\{0, 1, 4\}$. The graph of the set of ordered pairs is shown in Figure 6.

(c) $y = \sqrt{6 - x}$

SOLUTION For y to be a real number, $6 - x$ must be nonnegative. This happens only when $6 - x \geq 0$, or $6 \geq x$, making the domain $(-\infty, 6]$. The range is $[0, \infty)$ because $\sqrt{6 - x}$ is always nonnegative.

(d) $y = \sqrt{2x^2 + 5x - 12}$

SOLUTION The domain includes only those values of x satisfying $2x^2 + 5x - 12 \geq 0$. Using the methods for solving a quadratic inequality produces the domain

$$(-\infty, -4] \cup [3/2, \infty).$$

As in part (c), the range is $[0, \infty)$.

FIGURE 6

FOR REVIEW

Section R.5 demonstrates the method for solving a quadratic inequality. To solve $2x^2 + 5x - 12 \geq 0$, factor the quadratic to get $(2x - 3)(x + 4) \geq 0$. Setting each factor equal to 0 gives $x = 3/2$ or $x = -4$, leading to the intervals $(-\infty, -4]$, $[-4, 3/2]$, and $[3/2, \infty)$. Testing a number from each interval shows that the solution is $(-\infty, -4] \cup [3/2, \infty)$.

*The *union* of sets A and B, written $A \cup B$, is defined as the set of all elements in A or B or both.

(e) $y = \dfrac{2}{x^2 - 9}$

SOLUTION Since the denominator cannot be zero, $x \neq 3$ and $x \neq -3$. The domain is

$$(-\infty, -3) \cup (-3, 3) \cup (3, \infty).$$

Because the numerator can never be zero, $y \neq 0$. The denominator can take on any real number except for 0, allowing y to take on any value except for 0, so the range is $(-\infty, 0) \cup (0, \infty)$.   **TRY YOUR TURN 1** ▬▬

YOUR TURN 1 Find the domain and range for the function

$$y = \frac{1}{\sqrt{x^2 - 4}}.$$

To understand how a function works, think of a function f as a machine—for example, a calculator or computer—that takes an input x from the domain and uses it to produce an output $f(x)$ (which represents the y-value), as shown in Figure 7. In the Dow Jones example, when we put 4 into the machine, we get out 13,435.21, since $f(4) = 13{,}435.21$.

FIGURE 7

EXAMPLE 5 Evaluating Functions

Let $g(x) = -x^2 + 4x - 5$. Find the following.

(a) $g(3)$

SOLUTION Replace x with 3.

$$g(3) = -3^2 + 4 \cdot 3 - 5 = -9 + 12 - 5 = -2$$

(b) $g(a)$

SOLUTION Replace x with a to get $g(a) = -a^2 + 4a - 5$.

This replacement of one variable with another is important in later chapters.

(c) $g(x + h)$

SOLUTION Replace x with the expression $x + h$ and simplify.

$$
\begin{aligned}
g(x + h) &= -(x + h)^2 + 4(x + h) - 5 \\
&= -(x^2 + 2xh + h^2) + 4(x + h) - 5 &&\text{Square } (x + h) \\
&= -x^2 - 2xh - h^2 + 4x + 4h - 5 &&\text{Distributive Property}
\end{aligned}
$$

(d) $g\left(\dfrac{2}{r}\right)$

SOLUTION Replace x with $2/r$ and simplify.

$$g\left(\frac{2}{r}\right) = -\left(\frac{2}{r}\right)^2 + 4\left(\frac{2}{r}\right) - 5 = -\frac{4}{r^2} + \frac{8}{r} - 5$$

(e) Find all values of x such that $g(x) = -12$.

SOLUTION Set $g(x)$ equal to -12, and then add 12 to both sides to make one side equal to 0.

$$-x^2 + 4x - 5 = -12$$
$$-x^2 + 4x + 7 = 0$$

This equation does not factor, but can be solved with the quadratic formula, which says that if $ax^2 + bx + c = 0$, where $a \neq 0$, then

$$x = \frac{-b \pm \sqrt{b^2 - 4ac}}{2a}.$$

In this case, with $a = -1$, $b = 4$, and $c = 7$, we have

$$x = \frac{-4 \pm \sqrt{16 - 4(-1)7}}{2(-1)}$$

$$= \frac{-4 \pm \sqrt{44}}{-2} \qquad \sqrt{44} = \sqrt{4 \cdot 11} = \sqrt{4} \cdot \sqrt{11} = 2\sqrt{11}$$

$$= 2 \pm \sqrt{11} \qquad \text{Factor and simplify.}$$

$$\approx -1.317 \quad \text{or} \quad 5.317. \qquad \text{TRY YOUR TURN 2}$$

YOUR TURN 2 Given the function $f(x) = 2x^2 - 3x - 4$, find each of the following.
(a) $f(x + h)$ **(b)** All values of x such that $f(x) = -5$

TECHNOLOGY NOTE

We can use a graphing calculator to evaluate functions, as in Example 5(a). One method is to simply type in the expression, replacing x with the desired value. For Example 5(a), entering $-3^2 + 4 \cdot 3 - 5$ gives the solution -2.

However, if you are evaluating the function at more than one value, it may be convenient to first store the function. On the TI-84 Plus C calculator, select Y_1 and enter the function. Then type Y_1 on the home screen by selecting VARS, then Y-VARS, then FUNCTION, and then Y_1. To evaluate the function Y_1 at any value of x, enter the value, in parentheses, after Y_1. For Example 5(a), after $-X^2 + 4X - 5$ is entered for Y_1, the command $Y_1(3)$ yields -2.

A third method uses the graph of a function. First, enter the function into Y_1, choose an appropriate window, and graph the function. Use the `value` command in the CALC menu to evaluate the function at any x-value. In Figure 8(a), after graphing $Y_1 = -X^2 + 4X - 5$, we use the `value` command to evaluate the function at $x = 3$. The answer, -2, appears after Y = and the cursor marks the location on the graph.

The TI-84 Plus C calculator can also be used to verify the results of Example 5(e). We first graphed $Y_1 = -X^2 + 4X - 5$ and $Y_2 = -12$ on the same axes. In Figure 8(b), we used the `intersect` command in the CALC menu to find $x = 5.3166248$, which is one of our two answers to part (e). To find the second solution, use the `intersect` command again, but this time move the cursor toward the intersection on the other side of the y-axis. As in part (e), the other result is $x = -1.316625$.

FIGURE 8

CAUTION Notice from Example 5(c) that $g(x + h)$ is *not* the same as $g(x) + h$, which equals $-x^2 + 4x - 5 + h$. There is a significant difference between applying a function to the quantity $x + h$ and applying a function to x and adding h afterward.

If you tend to get confused when replacing x with $x + h$, as in Example 5(c), you might try replacing the x in the original function with a box, like this:

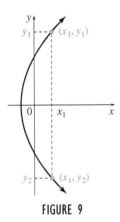

FIGURE 9

Then, to compute $g(x + h)$, just enter $x + h$ into the box:

$$g\left(\boxed{x + h}\right) = -\left(\boxed{x + h}\right)^2 + 4\left(\boxed{x + h}\right) - 5$$

and proceed as in Example 5(c).

Notice in the Dow Jones example that to find the value of the function for a given value of x, we drew a vertical line from the value of x and found where it intersected the graph. If a graph is to represent a function, each value of x from the domain must lead to exactly one value of y. In the graph in Figure 9, the domain value x_1 leads to *two* y-values, y_1 and y_2. Since the given x-value corresponds to two different y-values, this is not the graph of a function. This example suggests the **vertical line test** for the graph of a function.

Vertical Line Test

If a vertical line intersects a graph in more than one point, the graph is not the graph of a function.

A graph represents a function if and only if every vertical line intersects the graph in no more than one point.

EXAMPLE 6   Vertical Line Test

Use the vertical line test to determine which of the graphs in Example 3 represent functions.

SOLUTION   Every vertical line intersects the graphs in Figure 5(a), (c), and (d) in at most one point, so these are the graphs of functions. It is possible for a vertical line to intersect the graph in Figure 5(b) twice, so this is not a function.

A function f is called an **even function** if $f(-x) = f(x)$. This means that the graph is symmetric about the y-axis, so the left side is a mirror image of the right side. The function f is called an **odd function** if $f(-x) = -f(x)$. This means that the graph is symmetric about the origin, so the left side of the graph can be found by rotating the right side by $180°$ about the origin.

EXAMPLE 7   Even and Odd Functions

Determine whether each of the following functions is even, odd, or neither.

(a) $f(x) = x^4 - x^2$

SOLUTION   Calculate $f(-x) = (-x)^4 - (-x)^2 = x^4 - x^2 = f(x)$, so the function is even. Its graph, shown in Figure 10(a), is symmetric about the y-axis.

(b) $f(x) = \dfrac{x}{x^2 + 1}$

SOLUTION   Calculate $f(-x) = \dfrac{(-x)}{(-x)^2 + 1} = \dfrac{-x}{x^2 + 1} = -f(x)$, so the function is odd. Its graph, shown in Figure 10(b), is symmetric about the origin. You can see this by turning the book upside down and observing that the graph looks the same.

(c) $f(x) = x^4 - 4x^3$

SOLUTION   Calculate $f(-x) = (-x)^4 - 4(-x)^3 = x^4 + 4x^3$, which is equal neither to $f(x)$ nor to $-f(x)$. So the function is neither even nor odd. Its graph, shown in Figure 10(c), has no symmetry.

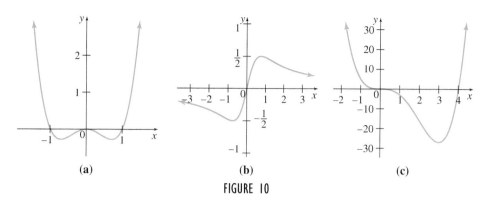

(a)   (b)   (c)

FIGURE 10

EXAMPLE 8 Delivery Charges

An overnight delivery service charges \$25 for a package weighing up to 2 lb. For each additional pound, or portion thereof, there is an additional charge of \$3. Let $D(w)$ represent the cost to send a package weighing w lb. Graph $D(w)$ for w in the interval $(0, 6]$.

SOLUTION For w in the interval $(0, 2]$, the shipping cost is $y = 25$. For w in $(2, 3]$, the shipping cost is $y = 25 + 3 = 28$. For w in $(3, 4]$, the shipping cost is $y = 28 + 3 = 31$, and so on. The graph is shown in Figure 11.

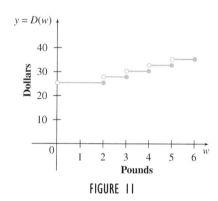

FIGURE 11

The function discussed in Example 8 is called a **step function**. Many real-life situations are best modeled by step functions. Additional examples are given in the exercises.

In Chapter 1 you saw several examples of linear models. In Example 9, we use a quadratic equation to model the area of a lot.

EXAMPLE 9 Area

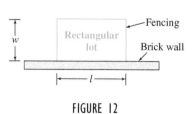

FIGURE 12

A fence is to be built against a brick wall to form a rectangular lot, as shown in Figure 12. Only three sides of the fence need to be built, because the wall forms the fourth side. The contractor will use 200 m of fencing. Let the length of the wall be l and the width w, as shown in Figure 12.

(a) Find the area of the lot as a function of the length l.

SOLUTION The area formula for a rectangle is area = length × width, or

$$A = lw.$$

We want the area as a function of the length only, so we must eliminate the width. We use the fact that the total amount of fencing is the sum of the three sections, one length and two widths, so $200 = l + 2w$. Solve this for w:

$$200 = l + 2w$$
$$200 - l = 2w \qquad \text{Subtract } l \text{ from both sides.}$$
$$100 - l/2 = w. \qquad \text{Divide both sides by 2.}$$

Substituting this expression into the formula for area gives

$$A = l(100 - l/2).$$

(b) Find the domain of the function in part (a).

SOLUTION The length cannot be negative, so $l \geq 0$. Similarly, the width cannot be negative, so $100 - l/2 \geq 0$, from which we find $l \leq 200$. Therefore, the domain is $[0, 200]$.

(c) Sketch a graph of the function in part (a).

SOLUTION The result from a graphing calculator is shown in Figure 13. Notice that at the endpoints of the domain, when $l = 0$ and $l = 200$, the area is 0. This makes sense: If the length or width is 0, the area will be 0 as well. In between, as the length increases from 0 to 100 m, the area increases, and seems to reach a peak of 5000 m² when $l = 100$ m. After that, the area decreases as the length continues to increase because the width is becoming smaller.

In the next section, we will study this type of function in more detail and determine exactly where the maximum occurs.

FIGURE 13

2.1 WARM-UP EXERCISES

Solve the following.

W1. $4x^2 - 9 = 0$ *(Sec. R.4)*

W2. $2x^2 + 9x - 5 = 0$ *(Sec. R.4)*

W3. $16 - x^2 \geq 0$ *(Sec. R.5)*

W4. $21 + 4x - x^2 \geq 0$ *(Sec. R.5)*

2.1 EXERCISES

Which of the following rules define y as a function of x?

1.

2.

3.

x	y
3	9
2	4
1	1
0	0
−1	1
−2	4
−3	9

4.

x	y
9	3
4	2
1	1
0	0
1	−1
4	−2
9	−3

5. $y = x^3 + 2$

6. $y = \sqrt{x}$

7. $x = |y|$

8. $x = y^2 + 4$

List the ordered pairs obtained from each equation, given $\{-2, -1, 0, 1, 2, 3\}$ as the domain. Graph each set of ordered pairs. Give the range.

9. $y = 2x + 3$

10. $y = -3x + 9$

11. $2y - x = 5$

12. $6x - y = -1$

13. $y = x(x + 2)$

14. $y = (x - 2)(x + 2)$

15. $y = x^2$

16. $y = -4x^2$

Give the domain of each function defined as follows.

17. $f(x) = 2x$

18. $f(x) = 2x + 3$

19. $f(x) = x^4$

20. $f(x) = (x + 3)^2$

21. $f(x) = \sqrt{4 - x^2}$

22. $f(x) = |3x - 6|$

23. $f(x) = (x - 3)^{1/2}$

24. $f(x) = (3x + 5)^{1/2}$

25. $f(x) = \dfrac{2}{1 - x^2}$

26. $f(x) = \dfrac{-8}{x^2 - 36}$

27. $f(x) = -\sqrt{\dfrac{2}{x^2 - 16}}$

28. $f(x) = -\sqrt{\dfrac{5}{x^2 + 36}}$

29. $f(x) = \sqrt{x^2 - 4x - 5}$

30. $f(x) = \sqrt{15x^2 + x - 2}$

31. $f(x) = \dfrac{1}{\sqrt{3x^2 + 2x - 1}}$

32. $f(x) = \sqrt{\dfrac{x^2}{3 - x}}$

Give the domain and the range of each function. Where arrows are drawn, assume the function continues in the indicated direction.

33.

34.

35.

36.

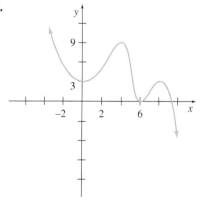

In Exercises 37–40, give the domain and range. Then, use each graph to find (a) $f(-2)$, (b) $f(0)$, (c) $f(1/2)$, and (d) any values of x such that $f(x) = 1$.

37.

38.

39.

40.

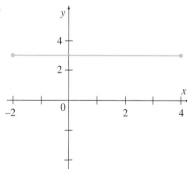

For each function, find (a) $f(4)$, (b) $f(-1/2)$, (c) $f(a)$, (d) $f(2/m)$, and (e) any values of x such that $f(x) = 1$.

41. $f(x) = 3x^2 - 4x + 1$ **42.** $f(x) = (x + 3)(x - 4)$

43. $f(x) = \begin{cases} \dfrac{2x+1}{x-4} & \text{if } x \neq 4 \\ 7 & \text{if } x = 4 \end{cases}$

44. $f(x) = \begin{cases} \dfrac{x-4}{2x+1} & \text{if } x \neq -\dfrac{1}{2} \\ 10 & \text{if } x = -\dfrac{1}{2} \end{cases}$

Let $f(x) = 6x^2 - 2$ and $g(x) = x^2 - 2x + 5$ to find the following values.

45. $f(t+1)$ **46.** $f(2-r)$

47. $g(r+h)$ **48.** $g(z-p)$

49. $g\left(\dfrac{3}{q}\right)$ **50.** $g\left(-\dfrac{5}{z}\right)$

For each function defined as follows, find (a) $f(x+h)$, (b) $f(x+h) - f(x)$, and (c) $[f(x+h) - f(x)]/h$.

51. $f(x) = 2x + 1$ **52.** $f(x) = x^2 - 3$

53. $f(x) = 2x^2 - 4x - 5$ **54.** $f(x) = -4x^2 + 3x + 2$

55. $f(x) = \dfrac{1}{x}$ **56.** $f(x) = -\dfrac{1}{x^2}$

Decide whether each graph represents a function.

57.   **58.**

59.   **60.**

61.   **62.**

Classify each of the functions in Exercises 63–70 as even, odd, or neither.

63. $f(x) = 3x$ **64.** $f(x) = 5x$

65. $f(x) = 2x^2$ **66.** $f(x) = x^2 - 3$

67. $f(x) = \dfrac{1}{x^2+4}$ **68.** $f(x) = x^3 + x$

69. $f(x) = \dfrac{x}{x^2-9}$ **70.** $f(x) = |x - 2|$

APPLICATIONS

Business and Economics

71. Price of Silver The figure below shows the average price of silver for recent years. Let $S(t)$ represent the average price (in dollars per ounce) of silver for year t. *Source: kitco.com.*

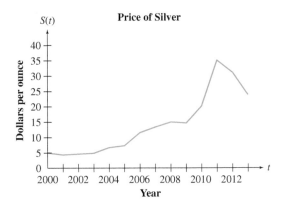

(a) Is $S(t)$ a linear function?

(b) What is the independent variable?

(c) What is the dependent variable?

(d) What is the domain of the function?

(e) Estimate the range of the function.

(f) What was the average price per ounce in 2008?

(g) In what year was the average price $35 per ounce?

72. APPLY IT Energy Consumption The U.S. Energy Information Administration has used recent consumption data to project the energy consumption (in quadrillion Btu) for the United States, China, and India until 2035. The following graph illustrates these data. *Source: U.S. Energy Information Administration.*

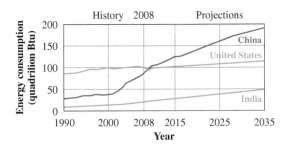

(a) Let $U(t)$ represent the energy consumption of the United States for year t. Is $U(t)$ a linear function?

(b) Let $I(t)$ represent the energy consumption of India for year t. What is the independent variable? What is the dependent variable?

(c) Let $C(t)$ represent the energy consumption of China for year t. What is the domain of the function? Estimate the range of the function.

(d) Estimate the energy consumption in 2015 for the United States. Repeat for China and India.

(e) In what year will China first consume 150 quadrillion Btu?

(f) In what year did the energy consumption of China equal the energy consumption of the United States?

73. Saw Rental A chain-saw rental firm charges $28 per day or fraction of a day to rent a saw, plus a fixed fee of $8 for re-sharpening the blade. Let $S(x)$ represent the cost of renting a saw for x days. Find the following.

(a) $S\left(\frac{1}{2}\right)$ **(b)** $S(1)$ **(c)** $S\left(1\frac{1}{4}\right)$

(d) $S\left(3\frac{1}{2}\right)$ **(e)** $S(4)$ **(f)** $S\left(4\frac{1}{10}\right)$

(g) What does it cost to rent a saw for $4\frac{9}{10}$ days?

(h) A portion of the graph of $y = S(x)$ is shown here. Explain how the graph could be continued.

(i) What is the independent variable?

(j) What is the dependent variable?

(k) Is S a linear function? Explain.

(l) Write a sentence or two explaining what part (f) and its answer represent.

(m) We have left $x = 0$ out of the graph. Discuss why it should or shouldn't be included. If it were included, how would you define $S(0)$?

74. Rental Car Cost The cost to rent a mid-size car is $54 per day or fraction of a day. If the car is picked up in Pittsburgh and dropped off in Cleveland, there is a fixed $44 drop-off charge. Let $C(x)$ represent the cost of renting the car for x days, taking it from Pittsburgh to Cleveland. Find the following.

(a) $C(3/4)$ **(b)** $C(9/10)$

(c) $C(1)$ **(d)** $C\left(1\frac{5}{8}\right)$

(e) Find the cost of renting the car for 2.4 days.

(f) Graph $y = C(x)$.

(g) Is C a function? Explain.

(h) Is C a linear function? Explain.

(i) What is the independent variable?

(j) What is the dependent variable?

75. Attorney Fees According to Massachusetts state law, the maximum amount of a jury award that attorneys can receive is:

40% of the first $150,000,

33.3% of the next $150,000,

30% of the next $200,000, and

24% of anything over $500,000.

Let $f(x)$ represent the maximum amount of money that an attorney in Massachusetts can receive for a jury award of size x. Find each of the following, and describe in a sentence what the answer tells you. *Source: The New Yorker.*

(a) $f(250,000)$ **(b)** $f(350,000)$ **(c)** $f(550,000)$

(d) Sketch a graph of $f(x)$.

76. Tax Rates In New York state in 2014, the income tax rates for a single person were as follows:

4% of the first $8200 earned,

4.5% of the next $3100 earned,

5.25% of the next $2050 earned,

5.9% of the next $7200 earned,

6.45% of the next $56,600 earned,

6.65% of the next $128,700 earned,

6.85% of the next $823,400 earned, and

8.82% of any amount earned over $1,029,250.

Let $f(x)$ represent the amount of tax owed on an income of x dollars. Find each of the following, and explain in a sentence what the answer tells you. *Source: New York State.*

(a) $f(10,000)$

(b) $f(12,000)$

(c) $f(18,000)$

(d) $f(50,000)$

(e) Find the domain and range of $f(x)$.

(f) Sketch a graph of $f(x)$ for $0 \le x \le 100,000$.

Life Sciences

77. Whales Diving The figure shows the depth of a diving sperm whale as a function of time, as recorded by researchers at the Woods Hole Oceanographic Institution in Massachusetts. *Source: Peter Tyack, Woods Hole Oceanographic Institution.*

Find the depth of the whale at the following times.

(a) 17 hours and 37 minutes

(b) 17 hours and 39 minutes

78. Metabolic Rate The basal metabolic rate (in kcal/day) for large anteaters is given by

$$y = f(x) = 19.7x^{0.753},$$

where x is the anteater's weight in kilograms.* *Source: Wildlife Feeding and Nutrition.*

(a) Find the basal metabolic rate for anteaters with the following weights.

(i) 5 kg **(ii)** 25 kg

*Technically, kilograms are a measure of mass, not weight. Weight is a measure of the force of gravity, which varies with the distance from the center of Earth. For objects on the surface of Earth, weight and mass are often used interchangeably, and we will do so in this text.

(b) Suppose the anteater's weight is given in pounds rather than kilograms. Given that 1 lb = 0.454 kg, find a function $x = g(z)$ giving the anteater's weight in kilograms if z is the animal's weight in pounds.

(c) Write the basal metabolic rate as a function of the weight in pounds in the form $y = az^b$ by calculating $f(g(z))$.[†]

79. Swimming Energy The energy expenditure (in kcal/km) for animals swimming at the surface of the water is given by

$$y = f(x) = 0.01x^{0.88},$$

where x is the animal's weight in grams. *Source: Wildlife Feeding and Nutrition*.

(a) Find the energy for the following animals swimming at the surface of the water.

(i) A muskrat weighing 800 g

(ii) A sea otter weighing 20,000 g

(b) Suppose the animal's weight is given in kilograms rather than grams. Given that 1 kg = 1000 g, find a function $x = g(z)$ giving the animal's weight in grams if z is the animal's weight in kilograms.

[†]The process of forming $f(g(z))$ is referred to as function composition, which we will study in Sec. 4.3.

(c) Write the energy expenditure as a function of the weight in kilograms in the form $y = az^b$ by calculating $f(g(z))$.[†]

General Interest

80. Area A rectangular field is to have a perimeter of 6000 ft.

(a) Write the area, A, of the field as a function of the width, w.

(b) Find the domain of the function in part (a).

(c) Use a graphing calculator to sketch the graph of the function in part (a).

(d) Describe what the graph found in part (c) tells you about how the area of the field varies with the width.

81. Perimeter A rectangular field is to have an area of 500 m².

(a) Write the perimeter, P, of the field as a function of the width, w.

(b) Find the domain of the function in part (a).

(c) Use a graphing calculator to sketch the graph of the function in part (a).

(d) Describe what the graph found in part (c) tells you about how the perimeter of the field varies with the width.

YOUR TURN ANSWERS

1. $(-\infty, -2) \cup (2, \infty), (0, \infty)$

2. (a) $2x^2 + 4xh + 2h^2 - 3x - 3h - 4$ **(b)** 1 and 1/2

2.2 Quadratic Functions; Translation and Reflection

In Example 7 in this section we will see how knowledge of quadratic functions *will help provide an answer to the question above.*

A linear function is defined by

$$f(x) = ax + b,$$

for real numbers a and b. In a *quadratic function* the independent variable is squared. A quadratic function is an especially good model for many situations with a maximum or a minimum function value. Quadratic functions also may be used to describe supply and demand curves; cost, revenue, and profit; as well as other quantities. Next to linear functions, they are the simplest type of function, and well worth studying thoroughly.

FOR REVIEW

In this section you will need to know how to solve a quadratic equation by factoring and by the quadratic formula, which are covered in Sections R.2 and R.4. Factoring is usually easiest; when a polynomial is set equal to zero and factored, then a solution is found by setting any one factor equal to zero. But factoring is not always possible. The quadratic formula will provide the solution to *any* quadratic equation.

Quadratic Function

A **quadratic function** is defined by

$$f(x) = ax^2 + bx + c,$$

where a, b, and c are real numbers, with $a \neq 0$.

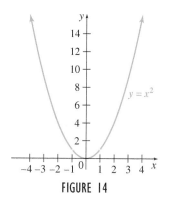

FIGURE 14

The simplest quadratic function has $f(x) = x^2$, with $a = 1, b = 0$, and $c = 0$. This function describes situations where the dependent variable y is proportional to the *square* of the independent variable x. The graph of this function is shown in Figure 14. This graph is called a **parabola**. Every quadratic function has a parabola as its graph. The lowest (or highest) point on a parabola is the **vertex** of the parabola. The vertex of the parabola in Figure 14 is $(0, 0)$.

If the graph in Figure 14 were folded in half along the y-axis, the two halves of the parabola would match exactly. This means that the graph of a quadratic function is *symmetric* with respect to a vertical line through the vertex; this line is the **axis of symmetry** of the parabola.

There are many real-world instances of parabolas. For example, cross sections of spotlight reflectors or radar dishes form parabolas. Also, a projectile thrown in the air follows a parabolic path. For such applications, we need to study more complicated quadratic functions than $y = x^2$, as in the next several examples.

EXAMPLE 1 Graphing a Quadratic Function

Graph $y = x^2 - 4$.

SOLUTION Each value of y will be 4 less than the corresponding value of y in $y = x^2$. The graph of $y = x^2 - 4$ has the same shape as that of $y = x^2$ but is 4 units lower. See Figure 15. The vertex of the parabola (on this parabola, the *lowest* point) is at $(0, -4)$. The x-intercepts can be found by letting $y = 0$ to get

$$0 = x^2 - 4,$$

from which $x = 2$ and $x = -2$ are the x-intercepts. The axis of symmetry of the parabola is the vertical line $x = 0$.

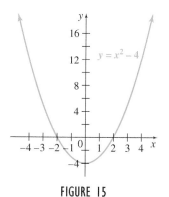

FIGURE 15

Example 1 suggests that the effect of c in $ax^2 + bx + c$ is to lower the graph if c is negative and to raise the graph if c is positive. This is true for any function; the movement up or down is referred to as a **vertical translation** of the function.

EXAMPLE 2 Graphing Quadratic Functions

Graph $y = ax^2$ with $a = -0.5, a = -1, a = -2$, and $a = -4$.

SOLUTION Figure 16 shows all four functions plotted on the same axes. We see that since a is negative, the graph opens downward. When a is between -1 and 1 (that is, when $a = -0.5$), the graph is wider than the original graph, because the values of y are smaller in magnitude. On the other hand, when a is greater than 1 or less than -1, the graph is steeper.

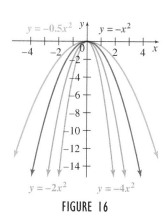

FIGURE 16

Example 2 shows that the sign of a in $ax^2 + bx + c$ determines whether the parabola opens upward or downward. Multiplying $f(x)$ by a negative number flips the graph of f upside down. This is called a **vertical reflection** of the graph. The magnitude of a determines how steeply the graph increases or decreases.

EXAMPLE 3 Graphing Quadratic Functions

Graph $y = (x - h)^2$ for $h = 3, 0$, and -4.

SOLUTION Figure 17 on the next page shows all three functions on the same axes. Notice that since the number is subtracted *before* the squaring occurs, the graph does not move up or down but instead moves left or right. Evaluating $f(x) = (x - 3)^2$ at $x = 3$ gives the same result as evaluating $f(x) = x^2$ at $x = 0$. Therefore, when we subtract the positive number 3 from x, the graph shifts 3 units to the right, so the vertex is at $(3, 0)$. Similarly, when we subtract the negative number -4 from x—in other words, when the function becomes $f(x) = (x + 4)^2$— the graph shifts to the left 4 units.

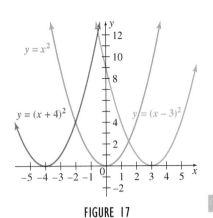

FIGURE 17

The left or right shift of the graph illustrated in Figure 17 is called a **horizontal translation** of the function.

If a quadratic equation is given in the form $ax^2 + bx + c$, we can identify the translations and any vertical reflection by rewriting it in the form

$$y = a(x - h)^2 + k.$$

In this form, we can identify the vertex as (h, k). A quadratic equation not given in this form can be converted by a process called **completing the square**. The next example illustrates the process.

EXAMPLE 4 Graphing a Quadratic Function (Completing the Square)

Graph $y = 3x^2 + 2x - 1$.

SOLUTION Rewrite the equation in the form $y = a(x - h)^2 + k$ by completing the square. First, factor 3 from the x-terms so the coefficient of x^2 is 1:

$$y = 3\left(x^2 + \frac{2}{3}x \quad\right) - 1.$$

Next, make the expression inside the parentheses a perfect square. Add the square of *half* the coefficient of x: $(\frac{1}{2} \cdot \frac{2}{3})^2 = \frac{1}{9}$. Since there is a factor of 3 outside the parentheses, we are actually adding $3 \cdot \frac{1}{9}$. To make sure that the value of the function is not changed, we must also subtract $3 \cdot \frac{1}{9}$ from the function. To summarize our steps,

$$y = 3\left(x^2 + \frac{2}{3}x \quad\right) - 1 \qquad \text{Factor out 3.}$$

$$= 3\left(x^2 + \frac{2}{3}x + \frac{1}{9}\right) - 1 - 3\left(\frac{1}{9}\right) \qquad \begin{array}{l}\text{Add and subtract 3 times} \\ (\frac{1}{2} \text{ the coefficient of } x)^2.\end{array}$$

$$= 3\left(x + \frac{1}{3}\right)^2 - \frac{4}{3}. \qquad \text{Factor and combine terms.}$$

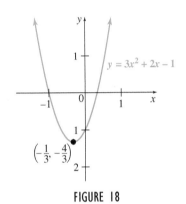

FIGURE 18

The function is now in the form $y = a(x - h)^2 + k$. Since $h = -1/3$ and $k = -4/3$, the vertex is $(-1/3, -4/3)$. The graph is the graph of the parabola $y = x^2$ translated 1/3 unit to the left and 4/3 unit downward. The parabola opens upward because $a = 3$ is positive. The 3 will cause the parabola to be stretched vertically by a factor of 3. These results are shown in Figure 18. **TRY YOUR TURN 1**

YOUR TURN 1 For the function $y = 2x^2 - 6x - 1$, **(a)** complete the square, **(b)** find the y-intercept, **(c)** find the x-intercepts, **(d)** find the vertex, and **(e)** sketch the graph.

Instead of completing the square to find the vertex of the graph of a quadratic function given in the form $y = ax^2 + bx + c$, we can develop a formula for the vertex. By the quadratic formula, if $ax^2 + bx + c = 0$, where $a \neq 0$, then

$$x = \frac{-b \pm \sqrt{b^2 - 4ac}}{2a}.$$

Notice that this is the same as

$$x = \frac{-b}{2a} \pm \frac{\sqrt{b^2 - 4ac}}{2a} = \frac{-b}{2a} \pm Q,$$

where $Q = \sqrt{b^2 - 4ac}/(2a)$. Since a parabola is symmetric with respect to its axis, the x-coordinate of the vertex is halfway between its two roots. Halfway between $x = -b/(2a) + Q$ and $x = -b/(2a) - Q$ is $x = -b/(2a)$. Once we have the x-coordinate of the vertex, we can easily find the y-coordinate by substituting the x-coordinate into the original equation.

Graph of the Quadratic Function

The graph of the quadratic function $f(x) = ax^2 + bx + c$ is a parabola with the following characteristics:

1. The parabola opens upward if $a > 0$ and downward if $a < 0$;
2. The y-intercept is at $(0, c)$;
3. The x-intercepts, if they exist, occur where $f(x) = 0$;
4. The vertex is at $(-b/(2a), f(-b/(2a)))$;
5. The axis of symmetry is $x = -b/(2a)$.

EXAMPLE 5 Graphing a Quadratic Function

Graph $f(x) = -4x^2 - 8x + 5$.

SOLUTION In this quadratic function, $a = -4$, $b = -8$, and $c = 5$. Since $a < 0$, the graph of the parabola opens downward.

The y-intercept is at $(0, 5)$. To find the x-intercepts, we set the equation equal to 0 and solve.

$$-4x^2 - 8x + 5 = 0$$
$$4x^2 + 8x - 5 = 0 \qquad \text{Multiply both sides by } -1.$$
$$(2x - 1)(2x + 5) = 0 \qquad \text{Factor.}$$
$$2x - 1 = 0 \quad \text{or} \quad 2x + 5 = 0$$
$$x = 1/2 \qquad\qquad x = -5/2$$

The x-intercepts are $(1/2, 0)$ and $(-5/2, 0)$.

The x-coordinate of the vertex is

$$x = \frac{-b}{2a} = \frac{-(-8)}{2(-4)} = -1.$$

The y-coordinate is then

$$y = f(-1) = -4(-1)^2 - 8(-1) + 5 = 9.$$

Therefore, the vertex is at $(-1, 9)$. Since the parabola opens downward, the vertex is the highest point on the parabola. The axis of symmetry is the vertical line through the vertex: $x = -1$. These results are shown in Figure 19. **TRY YOUR TURN 2**

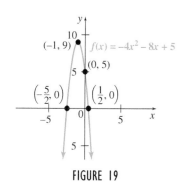

FIGURE 19

YOUR TURN 2 For the function $f(x) = x^2 - 4x - 5$ **(a)** find the y-intercept, **(b)** find the x-intercepts, **(c)** find the vertex, **(d)** find the axis of symmetry, and **(e)** sketch the graph.

Another situation that may arise is the absence of any x-intercepts, as in the next example.

EXAMPLE 6 Graphing a Quadratic Function

Graph $y = x^2 + 4x + 6$.

SOLUTION In this quadratic function, $a = 1$, $b = 4$, and $c = 6$. Since $a > 0$, the graph of the parabola opens upward.

The y-intercept is at $(0, 6)$. To find the x-intercepts, we solve the equation $x^2 + 4x + 6 = 0$. This does not appear to factor, so we'll try the quadratic formula.

$$x = \frac{-b \pm \sqrt{b^2 - 4ac}}{2a}$$
$$= \frac{-4 \pm \sqrt{4^2 - 4(1)(6)}}{2(1)}$$
$$= \frac{-4 \pm \sqrt{-8}}{2}$$

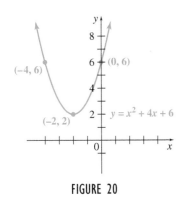

FIGURE 20

As soon as we see the negative value under the square root sign, we know that there are no real number solutions. Therefore, there are no x-intercepts.

The x-coordinate of the vertex is

$$x = \frac{-b}{2a} = \frac{-4}{2(1)} = -2.$$

Substituting this into the equation gives the y-coordinate:

$$y = (-2)^2 + 4(-2) + 6 = 2.$$

Therefore, the vertex is at $(-2, 2)$. Since the parabola opens upward, the vertex is the lowest point on the parabola. The axis of symmetry is the vertical line $x = -2$.

Since there are no x-intercepts, we only have two points (the vertex and y-intercept) to plot. Using the symmetry of the figure, we can also plot the mirror image of the y-intercept on the opposite side of the parabola's axis: at $x = -4$ (2 units to the left of the axis) and $y = 6$. Plotting the vertex, the y-intercept, and the point $(-4, 6)$ gives the graph in Figure 20. ▬

The concept of maximizing or minimizing a function is important in calculus, as we will see in future chapters. The following example illustrates how we can determine the maximum or minimum value of a quadratic function without graphing.

EXAMPLE 7 **Management Science**

When Power and Money, Inc., charges $600 for a seminar on management techniques, it attracts 1000 people. For each $20 decrease in the fee, an additional 100 people will attend the seminar. The managers are wondering how much to charge for the seminar to maximize their revenue.

SOLUTION Let x be the number of $20 decreases in the price. Then the price charged per person will be

$$\text{Price per person} = 600 - 20x,$$

and the number of people in the seminar will be

$$\text{Number of people} = 1000 + 100x.$$

The total revenue, $R(x)$, is given by the product of the price and the number of people attending, or

$$R(x) = (600 - 20x)(1000 + 100x)$$
$$= 600{,}000 + 40{,}000x - 2000x^2.$$

We see by the negative in the x^2-term that this defines a parabola opening downward, so the maximum revenue is at the vertex. The x-coordinate of the vertex is

$$x = \frac{-b}{2a} = \frac{-40{,}000}{2(-2000)} = 10.$$

YOUR TURN 3 Solve Example 7 with the following changes: a $1650 price attracts 900 people, and each $40 decrease in the price attracts an additional 80 people.
▬

The y-coordinate is then

$$y = 600{,}000 + 40{,}000(10) - 2000(10^2)$$
$$= 800{,}000.$$

Therefore, the maximum revenue is $800,000, which is achieved by charging $600 - 20x = 600 - 20(10) = \$400$ per person. **TRY YOUR TURN 3** ▬

Notice in this last example that the maximum revenue was achieved by charging less than the current price of $600, which was more than made up for by the increase in sales. This is typical of many applications. Mathematics is a powerful tool for solving such problems, since the answer is not always what one might have guessed intuitively.

In the next example, we show how the calculation of profit can involve a quadratic function.

EXAMPLE 8   Profit

A deli owner has found that his revenue from producing x pounds of vegetable cream cheese is given by $R(x) = -x^2 + 30x$, while the cost in dollars is given by $C(x) = 5x + 100$.

(a) Find the minimum break-even quantity.

SOLUTION   Notice from the graph in Figure 21 that the revenue function is a parabola opening downward and the cost function is a linear function that crosses the revenue function at two points. To find the minimum break-even quantity, we find where the two functions are equal.

$$R(x) = C(x)$$
$$-x^2 + 30x = 5x + 100$$
$$0 = x^2 - 25x + 100 \qquad \text{Subtract } -x^2 + 30x \text{ from both sides.}$$
$$= (x - 5)(x - 20) \qquad \text{Factor.}$$

The two graphs cross when $x = 5$ and $x = 20$. The minimum break-even point is at $x = 5$. The deli owner must sell at least 5 lb of cream cheese to break even.

(b) Find the maximum revenue.

SOLUTION   Since the revenue function is a quadratic with a negative x^2-term, the maximum is at the vertex. The vertex has a value of $x = -b/(2a) = -30/(-2) = 15$. The maximum revenue is $R(15) = -15^2 + 30(15) = 225$, or $225.

(c) Find the maximum profit.

SOLUTION   The profit is the difference between the revenue and the cost, or

$$P(x) = R(x) - C(x)$$
$$= (-x^2 + 30x) - (5x + 100)$$
$$= -x^2 + 25x - 100.$$

The value of x at the vertex is $x = -b/(2a) = -25/(-2) = 12.5$. The value of the function here is $P(12.5) = -12.5^2 + 25(12.5) - 100 = 56.25$. It is clear that this is a maximum, not only from Figure 21, but also because the profit function is a quadratic with a negative x^2-term. A maximum profit of $56.25 is achieved by selling 12.5 lb of cream cheese.   **TRY YOUR TURN 4**

FIGURE 21

YOUR TURN 4   Suppose the revenue in dollars is given by $R(x) = -x^2 + 40x$ and the cost is given by $C(x) = 8x + 192$. Find **(a)** the minimum break-even quantity, **(b)** the maximum revenue, and **(c)** the maximum profit.

In Section 1.1, we used two data points to determine the equation of a line that would model data that was approximately linear. In the next example, we have a set of data that would be better approximated with a quadratic model.

EXAMPLE 9   Elderly in the U.S. Labor Force

The table in the margin gives the labor force participation rate (LFPR) of men age 65 and older. (The LFPR for 2020 is projected.) *Source: Bureau of Labor Statistics.*

(a) Plot the data by letting $t = 50$ correspond to the year 1950. Would a linear or quadratic function model these data best?

Year	Participation Rate
1950	45.8
1960	33.1
1970	28.8
1980	19.0
1990	16.3
2000	17.7
2010	20.5
2020	26.7

SOLUTION

FIGURE 22

The scatterplot in Figure 22(a) suggests that a quadratic function with a positive value of a (so the graph opens upward) would be a reasonable model for the data. The minimum value (vertex) appears to be $(90, 16.3)$.

(b) Find the quadratic function defined by $f(t) = a(t - h)^2 + k$ that models the data. Use $(90, 16.3)$ as the vertex, (h, k). Then choose a second point, such as $(100, 17.7)$, to determine a.

SOLUTION Substitute $h = 90$, $k = 16.3$, $t = 100$, and $y = f(t) = 17.7$ into the function and solve for a.

$$f(t) = a(t - h)^2 + k$$
$$17.7 = a(100 - 90)^2 + 16.3$$
$$17.7 = 100a + 16.3 \qquad \text{\small$(100 - 90)^2 = 10^2 = 100$}$$
$$1.4 = 100a \qquad \text{\small Subtract 16.3 from both sides.}$$
$$a = 0.014 \qquad \text{\small Divide both sides by 100.}$$

Substituting a, h, and k into the model gives the quadratic function:

$$f(t) = 0.014(t - 90)^2 + 16.3.$$

Note that other choices for the second point will lead to slightly different equations.

TECHNOLOGY NOTE

Section 1.3 showed how the equation of a line that closely approximates a set of data points is found using linear regression. Some graphing calculators with statistics capability perform other types of regression. For example, **quadratic regression** gives the coefficients of a quadratic equation that models a given set of points.

On a T1-84 Plus C, enter the data into the calculator by selecting STAT and then EDIT, which brings up blank lists. (If the lists are not blank, clear them using the ClrList command.) Enter the year in L_1 (letting $t = 50$ correspond to 1950) and the rate in L_2.

Quit the editor, press STAT again, and then select the CALC menu. The command QuadReg L_1, L_2, Y_1 finds the regression equation for the data in L_1 and L_2 and stores the function in Y_1. The quadratic function (with rounded coefficients) is

$$f(t) = 0.0146t^2 - 2.76t + 147,$$

which models the data quite well, as shown in Figure 22(b). Verify that this function is close to the one found in Example 9(b).

On the next page we provide guidelines for sketching graphs that involve translations and reflections.

Translations and Reflections of Functions

Let f be any function, and let h and k be positive constants (Figure 23).

The graph of $y = f(x) + k$ is the graph of $y = f(x)$ translated upward k units (Figure 24).

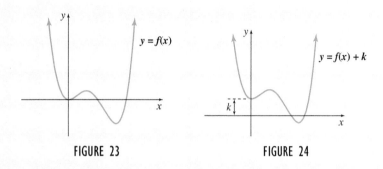

FIGURE 23   FIGURE 24

The graph of $y = f(x) - k$ is the graph of $y = f(x)$ translated downward by k units (Figure 25).

The graph of $y = f(x - h)$ is the graph of $y = f(x)$ translated to the right by h units (Figure 26).

The graph of $y = f(x + h)$ is the graph of $y = f(x)$ translated to the left by h units (Figure 27).

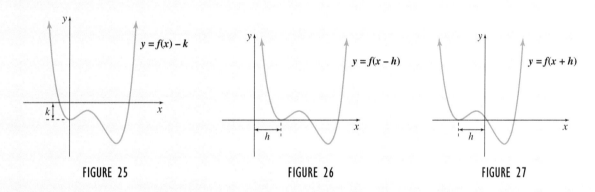

FIGURE 25   FIGURE 26   FIGURE 27

The graph of $y = -f(x)$ is the graph of $y = f(x)$ reflected vertically across the x-axis, that is, turned upside down (Figure 28).

The graph of $y = f(-x)$ is the graph of $y = f(x)$ reflected horizontally across the y-axis, that is, its mirror image (Figure 29).

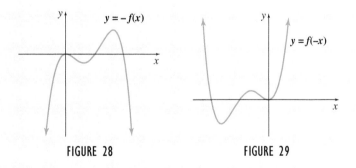

FIGURE 28   FIGURE 29

Notice in Figure 29 that another type of reflection, known as a **horizontal reflection**, is shown. Multiplying x or $f(x)$ by a constant a, to get $y = f(ax)$ or $y = a \cdot f(x)$, does not change the general appearance of the graph, except to compress or stretch it. When a is negative, it also causes a reflection, as shown in the last two figures in the summary for $a = -1$. Also see Exercises 39–46 in this section.

EXAMPLE 10 **Translations and Reflections of a Graph**

The graph of $y = \sqrt{x}$ is shown in Figure 30. (Note that this graph is the top half of the parabola $y = x^2$ lying on its side.) Use this graph, along with translations and reflections, to graph the following.

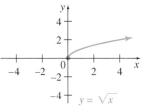

FIGURE 30

(a) $f(x) = \sqrt{x - 1} + 2$

SOLUTION The 1 under the square root sign is a translation of $y = \sqrt{x}$ to the right 1 unit. Adding the constant 2 raises the graph by 2 units. The graph of $f(x)$ is shown in Figure 31.

FIGURE 31

(b) $g(x) = -\sqrt{x + 2} - 1$

SOLUTION The effect of the negative sign in front of the radical is a vertical reflection of $y = \sqrt{x}$, as shown in Figure 32(a). Adding 2 under the square root shifts the graph to the left 2 units. Subtracting 1 lowers the graph by 1 unit. The graph of $g(x)$ is shown in Figure 32(b).

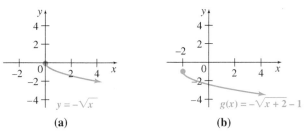

FIGURE 32

(c) $f(x) = \sqrt{4 - x}$

SOLUTION Consider the negative in front of the x, giving $f(x) = \sqrt{-x}$. This is a horizontal reflection of $y = \sqrt{x}$, as shown in Figure 33(a). Next, include the 4 under the square root sign. To get $4 - x$ into the form $f(x - h)$ or $f(x + h)$, we need to factor out the negative: $\sqrt{4 - x} = \sqrt{-(x - 4)}$. Now the 4 is subtracted, so this function is a translation to the right 4 units. The graph of $f(x)$ is shown in Figure 33(b).

YOUR TURN 5 Graph each of the following:
(a) $f(x) = \sqrt{x - 2} + 1$
(b) $g(x) = -\sqrt{x} - 1$

TRY YOUR TURN 5

(a)

(b)

FIGURE 33

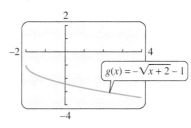

TECHNOLOGY NOTE

Graphing calculators can be used to graph most of the functions in this section. Be cautious, however, when selecting your viewing window to ensure that you capture the important features of the graph. For instance, a graphing calculator image of the function in Example 10(b) is shown in Figure 34. If you viewed this image, you might think that the function continues to go up and to the left. By realizing that $(-2, -1)$ is the vertex of the sideways parabola, we see that this is the leftmost point on the graph. Another approach is to find the domain of $g(x)$ by setting $x + 2 \geq 0$, from which we conclude that $x \geq -2$. This demonstrates the importance of knowing the algebraic techniques in order to interpret a graphing calculator image correctly.

FIGURE 34

2.2 WARM-UP EXERCISES

W1. Factor $4x^2 - 36x + 81$. *(Sec. R.2)*

W2. Factor $x^2 + \dfrac{5}{4}x + \dfrac{25}{64}$. *(Sec. R.2)*

W3. Solve $3x^2 - x - 10 = 0$. *(Sec. R.4)*

W4. Solve $2x^2 + 5x - 4 = 0$. *(Sec. R.4)*

2.2 EXERCISES

1. How does the value of a affect the graph of $y = ax^2$? Discuss the case for $a \geq 1$ and for $0 \leq a \leq 1$.

2. How does the value of a affect the graph of $y = ax^2$ if $a \leq 0$?

In Exercises 3–8, match the correct graph A–F to the function without using your calculator. Then, if you have a graphing calculator, use it to check your answers. Each graph in this group shows x and y in $[-10, 10]$.

3. $y = x^2 - 3$

4. $y = (x - 3)^2$

5. $y = (x - 3)^2 + 2$

6. $y = (x + 3)^2 + 2$

7. $y = -(3 - x)^2 + 2$

8. $y = -(x + 3)^2 + 2$

(A)

(B)

(C)

(D)

(E)

(F)

Complete the square and determine the vertex for each of the following.

9. $y = 3x^2 + 9x + 5$

10. $y = 4x^2 - 20x - 7$

11. $y = -2x^2 + 8x - 9$

12. $y = -5x^2 - 8x + 3$

In Exercises 13–24, graph each parabola and give its vertex, axis of symmetry, x-intercepts, and y-intercept.

13. $y = x^2 + 5x + 6$

14. $y = x^2 + 4x - 5$

15. $y = -2x^2 - 12x - 16$

16. $y = -3x^2 - 6x + 4$

17. $f(x) = 2x^2 + 8x - 8$

18. $f(x) = -x^2 + 6x - 6$

19. $f(x) = 2x^2 - 4x + 5$

20. $f(x) = \frac{1}{2}x^2 + 6x + 24$

21. $f(x) = -2x^2 + 16x - 21$

22. $f(x) = \frac{3}{2}x^2 - x - 4$

23. $f(x) = \frac{1}{3}x^2 - \frac{8}{3}x + \frac{1}{3}$

24. $f(x) = -\frac{1}{2}x^2 - x - \frac{7}{2}$

In Exercises 25–30, follow the directions for Exercises 3–8.

25. $y = \sqrt{x + 2} - 4$

26. $y = \sqrt{x - 2} - 4$

27. $y = \sqrt{-x + 2} - 4$

28. $y = \sqrt{-x - 2} - 4$

29. $y = -\sqrt{x + 2} - 4$

30. $y = -\sqrt{x - 2} - 4$

(A)

(B)

(C)

(D)

(E)

(F)

Given the following graph, sketch by hand the graph of the function described, giving the new coordinates for the three points labeled on the original graph.

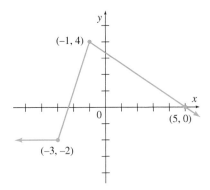

31. $y = -f(x)$

32. $y = f(x - 2) + 2$

33. $y = f(-x)$

34. $y = f(2 - x) + 2$

Use the ideas in this section to graph each function without a calculator.

35. $f(x) = \sqrt{x - 2} + 2$

36. $f(x) = \sqrt{x + 2} - 3$

37. $f(x) = -\sqrt{2 - x} - 2$

38. $f(x) = -\sqrt{2 - x} + 2$

Using the graph of $f(x)$ in Figure 23, show the graph of $f(ax)$ where a satisfies the given condition.

39. $0 < a < 1$ **40.** $1 < a$

41. $-1 < a < 0$ **42.** $a < -1$

Using the graph of $f(x)$ in Figure 23, show the graph of $af(x)$ where a satisfies the given condition.

43. $0 < a < 1$ **44.** $1 < a$

45. $-1 < a < 0$ **46.** $a < -1$

47. If r is an x-intercept of the graph of $y = f(x)$, what is an x-intercept of the graph of each of the following?

(a) $y = -f(x)$ (b) $y = f(-x)$

(c) $y = -f(-x)$

48. If b is the y-intercept of the graph of $y = f(x)$, what is the y-intercept of the graph of each of the following?

(a) $y = -f(x)$ (b) $y = f(-x)$

(c) $y = -f(-x)$

APPLICATIONS

Business and Economics

Profit In Exercises 49–52, let $C(x)$ be the cost to produce x batches of widgets, and let $R(x)$ be the revenue in thousands of dollars. For each exercise, (a) graph both functions, (b) find the minimum break-even quantity, (c) find the maximum revenue, and (d) find the maximum profit.

49. $R(x) = -x^2 + 8x$, $C(x) = 2x + 5$

50. $R(x) = -\dfrac{x^2}{2} + 5x$, $C(x) = \dfrac{3}{2}x + 3$

51. $R(x) = -\dfrac{4}{5}x^2 + 10x$, $C(x) = 2x + 15$

52. $R(x) = -4x^2 + 36x$, $C(x) = 16x + 24$

53. Maximizing Revenue The revenue of a charter bus company depends on the number of unsold seats. If the revenue $R(x)$ is given by

$$R(x) = 8000 + 70x - x^2,$$

where x is the number of unsold seats, find the maximum revenue and the number of unsold seats that corresponds to maximum revenue.

54. Maximizing Revenue A charter flight charges a fare of $200 per person plus $4 per person for each unsold seat on the plane. The plane holds 100 passengers. Let x represent the number of unsold seats.

(a) Find an expression for the total revenue received for the flight $R(x)$. (*Hint:* Multiply the number of people flying, $100 - x$, by the price per ticket.)

(b) Graph the expression from part (a).

(c) Find the number of unsold seats that will produce the maximum revenue.

(d) What is the maximum revenue?

(e) Some managers might be concerned about the empty seats, arguing that it doesn't make economic sense to leave any seats empty. Write a few sentences explaining why this is not necessarily so.

55. Income The manager of an 80-unit apartment complex is trying to decide what rent to charge. Experience has shown that at a rent of $800, all the units will be full. On the average, one additional unit will remain vacant for each $25 increase in rent.

(a) Let x represent the number of $25 increases. Find an expression for the rent for each apartment.

(b) Find an expression for the number of apartments rented.

(c) Find an expression for the total revenue from all rented apartments.

(d) What value of x leads to maximum revenue?

(e) What is the maximum revenue?

56. Revenue The manager of a peach orchard is trying to decide when to arrange for picking the peaches. If they are picked now, the average yield per tree will be 100 lb, which can be sold for 80¢ per pound. Past experience shows that the yield per tree will increase about 5 lb per week, while the price will decrease about 4¢ per pound per week.

(a) Let x represent the number of weeks that the manager should wait. Find the income per pound.

(b) Find the number of pounds per tree.

(c) Find the total revenue from a tree.

(d) When should the peaches be picked in order to produce maximum revenue?

(e) What is the maximum revenue?

57. Maximizing Revenue The demand for a certain type of cosmetic is given by

$$p = 500 - x,$$

where p is the price in dollars when x units are demanded.

(a) Find the revenue $R(x)$ that would be obtained at a price p. (*Hint:* Revenue = Demand × Price)

(b) Graph the revenue function $R(x)$.

(c) Find the price that will produce maximum revenue.

(d) What is the maximum revenue?

58. Advertising A study done by an advertising agency reveals that when x thousands of dollars are spent on advertising, it results in a sales increase in thousands of dollars given by the function

$$S(x) = -\frac{1}{4}(x - 10)^2 + 40, \quad \text{for } 0 \le x \le 10.$$

(a) Find the increase in sales when no money is spent on advertising.

(b) Find the increase in sales when $10,000 is spent on advertising.

(c) Sketch the graph of $S(x)$ without a calculator.

59. Coal Consumption The following table gives the amount of coal (in thousand short tons) consumed by U. S. utilities to produce electricity. *Source: U.S. Department of Energy.*

Year	Consumption
1995	850
2000	986
2005	1037
2010	934
2012	825

(a) Plot the data by letting t be the years since 1990. Would a linear or quadratic function model these data best?

(b) Find the quadratic function defined by $f(t) = a(t - h)^2 + k$ that models the data. Use the highest point as the vertex, (h, k). Then choose $(10, 986)$ as a second point to determine a.

(c) Use the quadratic regression feature on a graphing calculator to find the quadratic function that best fits the data. Graph this function on the same calculator window as the data.

(d) Graph the function from part (b) on the same calculator window as the data and function from part (c). Do the graphs of the two functions differ by much?

Life Sciences

60. Tooth Length The length (in mm) of the mesiodistal crown of the first molar for human fetuses can be approximated by

$$L(t) = -0.01t^2 + 0.788t - 7.048,$$

where t is the number of weeks since conception. *Source: American Journal of Physical Anthropology.*

(a) What does this formula predict for the length at 14 weeks? 24 weeks?

(b) What does this formula predict for the maximum length, and when does that occur? Explain why the formula does not make sense past that time.

61. Length of Life According to recent data from the Teachers Insurance and Annuity Association (TIAA), the survival function for life after 65 is approximately given by

$$S(x) = 1 - 0.058x - 0.076x^2,$$

where x is measured in decades. This function gives the probability that an individual who reaches the age of 65 will live at least x decades ($10x$ years) longer. *Source: Ralph DeMarr.*

(a) Find the median length of life for people who reach 65, that is, the age for which the survival rate is 0.50.

(b) Find the age beyond which virtually nobody lives. (There are, of course, exceptions.)

62. Cancer From 1992 to 2011, the age-adjusted incidence rate of invasive lung and bronchial cancer among women can be closely approximated by

$$f(t) = -0.0335t^2 + 0.490t + 48.4,$$

where t is the number of years since 1992. Based on this model, in what year did the incidence rate reach a maximum? On what years was the rate increasing? Decreasing? *Source: National Cancer Institute.*

63. Splenic Artery Resistance Blood flow to the fetal spleen is of research interest because several diseases are associated with increased resistance in the splenic artery (the artery that goes to the spleen). Researchers have found that the index of splenic artery resistance in the fetus can be described by the function

$$y = 0.057x - 0.001x^2,$$

where x is the number of weeks of gestation. *Source: American Journal of Obstetrics and Gynecology.*

(a) At how many weeks is the splenic artery resistance a maximum?

(b) What is the maximum splenic artery resistance?

(c) At how many weeks is the splenic artery resistance equal to 0, according to this formula? Is your answer reasonable for this function? Explain.

Social Sciences

64. Gender Ratio The number of males per 100 females, age 65 or over, in the United States for some recent years is shown in the following table. *Source: U.S. Census Bureau.*

Year	Males per 100 Females
1960	82.8
1970	72.1
1980	67.6
1990	67.2
2000	70.0
2010	77.0

(a) Plot the data, letting t be the years since 1900.

(b) Would a linear or quadratic function best model these data? Explain.

(c) If your graphing calculator has a quadratic regression feature, find the quadratic function that best fits the data. Graph this function on the same calculator window as the data.

(d) Choose the lowest point in the table above as the vertex and $(110, 77.0)$ as a second point to find a quadratic function defined by $f(t) = a(t - h)^2 + k$ that models the data.

(e) Graph the function from part (d) on the same calculator window as the data and function from part (c). Do the graphs of the two functions differ by much?

(f) Predict the number of males per 100 females in 2004 using the two functions from parts (c) and (d), and compare with the actual figure of 71.7.

65. Age of Marriage The following table gives the median age at their first marriage of women in the United States for some selected years. *Source: U.S. Census Bureau.*

Year	Age
1940	21.5
1950	20.3
1960	20.3
1970	20.8
1980	22.0
1990	23.9
2000	25.1
2010	26.5

(a) Plot the data using $t = 40$ for 1940, and so on.

(b) Would a linear or quadratic function best model this data? Explain.

(c) If your graphing calculator has a regression feature, find the quadratic function that best fits the data. Graph this function on the same calculator window as the data.

(d) Find a quadratic function defined by $f(t) = a(t - h)^2 + k$ that models the data using $(60, 20.3)$ as the vertex and then choosing $(110, 26.5)$ as a second point to determine the value of a.

(e) Graph the function from part (d) on the same calculator window as the data and function from part (c). Do the graphs of the two functions differ by much?

66. Accident Rate According to data from the National Highway Traffic Safety Administration, the accident rate as a function of the age of the driver in years x can be approximated by the function

$$f(x) = 60.0 - 2.28x + 0.0232x^2$$

for $16 \leq x \leq 85$. Find the age at which the accident rate is a minimum and the minimum rate. *Source: Ralph DeMarr.*

Physical Sciences

67. Maximizing the Height of an Object If an object is thrown upward with an initial velocity of 32 ft/second, then its height after t seconds is given by

$$h(t) = 32t - 16t^2.$$

(a) Find the maximum height attained by the object.

(b) Find the number of seconds it takes the object to hit the ground.

68. Stopping Distance According to data from the National Traffic Safety Institute, the stopping distance y in feet of a car traveling x mph can be described by the equation $y = 0.056057x^2 + 1.06657x$. *Source: National Traffic Safety Institute*.

(a) Find the stopping distance for a car traveling 25 mph.

(b) How fast can you drive if you need to be certain of stopping within 150 ft?

General Interest

69. Maximizing Area Glenview Community College wants to construct a rectangular parking lot on land bordered on one side by a highway. It has 380 ft of fencing to use along the other three sides. What should be the dimensions of the lot if the enclosed area is to be a maximum? (*Hint:* Let x represent the width of the lot, and let $380 - 2x$ represent the length.)

70. Maximizing Area What would be the maximum area that could be enclosed by the college's 380 ft of fencing if it decided to close the entrance by enclosing all four sides of the lot? (See Exercise 69.)

In Exercises 71 and 72, draw a sketch of the arch or culvert on coordinate axes, with the horizontal and vertical axes through the vertex of the parabola. Use the given information to label points on the parabola. Then give the equation of the parabola and answer the question.

71. Parabolic Arch An arch is shaped like a parabola. It is 30 m wide at the base and 15 m high. How wide is the arch 10 m from the ground?

72. Parabolic Culvert A culvert is shaped like a parabola, 18 ft across the top and 12 ft deep. How wide is the culvert 8 ft from the top?

YOUR TURN ANSWERS

1. (a) $y = 2(x - 3/2)^2 - 11/2$ (b) $(0, -1)$
   (c) $((3 + \sqrt{11})/2, 0), ((3 - \sqrt{11})/2, 0)$
   (d) $(3/2, -11/2)$
   (e)

2. (a) $(0, -5)$ (b) $(-1, 0), (5, 0)$
   (c) $(2, -9)$ (d) $x = 2$
   (e)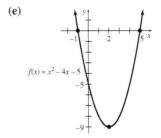

3. Charge $1050 for a maximum revenue of $2,205,000.

4. (a) 8   (b) $400   (c) $64

5. (a)   (b)

2.3 Polynomial and Rational Functions

APPLY IT How does the revenue collected by the government vary with the tax rate?

In Exercises 48–50 in this section, we will explore this question using polynomial *and* rational functions.

Polynomial Functions

Earlier, we discussed linear and quadratic functions and their graphs. Both of these functions are special types of *polynomial functions*.

Polynomial Function

A **polynomial function** of **degree** n, where n is a nonnegative integer, is defined by

$$f(x) = a_n x^n + a_{n-1}x^{n-1} + \cdots + a_1 x + a_0,$$

where $a_n, a_{n-1}, \ldots, a_1$, and a_0 are real numbers, called **coefficients**, with $a_n \neq 0$. The number a_n is called the **leading coefficient**.

For $n = 1$, a polynomial function takes the form

$$f(x) = a_1 x + a_0,$$

a linear function. A linear function, therefore, is a polynomial function of degree 1. (Note, however, that a linear function of the form $f(x) = a_0$ for a real number a_0 is a polynomial function of degree 0, the constant function.) A polynomial function of degree 2 is a quadratic function.

Accurate graphs of polynomial functions of degree 3 or higher require methods of calculus to be discussed later. Meanwhile, a graphing calculator is useful for obtaining such graphs, but care must be taken in choosing a viewing window that captures the significant behavior of the function.

The simplest polynomial functions of higher degree are those of the form $f(x) = x^n$. Such a function is known as a **power function**. Figure 35 below shows the graphs of $f(x) = x^3$ and $f(x) = x^5$, as well as tables of their values. These functions are simple enough that they can be drawn by hand by plotting a few points and connecting them with a smooth curve. An important property of all polynomials is that their graphs are smooth curves. Notice that $f(x) = x^3$ and $f(x) = x^5$ are odd functions, and that their graphs are symmetric about the origin.

$f(x) = x^3$		$f(x) = x^5$	
x	$f(x)$	x	$f(x)$
−2	−8	−1.5	−7.6
−1	−1	−1	−1
0	0	0	0
1	1	1	1
2	8	1.5	7.6

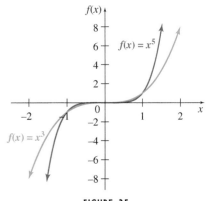

FIGURE 35

The graphs of $f(x) = x^4$ and $f(x) = x^6$, shown in Figure 36 along with tables of their values, can be sketched in a similar manner. These functions are even functions and their graphs have symmetry about the y-axis, as does the graph of $f(x) = ax^2$ for a nonzero real number a. As with the graph of $f(x) = ax^2$, the value of a in $f(x) = ax^n$ affects the direction of the graph. When $a > 0$, the graph has the same general appearance as the graph of $f(x) = x^n$. However, if $a < 0$, the graph is reflected vertically.

$f(x) = x^4$		$f(x) = x^6$	
x	$f(x)$	x	$f(x)$
-2	16	-1.5	11.4
-1	1	-1	1
0	0	0	0
1	1	1	1
2	16	1.5	11.4

FIGURE 36

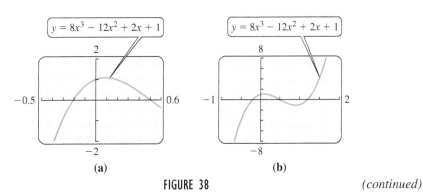

FIGURE 37

YOUR TURN 1 Graph $f(x) = 64 - x^6$.

EXAMPLE 1 Translations and Reflections

Graph $f(x) = -(x - 2)^3 + 3$.

SOLUTION Using the principles of translation and reflection from the previous section, we recognize that this is similar to the graph of $y = x^3$, but reflected vertically (because of the negative in front of $(x - 2)^3$), and with its center moved 2 units to the right and 3 units up. The result is shown in Figure 37. **TRY YOUR TURN 1**

A polynomial of degree 3, such as that in the previous example and in the next, is known as a **cubic polynomial**. A polynomial of degree 4, such as that in Example 3, is known as a **quartic polynomial**.

TECHNOLOGY

EXAMPLE 2 Graphing a Polynomial

Graph $f(x) = 8x^3 - 12x^2 + 2x + 1$.

SOLUTION Figure 38(a) shows the function graphed on the x- and y-intervals $[-0.5, 0.6]$ and $[-2, 2]$. In this view, it appears similar to a parabola opening downward. Zooming out to $[-1, 2]$ by $[-8, 8]$, we see in Figure 38(b) that the graph goes upward as x gets large. There are also two **turning points** near $x = 0$ and $x = 1$. (In a later chapter, we will introduce another term for such turning points: *relative extrema*.) By zooming in with the graphing calculator, we can find these turning points to be at approximately $(0.09175, 1.08866)$ and $(0.90825, -1.08866)$.

FIGURE 38 *(continued)*

$y = 8x^3 - 12x^2 + 2x + 1$

FIGURE 39

Zooming out still further, we see the function on $[-10, 10]$ by $[-300, 300]$ in Figure 39. From this viewpoint, we don't see the turning points at all, and the graph seems similar in shape to that of $y = x^3$. This is an important point: When x is large in magnitude, either positive or negative, $8x^3 - 12x^2 + 2x + 1$ behaves a lot like $8x^3$, because the other terms are small in comparison with the cubic term. So this viewpoint tells us something useful about the function, but it is less useful than the previous graph for determining the turning points.

After the previous example, you may wonder how to be sure you have the viewing window that exhibits all the important properties of a function. We will find an answer to this question in later chapters using the techniques of calculus. Meanwhile, let us consider one more example to get a better idea of how the graphs of polynomials look.

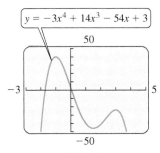

⊞ **TECHNOLOGY**

$y = -3x^4 + 14x^3 - 54x + 3$

FIGURE 40

EXAMPLE 3   **Graphing a Polynomial**

Graph $f(x) = -3x^4 + 14x^3 - 54x + 3$.

SOLUTION   Figure 40 shows a graphing calculator view on $[-3, 5]$ by $[-50, 50]$. If you have a graphing calculator, we recommend that you experiment with various viewpoints and verify for yourself that this viewpoint captures the important behavior of the function. Notice that it has three turning points. Notice also that as $|x|$ gets large, the graph turns downward. This is because as $|x|$ becomes large, the x^4-term dominates the other terms, which are small in comparison, and the x^4-term has a negative coefficient.

As suggested by the graphs above, the domain of a polynomial function is the set of all real numbers. The range of a polynomial function of odd degree is also the set of all real numbers. Some typical graphs of polynomial functions of odd and even degree are shown in Figure 41.

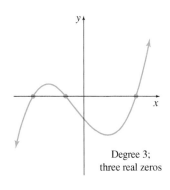

Degree 3;
three real zeros

Degree 3;
one real zero

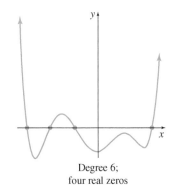

Degree 6;
four real zeros

FIGURE 41

The first two graphs suggest that for every polynomial f of odd degree, there is at least one real value of x for which $f(x) = 0$. Such a value of x is called a **real zero** or **root** of f. Notice that these values are the x-intercepts of the graph. As we saw in the previous section, polynomials of even degree (like the quadratic function) may or may not have real roots. When we graphed quadratic functions, we found the roots by factoring or by using the quadratic formula. With higher degree polynomials, determining roots can be more difficult.

TECHNOLOGY NOTE

We can use a graphing calculator to approximate the roots of higher degree polynomials. After the function has been graphed, select `zero` from the CALC menu. Because a function may have more than one root, we need to provide a range of x-values in which the calculator will search to find that particular root. To do this, we enter a "left" and "right" boundary for x. In Example 3, to find the smallest root, we enter -2 for the `Left Bound` and -1 for the `Right Bound`, and move the cursor toward the root for the `Guess?`. As shown in Figure 42, the calculator indicates that the root is approximately -1.7076. In a similar manner, the second root is found to be about 0.0556.

FIGURE 42

EXAMPLE 4   **Identifying the Degree of a Polynomial**

Identify the degree of the polynomial in each of the figures, and give the sign ($+$ or $-$) for the leading coefficient.

(a) Figure 43(a)

SOLUTION   Notice that the polynomial has a range $[k, \infty)$. This must be a polynomial of even degree, because if the highest power of x is an odd power, the polynomial can equal any real number, positive or negative. Notice also that the polynomial becomes a large positive number as x gets large in magnitude, either positive or negative, so the leading coefficient must be positive. Finally, notice that it has three turning points. Observe from the previous examples that a polynomial of degree n has at most $n - 1$ turning points. In a later chapter, we will use calculus to see why this is true. So the polynomial graphed in Figure 43(a) might be degree 4, although it could also be of degree 6, 8, etc. We can't be sure from the graph alone.

(b) Figure 43(b)

SOLUTION   Because the range is $(-\infty, \infty)$, this must be a polynomial of odd degree. Notice also that the polynomial becomes a large negative number as x becomes a large positive number, so the leading coefficient must be negative. Finally, notice that it has four turning points, so it might be degree 5, although it could also be of degree 7, 9, etc.

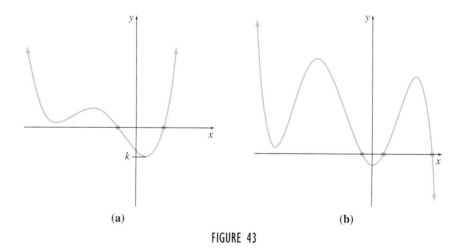

(a)   (b)

FIGURE 43

Properties of Polynomial Functions

1. A polynomial function of degree n can have at most $n - 1$ turning points. Conversely, if the graph of a polynomial function has n turning points, it must have degree at least $n + 1$.

2. In the graph of a polynomial function of even degree, both ends go up or both ends go down. For a polynomial function of odd degree, one end goes up and one end goes down.

3. If the graph goes up as x becomes a large positive number, the leading coefficient must be positive. If the graph goes down as x becomes a large positive number, the leading coefficient is negative.

⊞ TECHNOLOGY

Year	Percent
1980	26.1
1985	36.4
1990	42.1
1995	49.1
2000	34.7
2005	36.9
2010	62.9
2015	78.2

EXAMPLE 5 **Public Debt**

In economics, the federal debt held by the public is often given in relation to the size of the economy of that country. The gross domestic product (GDP), which is a measure of the economic output of a whole country or region, is used for comparison. The table in the margin gives the U.S. public debt as a percent of the GDP for some recent years. (The percent for 2015 is an estimate.) *Source: Budget of the U.S. Government.*

(a) Draw a scatterplot, letting $t = 0$ correspond to 1980. What type of polynomial function would approximate the data?

SOLUTION The scatterplot is shown in Figure 44(a). If we were to connect the data with a smooth curve, the beginning and end of the curve would head in opposite directions. This indicates that a polynomial of odd degree may be a good model. The data also appear to have two turning points, so the polynomial must be of at least degree three.

$$y = 0.00630t^3 - 0.289t^2 + 4.00t + 25.3$$

(a)   (b)

FIGURE 44

(b) Use the cubic regression feature of a graphing calculator to determine a cubic function that approximates the data. Graph the cubic function on the same window as the scatterplot.

SOLUTION The cubic equation is $y = 0.00630t^3 - 0.289t^2 + 4.00t + 25.3$, and is shown in Figure 44(b). The cubic function appears to be a good fit for the data. ▪

Rational Functions

Many situations require mathematical models that are quotients. A common model for such situations is a *rational function*.

Rational Function

A **rational function** is defined by

$$f(x) = \frac{p(x)}{q(x)},$$

where $p(x)$ and $q(x)$ are polynomial functions and $q(x) \neq 0$.

Since any values of x such that $q(x) = 0$ are excluded from the domain, a rational function often has a graph with one or more breaks.

EXAMPLE 6   Graphing a Rational Function

Graph $y = \dfrac{1}{x}$.

SOLUTION   This function is undefined for $x = 0$, since 0 is not allowed as the denominator of a fraction. For this reason, the graph of this function will not intersect the vertical line $x = 0$, which is the y-axis. Since x can equal any value except 0, the values of x can approach 0 as closely as desired from either side of 0.

Values of 1/x for Small x

				x approaches 0.				
x	-0.5	-0.2	-0.1	-0.01	0.01	0.1	0.2	0.5
$y = \dfrac{1}{x}$	-2	-5	-10	-100	100	10	5	2

$|y|$ gets larger and larger.

The table above suggests that as x gets closer and closer to 0, $|y|$ gets larger and larger. This is true in general: As the denominator gets smaller, the fraction gets larger. Thus, the graph of the function approaches the vertical line $x = 0$ (the y-axis) without ever touching it.

As $|x|$ gets larger and larger, $y = 1/x$ gets closer and closer to 0, as shown in the table below. This is also true in general: as the denominator gets larger, the fraction gets smaller.

Values of 1/x for Large $|x|$

x	-100	-10	-4	-1	1	4	10	100
$y = \dfrac{1}{x}$	-0.01	-0.1	-0.25	-1	1	0.25	0.1	0.01

The graph of the function approaches the horizontal line $y = 0$ (the x-axis). The information from both tables supports the graph in Figure 45.

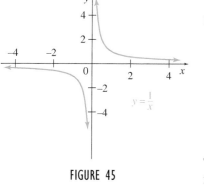

FIGURE 45

In Example 6, the vertical line $x = 0$ and the horizontal line $y = 0$ are *asymptotes*, defined as follows.

Asymptotes

If a function gets larger and larger in magnitude without bound as x approaches the number k, then the line $x = k$ is a **vertical asymptote**.

If the values of y approach a number k as $|x|$ gets larger and larger, the line $y = k$ is a **horizontal asymptote**.

There is an easy way to find any vertical asymptotes of a rational function. First, find the roots of the denominator. If a number k makes the denominator 0 but does not make the numerator 0, then the line $x = k$ is a vertical asymptote. If, however, a number k makes both the denominator and the numerator 0, then further investigation will be necessary, as we will see in the next example. In the next chapter we will show another way to find asymptotes using the concept of a *limit*.

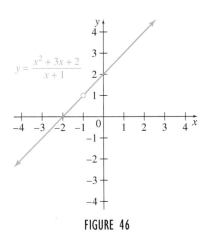

$y = \dfrac{x^2 + 3x + 2}{x + 1}$

FIGURE 46

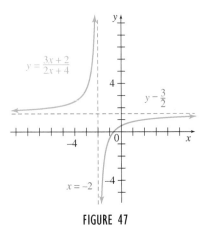

$y = \dfrac{3x + 2}{2x + 4}$

$y = \dfrac{3}{2}$

$x = -2$

FIGURE 47

YOUR TURN 2 Graph
$y = \dfrac{4x - 6}{x - 3}.$

EXAMPLE 7 Graphing a Rational Function

Graph the following rational functions:

(a) $y = \dfrac{x^2 + 3x + 2}{x + 1}.$

SOLUTION The value $x = -1$ makes the denominator 0, and so -1 is not in the domain of this function. Note that the value $x = -1$ also makes the numerator 0. In fact, if we factor the numerator and simplify the function, we get

$$y = \frac{x^2 + 3x + 2}{x + 1} = \frac{(x + 2)(x + 1)}{(x + 1)} = x + 2 \quad \text{for} \quad x \neq -1.$$

The graph of this function, therefore, is the graph of $y = x + 2$ with a hole at $x = -1$, as shown in Figure 46.

(b) $y = \dfrac{3x + 2}{2x + 4}.$

SOLUTION The value $x = -2$ makes the denominator 0, but not the numerator, so the line $x = -2$ is a vertical asymptote. To find a horizontal asymptote, let x get larger and larger, so that $3x + 2 \approx 3x$ because the 2 is very small compared with $3x$. Similarly, for x very large, $2x + 4 \approx 2x$. Therefore, $y = (3x + 2)/(2x + 4) \approx (3x)/(2x) = 3/2$. This means that the line $y = 3/2$ is a horizontal asymptote. (A more precise way of approaching this idea will be seen in the next chapter when limits at infinity are discussed.)

The intercepts should also be noted. Setting $x = 0$, we find that the y-intercept is $y = 2/4 = 1/2$. To find the x-intercept (also called the root), we set $y = 0$. Note that to make a fraction zero, the numerator must be zero. Solving $3x + 2 = 0$, we get the x-intercept $x = -2/3$. We can also use these values to determine where the function is positive and where it is negative. Using the techniques described in Chapter R, verify that the function is negative on $(-2, -2/3)$ and positive on $(-\infty, -2) \cup (-2/3, \infty)$. With this information, the two asymptotes to guide us, and the fact that there are only two intercepts, we suspect the graph is as shown in Figure 47. A graphing calculator can support this. **TRY YOUR TURN 2**

Rational functions occur often in practical applications. In many situations involving environmental pollution, much of the pollutant can be removed from the air or water at a fairly reasonable cost, but the last small part of the pollutant can be very expensive to remove. Cost as a function of the percentage of pollutant removed from the environment can be calculated for various percentages of removal, with a curve fitted through the resulting data points. This curve then leads to a mathematical model of the situation. Rational functions are often a good choice for these **cost-benefit models** because they rise rapidly as they approach a vertical asymptote.

EXAMPLE 8 Cost-Benefit Analysis

Suppose a cost-benefit model is given by

$$y = \frac{18x}{106 - x},$$

where y is the cost (in thousands of dollars) of removing x percent of a certain pollutant. The domain of x is the set of all numbers from 0 to 100 inclusive; any amount of pollutant from 0% to 100% can be removed. Find the cost to remove the following amounts of the pollutant: 100%, 95%, 90%, and 80%. Graph the function.

SOLUTION Removal of 100% of the pollutant would cost

$$y = \frac{18(100)}{106 - 100} = 300,$$

or $300,000. Check that 95% of the pollutant can be removed for $155,000, 90% for $101,000, and 80% for $55,000. Using these points, as well as others obtained from the function, gives the graph shown in Figure 48.

FIGURE 48

If a cost function has the form $C(x) = mx + b$, where x is the number of items produced, m is the marginal cost per item and b is the fixed cost, then the **average cost** per item is given by

$$\overline{C}(x) = \frac{C(x)}{x} = \frac{mx + b}{x}.$$

Notice that this is a rational function with a vertical asymptote at $x = 0$ and a horizontal asymptote at $y = m$. The vertical asymptote reflects the fact that, as the number of items produced approaches 0, the average cost per item becomes infinitely large, because the fixed costs are spread over fewer and fewer items. The horizontal asymptote shows that, as the number of items becomes large, the fixed costs are spread over more and more items, so most of the average cost per item is the marginal cost to produce each item. This is another example of how asymptotes give important information in real applications.

2.3 WARM-UP EXERCISES

Describe the graph of each of the following functions in terms of the graph of $f(x)$.   *(Sec. 2.2)*

W1. $y = f(x + 2) - 3$

W2. $y = -f(x - 3)$

W3. $y = f(-x) + 2$

W4. $y = f(2 - x)$

2.3 EXERCISES

1. Explain how translations and reflections can be used to graph $y = -(x - 1)^4 + 2$.

2. Describe an asymptote, and explain when a rational function will have (a) a vertical asymptote and (b) a horizontal asymptote.

Use the principles of the previous section with the graphs of this section to sketch a graph of the given function. Determine any roots of the polynomial.

3. $f(x) = (x - 2)^3 + 3$

4. $f(x) = (x + 1)^3 - 2$

5. $f(x) = -(x + 3)^4 + 1$

6. $f(x) = -(x - 1)^4 + 2$

In Exercises 7–15, match the correct graph A–I to the function without using your calculator. Then, after you have answered all of them, if you have a graphing calculator, use your calculator to check your answers. Each graph is plotted on $[-6, 6]$ by $[-50, 50]$.

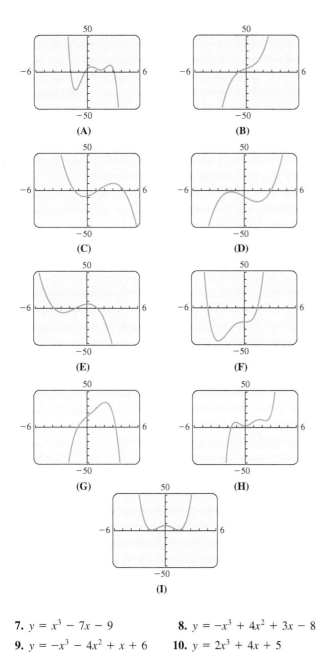

(A)

(B)

(C)

(D)

(E)

(F)

(G)

(H)

(I)

7. $y = x^3 - 7x - 9$ **8.** $y = -x^3 + 4x^2 + 3x - 8$

9. $y = -x^3 - 4x^2 + x + 6$ **10.** $y = 2x^3 + 4x + 5$

11. $y = x^4 - 5x^2 + 7$ **12.** $y = x^4 + 4x^3 - 20$

13. $y = -x^4 + 2x^3 + 10x + 15$

14. $y = 0.7x^5 - 2.5x^4 - x^3 + 8x^2 + x + 2$

15. $y = -x^5 + 4x^4 + x^3 - 16x^2 + 12x + 5$

In Exercises 16–20, match the correct graph A–E to the function without using your calculator. Then, after you have answered all of them, if you have a graphing calculator, use your calculator to check your answers. Each graph in this group is plotted on $[-6, 6]$ by $[-6, 6]$. Hint: Consider the asymptotes.

(A)

(B)

(C)

(D)

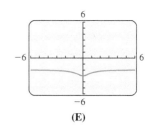

(E)

16. $y = \dfrac{2x^2 + 3}{x^2 - 1}$ **17.** $y = \dfrac{2x^2 + 3}{x^2 + 1}$

18. $y = \dfrac{-2x^2 - 3}{x^2 - 1}$ **19.** $y = \dfrac{-2x^2 - 3}{x^2 + 1}$

20. $y = \dfrac{2x^2 + 3}{x^3 - 1}$

Each of the following is the graph of a polynomial function. Give the possible values for the degree of the polynomial, and give the sign $(+$ or $-)$ for the leading coefficient.

21.

22.

23.

24.

25.

26.

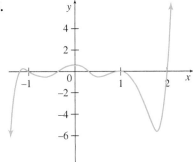

Find any horizontal and vertical asymptotes and any holes that may exist for each rational function. Draw the graph of each function, including any x- and y-intercepts.

27. $y = \dfrac{-4}{x + 2}$

28. $y = \dfrac{-1}{x + 3}$

29. $y = \dfrac{2}{3 + 2x}$

30. $y = \dfrac{8}{5 - 3x}$

31. $y = \dfrac{2x}{x - 3}$

32. $y = \dfrac{4x}{3 - 2x}$

33. $y = \dfrac{x + 1}{x - 4}$

34. $y = \dfrac{x - 4}{x + 1}$

35. $y = \dfrac{3 - 2x}{4x + 20}$

36. $y = \dfrac{6 - 3x}{4x + 12}$

37. $y = \dfrac{-x - 4}{3x + 6}$

38. $y = \dfrac{-2x + 5}{x + 3}$

39. $y = \dfrac{x^2 + 7x + 12}{x + 4}$

40. $y = \dfrac{9 - 6x + x^2}{3 - x}$

41. Write an equation that defines a rational function with a vertical asymptote at $x = 1$ and a horizontal asymptote at $y = 2$.

42. Write an equation that defines a rational function with a vertical asymptote at $x = -2$ and a horizontal asymptote at $y = 0$.

43. Consider the polynomial functions defined by $f(x) = (x - 1)(x - 2)(x + 3)$, $g(x) = x^3 + 2x^2 - x - 2$, and $h(x) = 3x^3 + 6x^2 - 3x - 6$.

  (a) What is the value of $f(1)$?

  (b) For what values, other than 1, is $f(x) = 0$?

  (c) Verify that $g(-1) = g(1) = g(-2) = 0$.

  (d) Based on your answer from part (c), what do you think is the factored form of $g(x)$? Verify your answer by multiplying these factors and comparing with $g(x)$.

  (e) Using your answer from part (d), what is the factored form of $h(x)$?

  (f) Based on what you have learned in this exercise, fill in the blank: If f is a polynomial and $f(a) = 0$ for some number a, then one factor of the polynomial is

  _____.

 44. Consider the function defined by

$$f(x) = \frac{x^7 - 4x^5 - 3x^4 + 4x^3 + 12x^2 - 12}{x^7}.$$

Source: The Mathematics Teacher.

  (a) Graph the function on $[-6, 6]$ by $[-6, 6]$. From your graph, estimate how many x-intercepts the function has and what their values are.

  (b) Now graph the function on $[-1.5, -1.4]$ by $[-10^{-4}, 10^{-4}]$ and on $[1.4, 1.5]$ by $[-10^{-5}, 10^{-5}]$. From your graphs, estimate how many x-intercepts the function has and what their values are.

  (c) From your results in parts (a) and (b), what advice would you give a friend on using a graphing calculator to find x-intercepts?

 45. Consider the function defined by

$$f(x) = \frac{1}{x^5 - 2x^3 - 3x^2 + 6}.$$

Source: The Mathematics Teacher.

  (a) Graph the function on $[-3.4, 3.4]$ by $[-3, 3]$. From your graph, estimate how many vertical asymptotes the function has and where they are located.

  (b) Now graph the function on $[-1.5, -1.4]$ by $[-10, 10]$ and on $[1.4, 1.5]$ by $[-1000, 1000]$. From your graphs, estimate how many vertical asymptotes the function has and where they are located.

  (c) From your results in parts (a) and (b), what advice would you give a friend on using a graphing calculator to find vertical asymptotes?

APPLICATIONS

Business and Economics

46. Average Cost Suppose the average cost per unit $\overline{C}(x)$, in dollars, to produce x units of yogurt is given by

$$\overline{C}(x) = \frac{600}{x + 20}.$$

(a) Find $\overline{C}(10), \overline{C}(20), \overline{C}(50), \overline{C}(75)$, and $\overline{C}(100)$.

(b) Which of the intervals $(0, \infty)$ and $[0, \infty)$ would be a more reasonable domain for $\overline{C}$? Why?

(c) Give the equations of any asymptotes on $(-\infty, \infty)$. Find any intercepts.

(d) Graph $y = \overline{C}(x)$ over all x for which $\overline{C}(x)$ is positive.

47. Cost Analysis In a recent year, the cost per ton, y, to build an oil tanker of x thousand deadweight tons was approximated by

$$\overline{C}(x) = \frac{220,000}{x + 475}$$

for $x > 0$.

(a) Find $\overline{C}(25), \overline{C}(50), \overline{C}(100), \overline{C}(200), \overline{C}(300)$, and $\overline{C}(400)$.

(b) Find any asymptotes on $(-\infty, \infty)$.

(c) Find any intercepts.

(d) Graph $y = \overline{C}(x)$ over all x for which $\overline{C}(x)$ is positive.

APPLY IT **Tax Rates** Exercises **48–50** refer to the *Laffer curve*, originated by the economist Arthur Laffer. An idealized version of this curve is shown here. According to this curve, decreasing a tax rate, say from x_2 percent to x_1 percent on the graph, can actually lead to an increase in government revenue. The theory is that people will work harder and earn more money if they are taxed at a lower rate, so the government ends up with more revenue than it would at a higher tax rate. All economists agree on the endpoints—0 revenue at tax rates of both 0% and 100%—but there is much disagreement on the location of the tax rate x_1 that produces the maximum revenue.

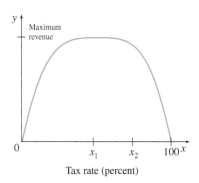

Tax rate (percent)

48. A function that might describe the entire Laffer curve is

$$y = x(100 - x)(x^2 + 500),$$

where y is government revenue in hundreds of thousands of dollars from a tax rate of x percent, with the function valid for $0 \le x \le 100$. Find the revenue from the following tax rates.

(a) 10%   (b) 40%   (c) 50%   (d) 80%

(e) Graph the function.

49. Find the equations of two quadratic functions that could describe the Laffer curve by having zeros at $x = 0$ and $x = 100$. Give the first a maximum of 100 and the second a maximum of 250, then multiply them together to get a new Laffer curve with a maximum of 25,000. Plot the resulting function.

50. An economist might argue that the models in the two previous exercises are unrealistic because they predict that a tax rate of 50% gives the maximum revenue, while the actual value is probably less than 50%. Consider the function

$$y = \frac{300x - 3x^2}{5x + 100},$$

where y is government revenue in millions of dollars from a tax rate of x percent, where $0 \le x \le 100$. *Source: Dana Lee Ling.*

(a) Graph the function, and discuss whether the shape of the graph is appropriate.

(b) Use a graphing calculator to find the tax rate that produces the maximum revenue. What is the maximum revenue?

51. Cost-Benefit Model Suppose a cost-benefit model is given by

$$y = \frac{6.7x}{100 - x},$$

where y is the cost in thousands of dollars of removing x percent of a given pollutant.

(a) Find the cost of removing each percent of pollutants: 50%; 70%; 80%; 90%; 95%; 98%; 99%.

(b) Is it possible, according to this function, to remove *all* the pollutant?

(c) Graph the function.

52. Cost-Benefit Model Suppose a cost-benefit model is given by

$$y = \frac{6.5x}{102 - x},$$

where y is the cost in thousands of dollars of removing x percent of a certain pollutant.

(a) Find the cost of removing each percent of pollutants: 0%; 50%; 80%; 90%; 95%; 99%; 100%.

(b) Graph the function.

53. Coal Consumption The table below gives U.S. coal consumption for selected years. *Source: U.S. Department of Energy.*

Year	Millions of Short Tons
1950	494.1
1960	398.1
1970	523.2
1980	702.7
1990	902.9
2000	1084.1
2010	1051.3
2012	890.5

(a) Draw a scatterplot, letting $t = 0$ represent 1950.

(b) Use the quadratic regression feature of a graphing calculator to get a quadratic function that approximates the data.

(c) Graph the function from part (b) on the same window as the scatterplot.

(d) Use cubic regression to get a cubic function that approximates the data.

(e) Graph the cubic from part (d) on the same window as the scatterplot.

(f) Which of the two functions in parts (b) and (d) appears to be a better fit for the data? Explain your reasoning.

Life Sciences

54. Cardiac Output A technique for measuring cardiac output depends on the concentration of a dye after a known amount is injected into a vein near the heart. In a normal heart, the concentration of the dye at time t (in seconds) is given by the function

$$g(t) = -0.006t^4 + 0.140t^3 - 0.053t^2 + 1.79t.$$

(a) Graph $g(t)$ on $[0, 6]$ by $[0, 20]$.

(b) In your graph from part (a), notice that the function initially increases. Considering the form of $g(t)$, do you think it can keep increasing forever? Explain.

(c) Write a short paragraph about the extent to which the concentration of dye might be described by the function $g(t)$.

55. Alcohol Concentration The polynomial function

$$A(x) = 0.003631x^3 - 0.03746x^2 + 0.1012x + 0.009$$

gives the approximate blood alcohol concentration in a 170-lb woman x hours after drinking 2 oz of alcohol on an empty stomach, for x in the interval $[0, 5]$. *Source: Medical Aspects of Alcohol Determination in Biological Specimens.*

(a) Graph $A(x)$ on $0 \le x \le 5$.

(b) Using the graph from part (a), estimate the time of maximum alcohol concentration.

(c) In many states, a person is legally drunk if the blood alcohol concentration exceeds 0.08%. Use the graph from part (a) to estimate the period in which this 170-lb woman is legally drunk.

56. Medical School For the years 2000 to 2013, the number of applicants to U.S. medical schools can be closely approximated by

$$A(t) = 6.110t^4 - 172.7t^3 + 1614t^2 - 4409t + 37,360,$$

where t is the number of years since 2000. *Source: Association of American Medical Colleges.*

(a) Graph the number of applicants on $0 \le t \le 13$.

(b) Based on the graph in part (a) during what year was the number of medical school applicants at a minimum?

57. Population Biology The function

$$f(x) = \frac{\lambda x}{1 + (ax)^b}$$

is used in population models to give the size of the next generation $(f(x))$ in terms of the current generation (x). *Source: Models in Ecology.*

(a) What is a reasonable domain for this function, considering what x represents?

(b) Graph this function for $\lambda = a = b = 1$.

(c) Graph this function for $\lambda = a = 1$ and $b = 2$.

(d) What is the effect of making b larger?

58. Growth Model The function

$$f(x) = \frac{Kx}{A + x}$$

is used in biology to give the growth rate of a population in the presence of a quantity x of food. This is called Michaelis-Menten kinetics. *Source: Mathematical Models in Biology.*

(a) What is a reasonable domain for this function, considering what x represents?

(b) Graph this function for $K = 5$ and $A = 2$.

(c) Show that $y = K$ is a horizontal asymptote.

(d) What do you think K represents?

(e) Show that A represents the quantity of food for which the growth rate is half of its maximum.

59. Brain Mass The mass (in grams) of the human brain during the last trimester of gestation and the first two years after birth can be approximated by the function

$$m(c) = \frac{c^3}{100} - \frac{1500}{c},$$

where c is the circumference of the head in centimeters. *Source: Early Human Development.*

(a) Find the approximate mass of brains with a head circumference of 30, 40, and 50 cm.

(b) Clearly the formula is invalid for any values of c yielding negative values of w. For what values of c is this true?

(c) Use a graphing calculator to sketch this graph on the interval $20 \le c \le 50$.

(d) Suppose an infant brain has mass of 700 g. Use features on a graphing calculator to find what the circumference of the head is expected to be.

60. Contact Lenses The strength of a contact lens is given in units known as diopters, as well as in mm of arc. The table on the next page is taken from a chart used by optometrists to convert diopters to mm of arc. *Source: Bausch & Lomb.*

(a) Notice that as the diopters increase, the mm of arc decrease. Find a value of k so the function $a = f(d) = k/d$ gives a, the mm of arc, as a function of d, the strength in diopters. (Round k to the nearest integer. For a more accurate answer, average all the values of k given by each pair of data.)

(b) An optometrist wants to order 40.50 diopter lenses for a patient. The manufacturer needs to know the strength in mm of arc. What is the strength in mm of arc?

Diopters	mm of Arc
36.000	9.37
36.125	9.34
36.250	9.31
36.375	9.27
36.500	9.24
36.625	9.21
36.750	9.18
36.875	9.15
37.000	9.12

Social Sciences

61. Head Start The enrollment in Head Start for some recent years is included in the table. *Source: Head Start.*

Year	Enrollment
1966	733,000
1970	477,400
1980	376,300
1990	540,930
1995	750,696
2000	857,664
2005	906,993
2010	904,118

(a) Plot the points from the table using 0 for 1960, and so on.

(b) Use the quadratic regression feature of a graphing calculator to get a quadratic function that approximates the data. Graph the function on the same window as the scatterplot.

(c) Use cubic regression to get a cubic function that approximates the data. Graph the function on the same window as the scatterplot.

(d) Which of the two functions in part (b) and (c) appears to be a better fit for the data? Explain your reasoning.

Physical Sciences

62. Length of a Pendulum A simple pendulum swings back and forth in regular time intervals. Grandfather clocks use pendulums to keep accurate time. The relationship between the length of a pendulum L and the period (time) T for one complete oscillation can be expressed by the function $L = kT^n$, where k is a constant and n is a positive integer to be determined. The data below were taken for different lengths of pendulums.* *Source: Gary Rockswold.*

T (sec)	L (ft)
1.11	1.0
1.36	1.5
1.57	2.0
1.76	2.5
1.92	3.0
2.08	3.5
2.22	4.0

(a) Find the value of k for $n = 1, 2$, and 3, using the data for the 4-ft pendulum.

(b) Use a graphing calculator to plot the data in the table and to graph the function $L = kT^n$ for the three values of k (and their corresponding values of n) found in part (a). Which function best fits the data?

(c) Use the best-fitting function from part (a) to predict the period of a pendulum having a length of 5 ft.

(d) If the length of pendulum doubles, what happens to the period?

(e) If you have a graphing calculator or computer program with a quadratic regression feature, use it to find a quadratic function that approximately fits the data. How does this answer compare with the answer to part (b)?

YOUR TURN ANSWERS

1.

$f(x) = 64 - x^6$

2.

$y = \dfrac{4x - 6}{x - 3}$

*See Exercise 24, Section 1.3.

2.4 Exponential Functions

Later in this section, in Example 5, we will see that determining the answer to this question depends on exponential functions.

In earlier sections we discussed functions involving expressions such as x^2, $(2x + 1)^3$, or x^{-1}, where the variable or variable expression is the base of an exponential expression, and the exponent is a constant. In an exponential function, the variable is in the exponent and the base is a constant.

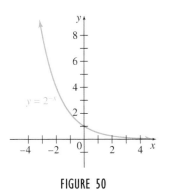

FIGURE 49

Exponential Function

An **exponential function** with base a is defined as

$$f(x) = a^x, \quad \text{where } a > 0 \text{ and } a \neq 1.$$

(If $a = 1$, the function is the constant function $f(x) = 1$.)

Exponential functions may be the single most important type of functions used in practical applications. They are used to describe growth and decay, which are important ideas in management, social science, and biology.

Figure 49 shows a graph of the exponential function defined by $f(x) = 2^x$. You could plot such a curve by hand by noting that $2^{-2} = 1/4$, $2^{-1} = 1/2$, $2^0 = 1$, $2^1 = 2$, and $2^2 = 4$, and then drawing a smooth curve through the points $(-2, 1/4)$, $(-1, 1/2)$, $(0, 1)$, $(1, 2)$, and $(2, 4)$. This graph is typical of the graphs of exponential functions of the form $y = a^x$, where $a > 1$. The y-intercept is $(0, 1)$. Notice that as x gets larger and larger, the function also gets larger. As x gets more and more negative, the function becomes smaller and smaller, approaching but never reaching 0. Therefore, the x-axis is a horizontal asymptote, but the function approaches only the left side of the asymptote. In contrast, rational functions approach both the left and right sides of the asymptote. The graph suggests that the domain is the set of all real numbers and the range is the set of all positive numbers.

> **FOR REVIEW**
>
> To review the properties of exponents used in this section, see Section R.6.

EXAMPLE 1 Graphing an Exponential Function

Graph $f(x) = 2^{-x}$.

SOLUTION The graph, shown in Figure 50, is the horizontal reflection of the graph of $f(x) = 2^x$ given in Figure 49. Since $2^{-x} = 1/2^x = (1/2)^x$, this graph is typical of the graphs of exponential functions of the form $y = a^x$ where $0 < a < 1$. The domain includes all real numbers and the range includes all positive numbers. The y-intercept is $(0, 1)$. Notice that this function, with $f(x) = 2^{-x} = (1/2)^x$, is decreasing over its domain. ▬

In the definition of an exponential function, notice that the base a is restricted to positive values, with negative or zero bases not allowed. For example, the function $y = (-4)^x$ could not include such numbers as $x = 1/2$ or $x = 1/4$ in the domain because the y-values would not be real numbers. The resulting graph would be at best a series of separate points having little practical use

EXAMPLE 2 Graphing an Exponential Function

Graph $f(x) = -2^x + 3$.

SOLUTION The graph of $y = -2^x$ is the vertical reflection of the graph of $y = 2^x$, so this is a decreasing function. (Notice that -2^x is not the same as $(-2)^x$. In -2^x, we raise 2 to the x power and then take the negative.) The 3 indicates that the graph should be translated

> **FOR REVIEW**
>
> Recall from Section 2.2 that the graph of $f(-x)$ is the reflection of the graph of $f(x)$ about the y-axis.

FIGURE 50

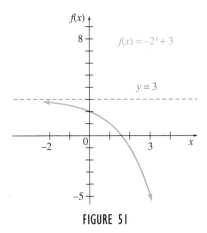

FIGURE 51

vertically 3 units, as compared to the graph of $y = -2^x$. Since $y = -2^x$ would have y-intercept $(0, -1)$, this function has y-intercept $(0, 2)$, which is up 3 units. For negative values of x, the graph approaches the line $y = 3$, which is a horizontal asymptote. The graph is shown in Figure 51. ▄

Exponential Equations

In Figures 49 and 50, which are typical graphs of exponential functions, a given value of x leads to exactly one value of a^x. Because of this, an equation with a variable in the exponent, called an **exponential equation**, often can be solved using the following property.

> If $a > 0, a \neq 1$, and $a^x = a^y$, then $x = y$.

The value $a = 1$ is excluded, since $1^2 = 1^3$, for example, even though $2 \neq 3$. To solve $2^{3x} = 2^7$ using this property, work as follows.

$$2^{3x} = 2^7$$
$$3x = 7$$
$$x = \frac{7}{3}$$

EXAMPLE 3 Solving Exponential Equations

(a) Solve $9^x = 27$.

SOLUTION First rewrite both sides of the equation so the bases are the same. Since $9 = 3^2$ and $27 = 3^3$,

$$9^x = 27$$
$$(3^2)^x = 3^3$$
$$3^{2x} = 3^3 \qquad \text{Multiply exponents.}$$
$$2x = 3 \qquad \text{Set exponents equal to each other.}$$
$$x = \frac{3}{2}.$$

(b) Solve $32^{2x-1} = 128^{x+3}$.

SOLUTION Since the bases must be the same, write 32 as 2^5 and 128 as 2^7, giving

$$32^{2x-1} = 128^{x+3}$$
$$(2^5)^{2x-1} = (2^7)^{x+3}$$
$$2^{10x-5} = 2^{7x+21}. \qquad \text{Multiply exponents.}$$

Now use the property from above to get

$$10x - 5 = 7x + 21$$
$$3x = 26$$
$$x = \frac{26}{3}.$$

Verify this solution in the original equation. TRY YOUR TURN 1 ▄

FOR REVIEW

Recall from Section R.6 that $(a^m)^n = a^{mn}$.

YOUR TURN 1 Solve $25^{x/2} = 125^{x+3}$.

Compound Interest

The calculation of compound interest is an important application of exponential functions. The cost of borrowing money or the return on an investment is called **interest**. The amount borrowed or invested is the **principal**, P. The **rate of interest** r is given as a percent per year, and t is the **time**, measured in years.

Simple Interest

The product of the principal P, rate r, and time t gives **simple interest**, I:

$$I = Prt.$$

With **compound interest**, interest is charged (or paid) on interest as well as on the principal. To find a formula for compound interest, first suppose that P dollars, the principal, is deposited at a rate of interest r per year. The interest earned during the first year is found using the formula for simple interest.

$$\text{First-year interest} = P \cdot r \cdot 1 = Pr.$$

At the end of one year, the amount on deposit will be the sum of the original principal and the interest earned, or

$$P + Pr = P(1 + r). \tag{1}$$

If the deposit earns compound interest, the interest earned during the second year is found from the total amount on deposit at the end of the first year. Thus, the interest earned during the second year (again found by the formula for simple interest), is

$$[P(1 + r)](r)(1) = P(1 + r)r, \tag{2}$$

so the total amount on deposit at the end of the second year is the sum of amounts from (1) and (2) above, or

$$P(1 + r) + P(1 + r)r = P(1 + r)(1 + r) = P(1 + r)^2.$$

In the same way, the total amount on deposit at the end of three years is

$$P(1 + r)^3.$$

After t years, the total amount on deposit, called the *compound amount*, is $P(1 + r)^t$.

When interest is compounded more than once a year, the compound amount formula is adjusted. For example, if interest is to be paid quarterly (four times a year), $1/4$ of the interest rate is used each time interest is calculated, so the rate becomes $r/4$, and the number of compounding periods in t years becomes $4t$. Generalizing from this idea gives the following formula.

Compound Amount

If P dollars is invested at a yearly rate of interest r per year, compounded m times per year for t years, the **compound amount** is

$$A = P\left(1 + \frac{r}{m}\right)^{tm} \text{ dollars.}$$

EXAMPLE 4   Compound Interest

Morgan Presley invests a bonus of $9000 at 6% annual interest compounded semiannually for 4 years. How much interest will she earn?

SOLUTION   Use the formula for compound amount with $P = 9000, r = 0.06, m = 2$ (since interest is compounded semiannually), and $t = 4$.

$$A = P\left(1 + \frac{r}{m}\right)^{tm}$$

$$= 9000\left(1 + \frac{0.06}{2}\right)^{4(2)}$$

$$= 9000(1.03)^8$$

$$\approx 11,400.93 \qquad \text{Use a calculator.}$$

YOUR TURN 2 Find the interest earned on $4400 at 3.25% interest compounded quarterly for 5 years.

The compound amount (investment plus the interest) is $11,400.93. The interest is $11,400.93 - $9000 = $2400.93. **TRY YOUR TURN 2**

NOTE When using a calculator to compute the compound amount, store each partial result in the calculator and avoid rounding off until the final answer.

The Number e

Perhaps the single most useful base for an exponential function is the number e, an irrational number that occurs often in practical applications. The famous Swiss mathematician Leonhard Euler (pronounced "oiler") (1707–1783) was the first person known to have referred to this number as e, and the notation has continued to this day. To see how the number e occurs in an application, begin with the formula for compound interest,

$$P\left(1 + \frac{r}{m}\right)^{tm}.$$

Suppose that a lucky investment produces annual interest of 100%, so that $r = 1.00 = 1$. Suppose also that you can deposit only \$1 at this rate, and for only one year. Then $P = 1$ and $t = 1$. Substituting these values into the formula for compound interest gives

$$P\left(1 + \frac{r}{m}\right)^{tm} = 1\left(1 + \frac{1}{m}\right)^{1(m)} = \left(1 + \frac{1}{m}\right)^{m}.$$

As interest is compounded more and more often, m gets larger and the value of this expression will increase. For example, if $m = 1$ (interest is compounded annually),

$$\left(1 + \frac{1}{m}\right)^{m} = \left(1 + \frac{1}{1}\right)^{1} = 2^{1} = 2,$$

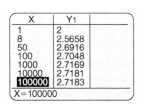

X	Y1
1	2
8	2.5658
50	2.6916
100	2.7048
1000	2.7169
10000	2.7181
100000	2.7183

X=100000

FIGURE 52

so that your \$1 becomes \$2 in one year. Using a graphing calculator, we produced Figure 52 (where m is represented by X and $(1 + 1/m)^{m}$ by Y_1) to see what happens as m becomes larger and larger. A spreadsheet can also be used to produce this table.

The table suggests that as m increases, the value of $(1 + 1/m)^{m}$ gets closer and closer to a fixed number, called e. As we shall see in the next chapter, this is an example of a limit.

Definition of e

As m becomes larger and larger, $\left(1 + \dfrac{1}{m}\right)^{m}$ becomes closer and closer to the number e,

whose approximate value is 2.718281828.

The value of e is approximated here to 9 decimal places. Euler approximated e to 18 decimal places. Many calculators give values of e^x, usually with a key labeled e^x. Some require two keys, either INV LN or 2nd LN. (We will define $\ln x$ in the next section.) In Figure 53, the functions $y = 2^x$, $y = e^x$, and $y = 3^x$ are graphed for comparison. Notice that the graph of e^x is between 2^x and 3^x, because e is between 2 and 3. For $x > 0$, the graphs show that $3^x > e^x > 2^x$. All three functions have y-intercept $(0, 1)$. It is difficult to see from the graph, but $3^x < e^x < 2^x$ when $x < 0$.

The number e is often used as the base in an exponential equation because it provides a good model for many natural, as well as economic, phenomena. In the exercises for this section, we will look at several examples of such applications.

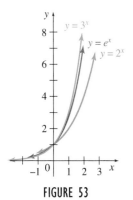

FIGURE 53

Continuous Compounding

In economics, the formula for **continuous compounding** is a good example of an exponential growth function. Recall the formula for compound amount

$$A = P\left(1 + \frac{r}{m}\right)^{tm},$$

where m is the number of times annually that interest is compounded. As m becomes larger and larger, the compound amount also becomes larger but not without bound. Recall that as m becomes larger and larger, $(1 + 1/m)^m$ becomes closer and closer to e. Similarly,

$$\left(1 + \frac{1}{(m/r)}\right)^{m/r}$$

becomes closer and closer to e. Let us rearrange the formula for compound amount to take advantage of this fact.

$$A = P\left(1 + \frac{r}{m}\right)^{tm}$$

$$= P\left(1 + \frac{1}{(m/r)}\right)^{tm} \qquad \frac{r}{m} = \frac{1}{(m/r)}$$

$$= P\left[\left(1 + \frac{1}{(m/r)}\right)^{m/r}\right]^{rt} \qquad \frac{m}{r} \cdot rt = tm$$

This last expression becomes closer and closer to Pe^{rt} as m becomes larger and larger, which describes what happens when interest is compounded continuously. Essentially, the number of times annually that interest is compounded becomes infinitely large. We thus have the following formula for the compound amount when interest is compounded continuously.

Continuous Compounding

If a deposit of P dollars is invested at a rate of interest r compounded continuously for t years, the compound amount is

$$A = Pe^{rt} \text{ dollars.}$$

EXAMPLE 5   Continuous Compound Interest

APPLY IT

If $600 is invested in an account earning 2.75% compounded continuously, how much would be in the account after 5 years?

SOLUTION In the formula for continuous compounding, let $P = 600, t = 5,$ and $r = 0.0275$ to get

$$A = 600e^{5(0.0275)} \approx 688.44,$$

or $688.44.

YOUR TURN 3 Find the amount after 4 years if $800 is invested in an account earning 3.15% compounded continuously.

TRY YOUR TURN 3

In situations that involve growth or decay of a population, the size of the population at a given time t often is determined by an exponential function of t. The next example illustrates a typical application of this kind.

EXAMPLE 6   Oxygen Consumption

Biologists studying salmon have found that the oxygen consumption of yearling salmon (in appropriate units) increases exponentially with the speed of swimming according to the function defined by

$$f(x) = 100e^{0.6x},$$

where x is the speed in feet per second. Find the following.

(a) The oxygen consumption when the fish are still

SOLUTION When the fish are still, their speed is 0. Substitute 0 for x:

$$f(0) = 100e^{(0.6)(0)} = 100e^0$$
$$= 100 \cdot 1 = 100. \qquad e^0 = 1$$

When the fish are still, their oxygen consumption is 100 units.

(b) The oxygen consumption at a speed of 2 ft per second

SOLUTION Find $f(2)$ as follows.

$$f(2) = 100e^{(0.6)(2)} = 100e^{1.2} \approx 332$$

At a speed of 2 ft per second, oxygen consumption is about 332 units rounded to the nearest integer. Because the function is only an approximation of the real situation, further accuracy is not realistic.

EXAMPLE 7 Food Surplus

A magazine article argued that the cause of the obesity epidemic in the United States is the decreasing cost of food (in real terms) due to the increasing surplus of food. As one piece of evidence, the following table was provided, which we have updated, showing U.S. corn production (in billions of bushels) for selected years. *Source: The New York Times Magazine.*

Year	Production (billions of bushels)
1930	1.757
1940	2.207
1950	2.764
1960	3.907
1970	4.152
1980	6.639
1990	7.934
2000	9.968
2010	12.447

14

0   80
0

FIGURE 54

$y = 1.757(1.0248)^t$

14

0   80
0

FIGURE 55

(a) Plot the data, letting $t = 0$ correspond to 1930. Does the production appear to grow linearly or exponentially?

SOLUTION Figure 54 shows a graphing calculator plot of the data, which suggests that corn production is growing exponentially.

(b) Find an exponential function in the form of $p(t) = p_0 a^t$ that models these data, where t is the year and $p(t)$ is the production of corn. Use the data $(0, 1.757)$ and $(80, 12.447)$.

SOLUTION Since $p(0) = p_0 a^0 = p_0$, we have $p_0 = 1.757$. Using $t = 80$, we have

$$p(80) = 1.757a^{80} = 12.447$$

$$a^{80} = \frac{12.447}{1.757} \qquad \text{Divide by 1.757.}$$

$$a = \left(\frac{12.447}{1.757}\right)^{1/80} \qquad \text{Take the 80th root.}$$

$$\approx 1.0248.$$

Thus $p(t) = 1.757(1.0248)^t$. Figure 55 shows that this function fits the data well.

(c) Determine the expected annual percentage increase in corn production during this time period.

SOLUTION   Since a is approximately 1.0248, the production of corn each year is 1.0248 times its value the previous year, for a rate of increase of about $0.0248 = 2.48\%$ per year.

(d) Graph $p(t)$ and estimate the year when corn production will be double what it was in 2010.

SOLUTION   Figure 56 shows the graphs of $p(t)$ and $y = 2 \cdot 12.447 = 24.894$ on the same coordinate axes. (Note that the scale in Figure 56 is different than the scale in Figures 54 and 55 so that larger values of t and $p(t)$ are visible.) Their graphs intersect at approximately $t = 108$, corresponding to 2038, which is the year when corn production will be double its 2010 level. In the next section, we will see another way to solve such problems that does not require the use of a graphing calculator.   ▬▬

FIGURE 56

Another way to check whether an exponential function fits the data is to see if points whose x-coordinates are equally spaced have y-coordinates with a constant ratio. This must be true for an exponential function because if $f(x) = a \cdot b^x$, then $f(x_1) = a \cdot b^{x_1}$ and $f(x_2) = a \cdot b^{x_2}$, so

$$\frac{f(x_2)}{f(x_1)} = \frac{a \cdot b^{x_2}}{a \cdot b^{x_1}} = b^{x_2 - x_1}.$$

This last expression is constant if $x_2 - x_1$ is constant, that is, if the x-coordinates are equally spaced.

In the previous example, all data points have t-coordinates 10 years apart, so we can compare the ratios of corn production for any of these first pairs of years. Here are the ratios for 1930–1940 and for 1990–2000:

$$\frac{2.207}{1.757} = 1.256$$

$$\frac{9.968}{7.934} = 1.256$$

These ratios are identical to 3 decimal places, so an exponential function fits the data very well. Not all ratios are this close; using the values at 1970 and 1980, we have $6.639/4.152 = 1.599$. From Figure 55, we can see that this is because the 1970 value is below the exponential curve and the 1980 value is above the curve.

⊞ TECHNOLOGY NOTE

Another way to find an exponential function that fits a set of data is to use a graphing calculator or computer program with an exponential regression feature. This fits an exponential function through a set of points using the least squares method, introduced in Section 1.3 for fitting a line through a set of points. On a TI-84 Plus C, for example, enter the year (with $t = 0$ corresponding to 1930) into the list L_1 and the corn production into L_2. Selecting ExpReg from the STAT CALC menu yields $y = 1.733(1.0253)^t$, which is close to the function we found in Example 7(b).

2.4 EXERCISES

A ream of 20-lb paper contains 500 sheets and is about 2 in. high. Suppose you take one sheet, fold it in half, then fold it in half again, continuing in this way as long as possible. *Source: The AMATYC Review.*

1. Complete the table.

Number of Folds	1	2	3	4	5	...	10	...	50
Layers of Paper									

2. After folding 50 times (if this were possible), what would be the height (in miles) of the folded paper?

For Exercises 3–11, match the correct graph A–F to the function without using your calculator. Notice that there are more functions than graphs; some of the functions are equivalent. After you have answered all of them, use a graphing calculator to check your answers. Each graph in this group is plotted on the window $[-2, 2]$ by $[-4, 4]$.

3. $y = 3^x$

4. $y = 3^{-x}$

5. $y = \left(\dfrac{1}{3}\right)^{1-x}$

6. $y = 3^{x+1}$

7. $y = 3(3)^x$

8. $y = \left(\dfrac{1}{3}\right)^x$

9. $y = 2 - 3^{-x}$

10. $y = -2 + 3^{-x}$

11. $y = 3^{x-1}$

(A)

(B)

(C)

(D)

(E)

(F)

12. In Exercises 3–11, there were more formulas for functions than there were graphs. Explain how this is possible.

Solve each equation.

13. $2^x = 32$

14. $4^x = 64$

15. $3^x = \dfrac{1}{81}$

16. $e^x = \dfrac{1}{e^5}$

17. $4^x = 8^{x+1}$

18. $25^x = 125^{x+2}$

19. $16^{x+3} = 64^{2x-5}$

20. $(e^3)^{-2x} = e^{-x+5}$

21. $e^{-x} = (e^4)^{x+3}$

22. $2^{|x|} = 8$

23. $5^{-|x|} = \dfrac{1}{25}$

24. $2^{x^2-4x} = \left(\dfrac{1}{16}\right)^{x-4}$

25. $5^{x^2+x} = 1$

26. $8^{x^2} = 2^{x+4}$

27. $27^x = 9^{x^2+x}$

28. $e^{x^2+5x+6} = 1$

Graph each of the following.

29. $y = 5e^x + 2$

30. $y = -2e^x - 3$

31. $y = -3e^{-2x} + 2$

32. $y = 4e^{-x/2} - 1$

33. In our definition of an exponential function, we ruled out negative values of a. The author of a textbook on mathematical economics, however, obtained a "graph" of $y = (-2)^x$ by plotting the following points and drawing a smooth curve through them.

x	−4	−3	−2	−1	0	1	2	3
y	1/16	−1/8	1/4	−1/2	1	−2	4	−8

The graph oscillates very neatly from positive to negative values of y. Comment on this approach. (This exercise shows the dangers of point plotting when drawing graphs.)

34. Explain why the exponential equation $4^x = 6$ cannot be solved using the method described in Example 3.

35. Explain why $3^x > e^x > 2^x$ when $x > 0$, but $3^x < e^x < 2^x$ when $x < 0$.

36. A friend claims that as x becomes large, the expression $1 + 1/x$ gets closer and closer to 1, and 1 raised to any power is still 1. Therefore, $f(x) = (1 + 1/x)^x$ gets closer and closer to 1 as x gets larger. Use a graphing calculator to graph f on $0.1 \leq x \leq 50$. How might you use this graph to explain to the friend why $f(x)$ does not approach 1 as x becomes large? What does it approach?

APPLICATIONS

Business and Economics

37. Interest Find the interest earned on $10,000 invested for 5 years at 4% interest compounded as follows.

(a) Annually   (b) Semiannually (twice a year)

(c) Quarterly   (d) Monthly   (e) Continuously

38. Interest Suppose $26,000 is borrowed for 4 years at 6% interest. Find the interest paid over this period if the interest is compounded as follows.

(a) Annually   (b) Semiannually

(c) Quarterly   (d) Monthly   (e) Continuously

39. Interest Lisa McNeil needs to choose between two investments: One pays 6% compounded annually, and the other pays 5.9% compounded monthly. If she plans to invest $18,000 for 2 years, which investment should she choose? How much extra interest will she earn by making the better choice?

40. Interest Find the interest rate required for an investment of $5000 to grow to $6100 in 5 years if interest is compounded as follows. (Round percent to nearest hundredth.)

(a) Annually   (b) Quarterly

41. Inflation Assuming continuous compounding, what will it cost to buy a $10 item in 3 years at the following inflation rates?

(a) 3%   (b) 4%   (c) 5%

42. Interest Janeli Bitor invests a $25,000 inheritance in a fund paying 5.5% per year compounded continuously. What will be the amount on deposit after each time period?

(a) 1 year   (b) 5 years   (c) 10 years

43. Interest Andrea Davis plans to invest $600 into a money market account. Find the interest rate that is needed for the money to grow to $1240 in 14 years if the interest is compounded quarterly.

44. Interest Lauren Snowden puts $10,500 into an account to save money to buy a car in 12 years. She expects the car of her dreams to cost $35,000 by then. Find the interest rate that is necessary if the interest is computed using the following methods.

(a) Compounded quarterly   (b) Compounded monthly

45. Inflation Inflation is generally described as the increase over time of the cost of a particular product or service. The rate of inflation depends on many factors and does not remain constant. Inflation causes the value of a dollar to decrease over time. Over the past ten years, the average rate of inflation was 2.4 %. *Source: Bureau of Labor Statistics.*

(a) Suppose a house costs $165,000 today. If the average rate of inflation continues at this same rate, estimate the cost of a similar house in 10 years.

(b) Find the cost of a $50 pair of jeans in 5 years.

(c) Explain why these estimates may or may not be accurate.

46. Interest On January 1, 2010, Jack deposited $1000 into Bank X to earn interest at the rate of j per annum compounded semiannually. On January 1, 2015, he transferred his account to Bank Y to earn interest at the rate of k per annum compounded quarterly. On January 1, 2018, the balance at Bank Y was $1990.76. If Jack could have earned interest at the rate of k per annum compounded quarterly from January 1, 2010, through January 1, 2018, his balance would have been $2203.76. Which of the following represents the ratio k/j? *Source: Society of Actuaries.*

(a) 1.25   (b) 1.30   (c) 1.35   (d) 1.40   (e) 1.45

47. Gross Domestic Product The U.S. Gross Domestic Product (GDP) is the output of goods and service produced by labor and property located in the United States. The following table gives the GDP (in billions of current dollars) for selected years. *Source: U.S. Department of Commerce.*

Year	GDP
1930	92
1940	103
1950	300
1960	543
1970	1076
1980	2863
1990	5980
2000	10,285
2010	14,965

(a) Plot the data, letting $t = 0$ correspond to 1930. Does the GDP appear to grow linearly or exponentially?

(b) Find an exponential function in the form of $f(t) = f_0 a^t$ that fits the data at 1930 and 2010, where t is the number of years since 1930 and $f(t)$ is the GDP.

(c) Approximate the average annual percentage increase in GDP during this time period.

(d) Graph $f(t)$ and estimate the first year when the GDP will be double what is was in 2010.

Life Sciences

48. Growth of Bacteria Salmonella bacteria, found on almost all chicken and eggs, grow rapidly in a nice warm place. If just a few hundred bacteria are left on the cutting board when a chicken is cut up, and they get into the potato salad, the population begins compounding. Suppose the number present in the potato salad after t hours is given by

$$f(t) = 500 \cdot 2^{3t}.$$

(a) If the potato salad is left out on the table, how many bacteria are present 1 hour later?

(b) How many were present initially?

(c) How often do the bacteria double?

(d) How quickly will the number of bacteria increase to 32,000?

49. Population Growth Since 1960, the growth in world population (in millions) closely fits the exponential function defined by

$$A(t) = 3100e^{0.0166t},$$

where t is the number of years since 1960. *Source: United Nations*.

(a) World population was about 3686 million in 1970. How closely does the function approximate this value?

(b) Use the function to approximate world population in 2000. (The actual 2000 population was about 6115 million.)

(c) Estimate world population in the year 2015.

50. Physician Demand The demand for physicians is expected to increase in the future, as shown in the table. *Source: Association of American Medical Colleges*.

Year	Demand for Physicians (in thousands)
2006	680.5
2015	758.6
2020	805.8
2025	859.3

(a) Plot the data, letting $t = 0$ correspond to 2000. Does fitting an exponential curve to the data seem reasonable?

(b) Use the data for 2006 and 2015 to find a function of the form $f(t) = Ce^{kt}$ that goes through these two points.

(c) Use your function from part (b) to predict the demand for physicians in 2020 and 2025. How well do these predictions fit the data?

 (d) If you have a graphing calculator or computer program with an exponential regression feature, use it to find an exponential function that approximately fits the data. How does this answer compare with the answer to part (b)?

51. Minority Population According to the U.S. Census Bureau, the United States is becoming more diverse. Based on U.S. Census population projections for 2000 to 2050, the projected Hispanic population (in millions) can be modeled by the exponential function

$$h(t) = 37.79(1.021)^t,$$

where $t = 0$ corresponds to 2000 and $0 \le t \le 50$. *Source: U.S. Census Bureau*.

(a) Find the projected Hispanic population for 2005. Compare this to the actual value of 42.69 million.

(b) The U.S. Asian population is also growing exponentially, and the projected Asian population (in millions) can be modeled by the exponential function

$$a(t) = 11.14(1.023)^t,$$

where $t = 0$ corresponds to 2000 and $0 \le t \le 50$. Find the projected Asian population for 2005, and compare this to the actual value of 12.69 million.

(c) Determine the expected annual percentage increase for Hispanics and for Asians. Which minority population, Hispanic or Asian, is growing at a faster rate?

(d) The U.S. black population is growing at a linear rate, and the projected black population (in millions) can be modeled by the linear function

$$b(t) = 0.5116t + 35.43,$$

where $t = 0$ corresponds to 2000 and $0 \le t \le 50$. Find the projected black population for 2005 and compare this projection to the actual value of 37.91 million.

 (e) Graph the projected population function for Hispanics and estimate when the Hispanic population will be double its actual value for 2005. Then do the same for the Asian and black populations. Comment on the accuracy of these numbers.

Physical Sciences

52. Radioactive Decay Suppose the quantity (in grams) of a radioactive substance present at time t is

$$Q(t) = 1000(5^{-0.3t}),$$

where t is measured in months.

(a) How much will be present in 6 months?

(b) How long will it take to reduce the substance to 8 g?

53. Atmospheric Pressure The atmospheric pressure (in millibars) at a given altitude (in meters) is listed in the table. *Source: Elements of Meteorology*.

Altitude	Pressure
0	1013
1000	899
2000	795
3000	701
4000	617
5000	541
6000	472
7000	411
8000	357
9000	308
10,000	265

(a) Find functions of the form $P = ae^{kx}$, $P = mx + b$, and $P = 1/(ax + b)$ that fit the data at $x = 0$ and $x = 10,000$, where P is the pressure and x is the altitude.

 (b) Plot the data in the table and graph the three functions found in part (a). Which function best fits the data?

(c) Use the best-fitting function from part (b) to predict pressure at 1500 m and 11,000 m. Compare your answers to the true values of 846 millibars and 227 millibars, respectively.

(d) If you have a graphing calculator or computer program with an exponential regression feature, use it to find an exponential function that approximately fits the data. How does this answer compare with the answer to part (b)?

54. Computer Chips The power of personal computers has increased dramatically as a result of the ability to place an increasing number of transistors on a single processor chip. The following table lists the number of transistors on some popular computer chips made by Intel. *Source: Intel.*

Year	Chip	Transistors (in millions)
1985	386	0.275
1989	486	1.2
1993	Pentium	3.1
1997	Pentium II	7.5
1999	Pentium III	9.5
2000	Pentium 4	42
2005	Pentium D	291
2007	Penryn	820
2009	Nehalem	1900

(a) Let t be the year, where $t = 0$ corresponds to 1985, and y be the number of transistors (in millions). Find functions of the form $y = mt + b$, $y = at^2 + b$, and $y = ab^t$ that fit the data at 1985 and 2009.

(b) Use a graphing calculator to plot the data in the table and to graph the three functions found in part (a). Which function best fits the data?

(c) Use the best-fitting function from part (b) to predict the number of transistors on a chip in the year 2015.

(d) If you have a graphing calculator or computer program with an exponential regression feature, use it to find an exponential function that approximately fits the data. How does this answer compare with the answer to part (b)?

(e) In 1965 Gordon Moore wrote a paper predicting how the power of computer chips would grow in the future. Moore's law says that the number of transistors that can be put on a chip doubles roughly every 18 months. Discuss the extent to which the data in this exercise confirms or refutes Moore's law.

55. Wind Energy The table in the next column gives the total world wind energy capacity (in megawatts) in recent years. *Source: World Wind Energy Association.*

(a) Let t be the number of years since 2000, and C the capacity (in MW). Find functions of the form $C = mt + b$, $C = at^2 + b$, and $C = ab^t$ that fit the data at 2001 and 2011.

(b) Use a graphing calculator to plot the data in the table and to graph the three functions found in part (a). Which function best fits the data?

(c) If you have a graphing calculator or computer program with an exponential regression feature, use it to find an exponential function that approximately fits the data in the table. How does this answer compare with the answer to part (a)?

Year	Capacity (MW)
2001	24,322
2002	31,181
2003	39,295
2004	47,681
2005	59,012
2006	74,112
2007	93,919
2008	120,913
2009	159,762
2010	196,986
2011	237,016

(d) Using the three functions from part (a) and the function from part (c), predict the total world wind capacity in 2012. Compare these with the World Wind Energy Association's prediction of 282,275.

56. Carbon Dioxide The table gives the estimated global carbon dioxide (CO_2) emissions from fossil-fuel burning, cement production, and gas flaring for selected years. The CO_2 estimates are expressed in millions of metric tons. *Source: Carbon Dioxide Information Analysis Center.*

Year	CO_2 Emissions (millions of metric tons)
1900	534
1910	819
1920	932
1930	1053
1940	1299
1950	1630
1960	2569
1970	4053
1980	5315
1990	6127
2000	6765
2010	9167

(a) Plot the data, letting $t = 0$ correspond to 1900. Do the emissions appear to grow linearly or exponentially?

(b) Find an exponential function in the form of $f(t) = f_0 a^t$ that fits these data at 1900 and 2010, where t is the number of years since 1900 and $f(t)$ is the CO_2 emissions.

(c) Approximate the average annual percentage increase in CO_2 emissions during this time period.

(d) Graph $f(t)$ and estimate the first year when emissions will be at least double what they were in 2010.

2.5 Logarithmic Functions

The number of years it will take for prices to double under given conditions is called the **doubling time**. For $1 to double (become $2) in t years, assuming 5% annual compounding, means that

$$A = P\left(1 + \frac{r}{m}\right)^{mt}$$

becomes

$$2 = 1\left(1 + \frac{0.05}{1}\right)^{1(t)}$$

or

$$2 = (1.05)^t.$$

This equation would be easier to solve if the variable were not in the exponent. **Logarithms** are defined for just this purpose. In Example 8, we will use logarithms to answer the question posed above.

> ## Logarithm
> For $a > 0, a \neq 1$, and $x > 0$,
>
> $$y = \log_a x \qquad \text{means} \qquad a^y = x.$$

(Read $y = \log_a x$ as "y is the logarithm of x to the base a.") For example, the exponential statement $2^4 = 16$ can be translated into the logarithmic statement $4 = \log_2 16$. Also, in the problem discussed above, $(1.05)^t = 2$ can be rewritten with this definition as $t = \log_{1.05} 2$. A logarithm is an exponent: $\log_a x$ **is the exponent used with the base a to get x.**

▢ EXAMPLE 1 Equivalent Expressions

This example shows the same statements written in both exponential and logarithmic forms.

Exponential Form	*Logarithmic Form*
(a) $3^2 = 9$	$\log_3 9 = 2$
(b) $(1/5)^{-2} = 25$	$\log_{1/5} 25 = -2$
(c) $10^5 = 100{,}000$	$\log_{10} 100{,}000 = 5$
(d) $4^{-3} = 1/64$	$\log_4(1/64) = -3$
(e) $2^{-4} = 1/16$	$\log_2(1/16) = -4$
(f) $e^0 = 1$	$\log_e 1 = 0$

YOUR TURN 1 Write the equation $5^{-2} = 1/25$ in logarithmic form.

TRY YOUR TURN 1 ▬

▢ EXAMPLE 2 Evaluating Logarithms

Evaluate each of the following logarithms.

(a) $\log_4 64$

SOLUTION We seek a number x such that $4^x = 64$. Since $4^3 = 64$, we conclude that $\log_4 64 = 3$.

(b) $\log_2(-8)$

SOLUTION We seek a number x such that $2^x = -8$. Since 2^x is positive for all real numbers x, we conclude that $\log_2(-8)$ is undefined. (Actually, $\log_2(-8)$ can be defined if we use complex numbers, but in this textbook, we restrict ourselves to real numbers.)

(c) $\log_5 80$

YOUR TURN 2 Evaluate $\log_3(1/81)$.

SOLUTION We know that $5^2 = 25$ and $5^3 = 125$, so $\log_5 25 = 2$ and $\log_5 125 = 3$. Therefore, $\log_5 80$ must be somewhere between 2 and 3. We will find a more accurate answer in Example 4. **TRY YOUR TURN 2**

Logarithmic Functions

For a given positive value of x, the definition of logarithm leads to exactly one value of y, so $y = \log_a x$ defines the *logarithmic function* of base a (the base a must be positive, with $a \neq 1$).

Logarithmic Function

If $a > 0$ and $a \neq 1$, then the **logarithmic function** of base a is defined by

$$f(x) = \log_a x$$

for $x > 0$.

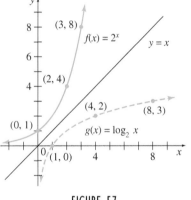

FIGURE 57

The graphs of the exponential function with $f(x) = 2^x$ and the logarithmic function with $g(x) = \log_2 x$ are shown in Figure 57. The graphs show that $f(3) = 2^3 = 8$, while $g(8) = \log_2 8 = 3$. Thus, $f(3) = 8$ and $g(8) = 3$. Also, $f(2) = 4$ and $g(4) = 2$. In fact, for any number m, if $f(m) = p$, then $g(p) = m$. Functions related in this way are called **inverse functions** of each other. The graphs also show that the domain of the exponential function (the set of real numbers) is the range of the logarithmic function. Also, the range of the exponential function (the set of positive real numbers) is the domain of the logarithmic function. Every logarithmic function is the inverse of some exponential function. This means that we can graph logarithmic functions by rewriting them as exponential functions using the definition of logarithm. The graphs in Figure 57 show a characteristic of a pair of inverse functions: Their graphs are mirror images about the line $y = x$. Therefore, since exponential functions pass through the point $(0, 1)$, logarithmic functions pass through the point $(1, 0)$. Notice that because the exponential function has the x-axis as a horizontal asymptote, the logarithmic function has the y-axis as a vertical asymptote. A more complete discussion of inverse functions is given in most standard intermediate algebra and college algebra books.

The graph of $\log_2 x$ is typical of logarithms with bases $a > 1$. When $0 < a < 1$, the graph is the vertical reflection of the logarithm graph in Figure 57. Because logarithms with bases less than 1 are rarely used, we will not explore them here.

| CAUTION | The domain of $\log_a x$ consists of all $x > 0$. In other words, you cannot take the logarithm of zero or a negative number. This also means that in a function such as $g(x) = \log_a(x - 2)$, the domain is given by $x - 2 > 0$, or $x > 2$. |

Properties of Logarithms

The usefulness of logarithmic functions depends in large part on the following **properties of logarithms**.

Properties of Logarithms

Let x and y be any positive real numbers and r be any real number. Let a be a positive real number, $a \neq 1$. Then

(a) $\log_a xy = \log_a x + \log_a y$

(b) $\log_a \dfrac{x}{y} = \log_a x - \log_a y$

(c) $\log_a x^r = r \log_a x$

(d) $\log_a a = 1$

(e) $\log_a 1 = 0$

(f) $\log_a a^r = r.$

To prove property (a), let $m = \log_a x$ and $n = \log_a y$. Then, by the definition of logarithm,

$$a^m = x \qquad \text{and} \qquad a^n = y.$$

Hence,

$$a^m a^n = xy.$$

By a property of exponents, $a^m a^n = a^{m+n}$, so

$$a^{m+n} = xy.$$

Now use the definition of logarithm to write

$$\log_a xy = m + n.$$

Since $m = \log_a x$ and $n = \log_a y$,

$$\log_a xy = \log_a x + \log_a y.$$

Proofs of properties (b) and (c) are left for the exercises. Properties (d) and (e) depend on the definition of a logarithm. Property (f) follows from properties (c) and (d).

EXAMPLE 3 Properties of Logarithms

If all the following variable expressions represent positive numbers, then for $a > 0, a \neq 1$, the statements in (a)–(c) are true.

YOUR TURN 3 Write the expression $\log_a \left(x^2 / y^3 \right)$ as a sum, difference, or product of simpler logarithms.

(a) $\log_a x + \log_a(x - 1) = \log_a x(x - 1)$

(b) $\log_a \dfrac{x^2 - 4x}{x + 6} = \log_a(x^2 - 4x) - \log_a(x + 6)$

(c) $\log_a(9x^5) = \log_a 9 + \log_a(x^5) = \log_a 9 + 5 \cdot \log_a x$ **TRY YOUR TURN 3**

Evaluating Logarithms

The invention of logarithms is credited to John Napier (1550–1617), who first called logarithms "artificial numbers." Later he joined the Greek words *logos* (ratio) and *arithmos* (number) to form the word used today. The development of logarithms was motivated by a need for faster computation. Tables of logarithms and slide rule devices were developed by Napier, Henry Briggs (1561–1631), Edmund Gunter (1581–1626), and others.

For many years logarithms were used primarily to assist in involved calculations. Current technology has made this use of logarithms obsolete, but logarithmic functions play an important role in many applications of mathematics. Since our number system has base 10, logarithms to base 10 were most convenient for numerical calculations and

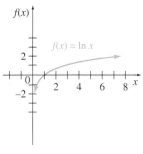

FIGURE 58

so base 10 logarithms were called **common logarithms**. Common logarithms are still useful in other applications. For simplicity,

$$\log_{10} x \text{ is abbreviated } \log x.$$

Most practical applications of logarithms use the number e as base. (Recall that to 7 decimal places, $e = 2.7182818$.) Logarithms to base e are called **natural logarithms**, and

$$\log_e x \text{ is abbreviated } \ln x$$

(read "el-en x"). A graph of $f(x) = \ln x$ is given in Figure 58.

NOTE   Keep in mind that $\ln x$ is a logarithmic function. Therefore, all of the properties of logarithms given previously are valid when a is replaced with e and $\log_e$ is replaced with $\ln$.

Although common logarithms may seem more "natural" than logarithms to base e, there are several good reasons for using natural logarithms instead. The most important reason is discussed later in Section 4.5 on Derivatives of Logarithmic Functions.

A calculator can be used to find both common and natural logarithms. For example, using a calculator and 4 decimal places, we get the following values.

$$\log 2.34 = 0.3692, \quad \log 594 = 2.7738, \quad \text{and} \quad \log 0.0028 = -2.5528.$$
$$\ln 2.34 = 0.8502, \quad \ln 594 = 6.3869, \quad \text{and} \quad \ln 0.0028 = -5.8781.$$

Notice that logarithms of numbers less than 1 are negative when the base is greater than 1. A look at the graph of $y = \log_2 x$ or $y = \ln x$ will show why.

Sometimes it is convenient to use logarithms to bases other than 10 or e. For example, some computer science applications use base 2. In such cases, the following theorem is useful for converting from one base to another.

Change-of-Base Theorem for Logarithms

If x is any positive number and if a and b are positive real numbers, $a \neq 1$, $b \neq 1$, then

$$\log_a x = \frac{\log_b x}{\log_b a}.$$

To prove this result, use the definition of logarithm to write $y = \log_a x$ as $x = a^y$ or $x = a^{\log_a x}$ (for positive x and positive a, $a \neq 1$). Now take base b logarithms of both sides of this last equation.

$$\log_b x = \log_b a^{\log_a x}$$
$$\log_b x = (\log_a x)(\log_b a), \qquad \log_a x^r = r \log_a x$$
$$\log_a x = \frac{\log_b x}{\log_b a} \qquad \text{Solve for } \log_a x.$$

If the base b is equal to e, then by the change-of-base theorem,

$$\log_a x = \frac{\log_e x}{\log_e a}.$$

Using $\ln x$ for $\log_e x$ gives the special case of the theorem using natural logarithms.

For any positive numbers a and x, $a \neq 1$,

$$\log_a x = \frac{\ln x}{\ln a}.$$

The change-of-base theorem for logarithms is useful when graphing $y = \log_a x$ on a graphing calculator for a base a other than e or 10. For example, to graph $y = \log_2 x$, let $y = \ln x/\ln 2$. The change-of-base theorem is also needed when using a calculator to evaluate a logarithm with a base other than e or 10.

> ░░░ **EXAMPLE 4** **Evaluating Logarithms**

Evaluate $\log_5 80$.

SOLUTION As we saw in Example 2, this number is between 2 and 3. Using the second form of the change-of-base theorem for logarithms with $x = 80$ and $a = 5$ gives

YOUR TURN 4 Evaluate $\log_3 50$.

$$\log_5 80 = \frac{\ln 80}{\ln 5} \approx \frac{4.3820}{1.6094} \approx 2.723.$$

To check, use a calculator to verify that $5^{2.723} \approx 80$.   **TRY YOUR TURN 4** ░░░

> **CAUTION** As mentioned earlier, when using a calculator, do not round off intermediate results. Keep all numbers in the calculator until you have the final answer. In Example 4, we showed the rounded intermediate values of $\ln 80$ and $\ln 5$, but we used the unrounded quantities when doing the division.

Logarithmic Equations
Equations involving logarithms are often solved by using the fact that exponential functions and logarithmic functions are inverses, so a logarithmic equation can be rewritten (with the definition of logarithm) as an exponential equation. In other cases, the properties of logarithms may be useful in simplifying a **logarithmic equation**.

> ░░░ **EXAMPLE 5** **Solving Logarithmic Equations**

Solve each equation.

(a) $\log_x \dfrac{8}{27} = 3$

SOLUTION Using the definition of logarithm, write the expression in exponential form. To solve for x, take the cube root on both sides.

$$x^3 = \frac{8}{27}$$

$$x = \frac{2}{3} \qquad \sqrt[3]{\frac{8}{27}} = \frac{\sqrt[3]{8}}{\sqrt[3]{27}} = \frac{2}{3}$$

(b) $\log_4 x = \dfrac{5}{2}$

SOLUTION In exponential form, the given statement becomes

$$4^{5/2} = x$$
$$\left(4^{1/2}\right)^5 = x \qquad a^{m/n} = \left(a^{1/n}\right)^m$$
$$2^5 = x \qquad 4^{1/2} = \sqrt{4} = 2$$
$$32 = x.$$

(c) $\log_2 x - \log_2 (x - 1) = 1$

SOLUTION Since $\log_a x - \log_a y = \log_a(x/y)$, we have

$$\log_2 x - \log_2(x - 1) = \log_2 \frac{x}{x - 1}.$$

The original equation becomes

$$\log_2 \frac{x}{x-1} = 1.$$

Now write this equation in exponential form.

$$\frac{x}{x-1} = 2^1 = 2$$

Solve this equation.

$$\frac{x}{x-1}(x-1) = 2(x-1) \qquad \text{Multiply both sides by } x-1.$$

$$x = 2(x-1)$$

$$x = 2x - 2 \qquad \text{Distributive property}$$

$$-x = -2 \qquad \text{Subtract } 2x \text{ from both sides.}$$

$$x = 2$$

(d) $\log x + \log (x-3) = 1$

SOLUTION Since $\log_a x + \log_a y = \log_a xy$, we have

$$\log x + \log (x-3) = \log [x(x-3)] = 1.$$

Since the logarithm base is 10, this means that

$$x(x-3) = 10 \qquad 10^1 = 10$$

$$x^2 - 3x - 10 = 0 \qquad \text{Distribute and subtract 10 from both sides.}$$

$$(x-5)(x+2) = 0. \qquad \text{Factor.}$$

YOUR TURN 5 Solve for x: $\log_2 x + \log_2 (x+2) = 3$.

This leads to two solutions: $x = 5$ and $x = -2$. But notice that -2 is not a valid value for x in the original equation, since the logarithm of a negative number is undefined. Therefore, the only solution is $x = 5$. **TRY YOUR TURN 5**

CAUTION It is important to check solutions when solving equations involving logarithms because $\log_a u$, where u is an expression in x, has domain given by $u > 0$.

Exponential Equations In the previous section exponential equations like $(1/3)^x = 81$ were solved by writing each side of the equation as a power of 3. That method cannot be used to solve an equation such as $3^x = 5$, however, since 5 cannot easily be written as a power of 3. Such equations can be solved approximately with a graphing calculator, but an algebraic method is also useful, particularly when the equation involves variables such as a and b rather than just numbers such as 3 and 5. A general method for solving these equations is shown in the following example.

EXAMPLE 6 **Solving Exponential Equations**

Solve each equation.

(a) $3^x = 5$

SOLUTION Taking natural logarithms (logarithms to any base could be used) on both sides gives

$$\ln 3^x = \ln 5$$

$$x \ln 3 = \ln 5 \qquad \ln u^r = r \ln u$$

$$x = \frac{\ln 5}{\ln 3} \approx 1.465. \qquad \text{Divide both sides by } \ln 3.$$

(b) $3^{2x} = 4^{x+1}$

SOLUTION Taking natural logarithms on both sides gives

$$\ln 3^{2x} = \ln 4^{x+1}$$

$$2x \ln 3 = (x + 1) \ln 4 \qquad \text{\small{ln u^r = r ln u}}$$

$$(2 \ln 3)x = (\ln 4)x + \ln 4$$

$$(2 \ln 3)x - (\ln 4)x = \ln 4 \qquad \text{\small{Subtract (ln 4) x from both sides.}}$$

$$(2 \ln 3 - \ln 4)x = \ln 4 \qquad \text{\small{Factor x.}}$$

$$x = \frac{\ln 4}{2 \ln 3 - \ln 4} \qquad \text{\small{Divide both sides by 2 ln 3 − ln 4.}}$$

$$x \approx 1.710$$

(c) $5e^{0.01x} = 9$

SOLUTION

$$e^{0.01x} = \frac{9}{5} = 1.8 \qquad \text{\small{Divide both sides by 5.}}$$

$$\ln e^{0.01x} = \ln 1.8 \qquad \text{\small{Take natural logarithms on both sides.}}$$

$$0.01x = \ln 1.8 \qquad \text{\small{ln e^u = u}}$$

$$x = \frac{\ln 1.8}{0.01} \approx 58.779 \qquad \text{\small{**TRY YOUR TURN 6**}}$$

YOUR TURN 6 Solve for x: $2^{x+1} = 3^x$.

Just as $\log_a x$ can be written as a base e logarithm, any exponential function $y = a^x$ can be written as an exponential function with base e. For example, there exists a real number k such that

$$2 = e^k.$$

Raising both sides to the power x gives

$$2^x = e^{kx},$$

so that powers of 2 can be found by evaluating appropriate powers of e. To find the necessary number k, solve the equation $2 = e^k$ for k by first taking logarithms on both sides.

$$2 = e^k$$

$$\ln 2 = \ln e^k$$

$$\ln 2 = k \ln e$$

$$\ln 2 = k \qquad \text{\small{ln e = 1}}$$

Thus, $k = \ln 2$. Later, in Section 4.4 on Derivatives of Exponential Functions, we will see why this change of base is useful. A general statement can be drawn from this example.

Change-of-Base Theorem for Exponentials

For every positive real number a,

$$a^x = e^{(\ln a)x}.$$

Another way to see why the change-of-base theorem for exponentials is true is to first observe that $e^{\ln a} = a$. Combining this with the fact that $e^{ab} = (e^a)^b$, we have $e^{(\ln a)x} = (e^{\ln a})^x = a^x$.

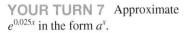 **EXAMPLE 7** **Change-of-Base-Theorem**

(a) Write 7^x using base e rather than base 7.

SOLUTION According to the change-of-base theorem,

$$7^x = e^{(\ln 7)x}.$$

Using a calculator to evaluate $\ln 7$, we could also approximate this as $e^{1.9459x}$.

(b) Approximate the function $f(x) = e^{2x}$ as $f(x) = a^x$ for some base a.

YOUR TURN 7 Approximate $e^{0.025x}$ in the form a^x.

SOLUTION We do not need the change-of-base theorem here. Just use the fact that

$$e^{2x} = (e^2)^x \approx 7.389^x,$$

where we have used a calculator to approximate e^2.   **TRY YOUR TURN 7** ▄▄▄

The final examples in this section illustrate the use of logarithms in practical problems.

 EXAMPLE 8 **Doubling Time**

With an inflation rate averaging 5% per year, how long will it take for prices to double?

▄▄▄ **APPLY IT**   **SOLUTION** Recall that if prices will double after t years at an inflation rate of 5%, compounded annually, t is given by the equation

$$A = P(1 + r)^t$$
$$2 = (1.05)^t. \qquad A = 2, P = 1, r = 0.05$$

We solve this equation by first taking natural logarithms on both sides.

$$\ln 2 = \ln(1.05)^t$$
$$\ln 2 = t \ln 1.05 \qquad \ln x^r = r \ln x$$
$$t = \frac{\ln 2}{\ln 1.05} \approx 14.2$$

It will take about 14 years for prices to double.   ▄▄▄

The problem solved in Example 8 can be generalized for the compound interest equation

$$A = P(1 + r)^t.$$

Solving for t as in Example 8 (with $A = 2$ and $P = 1$) gives the doubling time in years as

$$t = \frac{\ln 2}{\ln(1 + r)}.$$

If r is small, $\ln(1 + r) \approx r$, as Figure 59 shows, so that

$$t = \frac{\ln 2}{\ln(1 + r)} \approx \frac{\ln 2}{r} \approx \frac{0.693}{r}.$$

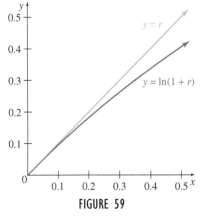

FIGURE 59

Notice from Figure 59 that the actual value of r is larger than $\ln(1 + r)$, causing the above formula to give a doubling time that is too small. By changing 0.693 to 0.70, the formula becomes reasonably accurate for $0.001 \le r < 0.05$. By increasing the numerator further to 0.72, the formula becomes reasonably accurate for $0.05 \le r < 0.12$. This leads to two useful rules for estimating the doubling time. The **rule of 70** says that for $0.001 \le r < 0.05$, the value of $70/100r$ gives a good approximation of t. The **rule of 72** says that for $0.05 \le r \le 0.12$, the value of $72/100r$ approximates t quite well.

Figure 60 on the next page shows the three functions $y = \ln 2/\ln(1 + r)$, $y = 70/100r$, and $y = 72/100r$ graphed on the same axes. Observe how close to each other they are. In an exercise we will ask you to explore the relationship between these functions further.

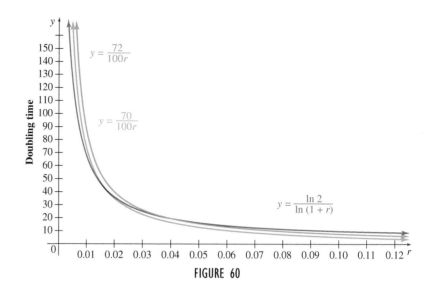

FIGURE 60

EXAMPLE 9 Rules of 70 and 72

Approximate the years to double at an interest rate of 6% using first the rule of 70, then the rule of 72.

SOLUTION By the rule of 70, money will double at 6% interest after

$$\frac{70}{100r} = \frac{70}{100(0.06)} = \frac{70}{6} = 11.67 \left(\text{or } 11\frac{2}{3} \right)$$

years.

Using the rule of 72 gives

$$\frac{72}{100r} = \frac{72}{6} = 12$$

years doubling time. Since a more precise answer is given by

$$\frac{\ln 2}{\ln(1 + r)} = \frac{\ln 2}{\ln(1.06)} \approx \frac{0.693}{0.058} \approx 11.9,$$

the rule of 72 gives a better approximation than the rule of 70. This agrees with the statement that the rule of 72 works well for values of r where $0.05 \le r \le 0.12$, since $r = 0.06$ falls into this category.

EXAMPLE 10 Index of Diversity

One measure of the diversity of the species in an ecological community is given by the **index of diversity** H, where

$$H = -[P_1 \ln P_1 + P_2 \ln P_2 + \cdots + P_n \ln P_n],$$

and $P_1, P_2, \ldots, P_n$ are the proportions of a sample belonging to each of n species found in the sample. *Source: Statistical Ecology.*

For example, in a community with two species, where there are 90 of one species and 10 of the other, $P_1 = 90/100 = 0.9$ and $P_2 = 10/100 = 0.1$, with

$$H = -[0.9 \ln 0.9 + 0.1 \ln 0.1] \approx 0.325.$$

Verify that $H \approx 0.673$ if there are 60 of one species and 40 of the other. As the proportions of n species get closer to $1/n$ each, the index of diversity increases to a maximum of $\ln n$.

2.5   EXERCISES

Write each exponential equation in logarithmic form.

1. $5^3 = 125$ **2.** $7^2 = 49$

3. $3^4 = 81$ **4.** $2^7 = 128$

5. $3^{-2} = \dfrac{1}{9}$ **6.** $\left(\dfrac{5}{4}\right)^{-2} = \dfrac{16}{25}$

Write each logarithmic equation in exponential form.

7. $\log_2 32 = 5$ **8.** $\log_3 81 = 4$

9. $\ln \dfrac{1}{e} = -1$ **10.** $\log_2 \dfrac{1}{8} = -3$

11. $\log 100{,}000 = 5$ **12.** $\log 0.001 = -3$

Evaluate each logarithm without using a calculator.

13. $\log_8 64$ **14.** $\log_9 81$

15. $\log_4 64$ **16.** $\log_3 27$

17. $\log_2 \dfrac{1}{16}$ **18.** $\log_3 \dfrac{1}{81}$

19. $\log_2 \sqrt[3]{\dfrac{1}{4}}$ **20.** $\log_8 \sqrt[4]{\dfrac{1}{2}}$

21. $\ln e$ **22.** $\ln e^3$

23. $\ln e^{5/3}$ **24.** $\ln 1$

25. Is the "logarithm to the base 3 of 4" written as $\log_4 3$ or $\log_3 4$?

26. Write a few sentences describing the relationship between e^x and $\ln x$.

Use the properties of logarithms to write each expression as a sum, difference, or product of simpler logarithms. For example,
$\log_2(\sqrt{3}x) = \frac{1}{2}\log_2 3 + \log_2 x.$

27. $\log_5 (3k)$ **28.** $\log_9(4m)$

29. $\log_3 \dfrac{3p}{5k}$ **30.** $\log_7 \dfrac{15p}{7y}$

31. $\ln \dfrac{3\sqrt{5}}{\sqrt[3]{6}}$ **32.** $\ln \dfrac{9\sqrt[3]{5}}{\sqrt[4]{3}}$

Suppose $\log_b 2 = a$ **and** $\log_b 3 = c$. **Use the properties of logarithms to find the following.**

33. $\log_b 32$ **34.** $\log_b 18$

35. $\log_b(72b)$ **36.** $\log_b(9b^2)$

Use natural logarithms to evaluate each logarithm to the nearest thousandth.

37. $\log_5 30$ **38.** $\log_{12} 210$

39. $\log_{1.2} 0.95$ **40.** $\log_{2.8} 0.12$

Solve each equation in Exercises 41–64. Round decimal answers to four decimal places.

41. $\log_x 36 = -2$ **42.** $\log_9 27 = m$

43. $\log_8 16 = z$ **44.** $\log_y 8 = \dfrac{3}{4}$

45. $\log_r 5 = \dfrac{1}{2}$ **46.** $\log_4(5x + 1) = 2$

47. $\log_5(9x - 4) = 1$

48. $\log_4 x - \log_4(x + 3) = -1$

49. $\log_9 m - \log_9(m - 4) = -2$

50. $\log(x + 5) + \log(x + 2) = 1$

51. $\log_3(x - 2) + \log_3(x + 6) = 2$

52. $\log_3(x^2 + 17) - \log_3(x + 5) = 1$

53. $\log_2(x^2 - 1) - \log_2(x + 1) = 2$

54. $\ln(5x + 4) = 2$

55. $\ln x + \ln 3x = -1$ **56.** $\ln (x + 1) - \ln x = 1$

57. $2^x = 6$ **58.** $5^x = 12$

59. $e^{k-1} = 6$ **60.** $e^{2y} = 15$

61. $3^{x+1} = 5^x$ **62.** $2^{x+1} = 6^{x-1}$

63. $5(0.10)^x = 4(0.12)^x$ **64.** $1.5(1.05)^x = 2(1.01)^x$

Write each expression using base e rather than base 10.

65. 10^{x+1} **66.** 10^{x^2}

Approximate each expression in the form a^x without using e.

67. e^{3x} **68.** e^{-4x}

Find the domain of each function.

69. $f(x) = \log(5 - x)$ **70.** $f(x) = \ln(x^2 - 9)$

71. Lucky Larry was faced with solving

$$\log(2x + 1) - \log(3x - 1) = 0.$$

Larry just dropped the logs and proceeded:

$$(2x + 1) - (3x - 1) = 0$$
$$-x + 2 = 0$$
$$x = 2.$$

Although Lucky Larry is wrong in dropping the logs, his procedure will always give the correct answer to an equation of the form

$$\log A - \log B = 0,$$

where A and B are any two expressions in x. Prove that this last equation leads to the equation $A - B = 0$, which is what you get when you drop the logs. *Source: The AMATYC Review.*

72. Find all errors in the following calculation.

$$(\log(x + 2))^2 = 2 \log(x + 2)$$
$$= 2(\log x + \log 2)$$
$$= 2(\log x + 100)$$

73. Prove: $\log_a \left(\dfrac{x}{y} \right) = \log_a x - \log_a y.$

74. Prove: $\log_a x^r = r \log_a x.$

APPLICATIONS

Business and Economics

75. Inflation Assuming annual compounding, find the time it would take for the general level of prices in the economy to double at the following annual inflation rates.

    (a) 3%   **(b)** 6%   **(c)** 8%

    (d) Check your answers using either the rule of 70 or the rule of 72, whichever applies.

76. Interest Allison Andrews invests $15,000 in an account paying 7% per year compounded annually.

    (a) How many years are required for the compound amount to at least double? (Note that interest is only paid at the end of each year.)

    (b) In how many years will the amount at least triple?

    (c) Check your answer to part (a) using either the rule of 70 or the rule of 72, whichever applies.

77. Interest Andrea Davis plans to invest $600 into a money market account. Find the interest rate that is needed for the money to grow to $1240 in 14 years if the interest is compounded continuously. (Compare with Exercise 43 in the previous section.)

78. Rule of 72 Complete the following table, and use the results to discuss when the rule of 70 gives a better approximation for the doubling time, and when the rule of 72 gives a better approximation.

r	0.001	0.02	0.05	0.08	0.12
$(\ln 2)/\ln(1 + r)$					
$70/100r$					
$72/100r$					

79. Gross Domestic Product In Exercise 47 of the previous section, the Gross Domestic Product (GDP) for the United States was modeled by the exponential function

$$f(t) = 92(1.0657)^t,$$

where $f(t)$ was the GDP (in billions of current dollars) and t was the number of years since 1930. In what year will the GDP reach 29,930?

80. Pay Increases You are offered two jobs starting July 1, 2016. Humongous Enterprises offers you $45,000 a year to start, with a raise of 4% every July 1. At Crabapple Inc. you start at $30,000, with an annual increase of 6% every July 1. On July 1 of what year would the job at Crabapple Inc. pay more than the job at Humongous Enterprises? Use the algebra of logarithms to solve this problem, and support your answer by using a graphing calculator to see where the two salary functions intersect.

Life Sciences

81. Physician's Demand Suppose that a quantity can be described by the exponential growth equation $y(t) = y_0 e^{kt}$.

    (a) Show that the quantity $\dfrac{1}{t} \ln \dfrac{y(t)}{y_0}$ is a constant.

    (b) For the physician's demand data in Exercise 50 of the previous section, let $t = 0$ correspond to 2006. Calculate the expression in part (a) for the other times. To what extent is this expression constant?

82. Insect Species An article in *Science* stated that the number of insect species of a given mass is proportional to $m^{-0.6}$, where m is the mass in grams. A graph accompanying the article shows the common logarithm of the mass on the horizontal axis and the common logarithm of the number of species on the vertical axis. Explain why the graph is a straight line. What is the slope of the line? *Source: Science.*

Index of Diversity For Exercises 83–85, refer to Example 10.

83. Suppose a sample of a small community shows two species with 50 individuals each.

    (a) Find the index of diversity H.

    (b) What is the maximum value of the index of diversity for two species?

    (c) Does your answer for part (a) equal $\ln 2$? Explain why.

84. A virgin forest in northwestern Pennsylvania has 4 species of large trees with the following proportions of each: hemlock, 0.521; beech, 0.324; birch, 0.081; maple, 0.074. Find the index of diversity H.

85. Find the value of the index of diversity for populations with n species and $1/n$ of each if

(a) $n = 3$;

(b) $n = 4$.

(c) Verify that your answers for parts (a) and (b) equal $\ln 3$ and $\ln 4$, respectively.

86. Allometric Growth   The allometric formula is used to describe a wide variety of growth patterns. It says that $y = nx^m$, where x and y are variables, and n and m are constants. For example, the famous biologist J. S. Huxley used this formula to relate the weight of the large claw of the fiddler crab to the weight of the body without the claw. Show that if x and y are given by the allometric formula, then $X = \log_b x$, $Y = \log_b y$, and $N = \log_b n$ are related by the linear equation

$$Y = mX + N.$$

Source: Problems of Relative Growth.

87. Drug Concentration   When a pharmaceutical drug is injected into the bloodstream, its concentration at time t can be approximated by $C(t) = C_0 e^{-kt}$, where C_0 is the concentration at $t = 0$. Suppose the drug is ineffective below a concentration C_1 and harmful above a concentration C_2. Then it can be shown that the drug should be given at intervals of time T, where

$$T = \frac{1}{k} \ln \frac{C_2}{C_1}.$$

A certain drug is harmful at a concentration five times the concentration below which it is ineffective. At noon an injection of the drug results in a concentration of 2 mg per liter of blood. Three hours later the concentration is down to 1 mg per liter. How often should the drug be given? *Source: Applications of Calculus to Medicine.*

The graph for Exercise 88 is plotted on a logarithmic scale where differences between successive measurements are not always the same. Data that do not plot in a linear pattern on the usual Cartesian axes often form a linear pattern when plotted on a logarithmic scale. Notice that on the horizontal scale, the distance from 5 to 10 is not the same as the distance from 10 to 15, and so on. This is characteristic of a graph drawn on logarithmic scales.

88. Metabolism Rate   The graph at the top of the next column shows the basal metabolism rate (in cm^3 of oxygen per gram per hour) for marsupial carnivores, which include the Tasmanian devil. This rate is inversely proportional to body mass raised to the power 0.25. *Source: The Quarterly Review of Biology.*

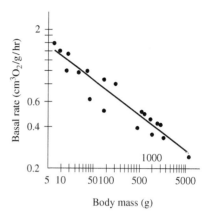

(a) Estimate the metabolism rate for a marsupial carnivore with body mass of 10 g. Do the same for one with body mass of 1000 g.

(b) Verify that if the relationship between x and y is of the form $y = ax^b$, then there will be a linear relationship between $\ln x$ and $\ln y$. (*Hint:* Apply $\ln$ to both sides of $y = ax^b$.)

(c) If a function of the form $y = ax^b$ contains the points (x_1, y_1) and (x_2, y_2), then values for a and b can be found by dividing $y_1 = ax_1^b$ by $y_2 = ax_2^b$, solving the resulting equation for b, and putting the result back into either equation to solve for a. Use this procedure and the results from part (a) to find an equation of the form $y = ax^b$ that gives the basal metabolism rate as a function of body mass.

(d) Use the result of part (c) to predict the basal metabolism rate of a marsupial carnivore whose body mass is 100 g.

89. Minority Population   The U.S. Census Bureau has reported that the United States is becoming more diverse. In Exercise 51 of the previous section, the projected Hispanic population (in millions) was modeled by the exponential function

$$h(t) = 37.79\,(1.021)^t$$

where $t = 0$ corresponds to 2000 and $0 \le t \le 50$. *Source: U.S. Census Bureau.*

(a) Estimate in what year the Hispanic population will double the 2005 population of 42.69 million. Use the algebra of logarithms to solve this problem.

(b) The projected U.S. Asian population (in millions) was modeled by the exponential function

$$a(t) = 11.14\,(1.023)^t,$$

where $t = 0$ corresponds to 2000 and $0 \le t \le 50$. Estimate in what year the Asian population will double the 2005 population of 12.69 million.

Social Sciences

90. Evolution of Languages   The number of years $N(r)$ since two independently evolving languages split off from a common ancestral language is approximated by

$$N(r) = -5000\,\ln r,$$

where r is the proportion of the words from the ancestral language that are common to both languages now. Find the following.

(a) $N(0.9)$

(b) $N(0.5)$

(c) $N(0.3)$

(d) How many years have elapsed since the split if 70% of the words of the ancestral language are common to both languages today?

(e) If two languages split off from a common ancestral language about 1000 years ago, find r.

Physical Sciences

91. **Communications Channel** According to the Shannon-Hartley theorem, the capacity of a communications channel in bits per second is given by

$$C = B \log_2\left(\frac{s}{n} + 1\right),$$

where B is the frequency bandwidth of the channel in hertz and s/n is its signal-to-noise ratio. It is physically impossible to exceed this limit. Solve the equation for the signal-to-noise ratio s/n. *Source: Scientific American.*

For Exercises 92–95, recall that log x represents the common (base 10) logarithm of x.

92. **Intensity of Sound** The loudness of sounds is measured in a unit called a *decibel*. To do this, a very faint sound, called the *threshold sound*, is assigned an intensity I_0. If a particular sound has intensity I, then the decibel rating of this louder sound is

$$10 \log \frac{I}{I_0}.$$

Find the decibel ratings of the following sounds having intensities as given. Round answers to the nearest whole number.

(a) Whisper, $115I_0$

(b) Busy street, $9,500,000I_0$

(c) Heavy truck, 20 m away, $1,200,000,000I_0$

(d) Rock music concert, $895,000,000,000I_0$

(e) Jetliner at takeoff, $109,000,000,000,000I_0$

(f) In a noise ordinance instituted in Stamford, Connecticut, the threshold sound I_0 was defined as 0.0002 microbars. Use this definition to express the sound levels in parts (c) and (d) in microbars. *Source: The New York Times.*

93. **Intensity of Sound** A story on the National Public Radio program *All Things Considered* discussed a proposal to lower the noise limit in Austin, Texas, from 85 decibels to 75 decibels. A manager for a restaurant was quoted as saying, "If you cut from 85 to 75, . . . you're basically cutting the sound down in half." Is this correct? If not, to what fraction of its original level is the sound being cut? *Source: National Public Radio.*

94. **Earthquake Intensity** The magnitude of an earthquake, measured on the Richter scale, is given by

$$R(I) = \log \frac{I}{I_0},$$

where I is the amplitude registered on a seismograph located 100 km from the epicenter of the earthquake, and I_0 is the amplitude of a certain small size earthquake. Find the Richter scale ratings of earthquakes with the following amplitudes.

(a) $1,000,000I_0$

(b) $100,000,000I_0$

(c) On June 15, 1999, the city of Puebla in central Mexico was shaken by an earthquake that measured 6.7 on the Richter scale. Express this reading in terms of I_0. *Source: Exploring Colonial Mexico.*

(d) On September 19, 1985, Mexico's largest recent earthquake, measuring 8.1 on the Richter scale, killed about 10,000 people. Express the magnitude of an 8.1 reading in terms of I_0. *Source: History.com.*

(e) Compare your answers to parts (c) and (d). How much greater was the force of the 1985 earthquake than the 1999 earthquake?

(f) The relationship between the energy E of an earthquake and the magnitude on the Richter scale is given by

$$R(E) = \frac{2}{3} \log\left(\frac{E}{E_0}\right),$$

where E_0 is the energy of a certain small earthquake. Compare the energies of the 1999 and 1985 earthquakes.

(g) According to a newspaper article, "Scientists say such an earthquake of magnitude 7.5 could release 15 times as much energy as the magnitude 6.7 trembler that struck the Northridge section of Los Angeles" in 1994. Using the formula from part (f), verify this quote by computing the magnitude of an earthquake with 15 times the energy of a magnitude 6.7 earthquake. *Source: The New York Times.*

95. **Acidity of a Solution** A common measure for the acidity of a solution is its pH. It is defined by pH $= -\log[H^+]$, where H^+ measures the concentration of hydrogen ions in the solution. The pH of pure water is 7. Solutions that are more acidic than pure water have a lower pH, while solutions that are less acidic (referred to as basic solutions) have a higher pH.

(a) Acid rain sometimes has a pH as low as 4. How much greater is the concentration of hydrogen ions in such rain than in pure water?

(b) A typical mixture of laundry soap and water for washing clothes has a pH of about 11, while black coffee has a pH of about 5. How much greater is the concentration of hydrogen ions in black coffee than in the laundry mixture?

96. **Music Theory** A music theorist associates the fundamental frequency of a pitch f with a real number defined by

$$p = 69 + 12 \log_2(f/440).$$

Source: Science.

(a) Standard concert pitch for an A is 440 cycles per second. Find the associated value of p.

(b) An A one octave higher than standard concert pitch is 880 cycles per second. Find the associated value of p.

YOUR TURN ANSWERS

1. $\log_5(1/25) = -2$ **2.** -4 **3.** $2\log_a x - 3\log_a y$

4. 3.561 **5.** 2 **6.** $(\ln 2)/\ln(3/2) \approx 1.7095$.

7. 1.0253^x

2.6 Applications: Growth and Decay; Mathematics of Finance

APPLY IT   What interest rate will cause $5000 to grow to $7250 in 6 years if money is compounded continuously?

This question, which will be answered in Example 7, is one of many situations that occur in biology, economics, and the social sciences, in which a quantity changes at a rate proportional to the amount of the quantity present.

In cases such as continuous compounding described above, the amount present at time t is a function of t, called the **exponential growth and decay function**. (The derivation of this equation is presented in a later section on Differential Equations.)

Exponential Growth and Decay Function

Let y_0 be the amount or number of some quantity present at time $t = 0$. The quantity is said to grow or decay exponentially if for some constant k, the amount present at time t is given by

$$y = y_0 e^{kt}.$$

If $k > 0$, then k is called the **growth constant**; if $k < 0$, then k is called the **decay constant**. A common example is the growth of bacteria in a culture. The more bacteria present, the faster the population increases.

EXAMPLE 1   Yeast Production

Yeast in a sugar solution is growing at a rate such that 1 g becomes 1.5 g after 20 hours. Find the growth function, assuming exponential growth.

SOLUTION   The values of y_0 and k in the exponential growth function $y = y_0 e^{kt}$ must be found. Since y_0 is the amount present at time $t = 0$, $y_0 = 1$. To find k, substitute $y = 1.5, t = 20$, and $y_0 = 1$ into the equation.

$$y = y_0 e^{kt}$$
$$1.5 = 1e^{k(20)}$$

Now take natural logarithms on both sides and use the power rule for logarithms.

$$1.5 = e^{20k}$$
$$\ln 1.5 = \ln e^{20k} \quad \text{Take ln of both sides.}$$
$$\ln 1.5 = 20k \quad \ln e^x = x$$
$$\frac{\ln 1.5}{20} = k \quad \text{Divide both sides by 20.}$$
$$k \approx 0.02 \text{ (to the nearest hundredth)}$$

YOUR TURN 1 Find the growth function if 5 g grows exponentially to 18 g after 16 hours.

The exponential growth function is $y = e^{0.02t}$, where y is the number of grams of yeast present after t hours.   TRY YOUR TURN 1

The decline of a population or decay of a substance may also be described by the exponential growth function. In this case the decay constant k is negative, since an increase in time leads to a decrease in the quantity present. Radioactive substances provide a good example of exponential decay. By definition, the **half-life** of a radioactive substance is the time it takes for exactly half of the initial quantity to decay.

EXAMPLE 2   Carbon Dating

Carbon-14 is a radioactive form of carbon that is found in all living plants and animals. After a plant or animal dies, the carbon-14 disintegrates. Scientists determine the age of the remains by comparing its carbon-14 with the amount found in living plants and animals. The amount of carbon-14 present after t years is given by the exponential equation

$$A(t) = A_0 e^{kt},$$

with $k = -[(\ln 2)/5600]$.

(a) Find the half-life of carbon-14.

SOLUTION   Let $A(t) = (1/2)A_0$ and $k = -[(\ln 2)/5600]$.

$$\frac{1}{2}A_0 = A_0 e^{-[(\ln 2)/5600]t}$$

$$\frac{1}{2} = e^{-[(\ln 2)/5600]t} \qquad \text{Divide by } A_0.$$

$$\ln \frac{1}{2} = \ln e^{-[(\ln 2)/5600]t} \qquad \text{Take ln of both sides.}$$

$$\ln \frac{1}{2} = -\frac{\ln 2}{5600}t \qquad \ln e^x = x$$

$$-\frac{5600}{\ln 2}\ln \frac{1}{2} = t \qquad \text{Multiply by } -\frac{5600}{\ln 2}.$$

$$-\frac{5600}{\ln 2}(\ln 1 - \ln 2) = t \qquad \ln \frac{x}{y} = \ln x - \ln y$$

$$-\frac{5600}{\ln 2}(-\ln 2) = t \qquad \ln 1 = 0$$

$$5600 = t$$

The half-life is 5600 years.

(b) Charcoal from an ancient fire pit on Java had 1/4 the amount of carbon-14 found in a living sample of wood of the same size. Estimate the age of the charcoal.

SOLUTION   Let $A(t) = (1/4)A_0$ and $k = -[(\ln 2)/5600]$.

$$\frac{1}{4}A_0 = A_0 e^{-[(\ln 2)/5600]t}$$

$$\frac{1}{4} = e^{-[(\ln 2)/5600]t} \qquad \text{Divide by } A_0.$$

$$\ln \frac{1}{4} = \ln e^{-[(\ln 2)/560]t} \qquad \text{Take ln of both sides.}$$

$$\ln \frac{1}{4} = -\frac{\ln 2}{5600}t \qquad \ln e^x = x$$

$$-\frac{5600}{\ln 2}\ln \frac{1}{4} = t \qquad \text{Multiply by } \frac{-5600}{\ln 2}.$$

$$t = 11,200$$

YOUR TURN 2 Estimate the age of a sample with 1/10 the amount of carbon-14 as a live sample.

The charcoal is about 11,200 years old.   TRY YOUR TURN 2 ▮

By following the steps in Example 2, we get the general equation giving the half-life T in terms of the decay constant k as

$$T = -\frac{\ln 2}{k}.$$

For example, the decay constant for potassium-40, where t is in billions of years, is approximately -0.5545 so its half-life is

$$T = -\frac{\ln 2}{(-0.5545)}$$
$$\approx 1.25 \text{ billion years.}$$

We can rewrite the growth and decay function as

$$y = y_0 e^{kt} = y_0 (e^k)^t = y_0 a^t,$$

where $a = e^k$. This is sometimes a helpful way to look at an exponential growth or decay function.

EXAMPLE 3   Radioactive Decay

Rewrite the function for radioactive decay of carbon-14 in the form $A(t) = A_0 a^{f(t)}$.

SOLUTION   From the previous example, we have

$$
\begin{aligned}
A(t) = A_0 e^{kt} &= A_0 e^{-[(\ln 2)/5600]t} \\
&= A_0 (e^{\ln 2})^{-t/5600} \qquad a^{mn} = (a^m)^n \\
&= A_0 2^{-t/5600} \qquad\quad e^{\ln a} = a \\
&= A_0 (2^{-1})^{t/5600} \\
&= A_0 \left(\frac{1}{2}\right)^{t/5600}. \qquad 2^{-1} = \frac{1}{2}
\end{aligned}
$$

This last expression shows clearly that every time t increases by 5600 years, the amount of carbon-14 decreases by a factor of $1/2$.

Effective Rate
We could use a calculator to see that \$1 at 8% interest (per year) compounded semiannually is $1(1.04)^2 = 1.0816$ or \$1.0816. The actual increase of \$0.0816 is 8.16% rather than the 8% that would be earned with interest compounded annually. To distinguish between these two amounts, 8% (the annual interest rate) is called the **nominal** or **stated** interest rate, and 8.16% is called the **effective** interest rate. We will continue to use r to designate the stated rate and we will use r_E for the effective rate.

Effective Rate for Compound Interest
If r is the annual stated rate of interest and m is the number of compounding periods per year, the effective rate of interest is

$$r_E = \left(1 + \frac{r}{m}\right)^m - 1.$$

Effective rate is sometimes called *annual yield*.

With continuous compounding, \$1 at 8% for 1 year becomes $(1)e^{1(0.08)} = e^{0.08} \approx 1.0833$. The increase is 8.33% rather than 8%, so a stated interest rate of 8% produces an effective rate of 8.33%.

Effective Rate for Continuous Compounding

If interest is compounded continuously at an annual stated rate of r, the effective rate of interest is

$$r_E = e^r - 1.$$

EXAMPLE 4   Effective Rate

Find the effective rate corresponding to each stated rate.

(a) 6% compounded quarterly

SOLUTION   Using the formula with $r = 0.06$ and $m = 4$, we get

$$r_E = \left(1 + \frac{0.06}{4}\right)^4 - 1 = (1.015)^4 - 1 \approx 0.0614.$$

The effective rate is 6.14%.

(b) 6% compounded continuously

SOLUTION   The formula for continuous compounding gives

$$r_E = e^{0.06} - 1 \approx 0.0618,$$

so the effective rate is 6.18%.   **TRY YOUR TURN 3**

YOUR TURN 3 Find the effective rate corresponding to each stated rate. **(a)** 4.25% compounded monthly **(b)** 3.75% compounded continuously

The formula for the compound amount when interest is compounded m times a year, $A = P(1 + r/m)^{tm}$, has five variables: $A, P, r, m,$ and t. If the values of any four are known, then the value of the fifth can be found.

EXAMPLE 5   Time of Investment

Susan Freilich has received a bonus of $25,000. She invests it in an account earning 7.2% compounded quarterly. Find how long it will take for her $25,000 investment to grow to $40,000.

SOLUTION   Here $P = \$25{,}000, r = 0.072,$ and $m = 4$. We also know the amount she wishes to end up with, $A = \$40{,}000$. Substitute these values into the compound amount formula and solve for time, t.

$$A = P\left(1 + \frac{r}{m}\right)^{tm}$$

$$40{,}000 = 25{,}000\left(1 + \frac{0.072}{4}\right)^{4t}$$

$$40{,}000 = 25{,}000(1.018)^{4t}$$

$$1.6 = 1.018^{4t} \qquad \text{Divide both sides by 25,000.}$$

$$\ln 1.6 = \ln(1.018)^{4t} \qquad \text{Take ln of both sides.}$$

$$\ln 1.6 = 4t \cdot \ln 1.018 \qquad \ln u^r = r \ln u$$

$$t = \frac{\ln 1.6}{4 \ln 1.018} \approx 6.586 \qquad \text{Divide both sides by 4 ln 1.018.}$$

YOUR TURN 4 Find the time needed for $30,000 to grow to $50,000 when invested in an account that pays 3.15% compounded quarterly.

Note that the interest is calculated quarterly and is added only at the *end* of each quarter. Therefore, we need to round up to the nearest quarter. She will have $40,000 in 6.75 years.   **TRY YOUR TURN 4**

| **CAUTION** | When calculating the time it takes for an investment to grow, take into account that interest is added only at the *end* of each compounding period. In Example 5, interest is added quarterly. At the end of the second quarter of the sixth year $(t = 6.5)$, she will have only $39,754.13, but at the end of the third quarter of that year $(t = 6.75)$, she will have $40,469.70. |

If A, the amount of money we wish to end up with, is given as well as r, m, and t, then P can be found using the formula for compounded interest. Here P is the amount that should be deposited today to produce A dollars in t years. The amount P is called the **present value** of A dollars.

NOTE
When computing the present value, you may have to round the principal up to the nearest cent to ensure that enough money is accumulated in the given amount of time. Always check your solution.

EXAMPLE 6 Present Value

Tom Shaffer has a balloon payment of $100,000 due in 5 years. What is the present value of that amount if the money earns interest at 4.5% annually?

SOLUTION Here P in the compound amount formula is unknown, with $A = 100,000$, $r = 0.045$, $t = 5$, and $m = 1$. Substitute the known values into the formula to get $100,000 = P(1.045)^5$. Solving for P yields.

$$P = \frac{100,000}{(1.045)^5} \approx 80,245.10465.$$

The present value of $100,000 in 5 years at 4.5% per year is rounded up to $80,245.11.

In general, to find the present value for an interest rate r compounded m times per year for t years, solve the equation

$$A = P\left(1 + \frac{r}{m}\right)^{tm}$$

for the variable P. To find the present value for an interest rate r compounded continuously for t years, solve the equation

$$A = Pe^{rt}$$

for the variable P.

EXAMPLE 7 Continuous Compound Interest Rate

Find the interest rate that will cause $5000 to grow to $7300 in 6 years if the money is compounded continuously.

APPLY IT

SOLUTION Use the formula for continuous compounding, $A = Pe^{rt}$, with $A = 7300$, $P = 5000$, and $t = 6$. Solve first for e^{rt}, then for r.

$$A = Pe^{rt}$$
$$7300 = 5000e^{6r}$$
$$1.46 = e^{6r} \qquad \text{Divide by 5000.}$$
$$\ln 1.46 = \ln e^{6r} \qquad \text{Take ln of both sides.}$$
$$\ln 1.46 = 6r \qquad \ln e^x = x$$
$$r = \frac{\ln 1.46}{6}$$
$$r \approx 0.0631$$

YOUR TURN 5 Find the interest rate that will cause $3200 to grow to $4500 in 7 years if the money is compounded continuously.

The required interest rate is 6.31%. **TRY YOUR TURN 5**

Limited Growth Functions

The exponential growth functions discussed so far all continued to grow without bound. More realistically, many populations grow exponentially for a while, but then the growth is slowed by some external constraint that eventually limits the growth. For example, an animal population may grow to the point where its habitat can no longer support the population and the growth rate begins to dwindle until a stable population size is reached. Models that reflect this pattern are called **limited growth functions**. The next example discusses a function of this type that occurs in industry.

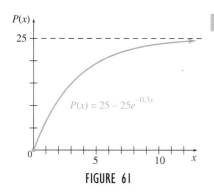

FIGURE 61

EXAMPLE 8 Employee Turnover

Assembly-line operations tend to have a high turnover of employees, forcing companies to spend much time and effort in training new workers. It has been found that a worker new to a task on the line will produce items according to the function defined by

$$P(x) = 25 - 25e^{-0.3x},$$

where $P(x)$ items are produced by the worker on day x.

(a) What happens to the number of items a worker can produce as x gets larger and larger?

SOLUTION As x gets larger, $e^{-0.3x}$ becomes closer to 0, so $P(x)$ approaches 25. This represents the limit on the number of items a worker can produce in a day. Note that this limit represents a horizontal asymptote on the graph of P, shown in Figure 61.

(b) How many days will it take for a new worker to produce at least 20 items in a day?

SOLUTION Let $P(x) = 20$ and solve for x.

$$P(x) = 25 - 25e^{-0.3x}$$
$$20 = 25 - 25e^{-0.3x}$$
$$-5 = -25e^{-0.3x} \qquad \text{Subtract 25 from both sides.}$$
$$0.2 = e^{-0.3x} \qquad \text{Divide by } -25.$$

Now take natural logarithms of both sides and use properties of logarithms.

$$\ln 0.2 = \ln e^{-0.3x}$$
$$\ln 0.2 = -0.3x \qquad \ln e^u = u$$
$$x = \frac{\ln 0.2}{-0.3} \approx 5.4$$

This means that 5 days are not quite enough; on the fifth day, a new worker produces $P(5) = 25 - 25e^{-0.3(5)} \approx 19.4$ items. It takes 6 days, and on the sixth day, a new worker produces $P(6) = 25 - 25e^{-0.3(5)} \approx 20.9$ items. ∎

Graphs such as the one in Figure 61 are called **learning curves**. According to such a graph, a new worker tends to learn quickly at first; then learning tapers off and approaches some upper limit. This is characteristic of the learning of certain types of skills involving the repetitive performance of the same task.

2.6 WARM-UP EXERCISES

Solve.

W1. $e^{2t} = 5$ *(Sec. 2.5)*

W2. $e^{k+2} = 3$ *(Sec. 2.5)*

W3. $10e^{3t} = 45$ *(Sec. 2.5)*

2.6 EXERCISES

 1. What is the difference between stated interest rate and effective interest rate?

2. In the exponential growth or decay function $y = y_0e^{kt}$, what does y_0 represent? What does k represent?

 3. In the exponential growth or decay function, explain the circumstances that cause k to be positive or negative.

 4. What is meant by the half-life of a quantity?

5. Show that if a radioactive substance has a half-life of T, then the corresponding constant k in the exponential decay function is given by $k = -(\ln 2)/T$.

6. Show that if a radioactive substance has a half-life of T, then the corresponding exponential decay function can be written as $y = y_0(1/2)^{(t/T)}$.

APPLICATIONS

Business and Economics

Effective Rate   Find the effective rate corresponding to each nominal rate of interest.

7. 4% compounded quarterly

8. 6% compounded monthly

9. 8% compounded continuously

10. 5% compounded continuously

Present Value   Find the present value of each amount.

11. $10,000 if interest is 6% compounded quarterly for 8 years

12. $45,678.93 if interest is 7.2% compounded monthly for 11 months

13. $7300 if interest is 5% compounded continuously for 3 years

14. $25,000 if interest is 4.6% compounded continuously for 8 years

15. Effective Rate   Vicki Kakounis bought a television set with money borrowed from the bank at 9% interest compounded semiannually. What effective interest rate did she pay?

16. Effective Rate   A firm deposits some funds in a special account at 6.2% compounded quarterly. What effective rate will they earn?

17. Effective Rate   Melissa Blum deposits $7500 of lottery winnings in an account paying 6% interest compounded monthly. What effective rate does the account earn?

18. Present Value   Frank Cronin must make a balloon payment of $20,000 in 4 years. Find the present value of the payment if it includes annual interest of 6.5% compounded monthly.

19. Present Value   A company must pay a $307,000 settlement in 3 years.

(a) What amount must be deposited now at 6% compounded semiannually to have enough money for the settlement?

(b) How much interest will be earned?

(c) Suppose the company can deposit only $200,000 now. How much more will be needed in 3 years?

(d) Suppose the company can deposit $200,000 now in an account that pays interest continuously. What interest rate would they need to accumulate the entire $307,000 in 3 years?

20. Present Value   A couple wants to have $40,000 in 5 years for a down payment on a new house.

(a) How much should they deposit today, at 6.4% compounded quarterly, to have the required amount in 5 years?

(b) How much interest will be earned?

(c) If they can deposit only $20,000 now, how much more will they need to complete the $40,000 after 5 years?

(d) Suppose they can deposit $20,000 now in an account that pays interest continuously. What interest rate would they need to accumulate the entire $40,000 in 5 years?

21. Interest   Christine O'Brien, who is self-employed, wants to invest $60,000 in a pension plan. One investment offers 8% compounded quarterly. Another offers 7.75% compounded continuously.

(a) Which investment will earn the most interest in 5 years?

(b) How much more will the better plan earn?

(c) What is the effective rate in each case?

(d) If Ms. O'Brien chooses the plan with continuous compounding, how long will it take for her $60,000 to grow to $80,000?

(e) How long will it take for her $60,000 to grow to at least $80,000 if she chooses the plan with quarterly compounding? (Be careful; interest is added to the account only every quarter. See Example 5.)

22. Interest   Greg Tobin wishes to invest a $5000 bonus check into a savings account that pays 6.3% interest. Find how many years it will take for the $5000 to grow to at least $11,000 if interest is compounded

(a) quarterly. (Be careful; interest is added to the account only every quarter. See Example 5.)

(b) continuously.

23. Sales   Sales of a new model of compact disc player are approximated by the function $S(x) = 1000 - 800e^{-x}$, where $S(x)$ is in appropriate units and x represents the number of years the disc player has been on the market.

(a) Find the sales during year 0.

(b) In how many years will sales reach 500 units?

(c) Will sales ever reach 1000 units?

(d) Is there a limit on sales for this product? If so, what is it?

24. Sales   Sales of a new model of digital camera are approximated by

$$S(x) = 5000 - 4000e^{-x},$$

where x represents the number of years that the digital camera has been on the market, and $S(x)$ represents sales in thousands of dollars.

(a) Find the sales in year 0.

(b) When will sales reach $4,500,000?

(c) Find the limit on sales.

Life Sciences

25. Population Growth   The population of the world in the year 1650 was about 500 million, and in the year 2010 was 6756 million. *Source: U.S. Census Bureau.*

(a) Assuming that the population of the world grows exponentially, find the equation for the population $P(t)$ in millions in the year t.

(b) Use your answer from part (a) to find the population of the world in the year 1.

(c) Is your answer to part (b) reasonable? What does this tell you about how the population of the world grows?

26. Giardia When a person swallows giardia cysts, stomach acids and pancreatic enzymes cause the cysts to release trophozoites, which divide every 12 hours. *Source: The New York Times*.

(a) Suppose the number of trophozoites at time $t = 0$ is y_0. Write a function in the form $y = y_0e^{kt}$ giving the number after t hours.

(b) Write the function from part (a) in the form $y = y_0 2^{f(t)}$.

(c) The article cited above said that a single trophozoite can multiply to a million in just 10 days and a billion in 15 days. Verify this fact.

27. Growth of Bacteria A culture contains 25,000 bacteria, with the population increasing exponentially. The culture contains 40,000 bacteria after 10 hours.

(a) Write a function in the form $y = y_0e^{kt}$ giving the number of bacteria after t hours.

(b) Write the function from part (a) in the form $y = y_0a^t$.

(c) How long will it be until there are 60,000 bacteria?

28. Decrease in Bacteria When an antibiotic is introduced into a culture of 50,000 bacteria, the number of bacteria decreases exponentially. After 9 hours, there are only 20,000 bacteria.

(a) Write an exponential equation to express the growth function y in terms of time t in hours.

(b) In how many hours will half the number of bacteria remain?

29. Growth of Bacteria The growth of bacteria in food products makes it necessary to time-date some products (such as milk) so that they will be sold and consumed before the bacteria count is too high. Suppose for a certain product that the number of bacteria present is given by

$$f(t) = 500e^{0.1t},$$

under certain storage conditions, where t is time in days after packing of the product and the value of $f(t)$ is in millions.

(a) If the product cannot be safely eaten after the bacteria count reaches 3000 million, how long will this take?

(b) If $t = 0$ corresponds to January 1, what date should be placed on the product?

30. Cancer Research An article on cancer treatment contains the following statement: A 37% 5-year survival rate for women with ovarian cancer yields an estimated annual mortality rate of 0.1989. The authors of this article assume that the number of survivors is described by the exponential decay function given at the beginning of this section, where y is the number of survivors and k is the mortality rate. Verify that the given survival rate leads to the given mortality rate. *Source: American Journal of Obstetrics and Gynecology*.

31. Chromosomal Abnormality The graph in the next column shows how the risk of chromosomal abnormality in a child rises with the age of the mother. *Source: downsyndrome.about.com*.

(a) Read from the graph the risk of chromosomal abnormality (per 1000) at ages 20, 35, 42, and 49.

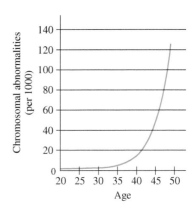

(b) Assuming the graph to be of the form $y = Ce^{kt}$, find k using $t = 20$ and $t = 35$.

(c) Still assuming the graph to be of the form $y = Ce^{kt}$, find k using $t = 42$ and $t = 49$.

(d) Based on your results from parts (a)–(c), is it reasonable to assume the graph is of the form $y = Ce^{kt}$? Explain.

(e) In situations such as parts (a)–(c), where an exponential function does not fit because different data points give different values for the growth constant k, it is often appropriate to describe the data using an equation of the form $y = Ce^{kt^n}$. Parts (b) and (c) show that $n = 1$ results in a smaller constant using the interval $[20, 35]$ than using the interval $[42, 49]$. Repeat parts (b) and (c) using $n = 2, 3$, etc., until the interval $[20, 35]$ yields a larger value of k than the interval $[42, 49]$, and then estimate what n should be.

Physical Sciences

32. Carbon Dating Refer to Example 2. A sample from a refuse deposit near the Strait of Magellan had 60% of the carbon-14 found in a contemporary living sample. How old was the sample?

Half-Life **Find the half-life of each radioactive substance. See Example 2.**

33. Plutonium-241; $A(t) = A_0e^{-0.053t}$

34. Radium-226; $A(t) = A_0e^{-0.00043t}$

35. Half-Life The half-life of plutonium-241 is approximately 13 years.

(a) How much of a sample weighing 4 g will remain after 100 years?

(b) How much time is necessary for a sample weighing 4 g to decay to 0.1 g?

36. Half-Life The half-life of radium-226 is approximately 1620 years.

(a) How much of a sample weighing 4 g will remain after 100 years?

(b) How much time is necessary for a sample weighing 4 g to decay to 0.1 g?

37. Radioactive Decay 500 g of iodine-131 is decaying exponentially. After 3 days 386 g of iodine-131 is left.

(a) Write a function in the form $y = y_0e^{kt}$ giving the number of grams of iodine-131 after t days.

(b) Write the function from part (a) in the form $y = y_0(386/500)^{f(t)}$.

(c) Use your answer from part (a) to find the half-life of iodine-131.

38. Radioactive Decay 25 g of polonium-210 is decaying exponentially. After 50 days 19.5 g of polonium-210 is left.

(a) Write a function in the form $y = y_0e^{kt}$ giving the number of grams of polonium-210 after t days.

(b) Write the function from part (a) in the form $y = y_0a^{t/50}$.

(c) Use your answer from part (a) to find the half-life of polonium-210.

39. Nuclear Energy Nuclear energy derived from radioactive isotopes can be used to supply power to space vehicles. The output of the radioactive power supply for a certain satellite is given by the function $y = 40e^{-0.004t}$, where y is in watts and t is the time in days.

(a) How much power will be available at the end of 180 days?

(b) How long will it take for the amount of power to be half of its original strength?

(c) Will the power ever be completely gone? Explain.

40. Botany A group of Tasmanian botanists have claimed that a King's holly shrub, the only one of its species in the world, is also the oldest living plant. Using carbon-14 dating of charcoal found along with fossilized leaf fragments, they arrived at an age of 43,000 years for the plant, whose exact location in southwest Tasmania is being kept a secret. What percent of the original carbon-14 in the charcoal was present? *Source: Science.*

41. Decay of Radioactivity A large cloud of radioactive debris from a nuclear explosion has floated over the Pacific Northwest, contaminating much of the hay supply. Consequently, farmers in the area are concerned that the cows who eat this hay will give contaminated milk. (The tolerance level for radioactive iodine in milk is 0.) The percent of the initial amount of radioactive iodine still present in the hay after t days is approximated by $P(t)$, which is given by the mathematical model

$$P(t) = 100e^{-0.1t}.$$

(a) Find the percent remaining after 4 days.

(b) Find the percent remaining after 10 days.

(c) Some scientists feel that the hay is safe after the percent of radioactive iodine has declined to 10% of the original amount. Solve the equation $10 = 100e^{-0.1t}$ to find the number of days before the hay may be used.

(d) Other scientists believe that the hay is not safe until the level of radioactive iodine has declined to only 1% of the original level. Find the number of days that this would take.

42. Chemical Dissolution The amount of chemical that will dissolve in a solution increases exponentially as the temperature is increased. At 0°C, 10 g of the chemical dissolves, and at 10°C, 11 g dissolves.

(a) Write an equation to express the amount of chemical dissolved, y, in terms of temperature, t, in degrees Celsius.

(b) At what temperature will 15 g dissolve?

Newton's Law of Cooling Newton's law of cooling says that the rate at which a body cools is proportional to the difference in temperature between the body and an environment into which it is introduced. This leads to an equation where the temperature $f(t)$ of the body at time t after being introduced into an environment having constant temperature T_0 is

$$f(t) = T_0 + Ce^{-kt},$$

where C and k are constants. Use this result in Exercises 43–45.

43. Find the temperature of an object when $t = 9$ if $T_0 = 18$, $C = 5$, and $k = 0.6$.

44. If $C = 100$, $k = 0.1$, and t is time in minutes, how long will it take a hot cup of coffee to cool to a temperature of 25°C in a room at 20°C?

45. If $C = -14.6$ and $k = 0.6$ and t is time in hours, how long will it take a frozen pizza to thaw to 10°C in a room at 18°C?

YOUR TURN ANSWERS
1. $y = 5e^{0.08t}$
2. 18,600 years old
3. **(a)** 4.33% **(b)** 3.82%
4. 16.5 years
5. 4.87%

CHAPTER REVIEW

SUMMARY

In this chapter we defined functions and studied some of their properties. In particular, we studied several families of functions including quadratic, polynomial, rational, exponential, and logarithmic functions. By knowing the properties of a family of functions, we can immediately apply that knowledge to any member of the family we encounter, giving us valuable information about the domain and the behavior of the function. Furthermore, this knowledge can help us to choose an appropriate function for an application. Exponential functions have so many important applications that we highlighted some of them in the last section of the chapter. In the next chapters, we see how calculus gives us even more information about the behavior of functions.

Function	A function is a rule that assigns to each element from one set exactly one element from another set.
Domain and Range	The set of all possible values of the independent variable in a function is called the domain of the function, and the resulting set of possible values of the dependent variable is called the range.
Vertical Line Test	A graph represents a function if and only if every vertical line intersects the graph in no more than one point.
Quadratic Function	A quadratic function is defined by $$f(x) = ax^2 + bx + c,$$ where a, b, and c are real numbers, with $a \neq 0$.
Graph of a Quadratic Function	The graph of the quadratic function $f(x) = ax^2 + bx + c$ is a parabola with the following characteristics: 1. The parabola opens upward if $a > 0$ and downward if $a < 0$; 2. The y-intercept is at $(0, c)$; 3. The x-intercepts, if they exist, occur where $f(x) = 0$; 4. The vertex is at $(-b/(2a), f(-b/(2a)))$; 5. The axis of symmetry is $x = -b/(2a)$.
Polynomial Function	A polynomial function of degree n, where n is a nonnegative integer, is defined by $$f(x) = a_n x^n + a_{n-1} x^{n-1} + \cdots + a_1 x + a_0,$$ where $a_n, a_{n-1}, \ldots, a_1$ and a_0 are real numbers, called coefficients, with $a_n \neq 0$. The number a_n is called the leading coefficient.
Properties of Polynomial Functions	1. A polynomial function of degree n can have at most $n - 1$ turning points. Conversely, if the graph of a polynomial function has n turning points, it must have degree at least $n + 1$. 2. In the graph of a polynomial function of even degree, both ends go up or both ends go down. For a polynomial function of odd degree, one end goes up and one end goes down. 3. If the graph goes up as x becomes large, the leading coefficient must be positive. If the graph goes down as x becomes large, the leading coefficient is negative.
Rational Function	A rational function is defined by $$f(x) = \frac{p(x)}{q(x)},$$ where $p(x)$ and $q(x)$ are polynomial functions and $q(x) \neq 0$.
Asymptotes	If a function gets larger and larger in magnitude without bound as x approaches the number k, then the line $x = k$ is a vertical asymptote. If the values of y approach a number k as $|x|$ gets larger and larger, the line $y = k$ is a horizontal asymptote.
Exponential Function	An exponential function with base a is defined as $$f(x) = a^x, \text{ where } a > 0 \text{ and } a \neq 1.$$
Simple Interest	If P dollars is invested at a yearly simple interest rate r per year for time t (in years), the interest I is given by $$I = Prt.$$

Math of Finance Formulas   If P is the principal or present value, r is the annual interest rate, t is time in years, and m is the number of compounding periods per year:

	Compounded m Times per Year	Compounded Continuously
Compound amount	$A = P\left(1 + \dfrac{r}{m}\right)^{tm}$	$A = Pe^{rt}$
Effective rate	$r_E = \left(1 + \dfrac{r}{m}\right)^{m} - 1$	$r_E = e^{r} - 1$

Definition of e   As m becomes larger and larger, $\left(1 + \dfrac{1}{m}\right)^{m}$ becomes closer and closer to the number e, whose approximate value is 2.718281828.

Logarithm   For $a > 0$, $a \neq 1$, and $x > 0$,

$$y = \log_a x \text{ means } a^y = x.$$

Logarithmic Function   If $a > 0$ and $a \neq 1$, then the logarithmic function of base a is defined by

$$f(x) = \log_a x,$$

for $x > 0$.

Properties of Logarithms   Let x and y be any positive real numbers and r be any real number. Let a be a positive real number, $a \neq 1$. Then

(a) $\log_a xy = \log_a x + \log_a y$

(b) $\log_a \dfrac{x}{y} = \log_a x - \log_a y$

(c) $\log_a x^r = r \log_a x$

(d) $\log_a a = 1$

(e) $\log_a 1 = 0$

(f) $\log_a a^r = r$.

Change-of-Base Theorem for Logarithms   If x is any positive number and if a and b are positive real numbers, $a \neq 1$, $b \neq 1$, then

$$\log_a x = \frac{\log_b x}{\log_b a} = \frac{\ln x}{\ln a}.$$

Change-of-Base Theorem for Exponentials   For every positive real number a

$$a^x = e^{(\ln a)x}.$$

Exponential Growth and Decay Function   Let y_0 be the amount or number of some quantity present at time $t = 0$. The quantity is said to grow or decay exponentially if, for some constant k, the amount present at any time t is given by

$$y = y_0 e^{kt}.$$

Graphs of Basic Functions

Quadratic

Absolute Value

Square Root

Rational

Exponential

Logarithmic

KEY TERMS

To understand the concepts presented in this chapter, you should know the meaning and use of the following terms. For easy reference, the section in the chapter where a word (or expression) was first used is provided.

2.1
function
domain
range
nonlinear function
constant function
vertical line test
even function
odd function
step function

2.2
quadratic function
parabola
vertex
axis of symmetry
vertical translation
vertical reflection
horizontal translation
completing the
 square

quadratic regression
horizontal reflection

2.3
polynomial function
degree
coefficient
leading coefficient
power function
cubic polynomial
quartic polynomial
turning point
real zero (or root)
rational function
vertical asymptote
horizontal asymptote
cost-benefit model
average cost

2.4
exponential function
exponential equation

interest
principal
rate of interest
time
simple interest
compound interest
compound amount
e
continuous compounding

2.5
doubling time
logarithm
logarithmic function
inverse function
properties of
 logarithms
common logarithms
natural logarithms
change-of-base theorem for
 logarithms

logarithmic equation
change-of-base theorem for
 exponentials
rule of 70
rule of 72
index of diversity

2.6
exponential growth and decay
 function
growth constant
decay constant
half-life
nominal (stated) rate
effective rate
present value
limited growth
 function
learning curve

REVIEW EXERCISES

CONCEPT CHECK

Determine whether each of the following statements is true or false, and explain why.

1. A linear function is an example of a polynomial function.

2. A rational function is an example of an exponential function.

3. The function $f(x) = 3x^2 - 75x + 2$ is a quadratic function.

4. The function $f(x) = x^2 - 6x + 4$ has a vertex at $x = 3$.

5. The function $g(x) = x^\pi$ is an exponential function.

6. The function $f(x) = \dfrac{1}{x-6}$ has a vertical asymptote at $y = 6$.

7. Since $3^{-2} = \dfrac{1}{9}$ we can conclude that $\log_3 \dfrac{1}{9} = -2$.

8. The domain of the function $f(x) = \dfrac{1}{x^2 - 4}$ includes all real numbers except $x = 2$.

9. The amount of money after two years if $2000 is invested in an account that is compounded monthly with an annual rate of 4% is $A = 2000\left(1 + \dfrac{4}{12}\right)^{24}$ dollars.

10. $\log_1 1 = 0$

11. $\ln(5 + 7) = \ln 5 + \ln 7$

12. $(\ln 3)^4 = 4 \ln 3$

13. $\log_{10} 0 = 1$

14. $e^{\ln 2} = 2$

15. $e^{\ln(-2)} = -2$

16. $\dfrac{\ln 4}{\ln 8} = \ln 4 - \ln 8$

17. The function $g(x) = e^x$ grows faster than the function $f(x) = \ln x$.

18. The half-life of a radioactive substance is the time required for half of the initial quantity to decay.

PRACTICE AND EXPLORATIONS

19. What is a function? A linear function? A quadratic function? A rational function?

20. How do you find a vertical asymptote? A horizontal asymptote?

21. What can you tell about the graph of a polynomial function of degree n before you plot any points?

22. Describe in words what a logarithm is.

List the ordered pairs obtained from the following if the domain of x for each exercise is $\{-3, -2, -1, 0, 1, 2, 3\}$. Graph each set of ordered pairs. Give the range.

23. $y = (2x - 1)(x + 1)$ **24.** $y = \dfrac{x}{x^2 + 1}$

25. Let $f(x) = 5x^2 - 3$ and $g(x) = -x^2 + 4x + 1$. Find the following.

(a) $f(-2)$ (b) $g(3)$ (c) $f(-k)$

(d) $g(3m)$ (e) $f(x + h)$ (f) $g(x + h)$

(g) $\dfrac{f(x + h) - f(x)}{h}$ (h) $\dfrac{g(x + h) - g(x)}{h}$

26. Let $f(x) = 2x^2 + 5$ and $g(x) = 3x^2 + 4x - 1$. Find the following.

(a) $f(-3)$ (b) $g(2)$ (c) $f(3m)$

(d) $g(-k)$ (e) $f(x + h)$ (f) $g(x + h)$

(g) $\dfrac{f(x + h) - f(x)}{h}$

(h) $\dfrac{g(x + h) - g(x)}{h}$

Find the domain of each function defined as follows.

27. $y = \dfrac{3x - 4}{x}$ **28.** $y = \dfrac{\sqrt{x - 2}}{2x + 3}$

29. $y = \ln(x + 7)$ **30.** $y = \ln(x^2 - 16)$

Graph the following by hand.

31. $y = 2x^2 + 3x - 1$ **32.** $y = -\dfrac{1}{4}x^2 + x + 2$

33. $y = -x^2 + 4x + 2$ **34.** $y = 3x^2 - 9x + 2$

35. $f(x) = x^3 - 3$ **36.** $f(x) = 1 - x^4$

37. $y = -(x - 1)^4 + 4$ **38.** $y = -(x + 2)^3 - 2$

39. $f(x) = \dfrac{8}{x}$ **40.** $f(x) = \dfrac{2}{3x - 6}$

41. $f(x) = \dfrac{4x - 2}{3x + 1}$ **42.** $f(x) = \dfrac{6x}{x + 2}$

43. $y = 4^x$ **44.** $y = 4^{-x} + 3$

45. $y = \left(\dfrac{1}{5}\right)^{2x-3}$ **46.** $y = \left(\dfrac{1}{2}\right)^{x-1}$

47. $y = \log_2(x - 1)$ **48.** $y = 1 + \log_3 x$

49. $y = -\ln(x + 3)$ **50.** $y = 2 - \ln x^2$

Solve each equation.

51. $2^{x+2} = \dfrac{1}{8}$ **52.** $\left(\dfrac{9}{16}\right)^x = \dfrac{3}{4}$

53. $9^{2y+3} = 27^y$ **54.** $\dfrac{1}{2} = \left(\dfrac{b}{4}\right)^{1/4}$

Write each equation using logarithms.

55. $3^5 = 243$ **56.** $5^{1/2} = \sqrt{5}$

57. $e^{0.8} \approx 2.22554$ **58.** $10^{1.07918} \approx 12$

Write each equation using exponents.

59. $\log_2 32 = 5$ **60.** $\log_9 3 = \dfrac{1}{2}$

61. $\ln 82.9 \approx 4.41763$ **62.** $\log 3.21 \approx 0.50651$

Evaluate each expression without using a calculator. Then support your work using a calculator and the change-of-base theorem for logarithms.

63. $\log_3 81$ **64.** $\log_{32} 16$

65. $\log_4 8$ **66.** $\log_{100} 1000$

Simplify each expression using the properties of logarithms.

67. $\log_5 3k + \log_5 7k^3$ **68.** $\log_3 2y^3 - \log_3 8y^2$

69. $4 \log_3 y - 2 \log_3 x$ **70.** $3 \log_4 r^2 - 2 \log_4 r$

Solve each equation. If necessary, round each answer to the nearest thousandth.

71. $6^p = 17$ **72.** $3^{z-2} = 11$

73. $2^{1-m} = 7$ **74.** $12^{-k} = 9$

75. $e^{-5-2x} = 5$ **76.** $e^{3x-1} = 14$

77. $\left(1 + \dfrac{m}{3}\right)^5 = 15$ **78.** $\left(1 + \dfrac{2p}{5}\right)^2 = 3$

79. $\log_k 64 = 6$

80. $\log_3(2x + 5) = 5$

81. $\log(4p + 1) + \log p = \log 3$

82. $\log_2(5m - 2) - \log_2(m + 3) = 2$

83. Give the following properties of the exponential function $f(x) = a^x; a > 0, a \neq 1$.

(a) Domain   (b) Range   (c) y-intercept

(d) Asymptote(s)

(e) Increasing if a is _____

(f) Decreasing if a is _____

84. Give the following properties of the logarithmic function $f(x) = \log_a x; a > 0, a \neq 1$.

(a) Domain   (b) Range   (c) x-intercept

(d) Asymptote(s)   (e) Increasing if a is _____

(f) Decreasing if a is _____

85. Compare your answers for Exercises 83 and 84. What similarities do you notice? What differences?

APPLICATIONS

Business and Economics

86. Car Rental To rent a mid-size car from one agency costs $60 per day or fraction of a day. If you pick up the car in Boston and drop it off in Utica, there is a fixed $40 charge. Let $C(x)$ represent the cost of renting the car for x days and taking it from Boston to Utica. Find the following.

(a) $C\left(\dfrac{3}{4}\right)$ (b) $C\left(\dfrac{9}{10}\right)$ (c) $C(1)$

(d) $C\left(1\dfrac{5}{8}\right)$ (e) $C\left(2\dfrac{1}{9}\right)$

(f) Graph the function defined by $y = C(x)$ for $0 < x \le 5$.

(g) What is the independent variable?

(h) What is the dependent variable?

87. Pollution The cost to remove x percent of a pollutant is

$$y = \frac{7x}{100 - x},$$

in thousands of dollars. Find the cost of removing the following percents of the pollutant.

(a) 80%   (b) 50%   (c) 90%

(d) Graph the function.

(e) Can all of the pollutant be removed?

Interest Find the amount of interest earned by each deposit.

88. $6902 if interest is 6% compounded semiannually for 8 years

89. $2781.36 if interest is 4.8% compounded quarterly for 6 years

Interest Find the compound amount if $12,104 is invested at 6.2% compounded continuously for each period.

90. 2 years   **91.** 4 years

Interest Find the compound amounts for the following deposits if interest is compounded continuously.

92. $1500 at 6% for 9 years

93. $12,000 at 5% for 8 years

94. How long will it take for $1000 deposited at 6% compounded semiannually to double? To triple?

95. How long will it take for $2100 deposited at 4% compounded quarterly to double? To triple?

Effective Rate Find the effective rate to the nearest hundredth for each nominal interest rate.

96. 7% compounded quarterly

97. 6% compounded monthly

98. 5% compounded continuously

Present Value Find the present value of each amount.

99. $2000 if interest is 6% compounded annually for 5 years

100. $10,000 if interest is 8% compounded semiannually for 6 years

101. Interest To help pay for college expenses, Julie Davis borrowed $10,000 at 7% interest compounded semiannually for 8 years. How much will she owe at the end of the 8-year period?

102. Inflation How long will it take for $1 to triple at an annual inflation rate of 8% compounded continuously?

103. Interest Find the interest rate needed for $6000 to grow to $8000 in 3 years with continuous compounding.

104. Present Value Frank Steed wants to open a camera shop. How much must he deposit now at 6% interest compounded monthly to have $25,000 at the end of 3 years?

105. Revenue A concert promoter finds she can sell 1000 tickets at $50 each. She will not sell the tickets for less than $50, but she finds that for every $1 increase in the ticket price above $50, she will sell 10 fewer tickets.

(a) Express n, the number of tickets sold, as a function of p, the price.

(b) Express R, the revenue, as a function of p, the price.

(c) Find the domain of the function found in part (b).

(d) Express R, the revenue, as a function of n, the number sold.

(e) Find the domain of the function found in part (d).

(f) Find the price that produces the maximum revenue.

(g) Find the number of tickets sold that produces the maximum revenue.

(h) Find the maximum revenue.

(i) Sketch the graph of the function found in part (b).

(j) Describe what the graph found in part (i) tells you about how the revenue varies with price.

106. Cost Suppose the cost in dollars to produce x posters is given by

$$C(x) = \frac{5x + 3}{x + 1}.$$

(a) Sketch a graph of $C(x)$.

(b) Find a formula for $C(x + 1) - C(x)$, the cost to produce an additional poster when x posters are already produced.

(c) Find a formula for $A(x)$, the average cost per poster.

(d) Find a formula for $A(x + 1) - A(x)$, the change in the average cost per poster when one additional poster is produced. (This quantity is approximately equal to the marginal average cost, which will be discussed in the chapter on the derivative.)

107. Cost Suppose the cost in dollars to produce x hundreds of nails is given by

$$C(x) = x^2 + 4x + 7.$$

(a) Sketch a graph of $C(x)$.

(b) Find a formula for $C(x + 1) - C(x)$, the cost to produce an additional hundred nails when x hundred are already produced. (This quantity is approximately equal to the marginal cost.)

(c) Find a formula for $A(x)$, the average cost per hundred nails.

(d) Find a formula for $A(x + 1) - A(x)$, the change in the average cost per nail when one additional batch of 100 nails is produced. (This quantity is approximately equal to the marginal average cost, which will be discussed in the chapter on the derivative.)

108. Consumer Price Index The U.S. consumer price index (CPI, or cost of living index) has risen over the years, as shown in the table below, using an index in which the average over the years 1982 to 1984 is set to 100. *Source: Bureau of Labor Statistics.*

Year	CPI
1960	29.6
1970	38.8
1980	82.4
1990	130.7
1995	152.4
2000	172.2
2005	195.3
2010	218.1

(a) Letting t be the years since 1960, write an exponential function in the form $f(t) = f_0 a^t$ that fits the data at 1960 and 2010.

(b) If your calculator has an exponential regression feature, find the best fitting exponential function for the data.

(c) Use a graphing calculator to plot the answers to parts (a) and (b) on the same axes as the data. Are the answers to parts (a) and (b) close to each other?

(d) If your calculator has a quadratic and cubic regression feature, find the best-fitting quadratic and cubic functions for the data.

(e) Use a graphing calculator to plot the answers to parts (b) and (d) on the same window as the data. Discuss the extent to which any one of these functions models the data better than the others.

Life Sciences

109. Fever A certain viral infection causes a fever that typically lasts 6 days. A model of the fever (in °F) on day x, $1 \le x \le 6$, is

$$F(x) = -\frac{2}{3}x^2 + \frac{14}{3}x + 96.$$

According to the model, on what day should the maximum fever occur? What is the maximum fever?

110. Sunscreen An article in a medical journal says that a sunscreen with a sun protection factor (SPF) of 2 provides 50% protection against ultraviolet B (UVB) radiation, an SPF of 4 provides 75% protection, and an SPF of 8 provides 87.5% protection (which the article rounds to 87%). *Source: Family Practice.*

(a) 87.5% protection means that 87.5% of the UVB radiation is screened out. Write as a fraction the amount of radiation that is let in, and then describe how this fraction, in general, relates to the SPF rating.

(b) Plot UVB percent protection (y) against x, where $x = 1/\text{SPF}$.

(c) Based on your graph from part (b), give an equation relating UVB protection to SPF rating.

(d) An SPF of 8 has double the chemical concentration of an SPF 4. Find the increase in the percent protection.

(e) An SPF of 30 has double the chemical concentration of an SPF 15. Find the increase in the percent protection.

(f) Based on your answers from parts (d) and (e), what happens to the increase in the percent protection as the SPF continues to double?

111. HIV in Infants The following table lists the reported number of cases of infants born in the United States with HIV in recent years because their mother was infected.* *Source: Centers for Disease Control and Prevention.*

Year	Cases
1995	295
1997	166
1999	109
2001	115
2003	94
2005	107
2007	79

(a) Plot the data on a graphing calculator, letting $t = 0$ correspond to the year 1995.

(b) Using the regression feature on your calculator, find a quadratic, a cubic, and an exponential function that models these data.

(c) Plot the three functions with the data on the same coordinate axes. Which function or functions best capture the behavior of the data over the years plotted?

(d) Find the number of cases predicted by all three functions for 2015. Which of these are realistic? Explain.

112. Respiratory Rate Researchers have found that the 95th percentile (the value at which 95% of the data are at or below) for respiratory rates (in breaths per minute) during the first 3 years of infancy is given by

$$y = 10^{1.82411 - 0.0125995x + 0.00013401x^2}$$

for awake infants and

$$y = 10^{1.72858 - 0.0139928x + 0.00017646x^2}$$

for sleeping infants, where x is the age in months. *Source: Pediatrics.*

(a) What is the domain for each function?

*These data include only those infants born in the 25 states with confidential name-based HIV infection reporting.

(b) For each respiratory rate, is the rate decreasing or increasing over the first 3 years of life? (*Hint:* Is the graph of the quadratic in the exponent opening upward or downward? Where is the vertex?)

(c) Verify your answer to part (b) using a graphing calculator.

(d) For a 1-year-old infant in the 95th percentile, how much higher is the waking respiratory rate than the sleeping respiratory rate?

113. Polar Bear Mass One formula for estimating the mass (in kg) of a polar bear is given by

$$m(g) = e^{0.02 + 0.062g - 0.000165g^2},$$

where g is the axillary girth in centimeters. It seems reasonable that as girth increases, so does the mass. What is the largest girth for which this formula gives a reasonable answer? What is the predicted mass of a polar bear with this girth? *Source: Wildlife Management.*

114. Population Growth A population of 15,000 small deer in a specific region has grown exponentially to 17,000 in 4 years.

(a) Write an exponential equation to express the population growth y in terms of time t in years.

(b) At this rate, how long will it take for the population to reach 45,000?

115. Population Growth In 1960 in an article in *Science* magazine, H. Van Forester, P. M. Mora, and W. Amiot predicted that world population would be infinite in the year 2026. Their projection was based on the rational function defined by

$$p(t) = \frac{1.79 \times 10^{11}}{(2026.87 - t)^{0.99}},$$

where $p(t)$ gives population in year t. This function has provided a relatively good fit to the population until very recently. *Source: Science.*

(a) Estimate world population in 2010 using this function, and compare it with the estimate of 6.909 billion. *Source: United Nations.*

(b) What does the function predict for world population in 2020? 2025?

(c) Discuss why this function is not realistic, despite its good fit to past data.

116. Intensity of Light The intensity of light (in appropriate units) passing through water decreases exponentially with the depth it penetrates beneath the surface according to the function

$$I(x) = 10e^{-0.3x},$$

where x is the depth in meters. A certain water plant requires light of an intensity of 1 unit. What is the greatest depth of water in which it will grow?

117. Drug Concentration The concentration of a certain drug in the bloodstream at time t (in minutes) is given by

$$c(t) = e^{-t} - e^{-2t}.$$

Use a graphing calculator to find the maximum concentration and the time when it occurs.

118. Glucose Concentration When glucose is infused into a person's bloodstream at a constant rate of c grams per minute, the glucose is converted and removed from the bloodstream at a rate proportional to the amount present. The amount of glucose in grams in the bloodstream at time t (in minutes) is given by

$$g(t) = \frac{c}{a} + \left(g_0 - \frac{c}{a}\right)e^{-at},$$

where a is a positive constant. Assume $g_0 = 0.08, c = 0.1$, and $a = 1.3$.

(a) At what time is the amount of glucose a maximum? What is the maximum amount of glucose in the bloodstream?

(b) When is the amount of glucose in the blood-stream 0.1 g?

(c) What happens to the amount of glucose in the blood-stream after a very long time?

119. Species Biologists have long noticed a relationship between the area of a piece of land and the number of species found there. The following data shows a sample of the British Isles and how many vascular plants are found on each. *Source: Journal of Biogeography.*

Isle	Area (km²)	Species
Ailsa	0.8	75
Fair	5.2	174
Iona	9.1	388
Man	571.6	765
N. Ronaldsay	7.3	131
Skye	1735.3	594
Stronsay	35.2	62
Wight	380.7	1008

(a) One common model for this relationship is logarithmic. Using the logarithmic regression feature on a graphing calculator, find a logarithmic function that best fits the data.

(b) An alternative to the logarithmic model is a power function of the form $S = b(A^c)$. Using the power regression feature on a graphing calculator, find a power function that best fits the data.

(c) Graph both functions from parts (a) and (b) along with the data. Give advantages and drawbacks of both models.

(d) Use both functions to predict the number of species found on the isle of Shetland, with an area of 984.2 km². Compare with the actual number of 421.

(e) Describe one or more situations in which being able to predict the number of species could be useful.

Physical Sciences

120. Oil Production The production of an oil well has decreased exponentially from 128,000 barrels per year 5 years ago to 100,000 barrels per year at present.

(a) Letting $t = 0$ represent the present time, write an exponential equation for production y in terms of time t in years.

(b) Find the time it will take for production to fall to 70,000 barrels per year.

121. Dating Rocks   Geologists sometimes measure the age of rocks by using "atomic clocks." By measuring the amounts of potassium-40 and argon-40 in a rock, the age t of the specimen (in years) is found with the formula

$$t = (1.26 \times 10^9) \frac{\ln[1 + 8.33(A/K)]}{\ln 2},$$

where A and K, respectively, are the numbers of atoms of argon-40 and potassium-40 in the specimen.

(a) How old is a rock in which $A = 0$ and $K > 0$?

(b) The ratio A/K for a sample of granite from New Hampshire is 0.212. How old is the sample?

(c) Let $A/K = r$. What happens to t as r gets larger? Smaller?

122. Planets   The following table contains the average distance D from the sun for the eight planets and their period P of revolution around the sun in years. *Source: The Natural History of the Universe.*

Planet	Distance (D)	Period (P)
Mercury	0.39	0.24
Venus	0.72	0.62
Earth	1	1
Mars	1.52	1.89
Jupiter	5.20	11.9
Saturn	9.54	29.5
Uranus	19.2	84.0
Neptune	30.1	164.8

The distances are given in astronomical units (A.U.); 1 A.U. is the average distance from Earth to the sun. For example,

since Jupiter's distance is 5.2 A.U., its distance from the sun is 5.2 times farther than Earth's.

(a) Find functions of the form $P = kD^n$ for $n = 1, 1.5$, and 2 that fit the data at Neptune.

(b) Use a graphing calculator to plot the data in the table and to graph the three functions found in part (a). Which function best fits the data?

(c) Use the best-fitting function from part (b) to predict the period of Pluto (which was removed from the list of planets in 2006), which has a distance from the sun of 39.5 A.U. Compare your answer to the true value of 248.5 years.

(d) If you have a graphing calculator or computer program with a power regression feature, use it to find a power function (a function of the form $P = kD^n$) that approximately fits the data. How does this answer compare with the answer to part (b)?

EXTENDED APPLICATION

POWER FUNCTIONS

In this chapter we have seen several applications of power functions, which have the general form $y = ax^b$. Power functions are so named because the independent variable is raised to a power. (These should not be confused with exponential functions, in which the independent variable appears in the power.) We explored some special cases of power functions, such as $b = 2$ (a simple quadratic function) and $b = 1/2$ (a square root function). But applications of power functions vary greatly and are not limited to these special cases.

For example, in Exercise 88 in Section 2.5, we saw that the basal metabolism rate of marsupial carnivores is a power function of the body mass. In that exercise we also saw a way to verify that empirical data can be modeled with a power function. By taking the natural logarithm of both sides of the equation for a power function,

$$y = ax^b, \tag{1}$$

we obtain the equation

$$\ln y = \ln a + b \ln x. \tag{2}$$

Letting $Y = \ln y$, $X = \ln x$, and $A = \ln a$ results in the linear equation

$$Y = A + bX. \qquad (3)$$

Plotting the logarithm of the original data reveals whether a straight line approximates the data well. If it does, then a power function is a good fit to the original data.

Here is another example. In an attempt to measure how the pace of city life is related to the population of the city, two researchers estimated the average speed of pedestrians in 15 cities by measuring the mean time it took them to walk 50 feet. Their results are shown in the table below. *Source: Nature.*

Figure 62(a) shows the speed (stored in the list L_2 on a TI-84 Plus C) plotted against the population (stored in L_1). The natural logarithm of the data was then calculated and stored using the commands $\ln(L_1) \to L_3$ and $\ln(L_2) \to L_4$. A plot of the data in L_3 and L_4 is shown in Figure 62(b). Notice that the data lie fairly closely along a straight line, confirming that a power function is an appropriate model for the original data. (These calculations and plots could also be carried out on a spreadsheet.)

A power function that best fits the data according to the least squares principle of Section 1.3 is found with the TI-84 Plus C command $\texttt{PwrReg } L_1, L_2, Y_1$. The result is

$$y = 1.363x^{0.09799}, \qquad (4)$$

with a correlation coefficient of $r = 0.9081$. (For more on the correlation coefficient, see Section 1.3.) Because this value is close to 1, it indicates a good fit. Similarly, we can find a line that fits the data in Figure 62(b) with the command $\texttt{LinReg(ax + b) } L_3, L_4, Y_2$.

The result is

$$Y = 0.30985 + 0.09799X, \qquad (5)$$

again with $r = 0.9081$. The identical correlation coefficient is not a surprise, since the two commands accomplish essentially the same

City	Population (x)	Speed (ft/sec) (y)
Brno, Czechoslovakia	341,948	4.81
Prague, Czechoslovakia	1,092,759	5.88
Corte, Corsica	5491	3.31
Bastia, France	49,375	4.90
Munich, Germany	1,340,000	5.62
Psychro, Crete	365	2.67
Itea, Greece	2500	2.27
Iráklion, Greece	78,200	3.85
Athens, Greece	867,023	5.21
Safed, Israel	14,000	3.70
Dimona, Israel	23,700	3.27
Netanya, Israel	70,700	4.31
Jerusalem, Israel	304,500	4.42
New Haven, U.S.A.	138,000	4.39
Brooklyn, U.S.A.	2,602,000	5.05

(a)

(b)

FIGURE 62

thing. Comparing Equations (4) and (5) with Equations (1), (2), and (3), notice that $b = 0.09799$ in both Equations (4) and (5), and that $A = 0.30985 \approx \ln a = \ln 1.363$. (The slight difference is due to rounding.) Equations (4) and (5) are plotted on the same window as the data in Figure 63(a) and (b), respectively.

(a)

(b)

FIGURE 63

These results raise numerous questions worth exploring. What does this analysis tell you about the connection between the pace of city life and the population of a city? What might be some reasons for this connection?

A third example was considered in Review Exercise 119, where we explored the relationship between the area of each of the British isles and the number of species of vascular plants on the isles. In Figure 64(a) we have plotted the complete set of data from the original article (except for the Isle of Britain, whose area is so large that it doesn't fit on a graph that shows the other data in detail). *Source: Journal of Biogeography*. In Figure 64(b) we have plotted the natural logarithm of the data (again leaving out Britain). Notice that despite the large amount of scatter in the data, there is a linear trend. Figure 64(a) includes the best-fitting power function

$$y = 125.9x^{0.2088}, \qquad (6)$$

while Figure 64 (b) includes the best-fitting line

$$Y = 4.836 + 0.2088X. \qquad (7)$$

(a)

(b)

FIGURE 64

(We have included the data for Britain in both of these calculations.) Notice as before that the exponent of the power function equals the slope of the linear function, and that $A = 4.836 \approx \ln a = \ln 125.9$. The correlation is 0.6917 for both, indicating that there is a trend, but that the data are somewhat scattered about the best-fitting curve or line.

There are many more examples of data that fit a power function. Bohorquez et al. have found that the frequency of attacks in a war is a power function of the number of people killed in the attacks. *Source: Nature*. Amazingly, this holds true for a wide variety of different wars, with the average value of the exponent as $b \approx 2.5$. For a fragmented, fluid enemy with more groups, the value of b tends to be larger, and for a robust, stronger enemy with fewer groups, the value of b tends to be smaller.* You will explore other applications of power functions in the exercises.

EXERCISES

1. Gwartney et al. listed the following data relating the price of a monthly cellular bill (in dollars) and the demand (in millions of subscribers).† *Source: Economics: Private and Public Choice*.

*For a TED video on this phenomenon, see www.ted.com/talks/sean_gourley_on_the_mathematics_of_war.html.
†The authors point out that these are actual prices and quantities annually for 1988 to 1994. If they could assume that other demand determinants, such as income, had remained constant during that period, this would give an accurate measurement of the demand function.

Quantity (in millions)	Price (in dollars)
2.1	123
3.5	107
5.3	92
7.6	79
11.0	73
16.0	63
24.1	56

(a) Using a graphing calculator, plot the natural logarithm of the price against the natural logarithm of the quantity. Does the relationship appear to be linear?

(b) Find the best-fitting line to the natural logarithm of the data, as plotted in part (a). Plot this line on the same axes as the data.

(c) Plot the price against the quantity. What is different about the trend in these data from the trend in Figures 62(a) and 64(a)? What does this tell you about the exponent of the best-fitting power function for these data? What conclusions can you make about how demand varies with the price?

(d) Find the best-fitting power function for the data plotted in part (c). Verify that this function is equivalent to the least squares line through the logarithm of the data found in part (b).

2. For many years researchers thought that the basal metabolic rate (BMR) in mammals was a power function of the mass, with disagreement on whether the power was 0.67 or 0.75. More recently, White et al. proposed that the power may vary for mammals with different evolutionary lineages. The following table shows a portion of their data containing the natural logarithm of the mass (in grams) and the natural logarithm of the BMR (in mL of oxygen per hour) for 12 species from the genus Peromyscus, or deer mouse. *Source: Evolution*.

Species	ln(mass)	ln(BMR)
Peromyscus boylii	3.1442	3.9943
Peromyscus californicus	3.8618	3.9506
Peromyscus crinitus	2.7663	3.2237
Peromyscus eremicus	3.0681	3.4998
Peromyscus gossypinus	3.0681	3.6104
Peromyscus leucopus	3.1355	3.8111
Peromyscus maniculatus	3.0282	3.6835
Peromyscus megalops	4.1927	4.5075
Peromyscus oreas	3.2019	3.7729
Peromyscus polionotus	2.4849	3.0671
Peromyscus sitkensis	3.3439	3.8447
Peromyscus truei	3.5041	4.0378

(a) Plot ln(BMR) against ln(mass) using a graphing calculator. Does the relationship appear to be linear?

(b) Find the least squares line for the data plotted in part (a). Plot the line on the same axes as the data.

(c) Calculate the mass and BMR for each species, and then find the best-fitting power function for these data. Plot this function on the same axes as the mass and BMR data.

(d) What would you conclude about whether the deer mouse BMR can be modeled as a power function of the mass? What seems to be an approximate value of the power?

DIRECTIONS FOR GROUP PROJECT

Go to the section on "build your own tables" of the Human Development Reports website at hdrstats.undp.org/en/buildtables. Select a group of countries, as well as two indicators that you think might be related by a power function. For example, you might choose "GDP per capita" and "Population not using an improved water source (%)"; click on "Display indicators in Row" and then "Show results." Then click on "Export to Excel." From the Excel spreadsheet, create a scatterplot of the original data, as well as a scatterplot of the natural logarithm of the data. Find data for which the natural logarithm is roughly a straight line, and find the least squares line. Then convert this to a power function modeling the original data. Present your results in a report, describing in detail what your analysis tells you about the countries under consideration and any other conclusions that you can make.

3

The Derivative

The population of the United States has been increasing since 1790, when the first census was taken. Over the past few decades, the population has not only been increasing, but the level of diversity has also been increasing. This fact is important to school districts, businesses, and government officials. Using examples in the third section of this chapter, we explore two rates of change related to the increase in minority population. In the first example, we calculate an average rate of change; in the second, we calculate the rate of change at a particular time. This latter rate is an example of a derivative, the subject of this chapter.

The algebraic problems considered in earlier chapters dealt with *static* situations:

- What is the revenue when 100 items are sold?
- How much interest is earned in three years?
- What is the equilibrium price?

Calculus, on the other hand, deals with *dynamic* situations:

- At what rate is the demand for a product changing?
- How fast is a car moving after 2 hours?
- When does the growth of a population begin to slow down?

The techniques of calculus allow us to answer these questions, which deal with rates of change.

The key idea underlying calculus is the concept of limit, so we will begin by studying limits.

3.1 Limits

APPLY IT What happens to the demand of an essential commodity as its price continues to increase?

We will find an answer to this question in Exercise 84 using the concept of limit.

The limit is one of the tools that we use to describe the behavior of a function as the values of x approach, or become closer and closer to, some particular number.

EXAMPLE 1 Finding a Limit

What happens to $f(x) = x^2$ when x is a number *very close* to (but not equal to) 2?

SOLUTION We can construct a table with x values getting closer and closer to 2 and find the corresponding values of $f(x)$.

	x approaches 2 from left				↓	x approaches 2 from right			
x	1.9	1.99	1.999	1.9999	2	2.0001	2.001	2.01	2.1
$f(x)$	3.61	3.9601	3.996001	3.99960001	4	4.00040001	4.004001	4.0401	4.41
	f(x) approaches 4				↑	f(x) approaches 4			

The table suggests that, as x gets closer and closer to 2 from either side, $f(x)$ gets closer and closer to 4. In fact, you can use a calculator to show the values of $f(x)$ can be made as close as you want to 4 by taking values of x close enough to 2. This is not surprising, since the value of the function at $x = 2$ is $f(x) = 4$. We can observe this fact by looking at the graph $y = x^2$, as shown in Figure 1. In such a case, we say "the limit of $f(x)$ as x approaches 2 is 4," which is written as

$$\lim_{x \to 2} f(x) = 4.$$

YOUR TURN 1

Find $\lim\limits_{x \to 1} (x^2 + 2)$.

TRY YOUR TURN 1

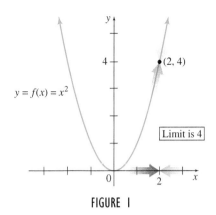

$y = f(x) = x^2$

Limit is 4

FIGURE 1

The phrase "x approaches 2 from the left" is written $x \to 2^-$. Similarly, "x approaches 2 from the right" is written $x \to 2^+$. These expressions are used to write **one-sided limits**. The **limit from the left** (as x approaches 2 from the negative direction) is written

$$\lim_{x \to 2^-} f(x) = 4,$$

and shown in red in Figure 1. The **limit from the right** (as x approaches 2 from the positive direction) is written

$$\lim_{x \to 2^+} f(x) = 4,$$

and shown in blue in Figure 1. A **two-sided limit**, such as

$$\lim_{x \to 2} f(x) = 4,$$

exists only if both one-sided limits exist and are the same; that is, if $f(x)$ approaches the same number as x approaches a given number from *either* side.

CAUTION | Notice that $\lim_{x \to a^-} f(x)$ does not mean to take negative values of x, nor does it mean to choose values of x to the right of a and then move in the negative direction. It means to use values less than a ($x < a$) that get closer and closer to a.

The previous example suggests the following informal definition.

Limit of a Function
Let f be a function and let a and L be real numbers. If

1. as x takes values closer and closer (but not equal) to a on both sides of a, the corresponding values of $f(x)$ get closer and closer (and perhaps equal) to L; and

2. the value of $f(x)$ can be made as close to L as desired by taking values of x close enough to a;

then L is the **limit** of $f(x)$ as x approaches a, written

$$\lim_{x \to a} f(x) = L.$$

This definition is informal because the expressions "closer and closer to" and "as close as desired" have not been defined. A more formal definition would be needed to prove the rules for limits given later in this section.*

---

*The limit is the key concept from which all the ideas of calculus flow. Calculus was independently discovered by the English mathematician Isaac Newton (1642–1727) and the German mathematician Gottfried Wilhelm Leibniz (1646–1716). For the next century, supporters of each accused the other of plagiarism, resulting in a lack of communication between mathematicians in England and on the European continent. Neither Newton nor Leibniz developed a mathematically rigorous definition of the limit (and we have no intention of doing so here). More than 100 years passed before the French mathematician Augustin-Louis Cauchy (1789–1857) accomplished this feat.

The definition of a limit describes what happens to $f(x)$ when x is near, but not equal to, the value a. It is not affected by how (or even whether) $f(a)$ is defined. Also the definition implies that the function values cannot approach two different numbers, so that if a limit exists, it is unique. These ideas are illustrated in the following examples.

> **CAUTION** Note that the limit is a value of y, not x.

EXAMPLE 2 Finding a Limit

Find $\lim_{x \to 2} g(x)$, where $g(x) = \dfrac{x^3 - 2x^2}{x - 2}$.

SOLUTION

Method 1
Using a Table

The function $g(x)$ is undefined when $x = 2$, since the value $x = 2$ makes the denominator 0. However, in determining the limit as x approaches 2 we are concerned only with the values of $g(x)$ when x is close to but *not equal to* 2. To determine if the limit exists, consider the value of g at some numbers close to but not equal to 2, as shown in the following table.

		x approaches 2 from left				x approaches 2 from right		
x	1.9	1.99	1.999	1.9999	2	2.0001	2.001	2.01	2.1
$g(x)$	3.61	3.9601	3.996001	3.99960001		4.00040001	4.004001	4.0401	4.41

f(x) approaches 4		f(x) approaches 4
	Undefined	

Notice that this table is almost identical to the previous table, except that g is undefined at $x = 2$. This suggests that $\lim_{x \to 2} g(x) = 4$, in spite of the fact that the function g does not exist at $x = 2$.

Method 2
Using Algebra

A second approach to this limit is to analyze the function. By factoring the numerator,

$$x^3 - 2x^2 = x^2(x - 2),$$

$g(x)$ simplifies to

$$g(x) = \frac{x^2(x - 2)}{x - 2} = x^2, \qquad \text{provided } x \neq 2.$$

The graph of $g(x)$, as shown in Figure 2, is almost the same as the graph of $y = x^2$, except that it is undefined at $x = 2$ (illustrated by the "hole" in the graph).

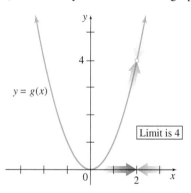

FIGURE 2

Since we are looking at the limit as x approaches 2, we look at values of the function for x close to but not equal to 2. Thus, the limit is

$$\lim_{x \to 2} g(x) = \lim_{x \to 2} x^2 = 4.$$

TRY YOUR TURN 2

YOUR TURN 2

Find $\lim_{x \to 2} \dfrac{x^2 - 4}{x - 2}$.

Method 3
Graphing Calculator

We can use the TRACE feature on a graphing calculator to determine the limit. Figure 3 shows the graph of the function in Example 2 drawn with a TI-84 Plus C. Notice that the function has a small gap at the point $(2, 4)$, which agrees with our previous observation that the function is undefined at $x = 2$, where the limit is 4. (Due to the limitations of the graphing calculator, this gap may vanish when the viewing window is changed very slightly.)

The result after pressing the TRACE key is shown in Figure 4. The cursor is already located at $x = 2$; if it were not, we could use the right or left arrow key to move the cursor there. The calculator does not give a y-value because the function is undefined at $x = 2$. Moving the cursor back a step gives $x = 1.9848485$, $y = 3.9396235$. Moving the cursor forward two steps gives $x = 2.0151515$, $y = 4.0608356$. It seems that as x approaches 2, y approaches 4, or at least something close to 4. Zooming in on the point $(2, 4)$ (such as using the window $[1.9, 2.1]$ by $[3.9, 4.1]$) allows the limit to be estimated more accurately and helps ensure that the graph has no unexpected behavior very close to $x = 2$.

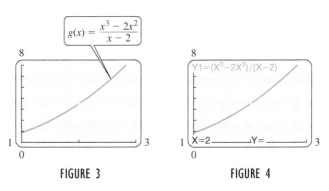

$$g(x) = \frac{x^3 - 2x^2}{x - 2}$$

FIGURE 3   **FIGURE 4**

The TABLE feature of a graphing calculator can also be used to investigate values of the function for values of x close to 2. See Example 10.

Note that a limit can be found in three ways:

1. algebraically;

2. using a graph (either drawn by hand or with a graphing calculator); and

3. using a table (either written out by hand or with a graphing calculator).

Which method you choose depends on the complexity of the function and the accuracy required by the application. Algebraic simplification gives the exact answer, but it can be difficult or even impossible to use in some situations. Calculating a table of numbers or tracing the graph may be easier when the function is complicated, but be careful, because the results could be inaccurate, inconclusive, or misleading. A graphing calculator does not tell us what happens between or beyond the points that are plotted.

EXAMPLE 3   **Finding a Limit**

Determine $\lim_{x \to 2} h(x)$ for the function h defined by

$$h(x) = \begin{cases} x^2, & \text{if } x \neq 2, \\ 1, & \text{if } x = 2. \end{cases}$$

SOLUTION   A function defined by two or more cases is called a **piecewise function**. The domain of h is all real numbers, and its graph is shown in Figure 5. Notice that $h(2) = 1$, but $h(x) = x^2$ when $x \neq 2$. To determine the limit as x approaches 2, we are concerned only with the values of $h(x)$ when x is close but not equal to 2. Once again,

$$\lim_{x \to 2} h(x) = \lim_{x \to 2} x^2 = 4.$$

TRY YOUR TURN 3

$y = h(x)$

Limit is 4

$\bullet (2, 1)$

FIGURE 5

YOUR TURN 3
Find $\lim_{x \to 3} f(x)$ if

$$f(x) = \begin{cases} 2x - 1 & \text{if } x \neq 3 \\ 1 & \text{if } x = 3. \end{cases}$$

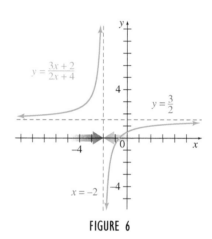

EXAMPLE 4 Finding a Limit

Find $\lim\limits_{x \to -2} f(x)$, where

$$f(x) = \frac{3x + 2}{2x + 4}.$$

SOLUTION The graph of the function is shown in Figure 6. A table with the values of $f(x)$ as x gets closer and closer to -2 is given below.

	x approaches −2 from left				*x* approaches −2 from right			
x	−2.1	−2.01	−2.001	−2.0001	−1.9999	−1.999	−1.99	−1.9
f(x)	21.5	201.5	2001.5	20,001.5	−19,998.5	−1998.5	−198.5	−18.5

FIGURE 6

Both the graph and the table suggest that as x approaches -2 from the left, $f(x)$ becomes larger and larger without bound. This happens because as x approaches -2, the denominator approaches 0, while the numerator approaches -4, and -4 divided by a smaller and smaller number becomes larger and larger. When this occurs, we say that "the limit as x approaches -2 from the left is infinity," and we write

$$\lim_{x \to -2^-} f(x) = \infty.$$

Because ∞ is not a real number, the limit in this case does not exist.

In the same way, the behavior of the function as x approaches -2 from the right is indicated by writing

$$\lim_{x \to -2^+} f(x) = -\infty,$$

since $f(x)$ becomes more and more negative without bound. Since there is no real number that $f(x)$ approaches as x approaches -2 (from either side), nor does $f(x)$ approach either ∞ or $-\infty$, we simply say

$$\lim_{x \to -2} \frac{3x + 2}{2x + 4} \text{ does not exist.}$$

TRY YOUR TURN 4

YOUR TURN 4

Find $\lim\limits_{x \to 0} \dfrac{2x - 1}{x}$.

NOTE In general, if both the limit from the left and from the right approach ∞, so that $\lim\limits_{x \to a} f(x) = \infty$, the limit would not exist because ∞ is not a real number. It is customary, however, to give ∞ as the answer, since it describes how the function is behaving near $x = a$. Likewise, if $\lim\limits_{x \to a} f(x) = -\infty$, we give $-\infty$ as the answer.

EXAMPLE 5 Finding a Limit

Find $\lim\limits_{x \to 0} \dfrac{|x|}{x}$.

SOLUTION

Method I
Algebraic Approach

The function $f(x) = |x|/x$ is not defined when $x = 0$. When $x > 0$, the definition of absolute value says that $|x| = x$, so $f(x) = |x|/x = x/x = 1$. When $x < 0$, then $|x| = -x$ and $f(x) = |x|/x = -x/x = -1$. Therefore,

$$\lim_{x \to 0^+} f(x) = 1 \qquad \text{and} \qquad \lim_{x \to 0^-} f(x) = -1.$$

Since the limits from the left and from the right are different, the limit does not exist.

Method 2
Graphing Calculator Approach

A calculator graph of f is shown in Figure 7.

As x approaches 0 from the right, x is always positive and the corresponding value of $f(x)$ is 1, so

$$\lim_{x \to 0^+} f(x) = 1.$$

But as x approaches 0 from the left, x is always negative and the corresponding value of $f(x)$ is -1, so

$$\lim_{x \to 0^-} f(x) = -1.$$

As in the algebraic approach, the limits from the left and from the right are different, so the limit does not exist.

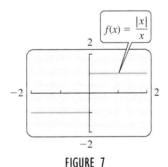

$$f(x) = \frac{|x|}{x}$$

FIGURE 7

The discussion up to this point can be summarized as follows.

Existence of Limits

The limit of f as x approaches a may not exist.

1. If $f(x)$ becomes infinitely large in magnitude (positive or negative) as x approaches the number a from either side, we write $\lim\limits_{x \to a} f(x) = \infty$ or $\lim\limits_{x \to a} f(x) = -\infty$. In either case, the limit does not exist.

2. If $f(x)$ becomes infinitely large in magnitude (positive) as x approaches a from one side and infinitely large in magnitude (negative) as x approaches a from the other side, then $\lim\limits_{x \to a} f(x)$ does not exist.

3. If $\lim\limits_{x \to a^-} f(x) = L$ and $\lim\limits_{x \to a^+} f(x) = M$, and $L \neq M$, then $\lim\limits_{x \to a} f(x)$ does not exist.

Figure 8 illustrates these three facts.

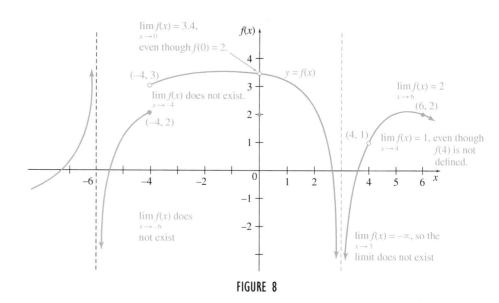

FIGURE 8

Rules for Limits

As shown by the preceding examples, tables and graphs can be used to find limits. However, it is usually more efficient to use the rules for limits given below. (Proofs of these rules require a formal definition of limit, which we have not given.)

Rules for Limits

Let a, A, and B be real numbers, and let f and g be functions such that

$$\lim_{x \to a} f(x) = A \quad \text{and} \quad \lim_{x \to a} g(x) = B.$$

1. If k is a constant, then $\lim_{x \to a} k = k$ and $\lim_{x \to a} [k \cdot f(x)] = k \cdot \lim_{x \to a} f(x) = k \cdot A$.

2. $\lim_{x \to a} [f(x) \pm g(x)] = \lim_{x \to a} f(x) \pm \lim_{x \to a} g(x) = A \pm B$
(The limit of a sum or difference is the sum or difference of the limits.)

3. $\lim_{x \to a} [f(x) \cdot g(x)] = [\lim_{x \to a} f(x)] \cdot [\lim_{x \to a} g(x)] = A \cdot B$
(The limit of a product is the product of the limits.)

4. $\lim_{x \to a} \dfrac{f(x)}{g(x)} = \dfrac{\lim_{x \to a} f(x)}{\lim_{x \to a} g(x)} = \dfrac{A}{B}$ if $B \neq 0$
(The limit of a quotient is the quotient of the limits, provided the limit of the denominator is not zero.)

5. If $p(x)$ is a polynomial, then $\lim_{x \to a} p(x) = p(a)$.

6. For any real number k, $\lim_{x \to a} [f(x)]^k = [\lim_{x \to a} f(x)]^k = A^k$, provided this limit exists.*

7. $\lim_{x \to a} f(x) = \lim_{x \to a} g(x)$ if $f(x) = g(x)$ for all $x \neq a$.

8. For any real number $b > 0$, $\lim_{x \to a} b^{f(x)} = b^{[\lim_{x \to a} f(x)]} = b^A$.

9. For any real number b such that $0 < b < 1$ or $1 < b$,
$\lim_{x \to a} [\log_b f(x)] = \log_b [\lim_{x \to a} f(x)] = \log_b A$ if $A > 0$.

*This limit does not exist, for example, when $A < 0$ and $k = 1/2$, or when $A = 0$ and $k \leq 0$.

This list may seem imposing, but these limit rules, once understood, agree with common sense. For example, Rule 3 says that if $f(x)$ becomes close to A as x approaches a, and if $g(x)$ becomes close to B, then $f(x) \cdot g(x)$ should become close to $A \cdot B$, which seems plausible.

EXAMPLE 6 Rules for Limits

Suppose $\lim_{x \to 2} f(x) = 3$ and $\lim_{x \to 2} g(x) = 4$. Use the limit rules to find the following limits.

(a) $\lim_{x \to 2} [f(x) + 5g(x)]$

SOLUTION

$$\lim_{x \to 2} [f(x) + 5g(x)] = \lim_{x \to 2} f(x) + \lim_{x \to 2} 5g(x) \qquad \text{Rule 2}$$
$$= \lim_{x \to 2} f(x) + 5 \lim_{x \to 2} g(x) \qquad \text{Rule 1}$$
$$= 3 + 5(4)$$
$$= 23$$

(b) $\lim_{x \to 2} \dfrac{[f(x)]^2}{\ln g(x)}$

SOLUTION

$$\lim_{x \to 2} \frac{[f(x)]^2}{\ln g(x)} = \frac{\lim_{x \to 2} [f(x)]^2}{\lim_{x \to 2} \ln g(x)} \qquad \text{Rule 4}$$
$$= \frac{[\lim_{x \to 2} f(x)]^2}{\ln[\lim_{x \to 2} g(x)]} \qquad \text{Rule 6 and Rule 9}$$
$$= \frac{3^2}{\ln 4}$$
$$\approx \frac{9}{1.38629} \approx 6.492 \qquad \text{TRY YOUR TURN 5}$$

YOUR TURN 5
Find $\lim_{x \to 2} [f(x) + g(x)]^2$.

EXAMPLE 7 Finding a Limit

Find $\lim_{x \to 3} \dfrac{x^2 - x - 1}{\sqrt{x + 1}}$.

SOLUTION

$$\lim_{x \to 3} \frac{x^2 - x - 1}{\sqrt{x + 1}} = \frac{\lim_{x \to 3} (x^2 - x - 1)}{\lim_{x \to 3} \sqrt{x + 1}} \qquad \text{Rule 4}$$
$$= \frac{\lim_{x \to 3} (x^2 - x - 1)}{\sqrt{\lim_{x \to 3} (x + 1)}} \qquad \text{Rule 6 } (\sqrt{a} = a^{1/2})$$
$$= \frac{3^2 - 3 - 1}{\sqrt{3 + 1}} \qquad \text{Rule 5}$$
$$= \frac{5}{\sqrt{4}}$$
$$= \frac{5}{2}$$

As Examples 6 and 7 suggest, the rules for limits actually mean that many limits can be found simply by evaluation. This process is valid for polynomials, rational functions, exponential functions, logarithmic functions, and roots and powers, as long as this does not involve an illegal operation, such as division by 0 or taking the logarithm of a negative number. Division by 0 presents particular problems that can often be solved by algebraic simplification, as the following example shows.

EXAMPLE 8 Finding a Limit

Find $\lim\limits_{x \to 2} \dfrac{x^2 + x - 6}{x - 2}$.

SOLUTION Rule 4 cannot be used here, since

$$\lim_{x \to 2} (x - 2) = 0.$$

The numerator also approaches 0 as x approaches 2, and 0/0 is meaningless. For $x \neq 2$, we can, however, simplify the function by rewriting the fraction as

$$\frac{x^2 + x - 6}{x - 2} = \frac{(x + 3)(x - 2)}{x - 2} = x + 3.$$

Now Rule 7 can be used.

YOUR TURN 6 Find
$\lim\limits_{x \to -3} \dfrac{x^2 - x - 12}{x + 3}$.

$$\lim_{x \to 2} \frac{x^2 + x - 6}{x - 2} = \lim_{x \to 2} (x + 3) = 2 + 3 = 5$$

TRY YOUR TURN 6

NOTE Mathematicians often refer to a limit that gives 0/0, as in Example 8, as an *indeterminate form*. This means that when the numerator and denominator are polynomials, they must have a common factor, which is why we factored the numerator in Example 8.

EXAMPLE 9 Finding a Limit

Find $\lim\limits_{x \to 4} \dfrac{\sqrt{x} - 2}{x - 4}$.

SOLUTION

Method 1
Rationalizing the Numerator

As $x \to 4$, the numerator approaches 0 and the denominator also approaches 0, giving the meaningless expression 0/0. In an expression such as this involving square roots, rather than trying to factor, you may find it simpler to use algebra to rationalize the numerator by multiplying both the numerator and the denominator by $\sqrt{x} + 2$. This gives

$$\frac{\sqrt{x} - 2}{x - 4} \cdot \frac{\sqrt{x} + 2}{\sqrt{x} + 2} = \frac{(\sqrt{x})^2 - 2^2}{(x - 4)(\sqrt{x} + 2)} \qquad (a - b)(a + b) = a^2 - b^2$$

$$= \frac{x - 4}{(x - 4)(\sqrt{x} + 2)} = \frac{1}{\sqrt{x} + 2}$$

if $x \neq 4$. Now use the rules for limits.

$$\lim_{x \to 4} \frac{\sqrt{x} - 2}{x - 4} = \lim_{x \to 4} \frac{1}{\sqrt{x} + 2} = \frac{1}{\sqrt{4} + 2} = \frac{1}{2 + 2} = \frac{1}{4}$$

Method 2
Factoring

Alternatively, we can take advantage of the fact that $x - 4 = (\sqrt{x})^2 - 2^2 = (\sqrt{x} + 2)(\sqrt{x} - 2)$ because of the factoring $a^2 - b^2 = (a + b)(a - b)$. Then

$$\lim_{x \to 4} \frac{\sqrt{x} - 2}{x - 4} = \lim_{x \to 4} \frac{\sqrt{x} - 2}{(\sqrt{x} + 2)(\sqrt{x} - 2)}$$

$$= \lim_{x \to 4} \frac{1}{\sqrt{x} + 2}$$

$$= \frac{1}{\sqrt{4} + 2} = \frac{1}{2 + 2} = \frac{1}{4}.$$

YOUR TURN 7

Find $\lim\limits_{x \to 1} \dfrac{\sqrt{x} - 1}{x - 1}$.

TRY YOUR TURN 7

| **CAUTION** | Simply because the expression in a limit is approaching 0/0, as in Examples 8 and 9, does *not* mean that the limit is 0 or that the limit does not exist. For such a limit, try to simplify the expression using the following principle: **To calculate the limit of $f(x)/g(x)$ as x approaches a, where $f(a) = g(a) = 0$, you should attempt to factor $x - a$ from both the numerator and the denominator.** |

EXAMPLE 10   Finding a Limit

Find $\lim\limits_{x \to 1} \dfrac{x^2 - 2x + 1}{(x - 1)^3}$.

SOLUTION

Method 1
Algebraic Approach

Again, Rule 4 cannot be used, since $\lim\limits_{x \to 1} (x - 1)^3 = 0$. If $x \neq 1$, the function can be rewritten as

$$\frac{x^2 - 2x + 1}{(x - 1)^3} = \frac{(x - 1)^2}{(x - 1)^3} = \frac{1}{x - 1}.$$

Then

$$\lim_{x \to 1} \frac{x^2 - 2x + 1}{(x - 1)^3} = \lim_{x \to 1} \frac{1}{x - 1}$$

by Rule 7. None of the rules can be used to find

$$\lim_{x \to 1} \frac{1}{x - 1},$$

but as x approaches 1, the denominator approaches 0 while the numerator stays at 1, making the result larger and larger in magnitude. If $x > 1$, both the numerator and denominator are positive, so $\lim\limits_{x \to 1^+} 1/(x - 1) = \infty$. If $x < 1$, the denominator is negative, so $\lim\limits_{x \to 1^-} 1/(x - 1) = -\infty$. Therefore,

$$\lim_{x \to 1} \frac{x^2 - 2x + 1}{(x - 1)^3} = \lim_{x \to 1} \frac{1}{x - 1} \text{ does not exist.}$$

Method 2
Graphing Calculator Approach

Using the TABLE feature on a TI-84 Plus C, we can produce the table of numbers shown in Figure 9, where Y_1 represents the function $y = 1/(x - 1)$. Figure 10 shows a graphing calculator view of the function on $[0, 2]$ by $[-10, 10]$. The behavior of the function indicates a vertical asymptote at $x = 1$, with the limit approaching $-\infty$ from the left and ∞ from the right, so

$$\lim_{x \to 1} \frac{x^2 - 2x + 1}{(x - 1)^3} = \lim_{x \to 1} \frac{1}{x - 1} \text{ does not exist.}$$

Both the table and the graph can be easily generated using a spreadsheet. Consult the *Graphing Calculator and Excel Spreadsheet Manual*, available with this text, for details.

FIGURE 9

FIGURE 10

NOTE Another way to understand the behavior of the function in the previous example near $x = 1$ is to recall from Section 2.3 on Polynomial and Rational Functions that a rational function often has a vertical asymptote at a value of x where the denominator is 0, although it may not if the numerator there is also 0. In this example, we see after simplifying that the function has a vertical asymptote at $x = 1$ because that would make the denominator of $1/(x - 1)$ equal to 0, while the numerator is 1.

Limits at Infinity

Sometimes it is useful to examine the behavior of the values of $f(x)$ as x gets larger and larger (or more and more negative). The phrase "x approaches infinity," written $x \rightarrow \infty$, expresses the fact that x becomes larger without bound. Similarly, the phrase "x approaches negative infinity" (symbolically, $x \rightarrow -\infty$) means that x becomes more and more negative without bound (such as $-10, -1000, -10,000$, etc.). The next example illustrates a **limit at infinity**.

EXAMPLE 11 Oxygen Concentration

Suppose a small pond normally contains 12 units of dissolved oxygen in a fixed volume of water. Suppose also that at time $t = 0$ a quantity of organic waste is introduced into the pond, with the oxygen concentration t weeks later given by

$$f(t) = \frac{12t^2 - 15t + 12}{t^2 + 1}.$$

As time goes on, what will be the ultimate concentration of oxygen? Will it return to 12 units?

SOLUTION After 2 weeks, the pond contains

$$f(2) = \frac{12 \cdot 2^2 - 15 \cdot 2 + 12}{2^2 + 1} = \frac{30}{5} = 6$$

units of oxygen, and after 4 weeks, it contains

$$f(4) = \frac{12 \cdot 4^2 - 15 \cdot 4 + 12}{4^2 + 1} \approx 8.5$$

units. Choosing several values of t and finding the corresponding values of $f(t)$, or using a graphing calculator or computer, leads to the table and graph in Figure 11.

The graph suggests that, as time goes on, the oxygen level gets closer and closer to the original 12 units. If so, the line $y = 12$ is a horizontal asymptote. The table suggests that

$$\lim_{t \to \infty} f(t) = 12.$$

Thus, the oxygen concentration will approach 12, but it will never be *exactly* 12.

t	$f(t)$
10	10.515
100	11.85
1000	11.985
10,000	11.9985
100,000	11.99985

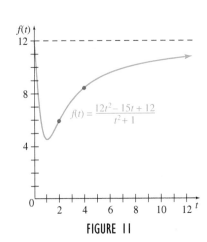

FIGURE 11

As we saw in the previous example, *limits at infinity or negative infinity, if they exist, correspond to horizontal asymptotes of the graph of the function.* In the previous chapter, we saw one way to find horizontal asymptotes. We will now show a more precise way, based upon some simple limits at infinity. The graphs of $f(x) = 1/x$ (in red) and $g(x) = 1/x^2$ (in blue) shown in Figure 12, as well as the table there, indicate that $\lim_{x \to \infty} 1/x = 0$, $\lim_{x \to -\infty} 1/x = 0$, $\lim_{x \to \infty} 1/x^2 = 0$, and $\lim_{x \to -\infty} 1/x^2 = 0$, suggesting the following rule.

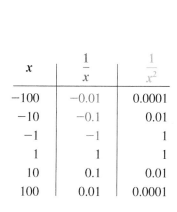

x	$\dfrac{1}{x}$	$\dfrac{1}{x^2}$
-100	-0.01	0.0001
-10	-0.1	0.01
-1	-1	1
1	1	1
10	0.1	0.01
100	0.01	0.0001

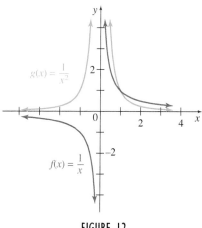

FIGURE 12

Limits at Infinity

For any positive real number n,

$$\lim_{x \to \infty} \frac{1}{x^n} = 0 \qquad \text{and} \qquad \lim_{x \to -\infty} \frac{1}{x^n} = 0.*$$

FOR REVIEW

In Section 2.3, we saw a way to find horizontal asymptotes by considering the behavior of the function as x (or t) gets large. For large t, $12t^2 - 15t + 12 \approx 12t^2$, because the t-term and the constant term are small compared with the t^2-term when t is large. Similarly, $t^2 + 1 \approx t^2$. Thus, for large t,

$$f(t) = \frac{12t^2 - 15t + 12}{t^2 + 1} \approx$$

$\dfrac{12t^2}{t^2} = 12$. Thus the function f has a horizontal asymptote at $y = 12$.

The rules for limits given earlier remain unchanged when a is replaced with ∞ or $-\infty$.

To evaluate the limit at infinity of a rational function, divide the numerator and denominator by the largest power of the variable that appears in the denominator, t^2 here, and then use these results. In the previous example, we find that

$$\lim_{t \to \infty} \frac{12t^2 - 15t + 12}{t^2 + 1} = \lim_{t \to \infty} \frac{\dfrac{12t^2}{t^2} - \dfrac{15t}{t^2} + \dfrac{12}{t^2}}{\dfrac{t^2}{t^2} + \dfrac{1}{t^2}}$$

$$= \lim_{t \to \infty} \frac{12 - 15 \cdot \dfrac{1}{t} + 12 \cdot \dfrac{1}{t^2}}{1 + \dfrac{1}{t^2}}.$$

---

*If x is negative, x^n does not exist for certain values of n, so the second limit is undefined for those values of n.

Now apply the limit rules and the fact that $\lim\limits_{t \to \infty} 1/t^n = 0$.

$$\frac{\lim\limits_{t \to \infty}\left(12 - 15 \cdot \dfrac{1}{t} + 12 \cdot \dfrac{1}{t^2}\right)}{\lim\limits_{t \to \infty}\left(1 + \dfrac{1}{t^2}\right)}$$

$$= \frac{\lim\limits_{t \to \infty} 12 - \lim\limits_{t \to \infty} 15 \cdot \dfrac{1}{t} + \lim\limits_{t \to \infty} 12 \cdot \dfrac{1}{t^2}}{\lim\limits_{t \to \infty} 1 + \lim\limits_{t \to \infty} \dfrac{1}{t^2}} \qquad \text{Rules 4 and 2}$$

$$= \frac{12 - 15\left(\lim\limits_{t \to \infty} \dfrac{1}{t}\right) + 12\left(\lim\limits_{t \to \infty} \dfrac{1}{t^2}\right)}{1 + \lim\limits_{t \to \infty} \dfrac{1}{t^2}} \qquad \text{Rule 1}$$

$$= \frac{12 - 15 \cdot 0 + 12 \cdot 0}{1 + 0} = 12. \qquad \text{Limits at infinity}$$

EXAMPLE 12   Limits at Infinity

Find each limit.

(a) $\lim\limits_{x \to \infty} \dfrac{8x + 6}{3x - 1}$

SOLUTION   We can use the rule $\lim\limits_{x \to \infty} 1/x^n = 0$ to find this limit by first dividing the numerator and denominator by x, as follows.

$$\lim_{x \to \infty} \frac{8x + 6}{3x - 1} = \lim_{x \to \infty} \frac{\dfrac{8x}{x} + \dfrac{6}{x}}{\dfrac{3x}{x} - \dfrac{1}{x}} = \lim_{x \to \infty} \frac{8 + 6 \cdot \dfrac{1}{x}}{3 - \dfrac{1}{x}} = \frac{8 + 0}{3 - 0} = \frac{8}{3}$$

(b) $\lim\limits_{x \to \infty} \dfrac{3x + 2}{4x^3 - 1} = \lim\limits_{x \to \infty} \dfrac{3 \cdot \dfrac{1}{x^2} + 2 \cdot \dfrac{1}{x^3}}{4 - \dfrac{1}{x^3}} = \dfrac{0 + 0}{4 - 0} = \dfrac{0}{4} = 0$

Here, the highest power of x in the denominator is x^3, which is used to divide each term in the numerator and denominator.

(c) $\lim\limits_{x \to \infty} \dfrac{3x^2 + 2}{4x - 3} = \lim\limits_{x \to \infty} \dfrac{3x + \dfrac{2}{x}}{4 - \dfrac{3}{x}}$

The highest power of x in the denominator is x (to the first power). There is a higher power of x in the numerator, but we don't divide by this. Notice that the denominator approaches 4, while the numerator becomes infinitely large, so

$$\lim_{x \to \infty} \frac{3x^2 + 2}{4x - 3} = \infty.$$

(d) $\lim\limits_{x \to \infty} \dfrac{5x^2 - 4x^3}{3x^2 + 2x - 1} = \lim\limits_{x \to \infty} \dfrac{5 - 4x}{3 + \dfrac{2}{x} - \dfrac{1}{x^2}}$

The highest power of x in the denominator is x^2. The denominator approaches 3, while the numerator becomes a negative number that is larger and larger in magnitude, so

$$\lim_{x \to \infty} \frac{2x^2 + 3x - 4}{6x^2 - 5x + 7}.$$

$$\lim_{x \to \infty} \frac{5x^2 - 4x^3}{3x^2 + 2x - 1} = -\infty.$$

TRY YOUR TURN 8

The method used in Example 12 is a useful way to rewrite expressions with fractions so that the rules for limits at infinity can be used.

Finding Limits at Infinity

If $f(x) = p(x)/q(x)$, for polynomials $p(x)$ and $q(x)$, $q(x) \neq 0$, $\lim_{x \to -\infty} f(x)$ and $\lim_{x \to \infty} f(x)$ can be found as follows.

1. Divide $p(x)$ and $q(x)$ by the highest power of x in $q(x)$.

2. Use the rules for limits, including the rules for limits at infinity,

$$\lim_{x \to \infty} \frac{1}{x^n} = 0 \quad \text{and} \quad \lim_{x \to -\infty} \frac{1}{x^n} = 0,$$

to find the limit of the result from Step 1.

For an alternate approach to finding limits at infinity, see Exercise 83.

3.1 WARM-UP EXERCISES

Factor each of the following expressions.   *(Sec. R.2)*

W1. $8x^2 + 22x + 15$

W2. $12x^2 - 7x - 12$

Simplify each of the following expressions.   *(Sec. R.3)*

W3. $\dfrac{3x^2 + x - 14}{x^2 - 4}$

W4. $\dfrac{2x^2 + x - 15}{x^2 - 9}$

3.1   EXERCISES

In Exercises 1–4, choose the best answer for each limit.

1. If $\lim_{x \to 2^-} f(x) = 5$ and $\lim_{x \to 2^+} f(x) = 6$, then $\lim_{x \to 2} f(x)$
   (a) is 5.
   (b) is 6.
   (c) does not exist.
   (d) is infinite.

2. If $\lim_{x \to 2^-} f(x) = \lim_{x \to 2^+} f(x) = -1$, but $f(2) = 1$, then $\lim_{x \to 2} f(x)$
   (a) is -1.
   (b) does not exist.
   (c) is infinite.
   (d) is 1.

3. If $\lim_{x \to 4^-} f(x) = \lim_{x \to 4^+} f(x) = 6$, but $f(4)$ does not exist, then $\lim_{x \to 4} f(x)$
   (a) does not exist.
   (b) is 6.
   (c) is $-\infty$.
   (d) is ∞.

4. If $\lim_{x \to 1^-} f(x) = -\infty$ and $\lim_{x \to 1^+} f(x) = -\infty$, then $\lim_{x \to 1} f(x)$
   (a) is ∞.
   (b) is $-\infty$.
   (c) does not exist.
   (d) is 1.

Decide whether each limit exists. If a limit exists, estimate its value.

5. (a) $\lim\limits_{x \to 3} f(x)$ **(b)** $\lim\limits_{x \to 0} f(x)$

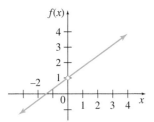

6. (a) $\lim\limits_{x \to 2} F(x)$ **(b)** $\lim\limits_{x \to -1} F(x)$

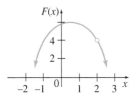

7. (a) $\lim\limits_{x \to 0} f(x)$ **(b)** $\lim\limits_{x \to 2} f(x)$

8. (a) $\lim\limits_{x \to 3} g(x)$ **(b)** $\lim\limits_{x \to 5} g(x)$

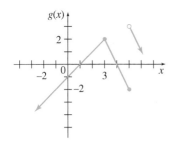

In Exercises 9 and 10, use the graph to find (i) $\lim\limits_{x \to a^-} f(x)$, (ii) $\lim\limits_{x \to a^+} f(x)$, (iii) $\lim\limits_{x \to a} f(x)$, and (iv) $f(a)$ if it exists.

9. (a) $a = -2$ **(b)** $a = -1$

10. (a) $a = 1$ **(b)** $a = 2$

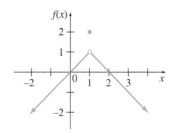

Decide whether each limit exists. If a limit exists, find its value.

11. $\lim\limits_{x \to \infty} f(x)$

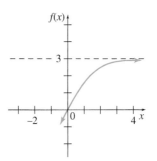

12. $\lim\limits_{x \to -\infty} g(x)$

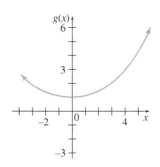

13. Explain why $\lim\limits_{x \to 2} F(x)$ in Exercise 6 exists, but $\lim\limits_{x \to -2} f(x)$ in Exercise 9 does not.

14. In Exercise 10, why does $\lim\limits_{x \to 1} f(x) = 1$, even though $f(1) = 2$?

15. Use the table of values to estimate $\lim\limits_{x \to 1} f(x)$.

x	0.9	0.99	0.999	0.9999	1.0001	1.001	1.01	1.1
$f(x)$	3.9	3.99	3.999	3.9999	4.0001	4.001	4.01	4.1

Complete the tables and use the results to find the indicated limits.

16. If $f(x) = 2x^2 - 4x + 7$, find $\lim\limits_{x \to 1} f(x)$.

x	0.9	0.99	0.999	1.001	1.01	1.1
$f(x)$			5.000002	5.000002		

17. If $k(x) = \dfrac{x^3 - 2x - 4}{x - 2}$, find $\lim\limits_{x \to 2} k(x)$.

x	1.9	1.99	1.999	2.001	2.01	2.1
$k(x)$						

18. If $f(x) = \dfrac{2x^3 + 3x^2 - 4x - 5}{x + 1}$, find $\lim\limits_{x \to -1} f(x)$.

x	-1.1	-1.01	-1.001	-0.999	-0.99	-0.9
$f(x)$						

19. If $h(x) = \dfrac{\sqrt{x} - 2}{x - 1}$, find $\lim\limits_{x \to 1} h(x)$.

x	0.9	0.99	0.999	1.001	1.01	1.1
$h(x)$						

20. If $f(x) = \dfrac{\sqrt{x} - 3}{x - 3}$, find $\lim\limits_{x \to 3} f(x)$.

x	2.9	2.99	2.999	3.001	3.01	3.1
$f(x)$						

Let $\lim\limits_{x \to 4} f(x) = 9$ and $\lim\limits_{x \to 4} g(x) = 27$. **Use the limit rules to find each limit.**

21. $\lim\limits_{x \to 4} [f(x) - g(x)]$

22. $\lim\limits_{x \to 4} [g(x) \cdot f(x)]$

23. $\lim\limits_{x \to 4} \dfrac{f(x)}{g(x)}$

24. $\lim\limits_{x \to 4} \log_3 f(x)$

25. $\lim\limits_{x \to 4} \sqrt{f(x)}$

26. $\lim\limits_{x \to 4} \sqrt[3]{g(x)}$

27. $\lim\limits_{x \to 4} 2^{f(x)}$

28. $\lim\limits_{x \to 4} [1 + f(x)]^2$

29. $\lim\limits_{x \to 4} \dfrac{f(x) + g(x)}{2g(x)}$

30. $\lim\limits_{x \to 4} \dfrac{5g(x) + 2}{1 - f(x)}$

Use the properties of limits to help decide whether each limit exists. If a limit exists, find its value.

31. $\lim\limits_{x \to 3} \dfrac{x^2 - 9}{x - 3}$

32. $\lim\limits_{x \to -2} \dfrac{x^2 - 4}{x + 2}$

33. $\lim\limits_{x \to 1} \dfrac{5x^2 - 7x + 2}{x^2 - 1}$

34. $\lim\limits_{x \to -3} \dfrac{x^2 - 9}{x^2 + x - 6}$

35. $\lim\limits_{x \to -2} \dfrac{x^2 - x - 6}{x + 2}$

36. $\lim\limits_{x \to 5} \dfrac{x^2 - 3x - 10}{x - 5}$

37. $\lim\limits_{x \to 0} \dfrac{1/(x + 3) - 1/3}{x}$

38. $\lim\limits_{x \to 0} \dfrac{-1/(x + 2) + 1/2}{x}$

39. $\lim\limits_{x \to 25} \dfrac{\sqrt{x} - 5}{x - 25}$

40. $\lim\limits_{x \to 36} \dfrac{\sqrt{x} - 6}{x - 36}$

41. $\lim\limits_{h \to 0} \dfrac{(x + h)^2 - x^2}{h}$

42. $\lim\limits_{h \to 0} \dfrac{(x + h)^3 - x^3}{h}$

43. $\lim\limits_{x \to \infty} \dfrac{3x}{7x - 1}$

44. $\lim\limits_{x \to -\infty} \dfrac{8x + 2}{4x - 5}$

45. $\lim\limits_{x \to -\infty} \dfrac{3x^2 + 2x}{2x^2 - 2x + 1}$

46. $\lim\limits_{x \to \infty} \dfrac{x^2 + 2x - 5}{3x^2 + 2}$

47. $\lim\limits_{x \to \infty} \dfrac{3x^3 + 2x - 1}{2x^4 - 3x^3 - 2}$

48. $\lim\limits_{x \to \infty} \dfrac{2x^2 - 1}{3x^4 + 2}$

49. $\lim\limits_{x \to \infty} \dfrac{2x^3 - x - 3}{6x^2 - x - 1}$

50. $\lim\limits_{x \to \infty} \dfrac{x^4 - x^3 - 3x}{7x^2 + 9}$

51. $\lim\limits_{x \to \infty} \dfrac{2x^2 - 7x^4}{9x^2 + 5x - 6}$

52. $\lim\limits_{x \to \infty} \dfrac{-5x^3 - 4x^2 + 8}{6x^2 + 3x + 2}$

53. Let $f(x) = \begin{cases} x^3 + 2 & \text{if } x \neq -1 \\ 5 & \text{if } x = -1 \end{cases}$. Find $\lim\limits_{x \to -1} f(x)$.

54. Let $g(x) = \begin{cases} 0 & \text{if } x = -2 \\ \frac{1}{2}x^2 - 3 & \text{if } x \neq -2 \end{cases}$. Find $\lim\limits_{x \to -2} g(x)$.

55. Let $f(x) = \begin{cases} x - 1 & \text{if } x < 3 \\ 2 & \text{if } 3 \leq x \leq 5 \\ x + 3 & \text{if } x > 5 \end{cases}$.

(a) Find $\lim\limits_{x \to 3} f(x)$.   (b) Find $\lim\limits_{x \to 5} f(x)$.

56. Let $g(x) = \begin{cases} 5 & \text{if } x < 0 \\ x^2 - 2 & \text{if } 0 \leq x \leq 3 \\ 7 & \text{if } x > 3 \end{cases}$.

(a) Find $\lim\limits_{x \to 0} g(x)$.   (b) Find $\lim\limits_{x \to 3} g(x)$.

57. Does a value of k exist such that the following limit exists?

$$\lim_{x \to 2} \frac{3x^2 + kx - 2}{x^2 - 3x + 2}$$

If so, find the value of k and the corresponding limit. If not, explain why not.

58. Repeat the instructions of Exercise 57 for the following limit.

$$\lim_{x \to 3} \frac{2x^2 + kx - 9}{x^2 - 4x + 3}$$

In Exercises 59–62, calculate the limit in the specified exercise, using a table such as in Exercises 15–20. Verify your answer by using a graphing calculator to zoom in on the point on the graph.

59. Exercise 31

60. Exercise 32

61. Exercise 33

62. Exercise 34

63. Let $F(x) = \dfrac{3x}{(x + 2)^3}$.

(a) Find $\lim\limits_{x \to -2} F(x)$.

(b) Find the vertical asymptote of the graph of $F(x)$.

(c) Compare your answers for parts (a) and (b). What can you conclude?

64. Let $G(x) = \dfrac{-6}{(x - 4)^2}$.

(a) Find $\lim\limits_{x \to 4} G(x)$.

(b) Find the vertical asymptote of the graph of $G(x)$.

(c) Compare your answers for parts (a) and (b). Are they related? How?

65. Describe how the behavior of the graph in Figure 10 near $x = 1$ can be predicted by the simplified expression for the function $y = 1/(x - 1)$.

66. A friend who is confused about limits wonders why you investigate the value of a function closer and closer to a point, instead of just finding the value of a function at the point. How would you respond?

67. Use a graph of $f(x) = e^x$ to answer the following questions.

(a) Find $\lim\limits_{x \to -\infty} e^x$.

(b) Where does the function e^x have a horizontal asymptote?

68. Use a graphing calculator to answer the following questions.

(a) From a graph of $y = xe^{-x}$, what do you think is the value of $\lim\limits_{x \to \infty} xe^{-x}$? Support this by evaluating the function for several large values of x.

(b) Repeat part (a), this time using the graph of $y = x^2e^{-x}$.

(c) Based on your results from parts (a) and (b), what do you think is the value of $\lim\limits_{x \to \infty} x^n e^{-x}$, where n is a positive integer? Support this by experimenting with other positive integers n.

69. Use a graph of $f(x) = \ln x$ to answer the following questions.

(a) Find $\lim\limits_{x \to 0^+} \ln x$.

(b) Where does the function $\ln x$ have a vertical asymptote?

70. Use a graphing calculator to answer the following questions.

(a) From a graph of $y = x \ln x$, what do you think is the value of $\lim\limits_{x \to 0^+} x \ln x$? Support this by evaluating the function for several small values of x.

(b) Repeat part (a), this time using the graph of $y = x(\ln x)^2$.

(c) Based on your results from parts (a) and (b), what do you think is the value of $\lim\limits_{x \to 0^+} x(\ln x)^n$, where n is a positive integer? Support this by experimenting with other positive integers n.

71. Explain in your own words why the rules for limits at infinity should be true.

72. Explain in your own words what Rule 4 for limits means.

Find each of the following limits (a) by investigating values of the function near the x-value where the limit is taken, and (b) using a graphing calculator to view the function near that value of x.

73. $\lim\limits_{x \to 1} \dfrac{x^4 + 4x^3 - 9x^2 + 7x - 3}{x - 1}$

74. $\lim\limits_{x \to 2} \dfrac{x^4 + x - 18}{x^2 - 4}$

75. $\lim\limits_{x \to -1} \dfrac{x^{1/3} + 1}{x + 1}$

76. $\lim\limits_{x \to 4} \dfrac{x^{3/2} - 8}{x + x^{1/2} - 6}$

Use a graphing calculator to graph the function. (a) Determine the limit from the graph. (b) Explain how your answer could be determined from the expression for $f(x)$.

77. $\lim\limits_{x \to \infty} \dfrac{\sqrt{9x^2 + 5}}{2x}$

78. $\lim\limits_{x \to -\infty} \dfrac{\sqrt{9x^2 + 5}}{2x}$

79. $\lim\limits_{x \to -\infty} \dfrac{\sqrt{36x^2 + 2x + 7}}{3x}$

80. $\lim\limits_{x \to \infty} \dfrac{\sqrt{36x^2 + 2x + 7}}{3x}$

81. $\lim\limits_{x \to \infty} \dfrac{(1 + 5x^{1/3} + 2x^{5/3})^3}{x^5}$

82. $\lim\limits_{x \to -\infty} \dfrac{(1 + 5x^{1/3} + 2x^{5/3})^3}{x^5}$

83. Explain why the following rules can be used to find $\lim\limits_{x \to \infty} [p(x)/q(x)]$:

(a) If the degree of $p(x)$ is less than the degree of $q(x)$, the limit is 0.

(b) If the degree of $p(x)$ is equal to the degree of $q(x)$, the limit is A/B, where A and B are the leading coefficients of $p(x)$ and $q(x)$, respectively.

(c) If the degree of $p(x)$ is greater than the degree of $q(x)$, the limit is ∞ or $-\infty$.

APPLICATIONS

Business and Economics

84. APPLY IT Consumer Demand When the price of an essential commodity (such as gasoline) rises rapidly, consumption drops slowly at first. If the price continues to rise, however, a "tipping" point may be reached, at which consumption takes a sudden substantial drop. Suppose the accompanying graph shows the consumption of gasoline, $G(t)$, in millions of gallons, in a certain area. We assume that the price is rising rapidly. Here t is time in months after the price began rising. Use the graph to find the following.

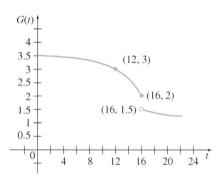

(a) $\lim\limits_{t \to 12} G(t)$

(b) $\lim\limits_{t \to 16} G(t)$

(c) $G(16)$

(d) The tipping point (in months)

85. Sales Tax Officials in California tend to raise the sales tax in years in which the state faces a budget deficit and then cut the tax when the state has a surplus. The graph below shows the California state sales tax in recent years. Let $T(x)$ represent the sales tax per dollar spent in year x. Find the following. *Source: California State.*

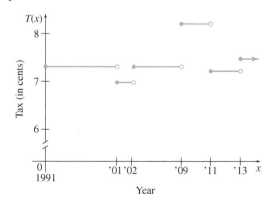

(a) $\lim_{x \to 94} T(x)$ (b) $\lim_{x \to 13^-} T(x)$
(c) $\lim_{x \to 13^+} T(x)$ (d) $\lim_{x \to 13} T(x)$
(e) $T(13)$

86. Postage The graph below shows how the postage required to mail a letter in the United States has changed in recent years. Let $C(t)$ be the cost to mail a letter in the year t. Find the following. *Source: United States Postal Service*.

(a) $\lim_{t \to 2014^-} C(t)$ (b) $\lim_{t \to 2014^+} C(t)$
(c) $\lim_{t \to 2014} C(t)$ (d) $C(2014)$

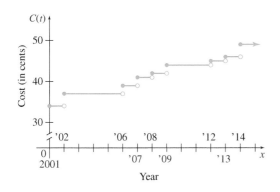

87. Average Cost The cost (in dollars) for manufacturing a particular DVD is

$$C(x) = 15,000 + 6x,$$

where x is the number of DVDs produced. Recall from the previous chapter that the average cost per DVD, denoted by $\overline{C}(x)$, is found by dividing $C(x)$ by x. Find and interpret $\lim_{x \to \infty} \overline{C}(x)$.

88. Average Cost In Chapter 1, we saw that the cost to fly x miles on American Airlines could be approximated by the equation

$$C(x) = 0.0417x + 167.55.$$

Recall from the previous chapter that the average cost per mile, denoted by $\overline{C}(x)$, is found by dividing $C(x)$ by x. Find and interpret $\lim_{x \to \infty} \overline{C}(x)$. *Source: American Airlines*.

89. Employee Productivity A company training program has determined that, on the average, a new employee produces $P(s)$ items per day after s days of on-the-job training, where

$$P(s) = \frac{63s}{s + 8}.$$

Find and interpret $\lim_{s \to \infty} P(s)$.

90. Preferred Stock In business finance, an annuity is a series of equal payments received at equal intervals for a finite period of time. The *present value* of an n-period annuity takes the form

$$P = R\left[\frac{1 - (1 + i)^{-n}}{i}\right],$$

where R is the amount of the periodic payment and i is the fixed interest rate per period. Many corporations raise money by issuing preferred stock. Holders of the preferred stock, called a *perpetuity*, receive payments that take the form of an annuity in that the amount of the payment never changes. However, normally the payments for preferred stock do not end but theoretically continue forever. Find the limit of this present value equation as n approaches infinity to derive a formula for the present value of a share of preferred stock paying a periodic dividend R. *Source: Robert D. Campbell*.

91. Growing Annuities For some annuities encountered in business finance, called *growing annuities*, the amount of the periodic payment is not constant but grows at a constant periodic rate. Leases with escalation clauses can be examples of growing annuities. The present value of a growing annuity takes the form

$$P = \frac{R}{i - g}\left[1 - \left(\frac{1 + g}{1 + i}\right)^n\right],$$

where

$\quad R$ = amount of the next annuity payment,

$\quad g$ = expected constant annuity growth rate,

$\quad i$ = required periodic return at the time the annuity is evaluated,

$\quad n$ = number of periodic payments.

A corporation's common stock may be thought of as a claim on a growing annuity where the annuity is the company's annual dividend. However, in the case of common stock, these payments have no contractual end but theoretically continue forever. Compute the limit of the expression above as n approaches infinity to derive the Gordon–Shapiro Dividend Model popularly used to estimate the value of common stock. Make the reasonable assumption that $i > g$. (*Hint:* What happens to a^n as $n \to \infty$ if $0 < a < 1$?) *Source: Robert D. Campbell*.

Life Sciences

92. Alligator Teeth Researchers have developed a mathematical model that can be used to estimate the number of teeth $N(t)$ at time t (days of incubation) for *Alligator mississippiensis*, where

$$N(t) = 71.8e^{-8.96e^{-0.0685t}}.$$

Source: Journal of Theoretical Biology.

(a) Find $N(65)$, the number of teeth of an alligator that hatched after 65 days.

(b) Find $\lim_{t \to \infty} N(t)$ and use this value as an estimate of the number of teeth of a newborn alligator. (*Hint:* See Exercise 67.) Does this estimate differ significantly from the estimate of part (a)?

93. Sediment To develop strategies to manage water quality in polluted lakes, biologists must determine the depths of sediments and the rate of sedimentation. It has been determined that the depth of sediment $D(t)$ (in centimeters) with respect to time (in years before 1990) for Lake Coeur d'Alene, Idaho, can be estimated by the equation

$$D(t) = 155(1 - e^{-0.0133t}).$$

Source: The Mathematics Teacher.

(a) Find $D(20)$ and interpret.

(b) Find $\lim_{t \to \infty} D(t)$ and interpret.

94. Drug Concentration The concentration of a drug in a patient's bloodstream h hours after it was injected is given by

$$A(h) = \frac{0.17h}{h^2 + 2}.$$

Find and interpret $\lim_{h \to \infty} A(h)$.

Social Sciences

95. Legislative Voting Members of a legislature often must vote repeatedly on the same bill. As time goes on, members may change their votes. Suppose that p_0 is the probability that an individual legislator favors an issue before the first roll call vote, and suppose that p is the probability of a change in position from one vote to the next. Then the probability that the legislator will vote "yes" on the nth roll call is given by

$$p_n = \frac{1}{2} + \left(p_0 - \frac{1}{2}\right)(1 - 2p)^n.$$

For example, the chance of a "yes" on the third roll call vote is

$$p_3 = \frac{1}{2} + \left(p_0 - \frac{1}{2}\right)(1 - 2p)^3.$$

Source: Mathematics in the Behavioral and Social Sciences.

Suppose that there is a chance of $p_0 = 0.7$ that Congressman Stephens will favor the budget appropriation bill before the first roll call, but only a probability of $p = 0.2$ that he will change his mind on the subsequent vote. Find and interpret the following.

(a) p_2 (b) p_4

(c) p_8 (d) $\lim_{n \to \infty} p_n$

YOUR TURN ANSWERS

1. 3   **2.** 4   **3.** 5   **4.** Does not exist.

5. 49   **6.** −7   **7.** 1/2   **8.** 1/3

3.2 Continuity

APPLY IT How does the average cost per day of a rental car change with the number of days the car is rented?

We will answer this question in Exercise 38.

In 2009, Congress passed legislation raising the federal minimum wage for the third time in three years. Figure 13 below shows how that wage has varied since it was instituted in 1938. We will denote this function by $f(t)$, where t is the year. *Source: U.S. Department of Labor.*

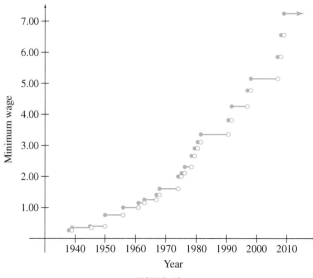

FIGURE 13

Notice from the graph that $\lim\limits_{t \to 1997^-} f(t) = 4.75$ and that $\lim\limits_{t \to 1997^+} f(t) = 5.15$, so that $\lim\limits_{t \to 1997} f(t)$ does not exist. Notice also that $f(1997) = 5.15$. A point such as this, where a function has a sudden sharp break, is a point where the function is *discontinuous*. In this case, the discontinuity is caused by the jump in the minimum wage from \$4.75 per hour to \$5.15 per hour in 1997.

Intuitively speaking, a function is *continuous* at a point if you can draw the graph of the function in the vicinity of that point without lifting your pencil from the paper. As we already mentioned, this would not be possible in Figure 13 if it were drawn correctly; there would be a break in the graph at $t = 1997$, for example. Conversely, a function is discontinuous at any x-value where the pencil *must* be lifted from the paper in order to draw the graph on both sides of the point. A more precise definition is as follows.

Continuity at $x = c$

A function f is **continuous** at $x = c$ if the following three conditions are satisfied:

1. $f(c)$ is defined,

2. $\lim\limits_{x \to c} f(x)$ exists, and

3. $\lim\limits_{x \to c} f(x) = f(c)$.

If f is not continuous at c, it is **discontinuous** there.

The following example shows how to check a function for continuity at a specific point. We use a three-step test, and if any step of the test fails, the function is not continuous at that point.

EXAMPLE 1 Continuity

Determine if each function is continuous at the indicated x-value.

(a) $f(x)$ in Figure 14 at $x = 3$

SOLUTION

> *Step 1* Does the function exist at $x = 3$?
>
> The open circle on the graph of Figure 14 at the point where $x = 3$ means that $f(x)$ does not exist at $x = 3$. Since the function does not pass the first test, it is discontinuous at $x = 3$, and there is no need to proceed to Step 2.

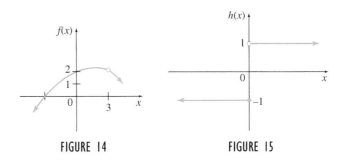

FIGURE 14   FIGURE 15

(b) $h(x)$ in Figure 15 at $x = 0$

SOLUTION

> *Step 1* Does the function exist at $x = 0$?
>
> According to the graph in Figure 15, $h(0)$ exists and is equal to -1.

Step 2 Does the limit exist at $x = 0$?

As x approaches 0 from the left, $h(x)$ is -1. As x approaches 0 from the right, however, $h(x)$ is 1. In other words,

$$\lim_{x \to 0^-} h(x) = -1,$$

while

$$\lim_{x \to 0^+} h(x) = 1.$$

Since no single number is approached by the values of $h(x)$ as x approaches 0, the limit $\lim_{x \to 0} h(x)$ does not exist. Since the function does not pass the second test, it is discontinuous at $x = 0$, and there is no need to proceed to Step 3.

(c) $g(x)$ in Figure 16 at $x = 4$

SOLUTION

Step 1 Is the function defined at $x = 4$?

In Figure 16, the heavy dot above 4 shows that $g(4)$ is defined. In fact, $g(4) = 1$.

Step 2 Does the limit exist at $x = 4$?

The graph shows that

$$\lim_{x \to 4^-} g(x) = -2, \text{ and } \lim_{x \to 4^+} g(x) = -2.$$

Therefore, the limit exists at $x = 4$ and

$$\lim_{x \to 4} g(x) = -2.$$

Step 3 Does $g(4) = \lim_{x \to 4} g(x)$?

Using the results of Step 1 and Step 2, we see that $g(4) \neq \lim_{x \to 4} g(x)$. Since the function does not pass the third test, it is discontinuous at $x = 4$.

FIGURE 16

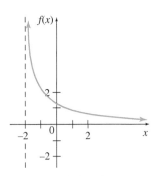

FIGURE 17

(d) $f(x)$ in Figure 17 at $x = -2$.

SOLUTION

Step 1 Does the function exist at $x = -2$?

The function f graphed in Figure 17 is not defined at $x = -2$. Since the function does not pass the first test, it is discontinuous at $x = -2$. (Function f is continuous at any value of x greater than -2, however.)

Notice that the function in part (a) of Example 1 could be made continuous simply by defining $f(3) = 2$. Similarly, the function in part (c) could be made continuous by redefining $g(4) = -2$. In such cases, when the function can be made continuous at a specific point simply by defining or redefining it at that point, the function is said to have a **removable discontinuity**.

A function is said to be **continuous on an open interval** if it is continuous at every x-value in the interval. Continuity on a closed interval is slightly more complicated because we must decide what to do with the endpoints. We will say that a function f is **continuous from the right** at $x = c$ if $\lim\limits_{x \to c^+} f(x) = f(c)$. A function f is **continuous from the left** at $x = c$ if $\lim\limits_{x \to c^-} f(x) = f(c)$. With these ideas, we can now define continuity on a closed interval.

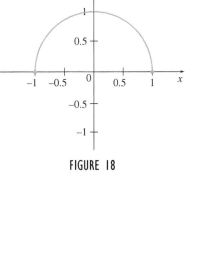

FIGURE 18

Continuity on a Closed Interval

A function is **continuous on a closed interval** $[a, b]$ if

1. it is continuous on the open interval (a, b),
2. it is continuous from the right at $x = a$, and
3. it is continuous from the left at $x = b$.

For example, the function $f(x) = \sqrt{1 - x^2}$, shown in Figure 18, is continuous on the closed interval $[-1, 1]$. By defining continuity on a closed interval in this way, we need not worry about the fact that $\sqrt{1 - x^2}$ does not exist to the left of $x = -1$ or to the right of $x = 1$. The table below lists some key functions and tells where each is continuous.

	Continuous Functions	
Type of Function	**Where It Is Continuous**	**Graphic Example**
Polynomial Function $y = a_n x^n + a_{n-1} x^{n-1} + \cdots + a_1 x + a_0$, where $a_n, a_{n-1}, \ldots, a_1, a_0$ are real numbers, not all 0	For all x	(graph)
Rational Function $y = \dfrac{p(x)}{q(x)}$, where $p(x)$ and $q(x)$ are polynomials, with $q(x) \neq 0$	For all x where $q(x) \neq 0$	(graph)
Root Function $y = \sqrt{ax + b}$, where a and b are real numbers, with $a \neq 0$ and $ax + b \geq 0$	For all x where $ax + b \geq 0$	(graph)

(continued)

Continuous Functions (cont.)		
Type of Function	**Where It Is Continuous**	**Graphic Example**
Exponential Function $y = a^x$ where $a > 0$	For all x	
Logarithmic Function $y = \log_a x$ where $a > 0$, $a \neq 1$	For all $x > 0$	

Continuous functions are nice to work with because finding $\lim_{x \to c} f(x)$ is simple if f is continuous: just evaluate $f(c)$.

When a function is given by a graph, any discontinuities are clearly visible. When a function is given by a formula, it is usually continuous at all x-values except those where the function is undefined or possibly where there is a change in the defining formula for the function, as shown in the following examples.

EXAMPLE 2   Continuity

Find all values $x = a$ where the function is discontinuous.

(a) $f(x) = \dfrac{4x - 3}{2x - 7}$

YOUR TURN 1   Find all values $x = a$ where the function is discontinuous.
$f(x) = \sqrt{5x + 3}$

SOLUTION   This rational function is discontinuous wherever the denominator is zero. There is a discontinuity when $a = 7/2$.

(b) $g(x) = e^{2x-3}$

SOLUTION   This exponential function is continuous for all x.   TRY YOUR TURN 1

EXAMPLE 3   Continuity

Find all values of x where the following piecewise function is discontinuous.

$$f(x) = \begin{cases} x + 1 & \text{if } x < 1 \\ x^2 - 3x + 4 & \text{if } 1 \leq x \leq 3. \\ 5 - x & \text{if } x > 3 \end{cases}$$

SOLUTION   Since each piece of this function is a polynomial, the only x-values where f might be discontinuous here are 1 and 3. We investigate at $x = 1$ first. From the left, where x-values are less than 1,

$$\lim_{x \to 1^-} f(x) = \lim_{x \to 1^-} (x + 1) = 1 + 1 = 2.$$

From the right, where x-values are greater than 1,

$$\lim_{x \to 1^+} f(x) = \lim_{x \to 1^+} (x^2 - 3x + 4) = 1^2 - 3 + 4 = 2.$$

Furthermore, $f(1) = 1^2 - 3 + 4 = 2$, so $\lim_{x \to 1} f(x) = f(1) = 2$. Thus f is continuous at $x = 1$, since $f(1) = \lim_{x \to 1} f(x)$.

Now let us investigate $x = 3$. From the left,

$$\lim_{x \to 3^-} f(x) = \lim_{x \to 3^-} (x^2 - 3x + 4) = 3^2 - 3(3) + 4 = 4.$$

From the right,

$$\lim_{x \to 3^+} f(x) = \lim_{x \to 3^+} (5 - x) = 5 - 3 = 2.$$

YOUR TURN 2 Find all values of x where the piecewise function is discontinuous.

$$f(x) = \begin{cases} 5x - 4 & \text{if } x < 0 \\ x^2 & \text{if } 0 \le x \le 3 \\ x + 6 & \text{if } x > 3 \end{cases}$$

Because $\lim_{x \to 3^-} f(x) \ne \lim_{x \to 3^+} f(x)$, the limit $\lim_{x \to 3} f(x)$ does not exist, so f is discontinuous at $x = 3$, regardless of the value of $f(3)$.

The graph of $f(x)$ can be drawn by considering each of the three parts separately. In the first part, the line $y = x + 1$ is drawn including only the section of the line to the left of $x = 1$. The other two parts are drawn similarly, as illustrated in Figure 19. We can see by the graph that the function is continuous at $x = 1$ and discontinuous at $x = 3$, which confirms our solution above. **TRY YOUR TURN 2**

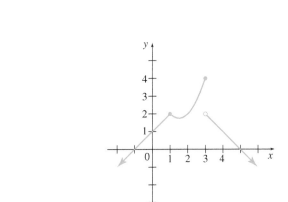

FIGURE 19

TECHNOLOGY NOTE

Some graphing calculators have the ability to draw piecewise functions. On the TI-84 Plus C, letting

$$Y_1 = (X + 1)(X < 1) + (X^2 - 3X + 4)(1 \le X)(X \le 3) + (5 - X)(X > 3)$$

produces the graph shown in Figure 20.

FIGURE 20

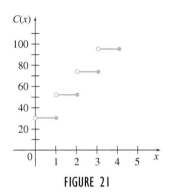

FIGURE 21

EXAMPLE 4 Cost Analysis

A trailer rental firm charges a flat $8 to rent a hitch. The trailer itself is rented for $22 per day or fraction of a day. Let $C(x)$ represent the cost of renting a hitch and trailer for x days.

(a) Graph C.

SOLUTION The charge for one day is $8 for the hitch and $22 for the trailer, or $30. In fact, if $0 < x \le 1$, then $C(x) = 30$. To rent the trailer for more than one day, but not more than two days, the charge is $8 + 2 \cdot 22 = 52$ dollars. For any value of x satisfying $1 < x \le 2$, the cost is $C(x) = 52$. Also, if $2 < x \le 3$, then $C(x) = 74$. These results lead to the graph in Figure 21.

(b) Find any values of x where C is discontinuous.

SOLUTION As the graph suggests, C is discontinuous at $x = 1, 2, 3, 4$, and all other positive integers.

One application of continuity is the **Intermediate Value Theorem**, which says that if a function is continuous on a closed interval $[a, b]$, the function takes on every value between $f(a)$ and $f(b)$. For example, if $f(1) = -3$ and $f(2) = 5$, then f must take on every value between -3 and 5 as x varies over the interval $[1, 2]$. In particular (in this case), there must be a value of x in the interval $(1, 2)$ such that $f(x) = 0$. If f were discontinuous, however, this conclusion would not necessarily be true. This is important because, if we are searching for a solution to $f(x) = 0$ in $[1, 2]$, we would like to know that a solution exists.

3.2 WARM-UP EXERCISES

Find each of the following limits. *(Sec. 3.1)*

W1. $\displaystyle\lim_{x \to 2} \frac{2x^2 - 11x + 14}{x^2 - 5x + 6}$

W2. $\displaystyle\lim_{x \to 4} \frac{3x^2 - 4x - 32}{x^2 - 6x + 8}$

Let $f(x) = \begin{cases} x + 1 & \text{if } x < 3 \\ 2x - 2 & \text{if } 3 \le x \le 5 \\ 4x + 1 & \text{if } x > 5 \end{cases}$. **Find each of the following limits.** *(Sec. 3.1)*

W3. $\displaystyle\lim_{x \to 3} f(x)$

W4. $\displaystyle\lim_{x \to 5} f(x)$

W5. $\displaystyle\lim_{x \to 6} f(x)$

3.2 EXERCISES

In Exercises 1–6, find all values $x = a$ where the function is discontinuous. For each point of discontinuity, give (a) $f(a)$ if it exists, (b) $\displaystyle\lim_{x \to a^-} f(x)$, (c) $\displaystyle\lim_{x \to a^+} f(x)$, (d) $\displaystyle\lim_{x \to a} f(x)$, and (e) identify which conditions for continuity are not met. Be sure to note when the limit doesn't exist.

1.

2.

3.

4.

5.

6.

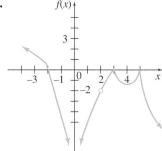

Find all values $x = a$ where the function is discontinuous. For each value of x, give the limit of the function as x approaches a. Be sure to note when the limit doesn't exist.

7. $f(x) = \dfrac{5 + x}{x(x - 2)}$

8. $f(x) = \dfrac{-2x}{(2x + 1)(3x + 6)}$

9. $f(x) = \dfrac{x^2 - 4}{x - 2}$ **10.** $f(x) = \dfrac{x^2 - 25}{x + 5}$

11. $p(x) = x^2 - 4x + 11$

12. $q(x) = -3x^3 + 2x^2 - 4x + 1$

13. $p(x) = \dfrac{|x + 2|}{x + 2}$ **14.** $r(x) = \dfrac{|5 - x|}{x - 5}$

15. $k(x) = e^{\sqrt{x-1}}$ **16.** $j(x) = e^{1/x}$

17. $r(x) = \ln\left|\dfrac{x}{x - 1}\right|$ **18.** $j(x) = \ln\left|\dfrac{x + 2}{x - 3}\right|$

In Exercises 19–24, (a) graph the given function, (b) find all values of x where the function is discontinuous, and (c) find the limit from the left and from the right at any values of x found in part (b).

19. $f(x) = \begin{cases} 1 & \text{if } x < 2 \\ x + 3 & \text{if } 2 \le x \le 4 \\ 7 & \text{if } x > 4 \end{cases}$

20. $f(x) = \begin{cases} x - 1 & \text{if } x < 1 \\ 0 & \text{if } 1 \le x \le 4 \\ x - 2 & \text{if } x > 4 \end{cases}$

21. $g(x) = \begin{cases} 11 & \text{if } x < -1 \\ x^2 + 2 & \text{if } -1 \le x \le 3 \\ 11 & \text{if } x > 3 \end{cases}$

22. $g(x) = \begin{cases} 0 & \text{if } x < 0 \\ x^2 - 5x & \text{if } 0 \le x \le 5 \\ 5 & \text{if } x > 5 \end{cases}$

23. $h(x) = \begin{cases} 4x + 4 & \text{if } x \le 0 \\ x^2 - 4x + 4 & \text{if } x > 0 \end{cases}$

24. $h(x) = \begin{cases} x^2 + x - 12 & \text{if } x \le 1 \\ 3 - x & \text{if } x > 1 \end{cases}$

In Exercises 25–28, find the value of the constant k that makes the function continuous.

25. $f(x) = \begin{cases} kx^2 & \text{if } x \le 2 \\ x + k & \text{if } x > 2 \end{cases}$

26. $g(x) = \begin{cases} x^3 + k & \text{if } x \le 3 \\ kx - 5 & \text{if } x > 3 \end{cases}$

27. $g(x) = \begin{cases} \dfrac{2x^2 - x - 15}{x - 3} & \text{if } x \ne 3 \\ kx - 1 & \text{if } x = 3 \end{cases}$

28. $h(x) = \begin{cases} \dfrac{3x^2 + 2x - 8}{x + 2} & \text{if } x \ne -2 \\ 3x + k & \text{if } x = -2 \end{cases}$

29. Explain in your own words what the Intermediate Value Theorem says and why it seems plausible.

30. Explain why $\lim\limits_{x \to 2}(3x^2 + 8x)$ can be evaluated by substituting $x = 2$.

In Exercises 31–32, (a) use a graphing calculator to tell where the rational function $P(x)/Q(x)$ is discontinuous, and (b) verify your answer from part (a) by using the graphing calculator to plot $Q(x)$ and determine where $Q(x) = 0$. You will need to choose the viewing window carefully.

31. $f(x) = \dfrac{x^2 + x + 2}{x^3 - 0.9x^2 + 4.14x - 5.4}$

32. $f(x) = \dfrac{x^2 + 3x - 2}{x^3 - 0.9x^2 + 4.14x + 5.4}$

33. Let $g(x) = \dfrac{x + 4}{x^2 + 2x - 8}$. Determine all values of x at which g is discontinuous, and for each of these values of x, define g in such a manner so as to remove the discontinuity, if possible. Choose one of the following. *Source: Society of Actuaries.*

(a) g is discontinuous only at -4 and 2.
   Define $g(-4) = -\frac{1}{6}$ to make g continuous at -4.
   $g(2)$ cannot be defined to make g continuous at 2.

(b) g is discontinuous only at -4 and 2.
   Define $g(-4) = -\frac{1}{6}$ to make g continuous at -4.
   Define $g(2) = 6$ to make g continuous at 2.

(c) g is discontinuous only at -4 and 2.
   $g(-4)$ cannot be defined to make g continuous at -4.
   $g(2)$ cannot be defined to make g continuous at 2.

(d) g is discontinuous only at 2.
   Define $g(2) = 6$ to make g continuous at 2.

(e) g is discontinuous only at 2.
   $g(2)$ cannot be defined to make g continuous at 2.

34. Tell at what values of x the function $f(x)$ in Figure 8 from the previous section is discontinuous. Explain why it is discontinuous at each of these values.

APPLICATIONS

Business and Economics

35. Production The graph shows the profit from the daily production of x thousand kilograms of an industrial chemical. Use the graph to find the following limits.

(a) $\lim\limits_{x \to 6} P(x)$

(b) $\lim\limits_{x \to 10^-} P(x)$

(c) $\lim\limits_{x \to 10^+} P(x)$

(d) $\lim\limits_{x \to 10} P(x)$

(e) Where is the function discontinuous? What might account for such a discontinuity?

(f) Use the graph to estimate the number of units of the chemical that must be produced before the second shift is as profitable as the first.

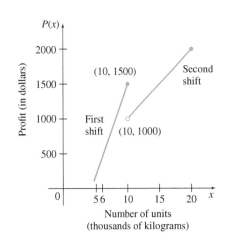

Number of units
(thousands of kilograms)

36. Cost Analysis The cost to transport a mobile home depends on the distance, x, in miles that the home is moved. Let $C(x)$ represent the cost to move a mobile home x miles. One firm charges as follows.

Cost per Mile	Distance in Miles
$4.00	$0 < x \le 150$
$3.00	$150 < x \le 400$
$2.50	$400 < x$

Find the cost to move a mobile home the following distances.

(a) 130 miles   (b) 150 miles   (c) 210 miles

(d) 400 miles   (e) 500 miles

(f) Where is C discontinuous?

37. Cost Analysis A company charges $1.25 per lb for a certain fertilizer on all orders 100 lb or less, and $1 per lb for orders over 100 lb. Let $F(x)$ represent the cost for buying x lb of the fertilizer. Find the cost of buying the following.

(a) 80 lb   (b) 150 lb   (c) 100 lb

(d) Where is F discontinuous?

38. APPLY IT   Car Rental Recently, a car rental firm charged $36 per day or portion of a day to rent a car for a period of 1 to 5 days. Days 6 and 7 were then free, while the charge for days 8 through 12 was again $36 per day. Let $A(t)$ represent the average cost to rent the car for t days, where $0 < t \le 12$. Find the average cost of a rental for the following number of days.

(a) 4   (b) 5   (c) 6   (d) 7   (e) 8

(f) Find $\lim\limits_{x \to 5^-} A(t)$.

(g) Find $\lim\limits_{x \to 5^+} A(t)$.

(h) Where is A discontinuous on the given interval?

39. Postage In 2014, it cost $0.98 to send a large envelope within the United States for the first ounce, and $0.21 for each additional ounce, or fraction thereof, up to 13 ounces. Let $C(x)$ be the cost to mail x ounces. Find the following. *Source: U.S. Postal Service.*

(a) $\lim\limits_{x \to 3^-} C(x)$

(b) $\lim\limits_{x \to 3^+} C(x)$

(c) $\lim\limits_{x \to 3} C(x)$

(d) $C(3)$

(e) $\lim\limits_{x \to 8.5^-} C(x)$

(f) $\lim\limits_{x \to 8.5^+} C(x)$

(g) $\lim\limits_{x \to 8.5} C(x)$

(h) $C(8.5)$

(i) Find all values on the interval $(0, 13)$ where the function C is discontinuous.

Life Sciences

40. Pregnancy A woman's weight naturally increases during the course of a pregnancy. When she delivers, her weight immediately decreases by the approximate weight of the child. Suppose that a 120-lb woman gains 27 lb during pregnancy, delivers a 7-lb baby, and then, through diet and exercise, loses the remaining weight during the next 20 weeks.

(a) Graph the weight gain and loss during the pregnancy and the 20 weeks following the birth of the baby. Assume that the pregnancy lasts 40 weeks, that delivery occurs immediately after this time interval, and that the weight gain/loss before and after birth is linear.

(b) Is this a continuous function? If not, then find the value(s) of t where the function is discontinuous.

41. Poultry Farming Researchers at Iowa State University and the University of Arkansas have developed a piecewise function that can be used to estimate the body weight (in grams) of a male broiler during the first 56 days of life according to

$$W(t) = \begin{cases} 48 + 3.64t + 0.6363t^2 + 0.00963t^3 & \text{if } 1 \le t \le 28, \\ -1004 + 65.8t & \text{if } 28 < t \le 56, \end{cases}$$

where t is the age of the chicken (in days). *Source: Poultry Science.*

(a) Determine the weight of a male broiler that is 25 days old.

(b) Is $W(t)$ a continuous function?

(c) Use a graphing calculator to graph $W(t)$ on $[1, 56]$ by $[0, 3000]$. Comment on the accuracy of the graph.

(d) Comment on why researchers would use two different types of functions to estimate the weight of a chicken at various ages.

YOUR TURN ANSWERS
1. Discontinuous when $a < -3/5$.
2. Discontinuous at $x = 0$.

3.3 Rates of Change

APPLY IT How does the manufacturing cost of a DVD change as the number of DVDs manufactured changes?

This question will be answered in Example 4 of this section as we develop a method for finding the rate of change of one variable with respect to a unit change in another variable.

Average Rate of Change One of the main applications of calculus is determining how one variable changes in relation to another. A marketing manager wants to know how profit changes with respect to the amount spent on advertising, while a physician wants to know how a patient's reaction to a drug changes with respect to the dose.

For example, suppose we take a trip from San Francisco driving south. Every half-hour we note how far we have traveled, with the following results for the first three hours.

Distance Traveled						
Time in Hours	0	0.5	1	1.5	2	2.5	3
Distance in Miles	0	30	55	80	104	124	138

If s is the function whose rule is

$$s(t) = \text{Distance from San Francisco at time } t,$$

then the table shows, for example, that $s(0) = 0$, $s(1) = 55$, $s(2.5) = 124$, and so on. The distance traveled during, say, the second hour can be calculated by $s(2) - s(1) = 104 - 55 = 49$ miles.

Distance equals time multiplied by rate (or speed); so the distance formula is $d = rt$. Solving for rate gives $r = d/t$, or

$$\text{Average speed} = \frac{\text{Distance}}{\text{Time}}.$$

For example, the average speed over the time interval from $t = 0$ to $t = 3$ is

$$\text{Average speed} = \frac{s(3) - s(0)}{3 - 0} = \frac{138 - 0}{3} = 46,$$

or 46 mph. We can use this formula to find the average speed for any interval of time during the trip, as shown below.

Average Speed		
Time Interval	**Average Speed**	$= \dfrac{\textbf{Distance}}{\textbf{Time}}$
$t = 0.5$ to $t = 1$	$\dfrac{s(1) - s(0.5)}{1 - 0.5} = \dfrac{25}{0.5}$	$= 50$
$t = 0.5$ to $t = 1.5$	$\dfrac{s(1.5) - s(0.5)}{1.5 - 0.5} = \dfrac{50}{1}$	$= 50$
$t = 1$ to $t = 2$	$\dfrac{s(2) - s(1)}{2 - 1} = \dfrac{49}{1}$	$= 49$
$t = 1$ to $t = 3$	$\dfrac{s(3) - s(1)}{3 - 1} = \dfrac{83}{2}$	$= 41.5$
$t = a$ to $t = b$	$\dfrac{s(b) - s(a)}{b - a}$	

The analysis of the average speed or *average rate of change* of distance s with respect to t can be extended to include any function defined by $f(x)$ to get a formula for the average rate of change of f with respect to x.

NOTE
The formula for the average rate of change is the same as the formula for the slope of the line through $(a, f(a))$ and $(b, f(b))$. This connection between slope and rate of change will be examined more closely in the next section.

Average Rate of Change
The **average rate of change** of $f(x)$ with respect to x for a function f as x changes from a to b is

$$\frac{f(b) - f(a)}{b - a}.$$

In Figure 22 we have plotted the distance vs. time for our trip from San Francisco, connecting the points with straight line segments. Because the change in y gives the change in distance, and the change in x gives the change in time, the slope of each line segment gives the average speed over that time interval:

$$\text{Slope} = \frac{\text{Change in } y}{\text{Change in } x} = \frac{\text{Change in distance}}{\text{Change in time}} = \text{Average speed.}$$

FIGURE 22

EXAMPLE 1   Minority Population

The United States population is becoming more diverse. Based on the U.S. Census population projections for 2000 to 2050, the projected Hispanic population (in millions) can be modeled by the exponential function

$$H(t) = 37.791(1.021)^t,$$

where $t = 0$ corresponds to 2000 and $0 \le t \le 50$. Use H to estimate the average rate of change in the Hispanic population from 2000 to 2010. *Source: U.S. Census Bureau.*

SOLUTION   On the interval from $t = 0$ (2000) to $t = 10$ (2010), the average rate of change was

$$\frac{H(10) - H(0)}{10 - 0} = \frac{37.791(1.021)^{10} - 37.791(1.021)^0}{10}$$

$$\approx \frac{46.521 - 37.791}{10} = \frac{8.73}{10}$$

$$= 0.873,$$

or 0.873 million. Based on this model, it is estimated that the Hispanic population in the United States increased, on average, at a rate of about 873,000 people per year between 2000 and 2010.   TRY YOUR TURN 1

YOUR TURN 1   The projected U.S. Asian population (in millions) for this same time period is $A(t) = 11.14(1.023)^t$. Use A to estimate the average rate of change from 2000 to 2010.

EXAMPLE 2   Education Spending

Figure 23 below shows the total amount appropriated annually (in billions of dollars) for the U.S. Department of Education in recent years. Find the average rate of change per year in Department of Education appropriation from 2009 to 2013. *Source: U.S. Department of Education.*

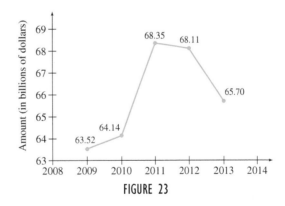

FIGURE 23

SOLUTION   Let $A(t)$ be the total amount appropriated for the Department of Education (in billions of dollars) in the year t. Then the average rate of change from 2009 to 2013 was

$$\frac{A(2013) - A(2009)}{2013 - 2009} = \frac{65.70 - 63.52}{4} = \frac{2.18}{4} = 0.545.$$

On average, the amount appropriated went up by 0.545 billion dollars, or \$545 million, per year during these years.   TRY YOUR TURN 2

YOUR TURN 2   In Example 2, find the average rate of change per year in the amount appropriated for the Department of Education from 2011 to 2013.

Instantaneous Rate of Change

Suppose a car is stopped at a traffic light. When the light turns green, the car begins to move along a straight road. Assume that the distance traveled by the car is given by the function

$$s(t) = 3t^2, \quad \text{for} \quad 0 \le t \le 15,$$

where t is the time in seconds and $s(t)$ is the distance in feet. We have already seen how to find the *average* speed of the car over any time interval. We now turn to a different problem, that of determining the exact speed of the car at a particular instant, say $t = 10$.

The intuitive idea is that the exact speed at $t = 10$ is very close to the average speed over a very short time interval near $t = 10$. If we take shorter and shorter time intervals near $t = 10$, the average speeds over these intervals should get closer and closer to the exact speed at $t = 10$. In other words, the exact speed at $t = 10$ is the limit of the average speeds over shorter and shorter time intervals near $t = 10$. The following chart illustrates this idea. The values in the chart are found using $s(t) = 3t^2$, so that, for example, $s(10) = 3(10)^2 = 300$ and $s(10.1) = 3(10.1)^2 = 306.03$.

Approximation of Speed at 10 Seconds	
Interval	**Average Speed**
$t = 10$ to $t = 10.1$	$\dfrac{s(10.1) - s(10)}{10.1 - 10} = \dfrac{306.03 - 300}{0.1} = 60.3$
$t = 10$ to $t = 10.01$	$\dfrac{s(10.01) - s(10)}{10.01 - 10} = \dfrac{300.6003 - 300}{0.01} = 60.03$
$t = 10$ to $t = 10.001$	$\dfrac{s(10.001) - s(10)}{10.001 - 10} = \dfrac{300.060003 - 300}{0.001} = 60.003$

The results in the chart suggest that the exact speed at $t = 10$ is 60 ft/sec. We can confirm this by computing the average speed from $t = 10$ to $t = 10 + h$, where h is a small, but nonzero, number that represents a small change in time. (The chart does this for $h = 0.1$, $h = 0.01$, and $h = 0.001$.) The average speed from $t = 10$ to $t = 10 + h$ is then

$$\frac{s(10 + h) - s(10)}{(10 + h) - 10} = \frac{3(10 + h)^2 - 3 \cdot (10)^2}{h}$$

$$= \frac{3(100 + 20h + h^2) - 300}{h}$$

$$= \frac{300 + 60h + 3h^2 - 300}{h}$$

$$= \frac{60h + 3h^2}{h}$$

$$= \frac{h(60 + 3h)}{h}$$

$$= 60 + 3h,$$

where h is not equal to 0. Saying that the time interval from 10 to $10 + h$ gets shorter and shorter is equivalent to saying that h gets closer and closer to 0. Therefore, the exact speed at $t = 10$ is the limit, as h approaches 0, of the average speed over the interval from $t = 10$ to $t = 10 + h$; that is,

$$\lim_{h \to 0} \frac{s(10 + h) - s(10)}{h} = \lim_{h \to 0}(60 + 3h)$$

$$= 60 \text{ ft/sec.}$$

This example can be easily generalized to any function f. Let a be a specific x-value, such as 10 in the example. Let h be a (small) number, which represents the distance between the two values of x, namely, a and $a + h$. The average rate of change of f as x changes from a to $a + h$ is

$$\frac{f(a + h) - f(a)}{(a + h) - a} = \frac{f(a + h) - f(a)}{h},$$

which is often called the **difference quotient**. Observe that the difference quotient is equivalent to the average rate of change formula, which can be verified by letting $b = a + h$ in the average rate of change formula. Furthermore, the exact rate of change of f at $x = a$, called the *instantaneous rate of change of f at x = a*, is the limit of this difference quotient.

Instantaneous Rate of Change

The **instantaneous rate of change** for a function f when $x = a$ is

$$\lim_{h \to 0} \frac{f(a + h) - f(a)}{h},$$

provided this limit exists.

CAUTION Remember that $f(x + h) \neq f(x) + f(h)$. To find $f(x + h)$, replace x with $(x + h)$ in the expression for $f(x)$. For example, if $f(x) = x^2$,

$$f(x + h) = (x + h)^2 = x^2 + 2xh + h^2,$$

but

$$f(x) + f(h) = x^2 + h^2.$$

In the example just discussed, with the car starting from the traffic light, we saw that the instantaneous rate of change gave the speed of the car. But speed is always positive, while instantaneous rate of change can be positive or negative. Therefore, we will refer to **velocity** when we want to consider not only how fast something is moving but also in what direction it is moving. In any motion along a straight line, one direction is arbitrarily labeled as positive, so when an object moves in the opposite direction, its velocity is negative. In general, velocity is the same as the instantaneous rate of change of a function that gives position in terms of time.

In Figure 24, we have plotted the function $s(t) = 3t^2$, giving distance as a function of time. We have also plotted in green a line through the points $(10, s(10))$ and $(15, s(15))$. As we observed earlier, the slope of this line is the same as the average speed between $t = 10$ and $t = 15$. Finally, in red, we have plotted the line that results when the second point, $(15, s(15))$, moves closer and closer to the first point until the two coincide. The slope of this line corresponds to the instantaneous velocity at $t = 10$. We will explore these ideas further in the next section. Meanwhile, you might think about how to calculate the equations of these lines.

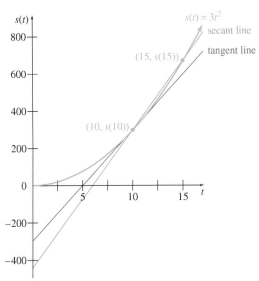

FIGURE 24

An alternate, but equivalent, approach is to let $a + h = b$ in the definition for instantaneous rate of change, so that $h = b - a$. This makes the instantaneous rate of change formula look more like the average rate of change formula.

Instantaneous Rate of Change (Alternate Form)

The **instantaneous rate of change** for a function f when $x = a$ can be written as

$$\lim_{b \to a} \frac{f(b) - f(a)}{b - a},$$

provided this limit exists.

EXAMPLE 3 Velocity

The distance in feet of an object from a starting point is given by $s(t) = 2t^2 - 5t + 40$, where t is time in seconds.

(a) Find the average velocity of the object from 2 seconds to 4 seconds.

SOLUTION The average velocity is

$$\frac{s(4) - s(2)}{4 - 2} = \frac{52 - 38}{2} = \frac{14}{2} = 7,$$

or 7 ft per second.

(b) Find the instantaneous velocity at 4 seconds.

SOLUTION

Method 1
Standard Form

For $t = 4$, the instantaneous velocity is

$$\lim_{h \to 0} \frac{s(4 + h) - s(4)}{h}$$

ft per second. We first calculate $s(4 + h)$ and $s(4)$, that is,

$$
\begin{aligned}
s(4 + h) &= 2(4 + h)^2 - 5(4 + h) + 40 \\
&= 2(16 + 8h + h^2) - 20 - 5h + 40 \\
&= 32 + 16h + 2h^2 - 20 - 5h + 40 \\
&= 2h^2 + 11h + 52,
\end{aligned}
$$

and

$$s(4) = 2(4)^2 - 5(4) + 40 = 52.$$

Therefore, the instantaneous velocity at $t = 4$ is

$$\lim_{h \to 0} \frac{(2h^2 + 11h + 52) - 52}{h} = \lim_{h \to 0} \frac{2h^2 + 11h}{h} = \lim_{h \to 0} \frac{h(2h + 11)}{h}$$
$$= \lim_{h \to 0} (2h + 11) = 11,$$

or 11 ft per second.

Method 2
Alternate Form

For $t = 4$, the instantaneous velocity is

$$\lim_{b \to 4} \frac{s(b) - s(4)}{b - 4}$$

ft per second. We first calculate $s(b)$ and $s(4)$, that is,

$$s(b) = 2b^2 - 5b + 40$$

and

$$s(4) = 2(4)^2 - 5(4) + 40 = 52.$$

The instantaneous rate of change is then

$$\lim_{b \to 4} \frac{2b^2 - 5b + 40 - 52}{b - 4} = \lim_{b \to 4} \frac{2b^2 - 5b - 12}{b - 4} \qquad \text{Simplify the numerator.}$$

$$= \lim_{b \to 4} \frac{(2b + 3)(b - 4)}{b - 4} \qquad \text{Factor the numerator.}$$

$$= \lim_{b \to 4} (2b + 3) \qquad \text{Cancel the } b - 4.$$

$$= 11, \qquad \text{Calculate the limit.}$$

YOUR TURN 3 For the function in Example 3, find the instantaneous velocity at 2 seconds.

or 11 ft per second.

TRY YOUR TURN 3

APPLY IT

EXAMPLE 4 Manufacturing

A company determines that the cost in dollars to manufacture x cases of the DVD "Mathematicians Caught in Embarrassing Moments" is given by

$$C(x) = 100 + 15x - x^2 \quad (0 \le x \le 7).$$

(a) Find the average change, per case, in the total cost if the number of cases manufactured changes from 1 to 5 cases.

SOLUTION Use the formula for average rate of change. The average rate of change of cost is

$$\frac{C(5) - C(1)}{5 - 1} = \frac{150 - 114}{4} = 9.$$

Thus, the total cost, on average, increases by $9 for each additional case when production increased from 1 to 5 cases.

(b) Find the additional cost when production is increased from 1 to 2 cases.

SOLUTION The additional cost can be found by calculating the cost to produce 2 cases, and subtracting the cost to produce 1 case; that is,

$$C(2) - C(1) = 126 - 114 = 12.$$

The additional cost to produce the second case is $12.

(c) Find the instantaneous rate of change of cost with respect to the number of cases produced when just one case is produced.

SOLUTION The instantaneous rate of change for $x = 1$ is given by

$$\lim_{h \to 0} \frac{C(1 + h) - C(1)}{h}$$

$$= \lim_{h \to 0} \frac{[100 + 15(1 + h) - (1 + h)^2] - [100 + 15(1) - 1^2]}{h}$$

$$= \lim_{h \to 0} \frac{100 + 15 + 15h - 1 - 2h - h^2 - 114}{h}$$

$$= \lim_{h \to 0} \frac{13h - h^2}{h} \qquad \text{Combine terms.}$$

$$= \lim_{h \to 0} \frac{h(13 - h)}{h} \qquad \text{Factor.}$$

$$= \lim_{h \to 0} (13 - h) \qquad \text{Divide by } h.$$

$$= 13. \qquad \text{Calculate the limit.}$$

When 1 case is manufactured, the cost is increasing at the rate of $13 per case. Notice that this is close to the value calculated in part (b). ■

As we mentioned in Chapter 1, economists sometimes define the marginal cost as the cost of producing one additional item and sometimes as the instantaneous rate of change of the cost function. These definitions are considered to be essentially equivalent. If a company (or an economy) produces millions of items, it makes little difference whether we let $h = 1$ or take the limit as h goes to 0, because 1 is very close to 0 when production is in the millions. The advantage of taking the instantaneous rate of change point of view is that it allows all the power of calculus to be used, including the Fundamental Theorem of Calculus, which is discussed later in this book.

Throughout this textbook, *we define the marginal cost to be the instantaneous rate of change of the cost function*. It can then be interpreted as the approximate cost of producing one additional item. For simplicity, we will make this interpretation even when production numbers are fairly small.

EXAMPLE 5 Manufacturing

For the cost function in the previous example, find the instantaneous rate of change of cost when 5 cases are made.

SOLUTION   The instantaneous rate of change for $x = 5$ is given by

$$\lim_{h \to 0} \frac{C(5 + h) - C(5)}{h}$$

$$= \lim_{h \to 0} \frac{[100 + 15(5 + h) - (5 + h)^2] - [100 + 15(5) - 5^2]}{h}$$

$$= \lim_{h \to 0} \frac{100 + 75 + 15h - 25 - 10h - h^2 - 150}{h}$$

$$= \lim_{h \to 0} \frac{5h - h^2}{h} \qquad \text{Combine terms.}$$

$$= \lim_{h \to 0} \frac{h(5 - h)}{h} \qquad \text{Factor.}$$

$$= \lim_{h \to 0} (5 - h) \qquad \text{Divide by } h.$$

$$= 5. \qquad \text{Calculate the limit.}$$

YOUR TURN 4   If the cost function is given by $C(x) = x^2 - 2x + 12$, find the instantaneous rate of change of cost when $x = 4$.

When 5 cases are manufactured, the cost is increasing at the rate of $5 per case; that is, the marginal cost when $x = 5$ is $5. Notice that as the number of items produced goes up, the marginal cost goes down, as might be expected.   TRY YOUR TURN 4 ■

EXAMPLE 6 Minority Population

Estimate the instantaneous rate of change in 2010 in the Hispanic population of the United States.

SOLUTION   We saw in Example 1 that the U.S. Hispanic population is approximately given by $H(t) = 37.791(1.021)^t$, where $t = 0$ corresponds to 2000. Unlike the previous example, in which the function was a polynomial, the function in this example is an exponential, making it harder to compute the limit directly using the formula for instantaneous rate of change at $t = 10$ (the year 2010):

$$\lim_{h \to 0} \frac{37.791(1.021)^{10+h} - 37.791(1.021)^{10}}{h}.$$

YOUR TURN 5   Estimate the instantaneous rate of change in 2010 in the Asian population of the United States. An estimate of the U.S. Asian population is given by $A(t) = 11.14(1.023)^t$, where $t = 0$ corresponds to 2000.

Instead, we will approximate the instantaneous rate of change by using smaller and smaller values of h. See the table on the next page. The limit seems to be approaching 0.96682 (million). Thus, the instantaneous rate of change in the U.S. Hispanic population was about 966,820 people per year in 2010.   TRY YOUR TURN 5 ■

	Limit Calculations
	$\dfrac{37.791(1.021)^{10+h} - 37.791(1.021)^{10}}{h}$
h	**h**
1	0.97693
0.1	0.96782
0.01	0.96692
0.001	0.96683
0.0001	0.96682
0.00001	0.96682

TECHNOLOGY NOTE

The table could be created using the TABLE feature on a TI-84 Plus C calculator by entering Y_1 as the function from Example 6, and $Y_2 = (Y_1)(10+X) - Y_1(10))/X$. (The calculator requires us to use X in place of h in the formula for instantaneous rate of change.) The result is shown in Figure 25. This table can also be generated using a spreadsheet.

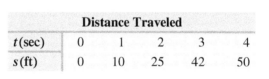

X	Y2	
1	.97693	
.1	.96782	
.01	.96692	
.001	.96683	
1E⁻4	.96682	
1E⁻5	.96682	
1E-6	.96682	
X=		

FIGURE 25

EXAMPLE 7   Velocity

One day Musk, the friendly pit bull, escaped from the yard and ran across the street to see a neighbor, who was 50 ft away. An estimate of the distance Musk ran as a function of time is given by the following table.

	Distance Traveled				
t (sec)	0	1	2	3	4
s (ft)	0	10	25	42	50

(a) Find Musk's average velocity during her 4-second trip.

SOLUTION The total distance she traveled is 50 ft, and the total time is 4 seconds, so her average velocity is $50/4 = 12.5$ ft per second.

(b) Estimate Musk's velocity at 2 seconds.

SOLUTION We could estimate her velocity by taking the short time interval from 2 to 3 seconds, for which the velocity is

$$\frac{42 - 25}{1} = 17 \text{ ft per second.}$$

Alternatively, we could estimate her velocity by taking the short time interval from 1 to 2 seconds, for which the velocity is

$$\frac{25 - 10}{1} = 15 \text{ ft per second.}$$

A better estimate is found by averaging these two values to get

$$\frac{17 + 15}{2} = 16 \text{ ft per second.}$$

Another way to get this same answer is to take the time interval from 1 to 3 seconds, for which the velocity is

$$\frac{42 - 10}{2} = 16 \text{ ft per second.}$$

This answer is reasonable if we assume Musk's velocity changes at a fairly steady rate and does not increase or decrease drastically from one second to the next. It is impossible to calculate Musk's exact velocity without knowing her position at times arbitrarily close to 2 seconds, or without a formula for her position as a function of time, or without a radar gun or speedometer on her. (In any case, she was very happy when she reached the neighbor.)

3.3 WARM-UP EXERCISES

For the function $f(x) = 2x^2 + 3x + 4$, find and simplify each of the following. *(Sec. 2.1)*

W1. $f(5 + h)$

W2. $\dfrac{f(2 + h) - f(2)}{h}$

For the function $f(x) = \dfrac{2}{x + 1}$, find and simplify each of the following. *(Sec. 2.1)*

W3. $f(3 + h)$

W4. $\dfrac{f(4 + h) - f(4)}{h}$

3.3 EXERCISES

Find the average rate of change for each function over the given interval.

1. $y = x^2 + 2x$ between $x = 1$ and $x = 3$

2. $y = -4x^2 - 6$ between $x = 2$ and $x = 6$

3. $y = -3x^3 + 2x^2 - 4x + 1$ between $x = -2$ and $x = 1$

4. $y = 2x^3 - 4x^2 + 6x$ between $x = -1$ and $x = 4$

5. $y = \sqrt{x}$ between $x = 1$ and $x = 4$

6. $y = \sqrt{3x - 2}$ between $x = 1$ and $x = 2$

7. $y = e^x$ between $x = -2$ and $x = 0$

8. $y = \ln x$ between $x = 2$ and $x = 4$

Suppose the position of an object moving in a straight line is given by $s(t) = t^2 + 5t + 2$. Find the instantaneous velocity at each time.

9. $t = 6$

10. $t = 1$

Suppose the position of an object moving in a straight line is given by $s(t) = 5t^2 - 2t - 7$. Find the instantaneous velocity at each time.

11. $t = 2$

12. $t = 3$

Suppose the position of an object moving in a straight line is given by $s(t) = t^3 + 2t + 9$. Find the instantaneous velocity at each time.

13. $t = 1$

14. $t = 4$

Find the instantaneous rate of change for each function at the given value.

15. $f(x) = x^2 + 2x$ at $x = 0$

16. $s(t) = -4t^2 - 6$ at $t = 2$

17. $g(t) = 1 - t^2$ at $t = -1$

18. $F(x) = x^2 + 2$ at $x = 0$

Use the formula for instantaneous rate of change, approximating the limit by using smaller and smaller values of h, to find the instantaneous rate of change for each function at the given value.

19. $f(x) = x^x$ at $x = 2$

20. $f(x) = x^x$ at $x = 3$

21. $f(x) = x^{\ln x}$ at $x = 2$

22. $f(x) = x^{\ln x}$ at $x = 3$

23. Explain the difference between the average rate of change of y as x changes from a to b, and the instantaneous rate of change of y at $x = a$.

24. If the instantaneous rate of change of $f(x)$ with respect to x is positive when $x = 1$, is f increasing or decreasing there?

APPLICATIONS

Business and Economics

25. Profit Suppose that the total profit in hundreds of dollars from selling x items is given by

$$P(x) = 2x^2 - 5x + 6.$$

Find the average rate of change of profit for the following changes in x.

(a) 2 to 4   **(b)** 2 to 3

(c) Find and interpret the instantaneous rate of change of profit with respect to the number of items produced when $x = 2$. (This number is called the *marginal profit* at $x = 2$.)

(d) Find the marginal profit at $x = 4$.

26. Revenue The revenue (in thousands of dollars) from producing x units of an item is

$$R(x) = 10x - 0.002x^2.$$

(a) Find the average rate of change of revenue when production is increased from 1000 to 1001 units.

(b) Find and interpret the instantaneous rate of change of revenue with respect to the number of items produced when 1000 units are produced. (This number is called the *marginal revenue* at $x = 1000$.)

(c) Find the additional revenue if production is increased from 1000 to 1001 units.

(d) Compare your answers for parts (a) and (c). What do you find? How do these answers compare with your answer to part (b)?

27. Demand Suppose customers in a hardware store are willing to buy $N(p)$ boxes of nails at p dollars per box, as given by

$$N(p) = 80 - 5p^2, \quad 1 \le p \le 4.$$

(a) Find the average rate of change of demand for a change in price from \$2 to \$3.

(b) Find and interpret the instantaneous rate of change of demand when the price is \$2.

(c) Find the instantaneous rate of change of demand when the price is \$3.

(d) As the price is increased from \$2 to \$3, how is demand changing? Is the change to be expected? Explain.

28. Interest If \$1000 is invested in an account that pays 3% compounded annually, the total amount, $A(t)$, in the account after t years is

$$A(t) = 1000(1.03)^t.$$

(a) Find the average rate of change per year of the total amount in the account for the first five years of the investment (from $t = 0$ to $t = 5$).

(b) Find the average rate of change per year of the total amount in the account for the second five years of the investment (from $t = 5$ to $t = 10$).

(c) Estimate the instantaneous rate of change for $t = 5$.

29. Interest If \$1000 is invested in an account that pays 3% compounded continuously, the total amount, $A(t)$, in the account after t years is

$$A(t) = 1000e^{0.03t}.$$

(a) Find the average rate of change per year of the total amount in the account for the first five years of the investment (from $t = 0$ to $t = 5$),

(b) Find the average rate of change per year of the total amount in the account for the second five years of the investment (from $t = 5$ to $t = 10$).

(c) Estimate the instantaneous rate of change for $t = 5$.

30. Unemployment The unemployment rates in the United States for the years 1994–2014 are shown in the graph below. Find the average change per year in unemployment for the following time periods. *Source: Bureau of Labor Statistics*.

(a) 1994 to 2000   **(b)** 2000 to 2006   **(c)** 2006 to 2012

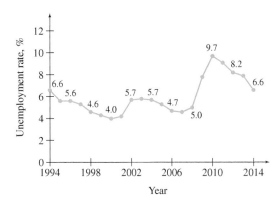

31. Gasoline Prices In 2013, the price of gasoline in the United States inexplicably spiked early and then gradually dropped. The average monthly price (in cents) per gallon of unleaded regular gasoline for 2013 is shown in the following chart. Find the average rate of change per month in the average price per gallon for each time period. *Source: U.S. Energy Information Administration*.

(a) From January to March (the peak)

(b) From March to December

(c) From January to December

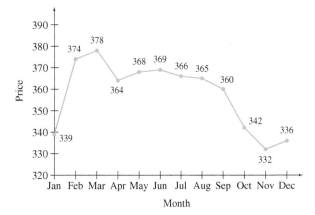

32. Medicare Trust Fund The graph shows the money remaining in the Medicare Trust Fund at the end of the fiscal year. *Source: Social Security Administration.*

Medicare Trust Fund

Using the Consumer Price Index for Urban Wage Earners and Clerical Workers

Find the approximate average rate of change in the trust fund for each time period.

(a) From 2008 to 2012   **(b)** From 2012 to 2020

Life Sciences

33. Minority Population Based on the U.S. Census population projections for 2000 to 2050, the projected Asian population (in millions) can be modeled by the exponential function

$$A(t) = 11.14(1.023)^t,$$

where $t = 0$ corresponds to 2000 and $0 \le t \le 50$. *Source: U.S. Census Bureau.*

(a) Use A to estimate the average rate of change in the Asian population from 2000 to 2015.

(b) Estimate the instantaneous rate of change in the Asian population in 2015.

34. World Population Growth The future size of the world population depends on how soon it reaches replacement-level fertility, the point at which each woman bears on average about 2.1 children. The graph shows projections for reaching that point in different years. *Source: Population Reference Bureau.*

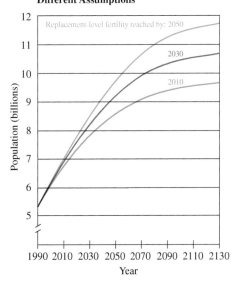

Ultimate World Population Size Under Different Assumptions

(a) Estimate the average rate of change in population for each projection from 1990 to 2050. Which projection shows the smallest rate of change in world population?

(b) Estimate the average rate of change in population from 2090 to 2130 for each projection. Interpret your answer.

35. HIV The estimated number of new HIV infections in recent years is shown in the graph. Find the average rate of change per year for each of the following time periods. *Source: AVERT.*

(a) 2006 to 2008   **(b)** 2008 to 2010   **(c)** 2007 to 2011

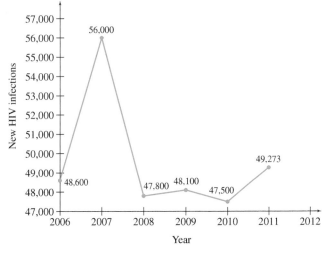

36. Molars The crown length (as shown below) of first molars in fetuses is related to the postconception age of the tooth as

$$L(t) = -0.01t^2 + 0.788t - 7.048,$$

where $L(t)$ is the crown length, in millimeters, of the molar t weeks after conception. *Source: American Journal of Physical Anthropology.*

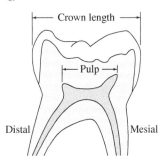

(a) Find the average rate of growth in crown length during weeks 22 through 28.

(b) Find the instantaneous rate of growth in crown length when the tooth is exactly 22 weeks of age.

 (c) Graph the function on $[0, 50]$ by $[0, 9]$. Does a function that increases and then begins to decrease make sense for this particular application? What do you suppose is happening during the first 11 weeks? Does this function accurately model crown length during those weeks?

 37. Thermic Effect of Food The metabolic rate of a person who has just eaten a meal tends to go up and then, after some time has passed, returns to a resting metabolic rate. This phenomenon is known as the thermic effect of food. Researchers have indicated that the thermic effect of food (in kJ/hr) for a particular person is

$$F(t) = -10.28 + 175.9te^{-t/1.3},$$

where t is the number of hours that have elapsed since eating a meal. *Source: American Journal of Clinical Nutrition.*

(a) Graph the function on $[0, 6]$ by $[-20, 100]$.

(b) Find the average rate of change of the thermic effect of food during the first hour after eating.

(c) Use a graphing calculator to find the instantaneous rate of change of the thermic effect of food exactly 1 hour after eating.

(d) Use a graphing calculator to estimate when the function stops increasing and begins to decrease.

38. Mass of Bighorn Yearlings The body mass of yearling bighorn sheep on Ram Mountain in Alberta, Canada, can be estimated by

$$M(t) = 27.5 + 0.3t - 0.001t^2$$

where $M(t)$ is measured in kilograms and t is days since May 25. *Source: Canadian Journal of Zoology.*

(a) Find the average rate of change of the weight of a bighorn yearling between 105 and 115 days past May 25.

(b) Find the instantaneous rate of change of weight for a bighorn yearling sheep whose age is 105 days past May 25.

(c) Graph the function $M(t)$ on $[5, 125]$ by $[25, 65]$.

(d) Does the behavior of the function past 125 days accurately model the mass of the sheep? Why or why not?

Social Sciences

39. Immigration The following graph shows how immigration (in thousands) to the United States has varied over the past century. *Source: Homeland Security.*

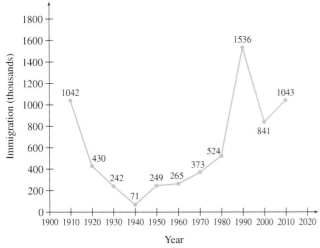

(a) Find the average annual rate of change in immigration for the first half of the century (from 1910 to 1960).

(b) Find the average annual rate of change in immigration for the second half of the century (from 1960 to 2010).

(c) Find the average annual rate of change in immigration for the entire century (from 1910 to 2010).

(d) Average your answers to parts (a) and (b), and compare the result with your answer from part (c). Will these always be equal for any two time periods?

(e) If the annual average rate of change for the last half-century continues, predict the number of immigrants in 2013. Compare your answer to the actual number of 990,553 immigrants.

40. Drug Use The following chart shows how the percentage of eighth graders, tenth graders, and twelfth graders who have used marijuana in their lifetime has varied in recent years. *Source: National Institutes of Health.*

(a) Find the average annual rate of change in the percent of eighth graders who have used marijuana in their lifetime over the three-year period 2007–2010 and the three-year period 2010–2013. Then calculate the annual rate of change for 2007–2013.

(b) Repeat part (a) for tenth graders.

(c) Repeat part (a) for twelfth graders.

(d) Discuss any similarities and differences between your answers to parts (a) through (c), as well as possible reasons for these differences and similarities.

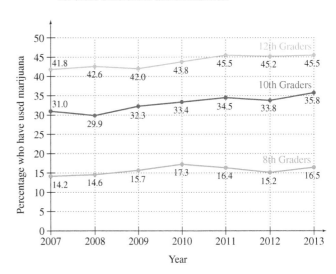

Physical Sciences

41. Temperature The graph shows the temperature T in degrees Celsius as a function of the altitude h in feet when an inversion layer is over Southern California. (An inversion layer is formed when air at a higher altitude, say 3000 ft, is warmer than air at sea level, even though air normally is cooler with increasing altitude.) Estimate and interpret the average rate of change in temperature for the following changes in altitude.

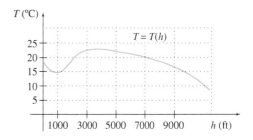

(a) 1000 to 3000 ft   (b) 1000 to 5000 ft

(c) 3000 to 9000 ft   (d) 1000 to 9000 ft

(e) At what altitude at or below 7000 ft is the temperature highest? Lowest? How would your answer change if 7000 ft is changed to 10,000 ft?

(f) At what altitude is the temperature the same as it is at 1000 ft?

42. Velocity A car is moving along a straight test track. The position in feet of the car, $s(t)$, at various times t is measured, with the following results.

t (sec)	0	2	4	6	8	10
$s(t)$ (ft)	0	10	14	20	30	36

Find and interpret the average velocities for the following changes in t.

(a) 0 to 2 seconds   (b) 2 to 4 seconds

(c) 4 to 6 seconds   (d) 6 to 8 seconds

(e) Estimate the instantaneous velocity at 4 seconds

   (i) by finding the average velocity between 2 and 6 seconds, and

   (ii) by averaging the answers for the average velocity in the two seconds before and the two seconds after (that is, the answers to parts (b) and (c)).

(f) Estimate the instantaneous velocity at 6 seconds using the two methods in part (e).

(g) Notice in parts (e) and (f) that your two answers are the same. Discuss whether this will always be the case, and why or why not.

43. Velocity Consider the example at the beginning of this section regarding the car traveling from San Francisco.

(a) Estimate the instantaneous velocity at 1 hour. Assume that the velocity changes at a steady rate from one half-hour to the next.

(b) Estimate the instantaneous velocity at 2 hours.

44. Velocity The distance of a particle from some fixed point is given by

$$s(t) = t^2 + 5t + 2,$$

where t is time measured in seconds. Find the average velocity of the particle over the following intervals.

(a) 4 to 6 seconds   (b) 4 to 5 seconds

(c) Find the instantaneous velocity of the particle when $t = 4$.

YOUR TURN ANSWERS

1. Increase, on average, by 284,000 people per year

2. Decrease, on average, of $1.325 billion per year

3. 3 ft per second   4. $6 per unit

5. About 0.318 million, or 318,000 people per year

3.4 Definition of the Derivative

APPLY IT How does the risk of chromosomal abnormality in a child change with the mother's age?

We will answer this question in Example 3, using the concept of the derivative.

In the previous section, the formula

$$\lim_{h \to 0} \frac{f(a + h) - f(a)}{h}$$

was used to calculate the instantaneous rate of change of a function f at the point where $x = a$. Now we will give a geometric interpretation of this limit.

The Tangent Line

In geometry, a *tangent line* to a circle is defined as a line that touches the circle at only one point, as at the point P in Figure 26 (which shows the top half of a circle). If you think of this half-circle as part of a curving road on which you are driving at night, then the tangent line indicates the direction of the light beam from your headlights as you pass through the point P. (We are not considering the new type of headlights on some cars that follow the direction of the curve.) Intuitively, the tangent line to an arbitrary curve at a point P on the curve should touch the curve at P, but not at any points nearby, and should indicate the direction of the curve. In Figure 27, for example, the lines through P_1 and P_3 are tangent lines, while the lines through P_2 and P_5 are not. The tangent lines just touch the curve and indicate the direction of the curve, while the other lines pass through the curve heading in some other direction. To decide about the line at P_4, we need to define the idea of a tangent line to the graph of a function more carefully.

FIGURE 26

FIGURE 27

To see how we might define the slope of a line tangent to the graph of a function f at a given point, let R be a fixed point with coordinates $(a, f(a))$ on the graph of a function $y = f(x)$, as in Figure 28 below. Choose a different point S on the graph and draw the line through R and S; this line is called a **secant line**. If S has coordinates $(a + h, f(a + h))$, then by the definition of slope, the slope of the secant line RS is given by

$$\text{Slope of secant} = \frac{\Delta y}{\Delta x} = \frac{f(a + h) - f(a)}{a + h - a} = \frac{f(a + h) - f(a)}{h}.$$

This slope corresponds to the average rate of change of y with respect to x over the interval from a to $a + h$. As h approaches 0, point S will slide along the curve, getting closer and closer to the fixed point R. See Figure 29, which shows successive positions S_1, S_2, S_3, and S_4 of the point S. If the slopes of the corresponding secant lines approach a limit as h approaches 0, then this limit is defined to be the slope of the tangent line at point R.

FIGURE 28

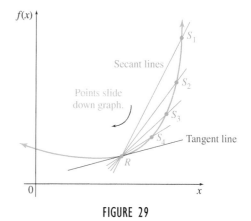

FIGURE 29

Slope of the Tangent Line

The **tangent line** of the graph of $y = f(x)$ at the point $(a, f(a))$ is the line through this point having slope

$$\lim_{h \to 0} \frac{f(a + h) - f(a)}{h},$$

provided this limit exists. If this limit does not exist, then there is no tangent at the point.

Notice that the definition of the slope of the tangent line is identical to that of the instantaneous rate of change discussed in the previous section and is calculated by the same procedure.

The slope of the tangent line at a point is also called the **slope of the curve** at the point and corresponds to the instantaneous rate of change of y with respect to x at the point. It indicates the direction of the curve at that point.

In Section 1.1, we saw that the equation of a line can be found with the point-slope form $y - y_1 = m(x - x_1)$, if the slope m and the coordinates (x_1, y_1) of a point on the line are known. Use the point-slope form to find the equation of the line with slope 3 that goes through the point $(-1, 4)$.

Let $m = 3, x_1 = -1, y_1 = 4$. Then

$$y - y_1 = m(x - x_1)$$
$$y - 4 = 3(x - (-1))$$
$$y - 4 = 3x + 3$$
$$y = 3x + 7.$$

EXAMPLE 1 Tangent Line

Consider the graph of $f(x) = x^2 + 2$.

(a) Find the slope and equation of the secant line through the points where $x = -1$ and $x = 2$.

SOLUTION Use the formula for slope as the change in y over the change in x, where y is given by $f(x)$. Since $f(-1) = (-1)^2 + 2 = 3$ and $f(2) = 2^2 + 2 = 6$, we have

$$\text{Slope of secant line} = \frac{f(2) - f(-1)}{2 - (-1)} = \frac{6 - 3}{3} = 1.$$

The slope of the secant line through $(-1, f(-1)) = (-1, 3)$ and $(2, f(2)) = (2, 6)$ is 1.

The equation of the secant line can be found with the point-slope form of the equation of a line from Chapter 1. We'll use the point $(-1, 3)$, although we could have just as well used the point $(2, 6)$.

$$y - y_1 = m(x - x_1)$$
$$y - 3 = 1[x - (-1)]$$
$$y - 3 = x + 1$$
$$y = x + 4$$

Figure 30 shows a graph of $f(x) = x^2 + 2$, along with a graph of the secant line (in green) through the points where $x = -1$ and $x = 2$.

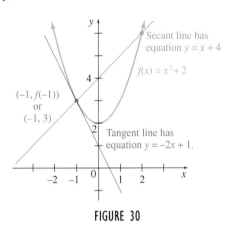

FIGURE 30

(b) Find the slope and equation of the tangent line at $x = -1$.

SOLUTION Use the definition given previously, with $f(x) = x^2 + 2$ and $a = -1$. The slope of the tangent line is given by

$$\text{Slope of tangent} = \lim_{h \to 0} \frac{f(a + h) - f(a)}{h}$$
$$= \lim_{h \to 0} \frac{[(-1 + h)^2 + 2] - [(-1)^2 + 2]}{h}$$
$$= \lim_{h \to 0} \frac{[1 - 2h + h^2 + 2] - [1 + 2]}{h}$$
$$= \lim_{h \to 0} \frac{-2h + h^2}{h}$$
$$= \lim_{h \to 0} \frac{h(-2 + h)}{h}$$
$$= \lim_{h \to 0} (-2 + h) = -2.$$

The slope of the tangent line at $(-1, f(-1)) = (-1, 3)$ is -2.

The equation of the tangent line can be found with the point-slope form of the equation of a line from Chapter 1.

YOUR TURN 1 For the graph of $f(x) = x^2 - x$, (a) find the equation of the secant line through the points where $x = -2$ and $x = 1$, and (b) find the equation of the tangent line at $x = -2$.

$$y - y_1 = m(x - x_1)$$
$$y - 3 = -2[x - (-1)]$$
$$y - 3 = -2(x + 1)$$
$$y - 3 = -2x - 2$$
$$y - -2x + 1$$

The tangent line at $x = -1$ is shown in red in Figure 30.   TRY YOUR TURN 1

Figure 31 shows the result of zooming in on the point $(-1, 3)$ in Figure 30. Notice that in this closeup view, the graph and its tangent line appear virtually identical. This gives us another interpretation of the tangent line. Suppose, as we zoom in on a function, the graph appears to become a straight line. Then this line is the tangent line to the graph at that point. In other words, the tangent line captures the behavior of the function very close to the point under consideration. (This assumes, of course, that the function when viewed close up is approximately a straight line. As we will see later in this section, this may not occur.)

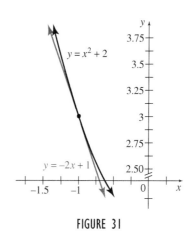

FIGURE 31

If it exists, the tangent line at $x = a$ is a good approximation of the graph of a function near $x = a$.

Consequently, another way to approximate the slope of the curve is to zoom in on the function using a graphing calculator until it appears to be a straight line (the tangent line). Then find the slope using any two points on that line.

TECHNOLOGY

EXAMPLE 2 **Slope (Using a Graphing Calculator)**

Use a graphing calculator to find the slope of the graph of $f(x) = x^x$ at $x = 1$.

SOLUTION The slope would be challenging to evaluate algebraically using the limit definition. Instead, using a graphing calculator on the window $[0, 2]$ by $[0, 2]$, we see the graph in Figure 32 on the next page. *(Continued)*

FIGURE 32   FIGURE 33

Zooming in gives the view in Figure 33. Using the TRACE key, we find two points on the line to be $(1, 1)$ and $(1.0015152, 1.0015174)$. Therefore, the slope is approximately

$$\frac{1.0015174 - 1}{1.0015152 - 1} \approx 1.$$

TECHNOLOGY NOTE

In addition to the method used in Example 2, there are other ways to use a graphing calculator to determine slopes of tangent lines and estimate instantaneous rates. Some alternate methods are listed below.

1. The `Tangent` command (under the `DRAW` menu) on a TI-84 Plus C allows the tangent line to be drawn on a curve, giving an easy way to generate a graph with its tangent line similar to Figure 30.

2. Rather than using a graph, we could use a TI-84 Plus C to create a table, as we did in the previous section, to estimate the instantaneous rate of change. Letting Y_1 = X^X and Y_2 = (Y$_1$(1 + X) − Y$_1$(1))/X, along with specific table settings, results in the table shown in Figure 34. Based on this table, we estimate that the slope of the graph of $f(x) = x^x$ at $x = 1$ is 1.

X	Y2
1	1.1053
.01	1.0101
.001	1.001
1E⁻4	1.0001
1E⁻5	1
1E⁻6	1

X=.1

FIGURE 34

3. An even simpler method on a TI-84 Plus C is to use the `dy/dx` command (under the `CALC` menu) or the `nDeriv` command (under the `MATH` menu). We will use this method in Example 4(b). But be careful, because sometimes these commands give erroneous results. For an example, see the Caution at the end of this section. For more details on the `dy/dx` command or the `nDeriv` command, see the *Graphing Calculator and Excel Spreadsheet Manual* available with this book.

EXAMPLE 3   **Genetics**

Figure 35 on the next page shows how the risk of chromosomal abnormality in a child increases with the age of the mother. Find the rate that the risk is rising when the mother is 40 years old. *Source: downsyndrome.about.com.*

APPLY IT   **SOLUTION**   In Figure 36 on the next page, we have added the tangent line to the graph at the point where the age of the mother is 40. At that point, the risk is approximately 15 per 1000. Extending the line, we estimate that when the age is 45, the y-coordinate of the line is roughly 35. Thus, the slope of the line is

$$\frac{35 - 15}{45 - 40} = \frac{20}{5} = 4.$$

FIGURE 35

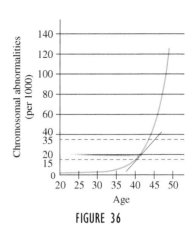

FIGURE 36

Therefore, at the age of 40, the risk of chromosomal abnormality in the child is increasing at the rate of about 4 per 1000 for each additional year of the mother's age.

The Derivative

If $y = f(x)$ is a function and a is a number in its domain, then we shall use the symbol $f'(a)$ to denote the special limit

$$\lim_{h \to 0} \frac{f(a + h) - f(a)}{h},$$

provided that it exists. This means that for each number a we can assign the number $f'(a)$ found by calculating this limit. This assignment defines an important new function.

Derivative

The **derivative** of the function f at x is defined as

$$f'(x) = \lim_{h \to 0} \frac{f(x + h) - f(x)}{h},$$

provided this limit exists.

The notation $f'(x)$ is read "f-prime of x." The function $f'(x)$ is called the derivative of f with respect to x. If x is a value in the domain of f and if $f'(x)$ exists, then f is **differentiable** at x. The process that produces f' is called **differentiation**.

NOTE The derivative is a *function of x*, since $f'(x)$ varies as x varies. This differs from both the slope of the tangent line and the instantaneous rate of change, either of which is represented by the number $f'(a)$ that corresponds to a number a. Otherwise, the formula for the derivative is identical to the formula for the slope of the tangent line given earlier in this section and to the formula for instantaneous rate of change given in the previous section.

The derivative function has several interpretations, two of which we have discussed.

1. The function $f'(x)$ represents the *instantaneous rate of change* of $y = f(x)$ with respect to x. This instantaneous rate of change could be interpreted as marginal cost, revenue, or profit (if the original function represented cost, revenue, or profit) or velocity (if the original function described displacement along a line). From now on we will use *rate of change* to mean *instantaneous* rate of change.

2. The function $f'(x)$ represents the *slope* of the graph of $f(x)$ at any point x. If the derivative is evaluated at the point $x = a$, then it represents the slope of the curve, or the slope of the tangent line, at that point.

The table below compares the different interpretations of the difference quotient and the derivative.

The Difference Quotient and the Derivative	
Difference Quotient	**Derivative**
$$\dfrac{f(x + h) - f(x)}{h}$$	$$\lim_{h \to 0} \dfrac{f(x + h) - f(x)}{h}$$
■ Slope of the secant line	■ Slope of the tangent line
■ Average rate of change	■ Instantaneous rate of change
■ Average velocity	■ Instantaneous velocity
■ Average rate of change in cost, revenue, or profit	■ Marginal cost, revenue, or profit

Just as we had an alternate definition in the previous section by using b instead of $a + h$, we now have an alternate definition by using b in place of $x + h$.

Derivative (Alternate Form)
The **derivative** of function f at x can be written as

$$f'(x) = \lim_{b \to x} \frac{f(b) - f(x)}{b - x},$$

provided this limit exists.

The next few examples show how to use the definition to find the derivative of a function by means of a four-step procedure.

EXAMPLE 4 **Derivative**

Let $f(x) = x^2$.

(a) Find the derivative.

SOLUTION

Method 1
Original Definition

By definition, for all values of x where the following limit exists, the derivative is given by

$$f'(x) = \lim_{h \to 0} \frac{f(x + h) - f(x)}{h}.$$

Use the following sequence of steps to evaluate this limit.

Step 1 Find $f(x + h)$.

Replace x with $x + h$ in the equation for $f(x)$. Simplify the result.

$$f(x) = x^2$$
$$f(x + h) = (x + h)^2$$
$$= x^2 + 2xh + h^2$$

(Note that $f(x + h) \neq f(x) + h$, since $f(x) + h = x^2 + h$.)

Step 2 Find $f(x + h) - f(x)$.
Since $f(x) = x^2$,

$$f(x + h) - f(x) = (x^2 + 2xh + h^2) - x^2 = 2xh + h^2.$$

Step 3 Find and simplify the quotient $\dfrac{f(x + h) - f(x)}{h}$. We find that

$$\frac{f(x + h) - f(x)}{h} = \frac{2xh + h^2}{h} = \frac{h(2x + h)}{h} = 2x + h,$$

except that $2x + h$ is defined for all real numbers h, while $[f(x + h) - f(x)]/h$ is not defined at $h = 0$. But this makes no difference in the limit, which ignores the value of the expression at $h = 0$.

Step 4 Finally, find the limit as h approaches 0. In this step, h is the variable and x is fixed.

$$f'(x) = \lim_{h \to 0} \frac{f(x + h) - f(x)}{h}.$$
$$= \lim_{h \to 0} (2x + h)$$
$$= 2x + 0 = 2x$$

Method 2
Alternate Form

Use

$$f(b) = b^2$$

and

$$f(x) = x^2.$$

We apply the alternate definition of the derivative as follows.

$$\lim_{b \to x} \frac{f(b) - f(x)}{b - x} = \lim_{b \to x} \frac{b^2 - x^2}{b - x}$$
$$= \lim_{b \to x} \frac{(b + x)(b - x)}{b - x} \qquad \text{Factor the numerator.}$$
$$= \lim_{b \to x} (b + x) \qquad \text{Divide by } b - x.$$
$$= x + x \qquad \text{Calculate the limit.}$$
$$= 2x$$

The alternate method appears shorter here because factoring $b^2 - x^2$ may seem simpler than calculating $f(x + h) - f(x)$. In other problems, however, factoring may be harder, in which case the first method may be preferable. Thus, from now on, we will use only the first method.

(b) Calculate and interpret $f'(3)$.

SOLUTION

Method 1
Algebraic Method

Since $f'(x) = 2x$, we have

$$f'(3) = 2 \cdot 3 = 6.$$

The number 6 is the slope of the tangent line to the graph of $f(x) = x^2$ at the point where $x = 3$, that is, at $(3, f(3)) = (3, 9)$. See Figure 37(a) on the next page.

Method 2
Graphing Calculator

As we mentioned earlier, some graphing calculators can calculate the value of the derivative at a given x-value. For example, the TI-84 Plus C uses the `nDeriv` command, with the variable, expression for $f(x)$, and the value of 3 entered to find $f'(3)$ for $f(x) = x^2$. The result is shown in Figure 37(b) on the next page.

YOUR TURN 2 Let $f(x) = x^2 - x$. Find the derivative, and then find $f'(-2)$.

TRY YOUR TURN 2

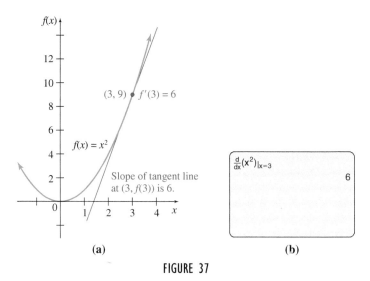

FIGURE 37

CAUTION 1. In Example 4(a) notice that $f(x + h)$ is *not* equal to $f(x) + h$. In fact,

$$f(x + h) = (x + h)^2 = x^2 + 2xh + h^2,$$

but

$$f(x) + h = x^2 + h.$$

2. In Example 4(b), do not confuse $f(3)$ and $f'(3)$. The value $f(3)$ is the y-value that corresponds to $x = 3$. It is found by substituting 3 for x in $f(x)$; $f(3) = 3^2 = 9$. On the other hand, $f'(3)$ is the slope of the tangent line to the curve at $x = 3$; as Example 4(b) shows, $f'(3) = 2 \cdot 3 = 6$.

Finding $f'(x)$ from the Definition of Derivative

The four steps used to find the derivative $f'(x)$ for a function $y = f(x)$ are summarized here.

1. Find $f(x + h)$.

2. Find and simplify $f(x + h) - f(x)$.

3. Divide by h to get $\dfrac{f(x + h) - f(x)}{h}$.

4. Let $h \to 0$; $f'(x) = \lim\limits_{h \to 0} \dfrac{f(x + h) - f(x)}{h}$, if this limit exists.

We now have four equivalent expressions for the change in x, but each has its uses, as the following box shows. We emphasize that these expressions all represent the same concept.

Equivalent Expressions for the Change in x

$x_2 - x_1$ Useful for describing the equation of a line through two points

$b - a$ A way to write $x_2 - x_1$ without the subscripts

Δx Useful for describing the change in x without referring to the individual points

h A way to write Δx with just one symbol

EXAMPLE 5 Derivative

Let $f(x) = 2x^3 + 4x$. Find $f'(x)$, $f'(2)$, and $f'(-3)$.

SOLUTION   Go through the four steps to find $f'(x)$.

Step 1 Find $f(x + h)$ by replacing x with $x + h$.

$$f(x + h) = 2(x + h)^3 + 4(x + h)$$
$$= 2(x^3 + 3x^2h + 3xh^2 + h^3) + 4(x + h)$$
$$= 2x^3 + 6x^2h + 6xh^2 + 2h^3 + 4x + 4h$$

Step 2 $f(x + h) - f(x) = 2x^3 + 6x^2h + 6xh^2 + 2h^3 + 4x + 4h$
$$-2x^3 - 4x$$
$$= 6x^2h + 6xh^2 + 2h^3 + 4h$$

Step 3 $\dfrac{f(x + h) - f(x)}{h} = \dfrac{6x^2h + 6xh^2 + 2h^3 + 4h}{h}$

$$= \dfrac{h(6x^2 + 6xh + 2h^2 + 4)}{h}$$

$$= 6x^2 + 6xh + 2h^2 + 4$$

Step 4 Now use the rules for limits to get

$$f'(x) = \lim_{h \to 0} \frac{f(x + h) - f(x)}{h}$$
$$= \lim_{h \to 0}(6x^2 + 6xh + 2h^2 + 4)$$
$$= 6x^2 + 6x(0) + 2(0)^2 + 4$$
$$f'(x) = 6x^2 + 4.$$

Use this result to find $f'(2)$ and $f'(-3)$.

$$f'(2) = 6 \cdot 2^2 + 4 = 28$$
$$f'(-3) = 6 \cdot (-3)^2 + 4 = 58 \qquad \text{TRY YOUR TURN 3}$$

YOUR TURN 3 Let $f(x) = x^3 - 1$. Find $f'(x)$ and $f'(-1)$.

TECHNOLOGY NOTE

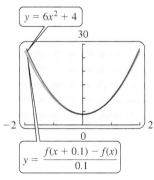

$y = 6x^2 + 4$

$y = \dfrac{f(x + 0.1) - f(x)}{0.1}$

FIGURE 38

One way to support the result in Example 5 is to plot $[f(x + h) - f(x)]/h$ on a graphing calculator with a small value of h. Figure 38 shows a graphing calculator screen of $y = [f(x + 0.1) - f(x)]/0.1$, where f is the function $f(x) = 2x^3 + 4x$, and $y = 6x^2 + 4$, which was just found to be the derivative of f. The two functions, plotted on the window $[-2, 2]$ by $[0, 30]$, appear virtually identical. If $h = 0.01$ had been used, the two functions would be indistinguishable.

EXAMPLE 6 Derivative

Let $f(x) = \dfrac{4}{x}$. Find $f'(x)$.

SOLUTION

Step 1 $f(x + h) = \dfrac{4}{x + h}$

Step 2 $f(x + h) - f(x) = \dfrac{4}{x + h} - \dfrac{4}{x}$

$$= \dfrac{4x - 4(x + h)}{x(x + h)} \qquad \text{Find a common denominator.}$$

$$= \dfrac{4x - 4x - 4h}{x(x + h)} \qquad \text{Simplify the numerator.}$$

$$= \dfrac{-4h}{x(x + h)}$$

Step 3 $\dfrac{f(x + h) - f(x)}{h} = \dfrac{\dfrac{-4h}{x(x + h)}}{h}$

$$= \dfrac{-4h}{x(x + h)} \cdot \dfrac{1}{h} \qquad \text{Invert and multiply.}$$

$$= \dfrac{-4}{x(x + h)}$$

Step 4 $f'(x) = \displaystyle\lim_{h \to 0} \dfrac{f(x + h) - f(x)}{h}$

$$= \lim_{h \to 0} \dfrac{-4}{x(x + h)}$$

$$= \dfrac{-4}{x(x + 0)}$$

$$f'(x) = \dfrac{-4}{x(x)} = \dfrac{-4}{x^2}$$

YOUR TURN 4 Let $f(x) = -\dfrac{2}{x}$. Find $f'(x)$.

TRY YOUR TURN 4

Notice that in Example 6 neither $f(x)$ nor $f'(x)$ is defined when $x = 0$. Look at a graph of $f(x) = 4/x$ to see why this is true.

EXAMPLE 7 Weight Gain

A mathematics professor found that, after introducing his dog Django to a new brand of food, Django's weight began to increase. After x weeks on the new food, Django's weight (in pounds) was approximately given by $w(x) = \sqrt{x} + 40$ for $0 \le x \le 6$. Find the rate of change of Django's weight after x weeks.

SOLUTION

Step 1 $w(x + h) = \sqrt{x + h} + 40$

Step 2 $w(x + h) - w(x) = \sqrt{x + h} + 40 - (\sqrt{x} + 40)$
$$= \sqrt{x + h} - \sqrt{x}$$

Step 3 $\dfrac{w(x + h) - w(x)}{h} = \dfrac{\sqrt{x + h} - \sqrt{x}}{h}$

In order to be able to divide by h, multiply both numerator and denominator by $\sqrt{x + h} + \sqrt{x}$; that is, rationalize the *numerator*.

$$\frac{w(x+h)-w(x)}{h}=\frac{\sqrt{x+h}-\sqrt{x}}{h}\cdot\frac{\sqrt{x+h}+\sqrt{x}}{\sqrt{x+h}+\sqrt{x}}$$ Rationalize the numerator

$$=\frac{(\sqrt{x+h})^2-(\sqrt{x})^2}{h(\sqrt{x+h}+\sqrt{x})}$$ $(a-b)(a+b)=a^2-b^2.$

$$=\frac{x+h-x}{h(\sqrt{x+h}+\sqrt{x})}$$

$$=\frac{h}{h(\sqrt{x+h}+\sqrt{x})}$$ Simplify.

$$=\frac{1}{\sqrt{x+h}+\sqrt{x}}$$ Divide by h.

Step 4 $w'(x)=\lim_{h\to0}\dfrac{1}{\sqrt{x+h}+\sqrt{x}}=\dfrac{1}{\sqrt{x}+\sqrt{x}}=\dfrac{1}{2\sqrt{x}}$

YOUR TURN 5 Let $f(x)=2\sqrt{x}$. Find $f'(x)$.

This tells us, for example, that after 4 weeks, when Django's weight is $w(4)=\sqrt{4}+40=42$ lb, her weight is increasing at a rate of $w'(4)=1/(2\sqrt{4})=1/4$ lb per week.   **TRY YOUR TURN 5**

EXAMPLE 8   **Cost Analysis**

The cost in dollars to manufacture x graphing calculators is given by $C(x)=-0.005x^2+20x+150$ when $0\le x\le2000$. Find the rate of change of cost with respect to the number manufactured when 100 calculators are made and when 1000 calculators are made.

SOLUTION   The rate of change of cost is given by the derivative of the cost function,

$$C'(x)=\lim_{h\to0}\frac{C(x+h)-C(x)}{h}.$$

Going through the steps for finding $C'(x)$ gives

$$C'(x)=-0.01x+20.$$

When $x=100$,

$$C'(100)=-0.01(100)+20=19.$$

This rate of change of cost per calculator gives the marginal cost at $x=100$, which means the approximate cost of producing the 101st calculator is $19.

YOUR TURN 6   If cost is given by $C(x)=10x-0.002x^2$, find the rate of change when $x=100$.

When 1000 calculators are made, the marginal cost is

$$C'(1000)=-0.01(1000)+20=10,$$

or $10.   **TRY YOUR TURN 6**

We can use the notation for the derivative to write the equation of the tangent line. Using the point-slope form, $y-y_1=m(x-x_1)$, and letting $y_1=f(x_1)$ and $m=f'(x_1)$, we have the following formula.

Equation of the Tangent Line

The tangent line to the graph of $y=f(x)$ at the point $(x_1,f(x_1))$ is given by the equation

$$y-f(x_1)=f'(x_1)(x-x_1),$$

provided $f'(x)$ exists.

EXAMPLE 9 **Tangent Line**

Find the equation of the tangent line to the graph of $f(x) = 4/x$ at $x = 2$.

SOLUTION From the answer to Example 6, we have $f'(x) = -4/x^2$, so $f'(x_1) = f'(2) = -4/2^2 = -1$. Also $f(x_1) = f(2) = 4/2 = 2$. Therefore the equation of the tangent line is

$$y - 2 = (-1)(x - 2),$$

or

$$y = -x + 4$$

after simplifying.

TRY YOUR TURN 7

YOUR TURN 7 Find the equation of the tangent line to the graph $f(x) = 2\sqrt{x}$ at $x = 4$.

Existence of the Derivative

The definition of the derivative included the phrase "provided this limit exists." If the limit used to define the derivative does not exist, then of course the derivative does not exist. For example, a derivative cannot exist at a point where the function itself is not defined. If there is no function value for a particular value of x, there can be no tangent line for that value. This was the case in Example 6—there was no tangent line (and no derivative) when $x = 0$.

Derivatives also do not exist at "corners" or "sharp points" on a graph. For example, the function graphed in Figure 39 is the *absolute value function*, defined previously as

$$f(x) = \begin{cases} x & \text{if } x \geq 0 \\ -x & \text{if } x < 0, \end{cases}$$

and written $f(x) = |x|$. By the definition of derivative, the derivative at any value of x is given by

$$f'(x) = \lim_{h \to 0} \frac{f(x + h) - f(x)}{h},$$

provided this limit exists. To find the derivative at 0 for $f(x) = |x|$, replace x with 0 and $f(x)$ with $|0|$ to get

$$f'(0) = \lim_{h \to 0} \frac{|0 + h| - |0|}{h} = \lim_{h \to 0} \frac{|h|}{h}.$$

In Example 5 in the first section of this chapter, we showed that

$$\lim_{h \to 0} \frac{|h|}{h} \text{ does not exist;}$$

therefore, the derivative does not exist at 0. However, the derivative does exist for all values of x other than 0.

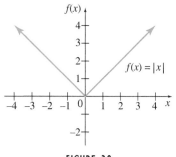

FIGURE 39

CAUTION The `nDeriv` command using the absolute value function at 0 on a TI-84 Plus C calculator gives the answer 0, which is wrong. It does this by investigating a point slightly to the left of 0 and slightly to the right of 0. Since the function has the same value at these two points, it assumes that the function must be flat around 0, which is false in this case because of the sharp corner at 0. Be careful about naively trusting your calculator; think about whether the answer is reasonable.

In Figure 40, we have zoomed in on the origin in Figure 39. Notice that the graph looks essentially the same. The corner is still sharp, and the graph does not resemble a straight line any more than it originally did. As we observed earlier, the derivative exists only at a point when the function more and more resembles a straight line as we zoom in on the point.

A graph of the function $f(x) = x^{1/3}$ is shown in Figure 41. As the graph suggests, the tangent line is vertical when $x = 0$. Since a vertical line has an undefined slope, the derivative

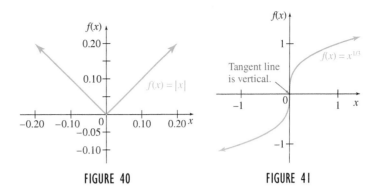

FIGURE 40   FIGURE 41

of $f(x) = x^{1/3}$ cannot exist when $x = 0$. Use the fact that $\lim\limits_{h \to 0} h^{1/3}/h = \lim\limits_{h \to 0} 1/h^{2/3}$ does not exist and the definition of the derivative to verify that $f'(0)$ does not exist for $f(x) = x^{1/3}$.

Figure 42 summarizes the various ways that a derivative can fail to exist. Notice in Figure 42 that at a point where the function is discontinuous, such as x_3, x_4, and x_6, the derivative does not exist. A function must be continuous at a point for the derivative to exist there. But just because a function is continuous at a point does not mean the derivative necessarily exists. For example, observe that the function in Figure 42 is continuous at x_1 and x_2, but the derivative does not exist at those values of x because of the sharp corners, making a tangent line impossible. This is exactly what happens with the function $f(x) = |x|$ at $x = 0$, as we saw in Figure 40. Also, the function is continuous at x_5, but the derivative doesn't exist there because the tangent line is vertical, and the slope of a vertical line is undefined.

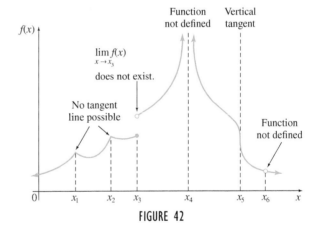

FIGURE 42

We summarize conditions for the derivative to exist or not exist below.

Existence of the Derivative

The derivative exists when a function f satisfies *all* of the following conditions at a point.

1. f is continuous,
2. f is smooth, and
3. f does not have a vertical tangent line.

The derivative does *not* exist when *any* of the following conditions are true for a function at a point.

1. f is discontinuous,
2. f has a sharp corner, or
3. f has a vertical tangent line.

EXAMPLE 10 **Astronomy**

In August 2013, Japanese amateur astronomer Koichi Itagaki discovered Nova Delphinus 2013. A nova is an exploding star whose brightness suddenly increases and then gradually fades. Astronomers use magnitudes of brightness to compare night sky objects. A lower star number indicates brighter objects. Prior to the explosion, the star was undetectable to the naked eye and had a brightness measure of $+17$, which is very dull. During a 24-hour period the brightness of the star increased to about $+4.5$, at which the explosion occurred. Gradually the brightness of the star has decreased. Suppose that the graph of the intensity of light emitted by Nova Delphinus 2013 is given as the function of time shown in Figure 43. Find where the function is not differentiable. *Source: Space.com.*

FIGURE 43

SOLUTION   Notice that although the graph is a continuous curve, it is not differentiable at the point of explosion.

3.4 WARM-UP EXERCISES

Find $\dfrac{f(x+h) - f(x)}{h}$ for each of the following functions.   *(Sec. 2.1)*

W1. $f(x) = 3x^2 - 2x - 5$

W2. $f(x) = \dfrac{3}{x - 2}$

Find the equation for a line satisfying the following conditions.
Put your answer in the form $y = mx + b$.   *(Sec. 1.1)*

W3. Through $(4, -1)$, parallel to $5x + 6y = 7$

W4. Through $(6, 2)$ and $(-2, 5)$

3.4   EXERCISES

1. By considering, but not calculating, the slope of the tangent line, give the derivative of the following.

   (a) $f(x) = 5$ (b) $f(x) = x$

   (c) $f(x) = -x$ (d) The line $x = 3$

   (e) The line $y = mx + b$

2. (a) Suppose $g(x) = \sqrt[3]{x}$. Use the graph of $g(x)$ to find $g'(0)$.

   (b) Explain why the derivative of a function does not exist at a point where the tangent line is vertical.

3. If $f(x) = \dfrac{x^2 - 1}{x + 2}$, where is f not differentiable?

4. If the rate of change of $f(x)$ is zero when $x = a$, what can be said about the tangent line to the graph of $f(x)$ at $x = a$?

Estimate the slope of the tangent line to each curve at the given point (x, y).

5.

6.

7.

8.

9.

10.

37.

38.

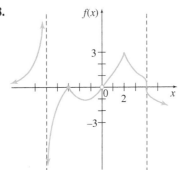

Using the definition of the derivative, find $f'(x)$. Then find $f'(-2)$, $f'(0)$, and $f'(3)$ when the derivative exists. (*Hint* for Exercises 17 and 18: See Example 9 in the first section of this chapter.)

11. $f(x) = 3x - 7$

12. $f(x) = -2x + 5$

13. $f(x) = -4x^2 + 9x + 2$

14. $f(x) = 6x^2 - 5x - 1$

15. $f(x) = 12/x$

16. $f(x) = 3/x$

17. $f(x) = \sqrt{x}$

18. $f(x) = -3\sqrt{x}$

19. $f(x) = 2x^3 + 5$

20. $f(x) = 4x^3 - 3$

For each function, find (a) the equation of the secant line through the points where x has the given values, and (b) the equation of the tangent line when x has the first value.

21. $f(x) = x^2 + 2x;$ $x = 3,$ $x = 5$

22. $f(x) = 6 - x^2;$ $x = -1,$ $x = 3$

23. $f(x) = 5/x;$ $x = 2,$ $x = 5$

24. $f(x) = -3/(x + 1);$ $x = 1,$ $x = 5$

25. $f(x) = 4\sqrt{x};$ $x = 9,$ $x = 16$

26. $f(x) = \sqrt{x};$ $x = 25,$ $x = 36$

 Use a graphing calculator to find $f'(2)$, $f'(16)$, and $f'(-3)$ for the following when the derivative exists.

27. $f(x) = -4x^2 + 11x$

28. $f(x) = 6x^2 - 4x$

29. $f(x) = e^x$

30. $f(x) = \ln|x|$

31. $f(x) = -\dfrac{2}{x}$

32. $f(x) = \dfrac{6}{x}$

33. $f(x) = \sqrt{x}$

34. $f(x) = -3\sqrt{x}$

Find the x-values where the following do not have derivatives.

35.

36.

39. For the function shown in the sketch, give the intervals or points on the x-axis where the rate of change of $f(x)$ with respect to x is

(**a**) positive;　　(**b**) negative;　　(**c**) zero.

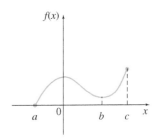

In Exercises 40 and 41, tell which graph, (a) or (b), represents velocity and which represents distance from a starting point. (*Hint*: Consider where the derivative is zero, positive, or negative.)

40. (a)

(b)

41. (a)

(b)

In Exercises 42–45, find the derivative of the function at the given point.

(a) Approximate the definition of the derivative with small values of *h*.

(b) Use a graphing calculator to zoom in on the function until it appears to be a straight line, and then find the slope of that line.

42. $f(x) = x^x$; $a = 2$

43. $f(x) = x^x$; $a = 3$

44. $f(x) = x^{1/x}$; $a = 2$

45. $f(x) = x^{1/x}$; $a = 3$

46. For each function in Column A, graph $[f(x + h) - f(x)]/h$ for a small value of *h* on the window $[-2, 2]$ by $[-2, 8]$. Then graph each function in Column B on the same window. Compare the first set of graphs with the second set to choose from Column B the derivative of each of the functions in Column A.

Column A	Column B
$\ln |x|$	e^x
e^x	$3x^2$
x^3	$\dfrac{1}{x}$

47. Explain why

$$\frac{f(x + h) - f(x - h)}{2h}$$

should give a reasonable approximation of $f'(x)$ when $f'(x)$ exists and *h* is small.

48. (a) For the function $f(x) = -4x^2 + 11x$, find the value of $f'(3)$, as well as the approximation using

$$\frac{f(x + h) - f(x)}{h}$$

and using the formula in Exercise 47 with $h = 0.1$.

(b) Repeat part (a) using $h = 0.01$.

(c) Repeat part (a) using the function $f(x) = -2/x$ and $h = 0.1$.

(d) Repeat part (c) using $h = 0.01$.

(e) Repeat part (a) using the function $f(x) = \sqrt{x}$ and $h = 0.1$.

(f) Repeat part (e) using $h = 0.01$.

(g) Using the results of parts (a) through (f), discuss which approximation formula seems to give better accuracy.

APPLICATIONS

Business and Economics

49. Demand Suppose the demand for a certain item is given by $D(p) = -2p^2 - 4p + 300$, where *p* represents the price of the item in dollars.

(a) Find the rate of change of demand with respect to price.

(b) Find and interpret the rate of change of demand when the price is $10.

50. Profit The profit (in thousands of dollars) from the expenditure of *x* thousand dollars on advertising is given by $P(x) = 1000 + 32x - 2x^2$. Find the marginal profit at the following expenditures. In each case, decide whether the firm should increase the expenditure.

(a) $8000 **(b)** $6000 **(c)** $12,000 **(d)** $20,000

51. Revenue The revenue in dollars generated from the sale of *x* picnic tables is given by $R(x) = 20x - \dfrac{x^2}{500}$.

(a) Find the marginal revenue when 1000 tables are sold.

(b) Estimate the revenue from the sale of the 1001st table by finding $R'(1000)$.

(c) Determine the actual revenue from the sale of the 1001st table.

(d) Compare your answers for parts (b) and (c). What do you find?

52. Cost The cost in dollars of producing *x* tacos is $C(x) = -0.00375x^2 + 1.5x + 1000$, for $0 \le x \le 180$.

(a) Find the marginal cost.

(b) Find and interpret the marginal cost at a production level of 100 tacos.

(c) Find the exact cost to produce the 101st taco.

(d) Compare the answers to parts (b) and (c). How are they related?

(e) Show that whenever $C(x) = ax^2 + bx + c$, $[C(x + 1) - C(x)] - C'(x) = a$. *Source: The College Mathematics Journal.*

(f) Show that whenever $C(x) = ax^2 + bx + c$,

$$C(x + 1) - C(x) = C'\left(x + \frac{1}{2}\right).$$

53. Social Security Assets The table on the next page gives actual and projected year-end assets in Social Security trust funds, in trillions of current dollars, where Year represents the number of years since 2000. The polynomial function defined by

$$f(x) = 0.0000329x^3 - 0.00450x^2 + 0.0613x + 2.34$$

models the data quite well. *Source: Social Security Administration.*

(a) To verify the fit of the model, find $f(10)$, $f(20)$, and $f(30)$.

(b) Use a graphing calculator with a command such as nDeriv to find the slope of the tangent line to the graph of *f* at the following *x*-values: 0, 10, 20, 30, and 35.

(c) Use your results in part (b) to describe the graph of f and interpret the corresponding changes in Social Security assets.

Year	Trillions of Dollars
10	2.45
20	2.34
30	1.03
40	−0.57
50	−1.86

Life Sciences

54. **Flight Speed** Many birds, such as cockatiels or the Arctic terns shown below, have flight muscles whose expenditure of power varies with the flight speed in a manner similar to the graph shown in the next column. The horizontal axis of the graph shows flight speed in meters per second, and the vertical axis shows power in watts per kilogram. *Source: Biolog-e: The Undergraduate Bioscience Research Journal.*

(a) The speed V_{mp} minimizes energy costs per unit of time. What is the slope of the line tangent to the curve at the point corresponding to V_{mp}? What is the physical significance of the slope at that point?

(b) The speed V_{mr} minimizes the energy costs per unit of distance covered. Estimate the slope of the curve at the point corresponding to V_{mr}. Give the significance of the slope at that point.

(c) By looking at the shape of the curve, describe how the power level decreases and increases for various speeds.

(d) Notice that the slope of the lines found in parts (a) and (b) represents the power divided by speed. Power is measured in energy per unit time per unit weight of the bird, and speed is distance per unit time, so the slope represents energy per unit distance per unit weight of the bird. If a line is drawn from the origin to a point on the graph,

at which point is the slope of the line (representing energy per unit distance per unit weight of the bird) smallest? How does this compare with your answers to parts (a) and (b)?

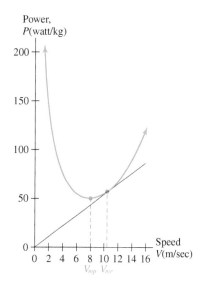

55. **Shellfish Population** In one research study, the population of a certain shellfish in an area at time t was closely approximated by the following graph. Estimate and interpret the derivative at each of the marked points.

56. **Eating Behavior** The eating behavior of a typical human during a meal can be described by

$$I(t) = 27 + 72t - 1.5t^2,$$

where t is the number of minutes since the meal began, and $I(t)$ represents the amount (in grams) that the person has eaten at time t. *Source: Appetite.*

(a) Find the rate of change of the intake of food for this particular person 5 minutes into a meal and interpret.

(b) Verify that the rate in which food is consumed is zero 24 minutes after the meal starts.

(c) Comment on the assumptions and usefulness of this function after 24 minutes. Given this fact, determine a logical domain for this function.

57. Quality Control of Cheese It is often difficult to evaluate the quality of products that undergo a ripening or maturation process. Researchers have successfully used ultrasonic velocity to determine the maturation time of Mahon cheese. The age can be determined by

$$M(v) = 0.0312443v^2 - 101.39v + 82{,}264, \quad v \geq 1620,$$

where $M(v)$ is the estimated age of the cheese (in days) for a velocity v (m per second). *Source: Journal of Food Science.*

(a) If Mahon cheese ripens in 150 days, determine the velocity of the ultrasound that one would expect to measure. (*Hint:* Set $M(v) = 150$ and solve for v.)

(b) Determine the derivative of this function when $v = 1700$ m per second and interpret.

Physical Sciences

58. Temperature The graph shows the temperature in degrees Celsius as a function of the altitude h in feet when an inversion layer is over Southern California. (See Exercise 41 in the previous section.) Estimate and interpret the derivatives of $T(h)$ at the marked points.

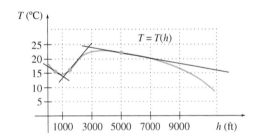

59. Oven Temperature The graph below shows the temperature one Christmas day of a Heat-Kit Bakeoven, a wood-burning oven for baking. The oven was lit at 8:30 A.M. Let $T(x)$ be the temperature x hours after 8:30 A.M. *Source: Heatkit.com.*

(a) Find and interpret $T'(0.5)$.

(b) Find and interpret $T'(3)$.

(c) Find and interpret $T'(4)$.

(d) At a certain time a Christmas turkey was put into the oven, causing the oven temperature to drop. Estimate when the turkey was put into the oven.

60. Baseball The graph shows how the velocity of a baseball that was traveling at 40 miles per hour when it was hit by a Little League baseball player varies with respect to the weight of the bat. *Source: Biological Cybernetics.*

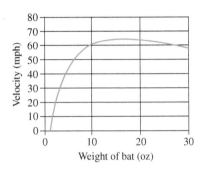

(a) Estimate and interpret the derivative for a 16-oz and 25-oz bat.

(b) What is the optimal bat weight for this player?

61. Baseball The graph shows how the velocity of a baseball that was traveling at 90 miles per hour when it was hit by a Major League baseball player varies with respect to the weight of the bat. *Source: Biological Cybernetics.*

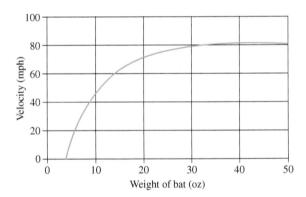

(a) Estimate and interpret the derivative for a 40-oz and 30-oz bat.

(b) What is the optimal bat weight for this player?

YOUR TURN ANSWERS

1. (a) $y = -2x + 2$

(b) $y = -5x - 4$

2. $f'(x) = 2x - 1$ and $f'(-2) = -5$.

3. $f'(x) = 3x^2$ and $f'(-1) = 3$.

4. $f'(x) = \dfrac{2}{x^2}$.

5. $f'(x) = \dfrac{1}{\sqrt{x}}$

6. $9.60

7. $y = \frac{1}{2}x + 2$

3.5 Graphical Differentiation

Given a graph of the production function, how can we find the graph of the marginal production function?

We will answer this question in Example 1 using graphical differentiation.

In the previous section, we estimated the derivative at various points of a graph by estimating the slope of the tangent line at those points. We will now extend this process to show how to sketch the graph of the derivative given the graph of the original function. This is important because, in many applications, a graph is all we have, and it is easier to find the derivative graphically than to find a formula that fits the graph and take the derivative of that formula.

EXAMPLE 1   Production of Landscape Mulch

In Figure 44(a), the graph shows total production (TP), measured in cubic yards of landscape mulch per week, as a function of labor used, measured in workers hired by a small business. The graph in Figure 44(b) shows the marginal production curve (MP_L), which is the derivative of the total production function. Verify that the graph of the marginal production curve (MP_L) is the graph of the derivative of the total production curve (TP).

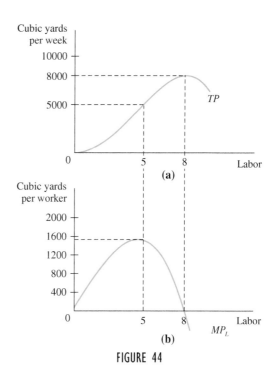

FIGURE 44

SOLUTION   Let q refer to the quantity of labor. We begin by choosing a point where estimating the derivative of TP is simple. Observe that when $q = 8$, TP has a horizontal tangent line, so its derivative is 0. This explains why the graph of MP_L equals 0 when $q = 8$.

Now, observe in Figure 44(a) that when $q < 8$, the tangent lines of TP have positive slope and the slope is steepest when $q = 5$. This means that the derivative should be positive for $q < 8$ and largest when $q = 5$. Verify that the graph of MP_L has this property.

Finally, as Figure 44(a) shows, the tangent lines of TP have negative slope when $q > 8$, so its derivative, represented by the graph of MP_L, should also be negative there. Verify that the graph of MP_L, in Figure 44(b), has this property as well.

In Example 1, we saw how the general shape of the graph of the derivative could be found from the graph of the original function. To get a more accurate graph of the derivative, we need to estimate the slope of the tangent line at various points, as we did in the previous section.

EXAMPLE 2 Temperature

Figure 45 gives the temperature in degrees Celsius as a function, $T(h)$, of the altitude h in feet when an inversion layer is over Southern California. Sketch the graph of the derivative of the function. (This graph appeared in Exercise 58 in the previous section.)

FIGURE 45

SOLUTION First, observe that when $h = 1000$ and $h = 3500$, $T(h)$ has horizontal tangent lines, so $T'(1000) = 0$ and $T'(3500) = 0$.

Notice that the tangent lines have a negative slope for $0 < h < 1000$. Thus, the graph of the derivative should be negative (below the x-axis) there. Then, for $1000 < h < 3500$, the tangent lines have positive slope, so the graph of the derivative should be positive (above the x-axis) there. Notice from the graph of $T(h)$ that the slope is largest when $h = 1500$. Finally, for $h > 3500$, the tangent lines would be negative again, forcing the graph of the derivative back down below the x-axis to take on negative values.

Now that we have a general shape of the graph, we can estimate the value of the derivative at several points to improve the accuracy of the graph. To estimate the derivative, find two points on each tangent line and compute the slope between the two points. The estimates at selected points are given in the table to the left. (Your answers may be slightly different, since estimation from a picture can be inexact.) Figure 46 shows a graph of these values of $T'(h)$.

Estimates of $T'(h)$

h	$T'(h)$
500	−0.005
1000	0
1500	0.008
3500	0
5000	−0.00125

YOUR TURN 1 Sketch the graph of the derivative of the function $f(x)$.

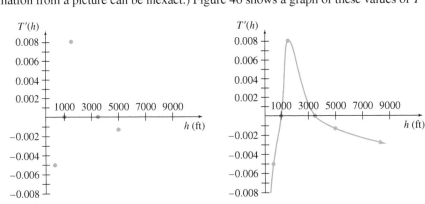

FIGURE 46   FIGURE 47

Using all of these facts, we connect the points in the graph $T'(h)$ smoothly, with the result shown in Figure 47.   TRY YOUR TURN 1 ▮

CAUTION Remember that when you graph the derivative, you are graphing the *slope* of the original function. Do not confuse the slope of the original function with the y-value of the original function. In fact, the slope of the original function is equal to the y-value of its derivative.

Sometimes the original function is not smooth or even continuous, so the graph of the derivative may also be discontinuous.

| EXAMPLE 3 | **Graphing a Derivative**

Sketch the graph of the derivative of the function shown in Figure 48.

SOLUTION Notice that when $x < -2$, the slope is 1, and when $-2 < x < 0$, the slope is -1. At $x = -2$, the derivative does not exist due to the sharp corner in the graph. The derivative also does not exist at $x = 0$ because the function is discontinuous there. Using this information, the graph of $f'(x)$ on $x < 0$ is shown in Figure 49.

FIGURE 48   FIGURE 49

FIGURE 50

YOUR TURN 2 Sketch the graph of the derivative of the function $g(x)$.

For $x > 0$, the derivative is positive. If you draw a tangent line at $x = 1$, you should find that the slope of this line is roughly 1. As x approaches 0 from the right, the derivative becomes larger and larger. As x approaches infinity, the tangent lines become more and more horizontal, so the derivative approaches 0. The resulting sketch of the graph of $y = f'(x)$ is shown in Figure 50.   TRY YOUR TURN 2

Finding the derivative graphically may seem difficult at first, but with practice you should be able to quickly sketch the derivative of any function graphed. Your answers to the exercises may not look exactly like those in the back of the book, because estimating the slope accurately can be difficult, but your answers should have the same general shape.

Figures 51(a), (b), and (c) on the next page show the graphs of $y = x^2$, $y = x^4$, and $y = x^{4/3}$ on a graphing calculator. When finding the derivative graphically, all three seem to have the same behavior: negative derivative for $x < 0$, 0 derivative at $x = 0$, and positive derivative for $x > 0$. Beyond these general features, however, the derivatives look quite different, as you can see from Figures 52(a), (b), and (c) on the next page, which show the graphs of the derivatives. When finding derivatives graphically, detailed information can be found only by very carefully measuring the slope of the tangent line at a large number of points.

TECHNOLOGY NOTE

On many calculators, the graph of the derivative can be plotted if a formula for the original function is known. For example, the graphs in Figure 52 were drawn on a TI-84 Plus C by using the `nDeriv` command. Define $Y_2 = \frac{d}{dx}(Y_1)\big|_{x\,=\,x}$ after entering the original function into Y_1. You can use this feature to practice finding the derivative graphically. Enter a function into Y_1, sketch the graph on the graphing calculator, and use it to draw by hand the graph of the derivative. Then use `nDeriv` to draw the graph of the derivative, and compare it with your sketch.

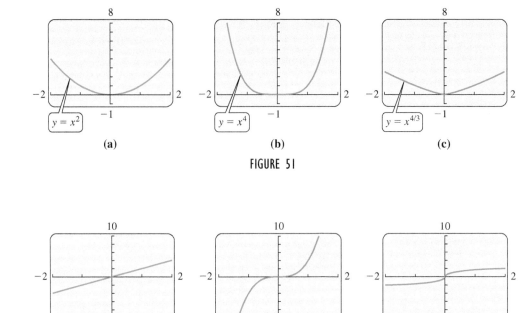

FIGURE 51

FIGURE 52

| EXAMPLE 4 | **Graphical Differentiation** |

Figure 53 shows the graph of a function f and its derivative function f'. Use slopes to decide which graph is that of f and which is the graph of f'.

SOLUTION   Look at the places where each graph crosses the x-axis; that is, the x-intercepts, since x-intercepts occur on the graph of f' whenever the graph of f has a horizontal tangent line or slope of zero. Also, a decreasing graph corresponds to negative slope or a negative derivative, while an increasing graph corresponds to positive slope or a positive derivative. Y_1 has zero slope near $x = 0$, $x = 1$, and $x = 5$; Y_2 has x-intercepts near these values of x. Y_1 decreases on $(-2, 0)$ and $(1, 5)$; Y_2 is negative on those intervals. Y_1 increases on $(0, 1)$ and $(5, 7)$; Y_2 is positive there. Thus, Y_1 is the graph of f and Y_2 is the graph of f'. ▬▬▬

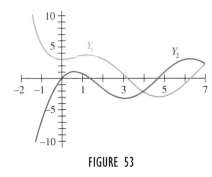

FIGURE 53

3.5 WARM-UP EXERCISES

Find the slope of the tangent line to each curve at the given point (x, y). *(Sec. 3.4)*

W1.

W2.

3.5   EXERCISES

1. Explain how to graph the derivative of a function given the graph of the function.

2. Explain how to graph a function given the graph of the derivative function. Note that the answer is not unique.

Each graphing calculator window shows the graph of a function $f(x)$ and its derivative function $f'(x)$. Decide which is the graph of the function and which is the graph of the derivative.

3.

4.

5.

6.

Sketch the graph of the derivative for each function shown.

7.

8.

9.

10.

11.

12.

13.

14.

15.

16.

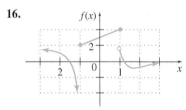

APPLICATIONS

Business and Economics

17. Consumer Demand When the price of an essential commodity rises rapidly, consumption drops slowly at first. If the price continues to rise, however, a "tipping" point may be reached, at which consumption takes a sudden substantial drop. Suppose the accompanying graph shows the consumption of gasoline, $G(t)$, in millions of gallons, in a certain area. We assume that the price is rising rapidly. Here t is the time in months after the price began rising. Sketch a graph of the rate of change in consumption as a function of time.

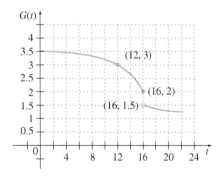

Life Sciences

18. Body Mass Index The graph at the top of the next column shows how the body mass index-for-age percentile for boys varies from the age of 2 to 20 years. *Source: Centers for Disease Control.*

(a) Sketch a graph of the rate of change of the 95th percentile as a function of age.

(b) Sketch a graph of the rate of change of the 50th percentile as a function of age.

Body Mass Index-for-Age Percentiles: Boys, 2 to 20 years

19. Shellfish Population In one research study, the population of a certain shellfish in an area at time t was closely approximated by the following graph. Sketch a graph of the growth rate of the population.

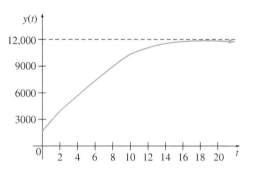

20. Flight Speed The graph below shows the relationship between the speed of a bird in flight and the required power expended by flight muscles. Sketch the graph of the rate of change of the power as a function of the speed. *Source: Biolog-e: The Undergraduate Bioscience Research Journal.*

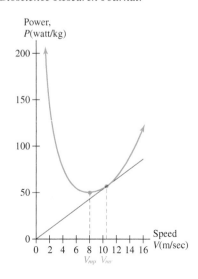

21. Human Growth The growth remaining in sitting height at consecutive skeletal age levels is indicated below for boys. Sketch a graph showing the rate of change of growth remaining for the indicated years. Use the graph and your sketch to estimate the remaining growth and the rate of change of remaining growth for a 14-year-old boy. *Source: Standards in Pediatric Orthopedics: Tables, Charts, and Graphs Illustrating Growth.*

22. Weight Gain The graph below shows the typical weight (in kilograms) of an English boy for his first 18 years of life. Sketch the graph of the rate of change of weight with respect to time. *Source: Human Growth After Birth.*

23. Creek Discharge The graph below shows the discharge of water (in cubic feet per second) from the Otter Creek in Middlebury, VT, during March 2013. Sketch a graph of the rate of change in the discharge with respect to time. *Source: usgs.gov.*

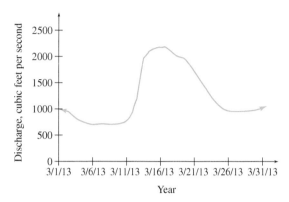

General Interest

24. Grades A group of MIT professors created a function for transforming student test scores of 0 to 100 into grades of 0 to 5. As the graph below illustrates, most of the scores transformed into grades of 3, 4, or 5. Draw a graph for the rate of change of the transformed grades with respect to the scores. *Source: MIT Faculty Newsletter.*

YOUR TURN ANSWERS

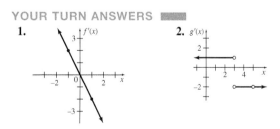

3

CHAPTER REVIEW

SUMMARY

In this chapter we introduced the ideas of limit and continuity of functions and then used these ideas to explore calculus. We saw that the difference quotient can represent

- the average rate of change,
- the slope of the secant line, and
- the average velocity.

We saw that the derivative can represent

- the instantaneous rate of change,
- the slope of the tangent line, and
- the instantaneous velocity.

We also learned how to estimate the value of the derivative using graphical differentiation. In the next chapter, we will take a closer look at the definition of the derivative to develop a set of rules to quickly and easily calculate the derivative of a wide range of functions without the need to directly apply the definition of the derivative each time.

Limit of a Function	Let f be a function and let a and L be real numbers. If **1.** as x takes values closer and closer (but not equal) to a on both sides of a, the corresponding values of $f(x)$ get closer and closer (and perhaps equal) to L; and **2.** the value of $f(x)$ can be made as close to L as desired by taking values of x close enough to a; then L is the limit of $f(x)$ as x approaches a, written $$\lim_{x \to a} f(x) = L.$$
Existence of Limits	The limit of f as x approaches a may not exist. **1.** If $f(x)$ becomes infinitely large in magnitude (positive or negative) as x approaches the number a from either side, we write $\lim_{x \to a} f(x) = \infty$ or $\lim_{x \to a} f(x) = -\infty$. In either case, the limit does not exist. **2.** If $f(x)$ becomes infinitely large in magnitude (positive) as x approaches a from one side and infinitely large in magnitude (negative) as x approaches a from the other side, then $\lim_{x \to a} f(x)$ does not exist. **3.** If $\lim_{x \to a^-} f(x) = L$ and $\lim_{x \to a^+} f(x) = M$, and $L \neq M$, then $\lim_{x \to a} f(x)$ does not exist.
Limits at Infinity	For any positive real number n, $$\lim_{x \to \infty} \frac{1}{x^n} = \lim_{x \to -\infty} \frac{1}{x^n} = 0.$$
Finding Limits at Infinity	If $f(x) = p(x)/q(x)$ for polynomials $p(x)$ and $q(x)$, $\lim_{x \to \infty} f(x)$ and $\lim_{x \to -\infty} f(x)$ can be found by dividing $p(x)$ and $q(x)$ by the highest power of x in $q(x)$.
Continuity	A function f is continuous at c if **1.** $f(c)$ is defined, **2.** $\lim_{x \to c} f(x)$ exists, and **3.** $\lim_{x \to c} f(x) = f(c)$.
Average Rate of Change	The average rate of change of $f(x)$ with respect to x as x changes from a to b is $$\frac{f(b) - f(a)}{b - a}.$$

Difference Quotient   The average rate of change can also be written as

$$\frac{f(x + h) - f(x)}{h}.$$

Derivative   The derivative of $f(x)$ with respect to x is

$$f'(x) = \lim_{h \to 0} \frac{f(x + h) - f(x)}{h}.$$

KEY TERMS

To understand the concepts presented in this chapter, you should know the meaning and use of the following terms.
For easy reference, the section in the chapter where a word (or expression) was first used is provided.

3.1
limit
limit from the left/right
one-/two-sided limit
piecewise function
limit at infinity

3.2
continuous
discontinuous
removable discontinuity
continuous on an open/closed
  interval
continuous from the right/left
Intermediate Value Theorem

3.3
average rate of change
difference quotient
instantaneous rate of change
velocity

3.4
secant line
tangent line
slope of the curve
derivative
differentiable
differentiation

REVIEW EXERCISES

CONCEPT CHECK

Determine whether each of the following statements is true or false, and explain why.

1. The limit of a product is the product of the limits when each of the limits exists.

2. The limit of a function may not exist at a point even though the function is defined there.

3. If a rational function has a polynomial in the denominator of higher degree than the polynomial in the numerator, then the limit at infinity must equal zero.

4. If the limit of a function exists at a point, then the function is continuous there.

5. A polynomial function is continuous everywhere.

6. A rational function is continuous everywhere.

7. The derivative gives the average rate of change of a function.

8. The derivative gives the instantaneous rate of change of a function.

9. The instantaneous rate of change is a limit.

10. The derivative is a function.

11. The slope of the tangent line gives the average rate of change.

12. The derivative of a function exists wherever the function is continuous.

PRACTICE AND EXPLORATIONS

13. Is a derivative always a limit? Is a limit always a derivative? Explain.

14. Is every continuous function differentiable? Is every differentiable function continuous? Explain.

15. Describe how to tell when a function is discontinuous at the real number $x = a$.

16. Give two applications of the derivative

$$f'(x) = \lim_{h \to 0} \frac{f(x + h) - f(x)}{h}.$$

Decide whether the limits in Exercises 17–34 exist. If a limit exists, find its value.

17. (a) $\displaystyle\lim_{x \to -3^-} f(x)$ **(b)** $\displaystyle\lim_{x \to -3^+} f(x)$ **(c)** $\displaystyle\lim_{x \to -3} f(x)$ **(d)** $f(-3)$

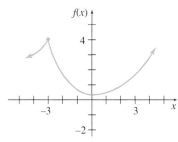

18. (a) $\displaystyle\lim_{x \to -1^-} g(x)$ **(b)** $\displaystyle\lim_{x \to -1^+} g(x)$ **(c)** $\displaystyle\lim_{x \to -1} g(x)$ **(d)** $g(-1)$

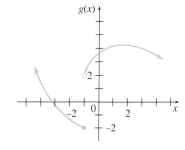

19. (a) $\lim\limits_{x \to 4^-} f(x)$ **(b)** $\lim\limits_{x \to 4^+} f(x)$ **(c)** $\lim\limits_{x \to 4} f(x)$ **(d)** $f(4)$

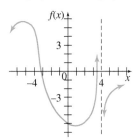

20. (a) $\lim\limits_{x \to 2^-} h(x)$ **(b)** $\lim\limits_{x \to 2^+} h(x)$ **(c)** $\lim\limits_{x \to 2} h(x)$ **(d)** $h(2)$

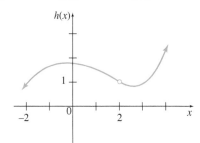

21. $\lim\limits_{x \to -\infty} g(x)$

22. $\lim\limits_{x \to \infty} f(x)$

23. $\lim\limits_{x \to 6} \dfrac{2x + 7}{x + 3}$

24. $\lim\limits_{x \to -3} \dfrac{2x + 5}{x + 3}$

25. $\lim\limits_{x \to 4} \dfrac{x^2 - 16}{x - 4}$

26. $\lim\limits_{x \to 2} \dfrac{x^2 + 3x - 10}{x - 2}$

27. $\lim\limits_{x \to -4} \dfrac{2x^2 + 3x - 20}{x + 4}$

28. $\lim\limits_{x \to 3} \dfrac{3x^2 - 2x - 21}{x - 3}$

29. $\lim\limits_{x \to 9} \dfrac{\sqrt{x} - 3}{x - 9}$

30. $\lim\limits_{x \to 16} \dfrac{\sqrt{x} - 4}{x - 16}$

31. $\lim\limits_{x \to \infty} \dfrac{2x^2 + 5}{5x^2 - 1}$

32. $\lim\limits_{x \to \infty} \dfrac{x^2 + 6x + 8}{x^3 + 2x + 1}$

33. $\lim\limits_{x \to -\infty} \left(\dfrac{3}{8} + \dfrac{3}{x} - \dfrac{6}{x^2} \right)$

34. $\lim\limits_{x \to -\infty} \left(\dfrac{9}{x^4} + \dfrac{10}{x^2} - 6 \right)$

Identify the x-values where f is discontinuous.

35.

36.

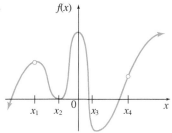

Find all x-values where the function is discontinuous. For each such value, give $f(a)$ and $\lim\limits_{x \to a} f(x)$ or state that it does not exist.

37. $f(x) = \dfrac{-5 + x}{3x(3x + 1)}$

38. $f(x) = \dfrac{7 - 3x}{(1 - x)(3 + x)}$

39. $f(x) = \dfrac{x - 6}{x + 5}$

40. $f(x) = \dfrac{x^2 - 9}{x + 3}$

41. $f(x) = x^2 + 3x - 4$

42. $f(x) = 2x^2 - 5x - 3$

In Exercises 43 and 44, (a) graph the given function, (b) find all values of x where the function is discontinuous, and (c) find the limit from the left and from the right at any values of x found in part (b).

43. $f(x) = \begin{cases} 1 - x & \text{if } x < 1 \\ 2 & \text{if } 1 \leq x \leq 2 \\ 4 - x & \text{if } x > 2 \end{cases}$

44. $f(x) = \begin{cases} 2 & \text{if } x < 0 \\ -x^2 + x + 2 & \text{if } 0 \leq x \leq 2 \\ 1 & \text{if } x > 2 \end{cases}$

Find each limit (a) by investigating values of the function near the point where the limit is taken and (b) by using a graphing calculator to view the function near the point.

45. $\lim\limits_{x \to 1} \dfrac{x^4 + 2x^3 + 2x^2 - 10x + 5}{x^2 - 1}$

46. $\lim\limits_{x \to -2} \dfrac{x^4 + 3x^3 + 7x^2 + 11x + 2}{x^3 + 2x^2 - 3x - 6}$

Find the average rate of change for the following on the given interval. Then find the instantaneous rate of change at the first x-value.

47. $y = 6x^3 + 2$ from $x = 1$ to $x = 4$

48. $y = -2x^3 - 3x^2 + 8$ from $x = -2$ to $x = 6$

49. $y = \dfrac{-6}{3x - 5}$ from $x = 4$ to $x = 9$

50. $y = \dfrac{x + 4}{x - 1}$ from $x = 2$ to $x = 5$

For each function, find (a) the equation of the secant line through the points where x has the given values, and (b) the equation of the tangent line when x has the first value.

51. $f(x) = 3x^2 - 5x + 7; x = 2, x = 4$

52. $f(x) = \dfrac{1}{x}; x = 1/2, x = 3$

53. $f(x) = \dfrac{12}{x-1}; x = 3, x = 7$

54. $f(x) = 2\sqrt{x-1}; x = 5, x = 10$

Use the definition of the derivative to find the derivative of the following.

55. $y = 4x^2 + 3x - 2$

56. $y = 5x^2 - 6x + 7$

In Exercises 57 and 58, find the derivative of the function at the given point (a) by approximating the definition of the derivative with small values of h and (b) by using a graphing calculator to zoom in on the function until it appears to be a straight line, and then finding the slope of that line.

57. $f(x) = (\ln x)^x; \quad x_0 = 3$

58. $f(x) = x^{\ln x}; \quad x_0 = 2$

Sketch the graph of the derivative for each function shown.

59.

60.

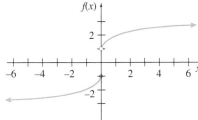

61. Let f and g be differentiable functions such that

$$\lim_{x \to \infty} f(x) = c$$

$$\lim_{x \to \infty} g(x) = d$$

where $c \neq d$. Determine

$$\lim_{x \to \infty} \frac{cf(x) - dg(x)}{f(x) - g(x)}.$$

(Choose one of the following.) *Source: Society of Actuaries.*

(a) 0   (b) $\dfrac{cf'(0) - dg'(0)}{f'(0) - g'(0)}$

(c) $f'(0) - g'(0)$ (d) $c - d$ (e) $c + d$

APPLICATIONS

Business and Economics

62. Revenue Waverly Products has found that its revenue is related to advertising expenditures by the function

$$R(x) = 5000 + 16x - 3x^2,$$

where $R(x)$ is the revenue in dollars when x hundred dollars are spent on advertising.

(a) Find the marginal revenue function.

(b) Find and interpret the marginal revenue when $1000 is spent on advertising.

63. Cost Analysis A company charges $1.50 per lb when a certain chemical is bought in lots of 125 lb or less, with a price per pound of $1.35 if more than 125 lb are purchased. Let $C(x)$ represent the cost of x lb. Find the cost for the following numbers of pounds.

(a) 100   (b) 125   (c) 140   (d) Graph $y = C(x)$.

(e) Where is C discontinuous?

Find the average cost per pound if the following number of pounds are bought.

(f) 100   (g) 125   (h) 140

Find and interpret the marginal cost (that is, the instantaneous rate of change of the cost) for the following numbers of pounds.

(i) 100   (j) 140

64. Marginal Analysis Suppose the profit (in cents) from selling x lb of potatoes is given by

$$P(x) = 15x + 25x^2.$$

Find the average rate of change in profit from selling each of the following amounts.

(a) 6 lb to 7 lb   (b) 6 lb to 6.5 lb   (c) 6 lb to 6.1 lb

Find the marginal profit (that is, the instantaneous rate of change of the profit) from selling the following amounts.

(d) 6 lb   (e) 20 lb   (f) 30 lb

(g) What is the domain of x?

(h) Is it possible for the marginal profit to be negative here? What does this mean?

(i) Find the average profit function. (Recall that average profit is given by total profit divided by the number produced, or $\overline{P}(x) = P(x)/x$.)

(j) Find the marginal average profit function (that is, the function giving the instantaneous rate of change of the average profit function).

(k) Is it possible for the marginal average profit to vary here? What does this mean?

(l) Discuss whether this function describes a realistic situation.

65. Average Cost The graph on the next page shows the total cost $C(x)$ to produce x tons of cement. (Recall that average cost is given by total cost divided by the number produced, or $\overline{C}(x) = C(x)/x$.)

(a) Draw a line through $(0, 0)$ and $(5, C(5))$. Explain why the slope of this line represents the average cost per ton when 5 tons of cement are produced.

(b) Find the value of x for which the average cost is smallest.

(c) What can you say about the marginal cost at the point where the average cost is smallest?

66. Tax Rates A simplified income tax considered in the U.S. Senate in 1986 had two tax brackets. Married couples earning $29,300 or less would pay 15% of their income in taxes. Those earning more than $29,300 would pay $4350 plus 27% of the income over $29,300 in taxes. Let $T(x)$ be the amount of taxes paid by someone earning x dollars in a year. *Source: Wall Street Journal.*

(a) Find $\lim\limits_{x \to 29,300^-} T(x)$.

(b) Find $\lim\limits_{x \to 29,300^+} T(x)$.

(c) Find $\lim\limits_{x \to 29,300} T(x)$.

(d) Sketch a graph of $T(x)$.

(e) Identify any x-values where T is discontinuous.

(f) Let $A(x) = T(x)/x$ be the average tax rate, that is, the amount paid in taxes divided by the income. Find a formula for $A(x)$. (*Note:* The formula will have two parts: one for $x \le 29,300$ and one for $x > 29,300$.)

(g) Find $\lim\limits_{x \to 29,300^-} A(x)$.

(h) Find $\lim\limits_{x \to 29,300^+} A(x)$.

(i) Find $\lim\limits_{x \to 29,300} A(x)$.

(j) Find $\lim\limits_{x \to \infty} A(x)$.

(k) Sketch the graph of $A(x)$.

67. Unemployment The unemployment rates in the United States for the years 1994–2014 are shown in the graph below. Sketch a graph showing the rate of change in the annual unemployment rates for this period. Use the graph and your sketch to estimate the annual unemployment rate and rate of change of the unemployment rate in 2012. *Source: Bureau of Labor Statistics.*

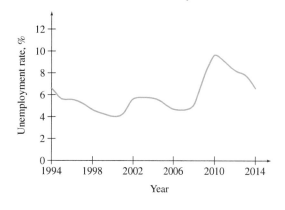

Life Sciences

68. Alzheimer's Disease The graph below shows the projected number of people aged 65 and over in the United States with Alzheimer's disease. Estimate and interpret the derivative in each of the following years. *Source: Alzheimer's Disease Facts and Figures.*

(a) 2000

(b) 2040

(c) Find the average rate of change between 2000 and 2040 in the number of people aged 65 and over in the United States with Alzheimer's disease.

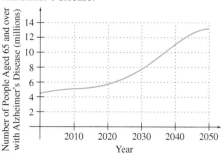

69. Spread of a Virus The spread of a virus is modeled by

$$V(t) = -t^2 + 6t - 4,$$

where $V(t)$ is the number of people (in hundreds) with the virus and t is the number of weeks since the first case was observed.

(a) Graph $V(t)$.

(b) What is a reasonable domain of t for this problem?

(c) When does the number of cases reach a maximum? What is the maximum number of cases?

(d) Find the rate of change function.

(e) What is the rate of change in the number of cases at the maximum?

(f) Give the sign (+ or −) of the rate of change up to the maximum and after the maximum.

70. Whales Diving The following figure, already shown in the section on Properties of Functions, shows the depth of a sperm whale as a function of time, recorded by researchers at the Woods Hole Oceanographic Institution in Massachusetts. *Source: Peter Tyack, Woods Hole Oceanographic Institution.*

(a) Find the rate that the whale was descending at the following times.

(i) 17 hours and 37 minutes

(ii) 17 hours and 39 minutes

(b) Sketch a graph of the rate the whale was descending as a function of time.

71. Body Mass Index   The following graph shows how the body mass index-for-age percentile for girls varies from the age of 2 to 20 years. *Source: Centers for Disease Control.*

(a) Sketch a graph of the rate of change of the 95th percentile as a function of age.

(b) Sketch a graph of the rate of change of the 50th percentile as a function of age.

**Body Mass Index-for-Age Percentiles:
Girls, 2 to 20 years**

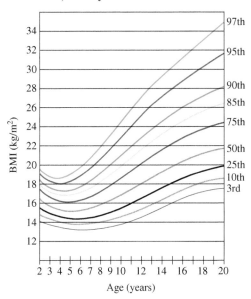

72. Human Growth   The growth remaining in sitting height for girls at consecutive skeletal age levels is indicated on the graph in the next column. Sketch a graph showing the rate of change of growth remaining for the indicated years. Use the graph and your sketch to estimate the remaining growth and the rate of change of remaining growth for a 10-year-old girl. *Source: Standards in Pediatric Orthopedics: Tables, Charts, and Graphs Illustrating Growth.*

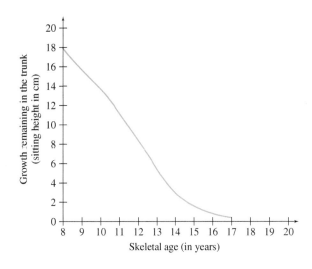

Physical Sciences

73. Temperature   Suppose a gram of ice is at a temperature of −100°C. The graph shows the temperature of the ice as increasing numbers of calories of heat are applied. It takes 80 calories to melt one gram of ice at 0°C into water, and 540 calories to boil one gram of water at 100°C into steam.

(a) Where is this graph discontinuous?

(b) Where is this graph not differentiable?

(c) Sketch the graph of the derivative.

A MODEL FOR DRUGS ADMINISTERED INTRAVENOUSLY

When a drug is administered intravenously it enters the bloodstream immediately, producing an immediate effect for the patient. The drug can be either given as a single rapid injection or given at a constant drip rate. The latter is commonly referred to as an intravenous (IV) infusion. Common drugs administered intravenously include morphine for pain, diazepam (or Valium) to control a seizure, and digoxin for heart failure.

SINGLE RAPID INJECTION

With a single rapid injection, the amount of drug in the bloodstream reaches its peak immediately and then the body eliminates the drug exponentially. The larger the amount of drug there is in the body, the faster the body eliminates it. If a lesser amount of drug is in the body, it is eliminated more slowly.

The amount of drug in the bloodstream t hours after a single rapid injection can be modeled using an exponential decay function, like those found in Chapter 2 on Nonlinear Functions, as follows:

$$A(t) = De^{kt},$$

where D is the size of the dose administered and k is the exponential decay constant for the drug.

EXAMPLE 1   Rapid Injection

The drug labetalol is used for the control of blood pressure in patients with severe hypertension. The half-life of labetalol is 4 hours. Suppose a 35-mg dose of the drug is administered to a patient by rapid injection.

(a) Find a model for the amount of drug in the bloodstream t hours after the drug is administered.

SOLUTION   Since $D = 35$ mg, the function has the form

$$A(t) = 35e^{kt}.$$

Recall from Chapter 2 on Nonlinear Functions that the general equation giving the half-life T in terms of the decay constant k was

$$T = -\frac{\ln 2}{k}.$$

Solving this equation for k, we get

$$k = -\frac{\ln 2}{T}.$$

Since the half-life of this drug is 4 hours,

$$k = -\frac{\ln 2}{4} \approx -0.17.$$

Therefore, the model is

$$A(t) = 35e^{-0.17t}.$$

The graph of $A(t)$ is given in Figure 54.

Rapid IV Injection

$A(t) = 35e^{-0.17t}$

Hours since dose was administered

FIGURE 54

(b) Find the average rate of change of drug in the bloodstream between $t = 0$ and $t = 2$. Repeat for $t = 4$ and $t = 6$.

SOLUTION   The average rate of change from $t = 0$ to $t = 2$ is

$$\frac{A(2) - A(0)}{2 - 0} \approx \frac{25 - 35}{2} = -5 \text{ mg/hr}.$$

The average rate of change from $t = 4$ to $t = 6$ is

$$\frac{A(6) - A(4)}{6 - 4} \approx \frac{13 - 18}{2} = -2.5 \text{ mg/hr}.$$

Notice that since the half-life of the drug is 4 hours, the average rate of change from $t = 4$ to $t = 6$ is half of the average rate of change from $t = 0$ to $t = 2$. What would the average rate of change be from $t = 8$ to $t = 10$?

(c) What happens to the amount of drug in the bloodstream as t increases? (i.e., What is the limit of the function as t approaches ∞?)

SOLUTION   Looking at the graph of $A(t)$, we can see that

$$\lim_{t \to \infty} A(t) = 0.$$

An advantage of an intravenous rapid injection is that the amount of drug in the body reaches a high level immediately. Suppose, however, that the effective level of this drug is between 30 mg and 40 mg. From the graph, we can see that it takes only an hour after the dose is given for the amount of drug in the body to fall below the effective level.

INTRAVENOUS INFUSION

With an IV infusion, the amount of drug in the bloodstream starts at zero and increases until the rate the drug is entering the body equals the rate the drug is being eliminated from the body. At this point, the amount of drug in the bloodstream levels off. This model is a limited growth function, like those from Chapter 2.

The amount of drug in the bloodstream t hours after an IV infusion begins can be modeled using a limited growth function, as follows.

$$A(t) = \frac{r}{-k}(1 - e^{kt}),$$

where r is the rate of infusion per hour and k is the exponential decay constant for the drug.

EXAMPLE 2 IV Infusion

The same drug used in Example 1 is given to a patient by IV infusion at a drip rate of 6 mg/hr. Recall that the half-life of this drug is 4 hours.

(a) Find a model for the amount of drug in the bloodstream t hours after the IV infusion begins.

SOLUTION Since $r = 6$ and $k = -0.17$, the function has the form

$$A(t) = 35(1 - e^{-0.17t}).$$

The graph of $A(t)$ is given in Figure 55.

IV Infusion

$A(t) = 35\left(1 - e^{-0.17t}\right)$

Hours since infusion began

FIGURE 55

(b) Find the average rate of change of drug in the bloodstream between $t = 0$ and $t = 2$. Repeat for $t = 4$ and $t = 6$.

SOLUTION The average rate of change from $t = 0$ to $t = 2$ is

$$\frac{A(2) - A(0)}{2 - 0} \approx \frac{10 - 0}{2} = 5 \text{ mg/hr.}$$

The average rate of change from $t = 4$ to $t = 6$ is

$$\frac{A(6) - A(4)}{6 - 4} \approx \frac{22 - 17}{2} = 2.5 \text{ mg/hr.}$$

Recall that the average rate of change from $t = 0$ to $t = 2$ for the rapid injection of this drug was -5 mg/hr and the average rate of change from $t = 4$ to $t = 6$ was -2.5 mg/hr. In fact, at

any given time, the rapid injection function is decreasing at the same rate the IV infusion function is increasing.

(c) What happens to the amount of drug in the bloodstream as t increases? (i.e., What is the limit of the function as t approaches ∞?)

SOLUTION Looking at the graph of $A(t)$ in Figure 55 and the formula for $A(t)$ in part (a), we can see that

$$\lim_{t \to \infty} A(t) = 35.$$

An advantage of an IV infusion is that a dose can be given such that the limit of $A(t)$ as t approaches ∞ is an effective level. Once the amount of drug has reached this effective level, it will remain there as long as the infusion continues. However, using this method of administration, it may take a while for the amount of drug in the body to reach an effective level. For our example, the effective level is between 30 mg and 40 mg. Looking at the graph, you can see that it takes about 11 hours to reach an effective level. If this patient were experiencing dangerously high blood pressure, you wouldn't want to wait 11 hours for the drug to reach an effective level.

SINGLE RAPID INJECTION FOLLOWED BY AN INTRAVENOUS INFUSION

Giving a patient a single rapid injection immediately followed by an intravenous infusion allows a patient to experience the advantages of both methods. The single rapid injection immediately produces an effective drug level in the patient's bloodstream. While the amount of drug in the bloodstream from the rapid infusion is decreasing, the amount of drug in the system from the IV infusion is increasing.

The amount of drug in the bloodstream t hours after the injection is given and infusion has started can be calculated by finding the sum of the two models.

$$A(t) = De^{kt} + \frac{r}{-k}(1 - e^{kt})$$

EXAMPLE 3 Combination Model

A 35-mg dose of labetalol is administered to a patient by rapid injection. Immediately thereafter, the patient is given an IV infusion at a drip rate of 6 mg/hr. Find a model for the amount of drug in the bloodstream t hours after the drug is administered.

SOLUTION Recall from Example 1, the amount of drug in the bloodstream t hours after the rapid injection was found to be

$$A(t) = 35e^{-0.17t}.$$

From Example 2, the amount of drug in the bloodstream t hours after the IV infusion began was found to be

$$A(t) = 35(1 - e^{-0.17t}).$$

Therefore, t hours after administering both the rapid injection and the IV infusion, the amount of drug in the bloodstream is

$$A(t) = 35e^{-0.17t} + 35(1 - e^{-0.17t})$$
$$= 35 \text{ mg.}$$

207

The graph of $A(t)$ is given in Figure 56.

Loading Dose Plus IV Infusion

FIGURE 56

Notice that the constant multiple of the rapid injection function, 35, is equal to the constant multiple of the IV infusion function. When this is the case, the sum of the two functions will be that constant.

EXAMPLE 4 Combination Model

A drug with a half-life of 3 hours is found to be effective when the amount of drug in the bloodstream is 58 mg. A 58-mg loading dose is given by rapid injection followed by an IV infusion. What should the rate of infusion be to maintain this level of drug in the bloodstream?

SOLUTION Recall that the amount of drug in the bloodstream t hours after both a rapid injection and IV infusion are administered is given by

$$A(t) = De^{kt} + \frac{r}{-k}(1 - e^{kt}).$$

The rapid injection dose, D, is 58 mg. The half-life of the drug is three hours; therefore,

$$k = -\frac{\ln 2}{3} \approx -0.23.$$

It follows that

$$A(t) = 58e^{-0.23t} + \frac{r}{0.23}(1 - e^{-0.23t}).$$

Since we want the sum of the rapid injection function and the IV infusion function to be 58 mg, it follows that

$$\frac{r}{0.23} = 58.$$

Solving for r, we get

$$r = 13.34 \text{ mg/hr}.$$

EXERCISES

1. A 500-mg dose of a drug is administered by rapid injection to a patient. The half-life of the drug is 9 hours.

   (a) Find a model for the amount of drug in the bloodstream t hours after the drug is administered.

   (b) Find the average rate of change of drug in the bloodstream between $t = 0$ and $t = 2$. Repeat for $t = 9$ and $t = 11$.

2. A drug is given to a patient by IV infusion at a drip rate of 350 mg/hr. The half-life of this drug is 3 hours.

   (a) Find a model of the amount of drug in the bloodstream t hours after the IV infusion begins.

   (b) Find the average rate of change of drug in the bloodstream between $t = 0$ and $t = 3$. Repeat for $t = 3$ and $t = 6$.

3. A drug with a half-life of 9 hours is found to be effective when the amount of drug in the bloodstream is 250 mg. A 250-mg loading dose is given by rapid injection followed by an IV infusion. What should the rate of infusion be to maintain this level of drug in the bloodstream?

4. Use the table feature on a graphing calculator or a spreadsheet to develop a table that shows how much of the drug is present in a patient's system at the end of each 1/2-hour time interval for 24 hours for the model found in Exercise 1(a). A chart such as this provides the health care worker with immediate information about patient drug levels.

5. Use the table feature on a graphing calculator or a spreadsheet to develop a table that shows how much of the drug is present in a patient's system at the end of each 1/2-hour time interval for 10 hours for the model found in Exercise 2(a). A chart such as this provides the health care worker with immediate information about patient drug levels.

6. Use the table feature on a graphing calculator or a spreadsheet to develop a table that shows how much of the drug is present in a patient's system at the end of each 1/2-hour time interval for 10 hours for the model found in Exercise 3. Are your results surprising?

DIRECTIONS FOR GROUP PROJECT

Choose a drug that is commonly prescribed by physicians for a common ailment. Develop an analysis for this drug that is similar to the analysis for labetalol in Examples 1 through 3. You can obtain information on the drug from the Internet or from advertisements found in various media. Once you complete the analysis, prepare a professional presentation that can be delivered at a public forum. The presentation should summarize the facts presented in this extended application but at a level that is understandable to a typical layperson.

4

Calculating the Derivative

By differentiating the function defining a mathematical model we can see how the model's output changes with the input. In an exercise in Section 2 we explore a rational-function model for the length of the rest period needed to recover from vigorous exercise such as riding a bike. The derivative indicates how the rest required changes with the work expended in kilocalories per minute.

n the previous chapter, we found the derivative to be a useful tool for describing the rate of change, velocity, and the slope of a curve. Taking the derivative by using the definition, however, can be difficult. To take full advantage of the power of the derivative, we need faster ways of calculating the derivative. That is the goal of this chapter.

4.1 Techniques for Finding Derivatives

APPLY IT How can a manager determine the best production level if the relationship between profit and production is known? How fast is the number of Americans who are expected to be over 100 years old growing?
These questions can be answered by finding the derivative of an appropriate function. We shall return to them at the end of this section in Examples 9 and 10.

Using the definition to calculate the derivative of a function is a very involved process even for simple functions. In this section we develop rules that make the calculation of derivatives much easier. Keep in mind that even though the process of finding a derivative will be greatly simplified with these rules, *the interpretation of the derivative will not change*. But first, a few words about notation are in order.

In addition to $f'(x)$, there are several other commonly used notations for the derivative.

Notations for the Derivative
The derivative of $y = f(x)$ may be written in any of the following ways:

$$f'(x), \quad \frac{dy}{dx}, \quad \frac{d}{dx}[f(x)], \quad \text{or} \quad D_x[f(x)].$$

The dy/dx notation for the derivative (read "the derivative of y with respect to x") is sometimes referred to as *Leibniz notation,* named after one of the co-inventors of calculus, Gottfried Wilhelm von Leibniz (1646–1716). (The other was Sir Isaac Newton, 1642–1727.)

With the above notation, the derivative of $y = f(x) = 2x^3 + 4x$, for example, which was found in Example 5 of Section 3.4 to be $f'(x) = 6x^2 + 4$, would be written

$$\frac{dy}{dx} = 6x^2 + 4$$

$$\frac{d}{dx}(2x^3 + 4x) = 6x^2 + 4$$

$$D_x(2x^3 + 4x) = 6x^2 + 4.$$

A variable other than x is often used as the independent variable. For example, if $y = f(t)$ gives population growth as a function of time, then the derivative of y with respect to t could be written

$$f'(t), \quad \frac{dy}{dt}, \quad \frac{d}{dt}[f(t)], \quad \text{or} \quad D_t[f(t)].$$

Other variables also may be used to name the function, as in $g(x)$ or $h(t)$.

Now we will use the definition

$$f'(x) = \lim_{h \to 0} \frac{f(x + h) - f(x)}{h}$$

to develop some rules for finding derivatives more easily than by the four-step process given in the previous chapter.

The first rule tells how to find the derivative of a constant function defined by $f(x) = k$, where k is a constant real number. Since $f(x + h)$ is also k, by definition $f'(x)$ is

$$f'(x) = \lim_{h \to 0} \frac{f(x + h) - f(x)}{h}$$

$$= \lim_{h \to 0} \frac{k - k}{h} = \lim_{h \to 0} \frac{0}{h} = \lim_{h \to 0} 0 = 0,$$

establishing the following rule.

Constant Rule

If $f(x) = k$, where k is any real number, then

$$f'(x) = \frac{d[k]}{dx} = 0.$$

(The derivative of a constant is 0.)

FIGURE 1

This rule is logical because the derivative represents rate of change, and a constant function, by definition, does not change. Figure 1 illustrates this constant rule geometrically; it shows a graph of the horizontal line $y = k$. At any point P on this line, the tangent line at P is the line itself. Since a horizontal line has a slope of 0, the slope of the tangent line is 0. This agrees with the result above: The derivative of a constant is 0.

EXAMPLE 1   Derivative of a Constant

(a) If $f(x) = 9$, then $f'(x) = 0$.

(b) If $h(t) = \pi$, then $D_t[h(t)] = 0$.

(c) If $y = 2^3$, then $dy/dx = 0$.

Functions of the form $y = x^n$, where n is a fixed real number, are very common in applications. To obtain a rule for finding the derivative of such a function, we can use the definition to work out the derivatives for various special values of n. This was done in Section 3.4 in Example 4 to show that for $f(x) = x^2$, $f'(x) = 2x$.

For $f(x) = x^3$, the derivative is found as follows.

$$f'(x) = \lim_{h \to 0} \frac{f(x + h) - f(x)}{h}$$

$$= \lim_{h \to 0} \frac{(x + h)^3 - x^3}{h}$$

$$= \lim_{h \to 0} \frac{(x^3 + 3x^2h + 3xh^2 + h^3) - x^3}{h}$$

The binomial theorem (discussed in most intermediate and college algebra texts) was used to expand $(x + h)^3$ in the last step. Now, the limit can be determined.

$$f'(x) = \lim_{h \to 0} \frac{3x^2h + 3xh^2 + h^3}{h}$$

$$= \lim_{h \to 0} (3x^2 + 3xh + h^2)$$

$$= 3x^2$$

The results in the table on the next page were found in a similar way, using the definition of the derivative. (These results are modifications of some of the examples and exercises from the previous chapter.)

Derivative of $f(x) = x^n$		
Function	n	Derivative
$f(x) = x$	1	$f'(x) = 1 = 1x^0$
$f(x) = x^2$	2	$f'(x) = 2x = 2x^1$
$f(x) = x^3$	3	$f'(x) = 3x^2$
$f(x) = x^4$	4	$f'(x) = 4x^3$
$f(x) = x^{-1}$	-1	$f'(x) = -1 \cdot x^{-2} = \dfrac{-1}{x^2}$
$f(x) = x^{1/2}$	$1/2$	$f'(x) = \dfrac{1}{2}x^{-1/2} = \dfrac{1}{2x^{1/2}}$

These results suggest the following rule.

Power Rule

If $f(x) = x^n$ for any real number n, then

$$f'(x) = \frac{d[x^n]}{dx} = nx^{n-1}.$$

(The derivative of $f(x) = x^n$ is found by multiplying by the exponent n and decreasing the exponent on x by 1.)

While the power rule is true for every real-number value of n, a proof is given here only for positive integer values of n. This proof follows the steps used above in finding the derivative of $f(x) = x^3$.

For any real numbers p and q, by the binomial theorem,

$$(p + q)^n = p^n + np^{n-1}q + \frac{n(n-1)}{2}p^{n-2}q^2 + \cdots + npq^{n-1} + q^n.$$

Replacing p with x and q with h gives

$$(x + h)^n = x^n + nx^{n-1}h + \frac{n(n-1)}{2}x^{n-2}h^2 + \cdots + nxh^{n-1} + h^n,$$

from which

$$(x + h)^n - x^n = nx^{n-1}h + \frac{n(n-1)}{2}x^{n-2}h^2 + \cdots + nxh^{n-1} + h^n.$$

Dividing each term by h yields

$$\frac{(x + h)^n - x^n}{h} = nx^{n-1} + \frac{n(n-1)}{2}x^{n-2}h + \cdots + nxh^{n-2} + h^{n-1}.$$

Use the definition of derivative, and the fact that each term except the first contains h as a factor and thus approaches 0 as h approaches 0, to get

$$f'(x) = \lim_{h \to 0} \frac{(x + h)^n - x^n}{h}$$

$$= nx^{n-1} + \frac{n(n-1)}{2}x^{n-2}0 + \cdots + nx0^{n-2} + 0^{n-1}$$

$$= nx^{n-1}.$$

This shows that the derivative of $f(x) = x^n$ is $f'(x) = nx^{n-1}$, proving the power rule for positive integer values of n.

EXAMPLE 2 Power Rule

(a) If $f(x) = x^6$, find $f'(x)$.

SOLUTION $f'(x) = 6x^{6-1} = 6x^5$

(b) If $y = t = t^1$, find $\dfrac{dy}{dt}$.

SOLUTION $\dfrac{dy}{dt} = 1t^{1-1} = t^0 = 1$

(c) If $y = 1/x^3$, find dy/dx.

SOLUTION Use a negative exponent to rewrite this equation as $y = x^{-3}$; then

$$\frac{dy}{dx} = -3x^{-3-1} = -3x^{-4} \quad \text{or} \quad \frac{-3}{x^4}.$$

(d) Find $D_x(x^{4/3})$.

SOLUTION $D_x(x^{4/3}) = \dfrac{4}{3}x^{4/3-1} = \dfrac{4}{3}x^{1/3}$

(e) If $y = \sqrt{z}$, find dy/dz.

SOLUTION Rewrite this as $y = z^{1/2}$; then

$$\frac{dy}{dz} = \frac{1}{2}z^{1/2-1} = \frac{1}{2}z^{-1/2} \quad \text{or} \quad \frac{1}{2z^{1/2}} \quad \text{or} \quad \frac{1}{2\sqrt{z}}.$$

TRY YOUR TURN 1

FOR REVIEW

At this point you may wish to turn back to Sections R.6 and R.7 for a review of negative exponents and rational exponents. The relationship between powers, roots, and rational exponents is explained there.

YOUR TURN 1 If $f(t) = \dfrac{1}{\sqrt{t}}$, find $f'(t)$.

The next rule shows how to find the derivative of the product of a constant and a function.

Constant Times a Function

Let k be a real number. If $g'(x)$ exists, then the derivative of $f(x) = k \cdot g(x)$ is

$$f'(x) = \frac{d}{dx}[k \cdot g(x)] = k \cdot g'(x).$$

(The derivative of a constant times a function is the constant times the derivative of the function.)

This rule is proved with the definition of the derivative and rules for limits.

$$
\begin{aligned}
f'(x) &= \lim_{h \to 0} \frac{kg(x + h) - kg(x)}{h} \\
&= \lim_{h \to 0} k\frac{[g(x + h) - g(x)]}{h} \qquad \text{Factor out } k. \\
&= k \lim_{h \to 0} \frac{g(x + h) - g(x)}{h} \qquad \text{Limit rule 1} \\
&= k \cdot g'(x) \qquad \text{Definition of derivative}
\end{aligned}
$$

EXAMPLE 3 Derivative of a Constant Times a Function

(a) If $y = 8x^4$, find $\dfrac{dy}{dx}$.

SOLUTION $\dfrac{dy}{dx} = 8(4x^3) = 32x^3$

(b) If $y = -\dfrac{3}{4}x^{12}$, find dy/dx.

SOLUTION $\dfrac{dy}{dx} = -\dfrac{3}{4}(12x^{11}) = -9x^{11}$

(c) Find $D_t(-8t)$.

SOLUTION $D_t(-8t) = -8(1) = -8$

(d) Find $D_p(10p^{3/2})$.

SOLUTION $D_p(10p^{3/2}) = 10\left(\dfrac{3}{2}p^{1/2}\right) = 15p^{1/2}$

(e) If $y = \dfrac{6}{x}$, find $\dfrac{dy}{dx}$.

SOLUTION Rewrite this as $y = 6x^{-1}$; then

YOUR TURN 2 If $y = 3\sqrt{x}$, find dy/dx.

$$\dfrac{dy}{dx} = 6(-1x^{-2}) = -6x^{-2} \quad \text{or} \quad \dfrac{-6}{x^2}.$$

TRY YOUR TURN 2

EXAMPLE 4 Beagles

Researchers have determined that the daily energy requirements of female beagles who are at least 1 year old change with respect to age according to the function

$$E(t) = 753t^{-0.1321},$$

where $E(t)$ is the daily energy requirements (in kJ/W$^{0.67}$) for a dog that is t years old. *Source: Journal of Nutrition.*

(a) Find $E'(t)$.

SOLUTION Using the rules of differentiation we find that

$$E'(t) = 753(-0.1321)t^{-0.1321-1} = -99.4713t^{-1.1321}.$$

(b) Determine the rate of change of the daily energy requirements of a 2-year-old female beagle.

SOLUTION $E'(2) = -99.4713(2)^{-1.1321} \approx -45.4$

Thus, the daily energy requirements of a 2-year-old female beagle are decreasing at the rate of 45.4 kJ/W$^{0.67}$ per year.

The final rule in this section is for the derivative of a function that is a sum or difference of terms.

Sum or Difference Rule

If $f(x) = u(x) \pm v(x)$, and if $u'(x)$ and $v'(x)$ exist, then

$$f'(x) = \dfrac{d}{dx}[u(x) \pm v(x)] = u'(x) \pm v'(x).$$

(The derivative of a sum or difference of functions is the sum or difference of the derivatives.)

The proof of the sum part of this rule is as follows: If $f(x) = u(x) + v(x)$, then

$$f'(x) = \lim_{h \to 0} \frac{[u(x + h) + v(x + h)] - [u(x) + v(x)]}{h}$$

$$= \lim_{h \to 0} \frac{[u(x + h) - u(x)] + [v(x + h) - v(x)]}{h}$$

$$= \lim_{h \to 0} \left[\frac{u(x + h) - u(x)}{h} + \frac{v(x + h) - v(x)}{h} \right]$$

$$= \lim_{h \to 0} \frac{u(x + h) - u(x)}{h} + \lim_{h \to 0} \frac{v(x + h) - v(x)}{h}$$

$$= u'(x) + v'(x).$$

A similar proof can be given for the difference of two functions.

EXAMPLE 5 Derivative of a Sum

Find the derivative of each function.

(a) $y = 6x^3 + 15x^2$

SOLUTION Let $u(x) = 6x^3$ and $v(x) = 15x^2$; then $y = u(x) + v(x)$. Since $u'(x) = 18x^2$ and $v'(x) = 30x$,

$$\frac{dy}{dx} = 18x^2 + 30x.$$

(b) $p(t) = 12t^4 - 6\sqrt{t} + \dfrac{5}{t}$

SOLUTION Rewrite $p(t)$ as $p(t) = 12t^4 - 6t^{1/2} + 5t^{-1}$; then

$$p'(t) = 48t^3 - 3t^{-1/2} - 5t^{-2}.$$

Also, $p'(t)$ may be written as $p'(t) = 48t^3 - \dfrac{3}{\sqrt{t}} - \dfrac{5}{t^2}.$

(c) $f(x) = \dfrac{x^3 + 3\sqrt{x}}{x}$

SOLUTION Rewrite $f(x)$ as $f(x) = \dfrac{x^3}{x} + \dfrac{3x^{1/2}}{x} = x^2 + 3x^{-1/2}.$ Then

$$D_x[f(x)] = 2x - \frac{3}{2}x^{-3/2},$$

or

$$D_x[f(x)] = 2x - \frac{3}{2\sqrt{x^3}}.$$

(d) $f(x) = (4x^2 - 3x)^2$

SOLUTION Rewrite $f(x)$ as $f(x) = 16x^4 - 24x^3 + 9x^2$ using the fact that $(a - b)^2 = a^2 - 2ab + b^2$; then

$$f'(x) = 64x^3 - 72x^2 + 18x.$$

YOUR TURN 3 If $h(t) = -3t^2 + 2\sqrt{t} + \dfrac{5}{t^4} - 7,$ find $h'(t)$.

TRY YOUR TURN 3

⊞ **TECHNOLOGY NOTE**

Some computer programs and calculators have built-in methods for taking derivatives symbolically, which is what we have been doing in this section, as opposed to approximating the derivative numerically by using a small number for h in the definition of the derivative. In the computer program Maple, we would do part (a) of Example 5 by entering

```
> diff(6*x^3+15*x^2,x);
```

where the x after the comma tells what variable the derivative is with respect to. Maple would respond with

```
18*x^2+30*x.
```

Similarly, on the TI-89, we would enter `d(6x^3+15x^2,x)` and the calculator would give "$18 \cdot x^2 + 30 \cdot x$."

Other graphing calculators, such as the TI-84 Plus C, do not have built-in methods for taking derivatives symbolically. As we saw in the last chapter, however, they do have the ability to calculate the derivative of a function at a particular point and to simultaneously graph a function and its derivative.

Recall that, on the TI-84 Plus C, we could use the `nDeriv` Command, as shown in Figure 2, to approximate the value of the derivative when $x = 1$. Figure 3(a) and Figure 3(b) indicate how to input the functions into the calculator and the corresponding graphs of both the function and its derivative. Consult the *Graphing Calculator and Excel Spreadsheet Manual*, available with this book, for assistance.

FIGURE 2

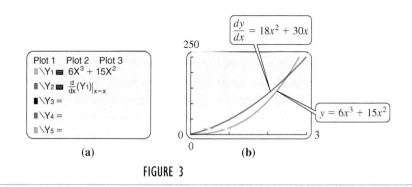

(a)   (b)

FIGURE 3

The rules developed in this section make it possible to find the derivative of a function more directly, so that applications of the derivative can be dealt with more effectively. The following examples illustrate some business applications.

Marginal Analysis

In previous sections we discussed the concepts of marginal cost, marginal revenue, and marginal profit. These concepts of **marginal analysis** are summarized here.

In business and economics the rates of change of such variables as cost, revenue, and profit are important considerations. Economists use the word *marginal* to refer to rates of change. For example, *marginal cost* refers to the rate of change of cost. Since the derivative of a function gives the rate of change of the function, a marginal cost (or revenue, or profit) function is found by taking the derivative of the cost (or revenue, or profit) function. Roughly speaking, the marginal cost at some level of production x is the cost to produce the $(x + 1)$st item. (Similar statements could be made for revenue or profit.)

To see why it is reasonable to say that the marginal cost function is approximately the cost of producing one more unit, look at Figure 4, where $C(x)$ represents the cost of producing x units of some item. Then the cost of producing $x + 1$ units is $C(x + 1)$. The cost of the $(x + 1)$st unit is, therefore, $C(x + 1) - C(x)$. This quantity is shown in the graph in Figure 4.

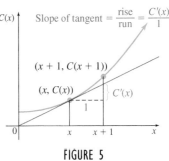

FIGURE 4   FIGURE 5

Now if $C(x)$ is the cost function, then the marginal cost $C'(x)$ represents the slope of the tangent line at any point $(x, C(x))$. The graph in Figure 5 shows the cost function $C(x)$ and the tangent line at a point $(x, C(x))$. Remember what it means for a line to have a given slope. If the slope of the line is $C'(x)$, then

$$\frac{\Delta y}{\Delta x} = C'(x) = \frac{C'(x)}{1},$$

and beginning at any point on the line and moving 1 unit to the right requires moving $C'(x)$ units up to get back to the line again. The vertical distance from the horizontal line to the tangent line shown in Figure 5 is therefore $C'(x)$.

Superimposing the graphs from Figures 4 and 5 as in Figure 6 shows that $C'(x)$ is indeed very close to $C(x + 1) - C(x)$. The two values are closest when x is very large, so that 1 unit is relatively small.

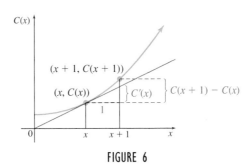

FIGURE 6

EXAMPLE 6   Marginal Cost

Suppose that the total cost in hundreds of dollars to produce x thousand barrels of a beverage is given by

$$C(x) = 4x^2 + 100x + 500.$$

Find the marginal cost for the following values of x.

(a) $x = 5$

SOLUTION   To find the marginal cost, first find $C'(x)$, the derivative of the total cost function.

$$C'(x) = 8x + 100$$

When $x = 5$,

$$C'(5) = 8(5) + 100 = 140.$$

After 5 thousand barrels of the beverage have been produced, the cost to produce one thousand more barrels will be *approximately* 140 hundred dollars, or $14,000.

The *actual* cost to produce one thousand more barrels is $C(6) - C(5)$:

$$C(6) - C(5) = (4 \cdot 6^2 + 100 \cdot 6 + 500) - (4 \cdot 5^2 + 100 \cdot 5 + 500)$$
$$= 1244 - 1100 = 144,$$

144 hundred dollars, or $14,400.

(b) $x = 30$

SOLUTION After 30 thousand barrels have been produced, the cost to produce one thousand more barrels will be approximately

$$C'(30) = 8(30) + 100 = 340,$$

or $34,000. Notice that the cost to produce an additional thousand barrels of beverage has increased by approximately $20,000 at a production level of 30,000 barrels compared to a production level of 5000 barrels. **TRY YOUR TURN 4**

YOUR TURN 4 If the cost function is given by $C(x) = 5x^3 - 10x^2 + 75$, find the marginal cost when $x = 100$.

Demand Functions The demand function, defined by $p = D(q)$, relates the number of units q of an item that consumers are willing to purchase to the price p. (Demand functions were also discussed in Chapter 1.) The total revenue $R(q)$ is related to price per unit and the amount demanded (or sold) by the equation

$$R(q) = qp = q \cdot D(q).$$

EXAMPLE 7 Marginal Revenue

The demand function (in dollars) for a certain product is given by

$$p = \frac{50,000 - q}{25,000}$$

for $0 \le q \le 50,000$. Find the marginal revenue when $q = 10,000$ units.

SOLUTION From the given function for p, the revenue function is given by

$$R(q) = qp$$
$$= q\left(\frac{50,000 - q}{25,000}\right)$$
$$= \frac{50,000q - q^2}{25,000}$$
$$= 2q - \frac{1}{25,000}q^2.$$

The marginal revenue is

$$R'(q) = 2 - \frac{2}{25,000}q.$$

When $q = 10,000$, the marginal revenue is

YOUR TURN 5 If the demand function is given by $p = 16 - 1.25q$, find the marginal revenue when $q = 5$.

$$R'(10,000) = 2 - \frac{2}{25,000}(10,000) = 1.2,$$

or $1.20 per unit. Thus, the next item sold (at sales of 10,000) will produce additional revenue of about $1.20. **TRY YOUR TURN 5**

The graphs of the demand function and marginal revenue function are shown in Figure 7(a). The graph of the revenue function is shown in Figure 7(b). Notice that when $q = 10,000$, the demand is $p = \$1.60$, the marginal revenue is $R'(10,000) = \$1.20$ per unit, and the revenue is $R(10,000) = \$16,000$.

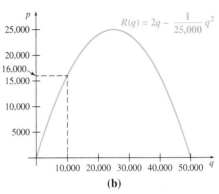

FIGURE 7

EXAMPLE 8   Marginal Revenue

Economists usually give demand functions with the quantity as the independent variable, as in Example 7. It might make more intuitive sense, however, to think of the quantity demanded as a function of the price. For example, suppose the demand function is given as

$$q = 50,000 - 25,000p$$

where p is the price in dollars and $0 \leq p \leq 2$. Find the marginal revenue when the price is \$1.60.

SOLUTION   When the quantity demanded is given as a function of the price, we must first solve the demand equation for p.

$$q = 50,000 - 25,000p$$
$$25,000p + q = 50,000 \qquad \text{Add 25,000}p \text{ to both sides.}$$
$$25,000p = 50,000 - q \qquad \text{Subtract } q \text{ from both sides.}$$
$$p = \frac{50,000 - q}{25,000} \qquad \text{Divide both sides by 25,000.}$$

Next, we must find the value of q corresponding to $p = 1.60$ by substituting this value of p into the original equation.

$$q = 50,000 - 25,000(1.6) = 10,000$$

We are now ready to proceed as we did in Example 7. In fact, notice that the rewritten demand function and the value of q correspond *exactly* to what we started with in Example 7, and so the solution is the same.

Management must be careful to keep track of marginal costs and revenue. If the marginal cost of producing an extra unit exceeds the marginal revenue received from selling it, then the company will lose money on that unit.

EXAMPLE 9   Marginal Profit

Suppose that the cost function for the product in Example 7 is given by

$$C(q) = 2100 + 0.25q, \qquad \text{where } 0 \leq q \leq 50,000.$$

Find the marginal profit from the production of the following numbers of units.

(a) 15,000

APPLY IT   **SOLUTION**   From Example 7, the revenue from the sale of q units is

$$R(q) = 2q - \frac{1}{25,000}q^2.$$

Since profit, P, is given by $P = R - C$,

$$P(q) = R(q) - C(q)$$

$$= \left(2q - \frac{1}{25,000}q^2\right) - (2100 + 0.25q)$$

$$= 2q - \frac{1}{25,000}q^2 - 2100 - 0.25q$$

$$= 1.75q - \frac{1}{25,000}q^2 - 2100. \qquad \text{See Figure 8(a).}$$

The marginal profit from the sale of q units is

$$P'(q) = 1.75 - \frac{2}{25,000}q = 1.75 - \frac{1}{12,500}q.$$

At $q = 15,000$ the marginal profit is

$$P'(15,000) = 1.75 - \frac{1}{12,500}(15,000) = 0.55,$$

or $0.55 per unit.

(b) 21,875

SOLUTION When $q = 21,875$, the marginal profit is

$$P'(21,875) = 1.75 - \frac{1}{12,500}(21,875) = 0.$$

(c) 25,000

SOLUTION When $q = 25,000$, the marginal profit is

$$P'(25,000) = 1.75 - \frac{1}{12,500}(25,000) = -0.25,$$

or $-$0.25 per unit.

As shown by parts (b) and (c), if more than 21,875 units are sold, the marginal profit is negative. This indicates that increasing production beyond that level will *reduce* profit.

Figure 8(a) shows a graph of the profit function. Figure 8(b) shows a graph of the marginal profit function. Notice that when $q = 15,000$, the profit is $P(15,000) = \$15,150$ and the marginal profit is $P'(15,000) = \$0.55$ per unit. When $q = 25,000$, the profit is $P(25,000) = \$16,650$ and the marginal profit is $P'(15,000) = -\$0.25$ per unit.

(a)

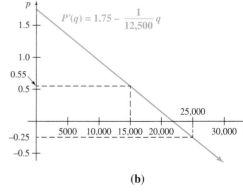

(b)

FIGURE 8

The final example shows an application of the derivative to a problem of demography.

EXAMPLE 10 Centenarians

The number of Americans (in thousands) who are expected to be over 100 years old can be approximated by the function

$$f(t) = 0.00943t^3 - 0.470t^2 + 11.085t + 23.441,$$

where t is the year, with $t = 0$ corresponding to 2000, and $0 \le t \le 50$. *Source: U.S. Census Bureau.*

(a) Find a formula giving the rate of change of the number of Americans over 100 years old.

APPLY IT

SOLUTION   Using the techniques for finding the derivative, we have

$$f'(t) = 0.02829t^2 - 0.940t + 11.085.$$

This tells us the rate of change in the number of Americans over 100 years old.

(b) Find the rate of change in the number of Americans who are expected to be over 100 years old in the year 2020.

SOLUTION   The year 2020 corresponds to $t = 20$.

$$f'(20) = 0.02829(20)^2 - 0.940(20) + 11.085 = 3.601$$

The number of Americans over 100 years old is expected to grow at a rate of about 3.6 thousand, or about 3600, per year in the year 2020.

4.1 WARM-UP EXERCISES

Find the equation for a line satisfying the following conditions. Put your answer in the form $y = mx + b$.   (Sec. 1.1)

W1.   Through $(2, 7)$, parallel to $4x + 2y = 9$

W2.   Through $(-3, 4)$, parallel to $2x - 5y = 3$

4.1   EXERCISES

Find the derivative of each function defined as follows.

1. $y = 12x^3 - 8x^2 + 7x + 5$

2. $y = 8x^3 - 5x^2 - \dfrac{x}{12}$

3. $y = 3x^4 - 6x^3 + \dfrac{x^2}{8} + 5$

4. $y = 5x^4 + 9x^3 + 12x^2 - 7x$

5. $f(x) = 6x^{3.5} - 10x^{0.5}$

6. $f(x) = -2x^{1.5} + 12x^{0.5}$

7. $y = 8\sqrt{x} + 6x^{3/4}$

8. $y = -100\sqrt{x} - 11x^{2/3}$

9. $y = 10x^{-3} + 5x^{-4} - 8x$

10. $y = 5x^{-5} - 6x^{-2} + 13x^{-1}$

11. $f(t) = \dfrac{7}{t} - \dfrac{5}{t^3}$

12. $f(t) = \dfrac{14}{t} + \dfrac{12}{t^4} + \sqrt{2}$

13. $y = \dfrac{6}{x^4} - \dfrac{7}{x^3} + \dfrac{3}{x} + \sqrt{5}$

14. $y = \dfrac{3}{x^6} + \dfrac{1}{x^5} - \dfrac{7}{x^2}$

15. $p(x) = -10x^{-1/2} + 8x^{-3/2}$

16. $h(x) = x^{-1/2} - 14x^{-3/2}$

17. $y = \dfrac{6}{\sqrt[4]{x}}$

18. $y = \dfrac{-2}{\sqrt[3]{x}}$

19. $f(x) = \dfrac{x^3 + 5}{x}$

20. $g(x) = \dfrac{x^3 - 4x}{\sqrt{x}}$

21. $g(x) = (8x^2 - 4x)^2$

22. $h(x) = (x^2 - 1)^3$

23. Which of the following describes the derivative function $f'(x)$ of a quadratic function $f(x)$?

   (a) Quadratic   **(b)** Linear

   (c) Constant   **(d)** Cubic (third degree)

24. Which of the following describes the derivative function $f'(x)$ of a cubic (third degree) function $f(x)$?

   (a) Quadratic   **(b)** Linear

   (c) Constant   **(d)** Cubic

25. Explain the relationship between the slope and the derivative of $f(x)$ at $x = a$.

26. Which of the following do *not* equal $\dfrac{d}{dx}(4x^3 - 6x^{-2})$?

   (a) $\dfrac{12x^2 + 12}{x^3}$ **(b)** $\dfrac{12x^5 + 12}{x^3}$

   (c) $12x^2 + \dfrac{12}{x^3}$ **(d)** $12x^3 + 12x^{-3}$

Find each derivative.

27. $D_x\left[9x^{-1/2} + \dfrac{2}{x^{3/2}}\right]$

28. $D_x\left[\dfrac{8}{\sqrt[4]{x}} - \dfrac{3}{\sqrt{x^3}}\right]$

29. $f'(-2)$ if $f(x) = \dfrac{x^4}{6} - 3x$

30. $f'(3)$ if $f(x) = \dfrac{x^3}{9} - 7x^2$

In Exercises 31–34, find the slope of the tangent line to the graph of the given function at the given value of x. Find the equation of the tangent line in Exercises 31 and 32.

31. $y = x^4 - 5x^3 + 2; \quad x = 2$

32. $y = -3x^5 - 8x^3 + 4x^2; \quad x = 1$

33. $y = -2x^{1/2} + x^{3/2}; \quad x = 9$

34. $y = -x^{-3} + x^{-2}; \quad x = 2$

35. Find all points on the graph of $f(x) = 9x^2 - 8x + 4$ where the slope of the tangent line is 0.

36. Find all points on the graph of $f(x) = x^3 + 9x^2 + 19x - 10$ where the slope of the tangent line is -5.

In Exercises 37–40, for each function find all values of x where the tangent line is horizontal.

37. $f(x) = 2x^3 + 9x^2 - 60x + 4$

38. $f(x) = x^3 + 15x^2 + 63x - 10$

39. $f(x) = x^3 - 4x^2 - 7x + 8$

40. $f(x) = x^3 - 5x^2 + 6x + 3$

41. At what points on the graph of $f(x) = 6x^2 + 4x - 9$ is the slope of the tangent line -2?

42. At what points on the graph of $f(x) = 2x^3 - 9x^2 - 12x + 5$ is the slope of the tangent line 12?

43. At what points on the graph of $f(x) = x^3 + 6x^2 + 21x + 2$ is the slope of the tangent line 9?

44. If $g'(5) = 12$ and $h'(5) = -3$, find $f'(5)$ for $f(x) = 3g(x) - 2h(x) + 3$.

45. If $g'(2) = 7$ and $h'(2) = 14$, find $f'(2)$ for
$$f(x) = \frac{1}{2}g(x) + \frac{1}{4}h(x).$$

46. Use the information given in the figure to find the following values.

   (a) $f(1)$ **(b)** $f'(1)$

   (c) The domain of f **(d)** The range of f

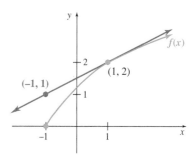

47. Explain the concept of marginal cost. How does it relate to cost? How is it found?

48. In Exercises 43–46 of Section 2.2, the effect of a when graphing $y = af(x)$ was discussed. Now describe how this relates to the fact that $D_x[af(x)] = af'(x)$.

49. Show that, for any constant k,
$$\frac{d}{dx}\left[\frac{f(x)}{k}\right] = \frac{f'(x)}{k}.$$

50. Use the differentiation feature on your graphing calculator to solve the problems (to 2 decimal places) below, where $f(x)$ is defined as follows:
$$f(x) = 1.25x^3 + 0.01x^2 - 2.9x + 1.$$

   (a) Find $f'(4)$.

   (b) Find all values of x where $f'(x) = 0$.

APPLICATIONS

Business and Economics

51. Revenue Assume that a demand equation is given by $q = 5000 - 100p$. Find the marginal revenue for the following production levels (values of q).

   (a) 1000 units   **(b)** 2500 units

   (c) 3000 units

52. Profit Suppose that for the situation in Exercise 51 the cost of producing q units is given by $C(q) = 3000 - 20q + 0.03q^2$. Find the marginal profit for the following production levels.

   (a) 500 units   **(b)** 815 units

   (c) 1000 units

53. Revenue If the price in dollars of a stereo system is given by

$$p(q) = \frac{1000}{q^2} + 1000,$$

where q represents the demand for the product, find the marginal revenue when the demand is 10.

54. Profit Suppose that for the situation in Exercise 53 the cost in dollars of producing q stereo systems is given by $C(q) = 0.2q^2 + 6q + 50$. Find the marginal profit when the demand is 10.

55. Marginal Product of Labor Economists have considered the output y of a manufacturing process as a function of the size of the labor force n using the function

$$y = kn^p,$$

where $0 < p < 1$. The marginal product of labor, defined as dy/dn, measures the rate that output increases with the size of the labor force, and is a measure of labor productivity. *Source: The Journal of Economic Education.*

(a) Show that

$$\frac{dy}{dn} = \frac{kp}{n^{1-p}}.$$

(b) How can you tell from your answer to part (a) that as the size of the labor force increases, the marginal product of labor gets smaller? This is a phenomenon known as the law of diminishing returns, discussed more in the next chapter.

56. Profit An analyst has found that a company's costs and revenues in dollars for one product are given by

$$C(x) = 2x \quad \text{and} \quad R(x) = 6x - \frac{x^2}{1000},$$

respectively, where x is the number of items produced.

(a) Find the marginal cost function.

(b) Find the marginal revenue function.

(c) Using the fact that profit is the difference between revenue and costs, find the marginal profit function.

(d) What value of x makes marginal profit equal 0?

(e) Find the profit when the marginal profit is 0.

(As we shall see in the next chapter, this process is used to find *maximum* profit.)

57. Postal Rates U.S. postal rates have steadily increased since 1932. Using data depicted in the table in the next column for the years 1932–2014, the cost in cents to mail a single letter can be modeled using a quadratic formula as follows:

$$C(t) = 0.007714t^2 - 0.03969t + 0.726,$$

where t is the number of years since 1932. *Source: U.S. Postal Service.*

(a) Find the predicted cost of mailing a letter in 1982 and 2002 and compare these estimates with the actual rates.

(b) Find the rate of change of the postage cost for the years 1982 and 2002 and interpret your results.

Year	Cost	Year	Cost
1932	3	1991	29
1958	4	1995	32
1963	5	1999	33
1968	6	2001	34
1971	8	2002	37
1974	10	2006	39
1975	13	2007	41
1978	15	2008	42
1981	18	2009	44
1981	20	2012	45
1985	22	2013	46
1988	25	2014	49

(c) Using the regression feature on a graphing calculator, find a cubic function that models these data, letting $t = 0$ correspond to the year 1932. Then use your answer to find the rate of change of the postage cost for the years 1982 and 2002.

(d) Discuss whether the quadratic or cubic function best describes the data. Do the answers from part (b) or from part (c) best describe the rate that postage was going up in the years 1982 and 2002?

(e) Explore other functions that could be used to model the data, using the various regression features on a graphing calculator, and discuss to what extent any of them are useful descriptions of the data.

58. Money The total amount of money in circulation for the years 1955–2014 can be closely approximated by

$$M(t) = 0.006934t^3 - 0.07837t^2 + 1.480t + 27.78,$$

where t represents the number of years since 1955 and $M(t)$ is in billions of dollars. Find the derivative of $M(t)$ and use it to find the rate of change of money in circulation in the following years. *Source: U.S. Treasury.*

(a) 1960   (b) 1980

(c) 2000   (d) 2010

(e) What do your answers to parts (a)–(d) tell you about the amount of money in circulation in those years?

Life Sciences

59. Cancer Insulation workers who were exposed to asbestos and employed before 1960 experienced an increased likelihood of lung cancer. If a group of insulation workers has a cumulative total of 100,000 years of work experience with their first date of employment t years ago, then the number of lung cancer cases occurring within the group can be modeled using the function

$$N(t) = 0.00437t^{3.2}.$$

Find the rate of growth of the number of workers with lung cancer in a group as described by the following first dates of employment. *Source: Observation and Inference: An Introduction to the Methods of Epidemiology.*

(a) 5 years ago   (b) 10 years ago

60. Blood Sugar Level Insulin affects the glucose, or blood sugar, level of some diabetics according to the function

$$G(x) = -0.2x^2 + 450,$$

where $G(x)$ is the blood sugar level 1 hour after x units of insulin are injected. (This mathematical model is only approximate, and it is valid only for values of x less than about 40.) Find the blood sugar level after the following numbers of units of insulin are injected.

(a) 0   (b) 25

Find the rate of change of blood sugar level after injection of the following numbers of units of insulin.

(c) 10   (d) 25

61. Bighorn Sheep The cumulative horn volume for certain types of bighorn rams, found in the Rocky Mountains, can be described by the quadratic function

$$V(t) = -2159 + 1313t - 60.82t^2,$$

where $V(t)$ is the horn volume (in cm³) and t is the year of growth, $2 \leq t \leq 9$. *Source: Conservation Biology.*

(a) Find the horn volume for a 3-year-old ram.

(b) Find the rate at which the horn volume of a 3-year-old ram is changing.

62. Brain Mass The brain mass of a human fetus during the last trimester can be accurately estimated from the circumference of the head by

$$m(c) = \frac{c^3}{100} - \frac{1500}{c},$$

where $m(c)$ is the mass of the brain (in grams) and c is the circumference (in centimeters) of the head. *Source: Early Human Development.*

(a) Estimate the brain mass of a fetus that has a head circumference of 30 cm.

(b) Find the rate of change of the brain mass for a fetus that has a head circumference of 30 cm and interpret your results.

63. Velocity of Marine Organism The typical velocity (in centimeters per second) of a marine organism of length l (in centimeters) is given by $v = 2.69l^{1.86}$. Find the rate of change of the velocity with respect to the length of the organism. *Source: Mathematical Topics in Population Biology Morphogenesis and Neurosciences.*

64. Heart The left ventricular length (viewed from the front of the heart) of a human fetus that is at least 18 weeks old can be estimated by

$$l(x) = -2.318 + 0.2356x - 0.002674x^2,$$

where $l(x)$ is the ventricular length (in centimeters) and x is the age (in weeks) of the fetus. *Source: American Journal of Cardiology.*

(a) Determine a meaningful domain for this function.

(b) Find $l'(x)$.

(c) Find $l'(25)$.

65. Track and Field In 1906 Kennelly developed a simple formula for predicting an upper limit on the fastest time that humans could ever run distances from 100 yards to 10 miles. His formula is given by

$$t = 0.0588s^{1.125},$$

where s is the distance in meters and t is the time to run that distance in seconds. *Source: Proceedings of the American Academy of Arts and Sciences.*

(a) Find Kennelly's estimate for the fastest mile. (*Hint:* 1 mile $\approx$ 1609 meters.)

(b) Find dt/ds when $s = 100$ and interpret your answer.

(c) Compare this and other estimates to the current world records. Have these estimates been surpassed?

66. Human Cough To increase the velocity of the air flowing through the trachea when a human coughs, the body contracts the windpipe, producing a more effective cough. Tuchinsky formulated that the velocity of air that is flowing through the trachea during a cough is

$$V = C(R_0 - R)R^2,$$

where C is a constant based on individual body characteristics, R_0 is the radius of the windpipe before the cough, and R is the radius of the windpipe during the cough. It can be shown that the maximum velocity of the cough occurs when $dV/dR = 0$. Find the value of R that maximizes the velocity.* *Source: COMAP, Inc.*

67. Body Mass Index The body mass index (BMI) is a number that can be calculated for any individual as follows: Multiply a person's weight by 703 and divide by the person's height squared. That is,

$$BMI = \frac{703w}{h^2},$$

where w is in pounds and h is in inches. The National Heart, Lung, and Blood Institute uses the BMI to determine whether a person is "overweight" $(25 \leq BMI < 30)$ or "obese" $(BMI \geq 30)$. *Source: The National Institutes of Health.*

(a) Calculate the BMI for Lebron James, basketball player for the Cleveland Cavaliers, who is 250 lb. and 6'8" tall.

(b) How much weight would Lebron James have to lose until he reaches a BMI of 24.9 and is no longer "overweight"? Comment on whether BMI cutoffs are appropriate for athletes with considerable muscle mass.

(c) For a 125-lb female, what is the rate of change of BMI with respect to height? (*Hint:* Take the derivative of the function: $f(h) = 703(125)/h^2$.)

(d) Calculate and interpret the meaning of $f'(65)$.

(e) Use the TABLE feature on your graphing calculator to construct a table for BMI for various weights and heights.

(f) Using the fact that 1 in. = 0.0254 m and 1 lb = 0.4536 kg, transform this formula to handle metric units.

---

*Interestingly, Tuchinsky also states that X-rays indicate that the body naturally contracts the windpipe to this radius during a cough.

Physical Sciences

Velocity We saw in the previous chapter that if a function $s(t)$ gives the position of an object at time t, the derivative gives the velocity, that is, $v(t) = s'(t)$. For each position function in Exercises 68–71, find (a) $v(t)$ and (b) the velocity when $t = 0$, $t = 5$, and $t = 10$.

68. $s(t) = 11t^2 + 4t + 2$

69. $s(t) = 18t^2 - 13t + 8$

70. $s(t) - 4t^3 + 8t^2 + t$

71. $s(t) = -3t^3 + 4t^2 - 10t + 5$

72. Velocity If a rock is dropped from a 144-ft building, its position (in feet above the ground) is given by $s(t) = -16t^2 + 144$, where t is the time in seconds since it was dropped.

  (a) What is its velocity 1 second after being dropped? 2 seconds after being dropped?

  (b) When will it hit the ground?

  (c) What is its velocity upon impact?

73. Velocity A ball is thrown vertically upward from the ground at a velocity of 64 ft per second. Its distance from the ground at t seconds is given by $s(t) = -16t^2 + 64t$.

  (a) How fast is the ball moving 2 seconds after being thrown? 3 seconds after being thrown?

  (b) How long after the ball is thrown does it reach its maximum height?

  (c) How high will it go?

74. Dead Sea Researchers who have been studying the alarming rate in which the level of the Dead Sea has been dropping have shown that the density $d(x)$ (in g per cm^3) of the Dead Sea brine during evaporation can be estimated by the function

$$d(x) = 1.66 - 0.90x + 0.47x^2,$$

where x is the fraction of the remaining brine, $0 \le x \le 1$. *Source: Geology*.

  (a) Estimate the density of the brine when 50% of the brine remains.

 (b) Find and interpret the instantaneous rate of change of the density when 50% of the brine remains.

75. Echoes Acoustical experts have found that clapping one's hands near the staircases of the Pyramid of Kukulkan at Chichen Itza results in an echo that sounds like a chirp. The initial frequency of the chirp f depends on the tread length of the stairs T according to the formula

$$f = \frac{c}{2T},$$

where c is the speed of sound, 340 m/sec. Find the initial frequency and the rate of change of the initial frequency with respect to tread length when the tread is 0.5 m. *Source: Acoustical Society of America*.

General Interest

76. AP Examinations The probability (as a percent) of scoring 3 or more on the Calculus AB Advanced Placement Examination can be very closely predicted as a function of a student's PSAT/NMSQT Score x by the function

$$P(x) = -0.00209x^3 + 0.3387x^2 - 15.15x + 208.6.$$

Source: The College Board.

  (a) Find the probability of scoring 3 or more for each of the following PSAT/NMSQT scores, as well as the rate that this probability is increasing with respect to the PSAT/NMSQT score.

    (i) 45   **(ii)** 75

  (b) Based on your results from part (a), what recommendations would you make about who should take the Calculus AB examination?

77. Dog's Human Age From the data printed in the following table from the *Minneapolis Star Tribune* a dog's age when compared to a human's age can be modeled using either a linear formula or a quadratic formula as follows:

$$y_1 = 4.13x + 14.63$$
$$y_2 = -0.033x^2 + 4.647x + 13.347,$$

where y_1 and y_2 represent a dog's human age for each formula and x represents a dog's actual age. *Source: The Mathematics Teacher*.

Dog Age	Human Age
1	16
2	24
3	28
5	36
7	44
9	52
11	60
13	68
15	76

  (a) Find y_1 and y_2 when $x = 5$.

  (b) Find dy_1/dx and dy_2/dx when $x = 5$ and interpret your answers.

  (c) If the first two points are eliminated from the table, find the equation of a line that perfectly fits the reduced set of data. Interpret your findings.

  (d) Of the three formulas, which do you prefer?

YOUR TURN ANSWERS

1. $f'(t) = -\dfrac{1}{2}t^{-\frac{3}{2}}$ or $f'(t) = -\dfrac{1}{2t^{\frac{3}{2}}}$ **2.** $\dfrac{dy}{dx} = \dfrac{3}{2}x^{-\frac{1}{2}}$ or $\dfrac{dy}{dx} = \dfrac{3}{2\sqrt{x}}$

3. $h'(t) = -6t + t^{-\frac{1}{2}} - 20t^{-5}$ or $h'(t) = -6t + \dfrac{1}{\sqrt{t}} - \dfrac{20}{t^5}$

4. \$148,000   **5.** \$3.50

4.2 Derivatives of Products and Quotients

We show how the derivative is used to solve a problem like this in Example 5, later in this section.

In the previous section we saw that the derivative of a sum of two functions is found from the sum of the derivatives. What about products? Is the derivative of a product equal to the product of the derivatives? For example, if

$$u(x) = 2x + 3 \quad \text{and} \quad v(x) = 3x^2,$$

then

$$u'(x) = 2 \quad \text{and} \quad v'(x) = 6x.$$

Let $f(x)$ be the product of u and v; that is, $f(x) = (2x + 3)(3x^2) = 6x^3 + 9x^2$. By the rules of the preceding section, $f'(x) = 18x^2 + 18x = 18x(x + 1)$. On the other hand, $u'(x) \cdot v'(x) = 2(6x) = 12x \neq f'(x)$. In this example, the derivative of a product is *not* equal to the product of the derivatives, nor is this usually the case.

The rule for finding derivatives of products is as follows.

Product Rule

If $f(x) = u(x) \cdot v(x)$, and if $u'(x)$ and $v'(x)$ both exist, then

$$f'(x) = \frac{d}{dx}[u(x) \cdot v(x)] = u(x) \cdot v'(x) + v(x) \cdot u'(x).$$

(The derivative of a product of two functions is the first function times the derivative of the second plus the second function times the derivative of the first.)

To sketch the method used to prove the product rule, let

$$f(x) = u(x) \cdot v(x).$$

Then $f(x + h) = u(x + h) \cdot v(x + h)$, and, by definition, $f'(x)$ is given by

$$f'(x) = \lim_{h \to 0} \frac{f(x + h) - f(x)}{h}$$

$$= \lim_{h \to 0} \frac{u(x + h) \cdot v(x + h) - u(x) \cdot v(x)}{h}.$$

Now subtract and add $u(x + h) \cdot v(x)$ in the numerator, giving

$$f'(x) = \lim_{h \to 0} \frac{u(x + h) \cdot v(x + h) - u(x + h) \cdot v(x) + u(x + h) \cdot v(x) - u(x) \cdot v(x)}{h}$$

$$= \lim_{h \to 0} \frac{u(x + h)[v(x + h) - v(x)] + v(x)[u(x + h) - u(x)]}{h}$$

$$= \lim_{h \to 0} u(x + h)\left[\frac{v(x + h) - v(x)}{h}\right] + \lim_{h \to 0} v(x)\left[\frac{u(x + h) - u(x)}{h}\right]$$

$$= \lim_{h \to 0} u(x + h) \cdot \lim_{h \to 0} \frac{v(x + h) - v(x)}{h} + \lim_{h \to 0} v(x) \cdot \lim_{h \to 0} \frac{u(x + h) - u(x)}{h}. \tag{1}$$

If $u'(x)$ and $v'(x)$ both exist, then

$$\lim_{h \to 0} \frac{u(x + h) - u(x)}{h} = u'(x) \qquad \text{and} \qquad \lim_{h \to 0} \frac{v(x + h) - v(x)}{h} = v'(x).$$

The fact that $u'(x)$ exists can be used to prove

$$\lim_{h \to 0} u(x + h) = u(x),$$

and since no h is involved in $v(x)$,

$$\lim_{h \to 0} v(x) = v(x).$$

Substituting these results into Equation (1) gives

$$f'(x) = u(x) \cdot v'(x) + v(x) \cdot u'(x),$$

the desired result.

To help see why the product rule is true, consider the special case in which u and v are positive functions. Then $u(x) \cdot v(x)$ represents the area of a rectangle, as shown in Figure 9. If we assume that u and v are increasing, then $u(x + h) \cdot v(x + h)$ represents the area of a slightly larger rectangle when h is a small positive number, as shown in the figure. The change in the area of the rectangle is given by the pink rectangle, with an area of $u(x)$ times the amount v has changed, plus the blue rectangle, with an area of $v(x)$ times the amount u has changed, plus the small green rectangle. As h becomes smaller and smaller, the green rectangle becomes negligibly small, and the change in the area is essentially $u(x)$ times the change in v plus $v(x)$ times the change in u.

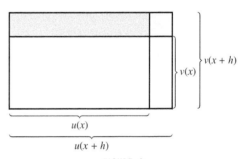

FIGURE 9

EXAMPLE 1 Product Rule

Let $f(x) = (2x + 3)(3x^2)$. Use the product rule to find $f'(x)$.

SOLUTION Here f is given as the product of $u(x) = 2x + 3$ and $v(x) = 3x^2$. By the product rule and the fact that $u'(x) = 2$ and $v'(x) = 6x$,

$$\begin{aligned}
f'(x) &= u(x) \cdot v'(x) + v(x) \cdot u'(x) \\
&= (2x + 3)(6x) + (3x^2)(2) \\
&= 12x^2 + 18x + 6x^2 = 18x^2 + 18x = 18x(x + 1).
\end{aligned}$$

This result is the same as that found at the beginning of the section.

EXAMPLE 2 Product Rule

Find the derivative of $y = (\sqrt{x} + 3)(x^2 - 5x)$.

SOLUTION Let $u(x) = \sqrt{x} + 3 = x^{1/2} + 3$, and $v(x) = x^2 - 5x$. Then

$$\begin{aligned}
\frac{dy}{dx} &= u(x) \cdot v'(x) + v(x) \cdot u'(x) \\
&= (x^{1/2} + 3)(2x - 5) + (x^2 - 5x)\left(\frac{1}{2}x^{-1/2}\right).
\end{aligned}$$

Simplify by multiplying and combining terms.

$$\frac{dy}{dx} = (2x)(x^{1/2}) + 6x - 5x^{1/2} - 15 + (x^2)\left(\frac{1}{2}x^{-1/2}\right) - (5x)\left(\frac{1}{2}x^{-1/2}\right)$$

$$= 2x^{3/2} + 6x - 5x^{1/2} - 15 + \frac{1}{2}x^{3/2} - \frac{5}{2}x^{1/2}$$

$$= \frac{5}{2}x^{3/2} + 6x - \frac{15}{2}x^{1/2} - 15$$

YOUR TURN 1 Find the derivative of $y = (x^3 + 7)(4 - x^2)$.

TRY YOUR TURN 1

We could have found the derivatives above by multiplying out the original functions. The product rule then would not have been needed. In the next section, however, we shall see products of functions where the product rule is essential.

What about *quotients* of functions? To find the derivative of the quotient of two functions, use the next rule.

Quotient Rule

If $f(x) = u(x)/v(x)$, if all indicated derivatives exist, and if $v(x) \neq 0$, then

$$f'(x) = \frac{d}{dx}\left[\frac{u(x)}{v(x)}\right] = \frac{v(x) \cdot u'(x) - u(x) \cdot v'(x)}{[v(x)]^2}.$$

(The derivative of a quotient is the denominator times the derivative of the numerator minus the numerator times the derivative of the denominator, all divided by the square of the denominator.)

The proof of the quotient rule is similar to that of the product rule and is left for the exercises. (See Exercises 37 and 38.)

FOR REVIEW

You may want to consult the Rational Expressions section of the Algebra Reference chapter (Section 3) to help you work with the fractions in the section.

CAUTION Just as the derivative of a product is *not* the product of the derivatives, the derivative of a quotient is *not* the quotient of the derivatives. If you are asked to take the derivative of a product or a quotient, it is essential that you recognize that the function contains a product or quotient and then use the appropriate rule.

EXAMPLE 3 Quotient Rule

Find $f'(x)$ if $f(x) = \dfrac{2x - 1}{4x + 3}$.

SOLUTION Let $u(x) = 2x - 1$, with $u'(x) = 2$. Also, let $v(x) = 4x + 3$, with $v'(x) = 4$. Then, by the quotient rule,

$$f'(x) = \frac{v(x) \cdot u'(x) - u(x) \cdot v'(x)}{[v(x)]^2}$$

$$= \frac{(4x + 3)(2) - (2x - 1)(4)}{(4x + 3)^2}$$

$$= \frac{8x + 6 - 8x + 4}{(4x + 3)^2}$$

YOUR TURN 2 Find $f'(x)$ if $f(x) = \dfrac{3x + 2}{5 - 2x}$.

$$= \frac{10}{(4x + 3)^2}.$$

TRY YOUR TURN 2

CAUTION In the second step of Example 3, we had the expression

$$\frac{(4x + 3)(2) - (2x - 1)(4)}{(4x + 3)^2}.$$

Students often incorrectly "cancel" the $4x + 3$ in the numerator with one factor of the denominator. Because the numerator is a *difference* of two products, however, you must multiply and combine terms *before* looking for common factors in the numerator and denominator.

EXAMPLE 4 Product and Quotient Rules

Find $D_x\left[\dfrac{(3 - 4x)(5x + 1)}{7x - 9}\right]$.

SOLUTION This function has a product within a quotient. Instead of multiplying the factors in the numerator first (which is an option), we can use the quotient rule together with the product rule, as follows. Use the quotient rule first to get

$$D_x\left[\frac{(3 - 4x)(5x + 1)}{7x - 9}\right] = \frac{(7x - 9)D_x[(3 - 4x)(5x + 1)] - [(3 - 4x)(5x + 1)D_x(7x - 9)]}{(7x - 9)^2}.$$

Now use the product rule to find $D_x[(3 - 4x)(5x + 1)]$ in the numerator.

$$= \frac{(7x - 9)[(3 - 4x)5 + (5x + 1)(-4)] - (3 + 11x - 20x^2)(7)}{(7x - 9)^2}$$

$$= \frac{(7x - 9)(15 - 20x - 20x - 4) - (21 + 77x - 140x^2)}{(7x - 9)^2}$$

$$= \frac{(7x - 9)(11 - 40x) - 21 - 77x + 140x^2}{(7x - 9)^2}$$

$$= \frac{-280x^2 + 437x - 99 - 21 - 77x + 140x^2}{(7x - 9)^2}$$

$$= \frac{-140x^2 + 360x - 120}{(7x - 9)^2}$$

YOUR TURN 3 Find $D_x\left[\dfrac{(5x - 3)(2x + 7)}{3x + 7}\right]$.

TRY YOUR TURN 3

Average Cost

Suppose $y = C(x)$ gives the total cost to manufacture x items. As mentioned earlier, the average cost per item is found by dividing the total cost by the number of items. The rate of change of average cost, called the *marginal average cost*, is the derivative of the average cost.

Marginal Average Cost

If the total cost to manufacture x items is given by $C(x)$, then the average cost per item is $\overline{C}(x) = C(x)/x$. The **marginal average cost** is the derivative of the average cost function, $\overline{C}'(x)$.

Similarly, the marginal average revenue function, $\overline{R}'(x)$, is defined as the derivative of the average revenue function, $\overline{R}(x) = R(x)/x$, and the marginal average profit function, $\overline{P}'(x)$, is defined as the derivative of the average profit function, $\overline{P}(x) = P(x)/x$.

A company naturally would be interested in making the average cost as small as possible. The next chapter will show that this can be done by using the derivative of $C(x)/x$. This derivative often can be found by means of the quotient rule, as in the next example.

| EXAMPLE 5 | **Minimum Average Cost**

Suppose the cost in dollars of manufacturing x hundred small motors is given by

$$C(x) = \frac{3x^2 + 120}{2x + 1}, \quad 10 \le x \le 200.$$

(a) Find the average cost per hundred motors.

SOLUTION The average cost is defined by

$$\overline{C}(x) = \frac{C(x)}{x} = \frac{3x^2 + 120}{2x + 1} \cdot \frac{1}{x} = \frac{3x^2 + 120}{2x^2 + x}.$$

(b) Find the marginal average cost.

SOLUTION The marginal average cost is given by

$$\frac{d}{dx}[\overline{C}(x)] = \frac{(2x^2 + x)(6x) - (3x^2 + 120)(4x + 1)}{(2x^2 + x)^2}$$

$$= \frac{12x^3 + 6x^2 - 12x^3 - 480x - 3x^2 - 120}{(2x^2 + x)^2}$$

$$= \frac{3x^2 - 480x - 120}{(2x^2 + x)^2}.$$

(c) As we shall see in the next chapter, average cost is generally minimized when the marginal average cost is zero. Find the level of production that minimizes average cost.

APPLY IT

SOLUTION Set the derivative $\overline{C}'(x) = 0$ and solve for x.

$$\frac{3x^2 - 480x - 120}{(2x^2 + x)^2} = 0$$

$$3x^2 - 480x - 120 = 0$$

$$3(x^2 - 160x - 40) = 0$$

YOUR TURN 4 Suppose cost is given by $C(x) = \dfrac{4x + 50}{x + 2}$. Find the marginal average cost.

Use the quadratic formula to solve this quadratic equation. Discarding the negative solution leaves $x = (160 + \sqrt{160^2 + 160})/2 \approx 160$ as the solution. Since x is in hundreds, production of 160 hundred or 16,000 motors will minimize average cost.

TRY YOUR TURN 4

4.2 WARM-UP EXERCISES

Find the derivative of each of the following functions. (*Sec. 4.1*)

W1. $f(x) = 3x^4 + 4x^3 - 5$

W2. $f(x) = \dfrac{2}{x^3} + 6\sqrt{x}$

W3. $f(x) = 9x^{2/3} + \dfrac{4}{\sqrt{x}}$

4.2 EXERCISES

Use the product rule to find the derivative of the following. (*Hint for Exercises 3–6:* Write the quantity as a product.)

1. $y = (3x^2 + 2)(2x - 1)$

2. $y = (5x^2 - 1)(4x + 3)$

3. $y = (2x - 5)^2$

4. $y = (7x - 6)^2$

5. $k(t) = (t^2 - 1)^2$

6. $g(t) = (3t^2 + 2)^2$

7. $y = (x + 1)(\sqrt{x} + 2)$

8. $y = (2x - 3)(\sqrt{x} - 1)$

9. $p(y) = (y^{-1} + y^{-2})(2y^{-3} - 5y^{-4})$

10. $q(x) = (x^{-2} - x^{-3})(3x^{-1} + 4x^{-4})$

Use the quotient rule to find the derivative of the following.

11. $f(x) = \dfrac{6x + 1}{3x + 10}$

12. $f(x) = \dfrac{8x \quad 11}{7x + 3}$

13. $y = \dfrac{5 - 3t}{4 + t}$

14. $y = \dfrac{9 - 7t}{1 - t}$

15. $y = \dfrac{x^2 + x}{x - 1}$

16. $y = \dfrac{x^2 - 4x}{x + 3}$

17. $f(t) = \dfrac{4t^2 + 11}{t^2 + 3}$

18. $y = \dfrac{-x^2 + 8x}{4x^2 - 5}$

19. $g(x) = \dfrac{x^2 - 4x + 2}{x^2 + 3}$

20. $k(x) = \dfrac{x^2 + 7x - 2}{x^2 - 2}$

21. $p(t) = \dfrac{\sqrt{t}}{t - 1}$

22. $r(t) = \dfrac{\sqrt{t}}{2t + 3}$

23. $y = \dfrac{5x + 6}{\sqrt{x}}$

24. $y = \dfrac{4x - 3}{\sqrt{x}}$

25. $h(z) = \dfrac{z^{2.2}}{z^{3.2} + 5}$

26. $g(y) = \dfrac{y^{1.4} + 1}{y^{2.5} + 2}$

27. $f(x) = \dfrac{(3x^2 + 1)(2x - 1)}{5x + 4}$

28. $g(x) = \dfrac{(2x^2 + 3)(5x + 2)}{6x - 7}$

29. If $g(3) = 4$, $g'(3) = 5$, $f(3) = 9$, and $f'(3) = 8$, find $h'(3)$ when $h(x) = f(x)g(x)$.

30. If $g(3) = 4$, $g'(3) = 5$, $f(3) = 9$, and $f'(3) = 8$, find $h'(3)$ when $h(x) = f(x)/g(x)$.

31. Find the error in the following work.

$$D_x\left(\frac{2x + 5}{x^2 - 1}\right) = \frac{(2x + 5)(2x) - (x^2 - 1)2}{(x^2 - 1)^2}$$

$$= \frac{4x^2 + 10x - 2x^2 + 2}{(x^2 - 1)^2}$$

$$= \frac{2x^2 + 10x + 2}{(x^2 - 1)^2}$$

32. Find the error in the following work.

$$D_x\left(\frac{x^2 - 4}{x^3}\right) = x^3(2x) - (x^2 - 4)(3x^2) = 2x^4 - 3x^4 + 12x^2$$

$$= -x^4 + 12x^2$$

33. Find an equation of the line tangent to the graph of $f(x) = x/(x - 2)$ at $(3, 3)$.

34. Find an equation of the line tangent to the graph of $f(x) = (2x - 1)(x + 4)$ at $(1, 5)$.

35. Consider the function

$$f(x) = \frac{3x^3 + 6}{x^{2/3}}.$$

    (a) Find the derivative using the quotient rule.

    (b) Find the derivative by first simplifying the function to

$$f(x) = \frac{3x^3}{x^{2/3}} + \frac{6}{x^{2/3}} = 3x^{7/3} + 6x^{-2/3}$$

    and using the rules from the previous section.

    (c) Compare your answers from parts (a) and (b) and explain any discrepancies.

36. What is the result of applying the product rule to the function

$$f(x) = kg(x),$$

where k is a constant? Compare with the rule for differentiating a constant times a function from the previous section.

37. Following the steps used to prove the product rule for derivatives, prove the quotient rule for derivatives.

38. Use the fact that $f(x) = u(x)/v(x)$ can be rewritten as $f(x)v(x) = u(x)$ and the product rule for derivatives to verify the quotient rule for derivatives. (*Hint:* After applying the product rule, substitute $u(x)/v(x)$ for $f(x)$ and simplify.)

For each function, find the value(s) of x in which $f'(x) = 0$, to 3 decimal places.

39. $f(x) = (x^2 - 2)(x^2 - \sqrt{2})$ **40.** $f(x) = \dfrac{x - 2}{x^2 + 4}$

41. Gottfried Leibniz, one of the inventors of calculus, initially thought that the derivative of $f(x)g(x)$ was $f'(x)g'(x)$, based on the example $f(x) = x^2 + bx$ and $g(x) = cx + d$. Besides neglecting to work out the details, which would have shown his error, he made an erroneous conclusion based on one example, rather than a mathematical proof. *Source: PRIMUS.*

    (a) Calculate $f'(x)g'(x)$.

    (b) Calculate $[f(x)g(x)]'$; and verify that your answer is different than your answer to part (a).

42. There are special cases in which the derivative of $f(x)g(x)$ *does* equal $f'(x)g'(x)$. Calculate both $[f(x)g(x)]'$ and $f'(x)g'(x)$ for $f(x) = a/(1 - x)$ and $g(x) = bx$, and verify that the two expressions are equal. *Source: PRIMUS*.

APPLICATIONS

Business and Economics

43. Average Cost The total cost (in hundreds of dollars) to produce x units of perfume is

$$C(x) = \frac{3x + 2}{x + 4}.$$

Find the average cost for each production level.

(a) 10 units   (b) 20 units   (c) x units

(d) Find the marginal average cost function.

44. Average Profit The total profit (in tens of dollars) from selling x self-help books is

$$P(x) = \frac{5x - 6}{2x + 3}.$$

Find the average profit from each sales level.

(a) 8 books   (b) 15 books   (c) x books

(d) Find the marginal average profit function.

(e) Is this a reasonable function for profit? Why or why not?

45. Employee Training A company that manufactures bicycles has determined that a new employee can assemble $M(d)$ bicycles per day after d days of on-the-job training, where

$$M(d) = \frac{100d^2}{3d^2 + 10}.$$

(a) Find the rate of change function for the number of bicycles assembled with respect to time.

(b) Find and interpret $M'(2)$ and $M'(5)$.

46. Marginal Revenue Suppose that the demand function is given by $p = D(q)$, where q is the quantity that consumers demand when the price is p. Show that the marginal revenue is given by

$$R'(q) = D(q) + qD'(q).$$

47. Marginal Average Cost Suppose that the average cost function is given by $\overline{C}(x) = C(x)/x$, where x is the number of items produced. Show that the marginal average cost function is given by

$$\overline{C}'(x) = \frac{xC'(x) - C(x)}{x^2}.$$

48. Revenue Suppose that at the beginning of the year, a Vermont maple syrup distributor found that the demand for maple syrup, sold at $15 a quart, was 500 quarts each month. At that time, the price was going up at a rate of $0.50 a month, but despite this, the demand was going up at a rate of 30 quarts a month due to increased advertising. How fast was the revenue increasing?

49. Average Cost A gasoline refinery found that the cost to produce 12,500 gallons of gasoline last month was $27,000.

At that time, the cost was going up at a rate of $1200 per month, while the number of gallons of gasoline the refinery produced was going up at a rate of 350 gallons per month. At what rate was the average cost to produce a gallon of gasoline increasing or decreasing last month?

Life Sciences

50. Muscle Reaction When a certain drug is injected into a muscle, the muscle responds by contracting. The amount of contraction, s (in millimeters), is related to the concentration of the drug, x (in milliliters), by

$$s(x) = \frac{x}{m + nx},$$

where m and n are constants.

(a) Find $s'(x)$.

(b) Find the rate of contraction when the concentration of the drug is 50 ml, $m = 10$, and $n = 3$.

51. Growth Models In Exercise 58 of Section 2.3, the formula for the growth rate of a population in the presence of a quantity x of food was given as

$$f(x) = \frac{Kx}{A + x}.$$

This was referred to as Michaelis-Menten kinetics.

(a) Find the rate of change of the growth rate with respect to the amount of food.

(b) The quantity A in the formula for $f(x)$ represents the quantity of food for which the growth rate is half of its maximum. Using your answer from part (a), find the rate of change of the growth rate when $x = A$.

52. Optimal Foraging Using data collected by zoologist Reto Zach, the work done by a crow to break open a whelk (large marine snail) can be estimated by the function

$$W = \left(1 + \frac{20}{H - 0.93}\right)H,$$

where H is the height (in meters) of the whelk when it is dropped. *Source: The Mathematics Teacher*.

(a) Find dW/dH.

(b) One can show that the amount of work is minimized when $dW/dH = 0$. Find the value of H that minimizes W.

(c) Interestingly, Zach observed the crows dropping the whelks from an average height of 5.23 m. What does this imply?

53. Work/Rest Cycles Murrell's formula for calculating the total amount of rest, in minutes, required after performing a particular type of work activity for 30 minutes is given by the formula

$$R(w) = \frac{30(w - 4)}{w - 1.5},$$

where w is the work expended in kilocalories per minute, kcal/min. *Source: Human Factors in Engineering and Design.*

(a) A value of 5 for w indicates light work, such as riding a bicycle on a flat surface at 10 mph. Find $R(5)$.

(b) A value of 7 for w indicates moderate work, such as mowing grass with a pushmower on level ground. Find $R(7)$.

(c) Find $R'(5)$ and $R'(7)$ and compare your answers. Explain whether these answers make sense.

Social Sciences

54. Memory Retention Some psychologists contend that the number of facts of a certain type that are remembered after t hours is given by

$$f(t) = \frac{90t}{99t - 90}.$$

Find the rate at which the number of facts remembered is changing after the following numbers of hours.

(a) 1   (b) 10

General Interest

55. Vehicle Waiting Time The average number of vehicles waiting in a line to enter a parking ramp can be modeled by the function

$$f(x) = \frac{x^2}{2(1 - x)},$$

where x is a quantity between 0 and 1 known as the traffic intensity. Find the rate of change of the number of vehicles in line with respect to the traffic intensity for the following values

of the intensity. *Source: Principles of Highway Engineering and Traffic Control.*

(a) $x = 0.1$ (b) $x = 0.6$

56. Baseball Some people have argued that the Pythagorean Theorem of Baseball is a better predictor of a team's ability than its winning percentage:

$$W = \frac{s^2}{s^2 + a^2},$$

where W is a predictor of what fraction of games a team is expected to win, s is the number of runs scored, and a is the number of runs allowed. *Source: baseball-reference.com.*

(a) The team with the best record in the 2013 season was the Boston Red Sox, with 97 wins and 65 losses. They scored 853 runs and allowed 656 runs. Calculate and compare their winning percentage (as a decimal) and their Pythagorean Theorem score W.

(b) Keeping their runs scored at 853 and considering their runs allowed as a variable a, find dW/da when $a = 656$.

(c) Interpreting your answer to part (b) as the approximate amount that W would change if a increased by one run, use your answer to approximate W when $a = 657$, and then calculate the actual value.

(d) Some people have argued that the exponent of s and a in the formula for W should be 1.81 rather than 2. Calculate W (with $s = 853$ and $a = 656$) and dW/da with this exponent.

YOUR TURN ANSWERS

1. $\dfrac{dy}{dx} = -5x^4 + 12x^2 - 14x$ 2. $f'(x) = \dfrac{19}{(5 - 2x)^2}$

3. $\dfrac{30x^2 + 140x + 266}{(3x + 7)^2}$ 4. $\dfrac{-4x^2 - 100x - 100}{(x^2 + 2x)^2}$

4.3 The Chain Rule

APPLY IT   Suppose we know how fast the radius of a circular oil slick is growing, and we know how much the area of the oil slick is growing per unit of change in the radius. How fast is the area growing?
We will answer this question in Example 4 using the chain rule for derivatives.

Before discussing the chain rule, we consider the composition of functions. Many of the most useful functions for modeling are created by combining simpler functions. Viewing complex functions as combinations of simpler functions often makes them easier to understand and use.

Composition of Functions
Suppose a function f assigns to each element x in set X some element $y = f(x)$ in set Y. Suppose also that a function g takes each element in set Y and assigns to it a value $z = g[f(x)]$ in set Z. By using both f and g, an element x in X is assigned to an element z in Z, as illustrated in Figure 10 on the next page. The result of this process is a new function called the *composition* of functions g and f and defined as follows.

FIGURE 10

Composite Function

Let f and g be functions. The **composite function**, or **composition**, of g and f is the function whose values are given by $g[f(x)]$ for all x in the domain of f such that $f(x)$ is in the domain of g. (Read $g[f(x)]$ as "g of f of x".)

The composition of g and f is also sometimes written as $(g \circ f)(x)$.

EXAMPLE 1 **Composite Functions**

Let $f(x) = 2x - 1$ and $g(x) = \sqrt{3x + 5}$. Find the following.

(a) $g[f(4)]$

SOLUTION Find $f(4)$ first.

$$f(4) = 2 \cdot 4 - 1 = 8 - 1 = 7$$

Then

$$g[f(4)] = g[7] = \sqrt{3 \cdot 7 + 5} = \sqrt{26}.$$

(b) $f[g(4)]$

SOLUTION Since $g(4) = \sqrt{3 \cdot 4 + 5} = \sqrt{17}$,

$$f[g(4)] = 2 \cdot \sqrt{17} - 1 = 2\sqrt{17} - 1.$$

YOUR TURN 1 For the functions in Example 1, find $f[g(0)]$ and $g[f(0)]$.

(c) $f[g(-2)]$

SOLUTION $f[g(-2)]$ does not exist, since -2 is not in the domain of g.

TRY YOUR TURN 1

EXAMPLE 2 **Composition of Functions**

Let $f(x) = 2x^2 + 5x$ and $g(x) = 4x + 1$. Find the following.

(a) $f[g(x)]$

SOLUTION Using the given functions, we have

$$\begin{aligned}
f[g(x)] &= f[4x + 1] \\
&= 2(4x + 1)^2 + 5(4x + 1) \\
&= 2(16x^2 + 8x + 1) + 20x + 5 \\
&= 32x^2 + 16x + 2 + 20x + 5 \\
&= 32x^2 + 36x + 7.
\end{aligned}$$

(b) $g[f(x)]$

SOLUTION By the definition above, with f and g interchanged,

YOUR TURN 2 Let $f(x) = 2x - 3$ and $g(x) = x^2 + 1$. Find $g[f(x)]$.

$$\begin{aligned}
g[f(x)] &= g[2x^2 + 5x] \\
&= 4(2x^2 + 5x) + 1 \\
&= 8x^2 + 20x + 1.
\end{aligned}$$

TRY YOUR TURN 2

As Example 2 shows, it is not always true that $f[g(x)] = g[f(x)]$. In fact, it is rare to find two functions f and g such that $f[g(x)] = g[f(x)]$. The domain of both composite functions given in Example 2 is the set of all real numbers.

EXAMPLE 3 Composition of Functions

Write each function as the composition of two functions f and g so that $h(x) = f[g(x)]$.

(a) $h(x) = 2(4x + 1)^2 + 5(4x + 1)$

   SOLUTION Let $f(x) = 2x^2 + 5x$ and $g(x) = 4x + 1$. Then $f[g(x)] = f(4x + 1) = 2(4x + 1)^2 + 5(4x + 1)$. Notice that $h(x)$ here is the same as $f[g(x)]$ in Example 2(a).

(b) $h(x) = \sqrt{1 - x^2}$

YOUR TURN 3 Write $h(x) = (2x - 3)^3$ as a composition of two functions f and g so that $h(x) = f[g(x)]$.

   SOLUTION One way to do this is to let $f(x) = \sqrt{x}$ and $g(x) = 1 - x^2$. Another choice is to let $f(x) = \sqrt{1 - x}$ and $g(x) = x^2$. Verify that with either choice, $f[g(x)] = \sqrt{1 - x^2}$. For the purposes of this section, the first choice is better; it is useful to think of f as being the function on the outer layer and g as the function on the inner layer. With this function h, we see a square root on the outer layer, and when we peel that away we see $1 - x^2$ on the inside.   **TRY YOUR TURN 3**

The Chain Rule Suppose $f(x) = x^2$ and $g(x) = 5x^3 + 2$. What is the derivative of $h(x) = f[g(x)] = (5x^3 + 2)^2$? At first you might think the answer is just $h'(x) = 2(5x^3 + 2) = 10x^3 + 4$ by using the power rule. You can check this answer by multiplying out $h(x) = (5x^3 + 2)^2 = 25x^6 + 20x^3 + 4$. Now calculate $h'(x) = 150x^5 + 60x^2$. The guess using the power rule was clearly wrong! The error is that the power rule applies to x raised to a power, not to some other function of x raised to a power.

How, then, could we take the derivative of $p(x) = (5x^3 + 2)^{20}$? This seems far too difficult to multiply out. Fortunately, there is a way. Notice from the previous paragraph that $h'(x) = 150x^5 + 60x^2 = 2(5x^3 + 2)15x^2$. So the original guess was almost correct, except it was missing the factor of $15x^2$, which just happens to be $g'(x)$. This is not a coincidence. To see why the derivative of $f[g(x)]$ involves taking the derivative of f and then multiplying by the derivative of g, let us consider a realistic example, the question from the beginning of this section.

EXAMPLE 4 Area of an Oil Slick

A leaking oil well off the Gulf Coast is spreading a circular film of oil over the water surface. At any time t (in minutes) after the beginning of the leak, the radius of the circular oil slick (in feet) is given by

$$r(t) = 4t.$$

Find the rate of change of the area of the oil slick with respect to time.

APPLY IT   **SOLUTION** We first find the rate of change in the radius over time by finding dr/dt:

$$\frac{dr}{dt} = 4.$$

This value indicates that the radius is increasing by 4 ft each minute.

   The area of the oil slick is given by

$$A(r) = \pi r^2, \quad \text{with} \quad \frac{dA}{dr} = 2\pi r.$$

The derivative, dA/dr, gives the rate of change in area per unit increase in the radius.

As these derivatives show, the radius is increasing at a rate of 4 ft/min, and for each foot that the radius increases, the area increases by $2\pi r$ ft^2. It seems reasonable, then, that the area is increasing at a rate of

$$2\pi r \text{ ft}^2/\text{ft} \times 4 \text{ ft/min} = 8\pi r \text{ ft}^2/\text{min}.$$

That is,

$$\frac{dA}{dt} = \frac{dA}{dr} \cdot \frac{dr}{dt} = 2\pi r \cdot 4 = 8\pi r.$$

Notice that because area (A) is a function of radius (r), which is a function of time (t), area as a function of time is a composition of two functions, written $A(r(t))$. The last step, then, can also be written as

$$\frac{dA}{dt} = \frac{d}{dt}A[r(t)] = A'[r(t)] \cdot r'(t) = 2\pi r \cdot 4 = 8\pi r.$$

Finally, we can substitute $r(t) = 4t$ to get the derivative in terms of t:

$$\frac{dA}{dt} = 8\pi r = 8\pi(4t) = 32\pi t.$$

The rate of change of the area of the oil slick with respect to time is $32\pi t$ ft^2/min. ▬

To check the result of Example 4, use the fact that $r = 4t$ and $A = \pi r^2$ to get the same result:

$$A = \pi(4t)^2 = 16\pi t^2, \qquad \text{with} \qquad \frac{dA}{dt} = 32\pi t.$$

The product used in Example 4,

$$\frac{dA}{dt} = \frac{dA}{dr} \cdot \frac{dr}{dt},$$

is an example of the **chain rule**, which is used to find the derivative of a composite function.

Chain Rule

If y is a function of u, say $y = f(u)$, and if u is a function of x, say $u = g(x)$, then $y = f(u) = f[g(x)]$, and

$$\frac{dy}{dx} = \frac{dy}{du} \cdot \frac{du}{dx}.$$

One way to remember the chain rule is to pretend that dy/du and du/dx are fractions, with du "canceling out." The proof of the chain rule requires advanced concepts and, therefore, is not given here.

EXAMPLE 5 Chain Rule

Find dy/dx if $y = (3x^2 - 5x)^{1/2}$.

SOLUTION Let $y = u^{1/2}$, and $u = 3x^2 - 5x$. Then

$$\frac{dy}{dx} = \frac{dy}{du} \cdot \frac{du}{dx}$$

$$= \frac{1}{2}u^{-1/2} \cdot (6x - 5).$$

Replacing u with $3x^2 - 5x$ gives

YOUR TURN 4 Find dy/dx
if $y = (5x^2 - 6x)^{-2}$.

$$\frac{dy}{dx} = \frac{1}{2}(3x^2 - 5x)^{-1/2}(6x - 5) = \frac{6x - 5}{2(3x^2 - 5x)^{1/2}}.$$

TRY YOUR TURN 4

The following alternative version of the chain rule is stated in terms of composite functions.

Chain Rule (Alternative Form)

If $y = f[g(x)]$, then

$$\frac{dy}{dx} = \frac{d}{dx}[f[g(x)]] = f'[g(x)] \cdot g'(x).$$

(To find the derivative of $f[g(x)]$, find the derivative of $f(u)$, replace each u with $g(x)$, and then multiply the result by the derivative of $g(x)$.)

In words, the chain rule tells us to first take the derivative of the outer function, then multiply it by the derivative of the inner function.

EXAMPLE 6 Chain Rule

Use the chain rule to find $D_x(x^2 + 5x)^8$.

SOLUTION As in Example 3(b), think of this as a function with layers. The outer layer is something being raised to the 8th power, so let $f(u) = u^8$. Once this layer is peeled away, we see that the inner layer is $x^2 + 5x$, so $g(x) = x^2 + 5x$. Then $(x^2 + 5x)^8 = f[g(x)]$ and

$$D_x(x^2 + 5x)^8 = f'[g(x)]g'(x).$$

Here $f'(u) = 8u^7$, with $f'[g(x)] = 8[g(x)]^7 = 8(x^2 + 5x)^7$ and $g'(x) = 2x + 5$.

$$D_x(x^2 + 5x)^8 = f'[g(x)]g'(x)$$

YOUR TURN 5
Find $D_x(x^2 - 7)^{10}$.

$$= 8[g(x)]^7 g'(x)$$

$$= 8(x^2 + 5x)^7(2x + 5)$$

TRY YOUR TURN 5

CAUTION (a) A common error is to forget to multiply by $g'(x)$ when using the chain rule. Remember, the derivative must involve a "chain," or product, of derivatives.

(b) Another common mistake is to write the derivative as $f'[g'(x)]$. Remember to leave $g(x)$ unchanged in $f'[g(x)]$ and then to multiply by $g'(x)$.

One way to avoid both of the errors described above is to remember that the chain rule is a two-step process. In Example 6, the first step was taking the derivative of the power, and the second step was multiplying by $g'(x)$. Forgetting to multiply by $g'(x)$ would be an erroneous one-step process. The other erroneous one-step process is to take the derivative inside the power, getting $f'[g'(x)]$, or $8(2x + 5)^7$ in Example 6.

Sometimes both the chain rule and either the product or quotient rule are needed to find a derivative, as the next examples show.

EXAMPLE 7 Derivative Rules

Find the derivative of $y = 4x(3x + 5)^5$.

SOLUTION Write $4x(3x + 5)^5$ as the product

$$(4x) \cdot (3x + 5)^5.$$

To find the derivative of $(3x + 5)^5$, let $g(x) = 3x + 5$, with $g'(x) = 3$. Now use the product rule and the chain rule.

$$\frac{dy}{dx} = 4x[5(3x + 5)^4 \cdot 3] + (3x + 5)^5(4)$$

Derivative of $(3x + 5)^5$ Derivative of $4x$

$$= 60x(3x + 5)^4 + 4(3x + 5)^5 \quad \text{Factor out the greatest common factor, } 4(3x + 5)^5.$$

$$= 4(3x + 5)^4[15x + (3x + 5)^1] \quad \text{Simplify inside brackets.}$$

$$= 4(3x + 5)^4(18x + 5) \quad \text{TRY YOUR TURN 6}$$

YOUR TURN 6 Find the derivative of $y = x^2(5x - 1)^3$.

EXAMPLE 8 Derivative Rules

Find $D_x\left[\dfrac{(3x + 2)^7}{x - 1}\right]$.

SOLUTION Use the quotient rule and the chain rule.

$$D_x\left[\frac{(3x + 2)^7}{x - 1}\right] = \frac{(x - 1)[7(3x + 2)^6 \cdot 3] - (3x + 2)^7(1)}{(x - 1)^2}$$

$$= \frac{21(x - 1)(3x + 2)^6 - (3x + 2)^7}{(x - 1)^2}$$

$$= \frac{(3x + 2)^6[21(x - 1) - (3x + 2)]}{(x - 1)^2} \quad \text{Factor out the greatest common factor, } (3x + 2)^6.$$

$$= \frac{(3x + 2)^6[21x - 21 - 3x - 2]}{(x - 1)^2} \quad \text{Simplify inside brackets.}$$

$$= \frac{(3x + 2)^6(18x - 23)}{(x - 1)^2} \quad \text{TRY YOUR TURN 7}$$

YOUR TURN 7 Find $D_x\left[\dfrac{(4x - 1)^3}{x + 3}\right]$.

Some applications requiring the use of the chain rule are illustrated in the next two examples.

EXAMPLE 9 City Revenue

The revenue realized by a small city from the collection of fines from parking tickets is given by

$$R(n) = \frac{8000n}{n + 2},$$

where n is the number of work-hours each day that can be devoted to parking patrol. At the outbreak of a flu epidemic, 30 work-hours are used daily in parking patrol, but during the epidemic that number is decreasing at the rate of 6 work-hours per day. How fast is revenue from parking fines decreasing at the outbreak of the epidemic?

SOLUTION We want to find dR/dt, the change in revenue with respect to time. By the chain rule,

$$\frac{dR}{dt} = \frac{dR}{dn} \cdot \frac{dn}{dt}.$$

First find dR/dn, using the quotient rule, as follows.

$$\frac{dR}{dn} = \frac{(n+2)(8000) - 8000n(1)}{(n+2)^2} = \frac{16,000}{(n+2)^2}$$

Since 30 work-hours were used at the outbreak of the epidemic, $n = 30$, so $dR/dn = 16,000/(30+2)^2 = 15.625$. Also, $dn/dt = -6$. Thus,

$$\frac{dR}{dt} = \frac{dR}{dn} \cdot \frac{dn}{dt} = (15.625)(-6) = -93.75.$$

Revenue is being lost at the rate of about \$94 per day at the outbreak of the epidemic. ▬

FIGURE 11

EXAMPLE 10 Compound Interest

Suppose a sum of \$500 is deposited in an account with an interest rate of r percent per year compounded monthly. At the end of 10 years, the balance in the account (as illustrated in Figure 11) is given by

$$A = 500\left(1 + \frac{r}{1200}\right)^{120}.$$

Find the rate of change of A with respect to r if $r = 5$ or 7.*

SOLUTION First find dA/dr using the chain rule.

$$\frac{dA}{dr} = (120)(500)\left(1 + \frac{r}{1200}\right)^{119}\left(\frac{1}{1200}\right)$$

$$= 50\left(1 + \frac{r}{1200}\right)^{119}$$

If $r = 5$,

$$\frac{dA}{dr} = 50\left(1 + \frac{5}{1200}\right)^{119}$$

$$\approx 82.01,$$

or \$82.01 per percentage point. If $r = 7$,

$$\frac{dA}{dr} = 50\left(1 + \frac{7}{1200}\right)^{119}$$

$$\approx 99.90,$$

or \$99.90 per percentage point. ▬

NOTE One lesson to learn from this section is that a derivative is always with respect to some variable. In the oil slick example, notice that the derivative of the area with respect to the radius is $2\pi r$, while the derivative of the area with respect to time is $8\pi r$. As another example, consider the velocity of a conductor walking at 2 mph on a train car. Her velocity with respect to the ground may be 50 mph, but the earth on which the train is running is moving about the sun at 1.6 million mph. The derivative of her position function might be 2, 50, or 1.6 million mph, depending on what variable it is with respect to.

---

*Notice that r is given here as an integer percent, rather than as a decimal, which is why the formula for compound interest has 1200 where you would expect to see 12. This leads to a simpler interpretation of the derivative.

4.3 WARM-UP EXERCISES

Find the derivative of each of the following functions. *(Sec. 4.2)*

W1. $f(x) = \dfrac{x^2}{x^3 + 1}$

W2. $f(x) = \dfrac{6\sqrt{x}}{x^4 + 1}$

W3. $f(x) = \dfrac{x^{2/3}}{x^{1/3} + 2}$

4.3 EXERCISES

Let $f(x) = 5x^2 - 2x$ and $g(x) = 8x + 3$. Find the following.

1. $f[g(2)]$ **2.** $f[g(-5)]$ **3.** $g[f(2)]$

4. $g[f(-5)]$ **5.** $f[g(k)]$ **6.** $g[f(5z)]$

In Exercises 7–14, find $f[g(x)]$ and $g[f(x)]$.

7. $f(x) = \dfrac{x}{8} + 7; \quad g(x) = 6x - 1$

8. $f(x) = -8x + 9; \quad g(x) = \dfrac{x}{5} + 4$

9. $f(x) = \dfrac{1}{x}; \quad g(x) = x^2$

10. $f(x) = \dfrac{2}{x^4}; \quad g(x) = 2 - x$

11. $f(x) = \sqrt{x + 2}; \quad g(x) = 8x^2 - 6$

12. $f(x) = 9x^2 - 11x; \quad g(x) = 2\sqrt{x + 2}$

13. $f(x) = \sqrt{x + 1}; \quad g(x) = \dfrac{-1}{x}$

14. $f(x) = \dfrac{8}{x}; \quad g(x) = \sqrt{3 - x}$

Write each function as the composition of two functions. (There may be more than one way to do this.)

15. $y = (5 - x^2)^{3/5}$ **16.** $y = (3x^2 - 7)^{2/3}$

17. $y = -\sqrt{13 + 7x}$ **18.** $y = \sqrt{9 - 4x}$

19. $y = (x^2 + 5x)^{1/3} - 2(x^2 + 5x)^{2/3} + 7$

20. $y = (x^{1/2} - 3)^2 + (x^{1/2} - 3) + 5$

Find the derivative of each function defined as follows.

21. $y = (8x^4 - 5x^2 + 1)^4$ **22.** $y = (2x^3 + 9x)^5$

23. $k(x) = -2(12x^2 + 5)^{-6}$ **24.** $f(x) = -7(3x^4 + 2)^{-4}$

25. $s(t) = 45(3t^3 - 8)^{3/2}$ **26.** $s(t) = 12(2t^4 + 5)^{3/2}$

27. $g(t) = -3\sqrt{7t^3 - 1}$ **28.** $f(t) = 8\sqrt{4t^2 + 7}$

29. $m(t) = -6t(5t^4 - 1)^4$ **30.** $r(t) = 4t(2t^5 + 3)^4$

31. $y = (3x^4 + 1)^4(x^3 + 4)$ **32.** $y = (x^3 + 2)(x^2 - 1)^4$

33. $q(y) = 4y^2(y^2 + 1)^{5/4}$ **34.** $p(z) = z(6z + 1)^{4/3}$

35. $y = \dfrac{-5}{(2x^3 + 1)^2}$ **36.** $y = \dfrac{1}{(3x^2 - 4)^5}$

37. $r(t) = \dfrac{(5t - 6)^4}{3t^2 + 4}$ **38.** $p(t) = \dfrac{(2t + 3)^3}{4t^2 - 1}$

39. $y = \dfrac{3x^2 - x}{(2x - 1)^5}$ **40.** $y = \dfrac{x^2 + 4x}{(3x^3 + 2)^4}$

41. In your own words explain how to form the composition of two functions.

42. The generalized power rule says that if $g(x)$ is a function of x and $y = [g(x)]^n$ for any real number n, then

$$\frac{dy}{dx} = n \cdot [g(x)]^{n-1} \cdot g'(x).$$

Explain why the generalized power rule is a consequence of the chain rule and the power rule.

Consider the following table of values of the functions f and g and their derivatives at various points.

x	1	2	3	4
$f(x)$	2	4	1	3
$f'(x)$	−6	−7	−8	−9
$g(x)$	2	3	4	1
$g'(x)$	2/7	3/7	4/7	5/7

Find the following using the table above.

43. (a) $D_x(f[g(x)])$ at $x = 1$ (b) $D_x(f[g(x)])$ at $x = 2$

44. (a) $D_x(g[f(x)])$ at $x = 1$ (b) $D_x(g[f(x)])$ at $x = 2$

In Exercises 45–48, find the equation of the tangent line to the graph of the given function at the given value of x.

45. $f(x) = \sqrt{x^2 + 16}; \quad x = 3$

46. $f(x) = (x^3 + 7)^{2/3}; \quad x = 1$

47. $f(x) = x(x^2 - 4x + 5)^4; \quad x = 2$

48. $f(x) = x^2\sqrt{x^4 - 12}; \quad x = 2$

In Exercises 49 and 50, find all values of x for the given function where the tangent line is horizontal.

49. $f(x) = \sqrt{x^3 - 6x^2 + 9x + 1}$

50. $f(x) = \dfrac{x}{(x^2 + 4)^4}$

51. Katie and Sarah are working on taking the derivative of

$$f(x) = \frac{2x}{3x + 4}.$$

Katie uses the quotient rule to get

$$f'(x) = \frac{(3x + 4)2 - 2x(3)}{(3x + 4)^2} = \frac{8}{(3x + 4)^2}.$$

Sarah converts it into a product and uses the product rule and the chain rule:

$$f(x) = 2x(3x + 4)^{-1}$$
$$f'(x) = 2x(-1)(3x + 4)^{-2}(3) + 2(3x + 4)^{-1}$$
$$= 2(3x + 4)^{-1} - 6x(3x + 4)^{-2}.$$

Explain the discrepancies between the two answers. Which procedure do you think is preferable?

52. Margy and Nate are working on taking the derivative of

$$f(x) = \frac{2}{(3x + 1)^4}.$$

Margy uses the quotient rule and chain rule as follows:

$$f'(x) = \frac{(3x + 1)^4 \cdot 0 - 2 \cdot 4(3x + 1)^3 \cdot 3}{(3x + 1)^8}$$
$$= \frac{-24(3x + 1)^3}{(3x + 1)^8} = \frac{-24}{(3x + 1)^5}.$$

Nate rewrites the function and uses the power rule and chain rule as follows:

$$f(x) = 2(3x + 1)^{-4}$$
$$f'(x) = (-4)2(3x + 1)^{-5} \cdot 3 = \frac{-24}{(3x + 1)^5}.$$

Compare the two procedures. Which procedure do you think is preferable?

APPLICATIONS

Business and Economics

53. Demand   Suppose the demand for a certain brand of vacuum cleaner is given by

$$D(p) = \frac{-p^2}{100} + 500,$$

where p is the price in dollars. If the price, in terms of the cost c, is expressed as

$$p(c) = 2c - 10,$$

find the demand in terms of the cost.

54. Revenue   Assume that the total revenue (in dollars) from the sale of x television sets is given by

$$R(x) = 24(x^2 + x)^{2/3}.$$

Find the marginal revenue when the following numbers of sets are sold.

(a) 100   (b) 200   (c) 300

(d) Find the average revenue from the sale of x sets.

(e) Find the marginal average revenue.

(f) Write a paragraph covering the following questions. How does the revenue change over time? What does the marginal revenue function tell you about the revenue function? What does the average revenue function tell you about the revenue function?

55. Interest   A sum of $1500 is deposited in an account with an interest rate of r percent per year, compounded daily. At the end of 5 years, the balance in the account is given by

$$A = 1500\left(1 + \frac{r}{36{,}500}\right)^{1825}.$$

Find the rate of change of A with respect to r for the following interest rates.

(a) 6%   (b) 8%   (c) 9%

56. Demand   Suppose a demand function is given by

$$q = D(p) = 30\left(5 - \frac{p}{\sqrt{p^2 + 1}}\right),$$

where q is the demand for a product and p is the price per item in dollars. Find the rate of change in the demand for the product per unit change in price (i.e., find dq/dp).

57. Depreciation   A certain truck depreciates according to the formula

$$V = \frac{60{,}000}{1 + 0.3t + 0.1t^2},$$

where V is the value of the truck (in dollars), t is time measured in years, and $t = 0$ represents the time of purchase (in years). Find the rate at which the value of the truck is changing at the following times.

(a) 2 years   (b) 4 years

58. Cost   Suppose the cost in dollars of manufacturing q items is given by

$$C = 2000q + 3500,$$

and the demand equation is given by

$$q = \sqrt{15{,}000 - 1.5p}.$$

In terms of the demand q,

(a) find an expression for the revenue R;

(b) find an expression for the profit P;

(c) find an expression for the marginal profit.

(d) Determine the value of the marginal profit when the price is $5000.

Life Sciences

59. Fish Population Suppose the population P of a certain species of fish depends on the number x (in hundreds) of a smaller fish that serves as its food supply, so that

$$P(x) = 2x^2 + 1.$$

Suppose, also, that the number of the smaller species of fish depends on the amount a (in appropriate units) of its food supply, a kind of plankton. Specifically,

$$x = f(a) = 3a + 2.$$

A biologist wants to find the relationship between the population P of the large fish and the amount a of plankton available, that is, $P[f(a)]$. What is the relationship?

60. Oil Pollution An oil well off the Gulf Coast is leaking, with the leak spreading oil over the surface as a circle. At any time t (in minutes) after the beginning of the leak, the radius of the circular oil slick on the surface is $r(t) = t^2$ feet. Let $A(r) = \pi r^2$ represent the area of a circle of radius r.

(a) Find and interpret $A[r(t)]$.

(b) Find and interpret $D_t A[r(t)]$ when $t = 100$.

61. Thermal Inversion When there is a thermal inversion layer over a city (as happens often in Los Angeles), pollutants cannot rise vertically but are trapped below the layer and must disperse horizontally. Assume that a factory smokestack begins emitting a pollutant at 8 A.M. Assume that the pollutant disperses horizontally, forming a circle. If t represents the time (in hours) since the factory began emitting pollutants ($t = 0$ represents 8 A.M.), assume that the radius of the circle of pollution is $r(t) = 2t$ miles. Let $A(r) = \pi r^2$ represent the area of a circle of radius r.

(a) Find and interpret $A[r(t)]$.

(b) Find and interpret $D_t A[r(t)]$ when $t = 4$.

62. Bacteria Population The total number of bacteria (in millions) present in a culture is given by

$$N(t) = 2t(5t + 9)^{1/2} + 12,$$

where t represents time (in hours) after the beginning of an experiment. Find the rate of change of the population of bacteria with respect to time for the following numbers of hours.

(a) 0   (b) 7/5   (c) 8

63. Calcium Usage To test an individual's use of calcium, a researcher injects a small amount of radioactive calcium into the person's bloodstream. The calcium remaining in the bloodstream is measured each day for several days. Suppose the amount of the calcium remaining in the bloodstream (in milligrams per cubic centimeter) t days after the initial injection is approximated by

$$C(t) = \frac{1}{2}(2t + 1)^{-1/2}.$$

Find the rate of change of the calcium level with respect to time for the following numbers of days.

(a) 0   (b) 4   (c) 7.5

(d) Is C always increasing or always decreasing? How can you tell?

64. Drug Reaction The strength of a person's reaction to a certain drug is given by

$$R(Q) = Q\left(C - \frac{Q}{3}\right)^{1/2},$$

where Q represents the quantity of the drug given to the patient and C is a constant.

(a) The derivative $R'(Q)$ is called the *sensitivity* to the drug. Find $R'(Q)$.

(b) Find the sensitivity to the drug if $C = 59$ and a patient is given 87 units of the drug.

(c) Is the patient's sensitivity to the drug increasing or decreasing when $Q = 87$?

General Interest

65. Candy The volume and surface area of a "jawbreaker" for any radius is given by the formulas

$$V(r) = \frac{4}{3}\pi r^3 \quad \text{and} \quad S(r) = 4\pi r^2,$$

respectively. Roger Guffey estimates the radius of a jawbreaker while in a person's mouth to be

$$r(t) = 6 - \frac{3}{17}t,$$

where $r(t)$ is in millimeters and t is in minutes. *Source: The Mathematics Teacher.*

(a) What is the life expectancy of a jawbreaker?

(b) Find dV/dt and dS/dt when $t = 17$ and interpret your answer.

(c) Construct an analogous experiment using some other type of food or verify the results of this experiment.

66. Zenzizenzizenzic Zenzizenzizenzic is an obsolete word with the distinction of containing the most z's of any word found in the *Oxford English Dictionary*. It was used in mathematics, before powers were written as superscript numbers, to represent the square of the square of the square of a number. In symbols, zenzizenzizenzic is written as $((x^2)^2)^2$. *Source: The Phrontistery.*

(a) Use the chain rule twice to find the derivative.

(b) Use the properties of exponents to first simplify the expression, and then find the derivative.

67. Zenzizenzicube Zenzizenzicube is another obsolete word (see Exercise 66) that represents the square of the square of a cube. In symbols, zenzizenzicube is written as $((x^3)^2)^2$. *Source: The Phrontistery.*

(a) Use the chain rule twice to find the derivative.

(b) Use the properties of exponents to first simplify the expression, and then find the derivative.

YOUR TURN ANSWERS

1. $2\sqrt{5} - 1; \sqrt{2}$ 2. $4x^2 - 12x + 10$

3. One possible answer is $g(x) = 2x - 3$ and $f(x) = x^3$

4. $\dfrac{dy}{dx} = \dfrac{-2(10x - 6)}{(5x^2 - 6x)^3}$ 5. $20x(x^2 - 7)^9$

6. $\dfrac{dy}{dx} = x(5x - 1)^2(25x - 2)$ 7. $\dfrac{(4x - 1)^2(8x + 37)}{(x + 3)^2}$

4.4 Derivatives of Exponential Functions

APPLY IT

Given a new product whose rate of growth is rapid at first and then slows, how can we find the rate of growth?
We will use a derivative to answer this question in Example 5 at the end of this section.

We can find the derivative of the exponential function by using the definition of the derivative. Thus

$$\frac{d(e^x)}{dx} = \lim_{h \to 0} \frac{e^{x+h} - e^x}{h}$$

$$= \lim_{h \to 0} \frac{e^x e^h - e^x}{h} \quad \text{Property 1 of exponents}$$

$$= e^x \lim_{h \to 0} \frac{e^h - 1}{h}. \quad \text{Property 1 of limits}$$

In the last step, since e^x does not involve h, we were able to bring e^x in front of the limit. The result says that the derivative of e^x is e^x times a constant, namely, $\lim_{h \to 0} (e^h - 1)/h$. To investigate this limit, we evaluate the expression for smaller and smaller values of h, as shown in the table in the margin. Based on the table, it appears that $\lim_{h \to 0} (e^h - 1)/h = 1$. This is proved in more advanced courses. We, therefore, have the following formula.

FOR REVIEW

Recall from Section 2.4 that e is a special irrational number whose value is approximately 2.718281828. It arises in many applications, such as continuously compounded interest, and it can be defined as

$$\lim_{m \to \infty} \left(1 + \frac{1}{m}\right)^m.$$

Approximation of $\lim\limits_{h \to 0} \dfrac{e^h - 1}{h}$	
h	$\dfrac{e^h - 1}{h}$
-0.1	0.9516
-0.01	0.9950
-0.001	0.9995
-0.0001	1.0000
0.00001	1.0000
0.0001	1.0001
0.001	1.0005
0.01	1.0050
0.1	1.0517

Derivative of e^x

$$\frac{d}{dx}(e^x) = e^x$$

To find the derivative of the exponential function with a base other than e, use the change-of-base theorem for exponentials to rewrite a^x as $e^{(\ln a)x}$. Thus, for any positive constant $a \neq 1$,

$$\frac{d(a^x)}{dx} = \frac{d[e^{(\ln a)x}]}{dx} \quad \text{Change-of-base theorem for exponentials}$$

$$= e^{(\ln a)x} \ln a \quad \text{Chain rule}$$

$$= (\ln a)a^x. \quad \text{Change-of-base theorem again}$$

Derivative of a^x

For any positive constant $a \neq 1$,

$$\frac{d}{dx}(a^x) = (\ln a)a^x.$$

(The derivative of an exponential function is the original function times the natural logarithm of the base.)

We now see why e is the best base to work with: It has the simplest derivative of all the exponential functions. Even if we choose a different base, e appears in the derivative anyway through the $\ln a$ term. (Recall that $\ln a$ is the logarithm of a to the base e.) In fact, of all the functions we have studied, e^x is the simplest to differentiate, because its derivative is just itself.*

The chain rule can be used to find the derivative of the more general exponential function $y = a^{g(x)}$. Let $y = f(u) = a^u$ and $u = g(x)$, so that $f[g(x)] = a^{g(x)}$. Then

$$f'[g(x)] = f'(u) = (\ln a)a^u = (\ln a)a^{g(x)},$$

and by the chain rule,

$$\frac{dy}{dx} = f'[g(x)] \cdot g'(x)$$

$$= (\ln a)a^{g(x)} \cdot g'(x).$$

As before, this formula becomes simpler when we use natural logarithms because $\ln e = 1$. We summarize these results next.

Derivative of $a^{g(x)}$ and $e^{g(x)}$

$$\frac{d}{dx}[a^{g(x)}] = (\ln a)a^{g(x)}g'(x)$$

and

$$\frac{d}{dx}[e^{g(x)}] = e^{g(x)}g'(x)$$

(The derivative of an exponential function is just the original exponential function times the natural logarithm of the base, which is not needed if the base is e. If the exponent is another function $g(x)$, the chain rule requires multiplying the answer by $g'(x)$.)

You need not memorize the previous two formulas. They are simply the result of applying the chain rule to the formula for the derivative of a^x.

CAUTION
Notice the difference between the derivative of a variable to a constant power, such as $D_x x^3 = 3x^2$, and a constant to a variable power, like $D_x 3^x = (\ln 3)3^x$. Remember, $D_x 3^x \neq x3^{x-1}$.

EXAMPLE 1 Derivatives of Exponential Functions

Find the derivative of each function.

(a) $y = e^{5x}$

SOLUTION Let $g(x) = 5x$, so $g'(x) = 5$. Then

$$\frac{dy}{dx} = 5e^{5x}.$$

(b) $s = 3^t$

SOLUTION $\frac{ds}{dt} = (\ln 3)3^t$

(c) $y = 10e^{3x^2}$

SOLUTION $\frac{dy}{dx} = 10(e^{3x^2})(6x) = 60xe^{3x^2}$

(d) $s = 8 \cdot 10^{1/t}$

SOLUTION $\frac{ds}{dt} = 8(\ln 10)10^{1/t}\left(\frac{-1}{t^2}\right)$

$$= \frac{-8(\ln 10)10^{1/t}}{t^2}$$

YOUR TURN 1 Find dy/dx for
(a) $y = 4^{3x}$,
(b) $y = e^{7x^3+5}$.

TRY YOUR TURN 1

*There is a joke about a deranged mathematician who frightened other inmates at an insane asylum by screaming at them, "I'm going to differentiate you!" But one inmate remained calm and simply responded, "I don't care; I'm e^x."

EXAMPLE 2   Derivative of an Exponential Function

Let $y = e^{x^2+1}\sqrt{5x + 2}$. Find $\dfrac{dy}{dx}$.

SOLUTION   Rewrite y as $y = e^{x^2+1}(5x + 2)^{1/2}$, and then use the product rule and the chain rule.

$$\frac{dy}{dx} = e^{x^2+1} \cdot \frac{1}{2}(5x + 2)^{-1/2} \cdot 5 + (5x + 2)^{1/2}e^{x^2+1} \cdot 2x$$

$$= e^{x^2+1}(5x + 2)^{-1/2}\left[\frac{5}{2} + (5x + 2) \cdot 2x\right] \quad \text{Factor out the greatest common factor, } e^{x^2+1}(5x + 2)^{-1/2}.$$

$$= e^{x^2+1}(5x + 2)^{-1/2}\left[\frac{5 + 4x(5x + 2)}{2}\right] \quad \text{Least common denominator.}$$

$$= e^{x^2+1}(5x + 2)^{-1/2}\left[\frac{5 + 20x^2 + 8x}{2}\right]$$

$$= \frac{e^{x^2+1}(20x^2 + 8x + 5)}{2\sqrt{5x + 2}} \quad \text{Simplify.}$$

TRY YOUR TURN 2

YOUR TURN 2 Let $y = (x^2 + 1)^2e^{2x}$. Find dy/dx.

EXAMPLE 3   Derivative of an Exponential Function

Let $f(x) = \dfrac{100{,}000}{1 + 100e^{-0.3x}}$. Find $f'(x)$.

SOLUTION   Use the quotient rule.

$$f'(x) = \frac{(1 + 100e^{-0.3x})(0) - 100{,}000(-30e^{-0.3x})}{(1 + 100e^{-0.3x})^2}$$

$$= \frac{3{,}000{,}000e^{-0.3x}}{(1 + 100e^{-0.3x})^2} \quad \text{TRY YOUR TURN 3}$$

YOUR TURN 3 Let $f(x) = \dfrac{100}{5 + 2e^{-0.01x}}$. Find $f'(x)$.

In the previous example, we could also have taken the derivative by writing $f(x) = 100{,}000(1 + 100e^{-0.3x})^{-1}$, from which we have

$$f'(x) = -100{,}000(1 + 100e^{-0.3x})^{-2}100e^{-0.3x}(-0.3).$$

This simplifies to the same expression as in Example 3.

EXAMPLE 4   Radioactivity

The amount in grams in a sample of uranium-239 after t years is given by

$$A(t) = 100e^{-0.362t}.$$

Find the rate of change of the amount present after 3 years.

SOLUTION   The rate of change is given by the derivative dA/dt.

$$\frac{dA}{dt} = 100(e^{-0.362t})(-0.362) = -36.2e^{-0.362t}$$

After 3 years $(t = 3)$, the rate of change is

$$\frac{dA}{dt} = -36.2e^{-0.362(3)} = -36.2e^{-1.086} \approx -12.2$$

YOUR TURN 4 The quantity (in grams) of a radioactive substance present after t years is $Q(t) = 100e^{-0.421t}$. Find the rate of change of the quantity present after 2 years.

grams per year.

TRY YOUR TURN 4

Frequently a population, or the sales of a certain product, will start growing slowly, then grow more rapidly, and then gradually level off. Such growth can often be approximated by a mathematical model known as the **logistic function**:

$$G(t) = \frac{mG_0}{G_0 + (m - G_0)e^{-kmt}},$$

where t represents time in appropriate units, G_0 is the initial number present, m is the maximum possible size of the population, k is a positive constant, and $G(t)$ is the population at time t. It is sometimes simpler to divide the numerator and denominator of the logistic function by G_0, writing the result as

$$G(t) = \frac{m}{1 + \left(\dfrac{m}{G_0} - 1\right)e^{-kmt}}.$$

Notice that

$$\lim_{t \to \infty} G(t) = \frac{m}{1 + 0} = m$$

because $\lim_{t \to \infty} e^{-kmt} = 0$.

EXAMPLE 5   Product Sales

A company sells 990 units of a new product in the first year and 3213 units in the fourth year. They expect that sales can be approximated by a logistic function, leveling off at around 100,000 in the long run.

(a) Find a formula $S(t)$ for the sales as a function of time.

SOLUTION   We already know that $S_0 = 990$ and $m = 100,000$, so

$$S(t) = \frac{100,000}{1 + \left(\dfrac{100,000}{990} - 1\right)e^{-k100,000t}}$$

$$= \frac{100,000}{1 + 100.01e^{-k100,000t}}.$$

To find k, use the fact that $S(4) = 3213$.

$$3213 = \frac{100,000}{1 + 100.01e^{-k100,000 \cdot 4}}$$

$$3213 = \frac{100,000}{1 + 100.01e^{-k400,000}}$$

$$3213(1 + 100.01e^{-k400,000}) = 100,000 \qquad \text{Cross multiply.}$$

$$3213 + 321,332e^{-k400,000} = 100,000$$

$$321,332e^{-k400,000} = 96,787 \qquad \text{Subtract 3213 from both sides.}$$

$$e^{-k400,000} \approx 0.3012 \qquad \text{Divide both sides by 321,332.}$$

$$-k400,000 = \ln 0.3012 \qquad \text{Take the natural logarithm of both sides.}$$

$$k = -\ln 0.3012/400,000$$

$$k \approx 3 \times 10^{-6}$$

Rounding 100.01 to 100 and simplifying $k100,000 = (3 \times 10^{-6})100,000 = 0.3$,

$$S(t) = \frac{100,000}{1 + 100e^{-k100,000t}}$$

$$= \frac{100,000}{1 + 100e^{-0.3t}}.$$

(b) Find the rate of change of sales after 4 years.

SOLUTION The derivative of this sales function, which gives the rate of change of sales, was found in Example 3. Using that derivative,

$$S'(4) = \frac{3{,}000{,}000e^{-0.3(4)}}{[1 + 100e^{-0.3(4)}]^2} = \frac{3{,}000{,}000e^{-1.2}}{(1 + 100e^{-1.2})^2}.$$

Using a calculator, $e^{-1.2} \approx 0.3012$, and

$$S'(4) \approx \frac{3{,}000{,}000(0.3012)}{[1 + 100(0.3012)]^2}$$

$$\approx \frac{903{,}600}{(1 + 30.12)^2}$$

$$\approx \frac{903{,}600}{968.5} \approx 933.$$

The rate of change of sales after 4 years is about 933 units per year. The positive number indicates that sales are increasing at this time.

The graph of the function in Example 5 is shown in Figure 12.

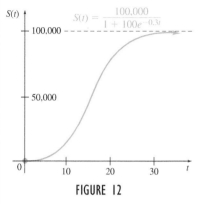

t	S(t)
0	990
5	4300
10	17,000
15	47,000
20	80,000
30	99,000

$$S(t) = \frac{100{,}000}{1 + 100e^{-0.3t}}$$

FIGURE 12

4.4 WARM-UP EXERCISES

Find the derivative of each of the following functions. *(Sec. 4.3)*

W1. $f(x) = \sqrt{x^6 + 5}$

W2. $f(x) = (4x^2 + 3x + 2)^{10}$

W3. $f(x) = \left(\dfrac{x^2 - 1}{x^2 + 1}\right)^{5/2}$

4.4 EXERCISES

Find derivatives of the functions defined as follows.

1. $y = e^{4x}$

2. $y = e^{-2x}$

3. $y = -8e^{3x}$

4. $y = 1.2e^{5x}$

5. $y = -16e^{2x+1}$

6. $y = -4e^{-0.3x}$

7. $y = e^{x^2}$

8. $y = e^{-x^2}$

9. $y = 3e^{2x^2}$

10. $y = -5e^{4x^3}$

11. $y = 4e^{2x^2-4}$

12. $y = -3e^{3x^2+5}$

13. $y = xe^x$

14. $y = x^2e^{-2x}$

15. $y = (x + 3)^2e^{4x}$

16. $y = (3x^3 - 4x)e^{-5x}$

17. $y = \dfrac{x^2}{e^x}$

18. $y = \dfrac{e^x}{2x + 1}$

19. $y = \dfrac{e^x + e^{-x}}{x}$

20. $y = \dfrac{e^x - e^{-x}}{x}$

21. $p = \dfrac{10{,}000}{9 + 4e^{-0.2t}}$

22. $p = \dfrac{500}{12 + 5e^{-0.5t}}$

23. $f(z) = (2z + e^{-z^2})^2$

24. $f(t) = (e^{t^2} + 5t)^3$

25. $y = 7^{3x+1}$

26. $y = 4^{-5x+2}$

27. $y = 3 \cdot 4^{x^2+2}$

28. $y = -10^{3x^2-4}$

29. $s = 2 \cdot 3^{\sqrt{t}}$

30. $s = 5 \cdot 2^{\sqrt{t-2}}$

31. $y = \dfrac{te^t + 2}{e^{2t} + 1}$

32. $y = \dfrac{t^2e^{2t}}{t + e^{3t}}$

33. $f(x) = e^{x\sqrt{3x+2}}$

34. $f(x) = e^{x^2/(x^3+2)}$

35. Prove that if $y = y_0e^{kt}$, where y_0 and k are constants, then $dy/dt = ky$. (This says that for exponential growth and decay, the rate of change of the population is proportional to the size of the population, and the constant of proportionality is the growth or decay constant.)

 36. Use a graphing calculator to sketch the graph of $y = [f(x + h) - f(x)]/h$ using $f(x) = e^x$ and $h = 0.0001$. Compare it with the graph of $y = e^x$ and discuss what you observe.

37. Use graphical differentiation to verify that $\dfrac{d}{dx}(e^x) = e^x$.

38. (a) Using a graphing calculator, sketch the graph of $y = 2^x$ on the interval $-1 \le x \le 1$, and have the calculator compute the derivative at $x = 0$. Compare the result with what you would predict based on the formula for the derivative of a^x.

(b) Repeat part (a), this time using the graph $y = 3^x$.

(c) You should have found that the answer to part (a) is less than 1 and the answer to part (b) is greater than 1. Using trial and error, find a function $y = a^x$ that gives a derivative of 1 at $x = 0$. What do you find about the value of a?

APPLICATIONS

Business and Economics

39. Cost The cost in dollars to produce x DVDs can be approximated by
$$C(x) = \sqrt{900 - 800 \cdot 1.1^{-x}}.$$
Find the marginal cost when the following quantities are made.

(a) 0   **(b)** 20

(c) What happens to the marginal cost as the number produced becomes larger and larger?

40. Sales The sales of a new personal computer (in thousands) are given by
$$S(t) = 100 - 90e^{-0.3t},$$
where t represents time in years. Find the rate of change of sales at each time.

(a) After 1 year   **(b)** After 5 years

(c) What is happening to the rate of change of sales as time goes on?

(d) Does the rate of change of sales ever equal zero?

41. Franchising Marketing researchers have studied the advertising fees that a franchisee (such as the owner of a fast food outlet) pays to the franchisor (the national fast food corporation). *Source: Journal of the Operational Research Society.*

(a) The researchers postulate that the anticipated sales would be a function of the level of local advertising A chosen by the franchisee of the form
$$S(A) = c - de^{-\theta A},$$
where c, d and θ are constants such that $c > d > 0$ and $\theta > 0$. Sketch the graph of this function, and discuss why such a shape seems plausible.

(b) The researchers propose hypothetical values $c = 60{,}000$ units, $d = 50{,}000$ units, and $\theta = 0.0006$. Find the values of $S(A)$ and $S'(A)$ when the level of local advertising is set at \$2500.

(c) Repeat part (b) using \$4000.

(d) If you were the franchisee, what decisions might you make on advertising based on the results of parts (b) and (c)?

42. Product Awareness After the introduction of a new product for tanning without sun exposure, the percent of the public that is aware of the product is approximated by
$$A(t) = 10t^2 2^{-t},$$
where t is the time in months. Find the rate of change of the percent of the public that is aware of the product after the following numbers of months.

(a) 2   **(b)** 4

(c) Notice that the answer to part (a) is positive and the answer to part (b) is negative. What does this tell you about how public awareness of the product has changed?

43. Internet Users The growth in the number (in millions) of Internet users in the United States between 1990 and 2015 can be approximated by a logistic function with $k = 0.0018$, where t is the number of years since 1990. In 1990 (when $t = 0$), the number of users was about 2 million, and the number is expected to level out around 250 million. *Source: World Bank.*

(a) Find the growth function $G(t)$ for the number of Internet users in the United States.

Estimate the number of Internet users in the United States and the rate of growth for the following years.

(b) 1995   **(c)** 2000   **(d)** 2010

(e) What happens to the rate of growth over time?

44. Investment The value of a particular investment changes over time according to the function
$$S(t) = 5000e^{0.1(e^{0.25t})},$$
where $S(t)$ is the value after t years. Calculate the rate at which the value of the investment is changing after 8 years. (Choose one of the following.) *Source: Society of Actuaries.*

(a) 618   **(b)** 1934   **(c)** 2011   **(d)** 7735   **(e)** 10,468

Life Sciences

45. Minority Population In Section 2.4, Exercise 51, we saw that the projected Hispanic population in the United States (in millions) can be approximated by the function
$$h(t) = 37.79(1.021)^t,$$
where $t = 0$ corresponds to 2000 and $0 \le t \le 50$. *Source: U.S. Census Bureau.*

(a) Estimate the Hispanic population in the United States for the year 2015.

(b) What is the instantaneous rate of change of the Hispanic population in the United States when $t = 15$? Interpret your answer.

46. Population Growth In Section 2.4, Exercise 49, the growth in world population (in millions) was approximated by the exponential function
$$A(t) = 3100e^{0.0166t},$$
where t is the number of years since 1960. Find the instantaneous rate of change in the world population at the following times. *Source: United Nations.*

(a) 2010   **(b)** 2015

47. Cactus Wrens The weight of cactus wrens in grams has been found to fit the logistic function with $k = 0.0125$ and t the age in days. The initial weight is 2.33 g, and the maximum size is 31.4 g. *Source: Ecology.*

(a) Find the growth function $G(t)$ for cactus wrens.

Find the weight and rate of growth at the following times.

(b) 1 day

(c) 5 days

(d) 10 days

(e) What happens to the rate of growth over time?

48. Whooping Cranes Based on data from the U.S. Fish and Wildlife Service, the population of whooping cranes in the Aransas-Wood Buffalo National Park can be approximated by a logistic function with $k = 6.01 \times 10^{-5}$, with a population in 1958 of 32 and a maximum population of 787. *Source: U.S. Fish and Wildlife Service.*

(a) Find the growth function $G(t)$ for the whooping crane population, where t is the time since 1938, when the park first started counting the cranes.

(b) Find the initial population G_0

Find the population and rate of growth in the following years.

(c) 1945

(d) 1985

(e) 2005

(f) What happens to the rate of growth over time?

49. Breast Cancer It has been observed that the following formula accurately models the relationship between the size of a breast tumor and the amount of time that it has been growing.

$$V(t) = 1100[1023e^{-0.02415t} + 1]^{-4},$$

where t is in months and $V(t)$ is measured in cubic centimeters. *Source: Cancer.*

(a) Find the tumor volume at 240 months.

(b) Assuming that the shape of a tumor is spherical, find the radius of the tumor from part (a). (*Hint:* The volume of a sphere is given by the formula $V = (4/3)\pi r^3$.)

(c) If a tumor of size 0.5 cm³ is detected, according to the formula, how long has it been growing? What does this imply?

(d) Find $\lim_{t \to \infty} V(t)$ and interpret this value. Explain whether this makes sense.

(e) Calculate the rate of change of tumor volume at 240 months and interpret.

50. Mortality The percentage of people of any particular age group that will die in a given year may be approximated by the formula

$$P(t) = 0.00239e^{0.0957t},$$

where t is the age of the person in years. *Source: U.S. Vital Statistics.*

(a) Find $P(25)$, $P(50)$, and $P(75)$.

(b) Find $P'(25)$, $P'(50)$, and $P'(75)$.

(c) Interpret your answers for parts (a) and (b). Are there any limitations of this formula?

51. Medical Literature It has been observed that there has been an increase in the proportion of medical research papers that use the word "novel" in the title or abstract, and that this proportion can be accurately modeled by the function

$$p(x) = 0.001131e^{0.1268x},$$

where x is the number of years since 1970. *Source: Nature.*

(a) Find $p(40)$.

(b) If this phenomenon continues, estimate the year in which every medical article will contain the word "novel" in its title or abstract.

(c) Estimate the rate of increase in the proportion of medical papers using this word in the year 2014.

(d) Explain some factors that may be contributing to researchers using this word.

52. Arctic Foxes The age/weight relationship of female Arctic foxes caught in Svalbard, Norway, can be estimated by the function

$$M(t) = 3102e^{-e^{-0.022(t-56)}},$$

where t is the age of the fox in days and $M(t)$ is the weight of the fox in grams. *Source: Journal of Mammalogy.*

(a) Estimate the weight of a female fox that is 200 days old.

(b) Use $M(t)$ to estimate the largest size that a female fox can attain. (*Hint:* Find $\lim_{t \to \infty} M(t)$.)

(c) Estimate the age of a female fox when it has reached 80% of its maximum weight.

(d) Estimate the rate of change in weight of an Arctic fox that is 200 days old. (*Hint:* Recall that $D_t[e^{f(t)}] = f'(t)e^{f(t)}$.)

(e) Use a graphing calculator to graph $M(t)$ and then describe the growth pattern.

(f) Use the table function on a graphing calculator or a spreadsheet to develop a chart that shows the estimated weight and growth rate of female foxes for days 50, 100, 150, 200, 250, and 300.

53. Beef Cattle Researchers have compared two models that are used to predict the weight of beef cattle of various ages,

$$W_1(t) = 509.7(1 - 0.941e^{-0.00181t})$$

and

$$W_2(t) = 498.4(1 - 0.889e^{-0.00219t})^{1.25},$$

where $W_1(t)$ and $W_2(t)$ represent the weight (in kilograms) of a t-day-old beef cow. *Source: Journal of Animal Science.*

(a) What is the maximum weight predicted by each function for the average beef cow? Is this difference significant?

(b) According to each function, find the age that the average beef cow reaches 90% of its maximum weight.

(c) Find $W_1{}'(750)$ and $W_2{}'(750)$. Compare your results.

(d) Graph the two functions on $[0, 2500]$ by $[0, 525]$ and comment on the differences in growth patterns for each of these functions.

(e) Graph the derivative of these two functions on $[0, 2500]$ by $[0, 1]$ and comment on any differences you notice between these functions.

54. Cholesterol Researchers have found that the risk of coronary heart disease rises as blood cholesterol increases. This risk may be approximated by the function

$$R(c) = 3.19(1.006)^c, \quad 100 \le c \le 300,$$

where R is the risk in terms of coronary heart disease incidence per 1000 per year, and c is the cholesterol in mg/dL. Suppose a person's cholesterol is 180 mg/dL and going up at a rate of 15 mg/dL per year. At what rate is the person's risk of coronary heart disease going up? *Source: Circulation.*

Social Sciences

55. Survival of Manuscripts Paleontologist John Cisne has demonstrated that the survival of ancient manuscripts can be modeled by the logistic equation. For example, the number of copies of the Venerable Bede's *De Temporum Ratione* was found to approach a limiting value over the five centuries after its publication in the year 725. Let $G(t)$ represent the proportion of manuscripts known to exist after t centuries out of the limiting value, so that $m = 1$. Cisne found that for Venerable Bede's *De Temporum Ratione, $k = 3.5$ and $G_0 = 0.00369$.* *Source: Science.*

(a) Find the growth function $G(t)$ for the proportion of copies of *De Temporum Ratione* found.

Find the proportion of manuscripts and their rate of growth after the following number of centuries.

(b) 1   (c) 2

(d) 3

(e) What happens to the rate of growth over time?

56. Habit Strength According to work by the psychologist C. L. Hull, the strength of a habit is a function of the number of times the habit is repeated. If N is the number of repetitions and $H(N)$ is the strength of the habit, then

$$H(N) = 1000(1 - e^{-kN}),$$

where k is a constant. Find $H'(N)$ if $k = 0.1$ and the number of times the habit is repeated is as follows.

(a) 10   (b) 100   (c) 1000

(d) Show that $H'(N)$ is always positive. What does this mean?

57. Online Learning The growth of the number of students taking at least one online course can be approximated by a logistic function with $k = 0.0440$, where t is the number of years since 2002. In 2002 (when $t = 0$), the number of students enrolled in at least one online course was 1.603 million. Assume that the number will level out at around 6.8 million students. *Source: The Sloan Consortium.*

(a) Find the growth function $G(t)$ for students enrolled in at least one online course.

Find the number of students enrolled in at least one online course and the rate of growth in the number in the following years.

(b) 2006   (c) 2010   (d) 2014

(e) What happens to the rate of growth over time?

Physical Sciences

58. Radioactive Decay The amount (in grams) of a sample of lead 214 present after t years is given by

$$A(t) = 500e^{-0.31t}.$$

Find the rate of change of the quantity present after each of the following years.

(a) 4   (b) 6   (c) 10

(d) What is happening to the rate of change of the amount present as the number of years increases?

(e) Will the substance ever be gone completely?

59. Electricity In a series resistance-capacitance DC circuit, the instantaneous charge Q on the capacitor as a function of time (where $t = 0$ is the moment the circuit is energized by closing a switch) is given by the equation

$$Q(t) = CV(1 - e^{-t/RC}),$$

where C, V, and R are constants. Further, the instantaneous charging current I_C is the rate of change of charge on the capacitor, or $I_C = dQ/dt$. *Source: Kevin Friedrich.*

(a) Find the expression for I_C as a function of time.

(b) If $C = 10^{-5}$ farads, $R = 10^7$ ohms, and $V = 10$ volts, what is the charging current after 200 seconds? (*Hint:* When placed into the function in part (a) the units can be combined into amps.)

60. Heat Index The heat index is a measure of how hot it really feels under different combinations of temperature and humidity. The heat index, in degrees Fahrenheit, can be approximated by

$$H(T) = T - 0.9971e^{0.02086T}[1 - e^{0.0445(D-57.2)}],$$

where the temperature T and dewpoint D are both expressed in degrees Fahrenheit. *Source: American Meteorological Society.*

(a) Assume the dewpoint is $D = 85°$ F. Find the function $H(T)$.

(b) Using the function you found in part (a), find the heat index when the temperature T is 80°F.

(c) Find the rate of change of the heat index when $T = 80°$ F.

61. Humidex In Canada, the humidex is used to measure the combined effect of temperature and humidity. It is defined by

$$H = T + \frac{5}{9}(6.11e^{5417.7530(1/273.16 - 1/D)} - 10),$$

where T is the temperature in degrees Celsius and D is the dewpoint (the temperature at which water vapor in air condenses into liquid water) in Kelvin (the temperature in degrees Celsius plus 273.15°). The dewpoint is less than the air temperature but approaches the air temperature as the humidity approaches 100%. Suppose $T = 30°$C (86°F). Find the humidex and dH/dD for the following dewpoints. *Source: Government of Canada.*

(a) 278 K (approximately 41°F)

(b) 290 K (approximately 62°F)

(c) 300 K (approximately 80°F)

(d) What do the answers to parts (a) through (c) tell you?

General Interest

62. Ballooning Suppose a person is going up in a hot air balloon. The surrounding air temperature in degrees Fahrenheit decreases with height according to the formula

$$T(h) = 80e^{-0.000065h},$$

where h is the height in feet. How fast is the temperature decreasing when the person is at a height of 1000 ft and rising at a height of 800 ft/hr?

63. The Gateway Arch The Gateway Arch in St. Louis, Missouri, is approximately 630 ft wide and 630 ft high. At first glance, the arch resembles a parabola, but its shape is actually known as a modified catenary. The height s (in feet) of the arch, measured from the ground to the middle of the arch, is approximated by the function

$$s(x) = 693.9 - 34.38(e^{0.01003x} + e^{-0.01003x}),$$

where $x = 0$ represents the center of the arch with $-299.2 \le x \le 299.2$. *Source: Notices of the AMS.*

(a) What is the height of the arch when $x = 0$?

(b) What is the slope of the line tangent to the curve when $x = 0$? Does this make sense?

(c) Find the rate of change of the height of the arc when $x = 150$ ft.

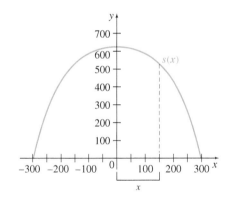

64. Dead Grandmother Syndrome Biology professor Mike Adams has found an increase over the years in the number of students claiming just before an exam that a relative (usually a grandmother) has died. Based on his data, he found that the average number of family deaths per 100 students could be approximated by

$$f(t) = 5.4572 \times 10^{-59}10^{2.923 \times 10^{-2}t},$$

where t is the year. Find the average number of deaths per 100 students in 2015 and the rate that the function was increasing in that year. *Source: Annals of Improbable Research.*

65. Track and Field In 1958, L. Lucy developed a method for predicting the world record for any given year that a human could run a distance of 1 mile. His formula is given as follows:

$$t(n) = 218 + 31(0.933)^n,$$

where $t(n)$ is the world record (in seconds) for the mile run in year $1950 + n$. Thus, $n = 5$ corresponds to the year 1955. *Source: Statistics in Sports.*

(a) Find the estimate for the world record in the year 2014.

(b) Calculate the instantaneous rate of change for the world record at the end of year 2014 and interpret.

(c) Find $\lim_{n \to \infty} t(n)$ and interpret. How does this compare with the current world record?

YOUR TURN ANSWERS

1. (a) $\dfrac{dy}{dx} = 3(\ln 4)4^{3x}$

   (b) $\dfrac{dy}{dx} = 21x^2e^{7x^3+5}$

2. $\dfrac{dy}{dx} = 2e^{2x}(x^2 + 1)(x + 1)^2$

3. $f'(x) = \dfrac{2e^{-0.01x}}{(5 + 2e^{-0.01x})^2}$

4. -18.1 grams per year

4.5 Derivatives of Logarithmic Functions

APPLY IT How does the average resale value of an automobile change with the age of the automobile?

We will use the derivative to answer this question in Example 4.

Recall that in the section on Logarithmic Functions, we showed that the logarithmic function and the exponential function are inverses of each other. In the last section we showed that the derivative of a^x is $(\ln a)a^x$. We can use this information and the chain rule to find the derivative of $\log_a x$. We begin by solving the general logarithmic function for x.

$$f(x) = \log_a x$$

$$a^{f(x)} = x \qquad \text{Definition of the logarithm}$$

Now consider the left and right sides of the last equation as functions of x that are equal, so their derivatives with respect to x will also be equal. Notice in the first step that we need to use the chain rule when differentiating $a^{f(x)}$.

$$(\ln a)a^{f(x)}f'(x) = 1 \qquad \text{Derivative of the exponential function}$$
$$(\ln a)xf'(x) = 1 \qquad \text{Substitute } a^{f(x)} = x.$$

Finally, divide both sides of this equation by $(\ln a)x$ to get

$$f'(x) = \frac{1}{(\ln a)x}.$$

Derivative of $\log_a x$

$$\frac{d}{dx}[\log_a x] = \frac{1}{(\ln a)x}$$

(The derivative of a logarithmic function is the reciprocal of the product of the variable and the natural logarithm of the base.)

As with the exponential function, this formula becomes particularly simple when we let $a = e$, because of the fact that $\ln e = 1$.

Derivative of $\ln x$

$$\frac{d}{dx}[\ln x] = \frac{1}{x}$$

This fact can be further justified geometrically. Notice what happens to the slope of the line $y = 2x + 4$ if the x-axis and y-axis are switched. That is, if we replace x with y and y with x, then the resulting line $x = 2y + 4$ or $y = x/2 - 2$ is a reflection of the line $y = 2x + 4$ across the line $y = x$, as seen in Figure 13. Furthermore, the slope of the new line is the reciprocal of the original line. In fact, the reciprocal property holds for all lines.

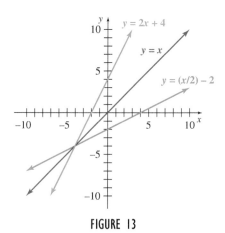

FIGURE 13

In the section on Logarithmic Functions, we showed that switching the x and y variables changes the exponential graph into a logarithmic graph, a defining property of functions that are inverses of each other. We also showed in the previous section that the slope of the tangent line of e^x at any point is e^x—that is, the y-coordinate itself. So, if we switch the x and y variables, the new slope of the tangent line will be $1/y$, except that it is no longer y, it is x. Thus, the slope of the tangent line of $y = \ln x$ must be $1/x$ and hence $D_x \ln x = 1/x$.

■■■■ **EXAMPLE 1** **Derivatives of Logarithmic Functions**

Find the derivative of each function.

(a) $f(x) = \ln 6x$

SOLUTION Use the properties of logarithms and the rules for derivatives.

$$f'(x) = \frac{d}{dx}(\ln 6x)$$

$$= \frac{d}{dx}(\ln 6 + \ln x)$$

$$= \frac{d}{dx}(\ln 6) + \frac{d}{dx}(\ln x) = 0 + \frac{1}{x} = \frac{1}{x}$$

(b) $y = \log x$

SOLUTION Recall that when the base is not specified, we assume that the logarithm is a common logarithm, which has a base of 10.

$$\frac{dy}{dx} = \frac{1}{(\ln 10)x}$$

YOUR TURN 1 Find the derivative of $f(x) = \log_3 x$.

TRY YOUR TURN 1 ■■■

Applying the chain rule to the formulas for the derivative of logarithmic functions gives us

$$\frac{d}{dx}\log_a g(x) = \frac{1}{\ln a} \cdot \frac{g'(x)}{g(x)}$$

and

$$\frac{d}{dx}\ln g(x) = \frac{g'(x)}{g(x)}.$$

■■■■ **EXAMPLE 2** **Derivatives of Logarithmic Functions**

Find the derivative of each function.

(a) $f(x) = \ln(x^2 + 1)$

SOLUTION Here $g(x) = x^2 + 1$ and $g'(x) = 2x$. Thus,

$$f'(x) = \frac{g'(x)}{g(x)} = \frac{2x}{x^2 + 1}.$$

(b) $y = \log_2(3x^2 - 4x)$

YOUR TURN 2 Find the derivative of

(a) $y = \ln(2x^3 - 3)$,
(b) $f(x) = \log_4(5x + 3x^3)$.

SOLUTION

$$\frac{dy}{dx} = \frac{1}{\ln 2} \cdot \frac{6x - 4}{3x^2 - 4x}$$

$$= \frac{6x - 4}{(\ln 2)(3x^2 - 4x)}$$

TRY YOUR TURN 2 ■■■

If $y = \ln(-x)$, where $x < 0$, the chain rule with $g(x) = -x$ and $g'(x) = -1$ gives

$$\frac{dy}{dx} = \frac{g'(x)}{g(x)} = \frac{-1}{-x} = \frac{1}{x}.$$

The derivative of $y = \ln(-x)$ is the same as the derivative of $y = \ln x$. For this reason, these two results can be combined into one rule using the absolute value of x. A similar situation holds true for $y = \ln[g(x)]$ and $y = \ln[-g(x)]$, as well as for $y = \log_a[g(x)]$ and $y = \log_a[-g(x)]$. These results are summarized as follows.

Derivative of $\log_a|x|$, $\log_a|g(x)|$, $\ln|x|$, and $\ln|g(x)|$

$$\frac{d}{dx}[\log_a|x|] = \frac{1}{(\ln a)x} \qquad \frac{d}{dx}[\log_a|g(x)|] = \frac{1}{\ln a} \cdot \frac{g'(x)}{g(x)}$$

$$\frac{d}{dx}[\ln|x|] = \frac{1}{x} \qquad \frac{d}{dx}[\ln|g(x)|] = \frac{g'(x)}{g(x)}$$

NOTE You need not memorize the previous four formulas. They are simply the result of the chain rule applied to the formula for the derivative of $y = \log_a x$, as well as the fact that when $\log_a x = \ln x$, so that $a = e$, then $\ln a = \ln e = 1$. An absolute value inside of a logarithm has no effect on the derivative, other than making the result valid for more values of x.

EXAMPLE 3   Derivatives of Logarithmic Functions

Find the derivative of each function.

(a) $y = \ln|5x|$

SOLUTION   Let $g(x) = 5x$, so that $g'(x) = 5$. From the previous formula,

$$\frac{dy}{dx} = \frac{g'(x)}{g(x)} = \frac{5}{5x} = \frac{1}{x}.$$

Notice that the derivative of $\ln|5x|$ is the same as the derivative of $\ln|x|$. Also notice that we would have found the exact same answer for the derivative of $y = \ln 5x$ (without the absolute value), but the result would not apply to negative values of x. Also, in Example 1, the derivative of $\ln 6x$ was the same as that for $\ln x$. This suggests that for any constant a,

$$\frac{d}{dx}\ln|ax| = \frac{d}{dx}\ln|x|$$

$$= \frac{1}{x}.$$

Exercise 46 asks for a proof of this result.

(b) $f(x) = 3x\ln x^2$

SOLUTION   This function is the product of the two functions $3x$ and $\ln x^2$, so use the product rule.

$$f'(x) = (3x)\left[\frac{d}{dx}\ln x^2\right] + (\ln x^2)\left[\frac{d}{dx}3x\right]$$

$$= 3x\left(\frac{2x}{x^2}\right) + (\ln x^2)(3)$$

$$= 6 + 3\ln x^2$$

By the power rule for logarithms,

$$f'(x) = 6 + \ln(x^2)^3$$

$$= 6 + \ln x^6.$$

Alternatively, write the answer as $f'(x) = 6 + 6\ln x$, except that this last form requires $x > 0$, while negative values of x are acceptable in $6 + \ln x^6$.

Another method would be to use a rule of logarithms to simplify the function to $f(x) = 3x \cdot 2\ln x = 6x\ln x$ and then to take the derivative. Note that this simplified function is valid only for $x > 0$, while the original function is valid for all $x \neq 0$.

(c) $s(t) = \dfrac{\log_8(t^{3/2} + 1)}{t}$

SOLUTION Use the quotient rule and the chain rule.

$$s'(t) = \dfrac{t \cdot \dfrac{1}{(t^{3/2} + 1)\ln 8} \cdot \dfrac{3}{2}t^{1/2} - \log_8(t^{3/2} + 1) \cdot 1}{t^2}$$

This expression can be simplified slightly by multiplying the numerator and the denominator by $2(t^{3/2} + 1)\ln 8$.

YOUR TURN 3 Find the derivative of each function.

(a) $y = \ln|2x + 6|$
(b) $f(x) = x^2 \ln 3x$
(c) $s(t) = \dfrac{\ln(t^2 - 1)}{t + 1}$

$$s'(t) = \dfrac{t \cdot \dfrac{1}{(t^{3/2} + 1)\ln 8} \cdot \dfrac{3}{2}t^{1/2} - \log_8(t^{3/2} + 1)}{t^2} \cdot \dfrac{2(t^{3/2} + 1)\ln 8}{2(t^{3/2} + 1)\ln 8}$$

$$= \dfrac{3t^{3/2} - 2(t^{3/2} + 1)(\ln 8)\log_8(t^{3/2} + 1)}{2t^2(t^{3/2} + 1)\ln 8} \qquad \text{TRY YOUR TURN 3}$$

EXAMPLE 4 Automobile Resale Value

Based on projections from the *Kelley Blue Book*, the resale value of a 2014 Toyota 4Runner SR5 can be approximated by the following function

$$f(t) = 30{,}781 - 24{,}277 \log(0.46t + 1),$$

where t is the number of years since 2014. Find and interpret $f(4)$ and $f'(4)$. *Source: Kelley Blue Book*.

APPLY IT **SOLUTION** Recognizing this function as a common (base 10) logarithm, we have

$$f(4) = 30{,}781 - 24{,}277 \log(0.46 \cdot 4 + 1) \approx 19{,}776.$$

The average resale value of a 2014 Toyota 4Runner in 2018 would be approximately $19,776. The derivative of $f(t)$ is

$$f'(t) = \dfrac{-24{,}277 \cdot 0.46}{(\ln 10)(0.46t + 1)},$$

so $f'(4) \approx -1708$. In 2018, the average resale value of a 2014 Toyota 4Runner is decreasing by $1708 per year.

4.5 WARM-UP EXERCISES

Find the derivative of each of the following functions. *(Sec. 4.4)*

W1. $f(x) = e^{x^2}$

W2. $f(x) = x^2 e^{4x}$

W3. $f(x) = \dfrac{e^{2x}}{e^{3x} + 1}$

4.5 EXERCISES

Find the derivative of each function.

1. $y = \ln(8x)$

2. $y = \ln(-4x)$

3. $y = \ln(8 - 3x)$

4. $y = \ln(1 + x^3)$

5. $y = \ln|4x^2 - 9x|$

6. $y = \ln|-8x^3 + 2x|$

7. $y = \ln\sqrt{x + 5}$

8. $y = \ln\sqrt{2x + 1}$

9. $y = \ln(x^4 + 5x^2)^{3/2}$

10. $y = \ln(5x^3 - 2x)^{3/2}$

11. $y = -5x \ln(3x + 2)$

12. $y = (3x + 7) \ln(2x - 1)$

13. $s = t^2 \ln|t|$

14. $y = x \ln|2 - x^2|$

15. $y = \dfrac{2 \ln(x + 3)}{x^2}$

16. $v = \dfrac{\ln u}{u^3}$

17. $y = \dfrac{\ln x}{4x + 7}$

18. $y = \dfrac{-2 \ln x}{3x - 1}$

19. $y = \dfrac{3x^2}{\ln x}$

20. $y = \dfrac{x^3 - 1}{2 \ln x}$

21. $y = (\ln |x + 1|)^4$

22. $y = \sqrt{\ln |x - 3|}$

23. $y = \ln |\ln x|$

24. $y = (\ln 4)(\ln |3x|)$

25. $y = e^{x^2} \ln x$

26. $y = e^{2x-1} \ln(2x - 1)$

27. $y = \dfrac{e^x}{\ln x}$

28. $p(y) = \dfrac{\ln y}{e^y}$

29. $g(z) = (e^{2z} + \ln z)^3$

30. $s(t) = \sqrt{e^{-t} + \ln 2t}$

31. $y = \log(6x)$

32. $y = \log(4x - 3)$

33. $y = \log |1 - x|$

34. $y = \log |3x|$

35. $y = \log_5 \sqrt{5x + 2}$

36. $y = \log_7 \sqrt{4x - 3}$

37. $y = \log_3(x^2 + 2x)^{3/2}$

38. $y = \log_2(2x^2 - x)^{5/2}$

39. $w = \log_8(2^p - 1)$

40. $z = 10^y \log y$

41. $f(x) = e^{\sqrt{x}} \ln(\sqrt{x} + 5)$

42. $f(x) = \ln(xe^{\sqrt{x}} + 2)$

43. $f(t) = \dfrac{\ln(t^2 + 1) + t}{\ln(t^2 + 1) + 1}$

44. $f(t) = \dfrac{2t^{3/2}}{\ln(2t^{3/2} + 1)}$

45. Why do we use the absolute value of x or of $g(x)$ in the derivative formulas for the natural logarithm?

46. Prove $\dfrac{d}{dx} \ln |ax| = \dfrac{d}{dx} \ln |x|$ for any constant a.

47. A friend concludes that because $y = \ln 6x$ and $y = \ln x$ have the same derivative, namely $dy/dx = 1/x$, these two functions must be the same. Explain why this is incorrect.

48. Use a graphing calculator to sketch the graph of $y = [f(x + h) - f(x)]/h$ using $f(x) = \ln |x|$ and $h = 0.0001$. Compare it with the graph of $y = 1/x$ and discuss what you observe.

49. Using the fact that

$$\ln[u(x)v(x)] = \ln u(x) + \ln v(x),$$

use the chain rule and the formula for the derivative of $\ln x$ to derive the product rule. In other words, find $[u(x)v(x)]'$ without assuming the product rule.

50. Using the fact that

$$\ln \dfrac{u(x)}{v(x)} = \ln u(x) - \ln v(x),$$

use the chain rule and the formula for the derivative of $\ln x$ to derive the quotient rule. In other words, find $[u(x)/v(x)]'$ without assuming the quotient rule.

51. Use graphical differentiation to verify that $\dfrac{d}{dx}(\ln x) = \dfrac{1}{x}$.

52. Use the fact that $d \ln x/dx = 1/x$, as well as the change-of-base theorem for logarithms, to prove that

$$\dfrac{d \log_a x}{dx} = \dfrac{1}{x \ln a}.$$

53. Let

$$h(x) = u(x)^{v(x)}.$$

(a) Using the fact that

$$\ln[u(x)^{v(x)}] = v(x) \ln u(x),$$

use the chain rule, the product rule, and the formula for the derivative of $\ln x$ to show that

$$\dfrac{d}{dx} \ln h(x) = \dfrac{v(x)u'(x)}{u(x)} + (\ln u(x)) v'(x).$$

(b) Use the result from part (a) and the fact that

$$\dfrac{d}{dx} \ln h(x) = \dfrac{h'(x)}{h(x)}$$

to show that

$$\dfrac{d}{dx} h(x) = u(x)^{v(x)} \left[\dfrac{v(x)u'(x)}{u(x)} + (\ln u(x)) v'(x) \right].$$

The idea of taking the logarithm of a function before differentiating is known as logarithmic differentiation.

Use the ideas from Exercise 53 to find the derivative of each of the following functions.

54. $h(x) = x^x$

55. $h(x) = (x^2 + 1)^{5x}$

56. This exercise shows another way to derive the formula for the derivative of the natural logarithm function using the definition of the derivative.

(a) Using the definition of the derivative, show that

$$\dfrac{d(\ln x)}{dx} = \lim_{h \to 0} \ln \left(1 + \dfrac{h}{x} \right)^{1/h}.$$

(b) Eliminate h from the result in part (a) using the substitution $h = x/m$ to show that

$$\dfrac{d(\ln x)}{dx} = \lim_{m \to \infty} \ln \left[\left(1 + \dfrac{1}{m} \right)^m \right]^{1/x}.$$

(c) What property should the function g have that would yield

$$\lim_{m \to \infty} g(h(m)) = g\left(\lim_{m \to \infty} h(m) \right)?$$

(d) Assuming that the natural logarithm has the property in part (c), and using a result about $\lim_{m \to \infty}(1+1/m)^m$ from Section 2.4, show that

$$\dfrac{d(\ln x)}{dx} = \ln e^{1/x} = \dfrac{1}{x}.$$

APPLICATIONS

Business and Economics

57. Revenue Suppose the demand function for q units of a certain item is

$$p = D(q) = 100 + \frac{50}{\ln q}, \quad q > 1,$$

where p is in dollars.

(a) Find the marginal revenue.

(b) Approximate the revenue from one more unit when 8 units are sold.

(c) How might a manager use the information from part (b)?

58. Profit If the cost function in dollars for q units of the item in Exercise 57 is $C(q) = 100q + 100$, find the following.

(a) The marginal cost

(b) The profit function $P(q)$

(c) The approximate profit from one more unit when 8 units are sold

(d) How might a manager use the information from part (c)?

59. Marginal Average Cost Suppose the cost in dollars to make x oboe reeds is given by

$$C(x) = 5 \log_2 x + 10.$$

Find the marginal average cost when the following numbers of reeds are sold.

(a) 10   (b) 20

60. Profit If the total revenue received from the sale of x items is given by

$$R(x) = 30 \ln(2x + 1),$$

while the total cost to produce x items is $C(x) = x/2$, find the following.

(a) The marginal revenue

(b) The profit function $P(x)$

(c) The marginal profit when $x = 60$

(d) Interpret the results of part (c).

Life Sciences

61. Bologna Sausage Scientists have developed a model to predict the growth of bacteria in bologna sausage at 32°C. The number of bacteria is given by

$$\ln\left(\frac{N(t)}{N_0}\right) = 9.8901e^{-e^{2.54197 - 0.2167t}},$$

where N_0 is the number of bacteria present at the beginning of the experiment and $N(t)$ is the number of bacteria present

at time t (in hours). *Source: Applied and Environmental Microbiology*.

(a) Use the properties of logarithms to find an expression for $N(t)$. Assume that $N_0 = 1000$.

(b) Use a graphing calculator to estimate the derivative of $N(t)$ when $t = 20$ and interpret.

(c) Let $S(t) = \ln(N(t)/N_0)$. Graph $S(t)$ on $[0, 35]$ by $[0, 12]$.

(d) Graph $N(t)$ on $[0, 35]$ by $[0, 20,000,000]$ and compare the graphs from parts (c) and (d).

(e) Find $\lim_{t \to \infty} S(t)$ and then use this limit to find $\lim_{t \to \infty} N(t)$.

62. Body Surface Area There is a mathematical relationship between an infant's weight and total body surface area (BSA), given by

$$A(w) = 4.688w^{0.8168 - 0.0154 \log w},$$

where w is the weight (in grams) and $A(w)$ is the BSA in square centimeters. *Source: British Journal of Cancer*.

(a) Find the BSA for an infant who weighs 4000 g.

(b) Find $A'(4000)$ and interpret your answer.

(c) Use a graphing calculator to graph $A(w)$ on $[2000, 10,000]$ by $[0, 6000]$.

63. Fruit Flies A study of the relation between the rate of reproduction in *Drosophila* (fruit flies) bred in bottles and the density of the mated population found that the number of imagoes (sexually mature adults) per mated female per day (y) can be approximated by

$$\log y = 1.54 - 0.008x - 0.658 \log x,$$

where x is the mean density of the mated population (measured as flies per bottle) over a 16-day period. *Source: Elements of Mathematical Biology*.

(a) Show that the above equation is equivalent to

$$y = 34.7(1.0186)^{-x}x^{-0.658}.$$

(b) Using your answer from part (a), find the number of imagoes per mated female per day when the density is

(i) 20 flies per bottle;

(ii) 40 flies per bottle.

(c) Using your answer from part (a), find the rate of change in the number of imagoes per mated female per day with respect to the density when the density is

(i) 20 flies per bottle;   (ii) 40 flies per bottle.

64. Pronghorn Fawns The field metabolic rate (FMR), or the total energy expenditure per day in excess of growth, can be calculated for pronghorn fawns using Nagy's formula,

$$F(x) = 0.774 + 0.727 \log x,$$

where x is the mass (in grams) of the fawn and $F(x)$ is the energy expenditure (in kJ/day). *Source: Animal Behavior*.

(a) Determine the total energy expenditure per day in excess of growth for a pronghorn fawn that weighs 25,000 g.

(b) Find $F'(25,000)$ and interpret the result.

(c) Graph the function on $[5000, 30,000]$ by $[3, 5]$.

65. Population Growth Suppose that the population of a certain collection of rare Brazilian ants is given by

$$P(t) = (t + 100) \ln(t + 2),$$

where t represents the time in days. Find the rates of change of the population on the second day and on the eighth day.

Social Sciences
66. Poverty The passage of the Social Security Amendments of 1965 resulted in the creation of the Medicare and Medicaid programs. Since then, the percent of persons 65 years and over with family income below the poverty level has declined. The percent can be approximated by the following function:

$$P(t) = 30.60 - 5.79 \ln t,$$

where t is the number of years since 1965. Find the percent of persons 65 years and over with family income below the poverty level and the rate of change in the following years. *Source: U.S. Census*.

(a) 1970   **(b)** 1990   **(c)** 2014

(d) What happens to the rate of change over time?

Physical Sciences
67. Richter Scale The Richter scale provides a measure of the magnitude of an earthquake. In fact, the largest Richter number M ever recorded for an earthquake was 8.9 from the 1933 earthquake in Japan. The following formula shows a relationship between the amount of energy released and the Richter number.

$$M = \frac{2}{3} \log \frac{E}{0.007},$$

where E is measured in kilowatt-hours. *Source: The Mathematics Teacher*.

(a) For the 1933 earthquake in Japan, what value of E gives a Richter number $M = 8.9$?

(b) If the average household uses 247 kWh per month, how many months would the energy released by an earthquake of this magnitude power 10 million households?

(c) Find the rate of change of the Richter number M with respect to energy when $E = 70,000$ kWh.

(d) What happens to dM/dE as E increases?

General Interest
68. Street Crossing Consider a child waiting at a street corner for a gap in traffic that is large enough so that he can safely cross the street. A mathematical model for traffic shows that if the expected waiting time for the child is to be at most 1 minute, then the maximum traffic flow, in cars per hour, is given by

$$f(x) = \frac{29,000(2.322 - \log x)}{x},$$

where x is the width of the street in feet. Find the maximum traffic flow and the rate of change of the maximum traffic flow with respect to street width for the following values of the street width. *Source: An Introduction to Mathematical Modeling*.

(a) 30 ft   **(b)** 40 ft

YOUR TURN ANSWERS

1. $f'(x) = \dfrac{1}{(\ln 3)x}$

2. (a) $\dfrac{dy}{dx} = \dfrac{6x^2}{2x^3 - 3}$ **(b)** $f'(x) = \dfrac{5 + 9x^2}{(\ln 4)(5x + 3x^3)}$

3. (a) $\dfrac{dy}{dx} = \dfrac{1}{x + 3}$ **(b)** $f'(x) = x + 2x \ln 3x$

    (c) $s'(t) = \dfrac{2t - (t - 1)\ln(t^2 - 1)}{(t - 1)(t + 1)^2}$

4 CHAPTER REVIEW

SUMMARY

In this chapter we used the definition of the derivative to develop techniques for finding derivatives of several types of functions. With the help of the rules that were developed, such as the power rule, product rule, quotient rule, and chain rule, we can now directly compute the derivative of a large variety of functions. In particular, we developed rules for finding derivatives of exponential and logarithmic functions. We also began to see the wide range of applications that these functions have in business, life sciences, social sciences, and the physical sciences. In the next chapter we will apply these techniques to study the behavior of certain functions, and we will learn that differentiation can be used to find maximum and minimum values of continuous functions.

Assume all indicated derivatives exist.

Constant Function   If $f(x) = k$, where k is any real number, then $f'(x) = \dfrac{d}{dx}[k] = 0$.

Power Rule   If $f(x) = x^n$, for any real number n, then $f'(x) = \dfrac{d}{dx}[x^n] = nx^{n-1}$.

Constant Times a Function   Let k be a real number. Then the derivative of $f(x) = k \cdot g(x)$ is $f'(x) = \dfrac{d}{dx}[k \cdot g(x)] = k \cdot g'(x)$.

Sum or Difference Rule   If $f(x) = u(x) \pm v(x)$, then $f'(x) = \dfrac{d}{dx}[u(x) \pm v(x)] = u'(x) \pm v'(x)$.

Product Rule   If $f(x) = u(x) \cdot v(x)$, then

$$f'(x) = \frac{d}{dx}[u(x) \cdot v(x)] = u(x) \cdot v'(x) + v(x) \cdot u'(x).$$

Quotient Rule   If $f(x) = \dfrac{u(x)}{v(x)}$, then

$$f'(x) = \frac{d}{dx}\left[\frac{u(x)}{v(x)}\right] = \frac{v(x) \cdot u'(x) - u(x) \cdot v'(x)}{[v(x)]^2}.$$

Chain Rule   If y is a function of u, say $y = f(u)$, and if u is a function of x, say $u = g(x)$, then $y = f(u) = f[g(x)]$, and

$$\frac{dy}{dx} = \frac{dy}{du} \cdot \frac{du}{dx}.$$

Chain Rule (Alternative Form)   Let $y = f[g(x)]$. Then $\dfrac{dy}{dx} = \dfrac{d}{dx}[f[g(x)]] = f'[g(x)] \cdot g'(x)$.

Exponential Functions $\dfrac{d}{dx}(e^x) = e^x$ $\dfrac{d}{dx}(a^x) = (\ln a)a^x$

$\dfrac{d}{dx}[e^{g(x)}] = e^{g(x)}g'(x)$ $\dfrac{d}{dx}[a^{g(x)}] = (\ln a)a^{g(x)}g'(x)$

Logarithmic Functions $\dfrac{d}{dx}[\ln |x|] = \dfrac{1}{x}$ $\dfrac{d}{dx}[\log_a |x|] = \dfrac{1}{(\ln a)x}$

$\dfrac{d}{dx}[\ln |g(x)|] = \dfrac{g'(x)}{g(x)}$ $\dfrac{d}{dx}[\log_a |g(x)|] = \dfrac{1}{\ln a} \cdot \dfrac{g'(x)}{g(x)}$

KEY TERMS

4.1
marginal analysis

4.2
marginal average cost

4.3
composite function
composition
chain rule

4.4
logistic function

REVIEW EXERCISES

CONCEPT CHECK

Determine whether each of the following statements is true or false, and explain why.

1. The derivative of π^3 is $3\pi^2$.

2. The derivative of a sum is the sum of the derivatives.

3. The derivative of a product is the product of the derivatives.

4. The marginal cost function is the derivative of the cost function.

5. The chain rule is used to take the derivative of a product of functions.

6. The only function that is its own derivative is e^x.

7. The derivative of 10^x is $x10^{x-1}$.

8. The derivative of $\ln |x|$ is the same as the derivative of $\ln x$.

9. The derivative of $\ln kx$ is the same as the derivative of $\ln x$.

10. The derivative of $\log x$ is the same as the derivative of $\ln x$.

PRACTICE AND EXPLORATIONS

Use the rules for derivatives to find the derivative of each function defined as follows.

11. $y = 5x^3 - 7x^2 - 9x + \sqrt{5}$

12. $y = 7x^3 - 4x^2 - 5x + \sqrt{2}$

13. $y = 9x^{8/3}$

14. $y = -4x^{-3}$

15. $f(x) = 3x^{-4} + 6\sqrt{x}$

16. $f(x) = 19x^{-1} - 8\sqrt{x}$

17. $k(x) = \dfrac{3x}{4x + 7}$

18. $r(x) = \dfrac{-8x}{2x + 1}$

19. $y = \dfrac{x^2 - x + 1}{x - 1}$

20. $y = \dfrac{2x^3 - 5x^2}{x + 2}$

21. $f(x) = (3x^2 - 2)^4$

22. $k(x) = (5x^3 - 1)^6$

23. $y = \sqrt{2t^7 - 5}$

24. $y = -3\sqrt{8t^4 - 1}$

25. $y = 3x(2x + 1)^3$

26. $y = 4x^2(3x - 2)^5$

27. $r(t) = \dfrac{5t^2 - 7t}{(3t + 1)^3}$

28. $s(t) = \dfrac{t^3 - 2t}{(4t - 3)^4}$

29. $p(t) = t^2(t^2 + 1)^{5/2}$

30. $g(t) = t^3(t^4 + 5)^{7/2}$

31. $y = -6e^{2x}$

32. $y = 8e^{0.5x}$

33. $y = e^{-2x^3}$

34. $y = -4e^{x^2}$

35. $y = 5xe^{2x}$

36. $y = -7x^2e^{-3x}$

37. $y = \ln(2 + x^2)$

38. $y = \ln(5x + 3)$

39. $y = \dfrac{\ln|3x|}{x - 3}$

40. $y = \dfrac{\ln|2x - 1|}{x + 3}$

41. $y = \dfrac{xe^x}{\ln(x^2 - 1)}$

42. $y = \dfrac{(x^2 + 1)e^{2x}}{\ln x}$

43. $s = (t^2 + e^t)^2$

44. $q = (e^{2p+1} - 2)^4$

45. $y = 3 \cdot 10^{-x^2}$

46. $y = 10 \cdot 2^{\sqrt{x}}$

47. $g(z) = \log_2(z^3 + z + 1)$

48. $h(z) = \log(1 + e^z)$

49. $f(x) = e^{2x}\ln(xe^x + 1)$

50. $f(x) = \dfrac{e^{\sqrt{x}}}{\ln(\sqrt{x} + 1)}$

Consider the following table of values of the functions f and g and their derivatives at various points.

x	1	2	3	4
$f(x)$	3	4	2	1
$f'(x)$	-5	-6	-7	-11
$g(x)$	4	1	2	3
$g'(x)$	2/9	3/10	4/11	6/13

Find the following using the table.

51. (a) $D_x(f[g(x)])$ at $x = 2$ **(b)** $D_x(f[g(x)])$ at $x = 3$

52. (a) $D_x(g[f(x)])$ at $x = 2$ **(b)** $D_x(g[f(x)])$ at $x = 3$

Find the slope of the tangent line to the given curve at the given value of x. Find the equation of each tangent line.

53. $y = x^2 - 6x;\quad x = 2$

54. $y = 8 - x^2;\quad x = 1$

55. $y = \dfrac{3}{x - 1};\quad x = -1$

56. $y = \dfrac{x}{x^2 - 1};\quad x = 2$

57. $y = \sqrt{6x - 2};\quad x = 3$

58. $y = -\sqrt{8x + 1};\quad x = 3$

59. $y = e^x;\quad x = 0$

60. $y = xe^x;\quad x = 1$

61. $y = \ln x;\quad x = 1$

62. $y = x \ln x;\quad x = e$

63. Consider the graphs of the function $y = \sqrt{2x - 1}$ and the straight line $y = x + k$. Discuss the number of points of intersection versus the change in the value of k. *Source: Japanese University Entrance Examination Problems in Mathematics.*

64. (a) Verify that

$$\frac{d \ln f(x)}{dx} = \frac{f'(x)}{f(x)}.$$

This expression is called the *relative rate of change*. It expresses the rate of change of f relative to the size of f. Stephen B. Maurer denotes this expression by $\hat{f}$ and notes that economists commonly work with relative rates of change. *Source: The College Mathematics Journal.*

(b) Verify that

$$\widehat{fg} = \hat{f} + \hat{g}.$$

Interpret this equation in terms of relative rates of change.

(c) In his article, Maurer uses the result of part (b) to solve the following problem:

"Last year, the population grew by 1% and the average income per person grew by 2%. By what approximate percent did the national income grow?"

Explain why the result from part (b) implies that the answer to this question is approximately 3%.

65. Suppose that the student body in your college grows by 2% and the tuition goes up by 3%. Use the result from the previous exercise to calculate the approximate amount that the total tuition collected goes up, and compare this with the actual amount.

66. Why is e a convenient base for exponential and logarithmic functions?

67. In Exercise 42 in Section 4.2, we saw an example for which the erroneous version of the product rule, $[f(x)g(x)]' = f'(x)g'(x)$, *does* hold. As a more general version of this example, calculate $[f(x)g(x)]'$ and $f'(x)g'(x)$ for $f(x) = a/(n - x)^n$ and $g(x) = bx^n$ for $n \neq 0$, and verify that they are equal. *Source: PRIMUS.*

68. Just as there are times when the erroneous version of the product rule holds (see the previous exercise), there are times when the erroneous version of the quotient rule holds. Calculate $[f(x)/g(x)]'$ and $f'(x)/g'(x)$ for $f(x) = ae^{k^2x/(k-1)}$ and $g(x) = be^{kx}$ for $k \neq 1$, and verify that they are equal. *Source: PRIMUS.*

APPLICATIONS

Business and Economics

Marginal Average Cost   Find the marginal average cost function of each function defined as follows.

69. $C(x) = \sqrt{x + 1}$

70. $C(x) = \sqrt{3x + 2}$

71. $C(x) = (x^2 + 3)^3$

72. $C(x) = (4x + 3)^4$

73. $C(x) = 10 - e^{-x}$

74. $C(x) = \ln(x + 5)$

75. Sales   The sales of a company are related to its expenditures on research by

$$S(x) = 1000 + 60\sqrt{x} + 12x,$$

where $S(x)$ gives sales in millions when x thousand dollars is spent on research. Find and interpret dS/dx if the following amounts are spent on research.

(a) $9000   (b) $16,000   (c) $25,000

(d) As the amount spent on research increases, what happens to sales?

76. Profit   Suppose that the profit (in hundreds of dollars) from selling x units of a product is given by

$$P(x) = \frac{x^2}{2x + 1}.$$

Find and interpret the marginal profit when the following numbers of units are sold.

(a) 4   (b) 12   (c) 20

(d) What is happening to the marginal profit as the number sold increases?

(e) Find and interpret the marginal average profit when 4 units are sold.

77. Costs   A company finds that its total costs are related to the amount spent on training programs by

$$T(x) = \frac{1000 + 60x}{4x + 5},$$

where $T(x)$ is costs in thousands of dollars when x hundred dollars are spent on training. Find and interpret $T'(x)$ if the following amounts are spent on training.

(a) $900   (b) $1900

(c) Are costs per dollar spent on training always increasing or decreasing?

78. Compound Interest   If a sum of $1000 is deposited into an account that pays $r\%$ interest compounded quarterly, the balance after 12 years is given by

$$A = 1000\left(1 + \frac{r}{400}\right)^{48}.$$

Find and interpret $\dfrac{dA}{dr}$ when $r = 5$.

79. Continuous Compounding   If a sum of $1000 is deposited into an account that pays $r\%$ interest compounded continuously, the balance after 12 years is given by

$$A = 1000e^{12r/100}.$$

Find and interpret $\dfrac{dA}{dr}$ when $r = 5$.

80. Doubling Time   If a sum of money is deposited into an account that pays $r\%$ interest compounded annually, the doubling time (in years) is given by

$$T = \frac{\ln 2}{\ln(1 + r/100)}.$$

Find and interpret dT/dr when $r = 5$.

81. U.S. Post Office   The number (in billions) of pieces of mail handled by the U.S. Post Office each year from 2004 through 2013 can be approximated by

$$P(t) = -0.01741t^4 + 0.6790t^3 - 7.141t^2 + 16.95t + 204.5,$$

where t is the number of years since 2004. Find and interpret the rate of change in the volume of mail for the following years. *Source: U.S. Postal Service.*

(a) 2005

(b) 2012

82. Elderly Employment   After declining over the last century, the percentage of men aged 65 and older in the workforce has begun to rise in recent years, as shown by the following table. (The last part of this data was modeled with a quadratic function in Section 2.4.) *Source: The World Almanac and Book of Facts 2014.*

Year	Percent of Men 65 and Older in Workforce
1900	63.1
1920	55.6
1930	54.0
1940	41.8
1950	45.8
1960	33.1
1970	26.8
1980	19.0
1990	16.3
2000	17.7
2010	20.5

(a) Using the regression feature on a graphing calculator, find a cubic and a quartic function that model this data, letting $t = 0$ correspond to the year 1900.

(b) Using each of your answers to part (a), find the rate that the percent of men aged 65 and older in the workforce was increasing in 2005.

(c) Discuss which model from part (a) best describes the data, as well as which answer from part (b) best describes the rate that the percent of men aged 65 and older in the workforce was increasing in 2005.

(d) Explore other functions that could be used to model the data, using the various regression features on a graphing calculator, and discuss to what extent any of them are useful descriptions of the data.

83. **Value of the Dollar** The U.S. dollar has been declining in value over the last century, except during the Great Depression, when it increased in value. The following table shows the number of dollars it took in various years to equal $1 in 1913. *Source: U.S. Bureau of Labor Statistics.*

Year	Number of Dollars It Took to Equal $1 in 1913
1913	1.00
1920	2.02
1930	1.69
1940	1.41
1950	2.53
1960	2.99
1970	3.92
1980	8.32
1990	13.20
2000	17.39
2010	22.02

(a) Using the regression feature on a graphing calculator, find a cubic and a quartic function that model these data, letting $t = 0$ correspond to the year 1900.

(b) Using each of your answers to part (a), find the rate that the number of dollars required to equal $1 in 1913 was increasing in 2005.

(c) Discuss which model from part (a) best describes the data, as well as which answer from part (b) best describes the rate that the number of dollars required to equal $1 in 1913 was increasing in 2005.

(d) Explore other functions that could be used to model the data, using the various regression features on a graphing calculator, and discuss to what extent any of them are useful descriptions of the data.

Life Sciences

84. **Exponential Growth** Suppose a population is growing exponentially with an annual growth constant $k = 0.05$. How fast is the population growing when it is 1,000,000? Use the derivative to calculate your answer, and then explain how the answer can be obtained without using the derivative.

85. **Logistic Growth** Suppose a population is growing logistically with $k = 5 \times 10^{-6}$, $m = 30,000$, and $G_0 = 2000$. Assume time is measured in years.

(a) Find the growth function $G(t)$ for this population.

(b) Find the population and rate of growth of the population after 6 years.

86. **Fish** The length of the monkeyface prickleback, a West Coast game fish, can be approximated by

$$L = 71.5(1 - e^{-0.1t})$$

and the weight by

$$W = 0.01289 \cdot L^{2.9},$$

where L is the length in centimeters, t is the age in years, and W is the weight in grams. *Source: California Fish and Game.*

(a) Find the approximate length of a 5-year-old monkeyface.

(b) Find how fast the length of a 5-year-old monkeyface is growing.

(c) Find the approximate weight of a 5-year-old monkeyface. (*Hint:* Use your answer from part (a).)

(d) Find the rate of change of the weight with respect to length for a 5-year-old monkeyface.

(e) Using the chain rule and your answers to parts (b) and (d), find how fast the weight of a 5-year-old monkeyface is growing.

87. **Arctic Foxes** The age/weight relationship of male Arctic foxes caught in Svalbard, Norway, can be estimated by the function

$$M(t) = 3583e^{-e^{-0.020(t-66)}},$$

where t is the age of the fox in days and $M(t)$ is the weight of the fox in grams. *Source: Journal of Mammology.*

(a) Estimate the weight of a male fox that is 250 days old.

(b) Use $M(t)$ to estimate the largest size that a male fox can attain. (*Hint:* Find $\lim_{t \to \infty} M(t)$.)

(c) Estimate the age of a male fox when it has reached 50% of its maximum weight.

(d) Estimate the rate of change in weight of a male Arctic fox that is 250 days old. (*Hint:* Recall that $D_t e^{f(t)} = f'(t)e^{f(t)}$.)

(e) Use a graphing calculator to graph $M(t)$ and then describe the growth pattern.

(f) Use the table function on a graphing calculator or a spreadsheet to develop a chart that shows the estimated weight and growth rate of male foxes for days 50, 100, 150, 200, 250, and 300.

88. **Hispanic Population** In Section 2.4, Exercise 51, we found that the projected Asian population in the United States, in millions, can be approximated by

$$a(t) = 11.14(1.023)^t,$$

where t is the years since 2000. Find the instantaneous rate of change in the projected Asian population in the United States in each of the following years. *Source: U.S. Census.*

(a) 2005

(b) 2025

Physical Sciences

89. Wind Energy In Section 2.4, Exercise 55, we found that the total world wind energy capacity (in megawatts) in recent years could be approximated by the function

$$C(t) = 19{,}259.86(1.2585)^t,$$

where t is the number of years since 2000. Find the rate of change in the energy capacity for the following years. *Source: World Wind Energy Association.*

(a) 2005   (b) 2010   (c) 2015

General Interest

90. Cats The distance from Lisa Wunderle's cat, Belmar, to a piece of string he is stalking is given in feet by

$$f(t) = \frac{8}{t+1} + \frac{20}{t^2+1},$$

where t is the time in seconds since he begins.

(a) Find Belmar's average velocity between 1 second and 3 seconds.

(b) Find Belmar's instantaneous velocity at 3 seconds.

91. Food Surplus In Section 2.4, Example 7, we found that the production of corn (in billions of bushels) in the United States since 1930 could be approximated by

$$p(t) = 1.757(1.0248)^t,$$

where t is the number of years since 1930. Find and interpret $p'(2000)$.

92. Dating a Language Over time, the number of original basic words in a language tends to decrease as words become obsolete or are replaced with new words. Linguists have used calculus to study this phenomenon and have developed a methodology for dating a language, called *glottochronology*. Experiments have indicated that a good estimate of the number of words that remain in use at a given time is given by

$$N(t) = N_0 e^{-0.217t},$$

where $N(t)$ is the number of words in a particular language, t is measured in the number of millennium, and N_0 is the original number of words in the language. *Source: The UMAP Journal.*

(a) In 1950, C. Feng and M. Swadesh established that of the original 210 basic ancient Chinese words from 950 A.D., 167 were still being used. Letting $t = 0$ correspond to 950, with $N_0 = 210$, find the number of words predicted to have been in use in 1950 A.D., and compare it with the actual number in use.

(b) Estimate the number of words that will remain in the year 2050 ($t = 1.1$).

(c) Find $N'(1.1)$ and interpret your answer.

93. Driving Fatalities A study by the National Highway Traffic Safety Administration found that driver fatalities rates were highest for the youngest and oldest drivers. The rates per 1000 licensed drivers for every 100 million miles may be approximated by the function

$$f(x) = k(x - 49)^6 + 0.8,$$

where x is the driver's age in years and k is the constant 3.8×10^{-9}. Find and interpret the rate of change of the fatality rate when the driver is

(a) 20 years old;   (b) 60 years old.

Source: National Highway Traffic Safety Administration.

EXTENDED APPLICATION

ELECTRIC POTENTIAL AND ELECTRIC FIELD

In physics, a major area of study is electricity, including such concepts as electric charge, electric force, and electric current. Two ideas that physicists use a great deal are electric potential and electric field. Electric potential is the same as voltage, such as for a battery. Electric field can be thought of in terms of a force field, which is often referred to in space movies as a deflector shield around a spaceship. An electric field causes an electric force to act on charged objects when they are in the electric field.

Both electric potential and electric field are produced by electric charges. It is important to physicists and electrical engineers to know what the electric field and the electric potential are near a charged object. Usually the problem involves finding the electric potential and taking the (negative) derivative of the electric potential to determine the electric field. More explicitly,

$$E = -\frac{dV}{dz}, \qquad (1)$$

where V is the electric potential (voltage) near the charged object, z is the distance from the object, and E is the electric field.

Let's look at an example. Suppose we have a charged disk of radius R and we want to determine the electric potential and electric field at a distance z along the axis of the disk (Figure 14).

FIGURE 14

Radius of Planet DISK = 10,000 m

FIGURE 15

Using some basic definitions and integral calculus (that you will learn about later in your calculus book), it can be shown that the electric potential on the axis is

$$V = k_1 \left(\sqrt{z^2 + R^2} - z \right), \tag{2}$$

where k_1 is a constant.* To determine the electric field, we apply Equation (1) and find

$$E = -k_1 \left(\frac{z}{\sqrt{z^2 + R^2}} - 1 \right). \tag{3}$$

You will provide the details for this example by working through Exercise 1 at the end of this section.

Physicists often like to see if complicated expressions such as these can be simplified when certain conditions apply. An example would be to imagine that the location z is very far away from the disk. In that case, the disk doesn't look much like a disk anymore, but more like a point. In Equations (2) and (3), we see that z will be much larger than R. If you reach the topic of Taylor series in calculus, you will learn how these series can be used to approximate square roots, so that for large values of z the voltage is inversely proportional to z and looks like this:

$$V = \frac{k_2}{z}, \tag{4}$$

where k_2 is a constant. To determine the electric field, we apply Equation (1) again to find

$$E = \frac{k_2}{z^2}. \tag{5}$$

You will be asked to prove this result in Exercise 2.

Thus we see that the exact functions for V and E of the disk become much simpler at locations far away from the disk. The voltage is a reciprocal function and the electric field is called an

*For those interested in the constants k_1 and k_2:

$$k_1 = \frac{Q}{2\pi\epsilon_0 R^2} \quad \text{and} \quad k_2 = \frac{1}{2}k_1 R^2 = \frac{Q}{4\pi\epsilon_0},$$

where Q is the charge, ϵ_0 is called the electric permittivity, and R is the radius of the disk.

inverse-square law. By the way, these functions are the same ones that would be used for a point charge, which is a charge that takes up very little space.

Now let's look at what happens when we are very close to the surface of the disk. You could imagine that an observer very close to the surface would see the disk as a large flat plane. If we apply the Taylor series once more to the exact function for the potential (Equation (2)), but this time with z much smaller than R, we find

$$V = k_1 \left(R - z + \frac{z^2}{2R} \right). \tag{6}$$

Notice that this is a quadratic function, or a parabola. Applying Equation (1) again, we see that the electric field is

$$E = k_1 \left(1 - \frac{z}{R} \right). \tag{7}$$

This is just a linear function that increases as we approach the surface (which would be $z = 0$) and decreases the farther away from the surface we move. Again, the details of this calculation are left for you to do in Exercise 3.

Here is a fun way of making sense of the equations listed above. Suppose you are in a spaceship approaching the planet DISK on a very important mission. Planet DISK is noted for the fact that there is always a sizeable amount of charge on it. You are approaching the planet from very far away along its axis. In Figure 15 the three voltage functions (Equations (2), (4), and (6)) are plotted, and in Figure 16 the three electric field functions (Equations (3), (5), and (7)) are plotted. The graphs were generated using $R = 10,000$, $k_1 = 100$, and $k_2 = 5 \times 10^9$.

Notice that it looks as though you can use the reciprocal function for the voltage (Equation (4)) and the inverse-square law function for the electric field (Equation (5)) when you are farther away than about 20,000 m because the exact functions and the approximate functions are almost exactly the same. Also, when you get close to the planet, say less than about 4000 m, you can use the quadratic function for the voltage (Equation (6)) and the linear function for the electric field (Equation (7)) because the exact functions and the approximate functions are nearly the same. This means that you should use the exact functions (Equations (2) and (3)) between about 20,000 m and about 4000 m because the other functions deviate substantially in that region.

Radius of Planet DISK = 10,000 m

Legend:
- Close to surface
- Exact function
- Far from planet

y-axis: Electric field (V/m)
x-axis: Distance from planet along z axis (m)

FIGURE 16

There are many other examples that could be studied, but they involve other functions that you haven't covered yet, especially trigonometric functions. But the process is still the same: If you can determine the electric potential in the region around a charged object, then the electric field is found by taking the negative derivative of the electric potential.

EXERCISES

1. Use Equation (1) to prove that the electric field of the disk (Equation (3)) is obtained from the voltage of the disk (Equation (2)). (*Hint:* It may help to write the square root in Equation (2) as a power.)

2. Apply Equation (1) to the voltage of a point charge (Equation (4)) to obtain the electric field of a point charge (Equation (5)).

3. Show that the electric field in Equation (7) results from the electric potential in Equation (6).

4. Sometimes for z very, very close to the disk, the third term in Equation (6) is so small that it can be dismissed. Show that the electric field is constant for this case.

5. Use a graphing calculator or Wolfram|Alpha (which can be found at www.wolframalpha.com) to recreate the graphs of the functions in Figures 15 and 16.

6. Use a spreadsheet to create a table of values for the functions displayed in Figures 15 and 16. Compare the three voltage functions and then compare the three electric field functions.

DIRECTIONS FOR GROUP PROJECT

Determine the electric potential and the electric field at various locations along the axis of a charged compact disc (CD). You will need to measure the radius of a typical CD and use the value for the electric permittivity (sometimes called the permittivity of free space), $\epsilon_0 = 8.85 \times 10^{-12} C^2/Nm^2$. To estimate the value for the charge on the CD, assume that a typical CD has about one mole of atoms (6.0×10^{23}, which is Avogadro's number) and that one out of every billion of these atoms loses an electron. The charge is found by multiplying 10^{-9} (one billionth) by Avogadro's number and by the charge of one electron (or proton), which is 1.6×10^{-19} C. With this information you can calculate the constants k_1 and k_2. Use appropriate graphing software, such as Microsoft Excel, to plot all three of the voltage functions (Equations (2), (4), and (6)) on one graph and all three electric field functions (Equations (3), (5), and (7)) on one graph.

5

Graphs and the Derivative

Derivatives provide useful information about the behavior of functions and the shapes of their graphs. The first derivative describes the rate of increase or decrease, while the second derivative indicates how the rate of increase or decrease is changing. In an exercise at the end of this chapter, we will see what changes in the sign of the second derivative tell us about the shape of the graph that shows a weightlifter's performance as a function of age.

The graph in Figure 1 shows the relationship between the number of sleep-related accidents and traffic density during a 24-hour period. The blue line indicates the hourly distribution of sleep-related accidents. The green line indicates the hourly distribution of traffic density. The red line indicates the relative risk of sleep-related accidents. For example, the relative risk graph shows us that a person is nearly seven times as likely to have an accident at 4:00 A.M. than at 10:00 P.M. *Source: Sleep*.

Given a graph like the one in Figure 1, we can often locate maximum and minimum values simply by looking at the graph. It is difficult to get *exact* values or *exact* locations of maxima and minima from a graph, however, and many functions are difficult to graph without the aid of technology. In Chapter 2, we saw how to find exact maximum and minimum values for quadratic functions by identifying the vertex. A more general approach is to use the derivative of a function to determine precise maximum and minimum values of the function. The procedure for doing this is described in this chapter, which begins with a discussion of increasing and decreasing functions.

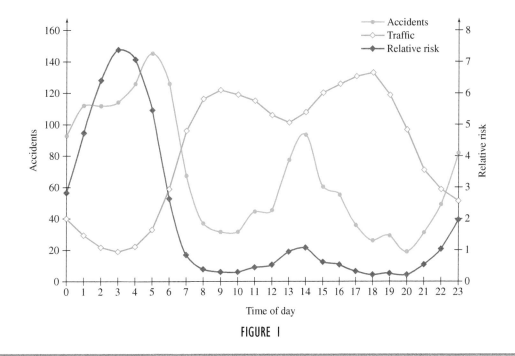

Time of day

FIGURE 1

5.1 Increasing and Decreasing Functions

APPLY IT   How long is it profitable to increase production?
We will answer this question in Example 5 after further investigating increasing and decreasing functions.

A function is *increasing* if the graph goes *up* from left to right and *decreasing* if its graph goes *down* from left to right. Examples of increasing functions are shown in Figures 2(a)–(c) on the following page, and examples of decreasing functions in Figures 2(d)–(f).

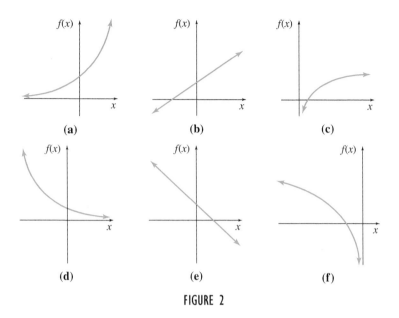

FIGURE 2

Increasing and Decreasing Functions

Let f be a function defined on some interval. Then for any two numbers x_1 and x_2 in the interval, f is **increasing** on the interval if

$$f(x_1) < f(x_2) \quad \text{whenever} \quad x_1 < x_2,$$

and f is **decreasing** on the interval if

$$f(x_1) > f(x_2) \quad \text{whenever} \quad x_1 < x_2.$$

EXAMPLE 1 Increasing and Decreasing

Where is the function graphed in Figure 3 increasing? Where is it decreasing?

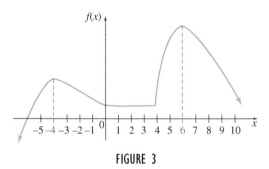

FIGURE 3

YOUR TURN 1 Find where the function is increasing and decreasing.

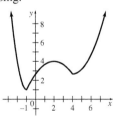

SOLUTION Moving from left to right, the function is increasing for x-values up to -4, then decreasing for x-values from -4 to 0, constant (neither increasing nor decreasing) for x-values from 0 to 4, increasing for x-values from 4 to 6, and decreasing for all x-values greater than 6. In interval notation, the function is increasing on $(-\infty, -4)$ and $(4, 6)$, decreasing on $(-4, 0)$ and $(6, \infty)$, and constant on $(0, 4)$. **TRY YOUR TURN 1**

How can we tell from the equation that defines a function where the graph increases and where it decreases? The derivative can be used to answer this question. Remember that the derivative of a function at a point gives the slope of the line tangent to the function at that point. Recall also that a line with a positive slope rises from left to right and a line with a negative slope falls from left to right.

The graph of a typical function, f, is shown in Figure 4. Think of the graph of f as a roller coaster track moving from left to right along the graph. Now, picture one of the cars on the roller coaster. As shown in Figure 5, when the car is on level ground or parallel to level ground, its floor is horizontal, but as the car moves up the slope, its floor tilts upward. When the car reaches a peak, its floor is again horizontal, but it then begins to tilt downward (very steeply) as the car rolls downhill. The floor of the car as it moves from left to right along the track represents the tangent line at each point. Using this analogy, we can see that the slope of the tangent line will be *positive* when the car travels uphill and f is *increasing*, and the slope of the tangent line will be *negative* when the car travels downhill and f *decreasing*. (In this case it is also true that the slope of the tangent line will be zero at "peaks" and "valleys.")

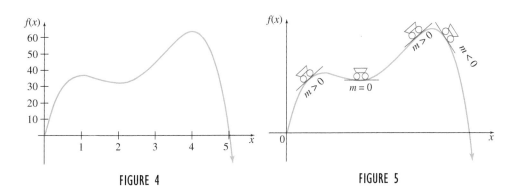

FIGURE 4   FIGURE 5

Thus, on intervals where $f'(x) > 0$, the function $f(x)$ will increase, and on intervals where $f'(x) < 0$, the function $f(x)$ will decrease. We can determine where $f(x)$ peaks by finding the intervals on which it increases and decreases.

Our discussion suggests the following test.

Test for Intervals Where $f(x)$ is Increasing and Decreasing

Suppose a function f has a derivative at each point in an open interval; then

if $f'(x) > 0$ for each x in the interval, f is *increasing* on the interval;   ↗

if $f'(x) < 0$ for each x in the interval, f is *decreasing* on the interval;   ↘

if $f'(x) = 0$ for each x in the interval, f is *constant* on the interval.   →

NOTE
The third condition must hold for an entire open interval, not a single point. It would not be correct to say that because $f'(x) = 0$ at a point, then $f(x)$ is constant at that point.

The derivative $f'(x)$ can change signs from positive to negative (or negative to positive) at points where $f'(x) = 0$ and at points where $f'(x)$ does not exist. The values of x where this occurs are called *critical numbers*.

Critical Numbers

The **critical numbers** for a function f are those numbers c in the domain of f for which $f'(c) = 0$ or $f'(c)$ does not exist. A **critical point** is a point whose x-coordinate is the critical number c and whose y-coordinate is $f(c)$.

It is shown in more advanced classes that if the critical numbers of a function are used to determine open intervals on a number line, then the sign of the derivative at any point in an interval will be the same as the sign of the derivative at any other point in the interval. This suggests that the test for increasing and decreasing functions be applied as follows (assuming that no open intervals exist where the function is constant).

Applying the Test for Increasing and Decreasing

1. Locate the critical numbers for f on a number line, as well as any points where f is undefined. These points determine several open intervals.
2. Choose a value of x in each of the intervals determined in Step 1. Use these values to determine whether $f'(x) > 0$ or $f'(x) < 0$ in that interval.
3. Use the test on the previous page to determine whether f is increasing or decreasing on the interval.

EXAMPLE 2   Increasing and Decreasing

Find the intervals in which the following function is increasing or decreasing. Locate all points where the tangent line is horizontal. Graph the function.

$$f(x) = x^3 + 3x^2 - 9x + 4$$

SOLUTION The derivative is $f'(x) = 3x^2 + 6x - 9$. To find the critical numbers, set this derivative equal to 0 and solve the resulting equation by factoring.

$$3x^2 + 6x - 9 = 0$$
$$3(x^2 + 2x - 3) = 0$$
$$3(x + 3)(x - 1) = 0$$
$$x = -3 \quad \text{or} \quad x = 1$$

The tangent line is horizontal at $x = -3$ or $x = 1$. Since there are no values of x where $f'(x)$ fails to exist, the only critical numbers are -3 and 1. To determine where the function is increasing or decreasing, locate -3 and 1 on a number line, as in Figure 6. (Be sure to place the values on the number line in numerical order.) These points determine three intervals: $(-\infty, -3)$, $(-3, 1)$, and $(1, \infty)$.

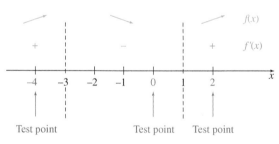

FIGURE 6

Now choose any value of x in the interval $(-\infty, -3)$. Choosing $x = -4$ and evaluating $f'(-4)$ using the factored form of $f'(x)$ gives

$$f'(-4) = 3(-4 + 3)(-4 - 1) = 3(-1)(-5) = 15,$$

which is positive. You could also substitute $x = -4$ in the unfactored form of $f'(x)$, but using the factored form makes it easier to see whether the result is positive or negative, depending upon whether you have an even or an odd number of negative factors. Since one value of x in this interval makes $f'(x) > 0$, all values will do so, and therefore, f is increasing on $(-\infty, -3)$. Selecting 0 from the middle interval gives $f'(0) = -9$, so f is decreasing on $(-3, 1)$. Finally, choosing 2 in the right-hand region gives $f'(2) = 15$, with f increasing on $(1, \infty)$. The arrows in each interval in Figure 6 indicate where f is increasing or decreasing.

$$f(x) = x^3 + 3x^2 - 9x + 4$$

FIGURE 7

YOUR TURN 2
Find the intervals in which
$f(x) = -x^3 - 2x^2 + 15x + 10$
is increasing or decreasing. Graph
the function.

We now have an additional tool for graphing functions: the test for determining where a function is increasing or decreasing. (Other tools are discussed in the next few sections.) To graph the function, plot a point at each of the critical numbers by finding $f(-3) = 31$ and $f(1) = -1$. Also plot points for $x = -4, 0,$ and 2, the test values of each interval. Use these points along with the information about where the function is increasing and decreasing to get the graph in Figure 7.   **TRY YOUR TURN 2**

CAUTION   Be careful to use $f(x)$, not $f'(x)$, to find the y-value of the points to plot.

Recall critical numbers are numbers c in the domain of f for which $f'(c) = 0$ or $f'(x)$ does not exist. In Example 2, there are no critical values, c, where $f'(c)$ fails to exist. The next example illustrates the case where a function has a critical number at c because the derivative does not exist at c.

EXAMPLE 3   **Increasing and Decreasing**

Find the critical numbers and decide where f is increasing and decreasing if $f(x) = (x - 1)^{2/3}$.

SOLUTION   We find $f'(x)$ first, using the power rule and the chain rule.

$$f'(x) = \frac{2}{3}(x - 1)^{-1/3}(1) = \frac{2}{3(x - 1)^{1/3}}$$

To find the critical numbers, we first find any values of x that make $f'(x) = 0$, but here $f'(x)$ is never 0. Next, we find any values of x where $f'(x)$ fails to exist. This occurs whenever the denominator of $f'(x)$ is 0, so set the denominator equal to 0 and solve.

$$3(x - 1)^{1/3} = 0 \qquad \text{Divide by 3.}$$
$$[(x - 1)^{1/3}]^3 = 0^3 \qquad \text{Raise both sides to the 3rd power.}$$
$$x - 1 = 0$$
$$x = 1$$

Since $f'(1)$ does not exist but $f(1)$ is defined, $x = 1$ is a critical number, the only critical number. This point divides the number line into two intervals: $(-\infty, 1)$ and $(1, \infty)$. Draw a number line for f', and use a test point in each of the intervals to find where f is increasing and decreasing.

$$f'(0) = \frac{2}{3(0 - 1)^{1/3}} = \frac{2}{-3} = -\frac{2}{3}$$

$$f'(2) = \frac{2}{3(2 - 1)^{1/3}} = \frac{2}{3}$$

Since f is defined for all x, these results show that f is decreasing on $(-\infty, 1)$ and increasing on $(1, \infty)$. The graph of f is shown in Figure 8. **TRY YOUR TURN 3**

YOUR TURN 3 Find where f is increasing and decreasing if $f(x) = (2x + 4)^{2/5}$. Graph the function.

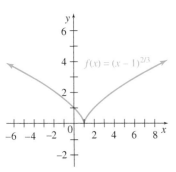

FIGURE 8

In Example 3, we found a critical number where $f'(x)$ failed to exist. This occurred when the denominator of $f'(x)$ was zero. Be on the alert for such values of x. Also be alert for values of x that would make the expression under a square root, or other even root, negative. For example, if $f(x) = \sqrt{x}$, then $f'(x) = 1/(2\sqrt{x})$. Notice that $f'(x)$ does not exist for $x \le 0$, but the values of $x < 0$ are not critical numbers because those values of x are not in the domain of f. The function $f(x) = \sqrt{x}$ does have a critical point at $x = 0$, because 0 is in the domain of f.

Sometimes a function may not have any critical numbers, but we are still able to determine where the function is increasing and decreasing, as shown in the next example.

EXAMPLE 4 Increasing and Decreasing (No Critical Numbers)

Find the intervals for which the following function increases and decreases. Graph the function.

$$f(x) = \frac{x - 1}{x + 1}.$$

SOLUTION Notice that the function f is undefined when $x = -1$, so -1 is not in the domain of f. To determine any critical numbers, first use the quotient rule to find $f'(x)$.

$$f'(x) = \frac{(x + 1)(1) - (x - 1)(1)}{(x + 1)^2}$$

$$= \frac{x + 1 - x + 1}{(x + 1)^2} = \frac{2}{(x + 1)^2}$$

This derivative is never 0, but it fails to exist at $x = -1$, where the function is undefined. Since -1 is not in the domain of f, there are no critical numbers for f.

We can still, however, apply the test for increasing and decreasing. The number -1 (where f is undefined) divides the number line into two intervals: $(-\infty, -1)$ and $(-1, \infty)$. Draw a number line for f', and use a test point in each of these intervals to find that $f'(x) > 0$ for all x except -1. (This can also be determined by observing that $f'(x)$ is the quotient of 2, which is positive, and $(x + 1)^2$, which is always positive when $x \ne -1$.) This means that the function f is increasing on both $(-\infty, -1)$ and $(-1, \infty)$.

To graph the function, we find any asymptotes. Since the value $x = -1$ makes the denominator 0 but not the numerator, the line $x = -1$ is a vertical asymptote. To find the horizontal asymptote, we find

$$\lim_{x \to \infty} \frac{x - 1}{x + 1} = \lim_{x \to \infty} \frac{1 - 1/x}{1 + 1/x}$$ Divide numerator and denominator by x.

$$= 1.$$

We get the same limit as x approaches $-\infty$, so the graph has the line $y = 1$ as a horizontal asymptote. Using this information, as well as the x-intercept $(1, 0)$ and the y-intercept $(0, -1)$, gives the graph in Figure 9. **TRY YOUR TURN 4**

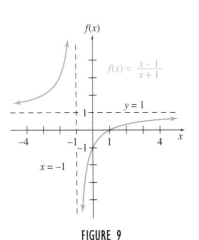

FIGURE 9

YOUR TURN 4 Find where f is increasing and decreasing if

$$f(x) = \frac{-2x}{x + 2}.$$

Graph the function.

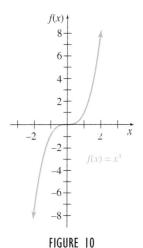

FIGURE 10

CAUTION It is important to note that the reverse of the test for increasing and decreasing functions is not true—it is possible for a function to be increasing on an interval even though the derivative is not positive at every point in the interval. A good example is given by $f(x) = x^3$, which is increasing on every interval, even though $f'(x) = 0$ when $x = 0$. See Figure 10.

Similarly, it is incorrect to assume that the sign of the derivative in regions separated by critical numbers must alternate between + and −. If this were always so, it would lead to a simple rule for finding the sign of the derivative: just check one test point, and then make the other regions alternate in sign. But this is not true if one of the factors in the derivative is raised to an even power. In the function $f(x) = x^3$ just considered, $f'(x) = 3x^2$ is positive on both sides of the critical number $x = 0$.

TECHNOLOGY NOTE

A graphing calculator can be used to find the derivative of a function at a particular x-value. The screen in Figure 11 supports our results in Example 2 for the test values, −4 and 2. The results are not exact because the calculator uses a numerical method to approximate the derivative at the given x-value.

Some graphing calculators can find where a function changes from increasing to decreasing by finding a maximum or minimum. The calculator windows in Figure 12 show this feature for the function in Example 2. Note that these, too, are approximations. This concept will be explored further in the next section.

FIGURE 11   FIGURE 12

Knowing the intervals where a function is increasing or decreasing can be important in applications, as shown by the next examples.

EXAMPLE 5 **Profit Analysis**

A company selling computers finds that the cost per computer decreases linearly with the number sold monthly, decreasing from \$1000 when none are sold to \$800 when 1000 are sold. Thus, the average cost function has a y-intercept of 1000 and a slope of $-200/1000 = -0.2$, so it is given by the formula

$$\overline{C}(x) = 1000 - 0.2x, \quad 0 \le x \le 1000,$$

where x is the number of computers sold monthly. Since $\overline{C}(x) = C(x)/x$, the cost function is given by

$$C(x) = x\overline{C}(x) = x(1000 - 0.2x)$$
$$= 1000x - 0.2x^2, \quad 0 \le x \le 1000.$$

Suppose the revenue function can be approximated by

$$R(x) = 0.0008x^3 - 2.4x^2 + 2400x, \quad 0 \le x \le 1000.$$

Determine any intervals on which the profit function is increasing.

APPLY IT **SOLUTION** First find the profit function $P(x)$.

$$P(x) = R(x) - C(x)$$
$$= (0.0008x^3 - 2.4x^2 + 2400x) - (1000x - 0.2x^2)$$
$$= 0.0008x^3 - 2.2x^2 + 1400x$$

To find any intervals where this function is increasing, set $P'(x) = 0$.

$$P'(x) = 0.0024x^2 - 4.4x + 1400 = 0$$

Solving this with the quadratic formula gives the approximate solutions $x = 409.8$ and $x = 1423.6$. The latter number is outside of the domain. Use $x = 409.8$ to determine two intervals on a number line, as shown in Figure 13. Choose $x = 0$ and $x = 1000$ as test points.

$$P'(0) = 0.0024(0^2) - 4.4(0) + 1400 = 1400$$
$$P'(1000) = 0.0024(1000^2) - 4.4(1000) + 1400 = -600$$

This means that when no computers are sold monthly, the profit is going up at a rate of $1400 per computer. When 1000 computers are sold monthly, the profit is going down at a rate of $600 per computer. The test points show that the function increases on $(0, 409.8)$ and decreases on $(409.8, 1000)$. See Figure 13. Thus, the profit is increasing when 409 computers or fewer are sold, and decreasing when 410 or more are sold, as shown in Figure 14.

FIGURE 13

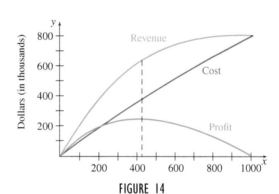

FIGURE 14

As the graph in Figure 14 shows, the profit will increase as long as the revenue function increases faster than the cost function. That is, increasing production will produce more profit as long as the marginal revenue is greater than the marginal cost.

EXAMPLE 6 Recollection of Facts

In the exercises in the previous chapter, the function

$$f(t) = \frac{90t}{99t - 90}$$

gave the number of facts recalled after t hours for $t > 10/11$. Find the intervals in which $f(t)$ is increasing or decreasing.

SOLUTION First use the quotient rule to find the derivative, $f'(t)$.

$$f'(t) = \frac{(99t - 90)(90) - 90t(99)}{(99t - 90)^2}$$
$$= \frac{8910t - 8100 - 8910t}{(99t - 90)^2} = \frac{-8100}{(99t - 90)^2}$$

Since $(99t - 90)^2$ is positive everywhere in the domain of the function and since the numerator is a negative constant, $f'(t) < 0$ for all t in the domain of $f(t)$. Thus $f(t)$ always decreases and, as expected, the number of words recalled decreases steadily over time.

5.1 EXERCISES

Find the open intervals where the functions graphed as follows are (a) increasing, or (b) decreasing.

1.

2.

3.

4.

5.

6.

7.

8.

For each of the exercises listed below, suppose that the function that is graphed is not $f(x)$, but $f'(x)$. Find the open intervals where $f(x)$ is (a) increasing or (b) decreasing.

9. Exercise 1

10. Exercise 2

11. Exercise 7

12. Exercise 8

For each function, find (a) the critical numbers; (b) the open intervals where the function is increasing; and (c) the open intervals where it is decreasing.

13. $y = 2.3 + 3.4x - 1.2x^2$

14. $y = 1.1 - 0.3x - 0.3x^2$

15. $f(x) = \dfrac{2}{3}x^3 - x^2 - 24x - 4$

16. $f(x) = \dfrac{2}{3}x^3 - x^2 - 4x + 2$

17. $f(x) = 4x^3 - 15x^2 - 72x + 5$

18. $f(x) = 4x^3 - 9x^2 - 30x + 6$

19. $f(x) = x^4 + 4x^3 + 4x^2 + 1$

20. $f(x) = 3x^4 + 8x^3 - 18x^2 + 5$

21. $y = -3x + 6$

22. $y = 6x - 9$

23. $f(x) = \dfrac{x + 2}{x + 1}$

24. $f(x) = \dfrac{x + 3}{x - 4}$

25. $y = \sqrt{x^2 + 1}$

26. $y = x\sqrt{9 - x^2}$

27. $f(x) = x^{2/3}$

28. $f(x) = (x + 1)^{4/5}$

29. $y = x - 4\ln(3x - 9)$

30. $f(x) = \ln \dfrac{5x^2 + 4}{x^2 + 1}$

31. $f(x) = xe^{-3x}$

32. $f(x) = xe^{x^2-3x}$

33. $f(x) = x^2 2^{-x}$

34. $f(x) = x2^{-x^2}$

35. $y = x^{2/3} - x^{5/3}$

36. $y = x^{1/3} + x^{4/3}$

37. A friend looks at the graph of $y = x^2$ and observes that if you start at the origin, the graph increases whether you go to the right or the left, so the graph is increasing everywhere. Explain why this reasoning is incorrect.

38. Use the techniques of this chapter to find the vertex and intervals where f is increasing and decreasing, given

$$f(x) = ax^2 + bx + c,$$

where we assume $a > 0$. Verify that this agrees with what we found in Chapter 2.

39. Repeat Exercise 38 under the assumption $a < 0$.

40. Where is the function defined by $f(x) = e^x$ increasing? Decreasing? Where is the tangent line horizontal?

41. Repeat Exercise 40 with the function defined by $f(x) = \ln x$.

42. (a) For the function in Exercise 15, find the average of the critical numbers.

(b) For the function in Exercise 15, use a graphing calculator to find the roots of the function, and then find the average of those roots.

(c) Compare your answers to parts (a) and (b). What do you notice?

(d) Repeat part (a) for the function in Exercise 17.

(e) Repeat part (b) for the function in Exercise 17.

(f) Compare your answers to parts (d) and (e). What do you notice?

It can be shown that the average of the roots of a polynomial (including the complex roots, if there are any) and the critical numbers of a polynomial (including complex roots of $f'(x) = 0$, if there are any) are always equal. *Source: The Mathematics Teacher*.

For each of the following functions, use a graphing calculator to find the open intervals where $f(x)$ is (a) increasing, or (b) decreasing.

43. $f(x) = e^{0.001x} - \ln x$

44. $f(x) = \ln(x^2 + 1) - x^{0.3}$

APPLICATIONS

Business and Economics

45. Housing Starts A county realty group estimates that the number of housing starts per year over the next three years will be

$$H(r) = \frac{300}{1 + 0.03r^2},$$

where r is the mortgage rate (in percent).

(a) Where is $H(r)$ increasing?

(b) Where is $H(r)$ decreasing?

46. Cost Suppose the total cost $C(x)$ (in dollars) to manufacture a quantity x of weed killer (in hundreds of liters) is given by

$$C(x) = x^3 - 2x^2 + 8x + 50.$$

(a) Where is $C(x)$ decreasing?

(b) Where is $C(x)$ increasing?

47. Profit A manufacturer sells video games with the following cost and revenue functions (in dollars), where x is the number of games sold, for $0 \le x \le 3300$.

$$C(x) = 0.32x^2 - 0.00004x^3$$
$$R(x) = 0.848x^2 - 0.0002x^3$$

Determine the interval(s) on which the profit function is increasing.

48. Profit A phone manufacturer has determined that the profit $P(x)$ (in thousands of dollars) is related to the quantity x of phones produced (in hundreds) per month by

$$P(x) = -(x - 4)e^x - 4, \qquad 0 < x \le 3.9.$$

(a) At what production levels is the profit increasing?

(b) At what levels is it decreasing?

49. Social Security Assets The projected year-end assets in the Social Security trust funds, in trillions of dollars, where t represents the number of years since 2000, can be approximated by

$$A(t) = 0.0000329t^3 - 0.00450t^2 + 0.0613t + 2.34,$$

where $0 \le t \le 50$. *Source: Social Security Administration*.

(a) Where is $A(t)$ increasing?

(b) Where is $A(t)$ decreasing?

50. Unemployment The annual unemployment rates of the U.S. civilian noninstitutional population for 1994–2014 are shown in the graph. When is the function increasing? Decreasing? Constant? *Source: Bureau of Labor Statistics*.

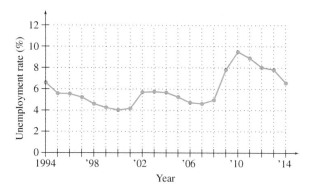

Life Sciences

51. Alcohol Concentration In Exercise 55 in the section on Polynomial and Rational Functions, we gave the function defined by

$$A(x) = 0.003631x^3 - 0.03746x^2 + 0.1012x + 0.009$$

as the approximate blood alcohol concentration in a 170-lb woman x hours after drinking 2 oz of alcohol on an empty stomach, for x in the interval $[0, 5]$. *Source: Medicolegal Aspects of Alcohol Determination in Biological Specimens*.

(a) On what time intervals is the alcohol concentration increasing?

(b) On what intervals is it decreasing?

52. Drug Concentration   The percent of concentration of a drug in the bloodstream t hours after the drug is administered is given by

$$K(t) = \frac{4t}{3t^2 + 27}$$

for $t \geq 0$.

(a) On what time intervals is the concentration of the drug increasing?

(b) On what intervals is it decreasing?

53. Drug Concentration   Suppose a certain drug is administered to a patient, with the percent of concentration of the drug in the bloodstream t hours later given by

$$K(t) = \frac{5t}{t^2 + 1}.$$

(a) On what time intervals is the concentration of the drug increasing?

(b) On what intervals is it decreasing?

54. Cardiology   The aortic pressure-diameter relation in a particular patient who underwent cardiac catheterization can be modeled by the polynomial

$$D(p) = 0.000002p^3 - 0.0008p^2 + 0.1141p + 16.683,$$

$$55 \leq p \leq 130,$$

where $D(p)$ is the aortic diameter (in millimeters) and p is the aortic pressure (in mm Hg). Determine where this function is increasing and where it is decreasing within the interval given above. *Source: Circulation.*

55. Thermic Effect of Food   The metabolic rate of a person who has just eaten a meal tends to go up and then, after some time has passed, returns to a resting metabolic rate. This phenomenon is known as the thermic effect of food. Researchers have indicated that the thermic effect of food for one particular person is

$$F(t) = -10.28 + 175.9te^{-t/1.3},$$

where $F(t)$ is the thermic effect of food (in kJ/hr) and t is the number of hours that have elapsed since eating a meal. *Source: American Journal of Clinical Nutrition.*

(a) Find $F'(t)$.

(b) Determine where this function is increasing and where it is decreasing. Interpret your answers.

56. Holstein Dairy Cattle   Researchers have developed the following function that can be used to accurately predict the weight of Holstein cows (females) of various ages:

$$W_1(t) = 619(1 - 0.905e^{-0.002t})^{1.2386},$$

where $W_1(t)$ is the weight of the Holstein cow (in kilograms) that is t days old. Where is this function increasing? *Source: Canadian Journal of Animal Science.*

57. Plant Growth   Researchers have found that the radius R to which a plant will grow is affected by the number of plants N in the area, as described by the equation

$$R = \sqrt{\frac{N_0 - N}{2\pi DN}},$$

where D is the density of the plants in an area and N_0 is the number of plants in the area initially. *Source: Ecology.*

(a) Find $\dfrac{dR}{dN}$.

(b) Based on the sign of your answer to part (a), what can you say about how the maximum radius of a plant is affected as the number of plants in the area increases?

(c) What happens to $\dfrac{dR}{dN}$ as N approaches 0? As N approaches N_0?

58. Spread of Infection   The number of people $P(t)$ (in hundreds) infected t days after an epidemic begins is approximated by

$$P(t) = \frac{10\ln(0.19t + 1)}{0.19t + 1}.$$

When will the number of people infected start to decline?

Social Sciences

59. Population   The standard normal probability function is used to describe many different populations. Its graph is the well-known normal curve. This function is defined by

$$f(x) = \frac{1}{\sqrt{2\pi}} e^{-x^2/2}.$$

Give the intervals where the function is increasing and decreasing.

60. Nuclear Arsenals   The figure shows estimated totals of nuclear weapons inventory for the United States and the former Soviet Union (and its successor states) from 1945 to 2010. *Source: Federation of American Scientists.*

(a) On what intervals were the total inventories of both countries increasing?

(b) On what intervals were the total inventories of both countries decreasing?

General Interest

61. Sports Cars   The graph on the next page shows the horsepower and torque as a function of the engine speed for a 1964 Ford Mustang. *Source: Online with Fuel Line Exhaust.*

(a) On what intervals is the power increasing with engine speed?

(b) On what intervals is the power decreasing with engine speed?

(c) On what intervals is the torque increasing with engine speed?

(d) On what intervals is the torque decreasing with engine speed?

1. Increasing on $(-1, 2)$ and $(4, \infty)$. Decreasing on $(-\infty, -1)$ and $(2, 4)$.

2. Increasing on $(-3, 5/3)$. Decreasing on $(-\infty, -3)$ and $(5/3, \infty)$

3. Increasing on $(-2, \infty)$ and decreasing on $(-\infty, -2)$.

4. Never increasing. Decreasing on $(-\infty, -2)$ and $(-2, \infty)$.

62. Automobile Mileage As a mathematics professor loads more weight in the back of his Subaru, the mileage goes down. Let x be the amount of weight (in pounds) that he adds, and let $y = f(x)$ be the mileage (in mpg).

(a) Is $f'(x)$ positive or negative? Explain.

(b) What are the units of $f'(x)$?

5.2 Relative Extrema

APPLY IT   In a 30-second commercial, when is the best time to present the sales message?

We will answer this question in Example 1 by investigating the idea of a relative maximum.

As we have seen throughout this text, the graph of a function may have peaks and valleys. It is important in many applications to determine where these points occur. For example, if the function represents the profit of a company, these peaks and valleys indicate maximum profits and losses. When the function is given as an equation, we can use the derivative to determine these points, as shown in the first example.

EXAMPLE 1   Maximizing Viewer's Attention

Suppose that the manufacturer of a diet soft drink is disappointed by sales after airing a new series of 30-second television commercials. The company's market research analysts hypothesize that the problem lies in the timing of the commercial's message, Drink Sparkling Light. Either it comes too early in the commercial, before the viewer has become involved; or it comes too late, after the viewer's attention has faded. After extensive experimentation, the research group finds that the percent of full attention that a viewer devotes to a commercial is a function of time (in seconds) since the commercial began, where

$$\text{Viewer's attention} = f(t) = -\frac{3}{20}t^2 + 6t + 20, \quad 0 \le t \le 30.$$

When is the best time to present the commercial's sales message?

SOLUTION Clearly, the message should be delivered when the viewer's attention is at a maximum. To find this time, find $f'(t)$.

$$f'(t) = -\frac{3}{10}t + 6 = -0.3t + 6$$

The derivative $f'(t)$ is greater than 0 when

$$-0.3t + 6 > 0,$$
$$3t > -60, \text{ or}$$
$$t < 20.$$

Similarly, $f'(t) < 0$ when $-0.3t + 6 < 0$, or $t > 20$. Thus, attention increases for the first 20 seconds and decreases for the last 10 seconds. The message should appear about 20 seconds into the commercial. At that time the viewer will devote $f(20) = 80\%$ of his attention to the commercial.

The maximum level of viewer attention (80%) in Example 1 is a *relative maximum*, defined as follows.

Relative Maximum or Minimum

Let c be a number in the domain of a function f. Then $f(c)$ is a **relative** (or **local**) **maximum** for f if there exists an open interval (a, b) containing c such that

$$f(x) \leq f(c)$$

for all x in (a, b).

Likewise, $f(c)$ is a **relative** (or **local**) **minimum** for f if there exists an open interval (a, b) containing c such that

$$f(x) \geq f(c)$$

for all x in (a, b).

A function has a **relative** (or **local**) **extremum** (plural: **extrema**) at c if it has either a relative maximum or a relative minimum there.

If c is an endpoint of the domain of f, we only consider x in the half-open interval that is in the domain.*

The intuitive idea is that a relative maximum is the greatest value of the function in some region right around the point, although there may be greater values elsewhere. For example, the highest value of the Dow Jones industrial average this week is a relative maximum, although the Dow may have reached a higher value earlier this year. Similarly, a relative minimum is the least value of a function in some region around the point.

A simple way to view these concepts is that a relative maximum is a peak, and a relative minimum is the bottom of a valley, although either a relative minimum or maximum can also occur at the endpoint of the domain.

NOTE Recall from Section 2.3 on Polynomials and Rational Functions that a relative extremum that is not an endpoint is also referred to as a turning point.

*There is disagreement on calling an endpoint a maximum or minimum. We define it this way because this is an applied calculus book, and in an application it would be considered a maximum or minimum value of the function.

EXAMPLE 2 Relative Extrema

Identify the x-values of all points where the graph in Figure 15 has relative extrema.

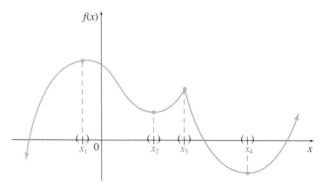

FIGURE 15

YOUR TURN 1 Identify the x-values of all points where the graph has relative extrema.

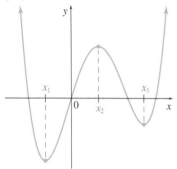

SOLUTION The parentheses around x_1 show an open interval containing x_1 such that $f(x) \leq f(x_1)$, so there is a relative maximum of $f(x_1)$ at $x = x_1$. Notice that many other open intervals would work just as well. Similar intervals around x_2, x_3, and x_4 can be used to find a relative maximum of $f(x_3)$ at $x = x_3$ and relative minima of $f(x_2)$ at $x = x_2$ and $f(x_4)$ at $x = x_4$.   TRY YOUR TURN 1

The function graphed in Figure 16 has relative maxima when $x = x_1$ or $x = x_3$ and relative minima when $x = x_2$ or $x = x_4$. The tangent lines at the points having x-values x_1 and x_2 are shown in the figure. Both tangent lines are horizontal and have slope 0. There is no single tangent line at the point where $x = x_3$.

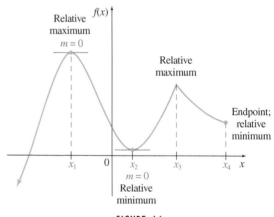

FIGURE 16

Since the derivative of a function gives the slope of a line tangent to the graph of the function, to find relative extrema we first identify all critical numbers and endpoints. A relative extremum *may* exist at a critical number. (A rough sketch of the graph of the function near a critical number often is enough to tell whether an extremum has been found.) These facts about extrema are summarized below.

If a function f has a relative extremum at c, then c is a critical number or c is an endpoint of the domain.

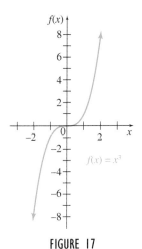

FIGURE 17

Be very careful not to get this result backward. It does *not* say that a function has relative extrema at all critical numbers of the function. For example, Figure 17 shows the graph of $f(x) = x^3$. The derivative, $f'(x) = 3x^2$, is 0 when $x = 0$, so that 0 is a critical number for that function. However, as suggested by the graph of Figure 17, $f(x) = x^3$ has neither a relative maximum nor a relative minimum at $x = 0$ (or anywhere else, for that matter). A critical number is a candidate for the location of a relative extremum, but only a candidate.

First Derivative Test

Suppose all critical numbers have been found for some function f. How is it possible to tell from the equation of the function whether these critical numbers produce relative maxima, relative minima, or neither? One way is suggested by the graph in Figure 18.

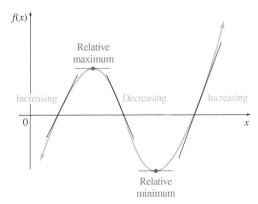

FIGURE 18

As shown in Figure 18, on the left of a relative maximum the tangent lines to the graph of a function have positive slopes, indicating that the function is increasing. At the relative maximum, the tangent line is horizontal. On the right of the relative maximum the tangent lines have negative slopes, indicating that the function is decreasing. Around a relative minimum the opposite occurs. As shown by the tangent lines in Figure 18, the function is decreasing on the left of the relative minimum, has a horizontal tangent at the minimum, and is increasing on the right of the minimum.

Putting this together with the methods from Section 1 for identifying intervals where a function is increasing or decreasing gives the following **first derivative test** for locating relative extrema.

First Derivative Test

Let c be a critical number for a function f. Suppose that f is continuous on (a, b) and differentiable on (a, b) except possibly at c, and that c is the only critical number for f in (a, b).

1. $f(c)$ is a relative maximum of f if the derivative $f'(x)$ is positive in the interval (a, c) and negative in the interval (c, b).

2. $f(c)$ is a relative minimum of f if the derivative $f'(x)$ is negative in the interval (a, c) and positive in the interval (c, b).

The sketches in the table on the next page show how the first derivative test works. Assume the same conditions on a, b, and c for the table as those given for the first derivative test.

Relative Extrema			
$f(x)$ has:	Sign of f' in (a, c)	Sign of f' in (c, b)	Sketches
Relative maximum	$+$	$-$	
Relative minimum	$-$	$+$	
No relative extrema	$+$	$+$	
No relative extrema	$-$	$-$	

EXAMPLE 3   Relative Extrema

Find all relative extrema for the following functions, as well as where each function is increasing and decreasing.

(a) $f(x) = 2x^3 - 3x^2 - 72x + 15$

SOLUTION

Method 1
First Derivative Test

The derivative is $f'(x) = 6x^2 - 6x - 72$. There are no points where $f'(x)$ fails to exist, so the only critical numbers will be found where the derivative equals 0. Setting the derivative equal to 0 gives

$$6x^2 - 6x - 72 = 0$$
$$6(x^2 - x - 12) = 0$$
$$6(x - 4)(x + 3) = 0 \quad \text{Factor.}$$
$$x - 4 = 0 \quad \text{or} \quad x + 3 = 0$$
$$x = 4 \quad \text{or} \quad x = -3.$$

As in the previous section, the critical numbers 4 and -3 are used to determine the three intervals $(-\infty, -3)$, $(-3, 4)$, and $(4, \infty)$ shown on the number line in Figure 19.

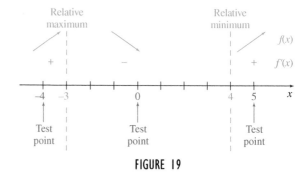

FIGURE 19

Any number from each of the three intervals can be used as a test point to find the sign of f' in each interval. Using -4, 0, and 5 gives the following information.

$$f'(-4) = 6(-8)(-1) > 0$$
$$f'(0) = 6(-4)(3) < 0$$
$$f'(5) = 6(1)(8) > 0$$

Thus, the derivative is positive on $(-\infty, -3)$, negative on $(-3, 4)$, and positive on $(4, \infty)$. By Part 1 of the first derivative test, this means that the function has a relative maximum of $f(-3) = 150$ when $x = -3$; by Part 2, f has a relative minimum of $f(4) = -193$ when $x = 4$. The function is increasing on $(-\infty, -3)$ and $(4, \infty)$ and decreasing on $(-3, 4)$. The graph is shown in Figure 20.   **TRY YOUR TURN 2**

YOUR TURN 2 Find all relative extrema of $f(x) = -x^3 - 2x^2 + 15x + 10$.

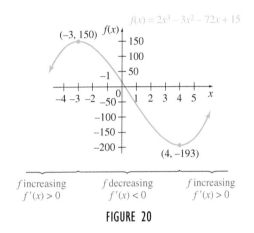

FIGURE 20

Method 2
Graphing Calculator

Many graphing calculators can locate a relative extremum when supplied with an interval containing the extremum. For example, after graphing the function $f(x) = 2x^3 - 3x^2 - 72x + 15$ on a TI-84 Plus C, we selected `maximum` from the CALC menu and entered a left bound of -4 and a right bound of 0. The calculator asks for an initial guess, but in this example it doesn't matter what we enter. The result of this process, as well as a similar process for finding the relative minimum, is shown in Figure 21.

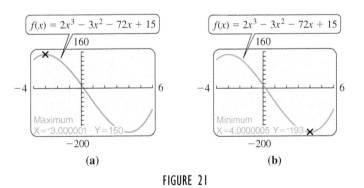

FIGURE 21

Another way to verify the extrema with a graphing calculator is to graph $y = f'(x)$ and find where the graph crosses the x-axis. Figure 22 shows the result of this approach for finding the relative minimum of the previous function.

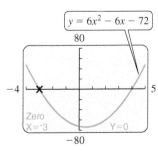

FIGURE 22

(b) $f(x) = 6x^{2/3} - 4x$

SOLUTION Find $f'(x)$.

$$f'(x) = 4x^{-1/3} - 4 = \frac{4}{x^{1/3}} - 4$$

The derivative fails to exist when $x = 0$, but the function itself is defined when $x = 0$, making 0 a critical number for f. To find other critical numbers, set $f'(x) = 0$.

$$f'(x) = 0$$
$$\frac{4}{x^{1/3}} - 4 = 0$$
$$\frac{4}{x^{1/3}} = 4$$
$$4 = 4x^{1/3} \qquad \text{Multiply both sides by } x^{1/3}.$$
$$1 = x^{1/3} \qquad \text{Divide both sides by 4.}$$
$$1 = x \qquad \text{Cube both sides.}$$

The critical numbers 0 and 1 are used to locate the intervals $(-\infty, 0)$, $(0, 1)$, and $(1, \infty)$ on a number line as in Figure 23. Evaluating $f'(x)$ at the test points -1, $1/2$, and 2 and using the first derivative test shows that f has a relative maximum at $x = 1$; the value of this relative maximum is $f(1) = 2$. Also, f has a relative minimum at $x = 0$; this relative minimum is $f(0) = 0$. The function is increasing on $(0, 1)$ and decreasing on $(-\infty, 0)$ and $(1, \infty)$. Notice that the graph, shown in Figure 24, has a sharp point at the critical number where the derivative does not exist. In the last section of this chapter we will show how to verify other features of the graph. **TRY YOUR TURN 3**

YOUR TURN 3 Find all relative extrema of $f(x) = x^{2/3} - x^{5/3}$.

FIGURE 23

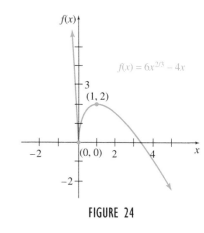

FIGURE 24

(c) $f(x) = x^3 e^x$

SOLUTION The derivative, found by using the product rule, is

$$f'(x) = x^3 e^x + e^x(3x^2)$$
$$= e^x x^2(x + 3). \qquad \text{Factor.}$$

This expression exists for all real numbers. Since e^x is always positive, the derivative is 0 only when $x = 0$ or $x = -3$. Using test points of -4, -1, and 1 gives the results shown in Figure 25. Notice that even though $x = 0$ is a critical number, it is neither a relative minimum nor a relative maximum because $f'(x) > 0$ (that is, the function is increasing) both to the left and to the right of 0. On the other hand, the function has a

YOUR TURN 4 Find all relative extrema of $f(x) = x^2 e^x$.

relative minimum at -3 of $f(-3) = (-3)^3 e^{-3} \approx -1.34$. It is decreasing on the interval $(-\infty, -3)$ and increasing on the interval $(-3, \infty)$. You might be surprised that the function is increasing at $x = 0$, where the derivative is 0, but notice that $f(x_1) < f(x_2)$ for any two numbers x_1 and x_2 in $(-3, \infty)$ with $x_1 < x_2$. See Figure 26.

TRY YOUR TURN 4

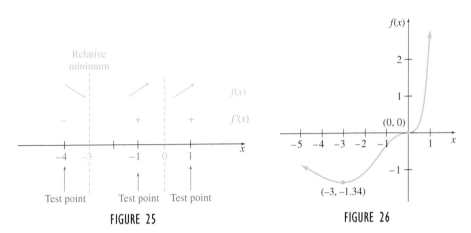

FIGURE 25   FIGURE 26

CAUTION A critical number must be in the domain of the function. For example, the derivative of $f(x) = x/(x - 4)$ is $f'(x) = -4/(x - 4)^2$, which fails to exist when $x = 4$. But $f(4)$ does not exist, so 4 is not a critical number, and the function has no relative extrema.

As mentioned at the beginning of this section, finding the maximum or minimum value of a quantity is important in applications of mathematics. The final example gives a further illustration.

EXAMPLE 4 Bicycle Sales

A small company manufactures and sells bicycles. The production manager has determined that the cost and demand functions for q $(q \geq 0)$ bicycles per week are

$$C(q) = 10 + 5q + \frac{1}{60}q^3 \quad \text{and} \quad p = D(q) = 90 - q,$$

where p is the price per bicycle.

(a) Find the maximum weekly revenue.

SOLUTION The revenue each week is given by

$$R(q) = qp = q(90 - q) = 90q - q^2.$$

To maximize $R(q) = 90q - q^2$, find $R'(q)$. Then find the critical numbers.

$$R'(q) = 90 - 2q = 0$$
$$90 = 2q$$
$$q = 45$$

Since $R'(q)$ exists for all q, 45 is the only critical number. To verify that $q = 45$ will produce a *maximum*, evaluate the derivative on both sides of $q = 45$.

$$R'(40) = 10 \quad \text{and} \quad R'(50) = -10$$

This shows that $R(q)$ is increasing up to $q = 45$, then decreasing, so there is a maximum value at $q = 45$ of $R(45) = 2025$. The maximum revenue will be \$2025 and will occur when 45 bicycles are produced and sold each week.

(b) Find the maximum weekly profit.

SOLUTION Since profit equals revenue minus cost, the profit is given by

$$P(q) = R(q) - C(q)$$

$$= (90q - q^2) - \left(10 + 5q + \frac{1}{60}q^3\right)$$

$$= -\frac{1}{60}q^3 - q^2 + 85q - 10.$$

Find the derivative and set it equal to 0 to find the critical numbers. (The derivative exists for all q.)

$$P'(q) = -\frac{1}{20}q^2 - 2q + 85 = 0$$

Solving this equation by the quadratic formula gives the solutions $q \approx 25.8$ and $q \approx -65.8$. Since q cannot be negative, the only critical number of concern is 25.8. Determine whether $q = 25.8$ produces a maximum by testing a value on either side of 25.8 in $P'(q)$.

$$P'(0) = 85 \quad \text{and} \quad P'(40) = -75$$

These results show that $P(q)$ increases to $q = 25.8$ and then decreases. Since q must be an integer, evaluate $P(q)$ at $q = 25$ and $q = 26$. Since $P(25) = 1229.58$ and $P(26) = 1231.07$, the maximum value occurs when $q = 26$. Thus, the maximum profit of $1231.07 occurs when 26 bicycles are produced and sold each week. Notice that this is not the same as the number that should be produced to yield maximum revenue.

(c) Find the price the company should charge to realize maximum profit.

SOLUTION As shown in part (b), 26 bicycles per week should be produced and sold to get the maximum profit of $1231.07 per week. Since the price is given by

$$p = 90 - q,$$

if $q = 26$, then $p = 64$. The manager should charge $64 per bicycle and produce and sell 26 bicycles per week to get the maximum profit of $1231.07 per week. Figure 27 shows the graphs of the functions used in this example. Notice that the slopes of the revenue and cost functions are the same at the point where the maximum profit occurs. Why is this true?

YOUR TURN 5 Find the maximum weekly profit and the price a company should charge to realize maximum profit if $C(q) = 100 + 10q$ and $p = D(q) = 50 - 2q$.

TRY YOUR TURN 5

FIGURE 27

CAUTION	Be careful to give the y-value of the point where an extremum occurs. Although we solve the equation $f'(x) = 0$ for x to find the extremum, the maximum or minimum value of the function is the corresponding y-value. Thus, in Example 4(a), we found that at $q = 45$, the maximum weekly revenue is $2025 (not $45).

The examples in this section involving the maximization of a quadratic function, such as the advertising example and the bicycle revenue example, could be solved by the methods described in Chapter 2 on Nonlinear Functions. But those involving more complicated functions, such as the bicycle profit example, are difficult to analyze without the tools of calculus.

Finding extrema for realistic problems requires an accurate mathematical model of the problem. In particular, it is important to be aware of restrictions on the values of the variables. For example, if $T(x)$ closely approximates the number of items that can be manufactured daily on a production line when x is the number of employees on the line, x must certainly be restricted to the positive integers or perhaps to a few common fractional values. (We can imagine half-time workers, but not 1/49-time workers.)

On the other hand, to apply the tools of calculus to obtain an extremum for some function, the function must be defined and be meaningful at every real number in some interval. Because of this, the answer obtained from a mathematical model might be a number that is not feasible in the actual problem.

Usually, the requirement that a continuous function be used, rather than one that can take on only certain selected values, is of theoretical interest only. In most cases, the methods of calculus give acceptable results as long as the assumptions of continuity and differentiability are not totally unreasonable. If they lead to the conclusion, say, that $80\sqrt{2}$ workers should be hired, it is usually only necessary to investigate acceptable values close to $80\sqrt{2}$. This was done in Example 4.

5.2 WARM-UP EXERCISES

Find the intervals where each function is increasing and decreasing. *(Sec. 5.1)*

W1. $f(x) = 2x^3 - 3x^2 - 36x + 4$

W2. $f(x) = 3x^4 - 10x^3 - 36x^2 - 72$

5.2   EXERCISES

Find the locations and values of all relative extrema for the functions with graphs as follows. Compare with Exercises 1–8 in the preceding section.

1.

2.

3.

4.

5.

6.

7.

8.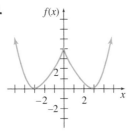

For each of the exercises listed below, suppose that the function that is graphed is not $f(x)$ but $f'(x)$. Find the locations of all relative extrema, and tell whether each extremum is a relative maximum or minimum.

9. Exercise 1   **10.** Exercise 2

11. Exercise 7   **12.** Exercise 8

Find the x-value of all points where the functions defined as follows have any relative extrema. Find the value(s) of any relative extrema.

13. $f(x) = x^2 - 10x + 33$ **14.** $f(x) = x^2 + 8x + 5$

15. $f(x) = x^3 + 6x^2 + 9x - 8$

16. $f(x) = x^3 + 3x^2 - 24x + 2$

17. $f(x) = -\frac{4}{3}x^3 - \frac{21}{2}x^2 - 5x + 8$

18. $f(x) = -\frac{2}{3}x^3 - \frac{1}{2}x^2 + 3x - 4$

19. $f(x) = x^4 - 18x^2 - 4$ **20.** $f(x) = x^4 - 8x^2 + 9$

21. $f(x) = 3 - (8 + 3x)^{2/3}$

22. $f(x) = \frac{(5 - 9x)^{2/3}}{7} + 1$

23. $f(x) = 2x + 3x^{2/3}$ **24.** $f(x) = 3x^{5/3} - 15x^{2/3}$

25. $f(x) = x - \frac{1}{x}$ **26.** $f(x) = x^2 + \frac{1}{x}$

27. $f(x) = \frac{x^2 - 2x + 1}{x - 3}$ **28.** $f(x) = \frac{x^2 - 6x + 9}{x + 2}$

29. $f(x) = x^2 e^x - 3$ **30.** $f(x) = 3xe^x + 2$

31. $f(x) = 2x + \ln x$ **32.** $f(x) = \frac{x^2}{\ln x}$

33. $f(x) = \frac{2^x}{x}$ **34.** $f(x) = x + 8^{-x}$

Use the derivative to find the vertex of each parabola.

35. $y = -2x^2 + 12x - 5$ **36.** $y = ax^2 + bx + c$

Graph each function on a graphing calculator, and then use the graph to find all relative extrema (to three decimal places). Then confirm your answer by finding the derivative and using the calculator to solve the equation $f'(x) = 0$.

37. $f(x) = x^5 - x^4 + 4x^3 - 30x^2 + 5x + 6$

38. $f(x) = -x^5 - x^4 + 2x^3 - 25x^2 + 9x + 12$

39. Graph $f(x) = 2|x + 1| + 4|x - 5| - 20$ with a graphing calculator in the window $[-10, 10]$ by $[-15, 30]$. Use the graph and the function to determine the x-values of all extrema.

40. Consider the function

$$g(x) = \frac{1}{x^{12}} - 2\left(\frac{1000}{x}\right)^6.$$

Source: The Mathematics Teacher.

  (a) Using a graphing calculator, try to find any local minima, or tell why finding a local minimum is difficult for this function.

  (b) Find any local minima using the techniques of calculus.

  (c) Based on your results in parts (a) and (b), describe circumstances under which relative extrema are easier to find using the techniques of calculus than using a graphing calculator.

In each figure, the graph of the derivative of $y = f(x)$ is shown. Find the locations of all relative extrema of $f(x)$.

41.

42.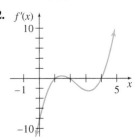

APPLICATIONS

Business and Economics

Profit In Exercises 43–46, find (a) the number, q, of units that produces maximum profit; (b) the price, p, per unit that produces maximum profit; and (c) the maximum profit, P.

43. $C(q) = 80 + 18q$; $p = 70 - 2q$

44. $C(q) = 25q + 5000$; $p = 90 - 0.02q$

45. $C(q) = 100 + 20qe^{-0.01q}$; $p = 40e^{-0.01q}$

46. $C(q) = 21.047q + 3$; $p = 50 - 5\ln(q + 10)$

47. Power On July 11, 2014, the power used in New York state (in thousands of megawatts) could be approximated by the function

$$P(t) = -0.006088t^3 + 0.1904t^2 - 1.076t + 18.39,$$

where t is the number of hours since midnight, for $0 \le t \le 24$. Find any relative extrema for power usage, as well as when they occurred. *Source: Current Energy.*

48. Profit The total profit $P(x)$ (in thousands of dollars) from the sale of x units of a certain prescription drug is given by

$$P(x) = \ln(-x^3 + 3x^2 + 72x + 1)$$

for x in $[0, 10]$.

(a) Find the number of units that should be sold in order to maximize the total profit.

(b) What is the maximum profit?

49. Revenue The demand equation for telephones at one store is

$$p = D(q) = 200e^{-0.1q},$$

where p is the price (in dollars) and q is the quantity of telephones sold per week. Find the values of q and p that maximize revenue.

50. Revenue The demand equation for one type of computer networking system is

$$p = D(q) = 500qe^{-0.0016q^2},$$

where p is the price (in dollars) and q is the quantity of servers sold per month. Find the values of q and p that maximize revenue.

51. Cost Suppose that the cost function for a product is given by $C(x) = 0.002x^3 + 9x + 6912$. Find the production level (i.e., value of x) that will produce the minimum average cost per unit $\overline{C}(x)$.

52. Unemployment The annual unemployment rates of the U.S. civilian noninstitutional population for 1994–2014 are shown in the graph. Identify the years where relative extrema occur, and estimate the unemployment rate at each of these years. *Source: Bureau of Labor Statistics*.

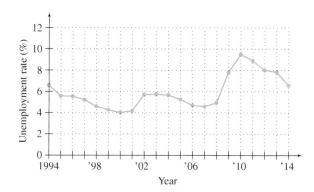

Life Sciences

53. Activity Level In the summer the activity level of a certain type of lizard varies according to the time of day. A biologist has determined that the activity level is given by the function

$$a(t) = 0.008t^3 - 0.288t^2 + 2.304t + 7,$$

where t is the number of hours after 12 noon. When is the activity level highest? When is it lowest?

54. Milk Consumption The average individual daily milk consumption for herds of Charolais, Angus, and Hereford calves can be described by the function

$$M(t) = 6.281t^{0.242}e^{-0.025t}, \qquad 1 \le t \le 26,$$

where $M(t)$ is the milk consumption (in kilograms) and t is the age of the calf (in weeks). *Source: Animal Production*.

(a) Find the time in which the maximum daily consumption occurs and the maximum daily consumption.

(b) If the general formula for this model is given by

$$M(t) = at^b e^{-ct},$$

find the time where the maximum consumption occurs and the maximum consumption. (*Hint:* Express your answer in terms of a, b, and c.)

55. Alaskan Moose The mathematical relationship between the age of a captive female moose and its mass can be described by the function

$$M(t) = 369(0.93)^t t^{0.36}, \quad t \le 12,$$

where $M(t)$ is the mass of the moose (in kilograms) and t is the age (in years) of the moose. Find the age at which the mass of a female moose is maximized. What is the maximum mass? *Source: Journal of Wildlife Management*.

56. Thermic Effect of Food As we saw in the last section, the metabolic rate after a person eats a meal tends to go up and then, after some time has passed, returns to a resting metabolic rate. This phenomenon is known as the thermic effect of food and can be described for a particular individual as

$$F(t) = -10.28 + 175.9te^{-t/1.3},$$

where $F(t)$ is the thermic effect of food (in kJ/hr), and t is the number of hours that have elapsed since eating a meal. Find the time after the meal when the thermic effect of the food is maximized. *Source: American Journal of Clinical Nutrition*.

Social Sciences

57. Attitude Change Social psychologists have found that as the discrepancy between the views of a speaker and those of an audience increases, the attitude change in the audience also increases to a point but decreases when the discrepancy becomes too large, particularly if the communicator is viewed by the audience as having low credibility. Suppose that the degree of change can be approximated by the function

$$D(x) = -x^4 + 8x^3 + 80x^2,$$

where x is the discrepancy between the views of the speaker and those of the audience, as measured by scores on a questionnaire. Find the amount of discrepancy the speaker should aim for to maximize the attitude change in the audience. *Source: Journal of Personality and Social Psychology*.

58. Film Length A group of researchers found that people prefer training films of moderate length; shorter films contain too little information, while longer films are boring. For a training film on the care of exotic birds, the researchers determined that the ratings people gave for the film could be approximated by

$$R(t) = \frac{20t}{t^2 + 100},$$

where t is the length of the film (in minutes). Find the film length that received the highest rating.

Physical Sciences

59. Height After a great deal of experimentation, two Atlantic Institute of Technology senior physics majors determined that when a bottle of French champagne is shaken several times, held upright, and uncorked, its cork travels according to

$$s(t) = -16t^2 + 40t + 3,$$

where s is its height (in feet) above the ground t seconds after being released.

(a) How high will it go?

(b) How long is it in the air?

5.3 Higher Derivatives, Concavity, and the Second Derivative Test

APPLY IT Just because the price of a stock is increasing, does that alone make it a good investment?

We will address this question in Example 1.

In the first section of this chapter, we used the derivative to determine intervals where a function is increasing or decreasing. For example, if the function represents the price of a stock, we can use the derivative to determine when the price is increasing. In addition, it would be important for us to know how the *rate of increase* is changing. We can determine how the *rate of increase* (or the *rate of decrease*) is changing by determining the rate of change of the derivative of the function. In other words, we can find the derivative of the derivative, called the **second derivative**, as shown in the following example.

EXAMPLE 1 Stock Prices

Suppose a friend is trying to get you to invest in the stock of a young company. The following function represents the price $P(t)$ of the company's stock since it became available two years ago:

$$P(t) = 17 + t^{1/2},$$

where t is the number of months since the stock became available. He claims that the price of the stock is always increasing and that you will make a fortune on it. Verify his claims. Is the price of the stock increasing? How fast? How much will you make if you invest now?

APPLY IT **SOLUTION** The derivative of $P(t)$,

$$P'(t) = \frac{1}{2}t^{-1/2} = \frac{1}{2\sqrt{t}},$$

is always positive because $\sqrt{t}$ is positive for $t > 0$. This means that the price function $P(t)$ is always increasing. But *how fast* is it increasing?

The derivative $P'(t) = 1/(2\sqrt{t})$ tells how fast the price is increasing at any number of months, t, since the stock became available. For example, when $t = 1$ month, $P'(t) = 1/2$, and the price is increasing at $1/2$ dollar, or 50 cents, per month. When $t = 4$ months, $P'(t) = 1/4$; the stock is increasing at 25 cents per month. By the time you buy in at $t = 24$ months, the price is increasing at 10 cents per month, and the *rate of increase* looks as though it will continue to decrease.

In general, the rate of increase in P' is given by the derivative of $P'(t)$, called the second derivative and denoted by $P''(t)$. Since $P'(t) = (1/2)t^{-1/2}$,

$$P''(t) = -\frac{1}{4}t^{-3/2} = -\frac{1}{4\sqrt{t^3}}.$$

$P''(t)$ is negative for $t > 0$ and, therefore, confirms the suspicion that the *rate* of increase in price does indeed decrease for all $t > 0$. The price of the company's stock will not drop, but the amount of return will certainly not be the fortune your friend predicts.

If you invest now, at $t = 24$ months, the price would be $21.90. A year later, it would be worth $23 a share. If you were rich enough to buy 100 shares for $21.90 each, the total investment would be worth $2300 in a year. The increase of $110 is about 5% of the investment. The only investors to make a lot of money on this stock would be those who bought early, when the rate of increase was much greater.

As mentioned earlier, the second derivative of a function f, written f'', gives the rate of change of the *derivative* of f. Before continuing to discuss applications of the second derivative, we need to introduce some additional terminology and notation.

Higher Derivatives

If a function f has a derivative f', then the derivative of f', if it exists, is the second derivative of f, written f''. The derivative of f'', if it exists, is called the **third derivative** of f, and so on. By continuing this process, we can find **fourth derivatives** and other higher derivatives. For example, if $f(x) = x^4 + 2x^3 + 3x^2 - 5x + 7$, then

$$f'(x) = 4x^3 + 6x^2 + 6x - 5, \qquad \text{First derivative of } f$$
$$f''(x) = 12x^2 + 12x + 6, \qquad \text{Second derivative of } f$$
$$f'''(x) = 24x + 12, \qquad \text{Third derivative of } f$$

and

$$f^{(4)}(x) = 24. \qquad \text{Fourth derivative of } f$$

Notation for Higher Derivatives

The second derivative of $y = f(x)$ can be written using any of the following notations:

$$f''(x), \quad \frac{d^2y}{dx^2}, \quad \text{or} \quad D_x^2[f(x)].$$

The third derivative can be written in a similar way. For $n \geq 4$, the nth derivative is written $f^{(n)}(x)$.

| CAUTION | Notice the difference in notation between $f^{(4)}(x)$, which indicates the fourth derivative of $f(x)$, and $f^4(x)$, which indicates $f(x)$ raised to the fourth power. |

EXAMPLE 2 Second Derivative

Let $f(x) = x^3 + 6x^2 - 9x + 8$.

(a) Find $f''(x)$.

SOLUTION To find the second derivative of $f(x)$, find the first derivative, and then take its derivative.

$$f'(x) = 3x^2 + 12x - 9$$
$$f''(x) = 6x + 12$$

(b) Find $f''(0)$.

YOUR TURN 1 Find $f''(1)$ if $f(x) = 5x^4 - 4x^3 + 3x$.

SOLUTION Since $f''(x) = 6x + 12$,

$$f''(0) = 6(0) + 12 = 12.$$ TRY YOUR TURN 1

EXAMPLE 3 Second Derivative

Find the second derivative for the functions defined as follows.

(a) $f(x) = (x^2 - 1)^2$

SOLUTION Here, using the chain rule,

$$f'(x) = 2(x^2 - 1)(2x) = 4x(x^2 - 1).$$

Use the product rule to find $f''(x)$.

$$\begin{aligned} f''(x) &= 4x(2x) + (x^2 - 1)(4) \\ &= 8x^2 + 4x^2 - 4 \\ &= 12x^2 - 4 \end{aligned}$$

(b) $g(x) = 4x(\ln x)$

SOLUTION Use the product rule.

$$g'(x) = 4x \cdot \frac{1}{x} + (\ln x) \cdot 4 = 4 + 4(\ln x)$$

$$g''(x) = 0 + 4 \cdot \frac{1}{x} = \frac{4}{x}$$

(c) $h(x) = \dfrac{x}{e^x}$

YOUR TURN 2 Find the second derivative for
(a) $f(x) = (x^3 + 1)^2$
(b) $g(x) = xe^x$
(c) $h(x) = \dfrac{\ln x}{x}$.

SOLUTION Here, we need the quotient rule.

$$h'(x) = \frac{e^x - xe^x}{(e^x)^2} = \frac{e^x(1 - x)}{(e^x)^2} = \frac{1 - x}{e^x}$$

$$h''(x) = \frac{e^x(-1) - (1 - x)e^x}{(e^x)^2} = \frac{e^x(-1 - 1 + x)}{(e^x)^2} = \frac{-2 + x}{e^x}$$

TRY YOUR TURN 2

Earlier, we saw that the first derivative of a function represents the rate of change of the function. The second derivative, then, represents the rate of change of the first derivative. If a function describes the position of a vehicle (along a straight line) at time t, then the first derivative gives the velocity of the vehicle. That is, if $y = s(t)$ describes the position (along a straight line) of the vehicle at time t, then $v(t) = s'(t)$ gives the velocity at time t.

We also saw that *velocity* is the rate of change of distance with respect to time. Recall, the difference between velocity and speed is that velocity may be positive or negative, whereas speed is always positive. A negative velocity indicates travel in a negative direction (backing up) with regard to the starting point; positive velocity indicates travel in the positive direction (going forward) from the starting point.

The instantaneous rate of change of velocity is called **acceleration**. Since instantaneous rate of change is the same as the derivative, acceleration is the derivative of velocity. Thus if $a(t)$ represents the acceleration at time t, then

$$a(t) = \frac{d}{dt}v(t) = s''(t).$$

If the velocity is positive and the acceleration is positive, the velocity is increasing, so the vehicle is speeding up. If the velocity is positive and the acceleration is negative, the vehicle is slowing down. A negative velocity and a positive acceleration mean the vehicle is backing up and slowing down. If both the velocity and acceleration are negative, the vehicle is speeding up in the negative direction.

EXAMPLE 4   Velocity and Acceleration

Suppose a car is moving in a straight line, with its position from a starting point (in feet) at time t (in seconds) given by

$$s(t) = t^3 - 2t^2 - 7t + 9.$$

Find the following.

(a) The velocity at any time t

   SOLUTION   The velocity is given by

$$v(t) = s'(t) = 3t^2 - 4t - 7$$

   feet per second.

(b) The acceleration at any time t

   SOLUTION   Acceleration is given by

$$a(t) = v'(t) = s''(t) = 6t - 4$$

   feet per second per second.

(c) The time intervals (for $t \geq 0$) when the car is going forward or backing up

   SOLUTION   We first find when the velocity is 0, that is, when the car is stopped.

$$v(t) = 3t^2 - 4t - 7 = 0$$
$$(3t - 7)(t + 1) = 0$$
$$t = 7/3 \quad \text{or} \quad t = -1$$

   We are interested in $t \geq 0$. Choose a value of t in each of the intervals $(0, 7/3)$ and $(7/3, \infty)$ to see that the velocity is negative in $(0, 7/3)$ and positive in $(7/3, \infty)$. The car is backing up for the first 7/3 seconds, then going forward.

(d) The time intervals (for $t \geq 0$) when the car is speeding up or slowing down

   SOLUTION   The car will speed up when the velocity and acceleration are the same sign and slow down when they have opposite signs. Here, the acceleration is positive when $6t - 4 > 0$, that is, $t > 2/3$ seconds, and negative for $t < 2/3$ seconds. Since the velocity is negative in $(0, 7/3)$ and positive in $(7/3, \infty)$, the car is speeding up for $0 < t < 2/3$ seconds, slowing down for $2/3 < t < 7/3$ seconds, and speeding up again for $t > 7/3$ seconds. See the sign graphs on the next page.

YOUR TURN 3 Find the velocity and acceleration of the car if the distance (in feet) is given by $s(t) = t^3 - 3t^2 - 24t + 10$, at time t (in seconds). When is the car going forward or backing up? When is the car speeding up or slowing down?

TRY YOUR TURN 3

Concavity of a Graph

The first derivative has been used to show where a function is increasing or decreasing and where the extrema occur. The second derivative gives the rate of change of the first derivative; it indicates *how fast* the function is increasing or decreasing. The rate of change of the derivative (the second derivative) affects the *shape* of the graph. Intuitively, we say that a graph is *concave upward* on an interval if it "holds water" and *concave downward* if it "spills water." See Figure 28.

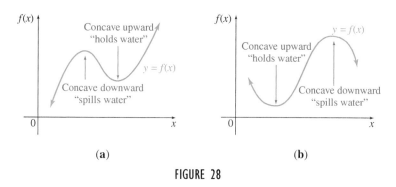

FIGURE 28

More precisely, a function is **concave upward** on an interval (a, b) if the graph of the function lies above its tangent line at each point of (a, b). A function is **concave downward** on (a, b) if the graph of the function lies below its tangent line at each point of (a, b). A point where a graph changes **concavity** is called an **inflection point**. See Figure 29.

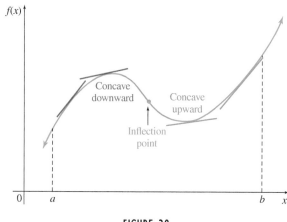

FIGURE 29

Users of soft contact lenses recognize concavity as the way to tell if a lens is inside out. As Figure 30 shows, a correct contact lens has a profile that is entirely concave upward. The profile of an inside-out lens has inflection points near the edges, where the profile begins to turn concave downward very slightly.

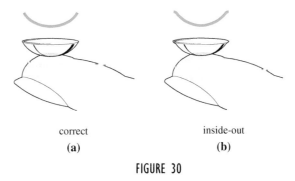

correct

(a)

inside-out

(b)

FIGURE 30

Just as a function can be either increasing or decreasing on an interval, it can be either concave upward or concave downward on an interval. Examples of various combinations are shown in Figure 31.

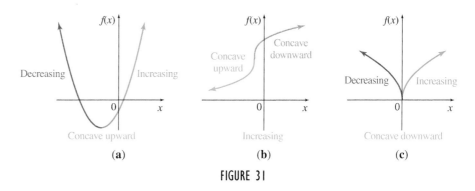

(a)   **(b)**   **(c)**

FIGURE 31

Figure 32 shows two functions that are concave upward on an interval (a, b) Several tangent lines are also shown. In Figure 32(a), the slopes of the tangent lines (moving from left to right) are first negative, then 0, and then positive. In Figure 32(b), the slopes are all positive, but they get larger.

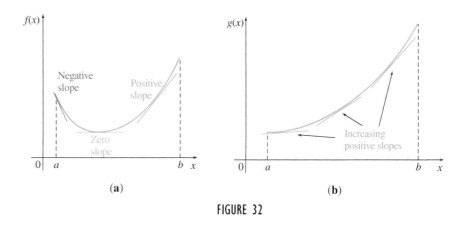

(a)   **(b)**

FIGURE 32

In both cases, the slopes are *increasing.* The slope at a point on a curve is given by the derivative. Since a function is increasing if its derivative is positive, its slope is increasing if the derivative of the slope function is positive. Since the derivative of a derivative is the second derivative, a function is concave upward on an interval if its second derivative is positive at each point of the interval.

A similar result is suggested by Figure 33 for functions whose graphs are concave downward. In both graphs, the slopes of the tangent lines are *decreasing* as we move from left to right. Since a function is decreasing if its derivative is negative, a function is concave downward on an interval if its second derivative is negative at each point of the interval. These observations suggest the following test.

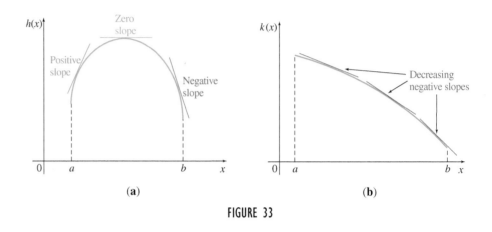

(a)   **(b)**

FIGURE 33

Test for Concavity

Let f be a function with derivatives f' and f'' existing at all points in an interval (a, b). Then f is concave upward on (a, b) if $f''(x) > 0$ for all x in (a, b) and concave downward on (a, b) if $f''(x) < 0$ for all x in (a, b).

FIGURE 34

An easy way to remember this test is by the faces shown in Figure 34. When the second derivative is positive at a point $(+ +)$, the graph is concave upward $(\smile)$. When the second derivative is negative at a point $(- -)$, the graph is concave downward $(\frown)$.

EXAMPLE 5 Concavity

Find all intervals where $f(x) = x^4 - 8x^3 + 18x^2$ is concave upward or downward, and find all inflection points.

SOLUTION The first derivative is $f'(x) = 4x^3 - 24x^2 + 36x$, and the second derivative is $f''(x) = 12x^2 - 48x + 36$. We factor $f''(x)$ as $12(x - 1)(x - 3)$, and then create a number line for $f''(x)$ as we did in the previous two sections for $f'(x)$.

We see from Figure 35 that $f''(x) > 0$ on the intervals $(-\infty, 1)$ and $(3, \infty)$, so f is concave upward on these intervals. Also, $f''(x) < 0$ on the interval $(1, 3)$, so f is concave downward on this interval.

FIGURE 35

Finally, we have inflection points where f'' changes sign, namely, at $x = 1$ and $x = 3$. Since $f(1) = 11$ and $f(3) = 27$, the inflection points are $(1, 11)$ and $(3, 27)$.

Although we were only seeking information about concavity and inflection points in this example, it is also worth noting that $f'(x) = 4x^3 - 24x^2 + 36x = 4x(x - 3)^2$, which has roots at $x = 0$ and $x = 3$. Verify that there is a relative minimum at $(0, 0)$, but that $(3, 27)$ is neither a relative minimum nor a relative maximum. The function is graphed in Figure 36.

YOUR TURN 4 Find all intervals where $f(x) = x^5 - 30x^3$ is concave upward or downward, and find all inflection points.

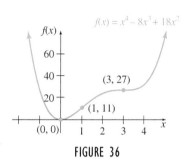

FIGURE 36   TRY YOUR TURN 4

Example 5 suggests the following result.

At an inflection point for a function f, the second derivative is 0 or does not exist.

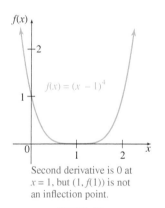

Second derivative is 0 at $x = 1$, but $(1, f(1))$ is not an inflection point.

FIGURE 37

CAUTION

1. Be careful with the previous statement. Finding a value of x where $f''(x) = 0$ does not mean that an inflection point has been located. For example, if $f(x) = (x - 1)^4$, then $f''(x) = 12(x - 1)^2$, which is 0 at $x = 1$. The graph of $f(x) = (x - 1)^4$ is always concave upward, however, so it has no inflection point. See Figure 37.

2. Note that the concavity of a function might change not only at a point where $f''(x) = 0$ but also where $f''(x)$ does not exist. For example, this happens at $x = 0$ for $f(x) = x^{1/3}$.

⊬ TECHNOLOGY NOTE

Most graphing calculators do not have a feature for finding inflection points. Nevertheless, a graphing calculator sketch can be useful for verifying that your calculations for finding inflection points and intervals where the function is concave up or down are correct.

Second Derivative Test

The idea of concavity can often be used to decide whether a given critical number produces a relative maximum or a relative minimum. This test, an alternative to the first derivative test, is based on the fact that a curve with a horizontal tangent at a point c and concave downward on an open interval containing c also has a relative maximum at c. A relative minimum occurs when a graph has a horizontal tangent at a point d and is concave upward on an open interval containing d. See Figure 38 on the next page.

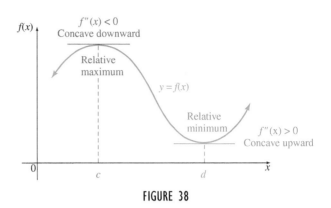

FIGURE 38

A function f is concave upward on an interval if $f''(x) > 0$ for all x in the interval, while f is concave downward on an interval if $f''(x) < 0$ for all x in the interval. These ideas lead to the **second derivative test** for relative extrema.

Second Derivative Test

Let f'' exist on some open interval containing c, (except possibly at c itself) and let $f'(c) = 0$.

1. If $f''(c) > 0$, then $f(c)$ is a relative minimum.

2. If $f''(c) < 0$, then $f(c)$ is a relative maximum.

3. If $f''(c) = 0$ or $f''(c)$ does not exist, then the test gives no information about extrema, so use the first derivative test.

NOTE
In Case 3 of the second derivative test (when $f''(c) = 0$ or does not exist), observe that if $f''(x)$ changes sign at c, there is an inflection point at $x = c$.

EXAMPLE 6 **Second Derivative Test**

Find all relative extrema and inflection points for each of the following functions.

(a) $f(x) = 4x^3 + 7x^2 - 10x + 8$.

SOLUTION First, find the points where the derivative is 0. Here $f'(x) = 12x^2 + 14x - 10$. Solve the equation $f'(x) = 0$ to get

$$12x^2 + 14x - 10 = 0$$
$$2(6x^2 + 7x - 5) = 0$$
$$2(3x + 5)(2x - 1) = 0$$

$$3x + 5 = 0 \quad \text{or} \quad 2x - 1 = 0$$
$$3x = -5 \qquad\qquad 2x = 1$$
$$x = -\frac{5}{3} \qquad\qquad x = \frac{1}{2}.$$

Now use the second derivative test. The second derivative is $f''(x) = 24x + 14$. Evaluate $f''(x)$ first at $-5/3$, getting

$$f''\left(-\frac{5}{3}\right) = 24\left(-\frac{5}{3}\right) + 14 = -40 + 14 = -26 < 0,$$

so that by Part 2 of the second derivative test, $-5/3$ leads to a relative maximum of $f(-5/3) = 691/27 \approx 25.59$. Also, when $x = 1/2$,

$$f''\left(\frac{1}{2}\right) = 24\left(\frac{1}{2}\right) + 14 = 12 + 14 = 26 > 0,$$

with $1/2$ leading to a relative minimum of $f(1/2) = 21/4 = 5.25$.

Notice also that $f''(x) = 24x + 14 = 0$ when $x = -14/24 = -7/12$. Since $f''(x) < 0$ when $x < -7/12$ and $f''(x) > 0$ when $x > -7/12$, there is an inflection point at $x = -7/12$, where $f(-7/12) \approx 15.42$. A graph of f is shown in Figure 39.

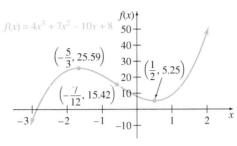

FIGURE 39

(b) $f(x) = 2x^5 - 5x^4 + 50$

SOLUTION   The derivative is $f'(x) = 10x^4 - 20x^3$. Find all points where $f'(x) = 0$.

$$10x^4 - 20x^3 = 0$$
$$10x^3(x - 2) = 0 \qquad \text{Factor.}$$
$$x = 0 \qquad \text{or} \qquad x = 2$$

To use the second derivative test, calculate $f''(x) = 40x^3 - 60x^2 = 20x^2(2x - 3)$. Notice that $f''(2) = 20(2)^2(2 \cdot 2 - 3) > 0$, so by the second derivative test, there is a relative minimum of $f(2) = 2(2)^5 - 5(2)^4 + 50 = 34$ at $x = 2$. Because $f''(0) = 0$, the second derivative test tells us nothing about the critical point at $x = 0$. Instead, we use the first derivative test to note that $f'(-1) = 10(-1)^3(-1 - 2) > 0$ and $f'(1) = 10(1)^3(1 - 2) < 0$. Because f is increasing to the left of 0 and decreasing to the right of 0, there is a relative maximum of $f(0) = 50$ at $x = 0$.

YOUR TURN 5

Find all relative extrema of $f(x) = -2x^3 + 3x^2 + 72x$.

Notice also that $f''(x) = 20x^2(2x - 3) = 0$ when $x = 0$ and $x = 3/2$. There is not an inflection point at $x = 0$ because $f''(x)$ is negative when x is slightly less than 0 or slightly greater than 0. There is, however, an inflection point at $x = 3/2$, where $f(3/2) = 39.875$. A graph of f is shown in Figure 40.   TRY YOUR TURN 5

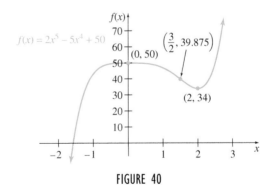

FIGURE 40

CAUTION   The second derivative test works only for those critical numbers c that make $f'(c) = 0$. This test does not work for critical numbers c for which $f'(c)$ does not exist (since $f''(c)$ would not exist either). Also, the second derivative test does not work for critical numbers c that make $f''(c) = 0$, such as in Example 6(b). In both of these cases, use the first derivative test.

The *law of diminishing returns* in economics is related to the idea of concavity. The function graphed in Figure 41 on the next page gives the output y from a given input x. If the input were advertising costs for some product, for example, the output might be the corresponding revenue from sales.

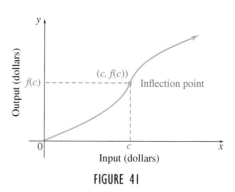

FIGURE 41

The graph in Figure 41 shows an inflection point at $(c, f(c))$. For $x < c$, the graph is concave upward, so the rate of change of the slope is increasing. This indicates that the output y is increasing at a faster rate with each additional dollar spent. When $x > c$, however, the graph is concave downward, the rate of change of the slope is decreasing, and the increase in y is smaller with each additional dollar spent. Thus, further input beyond c dollars produces diminishing returns. The inflection point at $(c, f(c))$ is called the **point of diminishing returns**. Beyond this point there is a smaller and smaller return for each dollar invested.

As another example of diminishing returns, consider a farm with a fixed amount of land, machinery, fertilizer, and so on. Adding workers increases production a lot at first, then less and less with each additional worker.

EXAMPLE 7   Point of Diminishing Returns

The revenue $R(x)$ generated from sales of a certain product is related to the amount x spent on advertising by

$$R(x) = \frac{1}{15,000}(600x^2 - x^3), \quad 0 \le x \le 600,$$

where x and $R(x)$ are in thousands of dollars. Is there a point of diminishing returns for this function? If so, what is it?

SOLUTION   Since a point of diminishing returns occurs at an inflection point, look for an x-value that makes $R''(x) = 0$. Write the function as

$$R(x) = \frac{600}{15,000}x^2 - \frac{1}{15,000}x^3 = \frac{1}{25}x^2 - \frac{1}{15,000}x^3.$$

Now find $R'(x)$ and then $R''(x)$.

$$R'(x) = \frac{2x}{25} - \frac{3x^2}{15,000} = \frac{2}{25}x - \frac{1}{5000}x^2$$

$$R''(x) = \frac{2}{25} - \frac{2x}{5000} = \frac{2}{25} - \frac{1}{2500}x$$

Set $R''(x)$ equal to 0 and solve for x.

$$\frac{2}{25} - \frac{1}{2500}x = 0$$

$$-\frac{1}{2500}x = -\frac{2}{25}$$

$$x = \frac{5000}{25} = 200$$

Test a number in the interval $(0, 200)$ to see that $R''(x)$ is positive there. Then test a number in the interval $(200, 600)$ to find that $R''(x)$ is negative in that interval. Since the sign of $R''(x)$ changes from positive to negative at $x = 200$, the graph changes from concave upward to concave downward at that point, and there is a point of diminishing returns at the inflection point $(200, 1066\frac{2}{3})$. Investments in advertising beyond \$200,000 return less and less for each dollar invested. Verify that $R'(200) = 8$. This means that when \$200,000 is invested, another \$1000 invested returns approximately \$8000 in additional revenue. Thus it may still be economically sound to invest in advertising beyond the point of diminishing returns. ▪

5.3 WARM-UP EXERCISES

Find all relative extrema for the following functions, as well as the x-values where they occur.   *(Sec. 5.2)*

W1. $f(x) = x^3 - 3x^2 - 72x + 20$

W2. $f(x) = x^4 - 8x^3 - 32x^2 + 10$

5.3 EXERCISES

Find $f''(x)$ for each function. Then find $f''(0)$ and $f''(2)$.

1. $f(x) = 5x^3 - 7x^2 + 4x + 3$

2. $f(x) = 4x^3 + 5x^2 + 6x - 7$

3. $f(x) = 4x^4 - 3x^3 - 2x^2 + 6$

4. $f(x) = -x^4 + 7x^3 - \dfrac{x^2}{2}$

5. $f(x) = 3x^2 - 4x + 8$

6. $f(x) = 8x^2 + 6x + 5$

7. $f(x) = \dfrac{x^2}{1 + x}$

8. $f(x) = \dfrac{-x}{1 - x^2}$

9. $f(x) = \sqrt{x^2 + 4}$

10. $f(x) = \sqrt{2x^2 + 9}$

11. $f(x) = 32x^{3/4}$

12. $f(x) = -6x^{1/3}$

13. $f(x) = 5e^{-x^2}$

14. $f(x) = 0.5e^{x^2}$

15. $f(x) = \dfrac{\ln x}{4x}$

16. $f(x) = \ln x + \dfrac{1}{x}$

Find $f'''(x)$, the third derivative of f, and $f^{(4)}(x)$, the fourth derivative of f, for each function.

17. $f(x) = 7x^4 + 6x^3 + 5x^2 + 4x + 3$

18. $f(x) = -2x^4 + 7x^3 + 4x^2 + x$

19. $f(x) = 5x^5 - 3x^4 + 2x^3 + 7x^2 + 4$

20. $f(x) = 2x^5 + 3x^4 - 5x^3 + 9x - 2$

21. $f(x) = \dfrac{x - 1}{x + 2}$

22. $f(x) = \dfrac{x + 1}{x}$

23. $f(x) = \dfrac{3x}{x - 2}$

24. $f(x) = \dfrac{x}{2x + 1}$

25. Let f be an nth degree polynomial of the form $f(x) = x^n + a_{n-1}x^{n-1} + \cdots + a_1x + a_0$.

(a) Find $f^{(n)}(x)$.

(b) Find $f^{(k)}(x)$ for $k > n$.

26. Let $f(x) = \ln x$.

(a) Compute $f'(x)$, $f''(x)$, $f'''(x)$, $f^{(4)}(x)$, and $f^{(5)}(x)$.

(b) Guess a formula for $f^{(n)}(x)$, where n is any positive integer.

27. For $f(x) = e^x$, find $f''(x)$ and $f'''(x)$. What is the nth derivative of f with respect to x?

28. For $f(x) = a^x$, find $f''(x)$ and $f'''(x)$. What is the nth derivative of f with respect to x?

In Exercises 29–50, find the open intervals where the functions are concave upward or concave downward. Find any inflection points.

29.

30.

31.   **32.**

33.

34.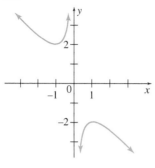

35. $f(x) = x^2 + 10x - 9$ **36.** $f(x) = 8 - 6x - x^2$

37. $f(x) = -2x^3 + 9x^2 + 168x - 3$

38. $f(x) = -x^3 - 12x^2 - 45x + 2$

39. $f(x) = \dfrac{3}{x - 5}$ **40.** $f(x) = \dfrac{-2}{x + 1}$

41. $f(x) = x(x + 5)^2$ **42.** $f(x) = -x(x - 3)^2$

43. $f(x) = 18x - 18e^{-x}$ **44.** $f(x) = 2e^{-x^2}$

45. $f(x) = x^{8/3} - 4x^{5/3}$ **46.** $f(x) = x^{7/3} + 56x^{4/3}$

47. $f(x) = \ln(x^2 + 1)$ **48.** $f(x) = x^2 + 8 \ln |x + 1|$

49. $f(x) = x^2 \log |x|$ **50.** $f(x) = 5^{-x^2}$

For each of the exercises listed below, suppose that the function that is graphed is not $f(x)$, but $f'(x)$. Find the open intervals where the original function is concave upward or concave downward, and find the location of any inflection points.

51. Exercise 29   **52.** Exercise 30

53. Exercise 31   **54.** Exercise 32

55. Give an example of a function $f(x)$ such that $f'(0) = 0$ but $f''(0)$ does not exist. Is there a relative minimum or maximum or an inflection point at $x = 0$?

 56. (a) Graph the two functions $f(x) = x^{7/3}$ and $g(x) = x^{5/3}$ on the window $[-2, 2]$ by $[-2, 2]$.

   (b) Verify that both f and g have an inflection point at $(0, 0)$.

   (c) How is the value of $f''(0)$ different from $g''(0)$?

   (d) Based on what you have seen so far in this exercise, is it always possible to tell the difference between a point where the second derivative is 0 or undefined based on the graph? Explain.

57. Describe the slope of the tangent line to the graph of $f(x) = e^x$ for the following.

   (a) $x \to -\infty$ **(b)** $x \to 0$

58. What is true about the slope of the tangent line to the graph of $f(x) = \ln x$ as $x \to \infty$? As $x \to 0$?

Find any critical numbers for f in Exercises 59–66 and then use the second derivative test to decide whether the critical numbers lead to relative maxima or relative minima. If $f''(c) = 0$ or $f''(c)$ does not exist for a critical number c, then the second derivative test gives no information. In this case, use the first derivative test instead.

59. $f(x) = -x^2 - 10x - 25$ **60.** $f(x) = x^2 - 12x + 36$

61. $f(x) = 3x^3 - 3x^2 + 1$ **62.** $f(x) = 2x^3 - 4x^2 + 2$

63. $f(x) = (x + 3)^4$ **64.** $f(x) = x^3$

65. $f(x) = x^{7/3} + x^{4/3}$ **66.** $f(x) = x^{8/3} + x^{5/3}$

 Sometimes the derivative of a function is known, but not the function. We will see more of this later in the book. For each function f' defined in Exercises 67–70, find $f''(x)$, then use a graphing calculator to graph f' and f'' in the indicated window. Use the graph to do the following.

(a) Give the (approximate) x-values where f has a maximum or minimum.

(b) By considering the sign of $f'(x)$, give the (approximate) intervals where $f(x)$ is increasing and decreasing.

(c) Give the (approximate) x-values of any inflection points.

(d) By considering the sign of $f''(x)$, give the intervals where f is concave upward or concave downward.

67. $f'(x) = x^3 - 6x^2 + 7x + 4; [-5, 5]$ by $[-5, 15]$

68. $f'(x) = 10x^2(x - 1)(5x - 3); [-1, 1.5]$ by $[-20, 20]$

69. $f'(x) = \dfrac{1 - x^2}{(x^2 + 1)^2}; [-3, 3]$ by $[-1.5, 1.5]$

70. $f'(x) = x^2 + x \ln x; [0, 1]$ by $[-2, 2]$

71. Suppose a friend makes the following argument. A function f is increasing and concave downward. Therefore, f' is positive and decreasing, so it eventually becomes 0 and then negative, at which point f decreases. Show that your friend is wrong by giving an example of a function that is always increasing and concave downward.

APPLICATIONS

Business and Economics

Point of Diminishing Returns   In Exercises 72–75, find the point of diminishing returns (x, y) for the given functions, where $R(x)$, represents revenue (in thousands of dollars) and x represents the amount spent on advertising (in thousands of dollars).

72. $R(x) = 10{,}000 - x^3 + 42x^2 + 800x, \quad 0 \le x \le 20$

73. $R(x) = \dfrac{4}{27}(-x^3 + 66x^2 + 1050x - 400), \quad 0 \le x \le 25$

74. $R(x) = -0.3x^3 + x^2 + 11.4x, \quad 0 \le x \le 6$

75. $R(x) = -0.6x^3 + 3.7x^2 + 5x, \quad 0 \le x \le 6$

76. Risk Aversion   In economics, an index of *absolute risk aversion* is defined as

$$I(M) = \frac{-U''(M)}{U'(M)},$$

where M measures how much of a commodity is owned and $U(M)$ is a *utility function,* which measures the ability of quantity M of a commodity to satisfy a consumer's wants. Find $I(M)$ for $U(M) = \sqrt{M}$ and for $U(M) = M^{2/3}$, and determine which indicates a greater aversion to risk.

77. Demand Function The authors of an article in an economics journal state that if $D(q)$ is the demand function, then the inequality

$$qD''(q) + D'(q) < 0$$

is equivalent to saying that the marginal revenue declines more quickly than does the price. Prove that this equivalence is true. *Source: Bell Journal of Economics.*

78. Social Security Assets As seen in the first section of this chapter, the projected year-end assets in the Social Security trust funds, in trillions of dollars, where t represents the number of years since 2000, can be approximated by

$$A(t) = 0.0000329t^3 - 0.00450t^2 + 0.0613t + 2.34,$$

where $0 \leq t \leq 50$. Find the value of t when Social Security assets will decrease most rapidly. Approximately when does this occur? *Source: Social Security Administration.*

79. Website Revenues A study on optimizing revenue from a website considered dividing customers into two groups based on a value x between 0 and 1, where x measures the proportion of the total bandwidth requested by a customer. Customers with a request less than x were considered low revenue, and those above x high revenue. The expected revenue from the low revenue customers was described by

$$R(x) = Cx(1 - e^{-kx}),$$

where C and k are positive constants based on attributes of the website and the customers. *Source: The Journal of the Operational Research Society.*

(a) Find $R'(x)$ and use it to find for what values of x in $[0, 1]$ the revenue is increasing.

(b) Find $R''(x)$ and use it to find for what values of x in $[0, 1]$ the revenue function is concave up.

80. Brand Choice A study showed that the ratio of the probability that consumers choose brand A over brand B can be given by

$$f(c) = \frac{a - c}{b - c},$$

where a is the number of features of brand A, b is the number of features of brand B, and c is the number of common features. Show the following. *Source: Marketing Science.*

(a) If $a > b > c$, then f is increasing and concave upward.

(b) If $b > a > c$, then f is decreasing and concave downward.

(c) Suppose you are the marketing manager for a company producing brand A. What marketing information could you derive from the results in parts (a) and (b)?

Life Sciences

81. Population Growth When a hardy new species is introduced into an area, the population often increases as shown in the graph in the next column. Explain the significance of the following function values on the graph.

(a) f_0 (b) $f(a)$ (c) f_M

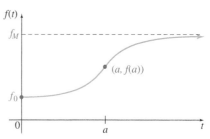

82. Ozone Depletion According to an article in *The New York Times,* "Government scientists reported last week that they had detected a slowdown in the rate at which chemicals that deplete the earth's protective ozone layer are accumulating in the atmosphere." Letting $c(t)$ be the amount of ozone-depleting chemicals at time t, what does this statement tell you about $c(t)$, $c'(t)$, and $c''(t)$? *Source: The New York Times.*

83. Drug Concentration The percent of concentration of a certain drug in the bloodstream t hours after the drug is administered is given by

$$K(t) = \frac{3t}{t^2 + 4}.$$

For example, after 1 hour the concentration is given by

$$K(1) = \frac{3(1)}{1^2 + 4} = \frac{3}{5}\% = 0.6\% = 0.006.$$

(a) Find the time at which concentration is a maximum.

(b) Find the maximum concentration.

84. Drug Concentration The percent of concentration of a drug in the bloodstream t hours after the drug is administered is given by

$$K(t) = \frac{4t}{3t^2 + 27}.$$

(a) Find the time at which the concentration is a maximum.

(b) Find the maximum concentration.

The next two exercises are a continuation of exercises first given in the section on Derivatives of Exponential Functions. Find the inflection point of the graph of each logistic function. This is the point at which the growth rate begins to decline.

85. Cactus Wrens The weight of cactus wrens in grams after t days is given by

$$G(t) = 31.4/(1 + 12.5e^{-0.393t}).$$

Source: Ecology.

86. Whooping Cranes The population of whooping cranes after t years is given by

$$G(t) = 787/(1 + 60.8e^{-0.0473t}).$$

Source: U.S. Fish and Wildlife Service.

Hints for Exercises 87 and 88: Leave B, c, and k as constants until you are ready to calculate your final answer.

87. Clam Growth Researchers used a version of the Gompertz curve to model the growth of razor clams during the first seven years of the clams' lives with the equation

$$L(t) = Be^{-ce^{-kt}},$$

where $L(t)$ gives the length (in centimeters) after t years, $B = 14.3032$, $c = 7.267963$, and $k = 0.670840$. Find the inflection point and describe what it signifies. *Source: Journal of Experimental Biology.*

88. Breast Cancer Growth Researchers used a version of the Gompertz curve to model the growth of breast cancer tumors with the equation

$$N(t) = e^{c(1-e^{-kt})},$$

where $N(t)$ is the number of cancer cells after t days, $c = 27.3$, and $k = 0.011$. Find the inflection point and describe what it signifies. *Source: Cancer Research.*

89. Popcorn Researchers have determined that the amount of moisture present in a kernel of popcorn affects the volume of the popped corn and can be modeled for certain sizes of kernels by the function

$$v(x) = -35.98 + 12.09x - 0.4450x^2,$$

where x is moisture content (%, wet basis) and $v(x)$ is the expansion volume (in cm^3/gram). Describe the concavity of this function. *Source: Cereal Chemistry.*

90. Alligator Teeth Researchers have developed a mathematical model that can be used to estimate the number of teeth $N(t)$ at time t (days of incubation) for *Alligator mississippiensis*, where

$$N(t) = 71.8e^{-8.96e^{-0.0685t}}.$$

Find the inflection point and describe its importance to this research. *Source: Journal of Theoretical Biology.*

91. Plant Growth Researchers have found that the probability P that a plant will grow to radius R can be described by the equation

$$P(R) = \frac{1}{1 + 2\pi DR^2},$$

where D is the density of the plants in an area. A graph in their publication shows an inflection point around $R = 0.022$. Find an expression for the value of R in terms of D at the inflection point, and find the value of D corresponding to an inflection point at $R = 0.022$. *Source: Ecology.*

Social Sciences

92. Crime In 1995, the rate of violent crimes in New York City continued to decrease, but at a slower rate than in previous years. Letting $f(t)$ be the rate of violent crime as a function of time, what does this tell you about $f(t)$, $f'(t)$, and $f''(t)$? *Source: The New York Times.*

Physical Sciences

93. Velocity and Acceleration When an object is dropped straight down, the distance (in feet) that it travels in t seconds is given by

$$s(t) = -16t^2.$$

Find the velocity at each of the following times.

(a) After 3 seconds   (b) After 5 seconds

(c) After 8 seconds

(d) Find the acceleration. (The answer here is a constant—the acceleration due to the influence of gravity alone near the surface of Earth.)

94. Velocity and Acceleration of a Car A car rolls down a hill. Its distance (in feet) from its starting point is given by $s(t) = 1.5t^2 + 4t$, where t is in seconds.

(a) How far will the car move in 10 seconds?

(b) What is the velocity at 5 seconds? At 10 seconds?

(c) How can you tell from $v(t)$ that the car will not stop?

(d) What is the acceleration at 5 seconds? At 10 seconds?

(e) What is happening to the velocity and the acceleration as t increases?

95. Height of a Ball If a cannonball is shot directly upward with a velocity of 256 ft per second, its height above the ground after t seconds is given by $s(t) = 256t - 16t^2$. Find the velocity and the acceleration after t seconds. What is the maximum height the cannonball reaches? When does it hit the ground?

96. Velocity and Acceleration A car is moving along a straight stretch of road. The acceleration of the car is given by the graph shown. Assume that the velocity of the car is always positive. At what time was the car moving most rapidly? Explain. *Source: Larry Taylor.*

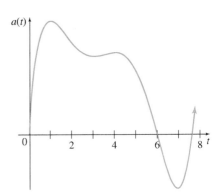

1. 36

2. (a) $f''(x) = 30x^4 + 12x$ (b) $g''(x) = 2e^x + xe^x$

(c) $h''(x) = \dfrac{-3 + 2\ln x}{x^3}$

3. $v(t) = 3t^2 - 6t - 24$ and $a(t) = 6t - 6$. Car backs up for the first 4 seconds and then goes forward. It speeds up for $0 < t < 1$, slows down for $1 < t < 4$, and then speeds up for $t > 4$.

4. Concave upward on $(-3, 0)$ and $(3, \infty)$; concave downward on $(-\infty, -3)$ and $(0, 3)$; inflection points are $(-3, 567)$, $(0, 0)$ and $(3, -567)$.

5. Relative maximum of $f(4) = 208$ at $x = 4$ and relative minimum of $f(-3) = -135$ at $x = -3$.

5.4 Curve Sketching

In the following examples, the test for concavity, the test for increasing and decreasing functions, and the concept of limits at infinity will help us sketch the graphs and describe the behavior of a variety of functions. This process, called **curve sketching**, has decreased somewhat in importance in recent years due to the widespread use of graphing calculators. We believe, however, that this topic is worth studying for the following reasons.

For one thing, a graphing calculator picture can be misleading, particularly if important points lie outside the viewing window. Even if all important features are within the viewing windows, there is still the problem that the calculator plots and connects points and misses what goes on between those points. As an example of the difficulty in choosing an appropriate window without a knowledge of calculus, see Exercise 40 in the second section of this chapter.

Furthermore, curve sketching may be the best way to learn the material in the previous three sections. You may feel confident that you understand what increasing and concave upward mean, but using those concepts in a graph will put your understanding to the test.

Curve sketching may be done with the following steps.

Curve Sketching

To sketch the graph of a function f:

1. Consider the domain of the function, and note any restrictions. (That is, avoid dividing by 0, taking a square root, or any even root, of a negative number, or taking the logarithm of 0 or a negative number.)

2. Find the y-intercept (if it exists) by substituting $x = 0$ into $f(x)$. Find any x-intercepts by solving $f(x) = 0$ if this is not too difficult.

3. (a) If f is a rational function, find any vertical asymptotes by investigating where the denominator is 0, and find any horizontal asymptotes by finding the limits as $x \to \infty$ and $x \to -\infty$.

   (b) If f is an exponential function, find any horizontal asymptotes; if f is a logarithmic function, find any vertical asymptotes.

4. Investigate symmetry. If $f(-x) = f(x)$, the function is even, so the graph is symmetric about the y-axis. If $f(-x) = -f(x)$, the function is odd, so the graph is symmetric about the origin.

5. Find $f'(x)$. Locate any critical points by solving the equation $f'(x) = 0$ and determining where $f'(x)$ does not exist, but $f(x)$ does. Find any relative extrema and determine where f is increasing or decreasing.

6. Find $f''(x)$. Locate potential inflection points by solving the equation $f''(x) = 0$ and determining where $f''(x)$ does not exist. Determine where f is concave upward or concave downward.

7. Plot the intercepts, the critical points, the inflection points, the asymptotes, and other points as needed. Take advantage of any symmetry found in Step 4.

8. Connect the points with a smooth curve using the correct concavity, being careful not to connect points where the function is not defined.

9. Check your graph using a graphing calculator. If the picture looks very different from what you've drawn, see in what ways the picture differs and use that information to help find your mistake.

There are four possible combinations for a function to be increasing or decreasing and concave up or concave down, as shown in the following table.

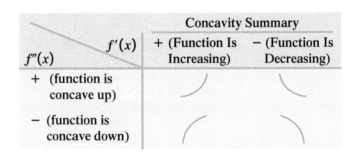

$f''(x)$ ╲ $f'(x)$	Concavity Summary	
	+ (Function Is Increasing)	− (Function Is Decreasing)
+ (function is concave up)		
− (function is concave down)		

EXAMPLE 1 Polynomial Function Graph

Graph $f(x) = 2x^3 - 3x^2 - 12x + 1$.

SOLUTION The domain is $(-\infty, \infty)$. The y-intercept is located at $y = f(0) = 1$. Finding the x-intercepts requires solving the equation $f(x) = 0$. But this is a third-degree equation; since we have not covered a procedure for solving such equations, we will skip this step. This is neither a rational nor an exponential nor a logarithmic function, so we also skip step 3. Observe that $f(-x) = 2(-x)^3 - 3(-x)^2 - 12(-x) + 1 = -2x^3 - 3x^2 + 12x + 1$, which is neither $f(x)$ nor $-f(x)$, so there is no symmetry about the y-axis or origin.

To find the intervals where the function is increasing or decreasing, find the first derivative.

$$f'(x) = 6x^2 - 6x - 12$$

This derivative is 0 when

$$6(x^2 - x - 2) = 0$$
$$6(x - 2)(x + 1) = 0$$
$$x = 2 \quad \text{or} \quad x = -1.$$

These critical numbers divide the number line in Figure 42 into three regions. Testing a number from each region in $f'(x)$ shows that f is increasing on $(-\infty, -1)$ and $(2, \infty)$ and decreasing on $(-1, 2)$. This is shown with the arrows in Figure 42. By the first derivative test, f has a relative maximum when $x = -1$ and a relative minimum when $x = 2$. The relative maximum is $f(-1) = 8$, while the relative minimum is $f(2) = -19$.

FIGURE 42

Now use the second derivative to find the intervals where the function is concave upward or downward. Here

$$f''(x) = 12x - 6,$$

which is 0 when $x = 1/2$. Testing a point with x less than $1/2$, and one with x greater than $1/2$, shows that f is concave downward on $(-\infty, 1/2)$ and concave upward on $(1/2, \infty)$.

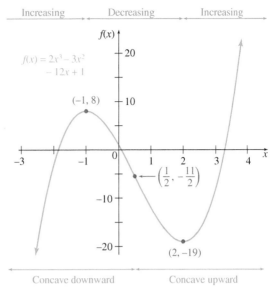

FIGURE 43

The graph has an inflection point at $(1/2, f(1/2))$, or $(1/2, -11/2)$. This information is summarized in the following table.

	Graph Summary			
Interval	$(-\infty, -1)$	$(-1, 1/2)$	$(1/2, 2)$	$(2, \infty)$
Sign of f'	+	−	−	+
Sign of f"	−	−	+	+
f Increasing or Decreasing	Increasing	Decreasing	Decreasing	Increasing
Concavity of f	Downward	Downward	Upward	Upward
Shape of Graph	⌢	⌢	⌣	⌣

Use this information and the critical points to get the graph shown in Figure 43. Notice that the graph appears to be symmetric about its inflection point. It can be shown that is always true for third-degree polynomials. In other words, if you put your pencil point at the inflection point and then spin the book 180° about the pencil point, the graph will appear to be unchanged. **TRY YOUR TURN 1**

YOUR TURN 1 Graph $f(x) = -x^3 + 3x^2 + 9x - 10$.

⊞ **TECHNOLOGY NOTE**

A graphing calculator picture of the function in Figure 43 on the arbitrarily chosen window $[-3, 3]$ by $[-7, 7]$ gives a misleading picture, as Figure 44(a) shows. Knowing where the turning points lie tells us that a better window would be $[-3, 4]$ by $[-20, 20]$, with the results shown in Figure 44(b).

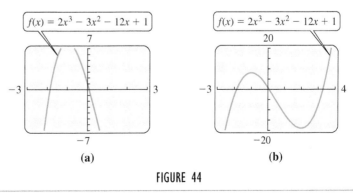

FIGURE 44

> **EXAMPLE 2** **Rational Function Graph**

Graph $f(x) = x + \dfrac{1}{x}$.

SOLUTION Notice that $x = 0$ is not in the domain of the function, so there is no y-intercept. To find the x-intercept, solve $f(x) = 0$.

$$x + \frac{1}{x} = 0$$

$$x = -\frac{1}{x}$$

$$x^2 = -1$$

Since x^2 is always positive, there is also no x-intercept.

The function is a rational function, but it is not written in the usual form of one polynomial over another. By getting a common denominator and adding the fractions, it can be rewritten in that form:

$$f(x) = x + \frac{1}{x} = \frac{x^2 + 1}{x}.$$

FOR REVIEW

Asymptotes were discussed in Section 2.3 on Polynomial and Rational Functions. You may wish to review that section. To review limits, refer to Section 3.1 in the chapter titled The Derivative.

Because $x = 0$ makes the denominator (but not the numerator) 0, the line $x = 0$ is a vertical asymptote. To find any horizontal asymptotes, we investigate

$$\lim_{x \to \infty} \frac{x^2 + 1}{x} = \lim_{x \to \infty} \left(\frac{x^2}{x} + \frac{1}{x} \right) = \lim_{x \to \infty} \left(x + \frac{1}{x} \right).$$

The second term, $1/x$, approaches 0 as $x \to \infty$, but the first term, x, becomes infinitely large, so the limit does not exist. Verify that $\lim_{x \to -\infty} f(x)$ also does not exist, so there are no horizontal asymptotes.

Observe that as x gets very large, the second term $(1/x)$ in $f(x)$ gets very small, so $f(x) = x + (1/x) \approx x$. The graph gets closer and closer to the straight line $y = x$ as x becomes larger and larger. This is what is known as an **oblique asymptote**.

Observe that

$$f(-x) = (-x) + \frac{1}{-x} = -\left(x + \frac{1}{x} \right) = -f(x),$$

so the graph is symmetric about the origin. This means that the left side of the graph can be found by rotating the right side $180°$ about the origin.

Here $f'(x) = 1 - (1/x^2)$, which is 0 when

$$\frac{1}{x^2} = 1$$

$$x^2 = 1$$

$$x^2 - 1 = (x - 1)(x + 1) = 0$$

$$x = 1 \quad \text{or} \quad x = -1.$$

The derivative fails to exist at 0, where the vertical asymptote is located. Evaluating $f'(x)$ in each of the regions determined by the critical numbers and the asymptote shows that f is increasing on $(-\infty, -1)$ and $(1, \infty)$ and decreasing on $(-1, 0)$ and $(0, 1)$. See Figure 45(a). By the first derivative test, f has a relative maximum of $y = f(-1) = -2$ when $x = -1$, and a relative minimum of $y = f(1) = 2$ when $x = 1$.

The second derivative is

$$f''(x) = \frac{2}{x^3},$$

which is never equal to 0 and does not exist when $x = 0$. (The function itself also does not exist at 0.) Because of this, there may be a change of concavity, but not an inflection point,

YOUR TURN 2 Graph $f(x) = 4x + \dfrac{1}{x}$.

when $x = 0$. The second derivative is negative when x is negative, making f concave downward on $(-\infty, 0)$. Also, $f''(x) > 0$ when $x > 0$, making f concave upward on $(0, \infty)$. See Figure 45(b).

Use this information, the asymptotes, and the critical points to get the graph shown in Figure 46.   **TRY YOUR TURN 2**

	Graph Summary			
Interval	$(-\infty, -1)$	$(-1, 0)$	$(0. 1)$	$(1, \infty)$
Sign of f'	$+$	$-$	$-$	$+$
Sign of f''	$-$	$-$	$+$	$+$
f Increasing or Decreasing	Increasing	Decreasing	Decreasing	Increasing
Concavity of f	Downward	Downward	Upward	Upward
Shape of Graph	⌢	⌢	⌣	⌣

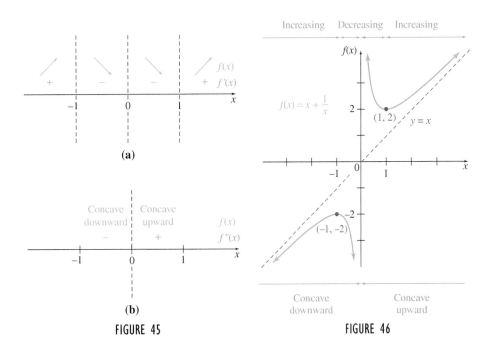

(a)

(b)

FIGURE 45

FIGURE 46

EXAMPLE 3 Rational Function Graph

Graph $f(x) = \dfrac{3x^2}{x^2 + 5}$.

SOLUTION The y-intercept is located at $y = f(0) = 0$. Verify that this is also the only x-intercept. There is no vertical asymptote, because $x^2 + 5 \neq 0$ for any value of x. Find any horizontal asymptote by calculating $\lim\limits_{x \to \infty} f(x)$ and $\lim\limits_{x \to -\infty} f(x)$. First, divide both the numerator and the denominator of $f(x)$ by x^2.

$$\lim_{x \to \infty} \frac{3x^2}{x^2 + 5} = \lim_{x \to \infty} \frac{\dfrac{3x^2}{x^2}}{\dfrac{x^2}{x^2} + \dfrac{5}{x^2}} = \frac{3}{1 + 0} = 3$$

Verify that the limit of $f(x)$ as $x \to -\infty$ is also 3. Thus, the horizontal asymptote is $y = 3$.

Observe that

$$f(-x) = \frac{3(-x)^2}{(-x)^2 + 5} = \frac{3x^2}{x^2 + 5} = f(x),$$

so the graph is symmetric about the y-axis. This means that the left side of the graph is the mirror image of the right side.

We now compute $f'(x)$:

$$f'(x) = \frac{(x^2 + 5)(6x) - (3x^2)(2x)}{(x^2 + 5)^2}.$$

Notice that $6x$ can be factored out of each term in the numerator:

$$f'(x) = \frac{(6x)[(x^2 + 5) - x^2]}{(x^2 + 5)^2}$$

$$= \frac{(6x)(5)}{(x^2 + 5)^2} = \frac{30x}{(x^2 + 5)^2}.$$

From the numerator, $x = 0$ is a critical number. The denominator is always positive. (Why?) Evaluating $f'(x)$ in each of the regions determined by $x = 0$ shows that f is decreasing on $(-\infty, 0)$ and increasing on $(0, \infty)$. By the first derivative test, f has a relative minimum when $x = 0$.

The second derivative is

$$f''(x) = \frac{(x^2 + 5)^2(30) - (30x)(2)(x^2 + 5)(2x)}{(x^2 + 5)^4}.$$

Factor $30(x^2 + 5)$ out of the numerator:

$$f''(x) = \frac{30(x^2 + 5)[(x^2 + 5) - (x)(2)(2x)]}{(x^2 + 5)^4}.$$

Divide a factor of $(x^2 + 5)$ out of the numerator and denominator, and simplify the numerator:

$$f''(x) = \frac{30[(x^2 + 5) - (x)(2)(2x)]}{(x^2 + 5)^3}$$

$$= \frac{30[(x^2 + 5) - (4x^2)]}{(x^2 + 5)^3}$$

$$= \frac{30(5 - 3x^2)}{(x^2 + 5)^3}.$$

The numerator of $f''(x)$ is 0 when $x = \pm \sqrt{5/3} \approx \pm 1.29$. Testing a point in each of the three intervals defined by these points shows that f is concave downward on $(-\infty, -1.29)$ and $(1.29, \infty)$, and concave upward on $(-1.29, 1.29)$. The graph has inflection points at $\left(\pm \sqrt{5/3}, f(\pm \sqrt{5/3}) \right) \approx (\pm 1.29, 0.75)$.

Use this information, the asymptote, the critical point, and the inflection points to get the graph shown in Figure 47. TRY YOUR TURN 3 ▊▊▊

YOUR TURN 3

Graph $f(x) = \dfrac{4x^2}{x^2 + 4}$.

Interval	Graph Summary			
	$(-\infty, -1.29)$	$(-1.29, 0)$	$(0, 1.29)$	$(1.29, \infty)$
Sign of f'	−	−	+	+
Sign of f"	−	+	+	−
f Increasing or Decreasing	Decreasing	Decreasing	Increasing	Increasing
Concavity of f	Downward	Upward	Upward	Downward
Shape of Graph	\	_	_/	⌐

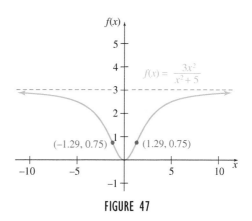

FIGURE 47

EXAMPLE 4   Graph with Logarithm

Graph $f(x) = \dfrac{\ln x}{x^2}$.

SOLUTION   The domain is $x > 0$, so there is no y-intercept. The x-intercept is 1, because $\ln 1 = 0$. We know that $y = \ln x$ has a vertical asymptote at $x = 0$, because $\lim_{x \to 0^+} \ln x = -\infty$. Dividing by x^2 when x is small makes $(\ln x)/x^2$ even more negative than $\ln x$. Therefore, $(\ln x)/x^2$ has a vertical asymptote at $x = 0$ as well. The first derivative is

$$f'(x) = \frac{x^2 \cdot \dfrac{1}{x} - 2x \ln x}{(x^2)^2} = \frac{x(1 - 2\ln x)}{x^4} = \frac{1 - 2\ln x}{x^3}$$

by the quotient rule. Setting the numerator equal to 0 and solving for x gives

$$1 - 2\ln x = 0$$
$$1 = 2\ln x$$
$$\ln x = 0.5$$
$$x = e^{0.5} \approx 1.65.$$

Since $f'(1)$ is positive and $f'(2)$ is negative, f increases on $(0, 1.65)$ then decreases on $(1.65, \infty)$, with a maximum value of $f(1.65) \approx 0.18$.

To find any inflection points, we calculate $f''(x)$.

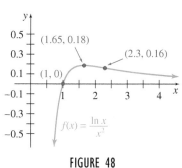

FIGURE 48

$$f''(x) = \frac{x^3\left(-2 \cdot \dfrac{1}{x}\right) - (1 - 2 \ln x) \cdot 3x^2}{(x^3)^2}$$

$$= \frac{-2x^2 - 3x^2 + 6x^2 \ln x}{x^6} = \frac{-5 + 6 \ln x}{x^4}$$

We then set $f''(x) = 0$.

$$\frac{-5 + 6 \ln x}{x^4} = 0$$

$-5 + 6 \ln x = 0$ Set the numerator equal to 0.

$6 \ln x = 5$ Add 5 to both sides.

$\ln x = 5/6$ Divide both sides by 6.

$x = e^{5/6} \approx 2.3$ $e^{\ln x} = x$.

YOUR TURN 4 Graph $f(x) = (x + 2)e^{-x}$. (Recall $\lim\limits_{x \to \infty} x^n e^{-x} = 0$)

There is an inflection point at $(2.3, f(2.3)) \approx (2.3, 0.16)$. Verify that $f''(1)$ is negative and $f''(3)$ is positive, so the graph is concave downward on $(0, 2.3)$ and upward on $(2.3, \infty)$. This information is summarized in the following table and could be used to sketch the graph. A graph of the function is shown in Figure 48.   **TRY YOUR TURN 4**

	Graph Summary		
Interval	$(0, 1.65)$	$(1.65, 2.3)$	$(2.3, \infty)$
Sign of f'	$+$	$-$	$-$
Sign of f''	$-$	$-$	$+$
f Increasing or Decreasing	Increasing	Decreasing	Decreasing
Concavity of f	Downward	Downward	Upward
Shape of Graph	⌢	⌢	⌣

As we saw earlier, a graphing calculator, when used with care, can be helpful in studying the behavior of functions. This section has illustrated that calculus is also a great help. The techniques of calculus show where the important points of a function, such as the relative extrema and the inflection points, are located. Furthermore, they tell how the function behaves between and beyond the points that are graphed, something a graphing calculator cannot always do.

5.4 WARM-UP EXERCISES

Find all inflection points for the following functions. *(Sec. 5.3)*

W1. $f(x) = x^4 - 2x^3 - 12x^2 + 4x + 13$

W2. $f(x) = x^3 - 21x^2 - 72x + 72x \ln x$

5.4 EXERCISES

1. By sketching a graph of the function or by investigating values of the function near 0, find $\lim_{x \to 0} x \ln |x|$. (This result will be useful in Exercise 21.)

2. Describe how you would find the equation of the horizontal asymptote for the graph of

$$f(x) = \frac{3x^2 - 2x}{2x^2 + 5}.$$

Graph each function, considering the domain, critical points, symmetry, relative extrema, regions where the function is increasing or decreasing, inflection points, regions where the function is concave upward or concave downward, intercepts where possible, and asymptotes where applicable. (*Hint*: In Exercise 21, use the result of Exercise 1. In Exercises 25–27, recall from Exercise 68 in Section 3.1 on Limits that $\lim_{x \to \infty} x^n e^{-x} = 0$.)

3. $f(x) = -2x^3 - 9x^2 + 108x - 10$

4. $f(x) = x^3 - \dfrac{15}{2}x^2 - 18x - 1$

5. $f(x) = -3x^3 + 6x^2 - 4x - 1$

6. $f(x) = x^3 - 6x^2 + 12x - 11$

7. $f(x) = x^4 - 24x^2 + 80$

8. $f(x) = -x^4 + 6x^2$

9. $f(x) = x^4 - 4x^3$

10. $f(x) = x^5 - 15x^3$

11. $f(x) = 2x + \dfrac{10}{x}$

12. $f(x) = 16x + \dfrac{1}{x^2}$

13. $f(x) = \dfrac{-x + 4}{x + 2}$

14. $f(x) = \dfrac{3x}{x - 2}$

15. $f(x) = \dfrac{1}{x^2 + 4x + 3}$

16. $f(x) = \dfrac{-8}{x^2 - 6x - 7}$

17. $f(x) = \dfrac{x}{x^2 + 1}$

18. $f(x) = \dfrac{1}{x^2 + 4}$

19. $f(x) = \dfrac{1}{x^2 - 9}$

20. $f(x) = \dfrac{-2x}{x^2 - 4}$

21. $f(x) = x \ln |x|$

22. $f(x) = x - \ln |x|$

23. $f(x) = \dfrac{\ln x}{x}$

24. $f(x) = \dfrac{\ln x^2}{x^2}$

25. $f(x) = xe^{-x}$

26. $f(x) = x^2 e^{-x}$

27. $f(x) = (x - 1)e^{-x}$

28. $f(x) = e^x + e^{-x}$

29. $f(x) = x^{2/3} - x^{5/3}$

30. $f(x) = x^{1/3} + x^{4/3}$

31. The default window on many calculators is $[-10, 10]$ by $[-10, 10]$. For the odd exercises between 3 and 15, tell which would give a poor representation in this window. (*Note:* Your answers may differ from ours, depending on what you consider "poor.")

32. Repeat Exercise 31 for the even exercises between 4 and 16.

33. Repeat Exercise 31 for the odd exercises between 17 and 29.

34. Repeat Exercise 31 for the even exercises between 18 and 30.

In Exercises 35–39, sketch the graph of a single function that has all of the properties listed.

35. (a) Continuous and differentiable everywhere except at $x = 1$, where it has a vertical asymptote
    (b) $f'(x) < 0$ everywhere it is defined
    (c) A horizontal asymptote at $y = 2$
    (d) $f''(x) < 0$ on $(-\infty, 1)$ and $(2, 4)$
    (e) $f''(x) > 0$ on $(1, 2)$ and $(4, \infty)$

36. (a) Continuous for all real numbers
    (b) $f'(x) < 0$ on $(-\infty, -6)$ and $(1, 3)$
    (c) $f'(x) > 0$ on $(-6, 1)$ and $(3, \infty)$
    (d) $f''(x) > 0$ on $(-\infty, -6)$ and $(3, \infty)$
    (e) $f''(x) < 0$ on $(-6, 3)$
    (f) A y-intercept at $(0, 2)$

37. (a) Continuous and differentiable for all real numbers
    (b) $f'(x) > 0$ on $(-\infty, -3)$ and $(1, 4)$
    (c) $f'(x) < 0$ on $(-3, 1)$ and $(4, \infty)$
    (d) $f''(x) < 0$ on $(-\infty, -1)$ and $(2, \infty)$
    (e) $f''(x) > 0$ on $(-1, 2)$
    (f) $f'(-3) = f'(4) = 0$
    (g) $f''(x) = 0$ at $(-1, 3)$ and $(2, 4)$

38. (a) Continuous for all real numbers
    (b) $f'(x) > 0$ on $(-\infty, -2)$ and $(0, 3)$
    (c) $f'(x) < 0$ on $(-2, 0)$ and $(3, \infty)$
    (d) $f''(x) < 0$ on $(-\infty, 0)$ and $(0, 5)$
    (e) $f''(x) > 0$ on $(5, \infty)$
    (f) $f'(-2) = f'(3) = 0$
    (g) $f'(0)$ doesn't exist
    (h) Differentiable everywhere except at $x = 0$
    (i) An inflection point at $(5, 1)$

39. (a) Continuous for all real numbers
    (b) Differentiable everywhere except at $x = 4$
    (c) $f(1) = 5$
    (d) $f'(1) = 0$ and $f'(3) = 0$
    (e) $f'(x) > 0$ on $(-\infty, 1)$ and $(4, \infty)$
    (f) $f'(x) < 0$ on $(1, 3)$ and $(3, 4)$
    (g) $\lim_{x \to 4^-} f'(x) = -\infty$ and $\lim_{x \to 4^+} f'(x) = \infty$
    (h) $f''(x) > 0$ on $(2, 3)$
    (i) $f''(x) < 0$ on $(-\infty, 2)$, $(3, 4)$, and $(4, \infty)$

YOUR TURN ANSWERS

1.

2.

3.

4.

5 Chapter Review

SUMMARY

In this chapter we have explored various concepts related to the graph of a function:

- increasing and decreasing,
- critical numbers (numbers c in the domain of f for which $f'(x) = 0$ or $f'(x)$ does not exist),
- critical points (whose x-coordinate is a critical number c and whose y-coordinate is $f(c)$),
- relative maxima and minima (together known as relative extrema),
- concavity, and
- inflection points (where the concavity changes).

The first and second derivative tests provide ways to locate relative extrema. The last section brings all these concepts together. Also, we investigated two applications of the second derivative:

- acceleration (the second derivative of the position function), and
- the point of diminishing returns (an inflection point on an input/output graph).

Test for Increasing/Decreasing	On any open interval,
	if $f'(x) > 0$, then f is increasing;
	if $f'(x) < 0$, then f is decreasing;
	if $f'(x) = 0$, then f is constant.

First Derivative Test	If c is a critical number for f on the open interval (a, b), f is continuous on (a, b), and f is differentiable on (a, b) (except possibly at c), then
	1. $f(c)$ is a relative maximum if $f'(x) > 0$ on (a, c) and $f'(x) < 0$ on (c, b);
	2. $f(c)$ is a relative minimum if $f'(x) < 0$ on (a, c) and $f'(x) > 0$ on (c, b).

Test for Concavity	On any open interval,
	if $f''(x) > 0$, then f is concave upward;
	if $f''(x) < 0$, then f is concave downward.

Second Derivative Test	Suppose f'' exists on an open interval containing c and $f'(c) = 0$.
	1. If $f''(c) > 0$, then $f(c)$ is a relative minimum.
	2. If $f''(c) < 0$, then $f(c)$ is a relative maximum.
	3. If $f''(c) = 0$ or $f'(c)$ does not exist, then the test gives no information about extrema, so use the first derivative test.

Curve Sketching   To sketch the graph of a function f:

1. Consider the domain of the function, and note any restrictions. (That is, avoid dividing by 0, taking a square root, or any even root, of a negative number, or taking the logarithm of 0 or a negative number.)

2. Find the y-intercept (if it exists) by substituting $x = 0$ into $f(x)$. Find any x-intercepts by solving $f(x) = 0$ if this is not too difficult.

3. (a) If f is a rational function, find any vertical asymptotes by investigating where the denominator is 0, and find any horizontal asymptotes by finding the limits as $x \to \infty$ and $x \to -\infty$.

   (b) If f is an exponential function, find any horizontal asymptotes; if f is a logarithmic function, find any vertical asymptotes.

4. Investigate symmetry. If $f(-x) = f(x)$, the function is even, so the graph is symmetric about the y-axis. If $f(-x) = -f(x)$, the function is odd, so the graph is symmetric about the origin.

5. Find $f'(x)$. Locate any critical points by solving the equation $f'(x) = 0$ and determining where $f'(x)$ does not exist, but $f(x)$ does. Find any relative extrema and determine where f is increasing or decreasing.

6. Find $f''(x)$. Locate potential inflection points by solving the equation $f''(x) = 0$ and determining where $f''(x)$ does not exist. Determine where f is concave upward or concave downward.

7. Plot the intercepts, the critical points, the inflection points, the asymptotes, and other points as needed. Take advantage of any symmetry found in Step 4.

8. Connect the points with a smooth curve using the correct concavity, being careful not to connect points where the function is not defined.

9. Check your graph using a graphing calculator. If the picture looks very different from what you've drawn, see in what ways the picture differs and use that information to help find your mistake.

KEY TERMS

5.1
increasing function
decreasing function
critical number
critical point

5.2
relative (or local) maximum

relative (or local) minimum
relative (or local) extremum
first derivative test

5.3
second derivative
third derivative

fourth derivative
acceleration
concave upward and downward
concavity
inflection point

second derivative test
point of diminishing returns

5.4
curve sketching
oblique asymptote

REVIEW EXERCISES

CONCEPT CHECK

For Exercises 1–12 determine whether each of the following statements is true or false, and explain why.

1. A critical number c is a number in the domain of a function f for which $f'(c) = 0$ or $f'(c)$ does not exist.

2. If $f'(x) > 0$ on an interval, the function is positive on that interval.

3. If c is a critical number, then the function must have a relative maximum or minimum at c.

4. If f is continuous on (a, b), $f'(x) < 0$ on (a, c), and $f'(x) > 0$ on (c, b), then f has a relative minimum at c.

5. If $f'(c)$ exists, $f''(c)$ also exists.

6. The acceleration is the second derivative of the position function.

7. If $f''(x) > 0$ on an interval, the function is increasing on that interval.

8. If $f''(c) = 0$, the function has an inflection point at c.

9. If $f''(c) = 0$, the function does not have a relative maximum or minimum at c.

10. Every rational function has either a vertical or a horizontal asymptote.

11. If an odd function has a y-intercept, it must pass through the origin.

12. If $f'(c) = 0$, where c is a value in interval (a, b), then f is a constant on the interval (a, b).

PRACTICE AND EXPLORATIONS

13. When given the equation for a function, how can you determine where it is increasing and where it is decreasing?

14. When given the equation for a function, how can you determine where the relative extrema are located? Give two ways to test whether a relative extremum is a minimum or a maximum.

15. Does a relative maximum of a function always have the largest y-value in the domain of the function? Explain your answer.

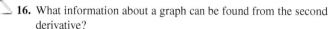 **16.** What information about a graph can be found from the second derivative?

Find the open intervals where f is increasing or decreasing.

17. $f(x) = x^2 + 9x + 8$ **18.** $f(x) = -2x^2 + 7x + 14$

19. $f(x) = -x^3 + 2x^2 + 15x + 16$

20. $f(x) = 4x^3 + 8x^2 - 16x + 11$

21. $f(x) = \dfrac{16}{9 - 3x}$ **22.** $f(x) = \dfrac{15}{2x + 7}$

23. $f(x) = \ln|x^2 - 1|$ **24.** $f(x) = 8xe^{-4x}$

Find the locations and values of all relative maxima and minima.

25. $f(x) = -x^2 + 4x - 8$ **26.** $f(x) = x^2 - 6x + 4$

27. $f(x) = 2x^2 - 8x + 1$ **28.** $f(x) = -3x^2 + 2x - 5$

29. $f(x) = 2x^3 + 3x^2 - 36x + 20$

30. $f(x) = 2x^3 + 3x^2 - 12x + 5$

31. $f(x) = \dfrac{xe^x}{x - 1}$ **32.** $f(x) = \dfrac{\ln(3x)}{2x^2}$

Find the second derivative of each function, and then find $f''(1)$ and $f''(-3)$.

33. $f(x) = 3x^4 - 5x^2 - 11x$ **34.** $f(x) = 9x^3 + \dfrac{1}{x}$

35. $f(x) = \dfrac{4x + 2}{3x} \quad \dfrac{}{6}$ **36.** $f(x) = \dfrac{1 - 2x}{4x + 5}$

37. $f(t) = \sqrt{t^2 + 1}$ **38.** $f(t) = -\sqrt{5 - t^2}$

Graph each function, considering the domain, critical points, symmetry, relative extrema, regions where the function is increasing or decreasing, inflection points, regions where the function is concave up or concave down, intercepts where possible, and asymptotes where applicable.

39. $f(x) = -2x^3 - \dfrac{1}{2}x^2 + x - 3$

40. $f(x) = -\dfrac{4}{3}x^3 + x^2 + 30x - 7$

41. $f(x) = x^4 - \dfrac{4}{3}x^3 - 4x^2 + 1$

42. $f(x) = -\dfrac{2}{3}x^3 + \dfrac{9}{2}x^2 + 5x + 1$

43. $f(x) = \dfrac{x - 1}{2x + 1}$ **44.** $f(x) = \dfrac{2x - 5}{x + 3}$

45. $f(x) = -4x^3 - x^2 + 4x + 5$

46. $f(x) = x^3 + \dfrac{5}{2}x^2 - 2x - 3$

47. $f(x) = x^4 + 2x^2$ **48.** $f(x) = 6x^3 - x^4$

49. $f(x) = \dfrac{x^2 + 4}{x}$ **50.** $f(x) = x + \dfrac{8}{x}$

51. $f(x) = \dfrac{2x}{3 - x}$ **52.** $f(x) = \dfrac{-4x}{1 + 2x}$

53. $f(x) = xe^{2x}$ **54.** $f(x) = x^2 e^{2x}$

55. $f(x) = \ln(x^2 + 4)$ **56.** $f(x) = x^2 \ln x$

57. $f(x) = 4x^{1/3} + x^{4/3}$ **58.** $f(x) = 5x^{2/3} + x^{5/3}$

In Exercises 59 and 60, sketch the graph of a single function that has all of the properties listed.

59. (a) Continuous everywhere except at $x = -4$, where there is a vertical asymptote

   (b) A y-intercept at $y = -2$

   (c) x-intercepts at $x = -3$, 1, and 4

   (d) $f'(x) < 0$ on $(-\infty, -5)$, $(-4, -1)$, and $(2, \infty)$

   (e) $f'(x) > 0$ on $(-5, -4)$ and $(-1, 2)$

   (f) $f''(x) > 0$ on $(-\infty, -4)$ and $(-4, -3)$

   (g) $f''(x) < 0$ on $(-3, -1)$ and $(-1, \infty)$

   (h) Differentiable everywhere except at $x = -4$ and $x = -1$

60. (a) Continuous and differentiable everywhere except at $x = -3$, where it has a vertical asymptote

   (b) A horizontal asymptote at $y = 1$

   (c) An x-intercept at $x = -2$

   (d) A y-intercept at $y = 4$

   (e) $f'(x) > 0$ on the intervals $(-\infty, -3)$ and $(-3, 2)$

   (f) $f'(x) < 0$ on the interval $(2, \infty)$

   (g) $f''(x) > 0$ on the intervals $(-\infty, -3)$ and $(4, \infty)$

   (h) $f''(x) < 0$ on the interval $(-3, 4)$

   (i) $f'(2) = 0$

   (j) An inflection point at $(4, 3)$

APPLICATIONS

Business and Economics

Stock Prices In Exercises 61 and 62, $P(t)$ is the price of a certain stock at time t during a particular day.

61. (a) If the price of the stock is falling faster and faster, are $P'(t)$ and $P''(t)$ positive or negative?

   (b) Explain your answer.

62. (a) When the stock reaches its highest price of the day, are $P'(t)$ and $P''(t)$ positive, zero, or negative?

   (b) Explain your answer.

63. Cat Brushes The cost function to produce q electric cat brushes is given by $C(q) = -10q^2 + 250q$. The demand equation is given by $p = -q^2 - 3q + 299$, where p is the price in dollars.

   (a) Find and simplify the profit function.

   (b) Find the number of brushes that will produce the maximum profit.

   (c) Find the price that produces the maximum profit.

   (d) Find the maximum profit.

   (e) Find the point of diminishing returns for the profit function

64. Gasoline Prices In 2013, the price of gasoline in the United States spiked and then dropped. The average monthly price (in cents per gallon) of unleaded regular gasoline for 2013 can be approximated by the function

$$p(t) = 0.1244t^3 - 3.001t^2 + 16.65t + 348, \quad \text{for} \quad 0 \le t \le 12,$$

where t is in months and $t = 0$ corresponds to the beginning of January 2013. *Source: U.S. Energy Information Administration.*

(a) Determine the interval(s) on which the price is increasing.

(b) Determine the interval(s) on which the price is decreasing.

(c) Find any relative extrema for the price of gasoline, as well as when they occurred.

65. College Presidents A survey of 185 public universities found that "the salaries and benefits of their presidents continued to rise, though at a slower rate than in years past." Let $f(t)$ represent the total salary and compensation of the average public university president in year t. What does this statement tell us about $f(t)$, $f'(t)$, and $f''(t)$? *Source: The New York Times.*

66. Inflation Mathematician Hugo Rossi wrote: "In the fall of 1972 President Nixon announced that the rate of increase of inflation was decreasing. This was the first time a sitting president used the third derivative to advance his case for reelection." Explain what President Nixon's announcement had to do with the third derivative. *Source: Notices of the American Mathematical Society.*

Life Sciences

67. Weightlifting An abstract for an article states, "We tentatively conclude that Olympic weightlifting ability in trained subjects undergoes a nonlinear decline with age, in which the second derivative of the performance versus age curve repeatedly changes sign." *Source: Medicine and Science in Sports and Exercise.*

(a) What does this quote tell you about the first derivative of the performance versus age curve?

(b) Describe what you know about the performance versus age curve based on the information in the quote.

68. Scaling Laws Many biological variables depend on body mass, with a functional relationship of the form

$$Y = Y_0 M^b,$$

where M represents body mass, b is a multiple of 1/4, and Y_0 is a constant. For example, when Y represents metabolic rate, $b = 3/4$. When Y represents heartbeat, $b = -1/4$. When Y represents life span, $b = 1/4$. *Source: Science.*

(a) Determine which of metabolic rate, heartbeat, and life span are increasing or decreasing functions of mass. Also determine which have graphs that are concave upward and which have graphs that are concave downward.

(b) Verify that all functions of the form given above satisfy the equation

$$\frac{dY}{dM} = \frac{b}{M} Y.$$

This means that the rate of change of Y is proportional to Y and inversely proportional to body mass.

69. Thoroughbred Horses The association between velocity during exercise and blood lactate concentration after submaximal 800-m exercise of thoroughbred racehorses on sand and grass tracks has been studied. The lactate-velocity relationship can be described by the functions

$$l_1(v) = 0.08e^{0.33v} \quad \text{and}$$

$$l_2(v) = -0.87v^2 + 28.17v - 211.41,$$

where $l_1(v)$ and $l_2(v)$ are the lactate concentrations (in mmol/L) and v is the velocity (in m/sec) of the horse during workout on sand and grass tracks, respectively. Sketch the graph of both functions for $13 \le v \le 17$. *Source: The Veterinary Journal.*

70. Neuron Communications In the FitzHugh-Nagumo model of how neurons communicate, the rate of change of the electric potential v with respect to time is given as a function of v by $f(v) = v(a - v)(v - 1)$, where a is a positive constant. Sketch a graph of this function when $a = 0.25$ and $0 \le v \le 1$. *Source: Mathematical Biology.*

71. Fruit Flies The number of imagoes (sexually mature adult fruit flies) per mated female per day (y) can be approximated by

$$y = 34.7(1.0186)^{-x} x^{-0.658},$$

where x is the mean density of the mated population (measured as flies per bottle) over a 16-day period. Sketch the graph of the function. *Source: Elements of Mathematical Biology.*

72. Blood Volume A formula proposed by Hurley for the red cell volume (RCV) in milliliters for males is

$$RCV = 1486S^2 - 4106S + 4514,$$

where S is the surface area (in square meters). A formula given by Pearson et al. is

$$RCV = 1486S - 825.$$

Source: Journal of Nuclear Medicine and British Journal of Haematology.

(a) For the value of S for which the RCV values given by the two formulas are closest, find the rate of change of RCV with respect to S for both formulas. What does this number represent?

(b) The formula for plasma volume for males given by Hurley is

$$PV = 995e^{0.6085S},$$

while the formula given by Pearson et al. is

$$PV = 1578S,$$

where PV is measured in milliliters and S in square meters. Find the value of S for which the PV values given by the two formulas are the closest. Then find the value of PV that each formula gives for this value of S.

(c) For the value of S found in part b, find the rate of change of PV with respect to S for both formulas. What does this number represent?

(d) Notice in parts (a) and (c) that both formulas give the same instantaneous rate of change at the value of S for which the function values are closest. Prove that if two functions f and g are differentiable and never cross but are closest together when $x = x_0$, then $f'(x_0) = g'(x_0)$.

Social Sciences

73. Learning Researchers used a version of the Gompertz curve to model the rate that children learn with the equation

$$y(t) = A^{c^t},$$

where $y(t)$ is the portion of children of age t years passing a certain mental test, $A = 0.3982 \times 10^{-291}$, and $c = 0.4252$. Find the inflection point and describe what it signifies. (*Hint:* Leave A and c as constants until you are ready to calculate your final answer. If A is too small for your calculator to handle, use common logarithms and properties of logarithms to calculate $(\log A)/(\log e)$.) *Source: School and Society*.

74. Population Under the scenario that the fertility rate in the European Union (EU) remains at 1.8 until 2020, when it rises to replacement level, the predicted population (in millions) of the 15 member countries of the EU can be approximated over the next century by

$$P(t) = 325 + 7.475(t + 10)e^{-(t+10)/20},$$

where t is the number of years since 2000. *Source: Science*.

(a) In what year is the population predicted to be largest? What is the population predicted to be in that year?

(b) In what year is the population declining most rapidly?

(c) What is the population approaching as time goes on?

75. Nuclear Weapons The graph shows the total inventory of nuclear weapons held by the United States and by the Soviet Union and its successor states from 1945 to 2010. (See Exercise 60 in the first section of this chapter.) *Source: Federation of American Scientists*.

(a) In what years was the U.S. total inventory of weapons at a relative maximum?

(b) When the U.S. total inventory of weapons was at the largest relative maximum, is the graph for the Soviet stockpile concave up or concave down? What does this mean?

Physical Sciences

76. Velocity and Acceleration A projectile is shot straight up with an initial velocity of 512 ft per second. Its height above the ground after t seconds is given by $s(t) = 512t - 16t^2$.

(a) Find the velocity and acceleration after t seconds.

(b) What is the maximum height attained?

(c) When does the projectile hit the ground and with what velocity?

General Interest

77. Grades As we saw in Chapter 3, a group of MIT professors created a function for transforming student test scores of 0 to 100 into scores from 0 to 5, with transformed scores concentrated around 3, 4, and 5. Transformed scores from 4.5 to 5 became grades of A, 3.5 to 4.5 became B, and 2.5 to 3.5 became C. They used the following function:

$$G(x) = \frac{3}{1 + e^{-c(x-50)}} + \frac{1}{1 + e^{-c(x-70)}} + \frac{1}{1 + e^{-c(x-90)}}$$

using the value $c = 0.7$. *Source: MIT Faculty Newsletter*.

(a) Show that $G'(x)$ is always positive, thus verifying that the function is increasing. Explain how you could have drawn this conclusion without any calculus by studying the original function.

(b) From a graphing calculator graph of the function on the interval $0 \leq x \leq 100$, estimate intervals where the derivative is close to 0.

(c) From the graphing calculator graph used in part (b), estimate all inflection points.

(d) To understand why the function is so flat when x is not close to 50, 70, or 90, show that the derivative can be rewritten as

$$G'(x) = \frac{3c}{[e^{c(x-50)/2} + e^{-c(x-50)/2}]^2}$$

$$+ \frac{c}{[e^{c(x-70)/2} + e^{-c(x-70)/2}]^2} + \frac{c}{[e^{c(x-90)/2} + e^{-c(x-90)/2}]^2}.$$

If x is much bigger or smaller than 50, what can you say about the first term? Make similar observations about the second and third terms.

EXTENDED APPLICATION

A DRUG CONCENTRATION MODEL FOR ORALLY ADMINISTERED MEDICATIONS

Finding a range for the concentration of a drug in the bloodstream that is both safe and effective is one of the primary goals in pharmaceutical research and development. This range is called the *therapeutic window*. When determining the proper dosage (both the size of the dose and the frequency of administration), it is important to understand the behavior of the drug once it enters the body. Using data gathered during research we can create a mathematical model that predicts the concentration of the drug in the bloodstream at any given time.

We will look at two examples that explore a mathematical model for the concentration of a particular drug in the bloodstream. We will find the maximum and minimum concentrations of the drug given the size of the dose and the frequency of administration. We will then determine what dose should be administered to maintain concentrations within a given therapeutic window.

The drug tolbutamide is used for the management of mild to moderately severe type 2 diabetes. Suppose a 1000-mg dose of this drug is taken every 12 hours for three days. The concentration of the drug in the bloodstream, t hours after the initial dose is taken, is shown in Figure 49.

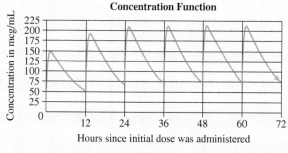

Concentration Function

Hours since initial dose was administered

FIGURE 49

Looking at the graph, you can see that after a few doses have been administered, the maximum values of the concentration function begin to level off. The function also becomes periodic, repeating itself between every dose. At this point, the concentration is said to be at steady-state. The time it takes to reach steady-state depends on the elimination half-life of the drug (the time it takes for half the dose to be eliminated from the body). The elimination half-life of the drug used for this function is about 7 hours. Generally speaking, we say that steady-state is reached after about 5 half-lives.

We will define the *steady-state concentration function*, $C_{ss}(t)$, to be the concentration of drug in the bloodstream t hours after a dose has been administered once steady-state has been reached.

The steady-state concentration function can be written as the difference of two exponential decay functions, or

$$C_{ss}(t) = c_1 e^{kt} - c_2 e^{k_a t}.$$

The constants c_1 and c_2 are influenced by several factors, including the size of the dose and how widely the particular drug disperses through the body. The constants k_a and k are decay constants reflecting the rate at which the drug is being absorbed into the bloodstream and eliminated from the bloodstream, respectively.

Consider the following steady-state concentration function:

$$C_{ss}(t) = 0.2473De^{-0.1t} - 0.1728De^{-2.8t} \text{ mcg/mL}$$

where D is the size of the dose (in milligrams) administered every 12 hours. The concentration is given in micrograms per milliliter.

If a single dose is 1000 mg, then the concentration of drug in the bloodstream is

$$C_{ss}(t) = 247.3e^{-0.1t} - 172.8e^{-2.8t} \text{ mcg/mL}.$$

The graph of $C_{ss}(t)$ is given below in Figure 50.

Steady-State Concentration Function

Hours since last dose was administered

FIGURE 50

EXAMPLE 1 **Drug Concentration**

Find the maximum and minimum concentrations for the steady-state concentration function

$$C_{ss}(t) = 247.3e^{-0.1t} - 172.8e^{-2.8t} \text{ mcg/mL}.$$

SOLUTION The maximum concentration occurs when $C'_{ss}(t) = 0$. Calculating the derivative, we get:

$$C'_{ss}(t) = 247.3(-0.1)e^{-0.1t} - 172.8(-2.8)e^{-2.8t}$$
$$= -24.73e^{-0.1t} + 483.84e^{-2.8t}.$$

If we factor out $e^{-0.1t}$, we can find where the derivative is equal to zero.

$$C'_{ss}(t) = e^{-0.1t}(-24.73 + 483.84e^{-2.7t}) = 0$$

$C'_{ss}(t) = 0$ when

$$-24.73 + 483.84e^{-2.7t} = 0.$$

Solving this equation for t, we get

$$t = \frac{\ln\left(\dfrac{24.73}{483.84}\right)}{-2.7} \approx 1.1 \text{ hours.}$$

Therefore, the maximum concentration is

$$C_{ss}(1.1) = 247.3e^{-0.1(1.1)} - 172.8e^{-2.8(1.1)} \approx 214 \text{ mcg/mL.}$$

Looking at the graph of $C_{ss}(t)$ in Figure 50, you can see that the minimum concentration occurs at the endpoints (when $t = 0$ and $t = 12$; immediately after a dose is administered and immediately before a next dose is to be administered, respectively).

319

Therefore, the minimum concentration is

$$C_{ss}(0) = 247.3e^{-0.1(0)} - 172.8e^{-2.8(0)} = 247.3 - 172.8$$
$$= 74.5 \text{ mcg/mL}.$$

Verify that $C_{ss}(12)$ gives the same value.

If the therapeutic window for this drug is 70–240 mcg/mL, then, once steady-state has been reached, the concentration remains safe and effective as long as treatment continues.

Suppose, however, that a new study found that this drug is effective only if the concentration remains between 100 and 400 mcg/mL. How could you adjust the dose so that the maximum and minimum steady-state concentrations fall within this range?

EXAMPLE 2 Therapeutic Window

Find a range for the size of doses such that the steady-state concentration remains within the therapeutic window of 100 to 400 mcg/mL.

SOLUTION Recall that the steady-state concentration function is

$$C_{ss}(t) = 0.2473De^{-0.1t} - 0.1728De^{-2.8t} \text{ mcg/mL},$$

where D is the size of the dose given (in milligrams) every 12 hours.

From Example 1, we found that the minimum concentration occurs when $t = 0$. Therefore, we want the minimum concentration, $C_{ss}(0)$, to be greater than or equal to 100 mcg/mL.

$$C_{ss}(0) = 0.2473De^{-0.1(0)} - 0.1728De^{-2.8(0)} \geq 100$$

or

$$0.2473D - 0.1728D \geq 100$$

Solving for D, we get

$$0.0745D \geq 100$$
$$D \geq 1342 \text{ mg}.$$

In Example 1, we also found that the maximum concentration occurs when $t = 1.1$ hours. If we change the size of the dose, the maximum concentration will change; however, the time it takes to reach the maximum concentration does not change. Can you see why this is true?

Since the maximum concentration occurs when $t = 1.1$, we want $C_{ss}(1.1)$, the maximum concentration, to be less than or equal to 400 mcg/mL.

$$C_{ss}(1.1) = 0.2473De^{-0.1(1.1)} - 0.1728De^{-2.8(1.1)} \leq 400$$

or

$$0.2215D - 0.0079D \leq 400.$$

Solving for D, we get

$$0.2136D \leq 400$$
$$D \leq 1873 \text{ mg}.$$

Therefore, if the dose is between 1342 mg and 1873 mg, the steady-state concentration remains within the new therapeutic window.

EXERCISES

Use the following information to answer Exercises 1–3.
A certain drug is given to a patient every 12 hours. The steady-state concentration function is given by

$$C_{ss}(t) = 1.99De^{-0.14t} - 1.62De^{-2.08t} \text{ mcg/mL},$$

where D is the size of the dose in milligrams.

1. If a 500-mg dose is given every 12 hours, find the maximum and minimum steady-state concentrations.

2. If the dose is increased to 1500 mg every 12 hours, find the maximum and minimum steady-state concentrations.

3. What dose should be given every 12 hours to maintain a steady-state concentration between 80 and 400 mcg/mL?

DIRECTIONS FOR GROUP PROJECT

Because of declining health, many elderly people rely on prescription medications to stabilize or improve their medical condition. Your group has been assigned the task of developing a brochure to be made available at senior citizens' centers and physicians' offices that describes drug concentrations in the body for orally administered medications. The brochure should summarize the facts presented in this extended application but at a level that is understandable to a typical layperson. The brochure should be designed to look professional with a marketing flair.

6

Applications of the Derivative

When several variables are related by a single equation, their rates of change are also related. For example, the height and horizontal distance of a kite are related to the length of the string holding the kite. In an exercise in Section 5 we differentiate this relationship to discover how fast the kite flier must let out the string to maintain the kite at a constant height and constant horizontal speed.

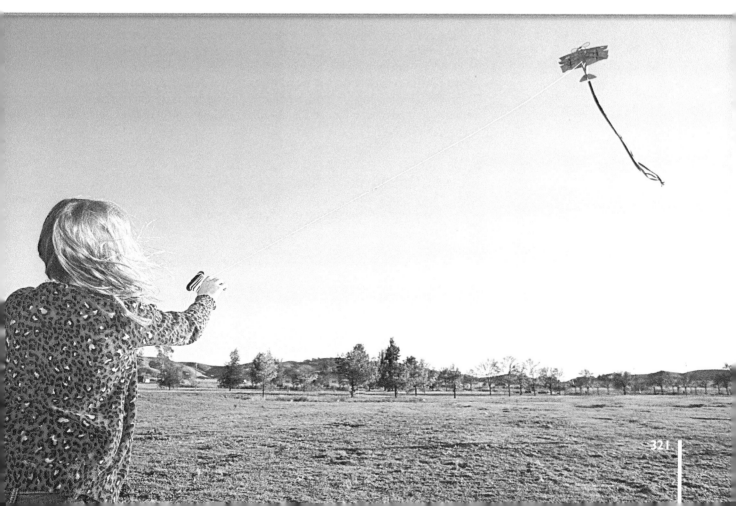

The previous chapter included examples in which we used the derivative to find the maximum or minimum value of a function. This problem is ubiquitous; consider the efforts people expend trying to maximize their income, or to minimize their costs or the time required to complete a task. In this chapter we will treat the topic of optimization in greater depth.

The derivative is applicable in far wider circumstances, however. In roughly 500 B.C., Heraclitus said, "Nothing endures but change," and his observation has relevance here. If change is continuous, rather than in sudden jumps, the derivative can be used to describe the rate of change. This explains why calculus has been applied to so many fields.

6.1 Absolute Extrema

APPLY IT During a 10-year period, when did the U.S. dollar reach a minimum exchange rate with the Canadian dollar, and how much was the U.S. dollar worth then?
We will answer this question in Example 3.

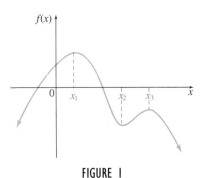

FIGURE 1

A function may have more than one relative maximum. It may be important, however, in some cases to determine if one function value is greater than any other. In other cases, we may want to know whether one function value is less than any other. For example, in Figure 1, $f(x_1) \geq f(x)$ for all x in the domain. There is no function value that is less than all others, however, because $f(x) \to -\infty$ as $x \to \infty$ or as $x \to -\infty$.

The greatest possible value of a function is called the *absolute maximum*, and the least possible value of a function is called the *absolute minimum*. As Figure 1 shows, one or both of these may not exist on the domain of the function, $(-\infty, \infty)$ here. Absolute extrema often coincide with relative extrema, as with $f(x_1)$ in Figure 1. Although a function may have several relative maxima or relative minima, it never has more than one *absolute maximum* or *absolute minimum*, although the absolute maximum or minimum might occur at more than one value of x.

Absolute Maximum or Minimum

Let f be a function defined on some interval. Let c be a number in the interval. Then $f(c)$ is the **absolute maximum** of f on the interval if

$$f(x) \leq f(c)$$

for every x in the interval, and $f(c)$ is the **absolute minimum** of f on the interval if

$$f(x) \geq f(c)$$

for every x in the interval.

A function has an **absolute extremum** (plural: **extrema**) at c if it has either an absolute maximum or an absolute minimum there.

| CAUTION | Notice that, just like a relative extremum, an absolute extremum is a y-value, not an x-value. |

Now look at Figure 2, which shows three functions defined on closed intervals. In each case there is an absolute maximum value and an absolute minimum value. These absolute extrema may occur at the endpoints or at relative extrema. As the graphs in Figure 2 show, an absolute extremum is either the largest or the smallest function value occurring on a closed interval, while a relative extremum is the largest or smallest function value in some (perhaps small) open interval.

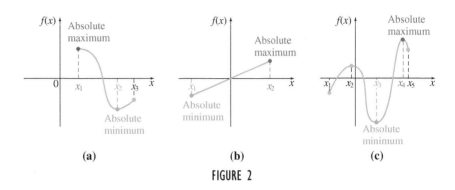

FIGURE 2

Although a function can have only one absolute minimum value and only one absolute maximum value, it can have many points where these values occur. (Note that the absolute maximum value and absolute minimum value are numbers, not points.) As an extreme example, consider the function $f(x) = 2$. The absolute minimum value of this function is clearly 2, as is the absolute maximum value. Both the absolute minimum and the absolute maximum occur at every real number x.

One reason for the importance of absolute extrema is given by the **extreme value theorem** (which is proved in more advanced courses).

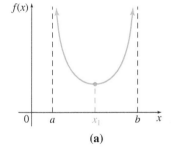

(a)

Extreme Value Theorem

A function f that is continuous on a closed interval $[a, b]$ will have both an absolute maximum and an absolute minimum on the interval.

A continuous function on an open interval may or may not have an absolute maximum or minimum. For example, the function in Figure 3(a) has an absolute minimum on the interval (a, b) at x_1, but it does not have an absolute maximum. Instead, it becomes arbitrarily large as x approaches a or b. Also, a discontinuous function on a closed interval may or may not have an absolute minimum or maximum. The function in Figure 3(b) has an absolute minimum at $x = a$, yet it has no absolute maximum. It may appear at first to have an absolute maximum at x_1, but notice that $f(x_1)$ has a smaller value than f at values of x less than x_1.

The extreme value theorem guarantees the existence of absolute extrema for a continuous function on a closed interval. To find these extrema, use the following steps.

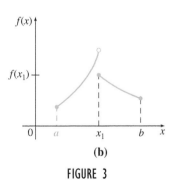

(b)

FIGURE 3

Finding Absolute Extrema

To find absolute extrema for a function f continuous on a closed interval $[a, b]$:

1. Find all critical numbers for f in (a, b).

2. Evaluate f for all critical numbers in (a, b).

3. Evaluate f for the endpoints a and b of the interval $[a, b]$.

4. The largest value found in Step 2 or 3 is the absolute maximum for f on $[a, b]$, and the smallest value found is the absolute minimum for f on $[a, b]$.

EXAMPLE 1 Absolute Extrema

Find the absolute extrema of the function

$$f(x) = x^{8/3} - 16x^{2/3}$$

on the interval $[-1, 8]$.

SOLUTION First look for critical numbers in the interval $(-1, 8)$.

$$f'(x) = \frac{8}{3}x^{5/3} - \frac{32}{3}x^{-1/3}$$

$$= \frac{8}{3}x^{-1/3}(x^2 - 4) \qquad \text{Factor.}$$

$$= \frac{8}{3}\left(\frac{x^2 - 4}{x^{1/3}}\right)$$

Set $f'(x) = 0$ and solve for x. Notice that $f'(x) = 0$ at $x = 2$ and $x = -2$, but -2 is not in the interval $(-1, 8)$, so we ignore it. The derivative is undefined at $x = 0$, but the function is defined there, so 0 is also a critical number.

Evaluate the function at the critical numbers and the endpoints.

	Extrema Candidates	
x-Value	Value of Function	
-1	-15	
0	0	
2	-19.05	← Absolute minimum
8	192	← Absolute maximum

The absolute maximum, 192, occurs when $x = 8$, and the absolute minimum, approximately -19.05, occurs when $x = 2$. A graph of f is shown in Figure 4.

TRY YOUR TURN 1

YOUR TURN 1 Find the absolute extrema of the function $f(x) = 3x^{2/3} - 3x^{5/3}$ on the interval $[0, 8]$.

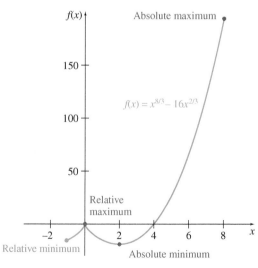

FIGURE 4

In Example 1, a graphing calculator that gives the maximum and minimum values of a function on an interval, such as the fMax or fMin feature of the TI-84 Plus C, could replace the table. Alternatively, we could first graph the function on the given interval and then select the feature that gives the maximum or minimum value of the graph of the function instead of completing the table.

EXAMPLE 2 Absolute Extrema

Find the locations and values of the absolute extrema, if they exist, for the function

$$f(x) = 3x^4 - 4x^3 - 12x^2 + 2.$$

SOLUTION In this example, the extreme value theorem does not apply, since the domain is an open interval, $(-\infty, \infty)$, which has no endpoints. Begin as before by finding any critical numbers.

$$f'(x) = 12x^3 - 12x^2 - 24x = 0$$
$$12x(x^2 - x - 2) = 0$$
$$12x(x + 1)(x - 2) = 0$$
$$x = 0 \quad \text{or} \quad x = -1 \quad \text{or} \quad x = 2$$

There are no values of x where $f'(x)$ does not exist. Evaluate the function at the critical numbers.

FIGURE 5

	Extrema Candidates	
x-Value	Value of Function	
−1	−3	
0	2	
2	−30	← Absolute minimum

For an open interval, rather than evaluating the function at the endpoints, we evaluate the limit of the function when the endpoints are approached. Because the positive x^4-term dominates the other terms as x becomes large,

$$\lim_{x \to \infty} (3x^4 - 4x^3 - 12x^2 + 2) = \infty.$$

The limit is also ∞ as x approaches $-\infty$. Since the function can be made arbitrarily large, it has no absolute maximum. The absolute minimum, -30, occurs at $x = 2$. This result can be confirmed with a graphing calculator, as shown in Figure 5. **TRY YOUR TURN 2**

YOUR TURN 2 Find the locations and values of the absolute extrema, if they exist, for the function $f(x) = -x^4 - 4x^3 + 8x^2 + 20$.

In many of the applied extrema problems in the next section, a continuous function on an open interval has just one critical number. In that case, we can use the following theorem, which also applies to closed intervals.

Critical Point Theorem

Suppose a function f is continuous on an interval I and that f has exactly one critical number in the interval I, located at $x = c$.

If f has a relative maximum at $x = c$, then this relative maximum is the absolute maximum of f on the interval I.

If f has a relative minimum at $x = c$, then this relative minimum is the absolute minimum of f on the interval I.

The critical point theorem is of no help in the previous two examples because they each had more than one critical point on the interval under consideration. But the theorem could be useful for some of Exercises 31–38 at the end of this section, and we will make good use of it in the next section.

EXAMPLE 3   U.S. and Canadian Dollar Exchange

The U.S. and Canadian exchange rate changes daily. The value of the U.S. dollar (in Canadian dollars) between 2010 and 2014 can be approximated by the function

$$f(t) = -0.002693t^3 + 0.03610t^2 - 0.1003t + 1.0645$$

where t is the number of years since 2010. Based on this approximation, in what year during this period did the value of the U.S. dollar reach its absolute minimum? What is the minimum value of the dollar during this period? *Source: The Federal Reserve.*

APPLY IT

SOLUTION   The function is defined on the interval $[0, 4]$. We first look for critical numbers in this interval. Here $f'(t) = -0.008079t^2 + 0.07220t - 0.1003$. We set this derivative equal to 0 and use the quadratic formula to solve for t.

$$-0.008079t^2 + 0.07220t - 0.1003 = 0$$

$$t = \frac{-0.07220 \pm \sqrt{(0.07220)^2 - 4(-0.008079)(-0.1003)}}{2(-0.008079)}$$

$$t = 1.72 \quad \text{or} \quad t = 7.2$$

Only the first value is in the interval $[0, 4]$. Now evaluate the function at the critical number and the endpoints 0 and 4.

	Extrema Candidates	
t-Value	Value of Function	
0	1.0645	
1.72	0.9851	← Absolute minimum
4	1.0685	

About 1.72 years after 2010, that is, in late 2011, the U.S. dollar was worth about $0.99 Canadian, which was an absolute minimum during this period. It is also worth noting the absolute maximum value of the U.S. dollar, which was about $1.07 Canadian, occurred approximately 4 years after 2010, or around early 2014.

Graphical Optimization

Figure 6 shows the production output for a family-owned business that produces landscape mulch. As the number of workers (measured in hours of labor) varies, the total production of mulch also varies. In Section 3.5, Example 1, we saw that maximum production occurs with 8 workers (corresponding to 320 hours of labor). A manager, however, may want to know how many hours of labor to use in order to maximize the output per hour of labor. For any point on the curve, the y-coordinate measures the output and the x-coordinate measures the hours of labor, so the y-coordinate divided by the x-coordinate gives the output per hour of labor. This quotient is also the slope of the line through the origin and the point on the curve. Therefore, to maximize the output per hour of labor, we need to find where this slope is greatest. As shown in Figure 6, this occurs when approximately 270 hours of labor are used. Notice that this is also where the line from the origin to the curve is tangent to the curve. Another way of looking at this is to say that the point on the curve where the tangent line passes through the origin is the point that maximizes the output per hour of labor.

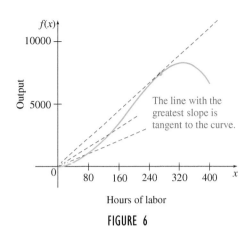

FIGURE 6

We can show that, in general, when $y = f(x)$ represents the output as a function of input, the maximum output per unit input occurs when the line from the origin to a point on the graph of the function is tangent to the function. Our goal is to maximize

$$g(x) = \frac{\text{output}}{\text{input}} = \frac{f(x)}{x}.$$

Taking the derivative and setting it equal to 0 gives

$$g'(x) = \frac{xf'(x) - f(x)}{x^2} = 0$$

$$xf'(x) = f(x)$$

$$f'(x) = \frac{f(x)}{x}.$$

Notice that $f'(x)$ gives the slope of the tangent line at the point, and $f(x)/x$ gives the slope of the line from the origin to the point. When these are equal, as in Figure 6, the output per input is maximized. In other examples, the point on the curve where the tangent line passes through the origin gives a minimum. For a life science example of this, see Exercise 54 in Section 3.4 on the Definition of the Derivative.

6.1 WARM-UP EXERCISES

Find all critical numbers for each of the following functions. *(Sec. 5.1)*

W1. $f(x) = 4x^3 - 7x^2 - 40x + 5$

W2. $f(x) = 15x^{7/2} - 7x^{9/2}$

6.1 EXERCISES

Find the locations of any absolute extrema for the functions with graphs as follows.

1.

2.

3.

4.

5.

6.

7.

8.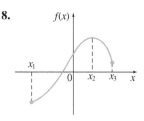

9. What is the difference between a relative extremum and an absolute extremum?

10. Can a relative extremum be an absolute extremum? Is a relative extremum necessarily an absolute extremum?

Find the absolute extrema if they exist, as well as all values of x where they occur, for each function, and specified domain. If you have one, use a graphing calculator to verify your answers.

11. $f(x) = x^3 - 6x^2 + 9x - 8;$ $[0, 5]$

12. $f(x) = x^3 - 3x^2 - 24x + 5;$ $[-3, 6]$

13. $f(x) = \dfrac{1}{3}x^3 + \dfrac{3}{2}x^2 - 4x + 1;$ $[-5, 2]$

14. $f(x) = \dfrac{1}{3}x^3 - \dfrac{1}{2}x^2 - 6x + 3;$ $[-4, 4]$

15. $f(x) = x^4 - 18x^2 + 1;$ $[-4, 4]$

16. $f(x) = x^4 - 32x^2 - 7;$ $[-5, 6]$

17. $f(x) = \dfrac{1 - x}{3 + x};$ $[0, 3]$ **18.** $f(x) = \dfrac{8 + x}{8 - x};$ $[4, 6]$

19. $f(x) = \dfrac{x - 1}{x^2 + 1};$ $[1, 5]$ **20.** $f(x) = \dfrac{x}{x^2 + 2};$ $[0, 4]$

21. $f(x) = (x^2 - 4)^{1/3};$ $[-2, 3]$

22. $f(x) = (x^2 - 16)^{2/3};$ $[-5, 8]$

23. $f(x) = 5x^{2/3} + 2x^{5/3};$ $[-2, 1]$

24. $f(x) = x + 3x^{2/3};$ $[-10, 1]$

25. $f(x) = x^2 - 8 \ln x;$ $[1, 4]$ **26.** $f(x) = \dfrac{\ln x}{x^2};$ $[1, 4]$

27. $f(x) = x + e^{-3x};$ $[-1, 3]$ **28.** $f(x) = x^2 e^{-0.5x};$ $[2, 5]$

Graph each function on the indicated domain, and use the capabilities of your calculator to find the location and value of the absolute extrema.

29. $f(x) = \dfrac{-5x^4 + 2x^3 + 3x^2 + 9}{x^4 - x^3 + x^2 + 7};$ $[-1, 1]$

30. $f(x) = \dfrac{x^3 + 2x + 5}{x^4 + 3x^3 + 10};$ $[-3, 0]$

Find the absolute extrema if they exist, as well as all values of x where they occur.

31. $f(x) = 2x + \dfrac{8}{x^2} + 1,$ $x > 0$

32. $f(x) = 12 - x - \dfrac{9}{x},$ $x > 0$

33. $f(x) = -3x^4 + 8x^3 + 18x^2 + 2$

34. $f(x) = x^4 - 4x^3 + 4x^2 + 1$

35. $f(x) = \dfrac{x - 1}{x^2 + 2x + 6}$ **36.** $f(x) = \dfrac{x}{x^2 + 1}$

37. $f(x) = \dfrac{\ln x}{x^3}$ **38.** $f(x) = x \ln x$

39. Find the absolute maximum and minimum of $f(x) = 2x - 3x^{2/3}$ **(a)** on the interval $[-1, 0.5]$; **(b)** on the interval $[0.5, 2]$.

40. Let $f(x) = e^{-2x}$. For $x > 0$, let $P(x)$ be the perimeter of the rectangle with vertices $(0, 0)$, $(x, 0)$, $(x, f(x))$ and $(0, f(x))$. Which of the following statements is true? *Source: Society of Actuaries*.

   (a) The function P has an absolute minimum but not an absolute maximum on the interval $(0, \infty)$.

   (b) The function P has an absolute maximum but not an absolute minimum on the interval $(0, \infty)$.

   (c) The function P has both an absolute minimum and an absolute maximum on the interval $(0, \infty)$.

   (d) The function P has neither an absolute maximum nor an absolute minimum on the interval $(0, \infty)$, but the graph of the function P does have an inflection point with positive x-coordinate.

   (e) The function P has neither an absolute maximum nor an absolute minimum on the interval $(0, \infty)$, and the graph of the function P does not have an inflection point with positive x-coordinate.

APPLICATIONS

Business and Economics

41. Bank Robberies   The number of bank robberies in the United States for the years 2000–2011 is given in the following figure. Consider the closed interval $[2000, 2011]$. *Source: FBI.*

Bank Robberies (2000–2011)

(a) Give all relative maxima and minima and when they occur on the interval.

(b) Give the absolute maxima and minima and when they occur on the interval. Interpret your results.

42. Bank Burglaries   The number of bank burglaries (entry into or theft from a bank during nonbusiness hours) in the United States for the years 2000–2011 is given in the figure. Consider the closed interval [2000, 2011]. *Source: FBI.*

(a) Give all relative maxima and minima and when they occur on the interval.

(b) Give the absolute maxima and minima and when they occur on the interval. Interpret your results.

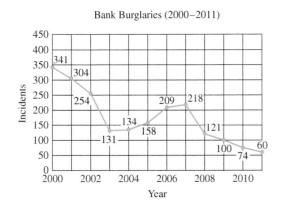

Bank Burglaries (2000–2011)

43. Profit   The total profit $P(x)$ (in thousands of dollars) from the sale of x hundred thousand automobile tires is approximated by

$$P(x) = -x^3 + 9x^2 + 120x - 400, \quad x \geq 5.$$

Find the number of hundred thousands of tires that must be sold to maximize profit. Find the maximum profit.

44. Profit   A company has found that its weekly profit from the sale of x units of an auto part is given by

$$P(x) = -0.02x^3 + 600x - 20{,}000.$$

Production bottlenecks limit the number of units that can be made per week to no more than 150, while a long-term contract requires that at least 50 units be made each week. Find the maximum possible weekly profit that the firm can make.

Average Cost   Find the minimum value of the average cost for the given cost function on the given intervals.

45. $C(x) = x^3 + 37x + 250$ on the following intervals.

   (a) $1 \leq x \leq 10$ (b) $10 \leq x \leq 20$

46. $C(x) = 81x^2 + 17x + 324$ on the following intervals.

   (a) $1 \leq x \leq 10$ (b) $10 \leq x \leq 20$

Cost   Each graph in the next column gives the cost as a function of production level. Use the method of graphical optimization to estimate the production level that results in the minimum cost per item produced.

47.

48.

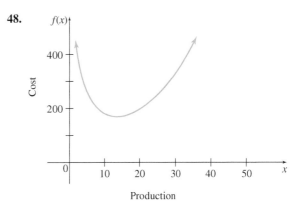

Profit   Each graph below gives the profit as a function of production level. Use graphical optimization to estimate the production level that gives the maximum profit per item produced.

49.

50.

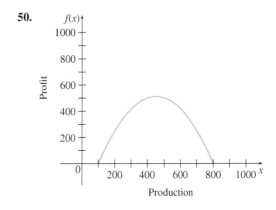

51. Cost A friend observes that the goal in Exercise 47 is to minimize the cost per item, and that the derivative $f'(x)$ has units of cost per item. To minimize the derivative would require setting its derivative, that is, the second derivative, equal to 0. Therefore, the sought-after point is the inflection point. Explain the error in this reasoning.

52. Administrative Intensity A study of administrative intensity (the number of managers divided by the number of workers) found that the profit P of a company depended on the administrative intensity i according to the formula

$$P(i) = (p - r)AS^a i^t k^b X^{1-c} - W_0 X - W_1 Xi - C,$$

where p is the selling price of a product; r is the nonwage cost of the product; A is a productivity scale factor; S is the number of occupational specialties; a, b, and t are positive numbers whose sum is less than 1 representing the marginal relative productivity of specialization, capital, and labor, respectively; k is the capital intensity per worker; X is the number of workers; $c = 1 - a - b - t$; W_0 is the average wage per worker; W_1 is the average salary per manager; and C is the fixed cost of production. Show that the profit is maximized when

$$i = [tA(p - r)/W_1]^{1/(1-t)} k^{b/(1-t)} S^{a/(1-t)} X^{-c/(1-t)}.$$

Source: Administrative Science Quarterly.

Life Sciences

53. Pollution A marshy region used for agricultural drainage has become contaminated with selenium. It has been determined that flushing the area with clean water will reduce the selenium for a while, but it will then begin to build up again. A biologist has found that the percent of selenium in the soil x months after the flushing begins is given by

$$f(x) = \frac{x^2 + 36}{2x}, \quad 1 \le x \le 12.$$

When will the selenium be reduced to a minimum? What is the minimum percent?

54. Salmon Spawning The number of salmon swimming upstream to spawn is approximated by

$$S(x) = -x^3 + 3x^2 + 360x + 5000, \quad 6 \le x \le 20,$$

where x represents the temperature of the water in degrees Celsius. Find the water temperature that produces the maximum number of salmon swimming upstream.

55. Molars Researchers have determined that the crown length of first molars in fetuses is related to the postconception age of the tooth as

$$L(t) = -0.01t^2 + 0.788t - 7.048,$$

where $L(t)$ is the crown length (in millimeters) of the molar t weeks after conception. Find the maximum length of the crown of first molars during weeks 22 through 28. *Source: American Journal of Physical Anthropology.*

56. Fungal Growth Because of the time that many people spend indoors, there is a concern about the health risk of being exposed to harmful fungi that thrive in buildings. The risk appears to increase in damp environments. Researchers have discovered that by controlling both the temperature and the relative humidity in a building, the growth of the fungus A.

versicolor can be limited. The relationship between temperature and relative humidity, which limits growth, can be described by

$$R(T) = -0.00007T^3 + 0.0401T^2 - 1.6572T + 97.086,$$

$$15 \le T \le 46,$$

where $R(T)$ is the relative humidity (in percent) and T is the temperature (in degrees Celsius). Find the temperature at which the relative humidity is minimized. *Source: Applied and Environmental Microbiology.*

Physical Sciences

57. Gasoline Mileage From information given in a recent business publication, we constructed the mathematical model

$$M(x) = -\frac{1}{45}x^2 + 2x - 20, \quad 30 \le x \le 65,$$

to represent the miles per gallon used by a certain car at a speed of x mph. Find the absolute maximum miles per gallon and the absolute minimum and the speeds at which they occur.

58. Gasoline Mileage For a certain sports utility vehicle,

$$M(x) = -0.015x^2 + 1.31x - 7.3, \quad 30 \le x \le 60,$$

represents the miles per gallon obtained at a speed of x mph. Find the absolute maximum miles per gallon and the absolute minimum and the speeds at which they occur.

General Interest

Area A piece of wire 12 ft long is cut into two pieces. (See the figure.) One piece is made into a circle and the other piece is made into a square. Let the piece of length x be formed into a circle. We allow x to equal 0 or 12, so all the wire may be used for the square or for the circle.

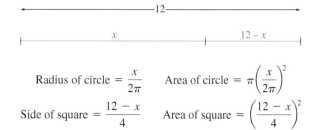

$$\text{Radius of circle} = \frac{x}{2\pi} \qquad \text{Area of circle} = \pi\left(\frac{x}{2\pi}\right)^2$$

$$\text{Side of square} = \frac{12 - x}{4} \qquad \text{Area of square} = \left(\frac{12 - x}{4}\right)^2$$

59. Where should the cut be made in order to minimize the sum of the areas enclosed by both figures?

60. Where should the cut be made in order to make the sum of the areas maximum? (*Hint:* Remember to use the endpoints of a domain when looking for absolute maxima and minima.)

61. For the solution to Exercise 59, show that the side of the square equals the diameter of the circle, that is, that the circle can be inscribed in the square.*

62. Information Content Suppose dots and dashes are transmitted over a telegraph line so that dots occur a fraction p of the time (where $0 < p < 1$) and dashes occur a fraction $1 - p$ of the time. The *information content* of the telegraph line is given by $I(p)$, where

$$I(p) = -p \ln p - (1 - p) \ln(1 - p).$$

*For a generalization of this phenomenon, see Cade, Pat and Russell A. Gordon, "An Apothem Apparently Appears," *The College Mathematics Journal,* Vol. 36, No. 1, Jan. 2005, pp. 52–55.

(a) Show that $I'(p) = -\ln p + \ln(1 - p)$.

(b) Set $I'(p) = 0$ and find the value of p that maximizes the information content.

(c) How might the result in part (b) be used?

6.2 Applications of Extrema

APPLY IT   How should boxes and cans be designed to minimize the material needed to construct them or to maximize the volume?

In Examples 3 and 4 we will use the techniques of calculus to find an answer to these questions.

In this section we give several examples showing applications of calculus to maximum and minimum problems. To solve these examples, go through the following steps.

Solving an Applied Extrema Problem

1. Read the problem carefully. Make sure you understand what is given and what is unknown.

2. If possible, sketch a diagram. Label the various parts.

3. Decide which variable must be maximized or minimized. Express that variable as a function of *one* other variable.

4. Find the domain of the function.

5. Find the critical points for the function from Step 3.

6. If the domain is a closed interval, evaluate the function at the endpoints and at each critical number to see which yields the absolute maximum or minimum. If the domain is an open interval, apply the critical point theorem when there is only one critical number. If there is more than one critical number, evaluate the function at the critical numbers and find the limit as the endpoints of the interval are approached to determine if an absolute maximum or minimum exists at one of the critical points.

CAUTION   Do not skip Step 6 in the preceding box. If a problem asks you to maximize a quantity and you find a critical point at Step 5, do not automatically assume the maximum occurs there, for it may occur at an endpoint, as in Exercise 60 of the previous section, or it may not exist at all.

An infamous case of such an error occurred in a 1945 study of "flying wing" aircraft designs similar to the Stealth bomber. In seeking to maximize the range of the aircraft (how far it can fly on a tank of fuel), the study's authors found that a critical point occurred when almost all of the volume of the plane was in the wing. They claimed that this critical point was a maximum. But another engineer later found that this critical point, in fact, *minimized* the range of the aircraft! *Source: Science.*

EXAMPLE 1   Maximization

Find two nonnegative numbers x and y for which $2x + y = 30$, such that xy^2 is maximized.

SOLUTION   Step 1, reading and understanding the problem, is up to you. Step 2 does not apply in this example; there is nothing to draw. We proceed to Step 3, in which we decide what is to be maximized and assign a variable to that quantity. Here, xy^2 is to be maximized, so let

$$M = xy^2.$$

According to Step 3, we must express M in terms of just *one* variable, which can be done using the equation $2x + y = 30$ by solving for either x or y. Solving for y gives

$$2x + y = 30$$
$$y = 30 - 2x.$$

Substitute $30 - 2x$ for y in the expression for M to get

$$M = x(30 - 2x)^2$$
$$= x(900 - 120x + 4x^2) \quad \text{Square } 30 - 2x.$$
$$= 900x - 120x^2 + 4x^3.$$

We are now ready for Step 4, when we find the domain of the function. Because of the nonnegativity requirement, x must be at least 0. Since y must also be at least 0, we require $30 - 2x \geq 0$, so $x \leq 15$. Thus x is confined to the interval $[0, 15]$.

Moving on to Step 5, we find the critical points for M by finding dM/dx, and then solving the equation $dM/dx = 0$ for x.

$$\frac{dM}{dx} = 900 - 240x + 12x^2 = 0$$
$$12(75 - 20x + x^2) = 0 \quad \text{Factor out the 12.}$$
$$12(5 - x)(15 - x) = 0 \quad \text{Factor the quadratic.}$$
$$x = 5 \quad \text{or} \quad x = 15$$

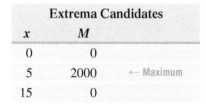

Extrema Candidates

x	M	
0	0	
5	2000	← Maximum
15	0	

YOUR TURN 1 Find two nonnegative numbers x and y for which $x + 3y = 30$, such that x^2y is maximized.

Finally, at Step 6, we find M for the critical numbers $x = 5$ and $x = 15$, as well as for $x = 0$, an endpoint of the domain. The other endpoint, $x = 15$, has already been included as a critical number. We see in the table that the maximum value of the function occurs when $x = 5$. Since $y = 30 - 2x = 30 - 2(5) = 20$, the values that maximize xy^2 are $x = 5$ and $y = 20$.   TRY YOUR TURN 1

NOTE A critical point is only a candidate for an absolute maximum or minimum. The absolute maximum or minimum might occur at a different critical point or at an endpoint.

EXAMPLE 2   Minimizing Time

A math professor participating in the sport of orienteering must get to a specific tree in the woods as fast as possible. He can get there by traveling east along the trail for 300 m and then north through the woods for 800 m. He can run 160 m per minute along the trail but only 70 m per minute through the woods. Running directly through the woods toward the tree minimizes the distance, but he will be going slowly the whole time. He could instead run 300 m along the trail before entering the woods, maximizing the total distance but minimizing the time in the woods. Perhaps the fastest route is a combination, as shown in Figure 7. Find the path that will get him to the tree in the minimum time.

SOLUTION As in Example 1, the first step is to read and understand the problem. If the statement of the problem is not clear to you, go back and reread it until you understand it before moving on.

We have already started Step 2 by providing Figure 7. Let x be the distance shown in Figure 7, so the distance he runs on the trail is $300 - x$. By the Pythagorean theorem, the distance he runs through the woods is $\sqrt{800^2 + x^2}$.

The first part of Step 3 is noting that we are trying to minimize the total amount of time, which is the sum of the time on the trail and the time through the woods. We must express this time as a function of x. Since time = distance/speed, the total time is

$$T(x) = \frac{300 - x}{160} + \frac{\sqrt{800^2 + x^2}}{70}.$$

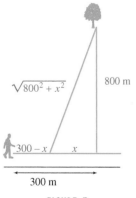

FIGURE 7

To complete Step 4, notice in this equation that $0 \leq x \leq 300$.

We now move to Step 5, in which we find the critical points by calculating the derivative and setting it equal to 0. Since $\sqrt{800^2 + x^2} = (800^2 + x^2)^{1/2}$,

$$T'(x) = -\frac{1}{160} + \frac{1}{70}\left(\frac{1}{2}\right)(800^2 + x^2)^{-1/2}(2x) = 0.$$

$$\frac{x}{70\sqrt{800^2 + x^2}} = \frac{1}{160}$$

	Extrema Candidates	
x	$T(x)$	
0	13.30	
300	12.21	← Minimum

$$16x = 7\sqrt{800^2 + x^2} \qquad \text{Cross multiply and divide by 10.}$$
$$256x^2 = 49(800^2 + x^2) = (49 \cdot 800^2) + 49x^2 \qquad \text{Square both sides.}$$
$$207x^2 = 49 \cdot 800^2 \qquad \text{Subtract } 49x^2 \text{ from both sides.}$$
$$x^2 = \frac{49 \cdot 800^2}{207}$$
$$x = \frac{7 \cdot 800}{\sqrt{207}} \approx 389 \qquad \text{Take the square root of both sides.}$$

YOUR TURN 2 Suppose the professor in Example 2 can only run 40 m per minute through the woods. Find the path that will get him to the tree in the minimum time.

Since 389 is not in the interval $[0, 300]$, the minimum time must occur at one of the endpoints.

We now complete Step 6 by creating a table with $T(x)$ evaluated at the endpoints. We see from the table that the time is minimized when $x = 300$, that is, when the professor heads straight for the tree.   TRY YOUR TURN 2

EXAMPLE 3   Maximizing Volume

A supplier of bolts wants to create open boxes for the bolts by cutting a square from each corner of a 12-in. by 12-in. piece of metal and then folding up the sides. What size square should be cut from each corner to produce a box of maximum volume?

APPLY IT

SOLUTION   Let x represent the length of a side of the square that is cut from each corner, as shown in Figure 8(a). The width of the box is $12 - 2x$, with the length also $12 - 2x$. As shown in Figure 8(b), the depth of the box will be x inches. The volume of the box is given by the product of the length, width, and height. In this example, the volume, $V(x)$, depends on x:

$$V(x) = x(12 - 2x)(12 - 2x) = 144x - 48x^2 + 4x^3.$$

Clearly, $0 \leq x$, and since neither the length nor the width can be negative, $0 \leq 12 - 2x$, so $x \leq 6$. Thus, the domain of V is the interval $[0, 6]$.

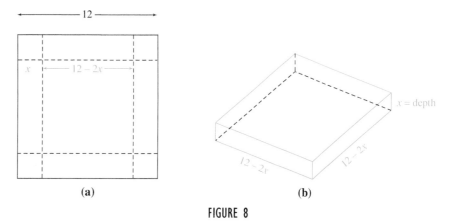

(a)   (b)

FIGURE 8

Extrema Candidates		
x	$V(x)$	
0	0	
2	128	← Maximum
6	0	

YOUR TURN 3 Repeat Example 3 using an 8-m by 8-m piece of metal.

The derivative is $V'(x) = 144 - 96x + 12x^2$. Set this derivative equal to 0.

$$12x^2 - 96x + 144 = 0$$
$$12(x^2 - 8x + 12) = 0$$
$$12(x - 2)(x - 6) = 0$$
$$x - 2 = 0 \quad \text{or} \quad x - 6 = 0$$
$$x = 2 \qquad\qquad x = 6$$

Find $V(x)$ for x equal to 0, 2, and 6 to find the depth that will maximize the volume. The table indicates that the box will have maximum volume when $x = 2$ and that the maximum volume will be 128 in^3.   **TRY YOUR TURN 3** ▬

EXAMPLE 4   Minimizing Area

A coffee company wants to manufacture cylindrical aluminum coffee cans with a volume of 1000 cm^3 (1 liter). What should the radius and height of the can be to minimize the amount of aluminum used?

APPLY IT

SOLUTION   The two variables in this problem are the radius and the height of the can, which we shall label r and h, as in Figure 9. Minimizing the amount of aluminum used requires minimizing the surface area of the can, which we will designate S. The surface area consists of a top and a bottom, each of which is a circle with an area πr^2, plus the side. If the side were sliced vertically and unrolled, it would form a rectangle with height h and width equal to the circumference of the can, which is $2\pi r$. Thus the surface area is given by

$$S = 2\pi r^2 + 2\pi rh.$$

FIGURE 9

The right side of the equation involves two variables. We need to get a function of a single variable. We can do this by using the information about the volume of the can:

$$V = \pi r^2 h = 1000.$$

(Here we have used the formula for the volume of a cylinder.) Solve this for h:

$$h = \frac{1000}{\pi r^2}.$$

(Solving for r would have involved a square root and a more complicated function.)

We now substitute this expression for h into the equation for S to get

$$S = 2\pi r^2 + 2\pi r\frac{1000}{\pi r^2} = 2\pi r^2 + \frac{2000}{r}.$$

There are no restrictions on r other than that it be a positive number, so the domain of S is $(0, \infty)$.

Find the critical points for S by finding dS/dr, then solving the equation $dS/dr = 0$ for r.

$$\frac{dS}{dr} = 4\pi r - \frac{2000}{r^2} = 0$$
$$4\pi r^3 = 2000 \qquad \text{Add } \tfrac{2000}{r^2} \text{ to both sides}$$
$$r^3 = \frac{500}{\pi} \qquad \text{and multiply by } r^2.$$

Take the cube root of both sides to get

$$r = \left(\frac{500}{\pi}\right)^{1/3} \approx 5.419$$

centimeters. Substitute this expression into the equation for h to get

$$h = \frac{1000}{\pi(5.419)^2} \approx 10.84$$

centimeters. Notice that the height of the can is twice its radius.

There are several ways to carry out Step 6 to verify that we have found the minimum. Because there is only one critical number, the critical point theorem applies.

Method 1
Critical Point Theorem
with First Derivative Test

Verify that when $r < 5.419$, then $dS/dr < 0$, and when $r > 5.419$, then $dS/dr > 0$. Since the function is decreasing before 5.419 and increasing after 5.419, there must be a relative minimum at $r = 5.419$ cm. By the critical point theorem, there is an absolute minimum there.

Method 2
Critical Point Theorem
with Second Derivative Test

We could also use the critical point theorem with the second derivative test.

$$\frac{d^2S}{dr^2} = 4\pi + \frac{4000}{r^3}$$

Notice that for positive r, the second derivative is always positive, so there is a relative minimum at $r = 5.419$ cm. By the critical point theorem, there is an absolute minimum there.

Method 3
Limits at Endpoints

We could also find the limit as the endpoints are approached.

$$\lim_{r \to 0} S = \lim_{r \to \infty} S = \infty$$

The surface area becomes arbitrarily large as r approaches the endpoints of the domain, so the absolute minimum surface area must be at the critical point.

YOUR TURN 4 Repeat Example 4 if the volume is to be 500 cm³.

The graphing calculator screen in Figure 10 confirms that there is an absolute minimum at $r = 5.419$ cm. **TRY YOUR TURN 4**

Notice that if the previous example had asked for the height and radius that maximize the amount of aluminum used, the problem would have no answer. There is no maximum for a function that can be made arbitrarily large.

$S = 2\pi r^2 + \frac{2000}{r}$

1600

Minimum
X=5.4192582 Y=553.58104

FIGURE 10

Maximum Sustainable Harvest

For most living things, reproduction is *seasonal*—it can take place only at selected times of the year. Large whales, for example, reproduce every two years during a relatively short time span of about two months. Shown on the time axis in Figure 11 are the reproductive periods. Let S = number of adults present during the reproductive period and let R = number of adults that return the next season to reproduce. *Source: Mathematics for the Biosciences.*

Reproductive period

S reproduce R return

FIGURE 11

If we find a relationship between R and S, $R = f(S)$, then we have formed a **spawner-recruit** function or **parent-progeny** function. These functions are notoriously hard to develop because of the difficulty of obtaining accurate counts and because of the many hypotheses that can be made about the life stages. We will simply suppose that the function f takes various forms.

If $R > S$, we can presumably harvest

$$H = R - S = f(S) - S$$

individuals, leaving S to reproduce. Next season, $R = f(S)$ will return and the harvesting process can be repeated, as shown in Figure 12.

Let S_0 be the number of spawners that will allow as large a harvest as possible without threatening the population with extinction. Then $H(S_0)$ is called the **maximum sustainable harvest**.

R return

S reproduce

$H = R - S$ caught

FIGURE 12

EXAMPLE 5 Maximum Sustainable Harvest

Suppose the spawner-recruit function for Idaho rabbits is $f(S) = 2.17\sqrt{S} \ln(S + 1)$, where S is measured in thousands of rabbits. Find S_0 and the maximum sustainable harvest, $H(S_0)$.

SOLUTION S_0 is the value of S that maximizes H. Since

$$H(S) = f(S) - S$$

$$= 2.17\sqrt{S} \ln(S + 1) - S,$$

$$H'(S) = 2.17\left(\frac{\ln(S + 1)}{2\sqrt{S}} + \frac{\sqrt{S}}{S + 1}\right) - 1.$$

Now we want to set this derivative equal to 0 and solve for S.

$$0 = 2.17\left(\frac{\ln(S + 1)}{2\sqrt{S}} + \frac{\sqrt{S}}{S + 1}\right) - 1.$$

$$H'(S) = 2.17\left(\frac{\ln(S + 1)}{2\sqrt{S}} + \frac{\sqrt{S}}{S + 1}\right) - 1$$

1.5

0   60

Zero
X=36.557775   Y=0

−0.5

FIGURE 13

This equation cannot be solved analytically, so we will graph $H'(S)$ with a graphing calculator and find any S-values where $H'(S)$ is 0. (An alternative approach is to use the equation solver some graphing calculators have.) The graph with the value where $H'(S)$ is 0 is shown in Figure 13.

From the graph we see that $H'(S) = 0$ when $S = 36.557775$, so the number of rabbits needed to sustain the population is about 36,600. A graph of H will show that this is a maximum. From the graph, using the capability of the calculator, we find that the harvest is $H(36.557775) \approx 11.015504$. These results indicate that after one reproductive season a population of 36,600 rabbits will have increased to 47,600. Of these, 11,000 may be harvested, leaving 36,600 to regenerate the population. Any harvest larger than 11,000 will threaten the future of the rabbit population, while a harvest smaller than 11,000 will allow the population to grow larger each season. Thus 11,000 is the maximum sustainable harvest for this population.

6.2 WARM-UP EXERCISES

Find all critical numbers for each of the following functions. *(Sec. 5.1)*

W1. $f(x) = 5x^4 - 18x^3 - 28x^2 + 12$

W2. $f(x) = \dfrac{2x + 1}{x^2 + 3}$

6.2 EXERCISES

In Exercises 1–4, use the steps shown in Exercise 1 to find non-negative numbers x and y that satisfy the given requirements. Give the optimum value of the indicated expression.

1. $x + y = 180$ and the product $P = xy$ is as large as possible.

   (a) Solve $x + y = 180$ for y.

   (b) Substitute the result from part (a) into $P = xy$, the equation for the variable that is to be maximized.

   (c) Find the domain of the function P found in part (b).

   (d) Find dP/dx. Solve the equation $dP/dx = 0$.

   (e) Evaluate P at any solutions found in part (d), as well as at the endpoints of the domain found in part (c).

   (f) Give the maximum value of P, as well as the two numbers x and y whose product is that value.

2. The sum of x and y is 140 and the sum of the squares of x and y is minimized.

3. $x + y = 90$ and $x^2 y$ is maximized.

4. $x + y = 105$ and xy^2 is maximized.

APPLICATIONS

Business and Economics

Average Cost In Exercises 5 and 6, determine the average cost function $\overline{C}(x) = C(x)/x$. To find where the average cost is smallest, first calculate $\overline{C}'(x)$, the derivative of the average cost function. Then use a graphing calculator to find where the derivative is 0. Check your work by finding the minimum from the graph of the function $\overline{C}(x)$.

5. $C(x) = \dfrac{1}{2}x^3 + 2x^2 - 3x + 35$

6. $C(x) = 10 + 20x^{1/2} + 16x^{3/2}$

7. **Revenue** If the price charged for a candy bar is $p(x)$ cents, then x thousand candy bars will be sold in a certain city, where

$$p(x) = 160 - \frac{x}{10}.$$

   (a) Find an expression for the total revenue from the sale of x thousand candy bars.

   (b) Find the value of x that leads to maximum revenue.

   (c) Find the maximum revenue.

8. **Revenue** The sale of compact discs of "lesser" performers is very sensitive to price. If a CD manufacturer charges $p(x)$ dollars per CD, where

$$p(x) = 12 - \frac{x}{8},$$

then x thousand CDs will be sold.

   (a) Find an expression for the total revenue from the sale of x thousand CDs.

   (b) Find the value of x that leads to maximum revenue.

   (c) Find the maximum revenue.

9. **Area** A campground owner has 1400 m of fencing. He wants to enclose a rectangular field bordering a river, with no fencing needed along the river. (See the sketch.) Let x represent the width of the field.

   (a) Write an expression for the length of the field.

   (b) Find the area of the field (area = length × width).

   (c) Find the value of x leading to the maximum area.

   (d) Find the maximum area.

10. **Area** Find the dimensions of the rectangular field of maximum area that can be made from 300 m of fencing material. (This fence has four sides.)

11. **Area** An ecologist is conducting a research project on breeding pheasants in captivity. She first must construct suitable pens. She wants a rectangular area with two additional fences across its width, as shown in the sketch. Find the maximum area she can enclose with 3600 m of fencing.

12. **Area** A farmer is constructing a rectangular pen with one additional fence across its width. Find the maximum area that can be enclosed with 2400 m of fencing.

13. **Cost with Fixed Area** A fence must be built in a large field to enclose a rectangular area of 25,600 m². One side of the area is bounded by an existing fence; no fence is needed there. Material for the fence costs $3 per meter for the two ends and $1.50 per meter for the side opposite the existing fence. Find the cost of the least expensive fence.

14. **Cost with Fixed Area** A fence must be built to enclose a rectangular area of 20,000 ft². Fencing material costs $2.50 per foot for the two sides facing north and south and $3.20 per foot for the other two sides. Find the cost of the least expensive fence.

15. **Revenue** A local club is arranging a charter flight to Hawaii. The cost of the trip is $1600 each for 90 passengers, with a refund of $10 per passenger for each passenger in excess of 90.

   (a) Find the number of passengers that will maximize the revenue received from the flight.

   (b) Find the maximum revenue.

16. Profit In planning a restaurant, it is estimated that a profit of $8 per seat will be made if the number of seats is no more than 50, inclusive. On the other hand, the profit on each seat will decrease by 10¢ for each seat above 50.

(a) Find the number of seats that will produce the maximum profit.

(b) What is the maximum profit?

17. Timing Income A local group of scouts has been collecting aluminum cans for recycling. The group has already collected 12,000 lb of cans, for which they could currently receive $7.50 per hundred pounds. The group can continue to collect cans at the rate of 400 lb per day. However, a glut in the aluminum market has caused the recycling company to announce that it will lower its price, starting immediately, by $0.15 per hundred pounds per day. The scouts can make only one trip to the recycling center. Find the best time for the trip. What total income will be received?

18. Pricing Decide what you would do if your assistant presented the following contract for your signature:

Your firm offers to deliver 250 tables to a dealer, at $160 per table, and to reduce the price per table on the entire order by 50¢ for each additional table over 250.

Find the dollar total involved in the largest possible transaction between the manufacturer and the dealer; then find the smallest possible dollar amount.

19. Packaging Design A television manufacturing firm needs to design an open-topped box with a square base. The box must hold 32 in³. Find the dimensions of the box that can be built with the minimum amount of materials. (See the figure.)

20. Packaging Design A company wishes to manufacture a box with a volume of 36 ft³ that is open on top and is twice as long as it is wide. Find the dimensions of the box produced from the minimum amount of material.

21. Container Design An open box will be made by cutting a square from each corner of a 3-ft by 8-ft piece of cardboard and then folding up the sides. What size square should be cut from each corner in order to produce a box of maximum volume?

22. Container Design Consider the problem of cutting corners out of a rectangle and folding up the sides to make a box. Specific examples of this problem are discussed in Example 3 and Exercise 21.

(a) In the solution to Example 3, compare the area of the base of the box with the area of the walls.

(b) Repeat part (a) for the solution to Exercise 21.

(c) Make a conjecture about the area of the base compared with the area of the walls for the box with the maximum volume.

23. Packaging Cost A closed box with a square base is to have a volume of 16,000 cm³. The material for the top and bottom of the box costs $3 per square centimeter, while the material for the sides costs $1.50 per square centimeter. Find the dimensions of the box that will lead to the minimum total cost. What is the minimum total cost?

24. Use of Materials A mathematics book is to contain 36 in² of printed matter per page, with margins of 1 in. along the sides and $1\frac{1}{2}$ in. along the top and bottom. Find the dimensions of the page that will require the minimum amount of paper. (See the figure.)

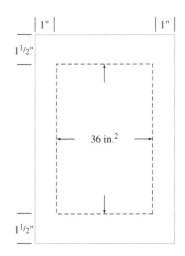

25. Can Design

(a) For the can problem in Example 4, the minimum surface area required that the height be twice the radius. Show that this is true for a can of arbitrary volume V.

(b) Do many cans in grocery stores have a height that is twice the radius? If not, discuss why this may be so.

26. Container Design Your company needs to design cylindrical metal containers with a volume of 16 cubic feet. The top and bottom will be made of a sturdy material that costs $2 per square foot, while the material for the sides costs $1 per square foot. Find the radius, height, and cost of the least expensive container.

27. Carpeting A company sells square carpets for $5 per square foot. It has a simplified manufacturing process for which all the carpets each week must be the same size, and the length must be a multiple of a half foot. It has found that it can sell 100 carpets in a week when the carpets are 2 ft by 2 ft, the minimum size. Beyond this, for each additional foot of length and width, the number sold goes down by 6. What size carpets should the company sell to maximize its revenue? What is the maximum weekly revenue?

28. Real Estate A real estate developer has determined that lots such as those described in Exercise 9 can be sold for $0.50 per square meter. The demand is proportional to the length of the property along the river; the number he can sell in a year is numerically half of that length. What property dimensions maximize his revenue, and how many would he sell in that case? (Remember that the number sold must be an integer.) What is the maximum revenue?

In Exercises 29–31, use a graphing calculator to determine where the derivative is equal to zero.

29. Can Design Modify the can problem in Example 4 so the cost must be minimized. Assume that aluminum costs 3¢ per square centimeter, and that there is an additional cost of 2¢ per cm times the perimeter of the top, and a similar cost for the bottom, to seal the top and bottom of the can to the side.

30. Can Design In this modification of the can problem in Example 4, the cost must be minimized. Assume that aluminum costs 3¢ per square centimeter, and that there is an additional cost of 1¢ per cm times the height of the can to make a vertical seam on the side.

31. Can Design This problem is a combination of Exercises 29 and 30. We will again minimize the cost of the can, assuming that aluminum costs 3¢ per square centimeter. In addition, there is a cost of 2¢ per cm to seal the top and bottom of the can to the side, plus 1¢ per cm to make a vertical seam.

32. Packaging Design A cylindrical box will be tied up with ribbon as shown in the figure. The longest piece of ribbon available is 130 cm long, and 10 cm of that are required for the bow. Find the radius and height of the box with the largest possible volume.

33. Cost A company wishes to run a utility cable from point A on the shore (see the figure below) to an installation at point B on the island. The island is 6 miles from the shore. Point A is 9 miles from Point C, the point on the shore closest to Point B. It costs $400 per mile to run the cable on land and $500 per mile underwater. Assume that the cable starts at A and runs along the shoreline, then angles and runs underwater to the island. Find the point at which the line should begin to angle in order to yield the minimum total cost.

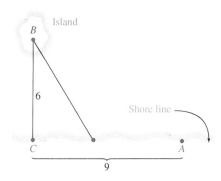

34. Cost Repeat Exercise 33, but make point A 7 miles from point C.

Life Sciences

35. Disease Epidemiologists have found a new communicable disease running rampant in College Station, Texas. They estimate that t days after the disease is first observed in the community, the percent of the population infected by the disease is approximated by

$$p(t) = \frac{20t^3 - t^4}{1000}$$

for $0 \le t \le 20$.

(a) After how many days is the percent of the population infected a maximum?

(b) What is the maximum percent of the population infected?

36. Disease Another disease hits the chronically ill town of College Station, Texas. This time the percent of the population infected by the disease t days after it hits town is approximated by $p(t) = 10te^{-t/8}$ for $0 \le t \le 40$.

(a) After how many days is the percent of the population infected a maximum?

(b) What is the maximum percent of the population infected?

Maximum Sustainable Harvest Find the maximum sustainable harvest in Exercises 37 and 38. See Example 5.

37. $f(S) = 12S^{0.25}$ **38.** $f(S) = \dfrac{25S}{S + 2}$

39. Pollution A lake polluted by bacteria is treated with an antibacterial chemical. After t days, the number N of bacteria per milliliter of water is approximated by

$$N(t) = 20\left(\frac{t}{12} - \ln\left(\frac{t}{12}\right)\right) + 30$$

for $1 \le t \le 15$.

(a) When during this time will the number of bacteria be a minimum?

(b) What is the minimum number of bacteria during this time?

(c) When during this time will the number of bacteria be a maximum?

(d) What is the maximum number of bacteria during this time?

40. Maximum Sustainable Harvest The population of salmon next year is given by $f(S) = Se^{r(1-S/P)}$, where S is this year's salmon population, P is the equilibrium population, and r is a constant that depends upon how fast the population grows. The number of salmon that can be fished next year while keeping the population the same is $H(S) = f(S) - S$. The maximum value of $H(S)$ is the maximum sustainable harvest. *Source: Journal of the Fisheries Research Board of Canada*.

(a) Show that the maximum sustainable harvest occurs when $f'(S) = 1$. (*Hint:* To maximize, set $H'(S) = 0$.)

(b) Let the value of S found in part (a) be denoted by S_0. Show that the maximum sustainable harvest is given by

$$S_0\left(\frac{1}{1 - rS_0/P} - 1\right).$$

(*Hint:* Set $f'(S_0) = 1$ and solve for $e^{r(1-S_0/P)}$. Then find $H(S_0)$ and substitute the expression for $e^{r(1-S_0/P)}$.)

Maximum Sustainable Harvest In Exercises 41 and 42, refer to Exercise 40. Find $f'(S_0)$ and solve the equation $f'(S_0) = 1$, using a calculator to find the intersection of the graphs of $f'(S_0)$ and $y = 1$.

41. Find the maximum sustainable harvest if $r = 0.1$ and $P = 100$.

42. Find the maximum sustainable harvest if $r = 0.4$ and $P = 500$.

43. **Pigeon Flight** Homing pigeons avoid flying over large bodies of water, preferring to fly around them instead. (One possible explanation is the fact that extra energy is required to fly over water because air pressure drops over water in the daytime.) Assume that a pigeon released from a boat 1 mile from the shore of a lake (point B in the figure) flies first to point P on the shore and then along the straight edge of the lake to reach its home at L. If L is 2 miles from point A, the point on the shore closest to the boat, and if a pigeon needs 4/3 as much energy per mile to fly over water as over land, find the location of point P, which minimizes energy used.

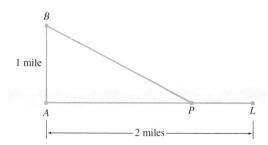

44. **Pigeon Flight** Repeat Exercise 43, but assume a pigeon needs 10/9 as much energy to fly over water as over land.

45. **Harvesting Cod** A recent article described the population $f(S)$ of cod in the North Sea next year as a function of this year's population S (in thousands of tons) by various mathematical models.

$$\text{Shepherd:} \quad f(S) = \frac{aS}{1 + (S/b)^c};$$

$$\text{Ricker:} \quad f(S) = aSe^{-bS};$$

$$\text{Beverton-Holt:} \quad f(S) = \frac{aS}{1 + (S/b)},$$

where a, b, and c are constants. *Source: Nature.*

(a) Find a replacement of variables in the Ricker model above that will make it the same as another form of the Ricker model described in Exercise 40 of this section, $f(S) = Se^{r(1-S/P)}$.

(b) Find $f'(S)$ for all three models.

(c) Find $f'(0)$ for all three models. From your answer, describe in words the geometric meaning of the constant a.

(d) The values of a, b, and c reported in the article for the Shepherd model are 3.026, 248.72, and 3.24, respectively. Find the value of this year's population that maximizes next year's population using the Shepherd model.

(e) The values of a and b reported in the article for the Ricker model are 4.151 and 0.0039, respectively. Find the value of this year's population that maximizes next year's population using the Ricker model.

(f) Explain why, for the Beverton–Holt model, there is no value of this year's population that maximizes next year's population.

46. **Harvesting Cod** For the Beverton–Holt model for cod in the previous exercise, consider harvesting at a rate proportional to each year's population, so that the population next year is given by

$$f(S) = \frac{aS}{1 + (S/b)} - rS,$$

where r is a positive constant of proportionality. Find the value of S that maximizes $f(S)$, and describe in words what this means.

47. **Cylindrical Cells** Cells in the human body are not necessarily spherical; some cancer cells are cylindrical. Suppose that the volume of such a cell is 65 μm^3. Find the radius and height of the cell with the minimum surface area. *Source: American Journal of Surgical Pathology.*

48. **Bird Migration** Suppose a migrating bird flies at a velocity v, and suppose the amount of time the bird can fly depends on its velocity according to the function $T(v)$. *Source: A Concrete Approach to Mathematical Modelling.*

(a) If E is the bird's initial energy, then the bird's effective power is given by kE/T, where k is the fraction of the power that can be converted into mechanical energy. According to principles of aerodynamics,

$$\frac{kE}{T} = aSv^3 + I,$$

where a is a constant, S is the wind speed, and I is the induced power, or rate of working against gravity. Using this result and the fact that distance is velocity multiplied by time, show that the distance that the bird can fly is given by

$$D(v) = \frac{kEv}{aSv^3 + I}.$$

(b) Show that the migrating bird can fly a maximum distance by flying at a velocity

$$v = \left(\frac{I}{2aS}\right)^{1/3}.$$

General Interest

49. **Travel Time** A hunter is at a point along a river bank. He wants to get to his cabin, located 3 miles north and 8 miles west. (See the figure.) He can travel 5 mph along the river but only 2 mph on this very rocky land. How far upriver should he go in order to reach the cabin in minimum time?

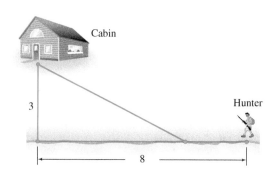

50. Travel Time   Repeat Exercise 49, but assume the cabin is 19 miles north and 8 miles west.

51. Postal Regulations   The U.S. Postal Service stipulates that any boxes sent through the mail must have a length plus girth totaling no more than 108 in. (See the figure.) Find the dimensions of the box with maximum volume that can be sent through the U.S. mail, assuming that the width and the height of the box are equal. ***Source: U.S. Postal Service.***

Length

FIRST CLASS   Girth

YOUR TURN ANSWERS

1. $x = 20$ and $y = 10/3$

2. Go 93 m along the trail and then head into the woods.

3. Box will have maximum volume when $x = 4/3$ m and the maximum volume is $1024/27$ m^3.

4. Radius is 4.3 cm and height is 8.6 cm.

6.3 Further Business Applications: Economic Lot Size; Economic Order Quantity; Elasticity of Demand

APPLY IT   How many batches of primer should a paint company produce per year to minimize its costs while meeting its customers' demand?

We will answer this question in Example 1 using the concept of economic lot size.

In this section we introduce three common business applications of calculus. The first two, *economic lot size* and *economic order quantity*, are related. A manufacturer must determine the production lot (or batch) size that will result in minimum production and storage costs, while a purchaser must decide what quantity of an item to order in an effort to minimize reordering and storage costs. The third application, *elasticity of demand*, deals with the sensitivity of demand for a product to changes in the price of the product.

Economic Lot Size
Suppose that a company manufactures a constant number of units of a product per year and that the product can be manufactured in several batches of equal size throughout the year. On the one hand, if the company were to manufacture one large batch every year, it would minimize setup costs but incur high warehouse costs. On the other hand, if it were to make many small batches, this would increase setup costs. Calculus can be used to find the number that should be manufactured in each batch in order to minimize the total cost. This number is called the **economic lot size**.

Figure 14 on the next page shows several possibilities for a product having an annual demand of 12,000 units. The top graph shows the results if all 12,000 units are made in one batch per year. In this case an average of 6000 units will be held in a warehouse. If 3000 units are made in each batch, four batches will be made at equal time intervals during the year, and the average number of units in the warehouse falls to only 1500. If 1000 units are made in each of twelve batches, an average of 500 units will be in the warehouse.

The variable in our discussion of economic lot size will be

$$q = \text{number of units in each batch.}$$

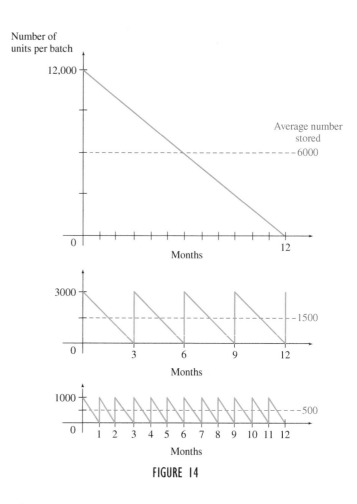

FIGURE 14

In addition, we have the following constants:

k = cost of storing one unit of the product for one year;

f = fixed setup cost to manufacture the product;

g = cost of manufacturing a single unit of the product;

M = total number of units produced annually.

The company has two types of costs: a cost associated with manufacturing the item and a cost associated with storing the finished product. Because q units are produced in each batch, and each batch has a fixed cost f and a variable cost g per unit, the manufacturing cost per batch is

$$f + gq.$$

The number of units produced in a year is M, so the number of batches per year must be M/q. Therefore, the total annual manufacturing cost is

$$(f + gq)\frac{M}{q} = \frac{fM}{q} + gM. \tag{1}$$

Since demand is constant, the inventory goes down linearly from q to 0, as in Figure 14, with an average inventory of $q/2$ units per year. The cost of storing one unit of the product for a year is k, so the total storage cost is

$$k\left(\frac{q}{2}\right) = \frac{kq}{2}. \tag{2}$$

The total production cost is the sum of the manufacturing and storage costs, or the sum of Equations (1) and (2). If $T(q)$ is the total cost of producing M units in batches of size q,

$$T(q) = \frac{fM}{q} + gM + \frac{kq}{2}.$$

In words, we have found that the total cost is equal to

$$\left(\text{fixed cost} + \frac{\text{cost}}{\text{unit}} \times \frac{\text{\# units}}{\text{batch}} \right) \frac{\text{\# batches}}{\text{year}} + \text{storage cost} \times \text{\# units in storage}.$$

Since the only constraint on q is that it be a positive number, the domain of T is $(0, \infty)$. To find the value of q that will minimize $T(q)$, remember that f, g, k, and M are constants and find $T'(q)$.

$$T'(q) = \frac{-fM}{q^2} + \frac{k}{2}$$

Set this derivative equal to 0.

$$\frac{-fM}{q^2} + \frac{k}{2} = 0$$

$$\frac{k}{2} = \frac{fM}{q^2} \qquad \text{Add } \tfrac{fM}{q^2} \text{ to both sides.}$$

$$q^2 \frac{k}{2} = fM \qquad \text{Multiply both sides by } q^2.$$

$$q^2 = \frac{2fM}{k} \qquad \text{Multiply both sides by } \tfrac{2}{k}.$$

$$q = \sqrt{\frac{2fM}{k}} \qquad \text{Take square root of both sides.} \qquad \textbf{(3)}$$

The critical point theorem can be used to show that $\sqrt{(2fM)/k}$ is the economic lot size that minimizes total production costs. (See Exercise 1.)

This application is referred to as the *inventory problem* and is treated in more detail in management science courses. Please note that Equation (3) was derived under very specific assumptions. If the assumptions are changed slightly, a different conclusion might be reached, and it would not necessarily be valid to use Equation (3).

In some examples Equation (3) may not give an integer value, in which case we must investigate the next integer smaller than q and the next integer larger to see which gives the minimum cost.

EXAMPLE 1   Lot Size

APPLY IT

A paint company has a steady annual demand for 24,500 cans of automobile primer. The comptroller for the company says that it costs $2 to store one can of paint for 1 year and $500 to set up the plant for the production of the primer. Find the number of cans of primer that should be produced in each batch, as well as the number of batches per year, in order to minimize total production costs.

SOLUTION   Use Equation (3), with $k = 2$, $M = 24,500$, and $f = 500$.

$$q = \sqrt{\frac{2fM}{k}} = \sqrt{\frac{2(500)(24,500)}{2}} = \sqrt{12,250,000} = 3500$$

The company should make 3500 cans of primer in each batch to minimize production costs. The number of batches per year is $M/q = 24,500/3500 = 7$.   TRY YOUR TURN 1

YOUR TURN 1 Suppose the annual demand in Example 1 is only 18,000 cans, the setup cost is $750, and the storage cost is $3 per can. Find the number of cans that should be produced in each batch and the number of batches per year to minimize production costs.

Economic Order Quantity

We can extend our previous discussion to the problem of reordering an item that is used at a constant rate throughout the year. Here, the company using a product must decide how often to order and how many units to request each time an order is placed; that is, it must identify the **economic order quantity**. In this case, the variable is

$$q = \text{number of units to order each time.}$$

We also have the following constants:

$$k = \text{cost of storing one unit for one year}$$
$$f = \text{fixed cost to place an order}$$
$$M = \text{total units needed per year}$$

The goal is to minimize the total cost of ordering over a year's time, where

$$\text{Total cost} = \text{Storage cost} + \text{Reorder cost.}$$

Again assume an average inventory of $q/2$, so the yearly storage cost is $kq/2$. The number of orders placed annually is M/q. The reorder cost is the product of this quantity and the cost per order, f. Thus, the reorder cost is fM/q, and the total cost is

$$T(q) = \frac{fM}{q} + \frac{kq}{2}.$$

This is almost the same formula we derived for the inventory problem, which also had a constant term gM. Since a constant does not affect the derivative, Equation (3) is also valid for the economic order quantity problem. As before, the number of orders placed annually is M/q. This illustrates how two different applications might have the same mathematical structure, so a solution to one applies to both.

EXAMPLE 2   Order Quantity

A large pharmacy has an annual need for 480 units of a certain antibiotic. It costs \$3 to store one unit for one year. The fixed cost of placing an order (clerical time, mailing, and so on) amounts to \$31. Find the number of units to order each time, and how many times a year the antibiotic should be ordered.

SOLUTION   Here $k = 3$, $M = 480$, and $f = 31$. We have

$$q = \sqrt{\frac{2fM}{k}} = \sqrt{\frac{2(31)(480)}{3}} = \sqrt{9920} \approx 99.6.$$

$T(99) = 298.803$ and $T(100) = 298.800$, so ordering 100 units of the drug each time minimizes the annual cost. The drug should be ordered $M/q = 480/100 = 4.8$ times a year, or about once every $2\frac{1}{2}$ months.   TRY YOUR TURN 2

YOUR TURN 2 Suppose the annual need in Example 2 is 320 units, the fixed cost amounts to \$30, and the storage cost is \$2 per unit. Find the number of units to order each time and how many times a year to order to minimize cost.

Elasticity of Demand

Anyone who sells a product or service is concerned with how a change in price affects demand. The sensitivity of demand to changes in price varies with different items. Luxury items tend to be more sensitive to price than essentials. For items such as milk, heating fuel, and light bulbs, relatively small percentage changes in price will not change the demand for the item much, so long as the price is not far from its normal range. For cars, home loans, jewelry, and concert tickets, however, small percentage changes in price can have a significant effect on demand.

One way to measure the sensitivity of demand to changes in price is by the relative change—the ratio of percent change in demand to percent change in price. If q represents the quantity demanded and p the price, this ratio can be written as

$$\frac{\Delta q/q}{\Delta p/p},$$

FOR REVIEW

Recall from Chapter 1 that the Greek letter Δ, pronounced *delta*, is used in mathematics to mean "change."

where Δq represents the change in q and Δp represents the change in p. This ratio is always negative, because q and p are positive, while Δq and Δp have opposite signs. (An *increase* in price causes a *decrease* in demand.) If the absolute value of this ratio is large, it suggests that a relatively small increase in price causes a relatively large drop (decrease) in demand.

This ratio can be rewritten as

$$\frac{\Delta q/q}{\Delta p/p} = \frac{\Delta q}{q} \cdot \frac{p}{\Delta p} = \frac{p}{q} \cdot \frac{\Delta q}{\Delta p}.$$

Suppose $q = f(p)$. (Note that this is the inverse of the way our demand functions have been expressed so far; previously we had $p = D(q)$.) Then $\Delta q = f(p + \Delta p) - f(p)$, and

$$\frac{\Delta q}{\Delta p} = \frac{f(p + \Delta p) - f(p)}{\Delta p}.$$

As $\Delta p \to 0$, this quotient becomes

$$\lim_{\Delta p \to 0} \frac{\Delta q}{\Delta p} = \lim_{\Delta p \to 0} \frac{f(p + \Delta p) - f(p)}{\Delta p} = \frac{dq}{dp},$$

and

$$\lim_{\Delta p \to 0} \frac{p}{q} \cdot \frac{\Delta q}{\Delta p} = \frac{p}{q} \cdot \frac{dq}{dp}.$$

The negative of this last quantity is called the *elasticity of demand* (E) and measures the instantaneous responsiveness of demand to price.*

Elasticity of Demand

Let $q = f(p)$, where q is demand at a price p. The **elasticity of demand** is

$$E = -\frac{p}{q} \cdot \frac{dq}{dp}.$$

Demand is inelastic if $E < 1$.

Demand is elastic if $E > 1$.

Demand has unit elasticity if $E = 1$.

For example, E has been estimated at 0.6 for physician services and at 2.3 for restaurant meals. The demand for medical care is much less responsive to price changes than is the demand for nonessential commodities, such as restaurant meals. ***Source: Economics: Private and Public Choice***.

If $E < 1$, the relative change in demand is less than the relative change in price, and the demand is called *inelastic*. If $E > 1$, the relative change in demand is greater than the relative change in price, and the demand is called *elastic*. In other words, inelastic means that a small change in price has little effect on demand, while elastic means that a small change in price has more effect on demand. When $E = 1$, the percentage changes in price and demand are relatively equal and the demand is said to have **unit elasticity**.

EXAMPLE 3   Elasticity

Terrence Wales described the demand for distilled spirits as

$$q = f(p) = -0.00375p + 7.87,$$

where p represents the retail price of a case of liquor in dollars per case. Here q represents the average number of cases purchased per year by a consumer. Calculate and interpret the elasticity of demand when $p = \$118.30$ per case. ***Source: The American Economic Review***.

---

*Economists often define elasticity as the negative of our definition, but give numerical values according to our definition.

SOLUTION From $q = -0.00375p + 7.87$, we determine $dq/dp = -0.00375$. Now we find E.

$$E = -\frac{p}{q} \cdot \frac{dq}{dp}$$

$$= -\frac{p}{-0.00375p + 7.87}(-0.00375)$$

$$= \frac{0.00375p}{-0.00375p + 7.87}$$

Let $p = 118.30$ to get

$$E = \frac{0.00375(118.30)}{-0.00375(118.30) + 7.87} \approx 0.0597.$$

Since $0.0597 < 1$, the demand is inelastic, and a percentage change in price will result in a smaller percentage change in demand. Thus an increase in price will increase revenue. For example, a 10% increase in price will cause an approximate decrease in demand of $(0.0597)(0.10) = 0.00597$ or about 0.6%. **TRY YOUR TURN 3**

YOUR TURN 3 Suppose the demand equation for a given commodity is $q = 24{,}000 - 3p^2$. Calculate and interpret E when $p = \$50$.

EXAMPLE 4 Elasticity

The demand for beer was modeled by Hogarty and Elzinga with the function given by $q = f(p) = 1/p$. The price was expressed in dollars per can of beer, and the quantity sold in cans per day per adult. Calculate and interpret the elasticity of demand. *Source: The Review of Economics and Statistics*.

SOLUTION Since $q = 1/p$,

$$\frac{dq}{dp} = \frac{-1}{p^2}, \quad \text{and}$$

$$E = -\frac{p}{q} \cdot \frac{dq}{dp} = -\frac{p}{1/p} \cdot \frac{-1}{p^2} = 1.$$

Here, the elasticity is 1, unit elasticity, at every (positive) price. As we will see shortly, this means that revenues remain constant when the price changes. **TRY YOUR TURN 4**

YOUR TURN 4 Suppose the demand equation for a given product is $q = 200e^{-0.4p}$. Calculate and interpret E when $p = \$100$.

Elasticity can be related to the total revenue, R, by considering the derivative of R. Since revenue is given by price times sales (demand),

$$R = pq.$$

Differentiate with respect to p using the product rule.

$$\frac{dR}{dp} = p \cdot \frac{dq}{dp} + q \cdot 1$$

$$= \frac{q}{q} \cdot p \cdot \frac{dq}{dp} + q \qquad \text{Multiply by } \tfrac{q}{q} \text{ (or 1)}.$$

$$= q\left(\frac{p}{q} \cdot \frac{dq}{dp}\right) + q$$

$$= q(-E) + q \qquad -E = \tfrac{p}{q} \cdot \tfrac{dq}{dp}.$$

$$= q(-E + 1) \qquad \text{Factor.}$$

$$= q(1 - E)$$

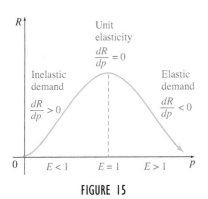

FIGURE 15

Total revenue R is increasing, optimized, or decreasing depending on whether $dR/dp > 0$, $dR/dp = 0$, or $dR/dp < 0$. These three situations correspond to $E < 1$, $E = 1$, or $E > 1$. See Figure 15.

In summary, total revenue is related to elasticity as follows.

Revenue and Elasticity

1. If the demand is inelastic, total revenue increases as price increases.
2. If the demand is elastic, total revenue decreases as price increases.
3. Total revenue is maximized at the price where demand has unit elasticity.

EXAMPLE 5 Elasticity

Assume that the demand for a product is $q = 216 - 2p^2$, where p is the price in dollars.

(a) Find the price intervals where demand is elastic and where demand is inelastic.

SOLUTION Since $q = 216 - 2p^2$, $dq/dp = -4p$, and

$$E = -\frac{p}{q} \cdot \frac{dq}{dp}$$

$$= -\frac{p}{216 - 2p^2}(-4p)$$

$$= \frac{4p^2}{216 - 2p^2}.$$

To decide where $E < 1$ or $E > 1$, solve the corresponding *equation*.

$$E = 1$$

$$\frac{4p^2}{216 - 2p^2} = 1 \qquad E = \frac{4p^2}{216 - 2p^2}$$

$$4p^2 = 216 - 2p^2 \qquad \text{Multiply both sides by } 216 - 2p^2.$$

$$6p^2 = 216 \qquad \text{Add } 2p^2 \text{ to both sides.}$$

$$p^2 = 36 \qquad \text{Divide both sides by 6.}$$

$$p = 6 \qquad \text{Take square root of both sides.}$$

Substitute a test number on either side of 6 in the expression for E to see which values make $E < 1$ and which make $E > 1$.

$$\text{Let } p = 1\text{: } E = \frac{4(1)^2}{216 - 2(1)^2} = \frac{4}{214} < 1.$$

$$\text{Let } p = 10\text{: } E = \frac{4(10)^2}{216 - 2(10)^2} = \frac{400}{216 - 200} > 1.$$

YOUR TURN 5 Suppose the demand function for a given product is $q = 3600 - 3p^2$. Find the price intervals where demand is elastic and where demand is inelastic. What price results in maximum revenue? What is the maximum revenue?

Demand is inelastic when $E < 1$. This occurs when $p < 6$. Demand is elastic when $E > 1$; that is, when $p > 6$.

(b) What price results in the maximum revenue? What is the maximum revenue?

SOLUTION Total revenue is maximized at the price where demand has unit elasticity. As we saw in part (a), this occurs when the price is set at $6 per item. The demand at this price is $q = 216 - 2(6)^2 = 144$. The maximum revenue is then $pq = 6 \cdot 144 = \$864$.

TRY YOUR TURN 5

6.3 WARM-UP EXERCISES

Find the derivative of each of the following functions. *(Sec. 4.1)*

W1. $f(x) = \dfrac{k}{x}$

W2. $f(x) = \dfrac{a}{x^2}$

6.3 EXERCISES

1. In the discussion of economic lot size, use the critical point theorem to show that $\sqrt{(2fM)/k}$ is the economic lot size that minimizes total production costs.

2. Why do you think that the cost g does not appear in the equation for q [Equation (3)]?

3. Choose the correct answer. *Source: American Institute of Certified Public Accountants.*

   The economic order quantity formula assumes that

   (a) Purchase costs per unit differ due to quantity discounts.

   (b) Costs of placing an order vary with quantity ordered.

   (c) Periodic demand for the goods is known.

   (d) Erratic usage rates are cushioned by safety stocks.

4. Describe elasticity of demand in your own words.

5. A Giffen good is a product for which the demand function is increasing. Economists debate whether such goods actually exist. What is true about the elasticity of a Giffen good? *Source: investopedia.com.*

6. What must be true about the demand function if $E = 0$?

7. Suppose that a demand function is linear—that is, $q = m - np$ for $0 \leq p \leq m/n$, where m and n are positive constants. Show that $E = 1$ at the midpoint of the demand curve on the interval $0 \leq p \leq m/n$; that is, at $p = m/(2n)$.

8. Suppose the demand function is of the form $q = Cp^{-k}$, where C and k are positive constants.

   (a) Find the elasticity E.

   (b) If $0 < k < 1$, what does your answer from part (a) say about how prices should be set to maximize the revenue?

   (c) If $k > 1$, what does your answer from part (a) say about how prices should be set to maximize the revenue?

   (d) If $k = 1$, what does your answer from part (a) tell you about setting prices to maximize revenue?

   (e) Based on your answers above, is a demand function of the form $q = Cp^{-k}$ realistic? Explain your answer.

APPLICATIONS

Business and Economics

9. **Lot Size** Suppose 100,000 lamps are to be manufactured annually. It costs $1 to store a lamp for 1 year, and it costs $500 to set up the factory to produce a batch of lamps. Find the number of lamps to produce in each batch.

10. **Lot Size** A manufacturer has a steady annual demand for 13,950 cases of sugar. It costs $9 to store 1 case for 1 year, $31 in setup cost to produce each batch, and $16 to produce each case. Find the number of cases per batch that should be produced.

11. **Lot Size** Find the number of batches of lamps that should be manufactured annually in Exercise 9.

12. **Lot Size** Find the number of batches of sugar that should be manufactured annually in Exercise 10.

13. **Order Quantity** A bookstore has an annual demand for 100,000 copies of a best-selling book. It costs $0.50 to store 1 copy for 1 year, and it costs $60 to place an order. Find the optimum number of copies per order.

14. **Order Quantity** A restaurant has an annual demand for 900 bottles of a California wine. It costs $1 to store 1 bottle for 1 year, and it costs $5 to place a reorder. Find the optimum number of bottles per order.

15. **Lot Size** Suppose that in the inventory problem, the storage cost depends on the maximum inventory size, rather than the average. This would be more realistic if, for example, the company had to build a warehouse large enough to hold the maximum inventory, and the cost of storage was the same no matter how full or empty the warehouse was. Show that in this case the number of units that should be ordered or manufactured to minimize the total cost is

$$q = \sqrt{\frac{fM}{k}}.$$

16. **Lot Size** A book publisher wants to know how many times a year a print run should be scheduled. Suppose it costs $1000 to set up the printing process, and the subsequent cost per book is so low it can be ignored. Suppose further that the annual warehouse cost is $6 times the maximum number of books stored. Assuming 5000 copies of the book are needed per year, how many books should be printed in each print run? (See Exercise 15.)

17. **Lot Size** Suppose that in the inventory problem, the storage cost is a combination of the cost described in the text and the cost described in Exercise 15. In other words, suppose there is an annual cost, k_1, for storing a single unit, plus an annual cost per unit, k_2, that must be paid for each unit up to the maximum number of units stored. Show that the number of units that should be ordered or manufactured to minimize the total cost in this case is

$$q = \sqrt{\frac{2fM}{k_1 + 2k_2}}.$$

18. **Lot Size** Every year, Erin D'Aquanni sells 30,000 cases of her Famous Spaghetti Sauce. It costs her $1 per year in electricity to store a case, plus she must pay annual warehouse fees of $2 per case for the maximum number of cases she will store. If it costs her $750 to set up a production run, plus $8 per case to manufacture a single case, how many production runs should she have each year to minimize her total costs? (See Exercise 17.)

Elasticity For each of the following demand functions, find (a) E, and (b) values of q (if any) at which total revenue is maximized.

19. $q = 50 - \dfrac{p}{4}$

20. $q = 25{,}000 - 50p$

21. $q = 37{,}500 - 5p^2$

22. $q = 48{,}000 - 10p^2$

23. $p = 400e^{-0.2q}$

24. $q = 10 - \ln p$

Elasticity   Find the elasticity of demand (E) for the given demand function at the indicated values of p. Is the demand elastic, inelastic, or neither at the indicated values? Interpret your results.

25. $q = 400 - 0.2p^2$

  (a) $p = \$20$ (b) $p = \$40$

26. $q = 300 - 2p$

  (a) $p = \$100$ (b) $p = \$50$

27. Elasticity of Crude Oil   The demand for crude oil in the United States was recently approximated by

$$q = f(p) = 2,431,129p^{-0.06},$$

where p represents the price of crude oil in dollars per barrel and q represents the per capita consumption of crude oil. Calculate and interpret the elasticity of demand when the price is $100 per barrel. *Source: OPEC Review.*

28. Elasticity of Rice   The demand for rice in Japan for a particular year was estimated by the general function

$$q = f(p) = Ap^{-0.13},$$

where p represents the price of a unit of rice, A represents a constant that can be calculated uniquely for a particular year, and q represents the annual per capita rice demand. Calculate and interpret the elasticity of demand. *Source: Agricultural Economics.*

29. Elasticity of Software   The Valve Corporation, a software entertainment company, recently ran a holiday sale on its popular Steam software program. Using data collected from the sale, it is possible to estimate the demand corresponding to various discounts in the price of the software. Assuming that the original price was $40, the demand for the software can be estimated by the function

$$q = 3,751,000p^{-2.826},$$

where p is the price and q is the demand. Calculate and interpret the elasticity of demand. *Source: codinghorror.com.*

30. Elasticity   The price of beef in the United States has been found to depend on the demand (measured by per capita consumption) according to the equation

$$q = \frac{342.5}{p^{0.5314}}.$$

Find the elasticity. Is the demand for beef elastic or inelastic? *Source: SAS Institute Inc.*

31. Elasticity   A study of the demand for air travel in Australia found that the demand for discount air travel from Sydney to Melbourne (in revenue passenger kilometer per capita, the product of the number of passengers traveling on a route and the distance of the route, divided by the populations of the host cities) depends on the airfare according to the equation

$$q = 55.2 - 0.022p.$$

Source: International Journal of Transport Economics.

  (a) Find the elasticity when the price is $166.10, the average discount airfare deflated by the consumer price index to 1989–1990 prices, according to the authors of the study.

  (b) Is the demand for airfare elastic or inelastic at this price?

  (c) Find the price that maximizes revenue.

32. Elasticity   The price along the West Coast of the United States for Japanese spruce logs (in dollars per cubic meter) based on the demand (in thousands of cubic meters per day) has been approximated by

$$p = 0.604q^2 - 20.16q + 263.067.$$

Source: Kamiak Econometrics.

  (a) Find the elasticity when the demand is 11 thousand cubic meters a day. (*Hint:* To find dq/dp when p is expressed in terms of q, you may use the fact that

$$\frac{dq}{dp} = \frac{1}{dp/dq}.$$

  Review the explanation of the derivative of the natural logarithm to see why this is true.)

  (b) Is the demand for spruce logs elastic or inelastic?

  (c) What happens to the elasticity as q approaches 16.6887? Discuss the limitations of this model for the price as a function of the demand.

33. Elasticity   A geometric interpretation of elasticity is as follows. Consider the tangent line to the demand curve $q = f(p)$ at the point $P_0 = (p_0, q_0)$. Let the point where the tangent line intersects the p-axis be called A, and the point where it intersects the q-axis be called B. Let P_0A and P_0B be the distances from P_0 to A and to B, respectively. Calculate the ratio P_0B/P_0A in terms of p_0, q_0, and $f'(p_0)$, and show that this ratio equals the elasticity. *Source: The AMATYC Review.*

34. Elasticity   Here's another way to identify the point of unit elasticity when the demand function is a straight line.

  (a) Suppose the demand function has the form $q = mp + b$, where $m < 0$ and $b > 0$. Find the points where the line intersects the p and q axes.

  (b) Find the midpoint of the two intercepts found in part (a).

  (c) Show that the elasticity equals 1 at the point found in part (b). In other words, the point of unit elasticity is midway between the two intercepts of a straight line demand function.

YOUR TURN ANSWERS

1. 6 batches per year with 3000 cans per batch

2. Order 98 units about every 3.675 months.

3. $E = 0.909$; the demand is inelastic.

4. $E = 40$; the demand is elastic.

5. Demand is inelastic when $p < 20$ and demand is elastic when $p > 20$. The maximum revenue is $48,000 when $p = \$20$.

6.4 Implicit Differentiation

APPLY IT How does demand for a certain commodity change with respect to price? *We will answer this question in Example 4.*

In almost all of the examples and applications so far, all functions have been defined in the form

$$y = f(x),$$

with y given **explicitly** in terms of x, or as an **explicit function** of x. For example,

$$y = 3x - 2, \qquad y = x^2 + x + 6, \qquad \text{and} \qquad y = -x^3 + 2$$

are all explicit functions of x. The equation $4xy - 3x = 6$ can be expressed as an explicit function of x by solving for y. This gives

$$4xy - 3x = 6$$
$$4xy = 3x + 6$$
$$y = \frac{3x + 6}{4x}.$$

On the other hand, some equations in x and y cannot be readily solved for y, and some equations cannot be solved for y at all. For example, while it would be possible (but tedious) to use the quadratic formula to solve for y in the equation $y^2 + 2yx + 4x^2 = 0$, it is not possible to solve for y in the equation $y^5 + 8y^3 + 6y^2x^2 + 2yx^3 + 6 = 0$. In equations such as these last two, y is said to be given **implicitly** in terms of x.

In such cases, it may still be possible to find the derivative dy/dx by a process called **implicit differentiation**. In doing so, we assume that there exists some function or functions f, which we may or may not be able to find, such that $y = f(x)$ and dy/dx exists. It is helpful to use dy/dx here rather than $f'(x)$ to make it clear which variable is independent and which is dependent.

EXAMPLE 1 Implicit Differentiation

Find dy/dx if $3xy + 4y^2 = 10$.

SOLUTION Differentiate with respect to x on both sides of the equation.

$$3xy + 4y^2 = 10$$
$$\frac{d}{dx}(3xy + 4y^2) = \frac{d}{dx}(10) \qquad (1)$$

FOR REVIEW
In Chapter 1, we pointed out that when y is given as a function of x, x is the independent variable and y is the dependent variable. We later defined the derivative dy/dx when y is a function of x. In an equation such as $3xy + 4y^2 = 10$, either variable can be considered the independent variable. If a problem asks for dy/dx, consider x the independent variable; if it asks for dx/dy, consider y the independent variable. A similar rule holds when other variables are used.

Now differentiate each term on the left side of the equation. Think of $3xy$ as the product $(3x)(y)$ and use the product rule and the chain rule. Since

$$\frac{d}{dx}(3x) = 3 \quad \text{and} \quad \frac{d}{dx}(y) = \frac{dy}{dx},$$

the derivative of $(3x)(y)$ is

$$(3x)\frac{dy}{dx} + (y)3 = 3x\frac{dy}{dx} + 3y.$$

To differentiate the second term, $4y^2$, use the chain rule, since y is assumed to be some function of x.

$$\frac{d}{dx}(4y^2) = 4(2y^1)\overset{\text{Derivative of } y^2}{\frac{dy}{dx}} = 8y\frac{dy}{dx}$$

On the right side of Equation (1), the derivative of 10 is 0. Taking the indicated derivatives in Equation (1) term by term gives

$$3x\frac{dy}{dx} + 3y + 8y\frac{dy}{dx} = 0.$$

Now solve this result for dy/dx.

$$(3x + 8y)\frac{dy}{dx} = -3y \qquad \text{Subtract } 3y \text{ from both sides and factor out } \tfrac{dy}{dx}.$$

$$\frac{dy}{dx} = \frac{-3y}{3x + 8y} \qquad \text{Divide by } 3x + 8y.$$

YOUR TURN 1 Find dy/dx if $x^2 + y^2 = xy$.

TRY YOUR TURN 1

NOTE Because we are treating y as a function of x, notice that each time an expression has y in it, we use the chain rule.

EXAMPLE 2 Implicit Differentiation

Find dy/dx for $x + \sqrt{x}\sqrt{y} = y^2$.

SOLUTION Take the derivative on both sides with respect to x.

$$\frac{d}{dx}\left(x + \sqrt{x}\sqrt{y}\right) = \frac{d}{dx}(y^2)$$

Since $\sqrt{x}\cdot\sqrt{y} = x^{1/2}\cdot y^{1/2}$, use the product rule and the chain rule as follows.

$$\underbrace{1}_{\substack{\text{Derivative}\\\text{of } x}} + \underbrace{x^{1/2}\left(\frac{1}{2}y^{-1/2}\cdot\frac{dy}{dx}\right) + y^{1/2}\left(\frac{1}{2}x^{-1/2}\right)}_{\substack{\text{Derivative}\\\text{of } x^{1/2}y^{1/2}}} = \underbrace{2y\frac{dy}{dx}}_{\substack{\text{Derivative}\\\text{of } y^2}}$$

$$1 + \frac{x^{1/2}}{2y^{1/2}}\cdot\frac{dy}{dx} + \frac{y^{1/2}}{2x^{1/2}} = 2y\frac{dy}{dx}$$

Multiply both sides by $2x^{1/2}\cdot y^{1/2}$.

$$2x^{1/2}\cdot y^{1/2} + x\frac{dy}{dx} + y = 4x^{1/2}\cdot y^{3/2}\cdot\frac{dy}{dx}$$

Combine terms and solve for dy/dx.

$$2x^{1/2}\cdot y^{1/2} + y = \left(4x^{1/2}\cdot y^{3/2} - x\right)\frac{dy}{dx}$$

YOUR TURN 2 Find dy/dx for $xe^y + x^2 = \ln y$.

$$\frac{dy}{dx} = \frac{2x^{1/2}\cdot y^{1/2} + y}{4x^{1/2}\cdot y^{3/2} - x} = \frac{2\sqrt{x}\sqrt{y} + y}{4y\sqrt{x}\sqrt{y} - x}$$

TRY YOUR TURN 2

EXAMPLE 3 Tangent Line

The graph of $x^3 + y^3 = 9xy$, shown in Figure 16, is a *folium of Descartes*.* Find the equation of the tangent line at the point $(2, 4)$, shown in Figure 16.

SOLUTION Since this is not the graph of a function, y is not a function of x, and dy/dx is not defined. But if we restrict the curve to the vicinity of $(2, 4)$, as shown in Figure 17,

*Information on this curve and others is available on the Famous Curves section of the MacTutor History of Mathematics Archive website at www-history.mcs.st-and.ac.uk/~history. See Exercises 33–36 for more curves.

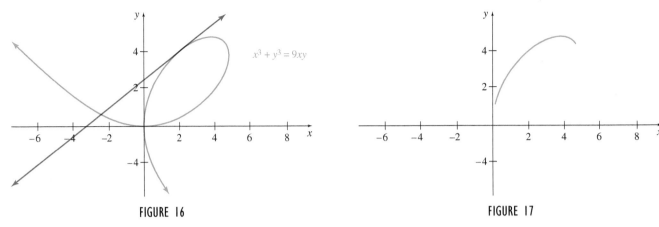

FIGURE 16   FIGURE 17

the curve does represent the graph of a function, and we can calculate dy/dx by implicit differentiation.

$$3x^2 + 3y^2 \cdot \frac{dy}{dx} = 9x\frac{dy}{dx} + 9y \qquad \text{Chain rule and product rule}$$

$$3y^2 \cdot \frac{dy}{dx} - 9x\frac{dy}{dx} = 9y - 3x^2 \qquad \text{Move all } \tfrac{dy}{dx} \text{ terms to the same side of the equation.}$$

$$\frac{dy}{dx}(3y^2 - 9x) = 9y - 3x^2 \qquad \text{Factor.}$$

$$\frac{dy}{dx} = \frac{9y - 3x^2}{3y^2 - 9x} \qquad \text{Divide by } 3y^2 - 9x.$$

$$= \frac{3(3y - x^2)}{3(y^2 - 3x)} = \frac{3y - x^2}{y^2 - 3x} \qquad \text{Factor and simplify.}$$

To find the slope of the tangent line at the point $(2, 4)$, let $x = 2$ and $y = 4$. The slope is

$$m = \frac{3y - x^2}{y^2 - 3x} = \frac{3(4) - 2^2}{4^2 - 3(2)} = \frac{8}{10} = \frac{4}{5}.$$

The equation of the tangent line is then found by using the point-slope form of the equation of a line.

$$y - y_1 = m(x - x_1)$$

$$y - 4 = \frac{4}{5}(x - 2)$$

$$y - 4 = \frac{4}{5}x - \frac{8}{5}$$

$$y = \frac{4}{5}x + \frac{12}{5}$$

The tangent line is graphed in Figure 16.   TRY YOUR TURN 3

NOTE
In Example 3, we could have substituted $x = 2$ and $y = 4$ immediately after taking the derivative implicitly. You may find that such a substitution makes solving the equation for dy/dx easier.

YOUR TURN 3 The graph of $y^4 - x^4 - y^2 + x^2 = 0$ is called the *devil's curve*. Find the equation of the tangent line at the point $(1, 1)$.

The steps used in implicit differentiation can be summarized as follows.

Implicit Differentiation
To find dy/dx for an equation containing x and y:

1. Differentiate on both sides of the equation with respect to x, keeping in mind that y is assumed to be a function of x.

2. Using algebra, place all terms with dy/dx on one side of the equals sign and all terms without dy/dx on the other side.

3. Factor out dy/dx, and then solve for dy/dx.

When an applied problem involves an equation that is not given in explicit form, implicit differentiation can be used to locate maxima and minima or to find rates of change.

EXAMPLE 4 Demand

The demand function for a certain commodity is given by

$$p = \frac{500{,}000}{2q^3 + 400q + 5000},$$

where p is the price in dollars and q is the demand in hundreds of units. Find the rate of change of demand with respect to price when $q = 100$ (that is, find dq/dp).

APPLY IT

SOLUTION Since we don't have q as a function of p, we will use implicit differentiation. Differentiate both sides with respect to p using the power rule (with a power of -1) and the chain rule.

$$1 = \frac{-500{,}000}{(2q^3 + 400q + 5000)^2}\left(6q^2\frac{dq}{dp} + 400\frac{dq}{dp}\right)$$

$$= \frac{-500{,}000}{(2q^3 + 400q + 5000)^2}(6q^2 + 400)\frac{dq}{dp}$$

Now substitute $q = 100$.

$$1 = \frac{-500{,}000}{(2 \cdot 100^3 + 400 \cdot 100 + 5000)^2}(6 \cdot 100^2 + 400)\frac{dq}{dp}$$

$$= \frac{-500{,}000 \cdot 60{,}400}{2{,}045{,}000^2} \cdot \frac{dq}{dp}$$

YOUR TURN 4 Find the rate of change of demand with respect to price when $q = 200$ if the demand function is given by $p = \dfrac{100{,}000}{q^2 + 100q}$.

Therefore,

$$\frac{dq}{dp} = -\frac{2{,}045{,}000^2}{500{,}000 \cdot 60{,}400} \approx -138.$$

This means that when demand (q) is 100 hundreds, or 10,000, demand is decreasing at the rate of 139 hundred, or 13,900, units per dollar change in price.   **TRY YOUR TURN 4**

6.4 WARM-UP EXERCISES

Find the derivative of each of the following functions.

W1. $f(x) = x \ln (x^2 + 1)$ *(Sec. 4.5)*

W2. $f(x) = \dfrac{e^{x^3}}{x^2}$ *(Sec. 4.4)*

6.4 EXERCISES

Find dy/dx by implicit differentiation for the following.

1. $6x^2 + 5y^2 = 36$

2. $7x^2 - 4y^2 = 24$

3. $8x^2 - 10xy + 3y^2 = 26$

4. $7x^2 = 5y^2 + 4xy + 1$

5. $5x^3 = 3y^2 + 4y$

6. $3x^3 - 8y^2 = 10y$

7. $3x^2 = \dfrac{2 - y}{2 + y}$

8. $2y^2 = \dfrac{5 + x}{5 - x}$

9. $2\sqrt{x} + 4\sqrt{y} = 5y$

10. $4\sqrt{x} - 8\sqrt{y} = 6y^{3/2}$

11. $x^4y^3 + 4x^{3/2} = 6y^{3/2} + 5$

12. $(xy)^{4/3} + x^{1/3} = y^6 + 1$

13. $e^{x^2y} = 5x + 4y + 2$

14. $x^2e^y + y = x^3$

15. $x + \ln y = x^2y^3$

16. $y \ln x + 2 = x^{3/2}y^{5/2}$

Find the equation of the tangent line at the given point on each curve.

17. $x^2 + y^2 = 25$; $(-3, 4)$

18. $x^2 + y^2 = 100$; $(8, -6)$

19. $x^2y^2 = 1$; $(-1, 1)$

20. $x^2y^3 = 8$; $(-1, 2)$

21. $2y^2 - \sqrt{x} = 4$; $(16, 2)$

22. $y + \dfrac{\sqrt{x}}{y} = 3$; $(4, 2)$

23. $e^{x^2+y^2} = xe^{5y} - y^2e^{5x/2}$; $(2, 1)$

24. $2xe^{xy} = e^{x^3} + ye^{x^2}$; $(1, 1)$

25. $\ln(x + y) = x^3y^2 + \ln(x^2 + 2) - 4$; $(1, 2)$

26. $\ln(x^2 + y^2) = \ln 5x + \dfrac{y}{x} - 2$; $(1, 2)$

In Exercises 27–32, find the equation of the tangent line at the given value of x on each curve.

27. $y^3 + xy - y = 8x^4$; $x = 1$

28. $y^3 + 2x^2y - 8y = x^3 + 19$; $x = 2$

29. $y^3 + xy^2 + 1 = x + 2y^2$; $x = 2$

30. $y^4(1 - x) + xy = 2$; $x = 1$

31. $2y^3(x - 3) + x\sqrt{y} = 3$; $x = 3$

32. $\dfrac{y}{18}(x^2 - 64) + x^{2/3}y^{1/3} = 12$; $x = 8$

Information on curves in Exercises 33–36, as well as many other curves, is available on the Famous Curves section of the MacTutor History of Mathematics Archive website at www-history.mcs.st-and.ac.uk/~history.

33. The graph of $x^{2/3} + y^{2/3} = 2$, shown in the figure, is an *astroid*. Find the equation of the tangent line at the point $(1, 1)$.

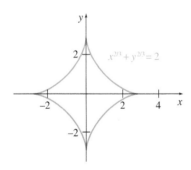

$x^{2/3} + y^{2/3} = 2$

34. The graph of $3(x^2 + y^2)^2 = 25(x^2 - y^2)$, shown in the figure, is a *lemniscate of Bernoulli*. Find the equation of the tangent line at the point $(2, 1)$.

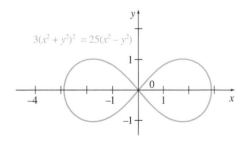

$3(x^2 + y^2)^2 = 25(x^2 - y^2)$

35. The graph of $y^2(x^2 + y^2) = 20x^2$, shown in the figure, is a *kappa curve*. Find the equation of the tangent line at the point $(1, 2)$.

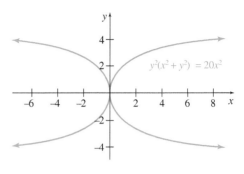

$y^2(x^2 + y^2) = 20x^2$

36. The graph of $2(x^2 + y^2)^2 = 25xy^2$, shown in the figure, is a *double folium*. Find the equation of the tangent line at the point $(2, 1)$.

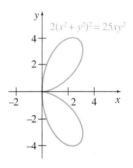

$2(x^2 + y^2)^2 = 25xy^2$

37. The graph of $x^2 + y^2 = 100$ is a circle having center at the origin and radius 10.

   (a) Write the equations of the tangent lines at the points where $x = 6$.

   (b) Graph the circle and the tangent lines.

38. Much has been written recently about elliptic curves because of their role in Andrew Wiles's 1995 proof of Fermat's Last Theorem. An elliptic curve can be written in the form

$$y^2 = x^3 + ax + b,$$

where a and b are constants, and the cubic function on the right has distinct roots. Find dy/dx for this curve.

39. Let $\sqrt{u} + \sqrt{2v + 1} = 5$. Find each derivative.

   (a) $\dfrac{du}{dv}$ **(b)** $\dfrac{dv}{du}$

   (c) Based on your answers to parts (a) and (b), what do you notice about the relationship between du/dv and dv/du?

40. Let $e^{u^2-v} - v = 1$. Find each derivative.

   (a) $\dfrac{du}{dv}$ **(b)** $\dfrac{dv}{du}$

   (c) Based on your answers to parts (a) and (b), what do you notice about the relationship between du/dv and dv/du?

41. Suppose $x^2 + y^2 + 1 = 0$. Use implicit differentiation to find dy/dx. Then explain why the result you got is meaningless. (*Hint*: Can $x^2 + y^2 + 1$ equal 0?)

42. The Lambert W function $w = f(x)$ is defined implicitly by the relation $x = we^w$ for $w \ge 1$. Among its many uses are finding the impact time for a projectile facing air resistance. Find dw/dx when $w = 1$. **Source: *The College Mathematics Journal*.**

APPLICATIONS

Business and Economics

43. Cost and Revenue For a certain product, cost C and revenue R are given as follows, where x is the number of units sold (in hundreds)

$$\text{Cost: } C^2 = x^2 + 100\sqrt{x} + 50$$

$$\text{Revenue: } 900(x - 5)^2 + 25R^2 = 22{,}500$$

(a) Find and interpret the marginal cost dC/dx at $x = 5$.

(b) Find and interpret the marginal revenue dR/dx at $x = 5$.

44. Demand The demand equation for a certain product is $2p^2 + q^2 = 1600$, where p is the price per unit in dollars and q is the number of units demanded.

(a) Find and interpret dq/dp.

(b) Find and interpret dp/dq.

45. Elasticity of Demand Researchers found the demand for cheese in Mexico for a particular year can be estimated by the implicit equation

$$\ln q = D - 0.44 \ln p,$$

where p represents the price of a unit of cheese and D represents a constant that can be calculated uniquely for a particular year. Here q represents the annual per capita cheese demand. *Source: Agricultural Economics.*

(a) Use implicit differentiation to calculate and interpret the elasticity of demand. (Recall from the previous section that elasticity of demand is $E = -(p/q) \cdot dq/dp$.)

(b) Solve the equation for q, then calculate the elasticity of demand.

46. Elasticity of Demand Researchers found the demand for milk in Mexico for a particular year can be estimated by the implicit equation

$$\ln q = C - 0.678 \ln p,$$

where p represents the price of a unit of fluid milk and C represents a constant that can be calculated uniquely for a particular year. Here q represents the annual per capita fluid milk demand. *Source: Agricultural Economics.*

(a) Use implicit differentiation to calculate and interpret the elasticity of demand. (Recall from the previous section that elasticity of demand is $E = -(p/q) \cdot dq/dp$.)

(b) Solve the equation for q, then calculate the elasticity of demand.

Life Sciences

47. Respiratory Rate Researchers have found a correlation between respiratory rate and body mass in the first three years of life. This correlation can be expressed by the function

$$\log R(w) = 1.83 - 0.43 \log(w),$$

where w is the body weight (in kilograms) and $R(w)$ is the respiratory rate (in breaths per minute). *Source: Archives of Disease in Children.*

(a) Find $R'(w)$ using implicit differentiation.

(b) Find $R'(w)$ by first solving the equation for $R(w)$.

(c) Discuss the two procedures. Is there a situation when you would want to use one method over another?

48. Species The relationship between the number of species in a genus (x) and the number of genera (y) comprising x species is given by

$$xy^a = k,$$

where a and k are constants. Find dy/dx. *Source: Elements of Mathematical Biology.*

49. Biochemical Reaction A simple biochemical reaction with three molecules has solutions that oscillate toward a steady state when positive constants a and b are below the curve $b - a = (b + a)^3$. Find the largest possible value of a for which the reaction has solutions that oscillate toward a steady state. (*Hint:* Find where $da/db = 0$. Derive values for $a + b$ and $a - b$, and then solve the equations in two unknowns.) *Source: Mathematical Biology.*

50. Phytoplankton Growth The equation

$$f(x)g(N) - m - s(x) = 0$$

describes the growth rate of phytoplankton at equilibrium, where x is the phytoplankton cell size, f is the maximum growth rate, N is the nutrient concentration, g represents the nutrient limitation experienced by phytoplankton, m is the mortality rate (a constant), and s is the loss due to sinking. Addressing the question of how phytoplankton evolution affects nutrient concentration requires finding the rate of change of N with respect to x. Using implicit differentiation, show that

$$\frac{dN}{dx} = \frac{s'(x) - f'(x)g(N)}{f(x)g'(N)}.$$

Source: The American Naturalist.

51. Foraging In studying how a predator forages a prey, an important concept is the functional response, defined as

$$f(R) = \frac{tCR}{1 + hCR},$$

where R is the prey population density, C is the attack rate constant, h is the handling time for a single prey, and t is the optimal foraging time, which is itself a function of R. A type-2 response, which is the type most commonly observed in the laboratory, is when the curve is concave down. Show that this occurs when

$$\frac{d^2t}{dR^2}R(1 + hCR)^2 + \frac{dt}{dR}2(1 + hCR) - 2thC < 0.$$

Source: Ecology.

Physical Sciences

Velocity The position of a particle at time t is given by s. Find the velocity ds/dt.

52. $2s^2 + \sqrt{st} - 4 = 3t$

53. $s^3 - 4st + 2t^3 - 5t = 0$

YOUR TURN ANSWERS

1. $dy/dx = (y - 2x)/(2y - x)$ **2.** $\dfrac{dy}{dx} = \dfrac{ye^y + 2xy}{1 - xye^y}$

3. $y = x$ **4.** $dq/dp = -72$

6.5 Related Rates

APPLY IT   When a skier's blood vessels contract because of the cold, how fast is the velocity of blood changing?
We use related rates to answer this question in Example 6 of this section.

It is common for variables to be functions of time; for example, sales of an item may depend on the season of the year, or a population of animals may be increasing at a certain rate several months after being introduced into an area. Time is often present implicitly in a mathematical model, meaning that derivatives with respect to time must be found by the method of implicit differentiation discussed in the previous section.

We start with a simple algebraic example.

EXAMPLE 1   Related Rates

Suppose that x and y are both functions of t, which can be considered to represent time, and that x and y are related by the equation

$$xy^2 + y = x^2 + 17.$$

Suppose further that when $x = 2$ and $y = 3$, then $dx/dt = 13$. Find the value of dy/dt at that moment.

SOLUTION   We start by taking the derivative of the relationship, using the product and chain rules. Keep in mind that both x and y are functions of t. The result is

$$x\left(2y\frac{dy}{dt}\right) + y^2\frac{dx}{dt} + \frac{dy}{dt} = 2x\frac{dx}{dt}.$$

Now substitute $x = 2$, $y = 3$, and $dx/dt = 13$ to get

$$2\left(6\frac{dy}{dt}\right) + 9(13) + \frac{dy}{dt} = 4(13),$$

$$12\frac{dy}{dt} + 117 + \frac{dy}{dt} = 52.$$

YOUR TURN 1 Suppose x and y are both functions of t and $x^3 + 2xy + y^2 = 1$. If $x = 1$, $y = -2$, and $dx/dt = 6$, then find dy/dt.

Solve this last equation for dy/dt to get

$$13\frac{dy}{dt} = -65,$$

$$\frac{dy}{dt} = -5.$$

TRY YOUR TURN 1

Our next example is typical of word problems involving related rates.

EXAMPLE 2   Area

A small rock is dropped into a lake. Circular ripples spread over the surface of the water, with the radius of each circle increasing at the rate of 3/2 ft per second. Find the rate of change of the area inside the circle formed by a ripple at the instant the radius is 4 ft.

SOLUTION   As shown in Figure 18, the area A and the radius r are related by

$$A = \pi r^2.$$

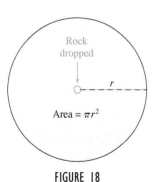

Rock dropped

r

Area $= \pi r^2$

FIGURE 18

Both A and r are functions of the time t in seconds. Take the derivative of both sides with respect to time.

$$\frac{d}{dt}(A) = \frac{d}{dt}(\pi r^2)$$

$$\frac{dA}{dt} = 2\pi r \cdot \frac{dr}{dt} \tag{1}$$

Since the radius is increasing at the rate of 3/2 ft per second,

$$\frac{dr}{dt} = \frac{3}{2}.$$

The rate of change of area at the instant $r = 4$ is given by dA/dt evaluated at $r = 4$. Substituting into Equation (1) gives

$$\frac{dA}{dt} = 2\pi \cdot 4 \cdot \frac{3}{2}$$

$$\frac{dA}{dt} = 12\pi \approx 37.7 \text{ ft}^2 \text{ per second.}$$

In Example 2, the derivatives (or rates of change) dA/dt and dr/dt are related by Equation (1); for this reason they are called **related rates**. As suggested by Example 2, four basic steps are involved in solving problems about related rates.

Solving a Related Rates Problem

1. Identify all given quantities, as well as the quantities to be found. Draw a sketch when possible.
2. Write an equation relating the variables of the problem.
3. Use implicit differentiation to find the derivative of both sides of the equation in Step 2 with respect to time.
4. Solve for the derivative giving the unknown rate of change and substitute the given values.

CAUTION
1. Differentiate *first*, and *then* substitute values for the variables. If the substitutions were performed first, differentiating would not lead to useful results.
2. Some students confuse related rates problems with applied extrema problems, perhaps because they are both word problems. There is an easy way to tell the difference. In applied extrema problems, you are always trying to maximize or minimize something, that is, make it as large or as small as possible. In related rates problems, you are trying to find how fast something is changing; time is always the independent variable.

EXAMPLE 3   **Sliding Ladder**

A 50-ft ladder is placed against a large building. The base of the ladder is resting on an oil spill, and it slips (to the right in Figure 19) at the rate of 3 ft per minute. Find the rate of change of the height of the top of the ladder above the ground at the instant when the base of the ladder is 30 ft from the base of the building.

SOLUTION   Starting with Step 1, let y be the height of the top of the ladder above the ground, and let x be the distance of the base of the ladder from the base of the building. We are trying to find dy/dt when $x = 30$. To perform Step 2, use the Pythagorean theorem to write

$$x^2 + y^2 = 50^2. \tag{2}$$

Ladder

50 ft

y

x

FIGURE 19

Both x and y are functions of time t (in minutes) after the moment that the ladder starts slipping. According to Step 3, take the derivative of both sides of Equation (2) with respect to time, getting

$$\frac{d}{dt}(x^2 + y^2) = \frac{d}{dt}(50^2)$$

$$2x\frac{dx}{dt} + 2y\frac{dy}{dt} = 0. \tag{3}$$

To complete Step 4, we need to find the values of x, y, and dx/dt. Once we find these, we can substitute them into Equation (3) to find dy/dt.

Since the base is sliding at the rate of 3 ft per minute,

$$\frac{dx}{dt} = 3.$$

Also, the base of the ladder is 30 ft from the base of the building, so $x = 30$. Use this to find y.

$$50^2 = 30^2 + y^2$$
$$2500 = 900 + y^2$$
$$1600 = y^2 \qquad \text{Subtract 900.}$$
$$y = 40 \qquad \text{Find the positive square root.}$$

In summary, $y = 40$ when $x = 30$. Also, the rate of change of x over time t is $dx/dt = 3$. Substituting these values into Equation (3) to find the rate of change of y over time gives

$$2(30)(3) + 2(40)\frac{dy}{dt} = 0$$

$$180 + 80\frac{dy}{dt} = 0$$

$$80\frac{dy}{dt} = -180$$

$$\frac{dy}{dt} = \frac{-180}{80} = \frac{-9}{4} = -2.25.$$

YOUR TURN 2 A 25-ft ladder is placed against a building. The base of the ladder is slipping away from the building at a rate of 3 ft per minute. Find the rate at which the top of the ladder is sliding down the building when the bottom of the ladder is 7 ft from the base of the building.

At the instant when the base of the ladder is 30 ft from the base of the building, the top of the ladder is sliding down the building at the rate of 2.25 ft per minute. (The minus sign shows that the ladder is sliding *down*, so the distance y is *decreasing*.)*   TRY YOUR TURN 2

EXAMPLE 4   Icicle

A cone-shaped icicle is dripping from the roof. The radius of the icicle is decreasing at a rate of 0.2 cm per hour, while the length is increasing at a rate of 0.8 cm per hour. If the icicle is currently 4 cm in radius and 20 cm long, is the volume of the icicle increasing or decreasing, and at what rate?

SOLUTION   For this problem we need the formula for the volume of a cone:

$$V = \frac{1}{3}\pi r^2 h, \tag{4}$$

where r is the radius of the cone and h is the height of the cone, which in this case is the length of the icicle, as in Figure 20.

FIGURE 20

---

*The model in Example 3 breaks down as the top of the ladder nears the ground. As y approaches 0, dy/dt becomes infinitely large. In reality, the ladder loses contact with the wall before y reaches 0.

In this problem, V, r, and h are functions of the time t in hours. Taking the derivative of both sides of Equation (4) with respect to time yields

$$\frac{dV}{dt} = \frac{1}{3}\pi\left[r^2\frac{dh}{dt} + (h)(2r)\frac{dr}{dt}\right].$$ (5)

Since the radius is decreasing at a rate of 0.2 cm per hour and the length is increasing at a rate of 0.8 cm per hour,

$$\frac{dr}{dt} = -0.2 \quad \text{and} \quad \frac{dh}{dt} = 0.8.$$

YOUR TURN 3 Suppose that in Example 4 the volume of the icicle is decreasing at a rate of 10 cm³ per hour and the radius is decreasing at a rate of 0.4 cm per hour. Find the rate of change of the length of the icicle when the radius is 4 cm and the length is 20 cm.

Substituting these, as well as $r = 4$ and $h = 20$, into Equation (5) yields

$$\frac{dV}{dt} = \frac{1}{3}\pi[4^2(0.8) + (20)(8)(-0.2)]$$

$$= \frac{1}{3}\pi(-19.2) \approx -20.$$

Because the sign of dV/dt is negative, the volume of the icicle is decreasing at a rate of 20 cm³ per hour. TRY YOUR TURN 3

EXAMPLE 5 **Revenue**

A company is increasing production of peanuts at the rate of 50 cases per day. All cases produced can be sold. The daily demand function is given by

$$p = 50 - \frac{q}{200},$$

where q is the number of units produced (and sold) and p is price in dollars. Find the rate of change of revenue with respect to time (in days) when the daily production is 200 units.

SOLUTION The revenue function,

$$R = qp = q\left(50 - \frac{q}{200}\right) = 50q - \frac{q^2}{200},$$

relates R and q. The rate of change of q over time (in days) is $dq/dt = 50$. The rate of change of revenue over time, dR/dt, is to be found when $q = 200$. Differentiate both sides of the equation

$$R = 50q - \frac{q^2}{200}$$

with respect to t.

$$\frac{dR}{dt} = 50\frac{dq}{dt} - \frac{1}{100}q\frac{dq}{dt} = \left(50 - \frac{1}{100}q\right)\frac{dq}{dt}$$

Now substitute the known values for q and dq/dt.

YOUR TURN 4 Repeat Example 5 using the daily demand function given by $p = 2000 - \dfrac{q^2}{100}$.

$$\frac{dR}{dt} = \left[50 - \frac{1}{100}(200)\right](50) = 2400$$

Thus revenue is increasing at the rate of $2400 per day. TRY YOUR TURN 4

FIGURE 21

APPLY IT

EXAMPLE 6 Blood Flow

Blood flows faster the closer it is to the center of a blood vessel. According to Poiseuille's laws, the velocity V of blood is given by

$$V = k(R^2 - r^2),$$

where R is the radius of the blood vessel, r is the distance of a layer of blood flow from the center of the vessel, and k is a constant, assumed here to equal 375. See Figure 21. Suppose that a skier's blood vessel has radius $R = 0.08$ mm and that cold weather is causing the vessel to contract at a rate of $dR/dt = -0.01$ mm per minute. How fast is the velocity of blood changing?

SOLUTION Find dV/dt. Treat r as a constant. Assume the given units are compatible.

$$V = 375(R^2 - r^2)$$

$$\frac{dV}{dt} = 375\left(2R\frac{dR}{dt} - 0\right) \qquad r \text{ is a constant.}$$

$$\frac{dV}{dt} = 750R\frac{dR}{dt}$$

Here $R = 0.08$ and $dR/dt = -0.01$, so

$$\frac{dV}{dt} = 750(0.08)(-0.01) = -0.6.$$

That is, the velocity of the blood is decreasing at a rate of -0.6 mm per minute each minute. The minus sign indicates that this is a deceleration (negative acceleration), since it represents a negative rate of change of velocity.

6.5 WARM-UP EXERCISES

Find dy/dx for each of the following. *(Sec. 6.4)*

W1. $x^3y + y^4x = 5$

W2. $x^2 + y^2 = 3xy^3$

6.5 EXERCISES

Assume x and y are functions of t. Evaluate dy/dt for each of the following.

1. $y^2 - 8x^3 = -55$; $\dfrac{dx}{dt} = -4, x = 2, y = 3$

2. $8y^3 + x^2 = 1$; $\dfrac{dx}{dt} = 2, x = 3, y = -1$

3. $2xy - 5x + 3y^3 = -51$; $\dfrac{dx}{dt} = -6, x = 3, y = -2$

4. $4x^3 - 6xy^2 + 3y^2 = 228$; $\dfrac{dx}{dt} = 3, x = -3, y = 4$

5. $\dfrac{x^2 + y}{x - y} = 9$; $\dfrac{dx}{dt} = 2, x = 4, y = 2$

6. $\dfrac{y^3 - 4x^2}{x^3 + 2y} = \dfrac{44}{31}$; $\dfrac{dx}{dt} = 5, x = -3, y = -2$

7. $xe^y = 2 - \ln 2 + \ln x$; $\dfrac{dx}{dt} = 6, x = 2, y = 0$

8. $y \ln x + xe^y = 1$; $\dfrac{dx}{dt} = 5, x = 1, y = 0$

APPLICATIONS

Business and Economics

9. Cost A manufacturer of handcrafted wine racks has determined that the cost to produce x units per month is given by $C = 0.2x^2 + 10,000$. How fast is cost per month changing when production is changing at the rate of 12 units per month and the production level is 80 units?

10. Cost/Revenue The manufacturer in Exercise 9 has found that the cost C and revenue R (in dollars) in one month are related by the equation

$$C = \frac{R^2}{450,000} + 12,000.$$

Find the rate of change of revenue with respect to time when the cost is changing by \$15 per month and the monthly revenue is \$25,000.

11. Revenue/Cost/Profit Given the revenue and cost functions $R = 50x - 0.4x^2$ and $C = 5x + 15$ (in dollars), where x is the daily production (and sales), find the following when 40 units are produced daily and the rate of change of production is 10 units per day.

(a) The rate of change of revenue with respect to time

(b) The rate of change of cost with respect to time

(c) The rate of change of profit with respect to time

12. Revenue/Cost/Profit Repeat Exercise 11, given that 80 units are produced daily and the rate of change of production is 12 units per day.

13. Demand The demand function for a certain product is determined by the fact that the product of the price and the quantity demanded equals 8000. The product currently sells for \$3.50 per unit. Suppose manufacturing costs are increasing over time at a rate of 15% and the company plans to increase the price p at this rate as well. Find the rate of change of demand over time.

14. Revenue A company is increasing production at the rate of 25 units per day. The daily demand function is determined by the fact that the price (in dollars) is a linear function of q. At a price of \$70, the demand is 0, and 100 items will be demanded at a price of \$60. Find the rate of change of revenue with respect to time (in days) when the daily production (and sales) is 20 items.

Life Sciences

15. Blood Velocity A cross-country skier has a history of heart problems. She takes nitroglycerin to dilate blood vessels, thus avoiding angina (chest pain) due to blood vessel contraction. Use Poiseuille's law with $k = 555.6$ to find the rate of change of the blood velocity when $R = 0.02$ mm and R is changing at 0.003 mm per minute. Assume r is constant. (See Example 6.)

16. Allometric Growth Suppose x and y are two quantities that vary with time according to the allometric formula $y = nx^m$. (See Exercise 86 in the section on Logarithmic Functions.) Show that the derivatives of x and y are related by the formula

$$\frac{1}{y}\frac{dy}{dt} = m\frac{1}{x}\frac{dx}{dt}.$$

(*Hint:* Take natural logarithms of both sides before taking the derivatives.)

17. Brain Mass The brain mass of a fetus can be estimated using the total mass of the fetus by the function

$$b = 0.22m^{0.87},$$

where m is the mass of the fetus (in grams) and b is the brain mass (in grams). Suppose the brain mass of a 25-g fetus is changing at a rate of 0.25 g per day. Use this to estimate the rate of change of the total mass of the fetus, dm/dt. *Source: Archives d'Anatomie, d'Histologie et d'Embryologie.*

18. Birds The energy cost of bird flight as a function of body mass is given by

$$E = 429m^{-0.35},$$

where m is the mass of the bird (in grams) and E is the energy expenditure (in calories per gram per hour). Suppose that the mass of a 10-g bird is increasing at a rate of 0.001 g per hour. Find the rate at which the energy expenditure is changing with respect to time. *Source: Wildlife Feeding and Nutrition.*

19. Metabolic Rate The average daily metabolic rate for captive animals from weasels to elk can be expressed as a function of mass by

$$r = 140.2m^{0.75},$$

where m is the mass of the animal (in kilograms) and r is the metabolic rate (in kcal per day). *Source: Wildlife Feeding and Nutrition.*

(a) Suppose that the mass of a weasel is changing with respect to time at a rate dm/dt. Find dr/dt.

(b) Determine dr/dt for a 250-kg elk that is gaining mass at a rate of 2 kg per day.

20. Lizards The energy cost of horizontal locomotion as a function of the body mass of a lizard is given by

$$E = 26.5m^{-0.34},$$

where m is the mass of the lizard (in kilograms) and E is the energy expenditure (in kcal/kg/km). Suppose that the mass of a 5-kg lizard is increasing at a rate of 0.05 kg per day. Find the rate at which the energy expenditure is changing with respect to time. *Source: Wildlife Feeding and Nutrition.*

Social Sciences

21. Crime Rate Sociologists have found that crime rates are influenced by temperature. In a midwestern town of 100,000 people, the crime rate has been approximated as

$$C = \frac{1}{10}(T - 60)^2 + 100,$$

where C is the number of crimes per month and T is the average monthly temperature in degrees Fahrenheit. The average temperature for May was 76°, and by the end of May the temperature was rising at the rate of 8° per month. How fast is the crime rate rising at the end of May?

22. Memorization Skills Under certain conditions, a person can memorize W words in t minutes, where

$$W(t) = \frac{-0.02t^2 + t}{t + 1}.$$

Find dW/dt when $t = 5$.

Physical Sciences

23. Sliding Ladder A 17-ft ladder is placed against a building. The base of the ladder is slipping away from the building at a rate of 9 ft per minute. Find the rate at which the top of the ladder is sliding down the building at the instant when the bottom of the ladder is 8 ft from the base of the building.

24. Distance

(a) One car leaves a given point and travels north at 30 mph. Another car leaves the same point at the same time and travels west at 40 mph. At what rate is the distance between the two cars changing at the instant when the cars have traveled 2 hours?

(b) Suppose that, in part (a), the second car left 1 hour later than the first car. At what rate is the distance between the two cars changing at the instant when the second car has traveled 1 hour?

25. Area   A rock is thrown into a still pond. The circular ripples move outward from the point of impact of the rock so that the radius of the circle formed by a ripple increases at the rate of 2 ft per minute. Find the rate at which the area is changing at the instant the radius is 4 ft.

26. Volume   A spherical snowball is placed in the sun. The sun melts the snowball so that its radius decreases 1/4 in. per hour. Find the rate of change of the volume with respect to time at the instant the radius is 4 in.

27. Ice Cube   An ice cube that is 3 cm on each side is melting at a rate of 2 cm^3 per min. How fast is the length of the side decreasing?

28. Volume   A sand storage tank used by the highway department for winter storms is leaking. As the sand leaks out, it forms a conical pile. The radius of the base of the pile increases at the rate of 0.75 in. per minute. The height of the pile is always twice the radius of the base. Find the rate at which the volume of the pile is increasing at the instant the radius of the base is 6 in.

29. Shadow Length   A man 6 ft tall is walking away from a lamp post at the rate of 50 ft per minute. When the man is 8 ft from the lamp post, his shadow is 10 ft long. Find the rate at which the length of the shadow is increasing when he is 25 ft from the lamp post. (See the figure.)

30. Water Level   A trough has a triangular cross section. The trough is 6 ft across the top, 6 ft deep, and 16 ft long. Water is being pumped into the trough at the rate of 4 ft^3 per minute.

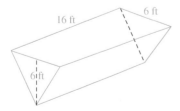

Find the rate at which the height of the water is increasing at the instant that the height is 4 ft.

31. Velocity   A pulley is on the edge of a dock, 8 ft above the water level. (See the figure below.) A rope is being used to pull in a boat. The rope is attached to the boat at water level. The rope is being pulled in at the rate of 1 ft per second. Find the rate at which the boat is approaching the dock at the instant the boat is 8 ft from the dock.

32. Kite Flying   Christine O'Brien is flying her kite in a wind that is blowing it east at a rate of 50 ft per minute. She has already let out 200 ft of string, and the kite is flying 100 ft above her hand. How fast must she let out string at this moment to keep the kite flying with the same speed and altitude?

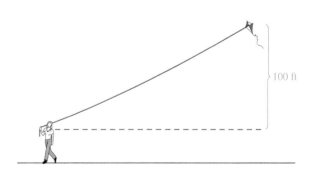

YOUR TURN ANSWERS

1. $dy/dt = -3$.

2. $-7/8$ ft/min

3. Increases 3.4 cm per hour

4. Revenue is increasing at the rate of \$40,000 per day.

6.6 Differentials: Linear Approximation

If the estimated sales of cellular telephones turn out to be inaccurate, approximately how much are profits affected?

Using differentials, we will answer this question in Example 4.

As mentioned earlier, the symbol Δx represents a change in the variable x. Similarly, Δy represents a change in y. An important problem that arises in many applications is to determine Δy given specific values of x and Δx. This quantity is often difficult to evaluate. In this section we show a method of approximating Δy that uses the derivative dy/dx. In essence, we use the tangent line at a particular value of x to approximate $f(x)$ for values close to x.

For values x_1 and x_2,

$$\Delta x = x_2 - x_1.$$

Solving for x_2 gives

$$x_2 = x_1 + \Delta x.$$

For a function $y = f(x)$, the symbol Δy represents a change in y:

$$\Delta y = f(x_2) - f(x_1).$$

Replacing x_2 with $x_1 + \Delta x$ gives

$$\Delta y = f(x_1 + \Delta x) - f(x_1).$$

If Δx is used instead of h, the derivative of a function f at x_1 could be defined as

$$\frac{dy}{dx} = \lim_{\Delta x \to 0} \frac{\Delta y}{\Delta x}.$$

If the derivative exists, then

$$\frac{dy}{dx} \approx \frac{\Delta y}{\Delta x}$$

as long as Δx is close to 0. Multiplying both sides by Δx (assume $\Delta x \neq 0$) gives

$$\Delta y \approx \frac{dy}{dx} \cdot \Delta x.$$

Until now, dy/dx has been used as a single symbol representing the derivative of y with respect to x. In this section, separate meanings for dy and dx are introduced in such a way that their quotient, when $dx \neq 0$, is the derivative of y with respect to x. These meanings of dy and dx are then used to find an approximate value of Δy.

To define dy and dx, look at Figure 22 on the next page, which shows the graph of a function $y = f(x)$. The tangent line to the graph has been drawn at the point P. Let Δx be any nonzero real number (in practical problems, Δx is a small number) and locate the point $x + \Delta x$ on the x-axis. Draw a vertical line through $x + \Delta x$. Let this vertical line cut the tangent line at M and the graph of the function at Q.

Define the new symbol dx to be the same as Δx. Define the new symbol dy to equal the length MR. The slope of PM is $f'(x)$. By the definition of slope, the slope of PM is also dy/dx, so that

$$f'(x) = \frac{dy}{dx},$$

or

$$dy = f'(x)dx.$$

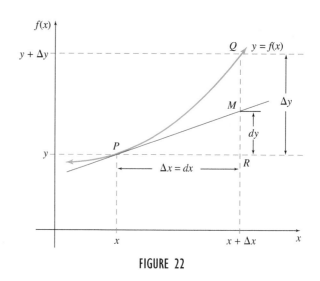

FIGURE 22

In summary, the definitions of the symbols dy and dx are as follows.

Differentials

For a function $y = f(x)$ whose derivative exists, the **differential** of x, written dx, is an arbitrary real number (usually small compared with x); the **differential** of y, written dy, is the product of $f'(x)$ and dx, or

$$dy = f'(x)dx.$$

The usefulness of the differential is suggested by Figure 22. As dx approaches 0, the value of dy gets closer and closer to that of Δy, so that for small nonzero values of dx

$$dy \approx \Delta y,$$

or

$$\Delta y \approx f'(x)dx.$$

EXAMPLE 1 Differential

Find dy for the following functions.

(a) $y = 6x^2$

SOLUTION The derivative is $dy/dx = 12x$ so

$$dy = 12x\,dx.$$

(b) $y = 800x^{-3/4}$, $x = 16$, and $dx = 0.01$

SOLUTION

$$\begin{aligned} dy &= -600x^{-7/4}dx & dy = f'(x)dx \\ &= -600(16)^{-7/4}(0.01) & x = 16 \text{ and } dx = 0.01 \\ &= -600\left(\frac{1}{2^7}\right)(0.01) = -0.046875 \end{aligned}$$

YOUR TURN 1 Find dy if $y = 300x^{-2/3}$, $x = 8$, and $dx = 0.05$.

This result tells us that when x increases from 16 to 16.01, y *decreases* by approximately 0.046875.

TRY YOUR TURN 1

Differentials can be used to approximate function values for a given x-value (in the absence of a calculator or computer). As discussed above,

$$\Delta y = f(x + \Delta x) - f(x).$$

For small nonzero values of Δx, $\Delta y \approx dy$, so that

$$dy \approx f(x + \Delta x) - f(x),$$

or

$$f(x) + dy \approx f(x + \Delta x).$$

Replacing dy with $f'(x)dx$ gives the following result.

Linear Approximation

Let f be a function whose derivative exists. For small nonzero values of Δx,

$$dy \approx \Delta y,$$

and

$$f(x + \Delta x) \approx f(x) + dy = f(x) + f'(x)dx.$$

EXAMPLE 2 Approximation

Approximate $\sqrt{50}$.

SOLUTION We will use the linear approximation

$$f(x + \Delta x) \approx f(x) + f'(x)dx$$

for a small value of Δx to form this estimate. We first choose a number x that is close to 50 for which we know its square root. Since $\sqrt{49} = 7$, we let $f(x) = \sqrt{x}, x = 49$, and $\Delta x = dx = 1$. Using this information, with the fact that when $f(x) = \sqrt{x} = x^{1/2}$,

$$f'(x) = \frac{1}{2}x^{-1/2} = \frac{1}{2x^{1/2}},$$

we have

$$f(x + \Delta x) \approx f(x) + f'(x)dx = \sqrt{x} + \frac{1}{2\sqrt{x}}dx.$$

Substituting $x = 49$ and $dx = 1$ into the preceding formula gives

$$f(50) = f(49 + 1) \approx \sqrt{49} + \frac{1}{2\sqrt{49}}(1)$$

$$= 7 + \frac{1}{14}$$

$$= 7\frac{1}{14}.$$

YOUR TURN 2

Approximate $\sqrt{99}$.

A calculator gives $7\frac{1}{14} \approx 7.07143$ and $\sqrt{50} \approx 7.07107$. Our approximation of $7\frac{1}{14}$ is close to the true answer and does not require a calculator. **TRY YOUR TURN 2**

While calculators have made differentials less important, the approximation of functions, including linear approximation, is still important in the branch of mathematics known as numerical analysis.

Marginal Analysis

Differentials are used to find an approximate value of the change in the dependent variable corresponding to a given change in the independent variable. When the concept of marginal cost (or profit or revenue) was used to approximate the change in cost for nonlinear functions, the same idea was developed. Thus the differential dy approximates Δy in much the same way as the marginal quantities approximate changes in functions.

For example, for a cost function $C(x)$,

$$dC = C'(x)dx = C'(x)\Delta x.$$

Since $\Delta C \approx dC$,

$$\Delta C \approx C'(x)\Delta x.$$

If the change in production, Δx, is equal to 1, then

$$\begin{aligned} C(x+1) - C(x) &= \Delta C \\ &\approx C'(x)\Delta x \\ &= C'(x), \end{aligned}$$

which shows that marginal cost $C'(x)$ approximates the cost of the next unit produced, as mentioned earlier.

EXAMPLE 3   Cost

Let $C(x) = 2x^3 + 300$.

(a) Use $C'(x)$ to approximate ΔC when $\Delta x = 1$ and $x = 3$. Then compare ΔC with $C'(3)$.

SOLUTION   Since $C(x) = 2x^3 + 300$, the derivative is

$$C'(x) = 6x^2$$

and the marginal cost approximation at $x = 3$ is

$$C'(3) = 6(3^2) = 54.$$

Now, the actual cost of the next unit produced is

$$\Delta C = C(4) - C(3) = 428 - 354 = 74.$$

Here, the approximation of $C'(3)$ for ΔC is poor, since $\Delta x = 1$ is large relative to $x = 3$.

(b) Use $C'(x)$ to approximate ΔC when $\Delta x = 1$ and $x = 50$. Then compare ΔC with $C'(50)$.

SOLUTION   The marginal cost approximation at $x = 50$ is

$$C'(50) = 6(50^2) = 15{,}000$$

and the actual cost of the next unit produced is

$$\Delta C = C(51) - C(50) = 265{,}602 - 250{,}300 = 15{,}302.$$

This approximation is quite good, since $\Delta x = 1$ is small compared to $x = 50$.   ▬

EXAMPLE 4   Profit

An analyst for a manufacturer of electronic devices estimates that the profit (in dollars) from the sale of x cellular telephones is given by

$$P(x) = 4000 \ln x.$$

In a report to management, the analyst projected sales for the coming year to be 10,000 phones, for a total profit of about \$36,840. He now realizes that his sales estimate may have been as much as 1000 phones too high. Approximately how far off is his profit estimate?

APPLY IT

SOLUTION   Differentials can be used to find the approximate change in P resulting from decreasing x by 1000. This change can be approximated by $dP = P'(x)dx$ where $x = 10,000$ and $dx = -1000$. Since $P'(x) = 4000/x$,

$$\Delta P \approx dP = \frac{4000}{x}dx$$

$$= \frac{4000}{10,000}(-1000)$$

$$= -400.$$

Thus the profit estimate may have been as much as \$400 too high. Computing the actual difference with a calculator gives $4000 \ln 9000 - 4000 \ln 10,000 \approx -421$, which is close to our approximation.

Error Estimation   The final example in this section shows how differentials are used to estimate errors that might enter into measurements of a physical quantity.

EXAMPLE 5   Error Estimation

In a precision manufacturing process, ball bearings must be made with a radius of 0.6 mm, with a maximum error in the radius of ± 0.015 mm. Estimate the maximum error in the volume of the ball bearing.

SOLUTION   The formula for the volume of a sphere is

$$V = \frac{4}{3}\pi r^3.$$

If an error of Δr is made in measuring the radius of the sphere, the maximum error in the volume is

$$\Delta V = \frac{4}{3}\pi(r + \Delta r)^3 - \frac{4}{3}\pi r^3.$$

Rather than calculating ΔV, approximate ΔV with dV, where

$$dV = 4\pi r^2 dr.$$

Replacing r with 0.6 and $dr = \Delta r$ with ± 0.015 gives

YOUR TURN 3   Repeat Example 5 for $r = 1.25$ mm with a maximum error in the radius of ± 0.025 mm.

$$dV = 4\pi(0.6)^2(\pm 0.015)$$

$$\approx \pm 0.0679.$$

The maximum error in the volume is about 0.07 mm^3.   TRY YOUR TURN 3

6.6 WARM-UP EXERCISES

Find the derivative of each of the following functions.

W1. $f(x) = \sqrt{x^4 + 2}$ *(Sec. 4.3)*

W2. $f(x) = \ln(e^{2x} + 1)$ *(Sec. 4.4, 4.5)*

6.6 EXERCISES

For Exercises 1–8, find dy for the given values of x and Δx.

1. $y = 2x^3 - 5x$; $\quad x = -2, \Delta x = 0.1$

2. $y = 4x^3 - 3x$; $\quad x = 3, \Delta x = 0.2$

3. $y = x^3 - 2x^2 + 3$; $\quad x = 1, \Delta x = -0.1$

4. $y = 2x^3 + x^2 - 4x$; $\quad x = 2, \Delta x = -0.2$

5. $y = \sqrt{3x + 2}$ $\quad x = 4, \Delta x = 0.15$

6. $y = \sqrt{4x - 1}$; $\quad x = 5, \Delta x = 0.08$

7. $y = \dfrac{2x - 5}{x + 1}$; $\quad x = 2, \Delta x = -0.03$

8. $y = \dfrac{6x - 3}{2x + 1}$; $\quad x = 3, \Delta x = -0.04$

Use the differential to approximate each quantity. Then use a calculator to approximate the quantity, and give the absolute value of the difference in the two results to 4 decimal places.

9. $\sqrt{145}$

10. $\sqrt{23}$

11. $\sqrt{0.99}$

12. $\sqrt{17.02}$

13. $e^{0.01}$

14. $e^{-0.002}$

15. $\ln 1.05$

16. $\ln 0.98$

APPLICATIONS

Business and Economics

17. Demand The demand for grass seed (in thousands of pounds) at a price of p dollars is

$$D(p) = -3p^3 - 2p^2 + 1500.$$

Use the differential to approximate the changes in demand for the following changes in p.

(a) \$2 to \$2.10   **(b)** \$6 to \$6.15

18. Average Cost The average cost (in dollars) to manufacture x dozen marking pencils is

$$A(x) = 0.04x^3 + 0.1x^2 + 0.5x + 6.$$

Use the differential to approximate the changes in the average cost for the following changes in x.

(a) 3 to 4   **(b)** 5 to 6

19. Revenue A company estimates that the revenue (in dollars) from the sale of x doghouses is given by

$$R(x) = 12{,}000 \ln(0.01x + 1).$$

Use the differential to approximate the change in revenue from the sale of one more doghouse when 100 doghouses are sold.

20. Profit The cost function for the company in Exercise 19 is

$$C(x) = 150 + 75x,$$

where x represents the demand for the product. Find the approximate change in profit for a 1-unit change in demand when demand is at a level of 100 doghouses. Use the differential.

21. Material Requirement A cube 4 in. on an edge is given a protective coating 0.1 in. thick. About how much coating should a production manager order for 1000 such cubes?

22. Material Requirement Beach balls 1 ft in diameter have a thickness of 0.03 in. How much material would be needed to make 5000 beach balls?

Life Sciences

23. Alcohol Concentration In Exercise 55 in the section on Polynomial and Rational Functions, we gave the function defined by

$$A(x) = 0.003631x^3 - 0.03746x^2 + 0.1012x + 0.009$$

as the approximate blood alcohol concentration in a 170-lb woman x hours after drinking 2 oz of alcohol on an empty stomach, for x in the interval $[0, 5]$. *Source: Medicolegal Aspects of Alcohol Determination in Biological Specimens*.

(a) Approximate the change in alcohol level from 1 to 1.2 hours.

(b) Approximate the change in alcohol level from 3 to 3.2 hours.

24. Drug Concentration The concentration of a certain drug in the bloodstream t hours after being administered is approximately

$$C(t) = \frac{5t}{9 + t^2}.$$

Use the differential to approximate the changes in concentration for the following changes in t.

(a) 1 to 1.5   **(b)** 2 to 2.25

25. Bacteria Population The population of bacteria (in millions) in a certain culture t hours after an experimental nutrient is introduced into the culture is

$$P(t) = \frac{25t}{8 + t^2}.$$

Use the differential to approximate the changes in population for the following changes in t.

(a) 2 to 2.5   **(b)** 3 to 3.25

26. Area of a Blood Vessel The radius of a blood vessel is 1.7 mm. A drug causes the radius to change to 1.6 mm. Find the approximate change in the area of a cross section of the vessel.

27. Volume of a Tumor A tumor is approximately spherical. If the radius of the tumor changes from 14 mm to 16 mm, find the approximate change in volume.

28. Area of an Oil Slick An oil slick is in the shape of a circle. Find the approximate increase in the area of the slick if its radius increases from 1.2 miles to 1.4 miles.

29. Area of a Bacteria Colony The shape of a colony of bacteria on a Petri dish is circular. Find the approximate increase in its area if the radius increases from 20 mm to 22 mm.

30. Gray Wolves Accurate methods of estimating the age of gray wolves are important to scientists who study wolf population dynamics. One method of estimating the age of a gray wolf is to measure the percent closure of the pulp cavity of a canine tooth and then estimate age by

$$A(p) = \frac{1.181p}{94.359 - p},$$

where p is the percent closure and $A(p)$ is the age of the wolf (in years). *Source: Journal of Wildlife Management.*

(a) What is a sensible domain for this function?

(b) Use differentials to estimate how long it will take for a gray wolf that first measures a 60% closure to obtain a 65% closure. Compare this with the actual value of about 0.55 year.

31. Pigs Researchers have observed that the mass of a female (gilt) pig can be estimated by the function

$$M(t) = -3.5 + 197.5e^{-e^{-0.01394(t - 108.4)}},$$

where t is the age of the pig (in days) and $M(t)$ is the mass of the pig (in kilograms). *Source: Animal Science.*

(a) If a particular gilt is 80 days old, use differentials to estimate how much it will gain before it is 90 days old.

(b) What is the actual gain in mass?

Physical Sciences

32. Volume A spherical balloon is being inflated. Find the approximate change in volume if the radius increases from 4 cm to 4.2 cm.

33. Volume A spherical snowball is melting. Find the approximate change in volume if the radius decreases from 3 cm to 2.8 cm.

34. Volume A cubical crystal is growing in size. Find the approximate change in the length of a side when the volume increases from 27 cubic mm to 27.1 cubic mm.

35. Volume An icicle is gradually increasing in length, while maintaining a cone shape with a length 15 times the radius.

Find the approximate amount that the volume of the icicle increases when the length increases from 13 cm to 13.2 cm.

General Interest

36. Measurement Error The edge of a square is measured as 3.45 in., with a possible error of ± 0.002 in. Estimate the maximum error in the area of the square.

37. Tolerance A worker is cutting a square from a piece of sheet metal. The specifications call for an area that is 16 cm^2 with an error of no more than 0.01 cm^2. How much error could be tolerated in the length of each side to ensure that the area is within the tolerance?

38. Measurement Error The radius of a circle is measured as 4.87 in., with a possible error of ± 0.040 in. Estimate the maximum error in the area of the circle.

39. Measurement Error A sphere has a radius of 5.81 in., with a possible error of ± 0.003 in. Estimate the maximum error in the volume of the sphere.

40. Tolerance A worker is constructing a cubical box that must contain 125 ft^3, with an error of no more than 0.3 ft^3. How much error could be tolerated in the length of each side to ensure that the volume is within the tolerance?

41. Measurement Error A cone has a known height of 7.284 in. The radius of the base is measured as 1.09 in., with a possible error of ± 0.007 in. Estimate the maximum error in the volume of the cone.

YOUR TURN ANSWERS
1. $-5/16$
2. 9.95
3. About 0.5 mm^3

CHAPTER REVIEW

SUMMARY

In this chapter, we began by discussing how to find an absolute maximum or minimum. In contrast to a relative extremum, which is the largest or smallest value of a function on some open interval about the point, an absolute extremum is the largest or smallest value of the function on the entire interval under consideration. We then studied various applications with maximizing or minimizing as the goal. Two more applications, economic lot size

and economic order quantity, were covered in a separate section, which also applied the derivative to the economic concept of elasticity of demand. Implicit differentiation is more of a technique than an application, but it underlies related rates problems, in which one or more rates are given and another is to be found. Finally, we studied the differential as a way to find linear approximations of functions.

Finding Absolute Extrema To find absolute extrema for a function f continuous on a closed interval $[a, b]$:

1. Find all critical numbers for f in (a, b).
2. Evaluate f for all critical numbers in (a, b).
3. Evaluate f for the endpoints a and b of the interval.
4. The largest value found in Step 2 or 3 is the maximum, and the smallest value is the minimum.

Solving an Applied Extrema Problem	1. Read the problem carefully. Make sure you understand what is given and what is unknown.
	2. If possible, sketch a diagram. Label the various parts.
	3. Decide which variable must be maximized or minimized. Express that variable as a function of *one* other variable.
	4. Find the domain of the function.
	5. Find the critical points for the function from Step 3.
	6. If the domain is a closed interval, evaluate the function at the endpoints and at each critical number to see which yields the absolute maximum or minimum. If the domain is an open interval, apply the critical point theorem when there is only one critical number. If there is more than one critical number, evaluate the function at the critical numbers and find the limit as the endpoints of the interval are approached to determine if an absolute maximum or minimum exists at one of the critical points.

Elasticity of Demand   Let $q = f(p)$, where q is demand at a price p.

$$E = -\frac{p}{q} \cdot \frac{dq}{dp}$$

Demand is inelastic if $E < 1$.
Demand is elastic if $E > 1$.
Demand has unit elasticity if $E = 1$.
Total revenue is maximized at the price where demand has unit elasticity.

Implicit Differentiation	To find dy/dx for an equation containing x and y:
	1. Differentiate on both sides of the equation with respect to x, keeping in mind that y is assumed to be a function of x.
	2. Place all terms with dy/dx on one side of the equals sign and all terms without dy/dx on the other side.
	3. Factor out dy/dx, and then solve for dy/dx.

Solving a Related Rates Problem	1. Identify all given quantities, as well as the quantities to be found. Draw a sketch when possible.
	2. Write an equation relating the variables of the problem.
	3. Use implicit differentiation to find the derivative of both sides of the equation in Step 2 with respect to time.
	4. Solve for the derivative, giving the unknown rate of change, and substitute the given values.

Differentials $dy = f'(x)dx$

Linear Approximation $f(x + \Delta x) \approx f(x) + dy = f(x) + f'(x)dx$

KEY TERMS

6.1
absolute maximum
absolute minimum
absolute extremum
 (or extrema)
extreme value theorem

critical point theorem
graphical optimization
6.2
spawner-recruit function
parent-progeny function
maximum sustainable
 harvest

6.3
economic lot size
economic order quantity
elasticity of demand
unit elasticity
6.4
explicit function

implicit differentiation
6.5
related rates
6.6
differential

REVIEW EXERCISES

CONCEPT CHECK

Determine whether each of the following statements is true or false, and explain why.

1. The absolute maximum of a function always occurs where the derivative has a critical number.

2. A continuous function on a closed interval has an absolute maximum and minimum.

3. A continuous function on an open interval does not have an absolute maximum or minimum.

4. Demand for a product is elastic if the elasticity is greater than 1.

5. Total revenue is maximized at the price where demand has unit elasticity.

6. Implicit differentiation can be used to find dy/dx when x is defined in terms of y.

7. In a related rates problem, all derivatives are with respect to time.

8. In a related rates problem, there can be more than two quantities that vary with time.

9. A differential is a real number.

10. When the change in x is small, the differential of y is approximately the change in y.

PRACTICE AND EXPLORATIONS

Find the absolute extrema if they exist, and all values of x where they occur on the given intervals.

11. $f(x) = -x^3 + 6x^2 + 1$; $[-1, 6]$

12. $f(x) = 4x^3 - 9x^2 - 3$; $[-1, 2]$

13. $f(x) = x^3 + 2x^2 - 15x + 3$; $[-4, 2]$

14. $f(x) = -2x^3 - 2x^2 + 2x - 1$; $[-3, 1]$

15. When solving applied extrema problems, why is it necessary to check the endpoints of the domain?

16. What is elasticity of demand (in words; no mathematical symbols allowed)? Why is the derivative used to describe elasticity?

17. Find the absolute maximum and minimum of $f(x) = \dfrac{2 \ln x}{x^2}$ on each interval.

(a) $[1, 4]$ (b) $[2, 5]$

18. Find the absolute maximum and minimum of $f(x) = \dfrac{e^{2x}}{x^2}$ on each interval.

(a) $[1/2, 2]$ (b) $[1, 3]$

19. When is it necessary to use implicit differentiation?

20. When a term involving y is differentiated in implicit differentiation, it is multiplied by dy/dx. Why? Why aren't terms involving x multiplied by dx/dx?

Find dy/dx.

21. $x^2 - 4y^2 = 3x^3y^4$

22. $x^2y^3 + 4xy = 2$

23. $2\sqrt{y-1} = 9x^{2/3} + y$

24. $9\sqrt{x} + 4y^3 = 2\sqrt{y}$

25. $\dfrac{6 + 5x}{2 - 3y} = \dfrac{1}{5x}$

26. $\dfrac{x + 2y}{x - 3y} = y^{1/2}$

27. $\ln(xy + 1) = 2xy^3 + 4$

28. $\ln(x + y) = 1 + x^2 + y^3$

29. Find the equation of the line tangent to the graph of $\sqrt{2y} - 4xy = -22$ at the point $(3, 2)$.

30. Find an equation of the line tangent to the graph of $8y^3 - 4xy^2 = 20$ at the point $(-3, 1)$.

31. What is the difference between a related rates problem and an applied extremum problem?

32. Why is implicit differentiation used in related rates problems?

Find dy/dt.

33. $y = 8x^3 - 7x^2$; $\dfrac{dx}{dt} = 4$, $x = 2$

34. $y = \dfrac{9 - 4x}{3 + 2x}$; $\dfrac{dx}{dt} = -1$, $x = -3$

35. $y = \dfrac{1 + \sqrt{x}}{1 - \sqrt{x}}$; $\dfrac{dx}{dt} = -4$, $x = 4$

36. $\dfrac{x^2 + 5y}{x - 2y} = 2$; $\dfrac{dx}{dt} = 1$, $x = 2$, $y = 0$

37. $y = xe^{3x}$; $\dfrac{dx}{dt} = -2$, $x = 1$

38. $y = \dfrac{1}{e^{x} + 1}$; $\dfrac{dx}{dt} = 3$, $x = 1$

39. What is a differential? What is it used for?

40. Describe when linear approximations are most accurate.

Evaluate dy.

41. $y = \dfrac{3x - 7}{2x + 1}$; $x = 2$, $\Delta x = 0.003$

42. $y = 8 - x^2 + x^3$; $x = -1$, $\Delta x = 0.02$

43. Suppose x and y are related by the equation

$$-12x + x^3 + y + y^2 = 4.$$

(a) Find all critical points on the curve.

(b) Determine whether the critical points found in part (a) are relative maxima or relative minima by taking values of x nearby and solving for the corresponding values of y.

(c) Is there an absolute maximum or minimum for x and y in the relationship given in part (a)? Why or why not?

44. In Exercise 43, implicit differentiation was used to find the relative extrema. The exercise was contrived to avoid various difficulties that could have arisen. Discuss some of the difficulties that might be encountered in such problems, and how these difficulties might be resolved.

APPLICATIONS

Business and Economics

45. Profit The total profit (in tens of dollars) from the sale of x hundred boxes of candy is given by

$$P(x) = -x^3 + 10x^2 - 12x.$$

(a) Find the number of boxes of candy that should be sold in order to produce maximum profit.

(b) Find the maximum profit.

46. Packaging Design The packaging department of a corporation is designing a box with a square base and no top. The volume is to be 32 m³. To reduce cost, the box is to have minimum surface area. What dimensions (height, length, and width) should the box have?

47. Packaging Design A company plans to package its product in a cylinder that is open at one end. The cylinder is to have a volume of 27π in³. What radius should the circular bottom of the cylinder have to minimize the cost of the material?

48. Packaging Design Fruit juice will be packaged in cylindrical cans with a volume of 40 in³ each. The top and bottom of the can cost 4¢ per in², while the sides cost 3¢ per in². Find the radius and height of the can of minimum cost.

49. Order Quantity A large camera store sells 20,000 batteries annually. It costs 15¢ to store 1 battery for 1 year and $12 to place a reorder. Find the number of batteries that should be ordered each time.

50. Order Quantity A store sells 180,000 cases of a product annually. It costs $12 to store 1 case for 1 year and $20 to place a reorder. Find the number of cases that should be ordered each time.

51. Lot Size A company produces 128,000 cases of soft drink annually. It costs $1 to store 1 case for 1 year and $10 to produce 1 lot. Find the number of lots that should be produced annually.

52. Lot Size In 1 year, a health food manufacturer produces and sells 240,000 cases of vitamins. It costs $2 to store 1 case for 1 year and $15 to produce each batch. Find the number of batches that should be produced annually.

53. Elasticity of Demand The demand for butter in Mexico for a particular year can be estimated by the general function

$$\ln q = D - 0.47 \ln p,$$

where p represents the price of a unit of butter and D represents a constant that can be calculated uniquely for a particular year. Here q represents the annual per capita butter demand. Calculate and interpret the elasticity of demand. *Source: Agricultural Economics.*

54. Elasticity Suppose the demand function for a product is given by $q = A/p^k$, where A and k are positive constants. For what values of k is the demand elastic? Inelastic?

Life Sciences

55. Pollution A circle of pollution is spreading from a broken underwater waste disposal pipe, with the radius increasing at the rate of 4 ft per minute. Find the rate of change of the area of the circle when the radius is 7 ft.

56. Logistic Growth Many populations grow according to the logistic equation

$$\frac{dx}{dt} = rx(N - x),$$

where r is a constant involving the rate of growth and N is the carrying capacity of the environment, beyond which the population decreases. Show that the graph of x has an inflection point where $x = N/2$. (*Hint:* Use implicit differentiation. Then set $d^2x/dt^2 = 0$, and factor.)

57. Dentin Growth The dentinal formation of molars in mice has been studied by researchers in Copenhagen. They determined that the growth curve that best fits dentinal formation for the first molar is

$$M(t) = 1.3386309 - 0.4321173t + 0.0564512t^2$$
$$-0.0020506t^3 + 0.0000315t^4 - 0.0000001785t^5,$$
$$5 \le t \le 51,$$

where t is the age of the mouse (in days), and $M(t)$ is the cumulative dentin volume (in 10^{-1} mm³). *Source: Journal of Craniofacial Genetics and Developmental Biology.*

(a) Use a graphing calculator to sketch the graph of this function on $[5, 51]$ by $[0, 7.5]$.

(b) Find the time in which the dentin formation is growing most rapidly. (*Hint:* Find the maximum value of the derivative of this function.)

58. Human Skin Surface The surface of the skin is made up of a network of intersecting lines that form polygons. Researchers have discovered a functional relationship between the age of a female and the number of polygons per area of skin according to

$$P(t) = 237.09 - 8.0398t + 0.20813t^2 - 0.0027563t^3$$
$$+ 0.000013016t^4, \quad 0 \le t \le 95,$$

where t is the age of the person (in years), and $P(t)$ is the number of polygons for a particular surface area of skin. *Source: Gerontology.*

(a) Use a graphing calculator to sketch a graph of $P(t)$ on $[0, 95]$ by $[0, 300]$.

(b) Find the maximum and minimum number of polygons per area predicted by the model.

(c) Discuss the accuracy of this model for older people.

Physical Sciences

59. Sliding Ladder A 50-ft ladder is placed against a building. The top of the ladder is sliding down the building at the rate of 2 ft per minute. Find the rate at which the base of the ladder is slipping away from the building at the instant that the base is 30 ft from the building.

60. Spherical Radius A large weather balloon is being inflated with air at the rate of 0.9 ft³ per minute. Find the rate of change of the radius when the radius is 1.7 ft.

61. Water Level   A water trough 2 ft across, 4 ft long, and 1 ft deep has ends in the shape of isosceles triangles. (See the figure.) It is being filled with 3.5 ft³ of water per minute. Find the rate at which the depth of water in the tank is changing when the water is 1/3 ft deep.

General Interest

62. Volume   Approximate the volume of coating on a sphere of radius 4 in. if the coating is 0.02 in. thick.

63. Area   A square has an edge of 9.2 in., with a possible error in the measurement of ± 0.04 in. Estimate the possible error in the area of the square.

64. Package Dimensions   UPS has the following rule regarding package dimensions. The length can be no more than 108 in., and the length plus the girth (twice the sum of the width and the height) can be no more than 130 in. If the width of a package is 4 in. more than its height and it has the maximum length plus girth allowed, find the length that produces maximum volume.

65. Pursuit   A boat moves north at a constant speed. A second boat, moving at the same speed, pursues the first boat in such a way that it always points directly at the first boat. When the first boat is at the point $(0, 1)$, the second boat is at the point $(6, 2.5)$, with the positive y-axis pointing north. It can then be shown that the curve traced by the second boat, known as a pursuit curve, is given by

$$y = \frac{x^2}{16} - 2 \ln x + \frac{1}{4} + 2 \ln 6.$$

Find the y-coordinate of the southernmost point of the second boat's path. *Source: Differential Equations: Theory and Applications.*

66. Playground Area   The city park department is planning an enclosed play area in a new park. One side of the area will be against an existing building, with no fence needed there. Find the dimensions of the maximum rectangular area that can be made with 900 m of fence.

67. Surfing   A mathematician is surfing in Long Beach, New York. He is standing on the shore and wants to paddle out to a spot 40 ft from shore; the closest point on the shore to that spot is 40 ft from where he is now standing. (See the figure.) If he can walk 5 ft per second along the shore and paddle 3 ft per second once he's in the water, how far along the shore should he walk before paddling toward the desired destination if he wants to complete the trip in the shortest possible time? What is the shortest possible time?

68. Repeat Exercise 67, but the closest point on the shore to the desired destination is now 25 ft from where he is standing.

EXTENDED APPLICATION

A TOTAL MODEL FOR A TRAINING PROGRAM

In this application, we set up a mathematical model for determining the total costs in setting up a training program. Then we use calculus to find the time interval between training programs that produces the minimum total cost. The model assumes that the demand for trainees is constant and that the fixed cost of training a batch of trainees is known. Also, it is assumed that people who are trained, but for whom no job is readily available, will be paid a fixed amount per month while waiting for a job to open up.

The model uses the following variables.

D = demand for trainees per month

N = number of trainees per batch

C_1 = fixed cost of training a batch of trainees

C_2 = marginal cost of training per trainee per month

C_3 = salary paid monthly to a trainee who has not yet been given a job after training

m = time interval in months between successive batches of trainees

t = length of training program in months

$Z(m)$ = total monthly cost of program

The total cost of training a batch of trainees is given by $C_1 + NtC_2$. However, $N = mD$, so that the total cost per batch is $C_1 + mDtC_2$.

After training, personnel are given jobs at the rate of D per month. Thus, $N - D$ of the trainees will not get a job the first month, $N - 2D$ will not get a job the second month, and so on. The $N - D$ trainees who do not get a job the first month produce total costs of $(N - D)C_3$, those not getting jobs during the second month produce costs of $(N - 2D)C_3$, and so on. Since $N = mD$, the costs during the first month can be written as

$$(N - D)C_3 = (mD - D)C_3 = (m - 1)DC_3,$$

while the costs during the second month are $(m - 2)DC_3$, and so on. The total cost for keeping the trainees without a job is thus

$$(m - 1)DC_3 + (m - 2)DC_3$$
$$+ (m - 3)DC_3 + \cdots + 2DC_3 + DC_3,$$

which can be factored to give

$$DC_3[(m - 1) + (m - 2) + (m - 3) + \cdots + 2 + 1].$$

The expression in brackets is the sum of the terms of an arithmetic sequence, discussed in most algebra texts. Using formulas for arithmetic sequences, the expression in brackets can be shown to equal $m(m - 1)/2$, so that we have

$$DC_3\left[\frac{m(m - 1)}{2}\right] \tag{1}$$

as the total cost for keeping jobless trainees.

The total cost per batch is the sum of the training cost per batch, $C_1 + mDtC_2$, and the cost of keeping trainees without a proper job, given by Equation (1). Since we assume that a batch of trainees is trained every m months, the total cost per month, $Z(m)$, is given by

$$Z(m) = \frac{C_1 + mDtC_2}{m} + \frac{DC_3\left[\dfrac{m(m - 1)}{2}\right]}{m}$$
$$= \frac{C_1}{m} + DtC_2 + DC_3\left(\frac{m - 1}{2}\right).$$

Source: P. L. Goyal and S. K. Goyal.

EXERCISES

1. Find $Z'(m)$.

2. Solve the equation $Z'(m) = 0$.

   As a practical matter, it is usually required that m be a whole number. If m does not come out to be a whole number, then m^+ and m^-, the two whole numbers closest to m, must be chosen. Calculate both $Z(m^+)$ and $Z(m^-)$; the smaller of the two provides the optimum value of Z.

3. Suppose a company finds that its demand for trainees is 3 per month, that a training program requires 12 months, that the fixed cost of training a batch of trainees is $15,000, that the marginal cost per trainee per month is $100, and that trainees are paid $900 per month after training but before going to work. Use your result from Exercise 2 and find m.

4. Since m is not a whole number, find m^+ and m^-.

5. Calculate $Z(m^+)$ and $Z(m^-)$.

6. What is the optimum time interval between successive batches of trainees? How many trainees should be in a batch?

7. The parameters of this model are likely to change over time; it is essential that such changes be incorporated into the model as they change. One way to anticipate this is to create a spreadsheet that gives the manager the total cost of training a batch of trainees for various scenarios. Using the data from Exercise 3 as a starting point, create a spreadsheet that varies these numbers and calculates the total cost of training a group of employees for each scenario. Graph the total cost of training with respect to changes in the various costs associated with training.

DIRECTIONS FOR GROUP PROJECT

Suppose you have read an article in the paper announcing that a new high-tech company is locating in your town. Given that the company is manufacturing very specialized equipment, you realize that it must develop a program to train all new employees. Because you would like to get an internship at this new company, use the information above to develop a hypothetical training program that optimizes the time interval between successive batches of trainees and the number of trainees that should be in each session. Assume that you know the new CEO because you and three of your friends have served her pizza at various times at the local pizza shop (your current jobs) and that she is willing to listen to a proposal that describes your training program. Prepare a presentation for your interview that will describe your training program. Use presentation software such as Microsoft PowerPoint.

7 Integration

Consumption of energy generated by the wind has increased from 70 trillion BTUs in 2001 to 1595 trillion BTUs in 2013. In this chapter, we will estimate wind energy consumption over this period using the definite integral, a mathematical concept that measures area under a curve. This same concept allows us to determine the distance a car has traveled, given its speed as a function of time; how much a culture of bacteria will grow; and how much consumers benefit by buying a product at the price determined by supply and demand.

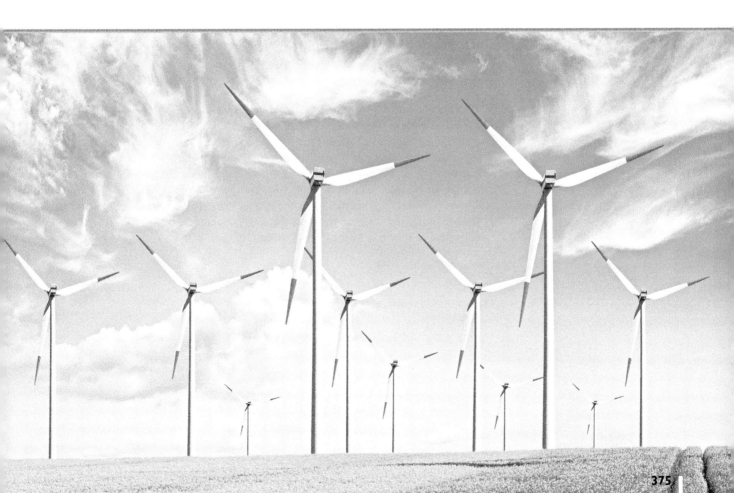

U p to this point in calculus you have solved problems such as

$$f(x) = x^5; \text{find } f'(x).$$

In this chapter you will be asked to solve problems that are the reverse of these, that is, problems of the form

$$f'(x) = 5x^4; \text{find } f(x).$$

The derivative and its applications, which you studied in previous chapters, are part of what is called *differential calculus*. The next two chapters are devoted to the other main branch of calculus, *integral calculus*. Integrals have many applications: finding areas; determining the lengths of curved paths; solving complicated probability problems; and calculating the location of an object (such as the distance of a space shuttle from Earth) when its velocity and initial position are known. The Fundamental Theorem of Calculus, presented later in this chapter, will reveal a surprisingly close connection between differential and integral calculus.

7.1 Antiderivatives

APPLY IT If an object is thrown from the top of the Willis Tower in Chicago, how fast is it going when it hits the ground?
Using antiderivatives, we will answer this question in Example 11.

Functions used in applications in previous chapters have provided information about a *total amount* of a quantity, such as cost, revenue, profit, temperature, gallons of oil, or distance. Derivatives of these functions provided information about the rate of change of these quantities and allowed us to answer important questions about the extrema of the functions. It is not always possible to find ready-made functions that provide information about the total amount of a quantity, but it is often possible to collect enough data to come up with a function that gives the *rate of change* of a quantity. We know that derivatives give the rate of change when the total amount is known. The reverse of finding a derivative is known as **antidifferentiation**. The goal is to find an *antiderivative*, defined as follows.

Antiderivative
If $F'(x) = f(x)$, then $F(x)$ is an **antiderivative** of $f(x)$.

EXAMPLE 1 Antiderivative

(a) If $F(x) = 10x$, then $F'(x) = 10$, so $F(x) = 10x$ is an antiderivative of $f(x) = 10$.

(b) For $F(x) = x^2$, $F'(x) = 2x$, making $F(x) = x^2$ an antiderivative of $f(x) = 2x$.

EXAMPLE 2 Antiderivative

Find an antiderivative of $f(x) = 5x^4$.

SOLUTION To find a function $F(x)$ whose derivative is $5x^4$, work backwards. Recall that the derivative of x^n is nx^{n-1}. If

$$nx^{n-1} \text{ is } 5x^4,$$

YOUR TURN 1 Find an antiderivative $f(x) = 8x^7$.

then $n - 1 = 4$ and $n = 5$, so x^5 is an antiderivative of $5x^4$. **TRY YOUR TURN 1**

EXAMPLE 3 Population

Suppose a population (in millions) is growing at a rate given by $f(x) = e^x$, where x is time in years from some initial date. Find a function giving the population at time x.

SOLUTION Let the population function be $F(x)$. Then

$$f(x) = F'(x) = e^x.$$

The derivative of the function defined by $F(x) = e^x$ is $F'(x) = e^x$, so one possible population function with the given growth rate is $F(x) = e^x$.

The function from Example 1(b), defined by $F(x) = x^2$, is not the only function whose derivative is $f(x) = 2x$. For example,

$$F(x) = x^2, \qquad G(x) = x^2 + 2, \qquad \text{and} \qquad H(x) = x^2 - 4$$

are all antiderivatives of $f(x) = 2x$, and any two of them differ only by a constant. These three functions, shown in Figure 1, have the same derivative, $f(x) = 2x$, and the slopes of their tangent lines at any particular value of x are the same. In fact, for any real number C, the function $F(x) = x^2 + C$ has $f(x) = 2x$ as its derivative. This means that there is a *family* or *class* of functions having $2x$ as a derivative. As the next theorem states, if two functions $F(x)$ and $G(x)$ are antiderivatives of $f(x)$, then $F(x)$ and $G(x)$ can differ only by a constant.

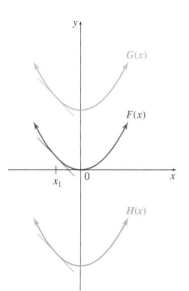

Slopes of the tangent lines
at $x = x_1$ are the same.

FIGURE 1

If $F(x)$ and $G(x)$ are both antiderivatives of a function $f(x)$ on an interval, then there is a constant C such that

$$F(x) - G(x) = C.$$

(Two antiderivatives of a function can differ only by a constant.) The arbitrary real number C is called an integration constant.

The family of all antiderivatives of the function f is indicated by

$$\int f(x)\, dx = F(x) + C.$$

The symbol $\int$ is the **integral sign**, $f(x)$ is the **integrand**, and $\int f(x)\, dx$ is called an **indefinite integral**, the most general antiderivative of f.

Indefinite Integral

If $F'(x) = f(x)$, then

$$\int f(x)\, dx = F(x) + C,$$

for any real number C.

For example, using this notation,

$$\int 2x\, dx = x^2 + C.$$

The dx in the indefinite integral indicates that $\int f(x)\, dx$ is the "integral of $f(x)$ *with respect to* x" just as the symbol dy/dx denotes the "derivative of y with respect to x." For example, in the indefinite integral $\int 2ax\, dx$, dx indicates that a is to be treated as a constant and x as the variable, so that

$$\int 2ax\, dx = \int a(2x)dx = ax^2 + C.$$

On the other hand,

$$\int 2ax\, da = a^2x + C = xa^2 + C.$$

A more complete interpretation of dx will be discussed later.

The symbol $\int f(x)\, dx$ was created by G. W. Leibniz (1646–1716) in the latter part of the seventeenth century. The $\int$ is an elongated S from *summa,* the Latin word for *sum.* The word *integral* as a term in the calculus was coined by Jakob Bernoulli (1654–1705), a Swiss mathematician who corresponded frequently with Leibniz. The relationship between sums and integrals will be clarified in a later section.

Finding an antiderivative is the reverse of finding a derivative. Therefore, each rule for derivatives leads to a rule for antiderivatives. For example, the power rule for derivatives tells us that

$$\frac{d}{dx}x^5 = 5x^4.$$

Consequently,

$$\int 5x^4\, dx = x^5 + C,$$

FOR REVIEW

Recall that $\dfrac{d}{dx}x^n = nx^{n-1}$.

the result found in Example 2. Note that the derivative of x^n is found by multiplying x by n and reducing the exponent on x by 1. To find an indefinite integral—that is, to undo what was done—*increase* the exponent by 1 and *divide* by the new exponent, $n + 1$.

Power Rule

For any real number $n \neq -1$,

$$\int x^n\, dx = \frac{x^{n+1}}{n + 1} + C.$$

(The antiderivative of $f(x) = x^n$ for $n \neq -1$ is found by increasing the exponent n by 1 and dividing x raised to the new power by the new value of the exponent.)

This rule can be verified by differentiating the expression on the right side of the power rule:

$$\frac{d}{dx}\left(\frac{x^{n+1}}{n + 1} + C\right) = \frac{n + 1}{n + 1}x^{(n+1)-1} + 0 = x^n.$$

(If $n = -1$, the expression in the denominator is 0, and the above rule cannot be used. Finding an antiderivative for this case is discussed later.)

EXAMPLE 9   Cost

Suppose a publishing company has found that the marginal cost at a level of production of x thousand books is given by

$$C'(x) = \frac{50}{\sqrt{x}}$$

and that the fixed cost (the cost before the first book can be produced) is $25,000. Find the cost function $C(x)$.

SOLUTION   Write $50/\sqrt{x}$ as $50/x^{1/2}$ or $50x^{-1/2}$, and then use the indefinite integral rules to integrate the function.

$$C(x) = \int \frac{50}{\sqrt{x}} \, dx = \int 50x^{-1/2} \, dx = 50(2x^{1/2}) + k = 100x^{1/2} + k$$

(Here k is used instead of C to avoid confusion with the cost function $C(x)$.) To find the value of k, use the fact that $C(0)$ is 25,000.

$$C(x) = 100x^{1/2} + k$$
$$25,000 = 100 \cdot 0 + k$$
$$k = 25,000$$

With this result, the cost function is $C(x) = 100x^{1/2} + 25,000$.

EXAMPLE 10   Demand

Suppose the marginal revenue from a product is given by $400e^{-0.1q} + 8$, where q is the number of products produced.

(a) Find the revenue function for the product.

SOLUTION   The marginal revenue is the derivative of the revenue function, so

$$R'(q) = 400e^{-0.1q} + 8$$
$$R(q) = \int \left(400e^{-0.1q} + 8 \right) dq \qquad \text{Integrate both sides.}$$
$$= 400 \frac{e^{-0.1q}}{-0.1} + 8q + C$$
$$= -4000e^{-0.1q} + 8q + C.$$

If $q = 0$, then $R = 0$ (no items sold means no revenue), so that

$$0 = -4000e^{-0.1(0)} + 8(0) + C$$
$$0 = -4000 + 0 + C$$
$$4000 = C.$$

Thus, the revenue function is

$$R(q) = -4000e^{-0.1q} + 8q + 4000.$$

(b) Find the demand function for this product.

SOLUTION   Recall that $R = qp$, where p is the demand function giving the price p as a function of q. Then

$$-4000e^{-0.1q} + 8q + 4000 = qp$$
$$\frac{-4000e^{-0.1q} + 8q + 4000}{q} = p. \qquad \text{Divide by } q.$$

The demand function is $p = \dfrac{-4000e^{-0.1q} + 8q + 4000}{q}$.

In the next example, integrals are used to find the position of a particle when the acceleration of the particle is given.

EXAMPLE 11 Velocity and Acceleration

Recall that if the function $s(t)$ gives the position of a particle at time t, then its velocity $v(t)$ and its acceleration $a(t)$ are given by

$$v(t) = s'(t) \quad \text{and} \quad a(t) = v'(t) = s''(t).$$

(a) Suppose the velocity of an object is $v(t) = 6t^2 - 8t$ and that the object is at 5 when time is 0. Find $s(t)$.

SOLUTION Since $v(t) = s'(t)$, the function $s(t)$ is an antiderivative of $v(t)$:

$$s(t) = \int v(t)\, dt = \int (6t^2 - 8t)\, dt$$
$$= 2t^3 - 4t^2 + C$$

for some constant C. Find C from the given information that $s = 5$ when $t = 0$.

$$s(t) = 2t^3 - 4t^2 + C$$
$$5 = 2(0)^3 - 4(0)^2 + C$$
$$5 = C$$
$$s(t) = 2t^3 - 4t^2 + 5$$

(b) Many experiments have shown that when an object is dropped, its acceleration (ignoring air resistance) is constant. This constant has been found to be approximately 32 ft per second every second; that is,

$$a(t) = -32.$$

The negative sign is used because the object is falling. Suppose an object is thrown down from the top of the 1100-ft-tall Willis Tower (formerly known as the Sears Tower) in Chicago. If the initial velocity of the object is -20 ft per second, find $s(t)$, the distance of the object from the ground at time t.

SOLUTION First find $v(t)$ by integrating $a(t)$:

$$v(t) = \int (-32)\, dt = -32t + k.$$

When $t = 0$, $v(t) = -20$:

$$-20 = -32(0) + k$$
$$-20 = k$$

and

$$v(t) = -32t - 20.$$

Be sure to evaluate the constant of integration k before integrating again to get $s(t)$. Now integrate $v(t)$ to find $s(t)$.

$$s(t) = \int (-32t - 20)\, dt = -16t^2 - 20t + C$$

Since $s(t) = 1100$ when $t = 0$, we can substitute these values into the equation for $s(t)$ to get $C = 1100$ and

$$s(t) = -16t^2 - 20t + 1100$$

as the distance of the object from the ground after t seconds.

(c) Use the equations derived in (b) to find the velocity of the object when it hit the ground and how long it took to strike the ground.

APPLY IT

SOLUTION When the object strikes the ground, $s = 0$, so

$$0 = -16t^2 - 20t + 1100.$$

To solve this equation for t, factor out the common factor of -4 and then use the quadratic formula.

$$0 = -4(4t^2 + 5t - 275)$$

$$t = \frac{-5 \pm \sqrt{25 + 4400}}{8} \approx \frac{-5 \pm 66.5}{8}$$

YOUR TURN 6 Repeat Example 11(b) and 11(c) for the Burj Khalifa in Dubai, which is the tallest building in the world, standing 2717 ft. The initial velocity is -20 ft per second.

Only the positive value of t is meaningful here: $t \approx 7.69$. It took the object about 7.69 seconds to strike the ground. From the velocity equation, with $t = 7.69$, we find

$$v(t) = -32t - 20$$
$$v(7.69) = -32(7.69) - 20 \approx -266,$$

so the object was falling (as indicated by the negative sign) at about 266 ft per second when it hit the ground. **TRY YOUR TURN 6**

EXAMPLE 12 Slope

Find a function f whose graph has slope $f'(x) = 6x^2 + 4$ and passes through the point $(1, 1)$.

SOLUTION Since $f'(x) = 6x^2 + 4$,

$$f(x) = \int (6x^2 + 4)\, dx = 2x^3 + 4x + C.$$

The graph of f passes through $(1, 1)$, so C can be found by substituting 1 for x and 1 for $f(x)$.

YOUR TURN 7 Find an equation of the curve whose tangent line has slope $f'(x) = 3x^{1/2} + 4$ and passes through the point $(1, -2)$.

$$1 = 2(1)^3 + 4(1) + C$$
$$1 = 6 + C$$
$$C = -5$$

Therefore, $f(x) = 2x^3 + 4x - 5.$ **TRY YOUR TURN 7**

7.1 WARM-UP EXERCISES

Find the derivative of the following functions.

W1. $f(x) = 5x^4 + 6\sqrt{x}$ *(Sec. 4.1)*

W2. $f(x) = 7e^{3x} - 8x^{5/2}$ *(Sec. 4.4)*

7.1 EXERCISES

1. What must be true of $F(x)$ and $G(x)$ if both are antiderivatives of $f(x)$?

2. How is the antiderivative of a function related to the function?

3. Explain what is wrong with the following use of the power rule:
$$\int \frac{5}{x^2}\, dx = \frac{5}{x^3/3} + C.$$

4. Explain why the restriction $n \neq -1$ is necessary in the rule
$$\int x^n\, dx = \frac{x^{n+1}}{n+1} + C.$$

Find the following.

5. $\int 6\, dk$

6. $\int 9\, dy$

7. $\int (2z + 3)\,dz$

8. $\int (3x - 5)\,dx$

9. $\int (6t^2 - 8t + 7)\,dt$

10. $\int (5x^2 - 6x + 3)\,dx$

11. $\int (4z^3 + 3z^2 + 2z - 6)\,dz$

12. $\int (16y^3 + 9y^2 - 6y + 3)\,dy$

13. $\int (5\sqrt{z} + \sqrt{2})\,dz$

14. $\int (t^{1/4} + \pi^{1/4})\,dt$

15. $\int 5x(x^2 - 8)\,dx$

16. $\int x^2(x^4 + 4x + 3)\,dx$

17. $\int (4\sqrt{v} - 3v^{3/2})\,dv$

18. $\int (15x\sqrt{x} + 2\sqrt{x})\,dx$

19. $\int (10u^{3/2} - 14u^{5/2})\,du$

20. $\int (56t^{5/2} + 18t^{7/2})\,dt$

21. $\int \left(\dfrac{7}{z^2}\right) dz$

22. $\int \left(\dfrac{4}{x^3}\right) dx$

23. $\int \left(\dfrac{\pi^3}{y^3} - \dfrac{\sqrt{\pi}}{\sqrt{y}}\right) dy$

24. $\int \left(\sqrt{u} + \dfrac{1}{u^2}\right) du$

25. $\int (-9t^{-2.5} - 2t^{-1})\,dt$

26. $\int (10x^{-3.5} + 4x^{-1})\,dx$

27. $\int \dfrac{1}{3x^2}\,dx$

28. $\int \dfrac{2}{3x^4}\,dx$

29. $\int 3e^{-0.2x}\,dx$

30. $\int -4e^{0.2v}\,dv$

31. $\int \left(\dfrac{-3}{x} + 4e^{-0.4x} + e^{0.1}\right) dx$

32. $\int \left(\dfrac{9}{x} - 3e^{-0.4x}\right) dx$

33. $\int \dfrac{1 + 2t^3}{4t}\,dt$

34. $\int \dfrac{2y^{1/2} - 3y^2}{6y}\,dy$

35. $\int (e^{2u} + 4u)\,du$

36. $\int (v^2 - e^{3v})\,dv$

37. $\int (x + 1)^2\,dx$

38. $\int (2y - 1)^2\,dy$

39. $\int \dfrac{\sqrt{x} + 1}{\sqrt[3]{x}}\,dx$

40. $\int \dfrac{1 - 2\sqrt[3]{z}}{\sqrt[3]{z}}\,dz$

41. $\int 10^x\,dx$

42. $\int 3^{2x}\,dx$

43. Find an equation of the curve whose tangent line has a slope of
$$f'(x) = x^{2/3},$$
given that the point $(1, 3/5)$ is on the curve.

44. The slope of the tangent line to a curve is given by
$$f'(x) = 6x^2 - 4x + 3.$$
If the point $(0, 1)$ is on the curve, find an equation of the curve.

APPLICATIONS

Business and Economics

Cost   **Find the cost function for each marginal cost function.**

45. $C'(x) = 4x - 5;$ fixed cost is $8

46. $C'(x) = 0.2x^2 + 5x;$ fixed cost is $10

47. $C'(x) = 0.03e^{0.01x};$ fixed cost is $8

48. $C'(x) = x^{1/2};$ 16 units cost $45

49. $C'(x) = x^{2/3} + 2;$ 8 units cost $58

50. $C'(x) = x + 1/x^2;$ 2 units cost $5.50

51. $C'(x) = 5x - 1/x;$ 10 units cost $94.20

52. $C'(x) = 1.2^x(\ln 1.2);$ 2 units cost $9.44

Demand   **Find the demand function for each marginal revenue function. Recall that if no items are sold, the revenue is 0.**

53. $R'(x) = 175 - 0.02x - 0.03x^2$

54. $R'(x) = 50 - 5x^{2/3}$

55. $R'(x) = 500 - 0.15\sqrt{x}$

56. $R'(x) = 600 - 5e^{0.0002x}$

57. Chinese Patents   The approximate rate of change in the number of patent applications received in China in recent years is given by
$$p'(t) = 43.14t - 143.5,$$
where t represents the number of years since 2000, and p represents the number of patents (in thousands). In 2008, 828,328 patent applications were received. *Source: State Intellectual Property Office of the P.R.C.*

(a) Find the function that gives the total number (in thousands) of patent applications received in China in year t.

(b) According to this function, how many patent applications were received in 2013? Compare this with the actual number of 2,377,061.

58. Profit   The marginal profit of a small fast-food stand is given, in thousands of dollars, by
$$P'(x) = \sqrt{x} + \dfrac{1}{2},$$
where x is the sales volume in thousands of hamburgers. The "profit" is $-\$1000$ when no hamburgers are sold. Find the profit function.

59. Profit   The marginal profit in dollars on Brie cheese sold at a cheese store is given by
$$P'(x) = x(50x^2 + 30x),$$
where x is the amount of cheese sold, in hundreds of pounds. The "profit" is $-\$40$ when no cheese is sold.

(a) Find the profit function.

(b) Find the profit from selling 200 lb of Brie cheese.

Life Sciences

60. Biochemical Excretion   If the rate of excretion of a biochemical compound is given by
$$f'(t) = 0.01e^{-0.01t},$$
the total amount excreted by time t (in minutes) is $f(t)$.

(a) Find an expression for $f(t)$.

(b) If 0 units are excreted at time $t = 0$, how many units are excreted in 10 minutes?

61. Flour Beetles A model for describing the population of adult flour beetles involves evaluating the integral

$$\int \frac{g(x)}{x} \, dx,$$

where $g(x)$ is the per-unit-abundance growth rate for a population of size x. The researchers consider the simple case in which $g(x) = a - bx$ for positive constants a and b. Find the integral in this case. *Source: Ecology.*

62. Concentration of a Solute According to Fick's law, the diffusion of a solute across a cell membrane is given by

$$c'(t) = \frac{kA}{V}[C - c(t)], \qquad \text{(1)}$$

where A is the area of the cell membrane, V is the volume of the cell, $c(t)$ is the concentration inside the cell at time t, C is the concentration outside the cell, and k is a constant. If c_0 represents the concentration of the solute inside the cell when $t = 0$, then it can be shown that

$$c(t) = (c_0 - C)e^{-kAt/V} + C. \qquad \text{(2)}$$

(a) Use the last result to find $c'(t)$.

(b) Substitute back into Equation (1) to show that (2) is indeed the correct antiderivative of (1).

63. Cell Growth Under certain conditions, the number of cancer cells $N(t)$ at time t increases at a rate

$$N'(t) = Ae^{kt},$$

where A is the rate of increase at time 0 (in cells per day) and k is a constant.

(a) Suppose $A = 50$, and at 5 days, the cells are growing at a rate of 250 per day. Find a formula for the number of cells after t days, given that 300 cells are present at $t = 0$.

(b) Use your answer from part (a) to find the number of cells present after 12 days.

64. Blood Pressure The rate of change of the volume $V(t)$ of blood in the aorta at time t is given by

$$V'(t) = -kP(t),$$

where $P(t)$ is the pressure in the aorta at time t and k is a constant that depends upon properties of the aorta. The pressure in the aorta is given by

$$P(t) = P_0 e^{-mt},$$

where P_0 is the pressure at time $t = 0$ and m is another constant. Letting V_0 be the volume at time $t = 0$, find a formula for $V(t)$.

Social Sciences

65. Bachelor's Degrees The number of bachelor's degrees conferred in the United States has been increasing steadily in recent decades. Based on data from the National Center for Education Statistics, the rate of change of the number of bachelor's degrees (in thousands) can be approximated by the function

$$B'(t) = 0.04764t^2 - 0.9148t + 13.20,$$

where t is the number of years since 1970. *Source: National Center for Education Statistics.*

(a) Find $B(t)$, given that 792,316 degrees were conferred in 1970 $(t = 0)$.

(b) Use the formula from part (a) to project the number of bachelor's degrees that will be conferred in 2020.

66. Degrees in Dentistry The number of degrees in dentistry (D.D.S. or D.M.D.) conferred to females in the United States has been increasing steadily in recent decades. Based on data from the National Center for Education Statistics, the rate of change of the number of dentistry degrees can be approximated by the function

$$D'(t) = 35.352e^{0.017676t},$$

where t is the number of years since 1980. *Source: National Center for Education Statistics.*

(a) Find $D(t)$, given that 1695 degrees in dentistry were conferred to females in 2001.

(b) Use the formula from part (a) to project the number of degrees in dentistry that will be conferred to females in 2020.

Physical Sciences

Exercises 67–71 refer to Example 11 in this section.

67. Velocity For a particular object, $a(t) = 5t^2 + 4$ and $v(0) = 6$. Find $v(t)$.

68. Distance Suppose $v(t) = 9t^2 - 3\sqrt{t}$ and $s(1) = 8$. Find $s(t)$.

69. Time An object is dropped from a small plane flying at 6400 ft. Assume that $a(t) = -32$ ft per second and $v(0) = 0$. Find $s(t)$. How long will it take the object to hit the ground?

70. Distance Suppose $a(t) = 18t + 8$, $v(1) = 15$, and $s(1) = 19$. Find $s(t)$.

71. Distance Suppose $a(t) = (15/2)\sqrt{t} + 3e^{-t}$, $v(0) = -3$, and $s(0) = 4$. Find $s(t)$.

72. Motion Under Gravity Show that an object thrown from an initial height h_0 with an initial velocity v_0 has a height at time t given by the function

$$h(t) = \tfrac{1}{2}gt^2 + v_0t + h_0,$$

where g is the acceleration due to gravity, a constant with value -32 ft/sec^2.

73. Rocket A small rocket was launched straight up from a platform. After 5 seconds, the rocket reached a maximum height of 412 ft. Find the initial velocity and height of the rocket. (*Hint:* See the previous exercise.)

74. Rocket Science In the 1999 movie *October Sky*, Homer Hickam was accused of launching a rocket that started a forest fire. Homer proved his innocence by showing that his rocket could not have flown far enough to reach where the fire started. He used the following reasoning.

(a) Using the fact that $a(t) = -32$ (see Example 11(b)), find $v(t)$ and $s(t)$, given $v(0) = v_0$ and $s(0) = 0$. (The initial velocity was unknown, and the initial height was 0 ft.)

(b) Homer estimated that the rocket was in the air for 14 seconds. Use $s(14) = 0$ to find v_0.

(c) If the rocket left the ground at a 45° angle, the velocity in the horizontal direction would be equal to v_0, the velocity in the vertical direction, so the distance traveled horizontally would be $v_0 t$. (The rocket left the ground at a steeper angle, so this would overestimate the distance from starting to landing point.) Find the distance the rocket would travel horizontally during its 14-second flight.

YOUR TURN ANSWERS ▬▬

1. x^8 or $x^8 + C$

2. $-\dfrac{1}{3t^3} + C$

3. $2x^3 + 4x^2 - 9x + C$

4. $\dfrac{2}{7}x^{7/2} - 4x^{1/2} + C$

5. $3 \ln|x| - \dfrac{1}{3}e^{-3x} + C$

6. $s(t) = -16t^2 - 20t + 2717$; 12.42 sec; 417 ft/sec

7. $f(x) = 2x^{3/2} + 4x - 8$

7.2 Substitution

APPLY IT If a formula for the marginal revenue is known, how can a formula for the total revenue be found?

Using the method of substitution, we will answer this question in Exercise 39.

In earlier chapters you learned all the rules for finding derivatives of elementary functions. By correctly applying those rules, you can take the derivative of any function involving powers of x, exponential functions, and logarithmic functions, combined in any way using the operations of arithmetic (addition, subtraction, multiplication, division, and exponentiation). By contrast, finding the antiderivative is much more complicated. There are a large number of techniques—more than we can cover in this book. Furthermore, for some functions all possible techniques fail. In the last section we saw how to integrate a few simple functions. In this section we introduce a technique known as *substitution* that will greatly expand the set of functions you can integrate.

The substitution technique depends on the idea of a differential, discussed in Chapter 6 on Applications of the Derivative. If $u = f(x)$, the *differential* of u, written du, is defined as

$$du = f'(x)\, dx.$$

For example, if $u = 2x^3 + 1$, then $du = 6x^2\, dx$. In this chapter we will only use differentials as a convenient notational device when finding an antiderivative such as

$$\int (2x^3 + 1)^4 6x^2\, dx.$$

FOR REVIEW

The chain rule, discussed in detail in Chapter 4 on Calculating the Derivative, states that

$$\frac{d}{dx}[f(g(x))] = f'(g(x)) \cdot g'(x).$$

The function $(2x^3 + 1)^4 6x^2$ might remind you of the result when using the chain rule to take the derivative. We will now use differentials and the chain rule in reverse to find the antiderivative. Let $u = 2x^3 + 1$; then $du = 6x^2\, dx$. Now substitute u for $2x^3 + 1$ and du for $6x^2\, dx$ in the indefinite integral.

$$\int (2x^3 + 1)^4 6x^2\, dx = \int \overset{u}{\overbrace{(2x^3 + 1)}}{}^4 \overset{du}{\overbrace{(6x^2\, dx)}}$$

$$= \int u^4\, du$$

With substitution we have changed a complicated integral into a simple one. This last integral can now be found by the power rule.

$$\int u^4\, du = \frac{u^5}{5} + C$$

$$= \frac{(2x^3 + 1)^5}{5} + C. \qquad \text{Substitute } u = 2x^3 + 1.$$

We can check the accuracy of this result by using the chain rule to take the derivative. We get

$$\frac{d}{dx}\left[\frac{(2x^3 + 1)^5}{5} + C\right] = \frac{1}{5} \cdot 5(2x^3 + 1)^4(6x^2) + 0$$

$$= (2x^3 + 1)^4 6x^2.$$

This method of integration is called **integration by substitution**. As shown above, it is simply the chain rule for derivatives in reverse. The results can always be verified by differentiation.

EXAMPLE 1 Substitution

Find $\int 6x(3x^2 + 4)^7 \, dx$.

SOLUTION If we choose $u = 3x^2 + 4$, then $du = 6x \, dx$ and the integrand can be written as the product of $(3x^2 + 4)^7$ and $6x \, dx$. Now substitute.

$$\int 6x(3x^2 + 4)^7 \, dx = \int (3x^2 + 4)^7(6x \, dx) = \int u^7 \, du$$

Find this last indefinite integral.

$$\int u^7 \, du = \frac{u^8}{8} + C$$

$$= \frac{(3x^2 + 4)^8}{8} + C \qquad \text{Substitute } u = 3x^2 + 4.$$

To verify this result, find the derivative.

YOUR TURN 1

Find $\int 8x(4x^2 + 8)^6 dx$.

$$\frac{d}{dx}\left[\frac{(3x^2 + 4)^8}{8} + C\right] = \frac{8}{8}(3x^2 + 4)^7(6x) + 0 = (3x^2 + 4)^7(6x)$$

The derivative is the original function, as required.   TRY YOUR TURN 1

EXAMPLE 2 Substitution

Find $\int x^2 \sqrt{x^3 + 1} \, dx$.

SOLUTION

Method 1
Modifying the Integral

An expression raised to a power is usually a good choice for u, so because of the square root or 1/2 power, let $u = x^3 + 1$; then $du = 3x^2 \, dx$. The integrand does not contain the constant 3, which is needed for du. To take care of this, multiply by 3/3, placing 3 inside the integral sign and 1/3 outside.

$$\int x^2 \sqrt{x^3 + 1} \, dx = \frac{1}{3}\int 3x^2 \sqrt{x^3 + 1} \, dx = \frac{1}{3}\int \sqrt{x^3 + 1} \, (3x^2 \, dx)$$

Now substitute u for $x^3 + 1$ and du for $3x^2 \, dx$, and then integrate.

$$\frac{1}{3}\int \sqrt{x^3 + 1} \, (3x^2 \, dx) = \frac{1}{3}\int \sqrt{u} \, du = \frac{1}{3}\int u^{1/2} \, du$$

$$= \frac{1}{3} \cdot \frac{u^{3/2}}{3/2} + C = \frac{2}{9}u^{3/2} + C$$

$$= \frac{2}{9}(x^3 + 1)^{3/2} + C. \qquad \text{Substitute } u = x^3 + 1.$$

Method 2
Eliminating the Constant

As in Method 1, we let $u = x^3 + 1$, so that $du = 3x^2\,dx$. Since there is no 3 in the integral, we divide the equation for du by 3 to get

$$\frac{1}{3}\,du = x^2\,dx.$$

We then substitute u for $x^3 + 1$ and $du/3$ for $x^2\,dx$ to get

$$\int \sqrt{u}\,\frac{1}{3}\,du = \frac{1}{3}\int u^{1/2}\,du$$

and proceed as we did in Method 1. The two methods are just slightly different ways of doing the same thing, but some people prefer one method over the other.

TRY YOUR TURN 2

YOUR TURN 2

Find $\int x^3\,\sqrt{3x^4 + 10}\,dx$.

The substitution method given in the examples above *will not always work*. For example, you might try to find

$$\int x^3\,\sqrt{x^3 + 1}\,dx$$

by substituting $u = x^3 + 1$, so that $du = 3x^2\,dx$. However, there is no *constant* that can be inserted inside the integral sign to give $3x^2$ alone. This integral, and a great many others, cannot be evaluated by substitution.

With practice, choosing u will become easy if you keep two principles in mind.

1. u should equal some expression in the integral that, when replaced with u, tends to make the integral simpler.

2. u must be an expression whose derivative—disregarding any constant multiplier, such as the 3 in $3x^2$—is also present in the integral.

The substitution should include as much of the integral as possible, as long as its derivative is still present. In Example 1, we could have chosen $u = 3x^2$, but $u = 3x^2 + 4$ is better, because it has the same derivative as $3x^2$ and captures more of the original integral. If we carry this reasoning further, we might try $u = (3x^2 + 4)^4$, but this is a poor choice, for $du = 4(3x^2 + 4)^3(6x)\,dx$, an expression not present in the original integral.

EXAMPLE 3 **Substitution**

Find $\displaystyle\int \frac{x + 3}{(x^2 + 6x)^2}\,dx$.

SOLUTION Let $u = x^2 + 6x$, so that $du = (2x + 6)\,dx = 2(x + 3)\,dx$. The integral is missing the 2, so multiply by $2 \cdot (1/2)$, putting 2 inside the integral sign and 1/2 outside.

$$\int \frac{x + 3}{(x^2 + 6x)^2}\,dx = \frac{1}{2}\int \frac{2(x + 3)}{(x^2 + 6x)^2}\,dx$$

$$= \frac{1}{2}\int \frac{du}{u^2} = \frac{1}{2}\int u^{-2}\,du$$

$$= \frac{1}{2}\cdot\frac{u^{-1}}{-1} + C = \frac{-1}{2u} + C$$

$$= \frac{-1}{2(x^2 + 6x)} + C. \qquad \text{Substitute } u = x^2 + 6x.$$

YOUR TURN 3

Find $\displaystyle\int \frac{x + 1}{(4x^2 + 8x)^3}\,dx$.

TRY YOUR TURN 3

In Example 3, the quantity $x^2 + 6x$ was raised to a power in the denominator. When such an expression is not raised to a power, the function can often be integrated using the fact that

$$\frac{d}{dx} \ln |f(x)| = \frac{1}{f(x)} \cdot f'(x).$$

This suggests that such integrals can be solved by letting u equal the expression in the denominator, as long as the derivative of the denominator is present in the numerator (disregarding any constant multiplier as usual). The next example illustrates this idea.

EXAMPLE 4 Substitution

Find $\displaystyle\int \frac{(2x - 3) \, dx}{x^2 - 3x}$.

SOLUTION Let $u = x^2 - 3x$, so that $du = (2x - 3) \, dx$. Then

$$\int \frac{(2x - 3) \, dx}{x^2 - 3x} = \int \frac{du}{u} = \ln |u| + C = \ln |x^2 - 3x| + C.$$

TRY YOUR TURN 4

YOUR TURN 4

Find $\displaystyle\int \frac{x + 3}{x^2 + 6x} \, dx$.

Recall that if $f(x)$ is a function, then by the chain rule, the derivative of the exponential function $y = e^{f(x)}$ is

$$\frac{d}{dx} e^{f(x)} = e^{f(x)} \cdot f'(x).$$

This suggests that the antiderivative of a function of the form $e^{f(x)}$ can be found by letting u be the exponent, as long as $f'(x)$ is also present in the integral (disregarding any constant multiplier as usual).

EXAMPLE 5 Substitution

Find $\int x^2 e^{x^3} \, dx$.

SOLUTION Let $u = x^3$, the exponent on e. Then $du = 3x^2 \, dx$. Multiplying by 3/3 gives

$$\int x^2 e^{x^3} \, dx = \frac{1}{3} \int e^{x^3} (3x^2 \, dx)$$

$$= \frac{1}{3} \int e^u \, du = \frac{1}{3} e^u + C = \frac{1}{3} e^{x^3} + C. \quad \text{TRY YOUR TURN 5}$$

YOUR TURN 5

Find $\int x^3 e^{x^4} \, dx$.

The techniques in the preceding examples can be summarized as follows.

Substitution

Each of the following forms can be integrated using the substitution $u = f(x)$.

Form of the Integral	Result
1. $\displaystyle\int [f(x)]^n f'(x) \, dx, \quad n \neq -1$	$\displaystyle\int u^n \, du = \frac{u^{n+1}}{n+1} + C = \frac{[f(x)]^{n+1}}{n+1} + C$
2. $\displaystyle\int \frac{f'(x)}{f(x)} \, dx$	$\displaystyle\int \frac{1}{u} \, du = \ln |u| + C = \ln |f(x)| + C$
3. $\displaystyle\int e^{f(x)} f'(x) \, dx$	$\displaystyle\int e^u \, du = e^u + C = e^{f(x)} + C$

The next example shows a more complicated integral in which none of the previous forms apply, but for which substitution still works.

EXAMPLE 6 Substitution

Find $\int x\sqrt{1-x}\,dx$.

SOLUTION Let $u = 1 - x$, so $du = -dx$. This example is different from the previous examples in this section because there is an extra x right after the integral sign. It would be a major error to rewrite this integral as $\int x\sqrt{u}(-du)$ because an integral should have only one variable. (We will modify this restriction in Chapter 9.) Instead we solve $u = 1 - x$ for x to get $x = 1 - u$ and substitute as follows.

$$\int x\sqrt{1-x}\,dx = \int (1-u)\sqrt{u}(-du)$$

$$= \int (u-1)u^{1/2}\,du$$

$$= \int (u^{3/2} - u^{1/2})\,du \qquad \text{Distribute.}$$

$$= \frac{2}{5}u^{5/2} - \frac{2}{3}u^{3/2} + C \qquad \text{Use power rule.}$$

$$= \frac{2}{5}(1-x)^{5/2} - \frac{2}{3}(1-x)^{3/2} + C \qquad \text{Substitute } u = 1 - x.$$

YOUR TURN 6

Find $\int x\sqrt{3+x}\,dx$

TRY YOUR TURN 6

The substitution method is useful if the integral can be written in one of the following forms, where u is some function of x.

Substitution Method

In general, for the types of problems in this book, there are three cases. We choose u to be one of the following:

1. the quantity under a root or raised to a power;
2. the quantity in the denominator;
3. the exponent on e.

Remember that some integrands may need to be rearranged to fit one of these cases.

TECHNOLOGY NOTE

FIGURE 2

Some calculators, such as the TI-89 and TI-Nspire CAS, can find indefinite integrals automatically. Many computer algebra systems, such as Maple, Matlab, and Mathematica, also do this. The website www.wolframalpha.com can also be used to symbolically determine indefinite integrals and derivatives of functions. Figure 2 shows the integral in Example 6 performed on a TI-89. The answer looks different but is algebraically equivalent to the answer found in Example 6.

EXAMPLE 7 Demand

The research department for a hardware chain has determined that at one store the marginal price of x boxes per week of a particular type of nails is

$$p'(x) = \frac{-4000}{(2x+15)^3}.$$

Find the demand equation if the weekly demand for this type of nails is 10 boxes when the price of a box of nails is $4.

SOLUTION To find the demand function, first integrate $p'(x)$ as follows.

$$p(x) = \int p'(x)\,dx = \int \frac{-4000}{(2x+15)^3}\,dx$$

Let $u = 2x + 15$. Then $du = 2\,dx$, and

$$p(x) = \frac{-4000}{2}\int (2x+15)^{-3}\,2\,dx \qquad \text{Multiply by 2/2.}$$

$$= -2000\int u^{-3}\,du \qquad\qquad \text{Substitute.}$$

$$= (-2000)\frac{u^{-2}}{-2} + C$$

$$= \frac{1000}{u^2} + C$$

$$p(x) = \frac{1000}{(2x+15)^2} + C. \qquad\qquad\qquad \textbf{(1)}$$

Find the value of C by using the given information that $p = 4$ when $x = 10$.

$$4 = \frac{1000}{(2\cdot 10 + 15)^2} + C$$

$$4 = \frac{1000}{35^2} + C$$

$$4 \approx 0.82 + C$$

$$3.18 \approx C$$

Replacing C with 3.18 in Equation (1) gives the demand function,

$$p(x) = \frac{1000}{(2x+15)^2} + 3.18. \qquad\qquad \rule{1.5em}{0.6em}$$

With a little practice, you will find you can skip the substitution step for integrals such as that shown in Example 7, in which the derivative of u is a constant. Recall from the chain rule that when you differentiate a function, such as $p(x) = 1000/(2x+15)^2 + 3.18$ in the previous example, you multiply by 2, the derivative of $(2x + 15)$. So when taking the antiderivative, simply divide by 2:

$$\int -4000(2x+15)^{-3}\,dx = \frac{-4000}{2}\cdot\frac{(2x+15)^{-2}}{-2} + C$$

$$= \frac{1000}{(2x+15)^2} + C.$$

| CAUTION | This procedure is valid because of the constant multiple rule presented in the previous section, which says that constant multiples can be brought into or out of integrals, just as they can with derivatives. This procedure is *not* valid with any expression other than a constant. |

EXAMPLE 8 Popularity Index

To determine the top 100 popular songs of each year since 1956, Jim Quirin and Barry Cohen developed a function that represents the rate of change on the charts of *Billboard* magazine required for a song to earn a "star" on the *Billboard* "Hot 100" survey. They developed the function

$$f(x) = \frac{A}{B + x},$$

where $f(x)$ represents the rate of change in position on the charts, x is the position on the "Hot 100" survey, and A and B are positive constants. The function

$$F(x) = \int f(x)\, dx$$

is defined as the "Popularity Index." Find $F(x)$. *Source: Chartmasters' Rock 100.*

SOLUTION Integrating $f(x)$ gives

$$F(x) = \int f(x)\, dx$$

$$= \int \frac{A}{B + x}\, dx$$

$$= A \int \frac{1}{B + x}\, dx.$$

Let $u = B + x$, so that $du = dx$. Then

$$F(x) = A \int \frac{1}{u}\, du = A \ln u + C$$

$$= A \ln(B + x) + C.$$

(The absolute value bars are not necessary, since $B + x$ is always positive here.) ▬

7.2 WARM-UP EXERCISES

Evaluate each of the following integrals. *(Sec. 7.1)*

W1. $\displaystyle\int 10x^4 + 6\sqrt{x}\, dx$

W2. $\displaystyle\int \frac{5}{x} + \frac{4}{x^3}\, dx$

W3. $\displaystyle\int e^{6x}\, dx$

7.2 EXERCISES

1. Integration by substitution is related to what differentiation method? What type of integrand suggests using integration by substitution?

2. The following integrals may be solved using substitution. Choose a function u that may be used to solve each problem. Then find du.

(a) $\displaystyle\int (3x^2 - 5)^4\, 2x\, dx$

(b) $\displaystyle\int \sqrt{1 - x}\, dx$

(c) $\displaystyle\int \frac{x^2}{2x^3 + 1}\, dx$

(d) $\displaystyle\int 4x^3 e^{x^4}\, dx$

Use substitution to find each indefinite integral.

3. $\displaystyle\int 4(2x + 3)^4\, dx$

4. $\displaystyle\int (-4t + 1)^3\, dt$

5. $\displaystyle\int \frac{2\, dm}{(2m + 1)^3}$

6. $\displaystyle\int \frac{3\, du}{\sqrt{3u - 5}}$

7. $\displaystyle\int \frac{2x + 2}{(x^2 + 2x - 4)^4}\, dx$

8. $\displaystyle\int \frac{6x^2\, dx}{(2x^3 + 7)^{3/2}}$

9. $\displaystyle\int z\sqrt{4z^2 - 5}\, dz$

10. $\displaystyle\int r\sqrt{5r^2 + 2}\, dr$

11. $\displaystyle\int 3x^2 e^{2x^3}\, dx$

12. $\displaystyle\int re^{-r^2}\, dr$

13. $\displaystyle\int (1 - t)e^{2t - t^2}\, dt$

14. $\displaystyle\int (x^2 - 1)e^{x^3 - 3x}\, dx$

15. $\displaystyle\int \frac{e^{1/z}}{z^2}\, dz$

16. $\displaystyle\int \frac{e^{\sqrt{y}}}{2\sqrt{y}}\, dy$

17. $\displaystyle\int \frac{t}{t^2 + 2}\, dt$

18. $\displaystyle\int \frac{-4x}{x^2 + 3}\, dx$

19. $\displaystyle\int \frac{x^3 + 2x}{x^4 + 4x^2 + 7}\, dx$

20. $\displaystyle\int \frac{t^2 + 2}{t^3 + 6t + 3}\, dt$

21. $\displaystyle\int \frac{2x + 1}{(x^2 + x)^3}\, dx$

22. $\displaystyle\int \frac{y^2 + y}{(2y^3 + 3y^2 + 1)^{2/3}}\, dy$

23. $\displaystyle\int p(p + 1)^5\, dp$

24. $\displaystyle\int 4r\sqrt{8 - r}\, dr$

25. $\displaystyle\int \frac{u}{\sqrt{u-1}}\, du$

26. $\displaystyle\int \frac{2x}{(x+5)^6}\, dx$

27. $\displaystyle\int (\sqrt{x^2+12x})(x+6)\, dx$

28. $\displaystyle\int (\sqrt{x^2-6x})(x-3)\, dx$

29. $\displaystyle\int \frac{(1+3\ln x)^2}{x}\, dx$

30. $\displaystyle\int \frac{\sqrt{2+\ln x}}{x}\, dx$

31. $\displaystyle\int \frac{e^{2x}}{e^{2x}+5}\, dx$

32. $\displaystyle\int \frac{1}{x(\ln x)}\, dx$

33. $\displaystyle\int \frac{\log x}{x}\, dx$

34. $\displaystyle\int \frac{(\log_2 (5x+1))^2}{5x+1}\, dx$

35. $\displaystyle\int x8^{3x^2+1}\, dx$

36. $\displaystyle\int \frac{10^{5\sqrt{x}+2}}{\sqrt{x}}\, dx$

37. Stan and Ollie work on the integral

$$\int 3x^2 e^{x^3}\, dx.$$

Stan lets $u = x^3$ and proceeds to get

$$\int e^u\, du = e^u + C = e^{x^3} + C.$$

Ollie tries $u = e^{x^3}$ and proceeds to get

$$\int du = u + C = e^{x^3} + C.$$

Discuss which procedure you prefer, and why.

38. Stan and Ollie work on the integral

$$\int 2x(x^2+2)\, dx.$$

Stan lets $u = x^2 + 2$ and proceeds to get

$$\int u\, du = \frac{u^2}{2} + C = \frac{(x^2+2)^2}{2} + C.$$

Ollie multiplies out the function under the integral and gets

$$\int (2x^3 + 4x)\, dx = \frac{x^4}{2} + 2x^2 + C.$$

How can they both be right?

APPLICATIONS

Business and Economics

39. APPLY IT   Revenue   The marginal revenue (in thousands of dollars) from the sale of x MP3 players is given by

$$R'(x) = 4x(x^2 + 27{,}000)^{-2/3}.$$

(a) Find the total revenue function if the revenue from 125 players is $29,591.

(b) How many players must be sold for a revenue of at least $40,000?

40. Debt   A company incurs debt at a rate of

$$D'(t) = 90(t+6)\sqrt{t^2+12t}$$

dollars per year, where t is the amount of time (in years) since the company began. By the fourth year the company had accumulated $16,260 in debt.

(a) Find the total debt function.

(b) How many years must pass before the total debt exceeds $40,000?

41. Cost   A company has found that the marginal cost (in thousands of dollars) to produce x central air conditioning units is

$$C'(x) = \frac{60x}{5x^2 + e},$$

where x is the number of units produced.

(a) Find the cost function, given that the company incurs a fixed cost of $10,000 even if no units are built.

(b) The company will seek a new source of investment income if the cost is more than $20,000 to produce 5 units. Should they seek this new source?

42. Profit   The rate of growth of the profit (in millions of dollars) from a new technology is approximated by

$$P'(t) = te^{-t^2},$$

where t represents time measured in years. The total profit in the third year that the new technology is in operation is $10,000.

(a) Find the total profit function.

(b) What happens to the total amount of profit in the long run?

43. Transportation   According to data from the Bureau of Transportation Statistics, the rate of change in the number of local transit vehicles (buses, light rail, etc.), in thousands, in the United States from 1970 to the present can be approximated by

$$f'(t) = 4.0674 \times 10^{-4} t(t - 1970)^{0.3},$$

where t is the year. *Source: National Transportation Statistics 2014.*

(a) Using the fact that in 1970 there were 61,298 such vehicles, find a formula giving the approximate number of local transit vehicles as a function of time.

(b) Use the answer to part (a) to forecast the number of local transit vehicles in the year 2020.

Life Sciences

44. Outpatient Visits   According to data from the American Hospital Association, the rate of change in the number of hospital outpatient visits, in millions, in the United States each year from 1980 to the present can be approximated by

$$f'(t) = 0.001483t(t - 1980)^{0.75},$$

where t is the year. *Source: Hospital Statistics.*

(a) Using the fact that in 1980 there were 262,951,000 outpatient visits, find a formula giving the approximate number of outpatient visits as a function of time.

(b) Use the answer to part (a) to forecast the number of outpatient visits in the year 2020.

45. Population Growth   A population grows at a rate

$$P'(t) = 500te^{-t^2/5},$$

where $P(t)$ is the population after t months.

(a) Find a formula for the population size after t months, given that the population is 2000 at $t = 0$.

(b) Use your answer from part (a) to find the population size after 3 months.

46. Epidemic An epidemic is growing in a region according to the rate

$$N'(t) = \frac{100t}{t^2 + 2},$$

where $N(t)$ is the number of people infected after t days.

(a) Find a formula for the number of people infected after t days, given that 37 people were infected at $t = 0$.

(b) Use your answer from part (a) to find the number of people infected after 21 days.

7.3 Area and the Definite Integral

We will answer this question in Example 3 using a method introduced in this section.

To calculate the areas of geometric figures such as rectangles, squares, triangles, and circles, we use specific formulas. In this section we consider the problem of finding the area of a figure or region that is bounded by curves, such as the shaded region in Figure 3.

The brilliant Greek mathematician Archimedes (about 287 B.C.–212 B.C.) is considered one of the greatest mathematicians of all time. His development of a rigorous method known as *exhaustion* to derive results was a forerunner of the ideas of integral calculus. Archimedes used a method that would later be verified by the theory of integration. His method involved viewing a geometric figure as a sum of other figures. For example, he thought of a plane surface area as a figure consisting of infinitely many parallel line segments. Among the results established by Archimedes' method was the fact that the area of a segment of a parabola (shown in color in Figure 3) is equal to 4/3 the area of a triangle with the same base and the same height.

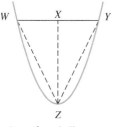

Area of parabolic segment

$= \frac{4}{3}$ (area of triangle WYZ)

FIGURE 3

EXAMPLE 1 **Approximation of Area**

Consider the region bounded by the y-axis, the x-axis, and the graph of $f(x) = \sqrt{4 - x^2}$, shown in Figure 4.

FIGURE 4

FIGURE 5

(a) Approximate the area of the region using two rectangles. Determine the height of the rectangle by the value of the function at the *left* endpoint.

SOLUTION A very rough approximation of the area of this region can be found by using two rectangles whose heights are determined by the value of the function at the left endpoints, as in Figure 5. The height of the rectangle on the left is $f(0) = 2$ and the height of the rectangle on the right is $f(1) = \sqrt{3}$. The width of each rectangle is 1, making the total area of the two rectangles

$$1 \cdot f(0) + 1 \cdot f(1) = 2 + \sqrt{3} \approx 3.7321 \text{ square units.}$$

Note that $f(x)$ is a decreasing function, and that we will overestimate the area when we evaluate the function at the left endpoint to determine the height of the rectangle in that interval.

(b) Repeat part (a) using the value of the function at the *right* endpoint to determine the height of the rectangle.

SOLUTION   Using the right endpoints, as in Figure 6, the area of the two rectangles is

$$1 \cdot f(1) + 1 \cdot f(2) = \sqrt{3} + 0 \approx 1.7321 \text{ square units.}$$

Note that we underestimate the area of this particular region when we use the right endpoints.

If the left endpoint gives an answer too big and the right endpoint an answer too small, it seems reasonable to average the two answers. This produces the method called the *trapezoidal rule*, discussed in more detail later in this chapter. In this example, we get

$$\frac{3.7321 + 1.7321}{2} = 2.7321 \text{ square units.}$$

(c) Repeat part (a) using the value of the function at the *midpoint* of each interval to determine the height of the rectangle.

SOLUTION   In Figure 7, the rectangles are drawn with height determined by the midpoint of each interval. This method is called the **midpoint rule**, and gives

$$1 \cdot f\left(\frac{1}{2}\right) + 1 \cdot f\left(\frac{3}{2}\right) = \frac{\sqrt{15}}{2} + \frac{\sqrt{7}}{2} \approx 3.2594 \text{ square units.}$$

(d) We can improve the accuracy of the previous approximations by increasing the number of rectangles. Repeat part (a) using four rectangles.

SOLUTION   Divide the interval from $x = 0$ to $x = 2$ into four equal parts, each of width 1/2. The height of each rectangle is given by the value of f at the left side of the rectangle, as shown in Figure 8. The area of each rectangle is the width, 1/2, multiplied by the height. The total area of the four rectangles is

$$\frac{1}{2} \cdot f(0) + \frac{1}{2} \cdot f\left(\frac{1}{2}\right) + \frac{1}{2} \cdot f(1) + \frac{1}{2} \cdot f\left(1\frac{1}{2}\right)$$

$$= \frac{1}{2}(2) + \frac{1}{2}\left(\frac{\sqrt{15}}{2}\right) + \frac{1}{2}\left(\sqrt{3}\right) + \frac{1}{2}\left(\frac{\sqrt{7}}{2}\right)$$

$$= 1 + \frac{\sqrt{15}}{4} + \frac{\sqrt{3}}{2} + \frac{\sqrt{7}}{4} \approx 3.4957 \text{ square units.}$$

This approximation looks better, but it is still greater than the actual area.

(e) Repeat part (a) using eight rectangles.

SOLUTION   Divide the interval from $x = 0$ to $x = 2$ into 8 equal parts, each of width 1/4 (see Figure 9). The total area of all of these rectangles is

$$\frac{1}{4} \cdot f(0) + \frac{1}{4} \cdot f\left(\frac{1}{4}\right) + \frac{1}{4} \cdot f\left(\frac{1}{2}\right) + \frac{1}{4} \cdot f\left(\frac{3}{4}\right) + \frac{1}{4} \cdot f(1)$$

$$+ \frac{1}{4} \cdot f\left(\frac{5}{4}\right) + \frac{1}{4} \cdot f\left(\frac{3}{2}\right) + \frac{1}{4} \cdot f\left(\frac{7}{4}\right)$$

$$\approx 3.3398 \text{ square units.}$$

FIGURE 6

FIGURE 7

FIGURE 8

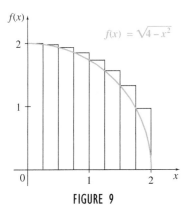

FIGURE 9

The process used in Example 1 of approximating the area under a curve by using more and more rectangles to get a better and better approximation can be generalized. To do this, divide the interval from $x = 0$ to $x = 2$ into n equal parts. Each of these n intervals has width

$$\frac{2 - 0}{n} = \frac{2}{n},$$

so each rectangle has width $2/n$ and height determined by the function value at the left side of the rectangle, or the right side, or the midpoint. We could also average the left and right side values as before. Using a computer or graphing calculator to find approximations to the area for several values of n gives the results in the following table.

	Approximations to the Area			
n	Left Sum	Right Sum	Trapezoidal	Midpoint
2	3.7321	1.7321	2.7321	3.2594
4	3.4957	2.4957	2.9957	3.1839
8	3.3398	2.8398	3.0898	3.1567
10	3.3045	2.9045	3.1045	3.1524
20	3.2285	3.0285	3.1285	3.1454
50	3.1783	3.0983	3.1383	3.1426
100	3.1604	3.1204	3.1404	3.1419
500	3.1455	3.1375	3.1415	3.1416

The numbers in the last four columns of this table represent approximations to the area under the curve, above the x-axis, and between the lines $x = 0$ and $x = 2$. As n becomes larger and larger, all four approximations become better and better, getting closer to the actual area. In this example, the exact area can be found by a formula from plane geometry. Write the given function as

$$y = \sqrt{4 - x^2},$$

then square both sides to get

$$y^2 = 4 - x^2$$
$$x^2 + y^2 = 4,$$

the equation of a circle centered at the origin with radius 2. The region in Figure 4 is the quarter of this circle that lies in the first quadrant. The actual area of this region is one-quarter of the area of the entire circle, or

$$\frac{1}{4}\pi(2)^2 = \pi \approx 3.1416.$$

As the number of rectangles increases without bound, the sum of the areas of these rectangles gets closer and closer to the actual area of the region, π. This can be written as

$$\lim_{n \to \infty} (\text{sum of areas of } n \text{ rectangles}) = \pi.$$

(The value of π was originally found by a process similar to this.)*

*The number π is the ratio of the circumference of a circle to its diameter. It is an example of an *irrational number*, and as such it cannot be expressed as a terminating or repeating decimal. Many approximations have been used for π over the years. A passage in the Bible (1 Kings 7:23) indicates a value of 3. The Egyptians used the value 3.16, and Archimedes showed that its value must be between 22/7 and 223/71. A Hindu writer, Brahmagupta, used $\sqrt{10}$ as its value in the seventh century. The search for the digits of π has continued into modern times. Shigeru Kondo, using a computer he built himself, recently computed the value to over 12 trillion digits.

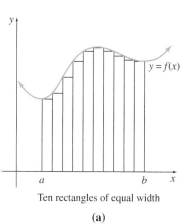

Ten rectangles of equal width

(a)

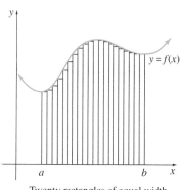

Twenty rectangles of equal width

(b)

FIGURE 10

Notice in the previous table that for a particular value of n, the midpoint rule gave the best answer (the one closest to the true value of $\pi \approx 3.1416$), followed by the trapezoidal rule, followed by the left and right sums. In fact, the midpoint rule with $n = 20$ gives a value (3.1454) that is slightly more accurate than the left sum with $n = 500$ (3.1455). It is usually the case that the midpoint rule gives a more accurate answer than either the left or the right sum.

Now we can generalize to get a method of finding the area bounded by the curve $y = f(x)$, the x-axis, and the vertical lines $x = a$ and $x = b$, as shown in Figure 10. To approximate this area, we could divide the region under the curve first into 10 rectangles (Figure 10(a)) and then into 20 rectangles (Figure 10(b)). The sum of the areas of the rectangles gives an approximation to the area under the curve when $f(x) \geq 0$. In the next section we will consider the case in which $f(x)$ might be negative.

To develop a process that would yield the *exact* area, begin by dividing the interval from a to b into n pieces of equal width, using each of these n pieces as the base of a rectangle (see Figure 11). Let x_1 be an arbitrary point in the first interval, x_2 be an arbitrary point in the second interval, and so on, up to the nth interval. In the graph of Figure 11, the symbol Δx is used to represent the width of each of the intervals. Since the length of the entire interval is $b - a$, each of the n pieces has length

$$\Delta x = \frac{b - a}{n}.$$

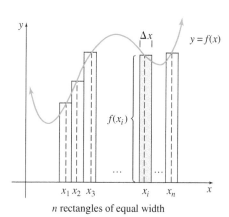

n rectangles of equal width

FIGURE 11

The pink rectangle is an arbitrary rectangle called the ith rectangle. Its area is the product of its length and width. Since the width of the ith rectangle is Δx and the length of the ith rectangle is given by the height $f(x_i)$,

$$\text{Area of the } i\text{th rectangle} = f(x_i) \cdot \Delta x.$$

The total area under the curve is approximated by the sum of the areas of all n of the rectangles. With sigma notation, the approximation of the total area becomes

$$\text{Area of all } n \text{ rectangles} = \sum_{i=1}^{n} f(x_i) \cdot \Delta x.$$

The exact area is defined to be the limit of this sum (if the limit exists) as the number of rectangles increases without bound:

$$\text{Exact area} = \lim_{n \to \infty} \sum_{i=1}^{n} f(x_i)\Delta x.$$

Whenever this limit exists, regardless of whether $f(x)$ is positive or negative, we will call it the *definite integral* of $f(x)$ from a to b. It is written as follows.

The Definite Integral

If f is defined on the interval $[a, b]$, the **definite integral** of f from a to b is given by

$$\int_a^b f(x)\, dx = \lim_{n \to \infty} \sum_{i=1}^{n} f(x_i)\Delta x,$$

provided the limit exists, where $\Delta x = (b - a)/n$ and x_i is *any* value of x in the ith interval.*

The definite integral can be approximated by

$$\sum_{i=1}^{n} f(x_i)\Delta x.$$

If $f(x) \geq 0$ on the interval $[a, b]$, the definite integral gives the area under the curve between $x = a$ and $x = b$. In the midpoint rule, x_i is the midpoint of the ith interval. We may also let x_i be the left endpoint, the right endpoint, or any other point in the ith interval.

In Example 1, the area bounded by the x-axis, the curve $y = \sqrt{4 - x^2}$, and the lines $x = 0$ and $x = 2$ could be written as the definite integral

$$\int_0^2 \sqrt{4 - x^2}\, dx = \pi.$$

NOTE Notice that unlike the indefinite integral, which is a set of *functions*, the definite integral represents a *number*. The next section will show how antiderivatives are used in finding the definite integral and, thus, the area under a curve.

Keep in mind that finding the definite integral of a function can be thought of as a mathematical process that gives the sum of an infinite number of individual parts (within certain limits). The definite integral represents area only if the function involved is *nonnegative* ($f(x) \geq 0$) at every x-value in the interval $[a, b]$. There are many other interpretations of the definite integral, and all of them involve this idea of approximation by appropriate sums. In the next section we will consider the definite integral when $f(x)$ might be negative.

As indicated in this definition, although the left endpoint of the ith interval has been used to find the height of the ith rectangle, any number in the ith interval can be used. (A more general definition is possible in which the rectangles do not necessarily all have the same width.) The b above the integral sign is called the **upper limit** of integration, and the a is the **lower limit** of integration. This use of the word *limit* has nothing to do with the limit of the sum; it refers to the limits, or boundaries, on x.

 **TECHNOLOGY NOTE**

Some calculators have a built-in function for evaluating the definite integral. For example, the TI-84 Plus C uses the `fnInt` command, found in the `MATH` menu, as shown in Figure 12(a). Figure 12(b) shows the command used for Example 1 and gives the answer 3.141593074, with an error of approximately 0.0000004.

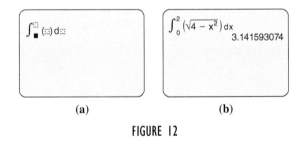

(a)   (b)

FIGURE 12

*The sum in the definition of the definite integral is an example of a Riemann sum, named for the German mathematician Georg Riemann (1826–1866), who at the age of 20 changed his field of study from theology and the classics to mathematics. Twenty years later he died of tuberculosis while traveling in Italy in search of a cure. The concepts of *Riemann sum* and *Riemann integral* are still studied in rigorous calculus textbooks.

EXAMPLE 2 **Approximation of Area**

Approximate $\int_0^4 2x\, dx$, the area of the region under the graph of $f(x) = 2x$, above the x-axis, and between $x = 0$ and $x = 4$, by using four rectangles of equal width whose heights are the values of the function at the midpoint of each subinterval.

SOLUTION

Method 1
Calculating by Hand

We want to find the area of the shaded region in Figure 13. The heights of the four rectangles given by $f(x_i)$ for $i = 1, 2, 3,$ and 4 are as follows.

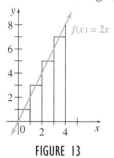

FIGURE 13

Rectangle Heights		
i	x_i	$f(x_i)$
1	$x_1 = 0.5$	$f(0.5) = 1.0$
2	$x_2 = 1.5$	$f(1.5) = 3.0$
3	$x_3 = 2.5$	$f(2.5) = 5.0$
4	$x_4 = 3.5$	$f(3.5) = 7.0$

The width of each rectangle is $\Delta x = (4 - 0)/4 = 1$. The sum of the areas of the four rectangles is

$$\sum_{i=1}^{4} f(x_i)\Delta x = f(x_1)\Delta x + f(x_2)\Delta x + f(x_3)\Delta x + f(x_4)\Delta x$$
$$= f(0.5)\Delta x + f(1.5)\Delta x + f(2.5)\Delta x + f(3.5)\Delta x$$
$$= 1(1) + 3(1) + 5(1) + 7(1)$$
$$= 16.$$

Using the formula for the area of a triangle, $A = (1/2)bh$, with b, the length of the base, equal to 4 and h, the height, equal to 8, gives

$$A = \frac{1}{2}bh = \frac{1}{2}(4)(8) = 16,$$

the exact value of the area. The approximation equals the exact area in this case because our use of the midpoints of each subinterval distributed the error evenly above and below the graph.

Method 2
Graphing Calculator

A graphing calculator can be used to organize the information in this example. For example, the `seq` feature in the LIST OPS menu of the TI-84 Plus C calculator can be used to store the values of i in the list L_1. Using the STAT EDIT menu, the entries for x_i can be generated by entering the formula $-.5 + L_1$ as the heading of L_2. Similarly, entering the formula for $f(x_i)$, $2 * L_2$, at the top of list L_3 will generate the values of $f(x_i)$ in L_3. (The entries are listed automatically when the formula is entered.) Then the `sum` feature in the LIST MATH menu can be used to add the values in L_3. The resulting screens are shown in Figure 14.

(a)   (b)

FIGURE 14

**Method 3
Spreadsheet**

YOUR TURN 1 Repeat
Example 1 to approximate

$$\int_1^5 4x\, dx.$$

The calculations in this example can also be done on a spreadsheet. In Microsoft Excel, for example, store the values of i in Column A. Put the command "$=A1-.5$" into B1; copying this formula into the rest of Column B gives the values of x_i. Similarly, use the formula for $f(x_i)$ to fill Column C. Column D is the product of Column C and Δx. Sum Column D to get the answer. For more details, see the *Graphing Calculator and Excel Spreadsheet Manual* available with this book.

TRY YOUR TURN 1

Total Change Suppose the function $f(x) = x^2 + 20$ gives the marginal cost of some item at a particular x-value. Then $f(2) = 24$ gives the rate of change of cost at $x = 2$. That is, a unit change in x (at this point) will produce a change of 24 units in the cost function. Also, $f(3) = 29$ means that each unit of change in x (when $x = 3$) will produce a change of 29 units in the cost function.

To find the *total* change in the cost function as x changes from 2 to 3, we could divide the interval from 2 to 3 into n equal parts, using each part as the base of a rectangle as we did above. The area of each rectangle would approximate the change in cost at the x-value that is the left endpoint of the base of the rectangle. Then the sum of the areas of these rectangles would approximate the net total change in cost from $x = 2$ to $x = 3$. The limit of this sum as $n \to \infty$ would give the exact total change.

This result produces another application of the definite integral: The area of the region under the graph of the marginal cost function $f(x)$ that is above the x-axis and between $x = a$ and $x = b$ gives the *net total change in the cost* as x goes from a to b.

Total Change in $F(x)$

If $f(x)$ gives the rate of change of $F(x)$ for x in $[a, b]$, then the **total change** in $F(x)$ as x goes from a to b is given by

$$\lim_{n \to \infty} \sum_{i=1}^{n} f(x_i)\Delta x = \int_a^b f(x)\, dx.$$

FIGURE 15

In other words, the total change in a quantity can be found from the function that gives the rate of change of the quantity, using the same methods used to approximate the area under a curve.

EXAMPLE 3 Oil Leakage

APPLY IT

Figure 15 shows the rate that oil is leaking from a machine in a large factory (in cubic centimeters per hour) with specific rates over a 12-hour period given in the table. Approximate the total amount of leakage over a 12-hour shift.

SOLUTION Use approximating rectangles, dividing the interval from 0 to 12 into 12 equal subdivisions. Each subinterval has width 1. Using the left endpoint of each subinterval and the table to determine the height of the rectangle, as shown, the approximation becomes

$$1 \cdot 15.2 + 1 \cdot 18.0 + 1 \cdot 18.8 + 1 \cdot 14.1 + 1 \cdot 9.5 + 1 \cdot 9.6 + 1 \cdot 13.1 + 1 \cdot 17.3$$
$$+ 1 \cdot 20.0 + 1 \cdot 19.2 + 1 \cdot 16.6 + 1 \cdot 16.4 = 187.8.$$

About 187.8 cubic centimeters of oil leak during this time. Mathematically, we could write

$$\int_0^{12} f(x)\, dx \approx 187.8,$$

where $f(x)$ is the function shown in Figure 15.

Oil Leakage (cc's per hour)			
x	y	x	y
0	15.2	6	13.1
1	18.0	7	17.3
2	18.8	8	20.0
3	14.1	9	19.2
4	9.5	10	16.6
5	9.6	11	16.4

Recall, velocity is the rate of change in distance from time a to time b. Thus the area under the velocity function defined by $v(t)$ from $t = a$ to $t = b$ gives the distance traveled in that time period.

EXAMPLE 4   Total Distance

A driver traveling on a business trip checks the speedometer each hour. The table shows the driver's velocity at several times.

Approximate the total distance traveled during the 3-hour period using the left endpoint of each interval, then the right endpoint.

Velocity				
Time (hr)	0	1	2	3
Velocity (mph)	0	52	58	60

SOLUTION   Using left endpoints, the total distance is

$$0 \cdot 1 + 52 \cdot 1 + 58 \cdot 1 = 110.$$

With right endpoints, we get

$$52 \cdot 1 + 58 \cdot 1 + 60 \cdot 1 = 170.$$

Again, left endpoints give a total that is too small, while right endpoints give a total that is too large. The average, 140 miles, is a better estimate of the total distance traveled.

TRY YOUR TURN 2 ▬▬

YOUR TURN 2 Repeat Example 4 for a driver traveling at the following velocities at various times.

Time (hr)	0	0.5	1	1.5	2
Velocity (mph)	0	50	56	40	48

Before discussing further applications of the definite integral, we need a more efficient method for evaluating it. This method will be developed in the next section.

7.3 EXERCISES

1. Explain the difference between an indefinite integral and a definite integral.

2. Complete the following statement.

$$\int_0^4 (x^2 + 3) \, dx = \lim_{n \to \infty} \underline{\qquad},$$ where $\Delta x =$ _____, and x_i is _____.

3. Let $f(x) = 2x + 5$, $x_1 = 0$, $x_2 = 2$, $x_3 = 4$, $x_4 = 6$, and $\Delta x = 2$.

   (a) Find $\sum_{i=1}^{4} f(x_i)\Delta x$.

   (b) The sum in part (a) approximates a definite integral using rectangles. The height of each rectangle is given by the value of the function at the left endpoint. Write the definite integral that the sum approximates.

4. Let $f(x) = 1/x$, $x_1 = 1/2$, $x_2 = 1$, $x_3 = 3/2$, $x_4 = 2$, and $\Delta x = 1/2$.

   (a) Find $\sum_{i=1}^{4} f(x_i)\Delta x$.

   (b) The sum in part (a) approximates a definite integral using rectangles. The height of each rectangle is given by the value of the function at the left endpoint. Write the definite integral that the sum approximates.

In Exercises 5–12, approximate the area under the graph of $f(x)$ and above the x-axis using the following methods with $n = 4$. (a) Use left endpoints. (b) Use right endpoints. (c) Average the answers in parts (a) and (b). (d) Use midpoints.

5. $f(x) = 2x + 5$ from $x = 2$ to $x = 4$

6. $f(x) = 3x + 2$ from $x = 1$ to $x = 3$

7. $f(x) = -x^2 + 4$ from $x = -2$ to $x = 2$

8. $f(x) = x^2$ from $x = 1$ to $x = 5$

9. $f(x) = e^x + 1$ from $x = -2$ to $x = 2$

10. $f(x) = e^x - 1$ from $x = 0$ to $x = 4$

11. $f(x) = \dfrac{2}{x}$ from $x = 1$ to $x = 9$

12. $f(x) = \dfrac{1}{x}$ from $x = 1$ to $x = 3$

13. Consider the region below $f(x) = x/2$, above the x-axis, and between $x = 0$ and $x = 4$. Let x_i be the midpoint of the ith subinterval.

   (a) Approximate the area of the region using four rectangles.

   (b) Find $\int_0^4 f(x) \, dx$ by using the formula for the area of a triangle.

14. Consider the region below $f(x) = 5 - x$, above the *x*-axis, and between $x = 0$ and $x = 5$. Let x_i be the midpoint of the *i*th subinterval.

(a) Approximate the area of the region using five rectangles.

(b) Find $\int_0^5 (5 - x)\, dx$ by using the formula for the area of a triangle.

15. Find $\int_0^4 f(x)\, dx$ for each graph of $y = f(x)$.

(a)

(b)

16. Find $\int_0^6 f(x)\, dx$ for each graph of $y = f(x)$, where $f(x)$ consists of line segments and circular arcs.

(a)

(b)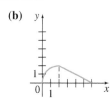

Find the exact value of each integral using formulas from geometry.

17. $\displaystyle\int_{-4}^{0} \sqrt{16 - x^2}\, dx$

18. $\displaystyle\int_{-3}^{3} \sqrt{9 - x^2}\, dx$

19. $\displaystyle\int_{2}^{5} (1 + 2x)\, dx$

20. $\displaystyle\int_{1}^{3} (5 + x)\, dx$

 21. In this exercise, we investigate the value of $\int_0^1 x^2\, dx$ using larger and larger values of *n* in the definition of the definite integral.

(a) First let $n = 10$, so $\Delta x = 0.1$. Fill a list on your calculator with values of x^2 as *x* goes from 0.1 to 1. (On a TI-84 Plus C, use the command `seq(X^2,X,.1,1,.1) → L1`.)

(b) Sum the values in the list formed in part (a), and multiply by 0.1, to estimate $\int_0^1 x^2\, dx$ with $n = 10$. (On a TI-84 Plus C, use the command `.1* sum(L1)`.)

(c) Repeat parts (a) and (b) with $n = 100$.

(d) Repeat parts (a) and (b) with $n = 500$.

(e) Based on your answers to parts (b) through (d), what do you estimate the value of $\int_0^1 x^2\, dx$ to be?

 22. Repeat Exercise 21 for $\int_0^1 x^3\, dx$.

 23. The booklet *All About Lawns*, published by Ortho Books, gives instructions for approximating the area of an irregularly shaped region. (See figure.) The booklet suggests measuring a long axis of the area, such as the vertical line in the figure. Then, every 10 feet along this axis, measure the width at right angles to the axis line. Sum these widths and multiply by 10. In the figure, for example, suppose the lengths of lines L_1, L_2 and L_3 are 34 ft, 48 ft, and 30 ft, respectively. Then

$$\text{Area} = (34 + 48 + 30)10 = 1120 \text{ square ft.}$$

How does this method relate to the discussion in this section? *Source: All About Lawns*.

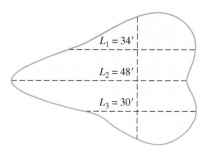

APPLICATIONS

In Exercises 24–30, estimate the area under each curve by summing the area of rectangles. Use the left endpoints, then the right endpoints, and then give the average of those answers.

Business and Economics

24. Electricity Consumption The following graph shows the rate of use of electrical energy (in millions of kilowatts) in a certain city on a very hot day. Estimate the total usage of electricity on that day. Let the width of each rectangle be 2 hours.

25. Wind Energy Consumption The following graph shows the U.S. wind energy consumption (trillion BTUs) for various years. Estimate the total consumption for the 12-year period from 2001 to 2013 using rectangles of width 3 years. *Source: Annual Energy Review*.

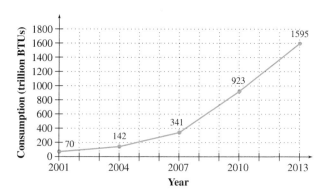

26. Revenue The total revenue for a product can be calculated as the area under the demand curve. Suppose that the demand curve for a certain wine (in dollars per liter) is given by

$$D(q) = \frac{1}{10}q^2 - 10q + 260$$

for $0 \le q \le 50$, where *q* is the demand in liters. Estimate the total revenue using rectangles of width 10 liters.

27. Revenue Repeat Exercise 26 using the demand curve for a fruit juice (in dollars per liter) given by

$$D(q) = \frac{2}{125}q^2 - \frac{4}{5}q + 15$$

for $0 \le q \le 25$, where q is the demand in liters. Estimate the total revenue using rectangles of width 5 liters.

Life Sciences

28. Oxygen Inhalation The graph below shows the rate of inhalation of oxygen (in liters per minute) by a person riding a bicycle very rapidly for 10 minutes. Estimate the total volume of oxygen inhaled in the first 20 minutes after the beginning of the ride. Use rectangles with widths of 1 minute.

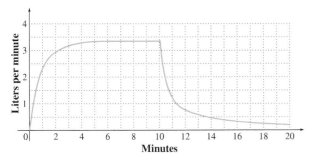

29. Foot-and-Mouth Epidemic In 2001, the United Kingdom suffered an epidemic of foot-and-mouth disease. The graph below shows the reported number of cattle (red) and pigs (blue) that were culled each month from mid-February through mid-October in an effort to stop the spread of the disease. *Source: Department of Environment, Food and Rural Affairs, United Kingdom.*

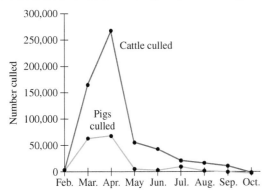

(a) Estimate the total number of cattle that were culled from mid-February through mid-October and compare this with 581,801, the actual number of cattle that were culled. Use rectangles that are one month in width, starting with mid-February.

(b) Estimate the total number of pigs that were culled from mid-February through mid-October and compare this with 146,145, the actual number of pigs that were culled. Use rectangles that are one month in width starting with mid-February.

Social Sciences

30. Automobile Accidents The graph at the top of the next column shows the number of fatal automobile accidents in California for various years. Estimate the total number of accidents in the 8-year period from 2003 to 2011 using rectangles of width 2 years. *Source: California Highway Patrol.*

Physical Sciences

Distance The next two graphs are from the Road & Track website. The curves show the velocity at t seconds after the car accelerates from a dead stop. To find the total distance traveled by the car in reaching 130 mph, we must estimate the definite integral

$$\int_0^T v(t)\,dt,$$

where T represents the number of seconds it takes for the car to reach 130 mph.

Use the graphs to estimate this distance by adding the areas of rectangles and using the midpoint rule. To adjust your answer to miles per hour, divide by 3600 (the number of seconds in an hour). You then have the number of miles that the car traveled in reaching 130 mph. Finally, multiply by 5280 ft per mile to convert the answer to feet. *Source: Road & Track.*

31. Estimate the distance traveled by the Lamborghini Gallardo LP560-4 using the graph below. Use rectangles with widths of 3 seconds, except for the last rectangle, which should have a width of 2 seconds. The circle marks the point where the car has gone a quarter mile. Does this seem correct?

32. Estimate the distance traveled by the Alfa Romeo 8C Competizione using the graph on the next page. Use rectangles with widths of 4 seconds, except for the last rectangle, which should have a width of 3.5 seconds. The circle marks the point where the car has gone a quarter mile. Does this seem correct?

Distance When data are given in tabular form, you may need to vary the size of the interval to calculate the area under the curve. The next two exercises include data from *Car and Driver* magazine. To estimate the total distance traveled by the car (in feet) during the time it took to reach its maximum velocity, estimate the area under the velocity versus time graph, as in the previous two exercises. Use the left endpoint for each time interval (the velocity at the beginning of that interval) and then the right endpoint (the velocity at the end of the interval). Finally, average the two answers together. Calculating and adding up the areas of the rectangles is most easily done on a spreadsheet or graphing calculator. As in the previous two exercises, you will need to multiply by a conversion factor of **5280/3600 = 22/15**, since the velocities are given in miles per hour, but the time is in seconds, and we want the answer in feet. *Source: Car and Driver.*

33. Using the table below, estimate the distance traveled by the Mercedes-Benz C63 AMG Edition 507.

Acceleration	Seconds
Zero to 30 mph	1.7
40 mph	2.3
50 mph	3.1
60 mph	3.9
70 mph	5.1
80 mph	6.3
90 mph	7.5
100 mph	9.2
110 mph	10.9
120 mph	12.9
130 mph	15.1
140 mph	18.2
150 mph	21.7
160 mph	27.0

34. Using the table below, estimate the distance traveled by the BMW M3.

Acceleration	Seconds
Zero to 30 mph	1.7
40 mph	2.3
50 mph	3.3
60 mph	4.1
70 mph	5.0
80 mph	6.5
90 mph	7.8
100 mph	9.2
110 mph	11.3
120 mph	13.3
130 mph	15.6
140 mph	18.4
150 mph	23.3
160 mph	27.8

Heat Gain The following graphs show the typical heat gain, in BTUs per hour per square foot, for windows (one with plain glass and one that is triple glazed) in Pittsburgh in June, one facing east and one facing south. The horizontal axis gives the time of the day. Estimate the total heat gain per square foot by summing the areas of rectangles. Use rectangles with widths of 2 hours, and let the function value at the midpoint of the subinterval give the height of the rectangle. *Source: Sustainable by Design.*

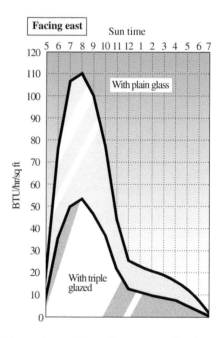

35. (a) Estimate the total heat gain per square foot for a plain glass window facing east.

(b) Estimate the total heat gain per square foot for a triple glazed window facing east.

36. **(a)** Estimate the total heat gain per square foot for a plain glass window facing south.

(b) Estimate the total heat gain per square foot for a triple glazed window facing south.

37. **Automobile Velocity** Two cars start from rest at a traffic light and accelerate for several minutes. The graph shows their velocities (in feet per second) as a function of time (in seconds). Car A is the one that initially has greater velocity. *Source: Stephen Monk.*

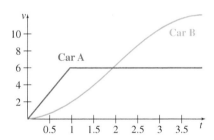

(a) How far has car A traveled after 2 seconds? (*Hint*: Use formulas from geometry.)

(b) When is car A farthest ahead of car B?

(c) Estimate the farthest that car A gets ahead of car B. For car A, use formulas from geometry. For car B, use $n = 4$ and the value of the function at the midpoint of each interval.

(d) Give a rough estimate of when car B catches up with car A.

38. **Distance** Musk the friendly pit bull has escaped again! Here is her velocity during the first 4 seconds of her romp.

t (sec)	0	1	2	3	4
v (ft/sec)	0	8	13	17	18

Give two estimates for the total distance Musk traveled during her 4-second trip, one using the left endpoint of each interval and one using the right endpoint.

39. **Distance** The speed of a particle in a test laboratory was noted every second for 3 seconds. The results are shown in the following table. Use the left endpoints and then the right endpoints to estimate the total distance the particle moved in the first three seconds.

t (sec)	0	1	2	3
v (ft/sec)	10	6.5	6	5.5

40. **Running** In 1987, Canadian Ben Johnson set a world record in the 100-m sprint. (The record was later taken away when he was found to have used an anabolic steroid to enhance his performance.) His speed at various times in the race is given in the following table.* *Source: Information Graphics.*

Time (sec)	Speed (mph)
0	0
1.84	12.9
3.80	23.8
6.38	26.3
7.23	26.3
8.96	26.0
9.83	25.7

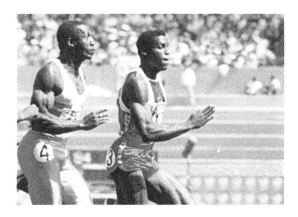

*The world record of 9.58 seconds is currently held by Usain Bolt of Jamaica.

(a) Use the information in the table and left endpoints to estimate the distance that Johnson ran in miles. You will first need to calculate Δt for each interval. At the end, you will need to divide by 3600 (the number of seconds in an hour), since the speed is in miles per hour.

(b) Repeat part (a), using right endpoints.

(c) Wait a minute; we know that the distance Johnson ran is 100 m. Divide this by 1609, the number of meters in a mile, to find how far Johnson ran in miles. Is your answer from part (a) or part (b) closer to the true answer? Briefly explain why you think this answer should be more accurate.

YOUR TURN ANSWERS

1. 48

2. Left endpoint estimate is 73 miles, right endpoint estimate is 97 miles, and average is 85 miles.

7.4 The Fundamental Theorem of Calculus

APPLY IT If we know how the rate of consumption of natural gas varies over time, how can we compute the total amount of natural gas used?
We will answer this question in Example 7.

In the first section of this chapter, you learned about antiderivatives. In the previous section, you learned about the definite integral. In this section, we connect these two separate topics and present one of the most powerful theorems of calculus.

We have seen that, if $f(x) \geq 0$,

$$\int_a^b f(x)\, dx$$

gives the area between the graph of $f(x)$ and the x-axis, from $x = a$ to $x = b$. In the previous section, the definite integral was defined and evaluated using the limit of a sum. In that section, we also saw that if $f(x)$ gives the rate of change of $F(x)$, the definite integral $\int_a^b f(x)\, dx$ gives the total change of $F(x)$ as x changes from a to b. If $f(x)$ gives the rate of change of $F(x)$, then $F(x)$ is an antiderivative of $f(x)$. Writing the total change in $F(x)$ from $x = a$ to $x = b$ as $F(b) - F(a)$ shows the connection between antiderivatives and definite integrals. This relationship is called the **Fundamental Theorem of Calculus**.

Fundamental Theorem of Calculus

Let f be continuous on the interval $[a, b]$, and let F be *any* antiderivative of f. Then

$$\int_a^b f(x)\, dx = F(b) - F(a) = F(x) \Big|_a^b.$$

The symbol $F(x)\big|_a^b$ is used to represent $F(b) - F(a)$. It is important to note that the Fundamental Theorem does not require $f(x) > 0$. The condition $f(x) > 0$ is necessary only when using the Fundamental Theorem to find area. Also, note that the Fundamental Theorem does not *define* the definite integral; it just provides a method for evaluating it.

EXAMPLE 1 **Fundamental Theorem of Calculus**

First find $\int 4t^3\, dt$ and then find $\int_1^2 4t^3\, dt$.

SOLUTION By the power rule given earlier, the indefinite integral is

$$\int 4t^3\, dt = t^4 + C.$$

By the Fundamental Theorem, the value of the definite integral $\int_1^2 4t^3\,dt$ is found by evaluating $t^4 \Big|_1^2$, with no constant C required.

YOUR TURN 1 Find $\int_1^3 3x^2\,dx.$

$$\int_1^2 4t^3\,dt = t^4 \Big|_1^2 = 2^4 - 1^4 = 15$$

TRY YOUR TURN 1

Example 1 illustrates the difference between the definite integral and the indefinite integral. A definite integral is a real number; an indefinite integral is a family of functions in which all the functions are antiderivatives of a function f.

NOTE No constant C is needed, as it is for the indefinite integral, because even if C were added to an antiderivative F, it would be eliminated in the final answer:

$$\int_a^b f(x)\,dx = (F(x) + C) \Big|_a^b$$
$$= (F(b) + C) - (F(a) + C)$$
$$= F(b) - F(a).$$

In other words, any antiderivative will give the same answer, so for simplicity, we choose the one with $C = 0$.

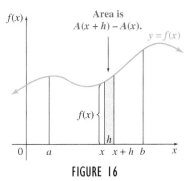

FIGURE 16

To see why the Fundamental Theorem of Calculus is true for $f(x) > 0$ when f is continuous, look at Figure 16. Define the function $A(x)$ as the area between the x-axis and the graph of $y = f(x)$ from a to x. We first show that A is an antiderivative of f; that is, $A'(x) = f(x)$.

To do this, let h be a small positive number. Then $A(x + h) - A(x)$ is the shaded area in Figure 16. This area can be approximated with a rectangle having width h and height $f(x)$. The area of the rectangle is $h \cdot f(x)$, and

$$A(x + h) - A(x) \approx h \cdot f(x).$$

Dividing both sides by h gives

$$\frac{A(x + h) - A(x)}{h} \approx f(x).$$

This approximation improves as h gets smaller and smaller. Taking the limit on the left as h approaches 0 gives an exact result.

$$\lim_{h \to 0} \frac{A(x + h) - A(x)}{h} = f(x)$$

This limit is simply $A'(x)$, so

$$A'(x) = f(x).$$

This result means that A is an antiderivative of f, as we set out to show.

$A(b)$ is the area under the curve from a to b, and $A(a) = 0$, so the area under the curve can be written as $A(b) - A(a)$. From the previous section, we know that the area under the curve is also given by $\int_a^b f(x)\,dx$. Putting these two results together gives

$$\int_a^b f(x)\,dx = A(b) - A(a)$$
$$= A(x) \Big|_a^b$$

where A is an antiderivative of f. From the note after Example 1, we know that any antiderivative will give the same answer, which proves the Fundamental Theorem of Calculus.

The Fundamental Theorem of Calculus certainly deserves its name, which sets it apart as the most important theorem of calculus. It is the key connection between differential calculus and integral calculus, which originally were developed separately without knowledge of this connection between them.

The variable used in the integrand does not matter; each of the following definite integrals represents the number $F(b) - F(a)$.

$$\int_a^b f(x)\,dx = \int_a^b f(t)\,dt = \int_a^b f(u)\,du$$

Key properties of definite integrals are listed below. Some of them are just restatements of properties from Section 1.

Properties of Definite Integrals

If all indicated definite integrals exist,

1. $\displaystyle\int_a^b k \cdot f(x)\,dx = k \cdot \int_a^b f(x)\,dx$ for any real constant k
(constant multiple of a function);

2. $\displaystyle\int_a^b [f(x) \pm g(x)]\,dx = \int_a^b f(x)\,dx \pm \int_a^b g(x)\,dx$
(sum or difference of functions);

3. $\displaystyle\int_a^a f(x)\,dx = 0;$

4. $\displaystyle\int_a^b f(x)\,dx = \int_a^c f(x)\,dx + \int_c^b f(x)\,dx$ for any real number c;

5. $\displaystyle\int_a^b f(x)\,dx = -\int_b^a f(x)\,dx.$

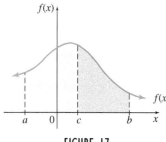

FIGURE 17

For $f(x) \geq 0$, since the distance from a to a is 0, the third property says that the "area" under the graph of f bounded by $x = a$ and $x = a$ is 0. Also, since $\int_a^c f(x)\,dx$ represents the blue region in Figure 17 and $\int_c^b f(x)\,dx$ represents the pink region,

$$\int_a^b f(x)\,dx = \int_a^c f(x)\,dx + \int_c^b f(x)\,dx,$$

as stated in the fourth property. While the figure shows $a < c < b$, the property is true for any value of c where both $f(x)$ and $F(x)$ are defined.

An algebraic proof is given here for the second property; proofs of the other properties are left for the exercises. If $F(x)$ and $G(x)$ are antiderivatives of $f(x)$ and $g(x)$, respectively,

$$\int_a^b [f(x) + g(x)]\,dx = [F(x) + G(x)]\Big|_a^b$$
$$= [F(b) + G(b)] - [F(a) + G(a)]$$
$$= [F(b) - F(a)] + [G(b) - G(a)]$$
$$= \int_a^b f(x)\,dx + \int_a^b g(x)\,dx.$$

EXAMPLE 2 Fundamental Theorem of Calculus

Find $\int_2^5 (6x^2 - 3x + 5)\, dx$.

SOLUTION Use the properties above and the Fundamental Theorem, along with properties from Section 1.

$$\int_2^5 (6x^2 - 3x + 5)\, dx = 6\int_2^5 x^2\, dx - 3\int_2^5 x\, dx + 5\int_2^5 dx$$

$$= 2x^3 \Big|_2^5 - \frac{3}{2}x^2 \Big|_2^5 + 5x \Big|_2^5$$

$$= 2(5^3 - 2^3) - \frac{3}{2}(5^2 - 2^2) + 5(5 - 2)$$

$$= 2(125 - 8) - \frac{3}{2}(25 - 4) + 5(3)$$

$$= 234 - \frac{63}{2} + 15 = \frac{435}{2}$$

TRY YOUR TURN 2

YOUR TURN 2

Find $\int_3^5 (2x^3 - 3x + 4)\, dx$.

EXAMPLE 3 Fundamental Theorem of Calculus

$$\int_1^2 \frac{dy}{y} = \ln |y| \Big|_1^2 = \ln |2| - \ln |1|$$

$$= \ln 2 - \ln 1 \approx 0.6931 - 0 = 0.6931$$

TRY YOUR TURN 3

YOUR TURN 3

Find $\int_1^3 \frac{2}{y}\, dy$.

EXAMPLE 4 Substitution

Evaluate $\int_0^5 x\sqrt{25 - x^2}\, dx$.

SOLUTION

Method 1
Changing the Limits

Use substitution. Let $u = 25 - x^2$, so that $du = -2x\, dx$. With a definite integral, the limits should be changed, too. The new limits on u are found as follows.

$$\text{If } x = 5, \text{ then } u = 25 - 5^2 = 0.$$
$$\text{If } x = 0, \text{ then } u = 25 - 0^2 = 25.$$

Then

$$\int_0^5 x\sqrt{25 - x^2}\, dx = -\frac{1}{2}\int_0^5 \sqrt{25 - x^2}\,(-2x\, dx) \qquad \text{Multiply by } -2/-2.$$

$$= -\frac{1}{2}\int_{25}^0 \sqrt{u}\, du \qquad \text{Substitute and change limits.}$$

$$= -\frac{1}{2}\int_{25}^0 u^{1/2}\, du \qquad \sqrt{u} = u^{1/2}$$

$$= -\frac{1}{2}\cdot\frac{u^{3/2}}{3/2} \Big|_{25}^0 \qquad \text{Use the power rule.}$$

$$= -\frac{1}{2}\cdot\frac{2}{3}[0^{3/2} - 25^{3/2}]$$

$$= -\frac{1}{3}(-125) = \frac{125}{3}.$$

Method 2
Evaluating the Antiderivative

An alternative method that some people prefer is to evaluate the antiderivative first and then calculate the definite integral. To evaluate the antiderivative in this example, ignore the limits on the original integral and use the substitution $u = 25 - x^2$, so that $du = -2x\,dx$. Then

$$\int x\sqrt{25 - x^2}\,dx = -\frac{1}{2}\int \sqrt{25 - x^2}(-2x\,dx)$$

$$= -\frac{1}{2}\int \sqrt{u}\,du$$

$$= -\frac{1}{2}\int u^{1/2}\,du$$

$$= -\frac{1}{2}\frac{u^{3/2}}{3/2} + C$$

$$= -\frac{u^{3/2}}{3} + C$$

$$= -\frac{(25 - x^2)^{3/2}}{3} + C.$$

We will ignore the constant C because it doesn't affect the answer, as we mentioned in the Note following Example 1.

Then, using the Fundamental Theorem of Calculus, we have

$$\int_0^5 x\sqrt{25 - x^2}\,dx = -\frac{(25 - x^2)^{3/2}}{3}\bigg|_0^5$$

$$= 0 - \left[-\frac{(25)^{3/2}}{3}\right]$$

$$= \frac{125}{3}.$$

YOUR TURN 4 Evaluate $\int_0^4 2x\sqrt{16 - x^2}\,dx$.

TRY YOUR TURN 4

CAUTION Don't confuse these two methods. In Method 1, we never return to the original variable or the original limits of integration. In Method 2, it is essential to return to the original variable and to not change the limits. When using Method 1, we recommend labeling the limits with the appropriate variable to avoid confusion, so the substitution in Example 4 becomes

$$\int_{x=0}^{x=5} x\sqrt{25 - x^2}\,dx = -\frac{1}{2}\int_{u=25}^{u=0} \sqrt{u}\,du.$$

The Fundamental Theorem of Calculus is a powerful tool, but it has a limitation. The problem is that not every function has an antiderivative in terms of the functions and operations you have seen so far. One example of an integral that cannot be evaluated by the Fundamental Theorem of Calculus for this reason is

$$\int_a^b e^{-x^2/2}\,dx,$$

yet this integral is crucial in probability and statistics. Such integrals may be evaluated by numerical integration, which is covered in the last section of this chapter. Fortunately for you, all the integrals in this section can be antidifferentiated using the techniques presented in the first two sections of this chapter.

Area
In the previous section we saw that, if $f(x) \geq 0$ in $[a, b]$, the definite integral $\int_a^b f(x)\,dx$ gives the area below the graph of the function $y = f(x)$, above the x-axis, and between the lines $x = a$ and $x = b$.

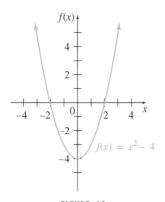

FIGURE 18

To see how to work around the requirement that $f(x) \geq 0$, look at the graph of $f(x) = x^2 - 4$ in Figure 18. The area bounded by the graph of f, the x-axis, and the vertical lines $x = 0$ and $x = 2$ lies below the x-axis. Using the Fundamental Theorem gives

$$\int_0^2 (x^2 - 4) \, dx = \left(\frac{x^3}{3} - 4x\right)\Big|_0^2$$

$$= \left(\frac{8}{3} - 8\right) - (0 - 0) = -\frac{16}{3}.$$

The result is a negative number because $f(x)$ is negative for values of x in the interval $[0, 2]$. Since Δx is always positive, if $f(x) < 0$ the product $f(x) \cdot \Delta x$ is negative, so $\int_0^2 f(x) \, dx$ is negative. Since area is nonnegative, the required area is given by $\left|-16/3\right|$ or $16/3$. Using a definite integral, the area could be written as

$$\left|\int_0^2 (x^2 - 4) \, dx\right| = \left|-\frac{16}{3}\right| = \frac{16}{3}.$$

EXAMPLE 5   Area

Find the area of the region between the x-axis and the graph of $f(x) = x^2 - 3x$ from $x = 1$ to $x = 3$.

SOLUTION   The region is shown in Figure 19. Since the region lies below the x-axis, the area is given by

$$\left|\int_1^3 (x^2 - 3x) \, dx\right|.$$

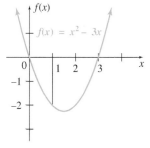

FIGURE 19

By the Fundamental Theorem,

$$\int_1^3 (x^2 - 3x) \, dx = \left(\frac{x^3}{3} - \frac{3x^2}{2}\right)\Big|_1^3 = \left(\frac{27}{3} - \frac{27}{2}\right) - \left(\frac{1}{3} - \frac{3}{2}\right) = -\frac{10}{3}.$$

The required area is $\left|-10/3\right| = 10/3$.

EXAMPLE 6   Area

Find the area between the x-axis and the graph of $f(x) = x^2 - 4$ from $x = 0$ to $x = 4$.

SOLUTION   Figure 20 shows the required region. Part of the region is below the x-axis. The definite integral over that interval will have a negative value. To find the area, integrate the negative and positive portions separately and take the absolute value of the first result before combining the two results to get the total area. Start by finding the point where the graph crosses the x-axis. This is done by solving the equation

$$x^2 - 4 = 0.$$

The solutions of this equation are 2 and -2. The only solution in the interval $[0, 4]$ is 2. The total area of the region in Figure 20 is

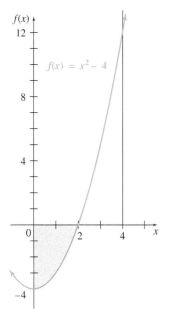

FIGURE 20

YOUR TURN 5   Repeat Example 6 for the function $f(x) = x^2 - 9$ from $x = 0$ to $x = 6$.

$$\left|\int_0^2 (x^2 - 4) \, dx\right| + \int_2^4 (x^2 - 4) \, dx = \left|\left(\frac{1}{3}x^3 - 4x\right)\Big|_0^2\right| + \left(\frac{1}{3}x^3 - 4x\right)\Big|_2^4$$

$$= \left|\frac{8}{3} - 8\right| + \left(\frac{64}{3} - 16\right) - \left(\frac{8}{3} - 8\right)$$

$$= 16.$$

TRY YOUR TURN 5

Incorrectly using one integral over the entire interval to find the area in Example 6 would have given

$$\int_0^4 (x^2 - 4)\, dx = \left(\frac{x^3}{3} - 4x\right)\Big|_0^4 = \left(\frac{64}{3} - 16\right) - 0 = \frac{16}{3},$$

which is not the correct area. This definite integral does not represent any area but is just a real number.

For instance, if $f(x)$ in Example 6 represents the annual rate of profit of a company, then 16/3 represents the total profit for the company over a 4-year period. The integral between 0 and 2 is $-16/3$; the negative sign indicates a loss for the first two years. The integral between 2 and 4 is 32/3, indicating a profit. The overall profit is $32/3 - 16/3 = 16/3$, although the total shaded area is $32/3 + |-16/3| = 16$.

Finding Area

In summary, to find the area bounded by $f(x)$, $x = a$, $x = b$, and the x-axis, use the following steps.

1. Sketch a graph.
2. Find any x-intercepts of $f(x)$ in $[a, b]$. These divide the total region into subregions.
3. The definite integral will be *positive* for subregions above the x-axis and *negative* for subregions below the x-axis. Use separate integrals to find the (positive) areas of the subregions.
4. The total area is the sum of the areas of all of the subregions.

In the last section, we saw that the area under a rate of change function $f'(x)$ from $x = a$ to $x = b$ gives the total value of $f(x)$ on $[a, b]$. Now we can use the definite integral to solve these problems.

EXAMPLE 7 Natural Gas Consumption

The yearly rate of consumption of natural gas (in trillions of cubic feet) for a certain city is

$$C'(t) = t + e^{0.01t},$$

where t is time in years and $t = 0$ corresponds to 2000. At this consumption rate, what was the total amount the city used in the 10-year period of 2005 to 2015?

APPLY IT **SOLUTION**   To find the consumption over the 10-year period from 2005 to 2015, use the definite integral.

$$\int_5^{15} (t + e^{0.01t})\, dt = \left(\frac{t^2}{2} + \frac{e^{0.01t}}{0.01}\right)\Big|_5^{15}$$
$$= (112.5 + 100e^{0.15}) - (12.5 + 100e^{0.05})$$
$$\approx 111$$

Therefore, a total of about 111 trillion ft³ of natural gas was used from 2005 to 2015 at this consumption rate.

Evaluate each of the following integrals.   *(Sec. 7.2)*

W1. $\displaystyle\int 6x\sqrt{x^2 + 4}\, dx$

W2. $\displaystyle\int \frac{4x}{x^2 + 2}\, dx$

W3. $\displaystyle\int 15x^2 e^{x^3}\, dx$

7.4   EXERCISES

Evaluate each definite integral.

1. $\displaystyle\int_{-2}^{4} (-3)\, dp$

2. $\displaystyle\int_{-4}^{1} \sqrt{2}\, dx$

3. $\displaystyle\int_{-1}^{2} (5t - 3)\, dt$

4. $\displaystyle\int_{-2}^{2} (4z + 3)\, dz$

5. $\displaystyle\int_{0}^{2} (5x^2 - 4x + 2)\, dx$

6. $\displaystyle\int_{-2}^{3} (-x^2 - 3x + 5)\, dx$

7. $\displaystyle\int_{0}^{2} 3\sqrt{4u + 1}\, du$

8. $\displaystyle\int_{3}^{9} \sqrt{2r - 2}\, dr$

9. $\displaystyle\int_{0}^{4} 2(t^{1/2} - t)\, dt$

10. $\displaystyle\int_{0}^{4} -(3x^{3/2} + x^{1/2})\, dx$

11. $\displaystyle\int_{1}^{4} (5y\sqrt{y} + 3\sqrt{y})\, dy$

12. $\displaystyle\int_{4}^{9} (4\sqrt{r} - 3r\sqrt{r})\, dr$

13. $\displaystyle\int_{4}^{6} \frac{2}{(2x - 7)^2}\, dx$

14. $\displaystyle\int_{1}^{4} \frac{-3}{(2p + 1)^2}\, dp$

15. $\displaystyle\int_{1}^{5} (6n^{-2} - n^{-3})\, dn$

16. $\displaystyle\int_{2}^{3} (3x^{-3} - 5x^{-4})\, dx$

17. $\displaystyle\int_{-3}^{-2} \left(2e^{-0.1y} + \frac{3}{y}\right) dy$

18. $\displaystyle\int_{-2}^{-1} \left(\frac{-2}{t} + 3e^{0.3t}\right) dt$

19. $\displaystyle\int_{1}^{2} \left(e^{4u} - \frac{1}{(u + 1)^2}\right) du$

20. $\displaystyle\int_{0.5}^{1} (p^3 - e^{4p})\, dp$

21. $\displaystyle\int_{-1}^{0} y(2y^2 - 3)^5\, dy$

22. $\displaystyle\int_{0}^{3} m^2(4m^3 + 2)^3\, dm$

23. $\displaystyle\int_{1}^{64} \frac{\sqrt{z} - 2}{\sqrt[3]{z}}\, dz$

24. $\displaystyle\int_{1}^{8} \frac{3 - y^{1/3}}{y^{2/3}}\, dy$

25. $\displaystyle\int_{1}^{2} \frac{\ln x}{x}\, dx$

26. $\displaystyle\int_{1}^{3} \frac{\sqrt{\ln x}}{x}\, dx$

27. $\displaystyle\int_{0}^{8} x^{1/3}\sqrt{x^{4/3} + 9}\, dx$

28. $\displaystyle\int_{1}^{2} \frac{3}{x(1 + \ln x)}\, dx$

29. $\displaystyle\int_{0}^{1} \frac{e^{2t}}{(3 + e^{2t})^2}\, dt$

30. $\displaystyle\int_{0}^{1} \frac{e^{2z}}{\sqrt{1 + e^{2z}}}\, dz$

In Exercises 31–40, use the definite integral to find the area between the x-axis and $f(x)$ over the indicated interval. Check first to see if the graph crosses the x-axis in the given interval.

31. $f(x) = 2x - 14;\quad [6, 10]$

32. $f(x) = 4x - 32;\quad [5, 10]$

33. $f(x) = 2 - 2x^2;\quad [0, 5]$

34. $f(x) = 9 - x^2;\quad [0, 6]$

35. $f(x) = x^3;\quad [-1, 3]$

36. $f(x) = x^3 - 2x;\quad [-2, 4]$

37. $f(x) = e^x - 1;\quad [-1, 2]$

38. $f(x) = 1 - e^{-x};\quad [-1, 2]$

39. $f(x) = \dfrac{1}{x} - \dfrac{1}{e};\quad [1, e^2]$

40. $f(x) = 1 - \dfrac{1}{x};\quad [e^{-1}, e]$

Find the area of each shaded region.

41.

42.

43.

44.

45. Assume $f(x)$ is continuous for $g \le x \le c$ as shown in the figure. Write an equation relating the three quantities

$$\int_a^b f(x)\,dx, \qquad \int_a^c f(x)\,dx, \qquad \int_b^c f(x)\,dx.$$

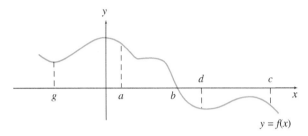

$y = f(x)$

46. Is the equation you wrote for Exercise 45 still true

(a) if b is replaced by d?

(b) if b is replaced by g?

47. The graph of $f(x)$, shown here, consists of two straight line segments and two quarter circles. Find the value of $\int_0^{16} f(x)\,dx$.

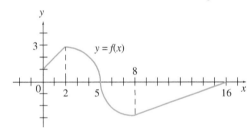

Use the Fundamental Theorem to show that the following are true.

48. $\displaystyle\int_a^b kf(x)\,dx = k\int_a^b f(x)\,dx$

49. $\displaystyle\int_a^b f(x)\,dx = \int_a^c f(x)\,dx + \int_c^b f(x)\,dx$

50. $\displaystyle\int_a^b f(x)\,dx = -\int_b^a f(x)\,dx$

51. Use Exercise 49 to find $\int_{-1}^4 f(x)\,dx$, given

$$f(x) = \begin{cases} 2x + 3 & \text{if } x \le 0 \\ -\dfrac{x}{4} - 3 & \text{if } x > 0. \end{cases}$$

52. You are given $\int_0^1 e^{x^2}\,dx = 1.46265$ and $\int_0^2 e^{x^2}\,dx = 16.45263$. Use this information to find

(a) $\displaystyle\int_{-1}^1 e^{x^2}\,dx;$ (b) $\displaystyle\int_1^2 e^{x^2}\,dx.$

53. Let $g(t) = t^4$ and define $f(x) = \int_c^x g(t)\,dt$ with $c = 1$.

(a) Find a formula for $f(x)$.

(b) Verify that $f'(x) = g(x)$. The fact that

$$\frac{d}{dx}\int_c^x g(t)\,dt = g(x)$$

is true for all continuous functions g is an alternative version of the Fundamental Theorem of Calculus.

(c) Let us verify the result in part (b) for a function whose antiderivative cannot be found. Let $g(t) = e^{t^2}$ and let $c = 0$. Use the integration feature on a graphing calculator to find $f(x)$ for $x = 1$ and $x = 1.01$. Then use the definition of the derivative with $h = 0.01$ to approximate $f'(1)$, and compare it with $g(1)$.

54. Consider the function $f(x) = x(x^2 + 3)^7$.

(a) Use the Fundamental Theorem of Calculus to evaluate

$$\int_{-5}^5 f(x)\,dx.$$

(b) Use symmetry to describe how the integral from part (a) could be evaluated without using substitution or finding an antiderivative.

APPLICATIONS

Business and Economics

55. Profit   Karla Harby Communications, a small company of science writers, found that its rate of profit (in thousands of dollars) after t years of operation is given by

$$P'(t) = (3t + 3)(t^2 + 2t + 2)^{1/3}.$$

(a) Find the total profit in the first three years.

(b) Find the profit in the fourth year of operation.

(c) What is happening to the annual profit over the long run?

56. Worker Efficiency   A worker new to a job will improve his efficiency with time so that it takes him fewer hours to produce an item with each day on the job, up to a certain point. Suppose the rate of change of the number of hours it takes a worker in a certain factory to produce the xth item is given by

$$H'(x) = 20 - 2x.$$

(a) What is the total number of hours required to produce the first 5 items?

(b) What is the total number of hours required to produce the first 10 items?

Life Sciences

57. Pollution   Pollution from a factory is entering a lake. The rate of concentration of the pollutant at time t is given by

$$P'(t) = 140t^{5/2},$$

where t is the number of years since the factory started introducing pollutants into the lake. Ecologists estimate that the lake can accept a total level of pollution of 4850 units before all the fish life in the lake ends. Can the factory operate for 4 years without killing all the fish in the lake?

58. Spread of an Oil Leak   An oil tanker is leaking oil at the rate given (in barrels per hour) by

$$L'(t) = \frac{80 \ln (t + 1)}{t + 1},$$

where t is the time (in hours) after the tanker hits a hidden rock (when $t = 0$).

(a) Find the total number of barrels that the ship will leak on the first day.

(b) Find the total number of barrels that the ship will leak on the second day.

(c) What is happening over the long run to the amount of oil leaked per day?

59. Tree Growth After long study, tree scientists conclude that a eucalyptus tree will grow at the rate of $0.6 + 4/(t + 1)^3$ ft per year, where t is time (in years).

(a) Find the number of feet that the tree will grow in the second year.

(b) Find the number of feet the tree will grow in the third year.

60. Growth of a Substance The rate at which a substance grows is given by

$$R'(x) = 150e^{0.2x},$$

where x is the time (in days). What is the total accumulated growth during the first 3.5 days?

61. Drug Reaction For a certain drug, the rate of reaction in appropriate units is given by

$$R'(t) = \frac{5}{t + 1} + \frac{2}{\sqrt{t + 1}},$$

where t is time (in hours) after the drug is administered. Find the total reaction to the drug over the following time periods.

(a) From $t = 1$ to $t = 12$

(b) From $t = 12$ to $t = 24$

62. Human Mortality If $f(x)$ is the instantaneous death rate for members of a population at time x, then the number of individuals who survive to age T is given by

$$F(T) = \int_0^T f(x)\, dx.$$

In 1825 the biologist Benjamin Gompertz proposed that $f(x) = kb^x$. Find a formula for $F(T)$. *Source: Philosophical Transactions of the Royal Society of London.*

63. Cell Division Let the expected number of cells in a culture that have an x percent probability of undergoing cell division during the next hour be denoted by $n(x)$.

(a) Explain why $\int_{20}^{30} n(x)\, dx$ approximates the total number of cells with a 20% to 30% chance of dividing during the next hour.

(b) Give an integral representing the number of cells that have less than a 60% chance of dividing during the next hour.

(c) Let $n(x) = \sqrt{5x + 1}$ give the expected number of cells (in millions) with x percent probability of dividing during the next hour. Find the number of cells with a 5 to 10% chance of dividing.

64. Bacterial Growth A population of *E. coli* bacteria will grow at a rate given by

$$w'(t) = (3t + 2)^{1/3},$$

where w is the weight (in milligrams) after t hours. Find the change in weight of the population from $t = 0$ to $t = 3$.

65. Blood Flow In an example from an earlier chapter, the velocity v of the blood in a blood vessel was given as

$$v = k(R^2 - r^2),$$

where R is the (constant) radius of the blood vessel, r is the distance of the flowing blood from the center of the blood vessel, and k is a constant. Total blood flow (in millimeters per minute) is given by

$$Q(R) = \int_0^R 2\pi vr\, dr.$$

(a) Find the general formula for Q in terms of R by evaluating the definite integral given above.

(b) Evaluate $Q(0.4)$.

66. Rams' Horns The average annual increment in the horn length (in centimeters) of bighorn rams born in recent years can be approximated by

$$y = 0.1762x^2 - 3.986x + 22.68,$$

where x is the ram's age (in years) for x between 3 and 9. Integrate to find the total increase in the length of a ram's horn during this time. *Source: Journal of Wildlife Management.*

67. Beagles The daily energy requirements of female beagles who are at least 1 year old change with respect to time according to the function

$$E(t) = 753t^{-0.1321},$$

where $E(t)$ is the daily energy requirement (in kJ/$W^{0.67}$), where W is the dog's weight (in kilograms) for a beagle that is t years old. *Source: Journal of Nutrition.*

(a) Assuming 365 days in a year, show that the energy requirement for a female beagle that is t days old is given by

$$E(t) = 1642t^{-0.1321}.$$

(b) Using the formula from part (a), determine the total energy requirements (in kJ/$W^{0.67}$) for a female beagle between her first and third birthday.

68. Sediment The density of sediment (in grams per cubic centimeter) at the bottom of Lake Coeur d'Alene, Idaho, is given by

$$p(x) = p_0 e^{0.0133x},$$

where x is the depth (in centimeters) and p_0 is the density at the surface. The total mass of a square-centimeter column of sediment above a depth of h cm is given by

$$\int_0^h p(x)\, dx.$$

If $p_0 = 0.85$ g per cm^3, find the total mass above a depth of 100 cm. *Source: The Mathematics Teacher.*

Social Sciences

69. Age Distribution The U.S. Census Bureau gives an age distribution that is approximately modeled (in millions) by the function

$$f(x) = 40.2 + 3.50x - 0.897x^2$$

where x varies from 0 to 9 decades. The population of a given age group can be found by integrating this function over the interval for that age group. *Source: Ralph DeMarr and the U.S. Census Bureau.*

(a) Find the integral of $f(x)$ over the interval $[0, 9]$. What does this integral represent?

(b) Baby boomers are those born between 1945 and 1965, that is, those in the range of 4.5 to 6.5 decades in 2010. Estimate the number of baby boomers.

70. Income Distribution Based on data from the U.S. Census Bureau, an approximate family income distribution for the United States is given by the function

$$f(x) = 0.00826x^3 - 0.211x^2 + 0.740x + 9.36,$$

where x is the annual income in units of $10,000, with $0 \leq x \leq 10$. For example, $x = 0.5$ represents an annual family income of $5000. (*Note:* This function does not give a good representation for family incomes over $100,000.) The percent of the families with an income in a given range can be found by integrating this function over that range. Find the percentage of families with an income between $35,000 and $60,000. *Source: Ralph DeMarr and the U.S. Census Bureau.*

Physical Sciences

71. Oil Consumption Suppose that the rate of consumption of a natural resource is $c'(t)$, where

$$c'(t) = ke^{rt}.$$

Here t is time in years, r is a constant, and k is the consumption in the year when $t = 0$. In 2010, an oil company sold 1.2 billion barrels of oil. Assume that $r = 0.04$.

(a) Write $c'(t)$ for the oil company, letting $t = 0$ represent 2010.

(b) Set up a definite integral for the amount of oil that the company will sell in the next 10 years.

(c) Evaluate the definite integral of part (b).

(d) The company has about 20 billion barrels of oil in reserve. To find the number of years that this amount will last, solve the equation

$$\int_0^T 1.2e^{0.04t}dt = 20.$$

(e) Rework part (d), assuming that $r = 0.02$.

72. Oil Consumption The rate of consumption of oil (in billions of barrels) by the company in Exercise 71 was given as

$$1.2e^{0.04t},$$

where $t = 0$ corresponds to 2010. Find the total amount of oil used by the company from 2010 to year T. At this rate, how much will be used in 5 years?

YOUR TURN ANSWERS

1. 26	**2.** 256	**3.** 2 ln 3 or ln 9
4. 128/3	**5.** 54	

7.5 The Area Between Two Curves

We will answer this question in *Example 4.*

Many important applications of integrals require finding the area between two graphs. The method used in previous sections to find the area between the graph of a function and the x-axis from $x = a$ to $x = b$ can be generalized to find such an area. For example, the area between the graphs of $f(x)$ and $g(x)$ from $x = a$ to $x = b$ in Figure 21(a) is the same as the area under the graph of $f(x)$, shown in Figure 21(b), minus the area under the graph of $g(x)$ (see Figure 21(c)). That is, the area between the graphs is given by

$$\int_a^b f(x)\, dx - \int_a^b g(x)\, dx,$$

(a)

(b)

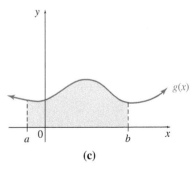

(c)

FIGURE 21

which can be written as

$$\int_a^b [f(x) - g(x)]\,dx.$$

Area Between Two Curves

If f and g are continuous functions and $f(x) \geq g(x)$ on $[a, b]$, then the area between the curves $f(x)$ and $g(x)$ from $x = a$ to $x = b$ is given by

$$\int_a^b [f(x) - g(x)]\,dx.$$

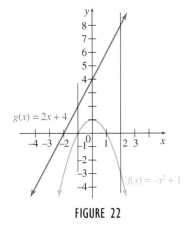

FIGURE 22

EXAMPLE 1   Area

Find the area bounded by $f(x) = -x^2 + 1$, $g(x) = 2x + 4$, $x = -1$, and $x = 2$.

SOLUTION   A sketch of the four equations is shown in Figure 22. In general, it is not necessary to spend time drawing a detailed sketch, but only to know whether the two functions intersect, and which function is greater between the intersections. To find out, set the two functions equal.

$$-x^2 + 1 = 2x + 4$$
$$0 = x^2 + 2x + 3$$

Verify by the quadratic formula that this equation has no real roots. Since the graph of f is a parabola opening downward that does not cross the graph of g (a line), the parabola must be entirely under the line, as shown in Figure 22. Therefore $g(x) \geq f(x)$ for x in the interval $[-1, 2]$, and the area is given by

$$\int_{-1}^2 [g(x) - f(x)]\,dx = \int_{-1}^2 [(2x + 4) - (-x^2 + 1)]\,dx$$

$$= \int_{-1}^2 (2x + 4 + x^2 - 1)\,dx$$

$$= \int_{-1}^2 (x^2 + 2x + 3)\,dx$$

$$= \frac{x^3}{3} + x^2 + 3x \Big|_{-1}^2$$

$$= \left(\frac{8}{3} + 4 + 6\right) - \left(\frac{-1}{3} + 1 - 3\right)$$

$$= \frac{8}{3} + 10 + \frac{1}{3} + 2$$

$$= 15.$$

YOUR TURN 1 Repeat Example 1 for $f(x) = 4 - x^2$, $g(x) = x + 2$, $x = -2$, and $x = 1$.

TRY YOUR TURN 1

NOTE   It is not necessary to draw the graphs to determine which function is greater. Since the functions in the previous example do not intersect, we can evaluate them at *any* point to make this determination. For example, $f(0) = 1$ and $g(0) = 4$. Because $g(x) > f(x)$ at $x = 4$, and the two functions are continuous and never intersect, $g(x) > f(x)$ for all x.

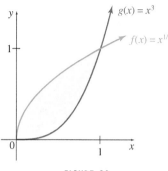

FIGURE 23

YOUR TURN 2 Repeat
Example 2 for $y = x^{1/4}$ and $y = x^2$.

EXAMPLE 2 Area

Find the area between the curves $y = x^{1/2}$ and $y = x^3$.

SOLUTION Let $f(x) = x^{1/2}$ and $g(x) = x^3$. As before, set the two equal to find where they intersect.

$$x^{1/2} = x^3$$
$$0 = x^3 - x^{1/2}$$
$$0 = x^{1/2}(x^{5/2} - 1)$$

The only solutions are $x = 0$ and $x = 1$. Verify that the graph of f is concave downward, while the graph of g is concave upward, so the graph of f must be greater between 0 and 1. (This may also be verified by taking a point between 0 and 1, such as 0.5, and verifying that $0.5^{1/2} > 0.5^3$.) The graph is shown in Figure 23.

The area between the two curves is given by

$$\int_a^b [f(x) - g(x)]\, dx = \int_0^1 (x^{1/2} - x^3)\, dx.$$

Using the Fundamental Theorem,

$$\int_0^1 (x^{1/2} - x^3)\, dx = \left(\frac{x^{3/2}}{3/2} - \frac{x^4}{4}\right)\Bigg|_0^1$$

$$= \left(\frac{2}{3}x^{3/2} - \frac{x^4}{4}\right)\Bigg|_0^1$$

$$= \frac{2}{3}(1) - \frac{1}{4}$$

$$= \frac{5}{12}.$$ **TRY YOUR TURN 2**

TECHNOLOGY NOTE

A graphing calculator is very useful in approximating solutions of problems involving the area between two curves. First, it can be used to graph the functions and identify any intersection points. Then it can be used to approximate the definite integral that represents the area. (A function that gives a numerical approximation to the integral is located in the MATH menu of a TI-84 Plus C calculator.) Figure 24 shows the results of using these steps for Example 2. The second window shows that the area closely approximates 5/12.

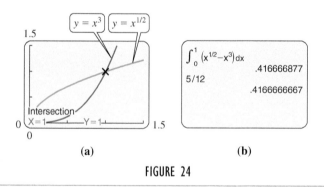

(a)   **(b)**

FIGURE 24

The difference between two integrals can be used to find the area between the graphs of two functions even if one graph lies below the x-axis. In fact, if $f(x) \geq g(x)$ for all values of x in the interval $[a, b]$, then the area between the two graphs is always given by

$$\int_a^b [f(x) - g(x)]\, dx.$$

To see this, look at the graphs in Figure 25(a), where $f(x) \geq g(x)$ for x in $[a, b]$. Suppose a constant C is added to both functions, with C large enough so that both graphs lie above the x-axis, as in Figure 25(b). The region between the graphs is not changed. By the work above, this area is given by $\int_a^b [f(x) - g(x)] \, dx$ regardless of where the graphs of $f(x)$ and $g(x)$ are located. As long as $f(x) \geq g(x)$ on $[a, b]$, then the area between the graphs from $x = a$ to $x = b$ will equal $\int_a^b [f(x) - g(x)] \, dx$.

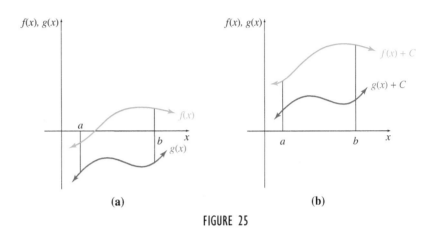

FIGURE 25

EXAMPLE 3 Area

Find the area of the region enclosed by $y = x^2 - 2x$ and $y = x$ on $[0, 4]$.

SOLUTION Verify that the two graphs cross at $x = 0$ and $x = 3$. Because the first graph is a parabola opening upward, the parabola must be below the line between 0 and 3 and above the line between 3 and 4. See Figure 26. (The greater function could also be identified by checking a point between 0 and 3, such as 1, and a point between 3 and 4, such as 3.5. For each of these values of x, we could calculate the corresponding value of y for the two functions and see which is greater.) Because the graphs cross at $x = 3$, the area is found by taking the sum of two integrals as follows.

$$\text{Area} = \int_0^3 [x - (x^2 - 2x)] \, dx + \int_3^4 [(x^2 - 2x) - x] \, dx$$

$$= \int_0^3 (-x^2 + 3x) \, dx + \int_3^4 (x^2 - 3x) \, dx$$

$$= \left(\frac{-x^3}{3} + \frac{3x^2}{2} \right) \Big|_0^3 + \left(\frac{x^3}{3} - \frac{3x^2}{2} \right) \Big|_3^4$$

$$= \left(-9 + \frac{27}{2} - 0 \right) + \left(\frac{64}{3} - 24 - 9 + \frac{27}{2} \right)$$

$$= \frac{19}{3}$$

TRY YOUR TURN 3

YOUR TURN 3 Repeat Example 3 for $y = x^2 - 3x$ and $y = 2x$ on $[0, 6]$.

FIGURE 26

In the remainder of this section we will consider some typical applications that require finding the area between two curves.

EXAMPLE 4 Savings

A company is considering a new manufacturing process in one of its plants. The new process provides substantial initial savings, with the savings declining with time t (in years) according to the rate-of-savings function

$$S'(t) = 100 - t^2,$$

where $S'(t)$ is in thousands of dollars per year. At the same time, the cost of operating the new process increases with time t (in years), according to the rate-of-cost function (in thousands of dollars per year)

$$C'(t) = t^2 + \frac{14}{3}t.$$

(a) For how many years will the company realize savings?

SOLUTION Figure 27 shows the graphs of the rate-of-savings and rate-of-cost functions. The rate of cost (marginal cost) is increasing, while the rate of savings (marginal savings) is decreasing. The company should use this new process until the difference between these quantities is zero; that is, until the time at which these graphs intersect. The graphs intersect when

$$C'(t) = S'(t),$$

or

$$t^2 + \frac{14}{3}t = 100 - t^2.$$

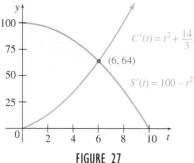

FIGURE 27

Solve this equation as follows.

$$0 = 2t^2 + \frac{14}{3}t - 100$$

$$0 = 3t^2 + 7t - 150 \qquad \text{Multiply by } \tfrac{3}{2}.$$

$$= (t - 6)(3t + 25) \qquad \text{Factor.}$$

Set each factor equal to 0 to get

$$t = 6 \qquad \text{or} \qquad t = -25/3.$$

Only 6 is a meaningful solution here. The company should use the new process for 6 years.

(b) What will be the net total savings during this period?

SOLUTION Since the total savings over the 6-year period is given by the area under the rate-of-savings curve and the total additional cost is given by the area under the rate-of-cost curve, the net total savings over the 6-year period is given by the area between the rate-of-cost and the rate-of-savings curves and the lines $t = 0$ and $t = 6$. This area can be evaluated with a definite integral as follows.

$$\text{Total savings} = \int_0^6 \left[(100 - t^2) - \left(t^2 + \frac{14}{3}t \right) \right] dt$$

$$= \int_0^6 \left(100 - \frac{14}{3}t - 2t^2 \right) dt$$

$$= \left(100t - \frac{7}{3}t^2 - \frac{2}{3}t^3 \right) \Big|_0^6$$

$$= 100(6) - \frac{7}{3}(36) - \frac{2}{3}(216) = 372$$

The company will save a total of $372,000 over the 6-year period.

The answer to a problem will not always be an integer. Suppose in solving the quadratic equation in Example 4 we found the solutions to be $t = 6.7$ and $t = -7.3$. It may not be realistic to use a new process for 6.7 years; it may be necessary to choose between 6 years and 7 years. Since the mathematical model produces a result that is not in the domain of the function in this case, it is necessary to find the total savings after 6 years and after 7 years and then select the better result.

Consumers' Surplus

Have you ever entered a store willing to pay, say, $10 for an item, only to discover that the item sells for less, perhaps just $7? The $3 that you didn't spend, plus the sum of all the money saved by all the other customers who would have been willing to pay a price higher than $7, is known as the **consumers' surplus**. According to supply and demand, an item will sell at the equilibrium price, where the demand and supply curves intersect. But some customers may be willing to pay more, as the demand curve indicates.

To calculate the total amount that consumers would be willing to pay for q_0 items, first consider the simple case in which everyone is willing to pay exactly p_0, the equilibrium price. Then the total amount everyone would pay would be the price times the quantity, or $p_0 q_0$, which is the green area in Figure 28. In fact, this is the exact total that everyone together pays when the item sells for p_0. Now divide the interval from 0 to q_0 into n intervals, each of width $\Delta q = q_0/n$. Each interval represents Δq people. Specifically, we will assume that the people represented by the ith interval, where i is a number between 1 and n, are those who are willing to pay a price $p_i = D(q_i)$ for the item, where q_i is some quantity on that interval. Then the total amount those people would be willing to pay would be $D(q_i)\Delta q$. The total amount that everyone together would be willing to pay is

$$\sum_{i=1}^{n} D(q_i)\Delta q.$$

In a more realistic situation, the demand curve changes continuously, so we find the total amount that everyone would be willing to pay by taking the limit as n goes to infinity:

$$\lim_{n \to \infty} \sum_{i=1}^{n} D(q_i)\Delta q = \int_0^{q_0} D(q)\,dq.$$

This quantity, which represents the total amount consumers are willing to spend for q_0 items, is the area under the demand curve in Figure 28. The pink shaded area represents the difference between what consumers would be willing to pay and what they actually pay, or the consumers' surplus.

As the figure suggests, the consumers' surplus is given by an area between the two curves $p = D(q)$ and $p = p_0$, so its value can be found with a definite integral as follows.

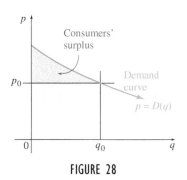

FIGURE 28

Consumers' Surplus

If $D(q)$ is a demand function with equilibrium price p_0 and equilibrium demand q_0, then

$$\textbf{Consumers' Surplus} = \int_0^{q_0} [D(q) - p_0]\,dq.$$

Similarly, if some manufacturers would be willing to supply a product at a price *lower* than the equilibrium price p_0, the total of the differences between the equilibrium price and the lower prices at which the manufacturers would sell the product is considered added income for the manufacturers and is called the **producers' surplus**. Figure 29 shows the (green shaded) total area under the supply curve from $q = 0$ to $q = q_0$, which is the minimum total amount the manufacturers are willing to realize from the sale of q_0 items. The total area under the line $p = p_0$ is the amount actually realized. The difference between these two areas, the producers' surplus, is also given by a definite integral.

Producers' Surplus

If $S(q)$ is a supply function with equilibrium price p_0 and equilibrium supply q_0, then

$$\textbf{Producer's Surplus} = \int_0^{q_0} [p_0 - S(q)]\,dq.$$

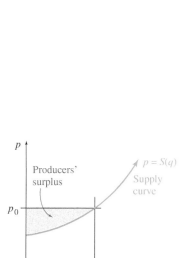

FIGURE 29

EXAMPLE 5   **Consumers' and Producers' Surplus**

Suppose the price (in dollars per ton) for oat bran is

$$D(q) = 344 - 3e^{q/2}$$

when the demand for the product is q tons. Also, suppose the function

$$S(q) = e^{q/2} - 1$$

describes the price when the supply is q tons. Find the consumers' surplus and the producers' surplus.

SOLUTION   Begin by finding the equilibrium quantity. Do this by setting the two equations equal.

$$e^{q/2} - 1 = 344 - 3e^{q/2}$$

$$4e^{q/2} = 345 \qquad\qquad \text{Add } 3e^{q/2} \text{ and 1 to both sides.}$$

$$e^{q/2} = \frac{345}{4} = 86.25 \qquad\qquad \text{Divide both sides by 4.}$$

$$\frac{q}{2} = \ln 86.25 \qquad\qquad \text{Take the natural logarithm of both sides.}$$

$$q = 2\ln 86.25 \approx 8.9145 \qquad \text{Multiply both sides by 2.}$$

At the equilibrium point where the supply and demand are both 8.9145 tons, the price is

$$S(8.9145) = e^{8.9145/2} - 1 \approx 85.25,$$

or $85.25. Verify that this same answer is found by computing $D(8.9145)$. The consumers' surplus, represented by the area shown in Figure 30, is

$$\int_0^{8.9145} \left[(344 - 3e^{q/2}) - 85.25 \right] dq = \int_0^{8.9145} (258.75 - 3e^{q/2})\, dq.$$

Evaluating the definite integral gives

$$(258.75q - 3(2)e^{q/2})\Big|_0^{8.9145} = (258.75(8.9145) - 6e^{8.9145/2}) - (0 - 6)$$

$$\approx 1795.13.$$

Here the consumers' surplus is $1795.13. The producers' surplus, also shown in Figure 30, is given by

$$\int_0^{8.9145} \left[85.25 - (e^{q/2} - 1) \right] dq = \int_0^{8.9145} (86.25 - e^{q/2})\, dq$$

$$= (86.25q - 2e^{q/2})\Big|_0^{8.9145}$$

$$= (86.25(8.9145) - 2e^{8.9145/2}) - (0 - 2)$$

$$\approx 598.38.$$

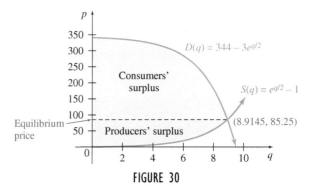

FIGURE 30

YOUR TURN 4 Repeat
Example 5 when
$D(q) = 600 - e^{q/3}$ and
$S(q) = e^{q/3} - 100$.

The producers' surplus of $598.38 is much smaller than the consumers' surplus in this case because the demand curve is more elastic than the supply curve. Notice in Figure 30 how steep the demand curve is at $q = 8.9145$ compared with the supply curve. Consumers have a greater surplus when they are sensitive to changes in the price.

TRY YOUR TURN 4 ▮▮▮▮

7.5 WARM-UP EXERCISES

Evaluate each of the following integrals. *(Sec. 7.4)*

W1. $\displaystyle\int_1^8 (x^{1/3} + x^{4/3})\, dx$

W2. $\displaystyle\int_0^3 e^{2x}\, dx$

7.5 EXERCISES

Find the area between the curves in Exercises 1–24.

1. $x = -2, \quad x = 1, \quad y = 2x^2 + 5, \quad y = 0$

2. $x = 1, \quad x = 2, \quad y = 3x^3 + 2, \quad y = 0$

3. $x = -3, \quad x = 1, \quad y = x^3 + 1, \quad y = 0$

4. $x = -3, \quad x = 0, \quad y = 1 - x^2, \quad y = 0$

5. $x = -2, \quad x = 1, \quad y = 2x, \quad y = x^2 - 3$

6. $x = 0, \quad x = 6, \quad y = 5x, \quad y = 3x + 10$

7. $y = x^2 - 30, \quad y = 10 - 3x$

8. $y = x^2 - 18, \quad y = x - 6$

9. $y = x^2, \quad y = 2x$

10. $y = x^2, \quad y = x^3$

11. $x = 1, \quad x = 6, \quad y = \dfrac{1}{x}, \quad y = \dfrac{1}{2}$

12. $x = 0, \quad x = 4, \quad y = \dfrac{1}{x+1}, \quad y = \dfrac{x-1}{2}$

13. $x = -1, \quad x = 1, \quad y = e^x, \quad y = 3 - e^x$

14. $x = -1, \quad x = 2, \quad y = e^{-x}, \quad y = e^x$

15. $x = -1, \quad x = 2, \quad y = 2e^{2x}, \quad y = e^{2x} + 1$

16. $x = 2, \quad x = 4, \quad y = \dfrac{x-1}{4}, \quad y = \dfrac{1}{x-1}$

17. $y = x^3 - x^2 + x + 1, \quad y = 2x^2 - x + 1$

18. $y = 2x^3 + x^2 + x + 5, \quad y = x^3 + x^2 + 2x + 5$

19. $y = x^4 + \ln(x + 10), \quad y = x^3 + \ln(x + 10)$

20. $y = x^5 - 2\ln(x + 5), \quad y = x^3 - 2\ln(x + 5)$

21. $y = x^{4/3}, \quad y = 2x^{1/3}$

22. $y = \sqrt{x}, \quad y = x\sqrt{x}$

23. $x = 0, \quad x = 3, \quad y = 2e^{3x}, \quad y = e^{3x} + e^6$

24. $x = 0, \quad x = 3, \quad y = e^x, \quad y = e^{4-x}$

In Exercises 25 and 26, use a graphing calculator to find the values of x where the curves intersect and then to find the area between the two curves.

25. $y = e^x, \quad y = -x^2 - 2x$

26. $y = \ln x, \quad y = x^3 - 5x^2 + 6x - 1$

APPLICATIONS

Business and Economics

27. **Net Savings** Suppose a company wants to introduce a new machine that will produce a rate of annual savings (in dollars) given by
$$S'(t) = 150 - t^2,$$
where t is the number of years of operation of the machine, while producing a rate of annual costs (in dollars) of
$$C'(t) = t^2 + \frac{11}{4}t.$$

   (a) For how many years will it be profitable to use this new machine?

   (b) What are the net total savings during the first year of use of the machine?

   (c) What are the net total savings over the entire period of use of the machine?

28. **Net Savings** A new smog-control device will reduce the output of sulfur oxides from automobile exhausts. It is estimated that the rate of savings (in millions of dollars per year) to the community from the use of this device will be approximated by
$$S'(t) = -t^2 + 4t + 8,$$
after t years of use of the device. The new device cuts down on the production of sulfur oxides, but it causes an increase in the production of nitrous oxides. The rate of additional costs

(in millions of dollars per year) to the community after t years is approximated by

$$C'(t) = \frac{3}{25}t^2.$$

(a) For how many years will it pay to use the new device?

(b) What will be the net savings over this period of time?

29. Profit Perdaris Enterprises had an expenditure rate of $E'(x) = e^{0.1x}$ dollars per day and an income rate of $I'(x) = 98.8 - e^{0.1x}$ dollars per day on a particular job, where x was the number of days from the start of the job. The company's profit on that job will equal total income less total expenditures. Profit will be maximized if the job ends at the optimum time, which is the point where the two curves meet. Find the following.

(a) The optimum number of days for the job to last

(b) The total income for the optimum number of days

(c) The total expenditures for the optimum number of days

(d) The maximum profit for the job

30. Net Savings A factory of Starr Smith Industries has installed a new process that will produce an increased rate of revenue (in thousands of dollars per year) of

$$R'(t) = 104 - 0.4e^{t/2},$$

where t is time measured in years. The new process produces additional costs (in thousands of dollars per year) at the rate of

$$C'(t) = 0.3e^{t/2}.$$

(a) When will it no longer be profitable to use this new process?

(b) Find the net total savings.

31. Producers' Surplus Find the producers' surplus if the supply function for pork bellies is given (in dollars) by

$$S(q) = q^{5/2} + 2q^{3/2} + 50.$$

Assume supply and demand are in equilibrium at $q = 16$.

32. Producers' Surplus Suppose the supply function for concrete is given (in dollars) by

$$S(q) = 100 + 3q^{3/2} + q^{5/2},$$

and that supply and demand are in equilibrium at $q = 9$. Find the producers' surplus.

33. Consumers' Surplus Find the consumers' surplus if the demand function for grass seed is given (in dollars) by

$$D(q) = \frac{200}{(3q + 1)^2},$$

assuming supply and demand are in equilibrium at $q = 3$.

34. Consumers' Surplus Find the consumers' surplus if the demand function for extra virgin olive oil is given (in dollars) by

$$D(q) = \frac{32,000}{(2q + 8)^3},$$

and if supply and demand are in equilibrium at $q = 6$.

35. Consumers' and Producers' Surplus Suppose the supply function for oil is given (in dollars) by

$$S(q) = q^2 + 10q,$$

and the demand function is given (in dollars) by

$$D(q) = 900 - 20q - q^2.$$

(a) Graph the supply and demand curves.

(b) Find the point at which supply and demand are in equilibrium.

(c) Find the consumers' surplus.

(d) Find the producers' surplus.

36. Consumers' and Producers' Surplus Suppose the supply function for a certain item is given (in dollars) by

$$S(q) = (q + 1)^2,$$

and the demand function is given by

$$D(q) = \frac{1000}{q + 1}.$$

(a) Graph the supply and demand curves.

(b) Find the point at which supply and demand are in equilibrium.

(c) Find the consumers' surplus.

(d) Find the producers' surplus.

37. Consumers' and Producers' Surplus Suppose that with the supply and demand for oil as in Exercise 35, the government sets the price at $264 per unit.

(a) Use the supply function to calculate the quantity that will be produced at the new price.

(b) Find the consumers' surplus for the new price, using the quantity found in part (a) in place of the equilibrium quantity. How much larger is this than the consumers' surplus in Exercise 35?

(c) Find the producers' surplus for the new price, using the quantity found in part (a) in place of the equilibrium quantity. How much smaller is this than the producers' surplus in Exercise 35?

(d) Calculate the difference between the total of the consumers' and producers' surplus under the equilibrium price and under the government price. Economists refer to this loss as the *welfare cost* of the government's setting the price.

(e) Because of the welfare cost calculated in part (d), many economists argue that it is bad economics for the government to set prices. Others point to the increase in the consumers' surplus, calculated in part (b), as a justification for such government action. Discuss the pros and cons of this issue.

38. Income Inequality For a family with an income in the lowest fifth of the U.S. population, the mean household income for the years 1970 to 2012 can be approximated (in thousands of 2012 dollars) by

$$f(t) = -0.0001844t^3 + 0.009938t^2 - 0.08515t + 11.33,$$

where t is the number of years since 1970. The mean household income for a family in the top 5% can be approximated by

$$g(t) = -0.01107t^3 + 0.6854t^2 - 6.587t + 193.3.$$

Source: U.S. Census Bureau.

(a) Find the area under $f(t)$ and $g(t)$, as well as the area between the two curves, over the interval $0 \le t \le 42$.

 (b) Interpret the answers from part (a) and discuss what they might tell us about income inequality.

Life Sciences

 39. **Pollution** Pollution begins to enter a lake at time $t = 0$ at a rate (in gallons per hour) given by the formula

$$f(t) = 10(1 - e^{-0.5t}),$$

where t is the time (in hours). At the same time, a pollution filter begins to remove the pollution at a rate

$$g(t) = 0.4t$$

as long as pollution remains in the lake.

(a) How much pollution is in the lake after 12 hours?

(b) Use a graphing calculator to find the time when the rate that pollution enters the lake equals the rate the pollution is removed.

(c) Find the amount of pollution in the lake at the time found in part (b).

(d) Use a graphing calculator to find the time when all the pollution has been removed from the lake.

40. **Pollution** Repeat the steps of Exercise 39, using the functions

$$f(t) = 15(1 - e^{-0.05t})$$

and

$$g(t) = 0.3t.$$

Social Sciences

41. **Distribution of Income** Suppose that all the people in a country are ranked according to their incomes, starting at the bottom. Let x represent a fraction of the country $(0 \le x \le 1)$.

Let $I(x)$ represent the proportion of the total income earned by the lowest x of all people. Thus, $I(0.4)$ represents the fraction of total income earned by the lowest 40% of the population. The curve described by this function is known as a **Lorenz curve**. Suppose

$$I(x) = 0.9x^2 + 0.1x.$$

Find and interpret the following.

(a) $I(0.1)$ (b) $I(0.4)$

If income were distributed uniformly, we would have $I(x) = x$. The area under this line of complete equality is $1/2$. As $I(x)$ dips further below $y = x$, there is less equality of income distribution. This inequality can be quantified by the ratio of the area between $I(x)$ and $y = x$ to $1/2$. This ratio is called the **Gini index of income inequality** and equals $2 \int_0^1 [x - I(x)] \, dx$.

(c) Graph $I(x) = x$ and $I(x) = 0.9x^2 + 0.1x$, for $0 \le x \le 1$, on the same axes.

(d) Find the area between the curves.

 (e) For U.S. families, the Gini index was 0.386 in 1968 and 0.477 in 2012. Describe how the distribution of family incomes has changed over this time. *Source: U.S. Census.*

Physical Sciences

42. **Metal Plate** A worker sketches the curves $y = \sqrt{x}$ and $y = x/2$ on a sheet of metal and cuts out the region between the curves to form a metal plate. Find the area of the plate.

7.6 Numerical Integration

APPLY IT   If the velocity of a vehicle is known only at certain points in time, how can the total distance traveled by the vehicle be estimated?

Using numerical integration, we will answer this question in Example 3 of this section.

     Some integrals cannot be evaluated by any technique. One solution to this problem was presented in Section 3 of this chapter, in which the area under a curve was approximated by summing the areas of rectangles. This method is seldom used in practice because better methods exist that are more accurate for the same amount of work. These methods are referred to as **numerical integration** methods. We discuss two such methods here: the trapezoidal rule and Simpson's rule.

Trapezoidal Rule   Recall, the trapezoidal rule was mentioned briefly in Section 3, where we found approximations with it by averaging the sums of rectangles found by using left endpoints and then using right endpoints. In this section we derive an explicit formula for the trapezoidal rule in terms of function values.* To illustrate the derivation of the trapezoidal rule, consider the integral

$$\int_1^5 \frac{1}{x}\, dx.$$

The shaded region in Figure 31 shows the area representing that integral, the area under the graph $f(x) = 1/x$, above the x-axis, and between the lines $x = 1$ and $x = 5$.

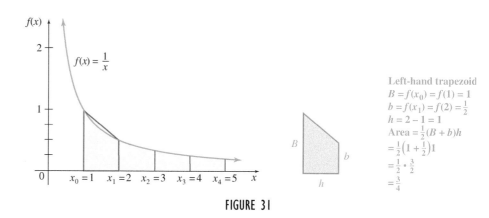

FIGURE 31

Note that this function can be integrated using the Fundamental Theorem of Calculus. Since $\int (1/x)\, dx = \ln |x| + C$,

$$\int_1^5 \frac{1}{x}\, dx = \ln |x| \Big|_1^5 = \ln 5 - \ln 1 = \ln 5 - 0 = \ln 5 \approx 1.609438.$$

We can also approximate the integral using numerical integration. As shown in the figure, if the area under the curve is approximated with trapezoids rather than rectangles, the approximation should be improved.

As in earlier work, to approximate this area we divide the interval $[1, 5]$ into subintervals of equal widths. To get a first approximation to ln 5 by the trapezoidal rule, find the sum of the areas of the four trapezoids shown in Figure 31. From geometry, the area of a trapezoid is half the product of the sum of the bases and the altitude. Each of the trapezoids in Figure 31 has altitude 1. (In this case, the bases of the trapezoid are vertical and the altitudes are horizontal.) Adding the areas gives

$$\ln 5 = \int_1^5 \frac{1}{x}\, dx \approx \frac{1}{2}\left(\frac{1}{1} + \frac{1}{2}\right)(1) + \frac{1}{2}\left(\frac{1}{2} + \frac{1}{3}\right)(1) + \frac{1}{2}\left(\frac{1}{3} + \frac{1}{4}\right)(1) + \frac{1}{2}\left(\frac{1}{4} + \frac{1}{5}\right)(1)$$

$$= \frac{1}{2}\left(\frac{3}{2} + \frac{5}{6} + \frac{7}{12} + \frac{9}{20}\right) \approx 1.68333.$$

To get a better approximation, divide the interval $[1, 5]$ into more subintervals. Generally speaking, the larger the number of subintervals, the better the approximation. The results for selected values of n are shown to 5 decimal places. When $n = 1000$, the approximation agrees with the true value of ln 5 ≈ 1.609438 to 5 decimal places.

Generalizing from this example, let f be a continuous function on an interval $[a, b]$. Divide the interval from a to b into n equal subintervals by the points $a = x_0, x_1, x_2, \ldots, x_n = b$ as shown in Figure 32 on the next page. Use the subintervals to make trapezoids that

Approximations to $\int_1^5 \frac{1}{x}\, dx$	
n	Trapezoidal Approximation
6	1.64360
8	1.62897
10	1.62204
20	1.61262
100	1.60957
1000	1.60944

*In American English a trapezoid is a four-sided figure with two parallel sides, contrasted with a trapezium, which has no parallel sides. In British English, however, it is just the opposite. What Americans call a trapezoid is called a trapezium in Great Britain.

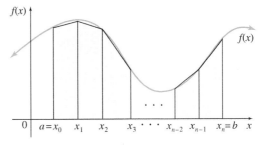

FIGURE 32

approximately fill in the region under the curve. The approximate value of the definite integral $\int_a^b f(x)\,dx$ is given by the sum of the areas of the trapezoids, or

$$\int_a^b f(x)\,dx \approx \frac{1}{2}[f(x_0) + f(x_1)]\left(\frac{b-a}{n}\right) + \frac{1}{2}[f(x_1) + f(x_2)]\left(\frac{b-a}{n}\right)$$

$$+ \cdots + \frac{1}{2}[f(x_{n-1}) + f(x_n)]\left(\frac{b-a}{n}\right)$$

$$= \left(\frac{b-a}{n}\right)\left[\frac{1}{2}f(x_0) + \frac{1}{2}f(x_1) + \frac{1}{2}f(x_1) + \frac{1}{2}f(x_2) + \frac{1}{2}f(x_2) + \cdots + \frac{1}{2}f(x_{n-1}) + \frac{1}{2}f(x_n)\right]$$

$$= \left(\frac{b-a}{n}\right)\left[\frac{1}{2}f(x_0) + f(x_1) + f(x_2) + \cdots + f(x_{n-1}) + \frac{1}{2}f(x_n)\right].$$

This result gives the following rule.

Trapezoidal Rule

Let f be a continuous function on $[a, b]$ and let $[a, b]$ be divided into n equal subintervals by the points $a = x_0, x_1, x_2, \ldots, x_n = b$. Then, by the **trapezoidal rule**,

$$\int_a^b f(x)\,dx \approx \left(\frac{b-a}{n}\right)\left[\frac{1}{2}f(x_0) + f(x_1) + \cdots + f(x_{n-1}) + \frac{1}{2}f(x_n)\right].$$

EXAMPLE 1   Trapezoidal Rule

Use the trapezoidal rule with $n = 4$ to approximate

$$\int_0^2 \sqrt{x^2 + 1}\,dx.$$

SOLUTION

Method I
Calculating by Hand

Here $a = 0$, $b = 2$, and $n = 4$, with $(b - a)/n = (2 - 0)/4 = 1/2$ as the altitude of each trapezoid. Then $x_0 = 0$, $x_1 = 1/2$, $x_2 = 1$, $x_3 = 3/2$, and $x_4 = 2$. Now find the corresponding function values. The work can be organized into a table, as follows.

Calculations for Trapezoidal Rule

i	x_i	$f(x_i)$
0	0	$\sqrt{0^2 + 1} = 1$
1	1/2	$\sqrt{(1/2)^2 + 1} \approx 1.11803$
2	1	$\sqrt{1^2 + 1} \approx 1.41421$
3	3/2	$\sqrt{(3/2)^2 + 1} \approx 1.80278$
4	2	$\sqrt{2^2 + 1} \approx 2.23607$

Substitution into the trapezoidal rule gives

$$\int_0^2 \sqrt{x^2 + 1}\, dx$$

$$\approx \frac{2 - 0}{4}\left[\frac{1}{2}(1) + 1.11803 + 1.41421 + 1.80278 + \frac{1}{2}(2.23607)\right]$$

$$\approx 2.97653.$$

The approximation 2.97653 found above using the trapezoidal rule with $n = 4$ differs from the true value of 2.95789 by 0.01864. As mentioned above, this error would be reduced if larger values were used for n. For example, if $n = 8$, the trapezoidal rule gives an answer of 2.96254, which differs from the true value by 0.00465. Techniques for estimating such errors are considered in more advanced courses.

**Method 2
Graphing Calculator**

Just as we used a graphing calculator to approximate area using rectangles, we can also use it for the trapezoidal rule. As before, put the values of i in L_1 and the values of x_i in L_2. In the heading for L_3, put $\sqrt{(L_2^2 + 1)}$. Using the fact that $(b - a)/n = (2 - 0)/4 = 0.5$, the command .5*(.5*L₃(1)+sum(L₃,2,4)+.5*L₃(5)) gives the result 2.976528589. For more details, see the *Graphing Calculator and Excel Spreadsheet Manual* available with this book.

**Method 3
Spreadsheet**

YOUR TURN 1 Use the trapezoidal rule with $n = 4$ to approximate $\int_1^3 \sqrt{x^2 + 3}\, dx$.

The trapezoidal rule can also be done on a spreadsheet. In Microsoft Excel, for example, store the values of 0 through n in column A. After putting the left endpoint in E1 and Δx in E2, put the command "=E1+A1*E2" into B1; copying this formula into the rest of column B gives the values of x_i. Similarly, use the formula for $f(x_i)$ to fill column C. Using the fact that $n = 5$ in this example, the command "E2*(.5*C1+sum(C2:C4)+.5*C5)" gives the result 2.976529. For more details, see the *Graphing Calculator and Excel Spreadsheet Manual* available with this book.

TRY YOUR TURN 1

The trapezoidal rule is not widely used because its results are not very accurate. In fact, the midpoint rule discussed earlier in this chapter is usually more accurate than the trapezoidal rule. We will now consider a method that usually gives more accurate results than either the trapezoidal or midpoint rule.

Simpson's Rule

Another numerical method, *Simpson's rule*, approximates consecutive portions of the curve with portions of parabolas rather than the line segments of the trapezoidal rule. Simpson's rule usually gives a better approximation than the trapezoidal rule for the same number of subintervals.* As shown in Figure 33 on the next page, one parabola is fitted through points *A*, *B*, and *C*; another through *C*, *D*, and *E*; and so on. Then the sum of the areas under these parabolas will approximate the area under the graph of the function. Because of the way the parabolas overlap, it is necessary to have an even number of intervals, and therefore an odd number of points, to apply Simpson's rule.

If h, the length of each subinterval, is $(b - a)/n$, the area under the parabola through points *A*, *B*, and *C* can be found by a definite integral. The details are omitted; the result is

$$\frac{h}{3}[f(x_0) + 4f(x_1) + f(x_2)].$$

---

*Although the trapezoidal rule is usually less accurate than Simpson's rule, there are cases for which the trapezoidal rule gives exceptional accuracy. See Trefethen, Lloyd N. and J. A. C. Weideman, "The Exponentially Convergent Trapezoidal Rule," *SIAM Review*, Vol. 56, Sept. 2014, pp. 385–458.

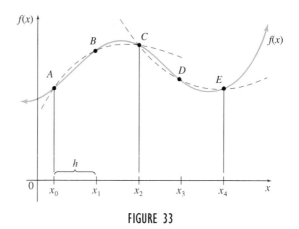

FIGURE 33

Similarly, the area under the parabola through points C, D, and E is

$$\frac{h}{3}[f(x_2) + 4f(x_3) + f(x_4)].$$

When these expressions are added, the last term of one expression equals the first term of the next. For example, the sum of the two areas given above is

$$\frac{h}{3}[f(x_0) + 4f(x_1) + 2f(x_2) + 4f(x_3) + f(x_4)].$$

This illustrates the origin of the pattern of the terms in the following rule.

Simpson's Rule

Let f be a continuous function on $[a, b]$ and let $[a, b]$ be divided into an even number n of equal subintervals by the points $a = x_0, x_1, x_2, \ldots, x_n = b$. Then by **Simpson's rule**,

$$\int_a^b f(x)\, dx \approx \left(\frac{b-a}{3n}\right)[f(x_0) + 4f(x_1) + 2f(x_2) + 4f(x_3) + \cdots$$
$$+ 2f(x_{n-2}) + 4f(x_{n-1}) + f(x_n)].$$

Thomas Simpson (1710–1761), a British mathematician, wrote texts on many branches of mathematics. Some of these texts went through as many as ten editions. His name became attached to this numerical method of approximating definite integrals even though the method preceded his work.

| CAUTION | In Simpson's rule, n (the number of subintervals) must be even.

EXAMPLE 2 Simpson's Rule

Use Simpson's rule with $n = 4$ to approximate

$$\int_0^2 \sqrt{x^2 + 1}\, dx,$$

which was approximated by the trapezoidal rule in Example 1.

SOLUTION As in Example 1, $a = 0, b = 2$, and $n = 4$, and the endpoints of the four intervals are $x_0 = 0, x_1 = 1/2, x_2 = 1, x_3 = 3/2$, and $x_4 = 2$. The table of values is also the same.

Calculations for Simpson's Rule

i	x_i	$f(x_i)$
0	0	1
1	1/2	1.11803
2	1	1.41421
3	3/2	1.80278
4	2	2.23607

Since $(b - a)/(3n) = 2/12 = 1/6$, substituting into Simpson's rule gives

YOUR TURN 2 Use Simpson's rule with $n = 4$ to approximate

$$\int_1^3 \sqrt{x^2 + 3}\, dx.$$

$$\int_0^2 \sqrt{x^2 + 1}\, dx \approx \frac{1}{6}[1 + 4(1.11803) + 2(1.41421) + 4(1.80278) + 2.23607] \approx 2.95796.$$

This differs from the true value by 0.00007, which is less than the trapezoidal rule with $n = 8$. If $n = 8$ for Simpson's rule, the approximation is 2.95788, which differs from the true value by only 0.00001. **TRY YOUR TURN 2**

NOTE

1. Just as we can use a graphing calculator or a spreadsheet for the trapezoidal rule, we can also use such technology for Simpson's rule. For more details, see the *Graphing Calculator and Excel Spreadsheet Manual* available with this book.

2. Let M represent the midpoint rule approximation and T the trapezoidal rule approximation, using n subintervals in each. Then the formula $S = (2M + T)/3$ gives the Simpson's rule approximation with $2n$ subintervals.

Numerical methods make it possible to approximate

$$\int_a^b f(x)\, dx$$

even when $f(x)$ is not known. The next example shows how this is done.

EXAMPLE 3 Total Distance

As mentioned earlier, the velocity $v(t)$ gives the rate of change of distance $s(t)$ with respect to time t. Suppose a vehicle travels an unknown distance. The passengers keep track of the velocity at 10-minute intervals (every 1/6 of an hour) with the following results.

Velocity of a Vehicle

Time in Hours, t	1/6	2/6	3/6	4/6	5/6	1	7/6
Velocity in Miles per Hour, $v(t)$	45	55	52	60	64	58	47

What is the total distance traveled in the 60-minute period from $t = 1/6$ to $t = 7/6$?

APPLY IT **SOLUTION** The distance traveled in t hours is $s(t)$, with $s'(t) = v(t)$. The total distance traveled between $t = 1/6$ and $t = 7/6$ is given by

$$\int_{1/6}^{7/6} v(t)\, dt.$$

Even though this integral cannot be evaluated because we do not have an expression for $v(t)$, either the trapezoidal rule or Simpson's rule can be used to approximate its value and give the total distance traveled. In either case, let $n = 6$, $a = t_0 = 1/6$, and $b = t_6 = 7/6$. By the trapezoidal rule,

$$\int_{1/6}^{7/6} v(t)\, dt \approx \frac{7/6 - 1/6}{6}\left[\frac{1}{2}(45) + 55 + 52 + 60 + 64 + 58 + \frac{1}{2}(47)\right]$$

$$\approx 55.83.$$

By Simpson's rule,

$$\int_{1/6}^{7/6} v(t)\, dt \approx \frac{7/6 - 1/6}{3(6)}[45 + 4(55) + 2(52) + 4(60) + 2(64) + 4(58) + 47]$$

$$= \frac{1}{18}(45 + 220 + 104 + 240 + 128 + 232 + 47) \approx 56.44.$$

The distance traveled in the 1-hour period was about 56 miles.

As already mentioned, Simpson's rule generally gives a better approximation than the trapezoidal rule. As n increases, the two approximations get closer and closer. For the same accuracy, however, a smaller value of n generally can be used with Simpson's rule so that less computation is necessary. Simpson's rule is the method used by many calculators that have a built-in integration feature.

The branch of mathematics that studies methods of approximating definite integrals (as well as many other topics) is called *numerical* analysis. Numerical integration is useful even with functions whose antiderivatives can be determined if the antidifferentiation is complicated and a computer or calculator programmed with Simpson's rule is handy. You may want to program your calculator for both the trapezoidal rule and Simpson's rule. For some calculators, these programs are in the *Graphing Calculator and Excel Spreadsheet Manual* available with this book.

7.6 EXERCISES

In Exercises 1–10, use $n = 4$ to approximate the value of the given integrals by the following methods: (a) the trapezoidal rule, and (b) Simpson's rule. (c) Find the exact value by integration.

1. $\int_0^2 (3x^2 + 2)\, dx$

2. $\int_0^2 (2x^2 + 1)\, dx$

3. $\int_{-1}^3 \frac{3}{5 - x}\, dx$

4. $\int_1^5 \frac{6}{2x + 1}\, dx$

5. $\int_{-1}^2 (2x^3 + 1)\, dx$

6. $\int_0^3 (2x^3 + 1)\, dx$

7. $\int_1^5 \frac{1}{x^2}\, dx$

8. $\int_2^4 \frac{1}{x^3}\, dx$

9. $\int_0^1 4xe^{-x^2}\, dx$

10. $\int_0^4 x\sqrt{2x^2 + 1}\, dx$

11. Find the area under the semicircle $y = \sqrt{4 - x^2}$ and above the x-axis by using $n = 8$ with the following methods.

(a) The trapezoidal rule

(b) Simpson's rule

(c) Compare the results with the area found by the formula for the area of a circle. Which of the two approximation techniques was more accurate?

12. Find the area between the x-axis and the upper half of the ellipse $4x^2 + 9y^2 = 36$ by using $n = 12$ with the following methods.

(a) The trapezoidal rule

(b) Simpson's rule

(*Hint:* Solve the equation for y and find the area of the semiellipse.)

(c) Compare the results with the actual area, $3\pi \approx 9.4248$ (which can be found by methods not considered in this text). Which approximation technique was more accurate?

13. Suppose that $f(x) > 0$ and $f''(x) > 0$ for all x between a and b, where $a < b$. Which of the following cases is true of a trapezoidal approximation T for the integral $\int_a^b f(x)\, dx$? Explain.

(a) $T < \int_a^b f(x)\, dx$

(b) $T > \int_a^b f(x)\, dx$

(c) Can't say which is larger

14. Refer to Exercise 13. Which of the three cases applies to these functions?

(a) $f(x) = x^2$; $[0, 3]$ (b) $f(x) = \sqrt{x}$; $[0, 9]$

(c)

Exercises 15–18 require both the trapezoidal rule and Simpson's rule. They can be worked without calculator programs if such programs are not available, although they require more calculation than the other problems in this exercise set.

Error Analysis The difference between the true value of an integral and the value given by the trapezoidal rule or Simpson's rule is known as the error. In numerical analysis, the error is studied to determine how large n must be for the error to be smaller than some specified amount. For both rules, the error is inversely proportional to a power of n, the number of subdivisions. In other words, the error is roughly k/n^p, where k is a constant that depends on the function and the interval, and p is a power that depends only on the method used. With a little experimentation, you can find out what the power p is for the trapezoidal rule and for Simpson's rule.

15. (a) Find the exact value of $\int_0^1 x^4\, dx$.

(b) Approximate the integral in part (a) using the trapezoidal rule with $n = 4, 8, 16$, and 32. For each of these answers, find the absolute value of the error by subtracting the trapezoidal rule answer from the exact answer found in part (a).

(c) If the error is k/n^p, then the error times n^p should be approximately a constant. Multiply the errors in part (b) times n^p for $p = 1, 2$, etc., until you find a power p yielding the same answer for all four values of n.

16. Based on the results of Exercise 15, what happens to the error in the trapezoidal rule when the number of intervals is doubled?

17. Repeat Exercise 15 using Simpson's rule.

18. Based on the results of Exercise 17, what happens to the error in Simpson's rule when the number of intervals is doubled?

19. For the integral in Exercise 7, apply the midpoint rule with $n = 4$ and Simpson's rule with $n = 8$ to verify the formula $S = (2M + T)/3$.

20. Repeat the instructions of Exercise 19 using the integral in Exercise 8.

APPLICATIONS

Business and Economics

21. Wind Energy In Exercise 25 of Section 7.3, we estimated the total wind energy consumption (in trillion BTUs) for the 12-year period from 2001 to 2013 using rectangles and the following data. *Source: Annual Energy Review.*

Year	Consumption
2001	70
2004	142
2007	341
2010	923
2013	1595

(a) Approximate the total wind energy consumption using the trapezoidal rule.

(b) Approximate the total wind energy consumption using Simpson's rule.

22. Total Cost A company's marginal costs (in hundreds of dollars per year) were as follows over a certain period.

Year, x	1	2	3	4	5	6	7
Marginal Cost, $f(x)$	9.0	9.2	9.5	9.4	9.8	10.1	10.5

Repeat parts (a) and (b) of Exercise 21 for these data to find the total cost over the given period.

Life Sciences

23. Drug Reaction Rate The reaction rate to a new drug is given by

$$y = e^{-t^2} + \frac{1}{t + 1},$$

where t is time (in hours) after the drug is administered. Find the total reaction to the drug from $t = 1$ to $t = 9$ by letting $n = 8$ and using the following methods.

(a) The trapezoidal rule   **(b)** Simpson's rule

24. Growth Rate The growth rate of a certain tree (in feet) is given by

$$y = \frac{2}{t + 2} + e^{-t^2/2},$$

where t is time (in years). Find the total growth from $t = 1$ to $t = 7$ by using $n = 12$ with the following methods.

(a) The trapezoidal rule   **(b)** Simpson's rule

Blood Level Curves In the study of bioavailability in pharmacy, a drug is given to a patient. The level of concentration of the drug is then measured periodically, producing blood level curves such as the ones shown in the figure.

The areas under the curves give the total amount of the drug available to the patient for each milliliter of blood. Use the trapezoidal rule with $n = 10$ to find the following areas. *Source: Basics of Bioavailability.*

25. Find the total area under the curve for Formulation A. What does this area represent?

26. Find the total area under the curve for Formulation B. What does this area represent?

27. Find the area between the curve for Formulation A and the minimum effective concentration line. What does your answer represent?

28. Find the area between the curve for Formulation B and the minimum effective concentration line. What does this area represent?

29. Calves The daily milk consumption (in kilograms) for calves can be approximated by the function

$$y = b_0 w^{b_1} e^{-b_2 w},$$

where w is the age of the calf (in weeks) and b_0, b_1, and b_2 are constants. *Source: Animal Production.*

(a) The age in days is given by $t = 7w$. Use this fact to convert the function above to a function in terms of t.

(b) For a group of Angus calves, $b_0 = 5.955$, $b_1 = 0.233$, and $b_2 = 0.027$. Use the trapezoidal rule with $n = 10$, and then Simpson's rule with $n = 10$, to find the total amount of milk consumed by one of these calves over the first 25 weeks of life.

(c) For a group of Nelore calves, $b_0 = 8.409$, $b_1 = 0.143$, and $b_2 = 0.037$. Use the trapezoidal rule with $n = 10$, and then Simpson's rule with $n = 10$, to find the total amount of milk consumed by one of these calves over the first 25 weeks of life.

30. Foot-and-Mouth Epidemic In 2001, the United Kingdom suffered an epidemic of foot-and-mouth disease. The graph below shows the reported number of cattle (red) and pigs (blue) that were culled each month from mid-February through mid-October in an effort to stop the spread of the disease. In Section 7.3 on Area and the Definite Integral we estimated the number of cattle and pigs that were culled using rectangles. *Source: Department of Environment, Food and Rural Affairs, United Kingdom.*

(a) Estimate the total number of cattle that were culled from mid-February through mid-October and compare this with 581,801, the actual number of cattle that were culled. Use Simpson's rule with interval widths of one month starting with mid-February.

(b) Estimate the total number of pigs that were culled from mid-February through mid-October and compare this with 146,145, the actual number of pigs that were culled. Use Simpson's rule with interval widths of one month starting with mid-February.

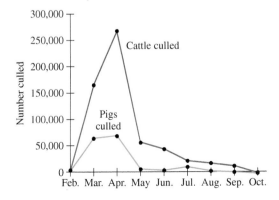

Social Sciences

31. Educational Psychology The results from a research study in psychology were as follows.

Number of Hours of Study, x	1	2	3	4	5	6	7
Rate of Extra Points Earned on a Test, $f(x)$	4	7	11	9	15	16	23

Repeat parts (a) and (b) of Exercise 21 for these data.

Physical Sciences

32. Chemical Formation The following table shows the results from a chemical experiment.

Concentration of Chemical A, x	1	2	3	4	5	6	7
Rate of Formation of Chemical B, $f(x)$	12	16	18	21	24	27	32

Repeat parts (a) and (b) of Exercise 21 for these data.

If you have a program for Simpson's rule in your graphing calculator, use it with $n = 20$ for Exercises 33–35.

33. Total Revenue An electronics company analyst has determined that the rate per month at which revenue comes in from the calculator division is given by

$$R(x) = 105e^{0.01x} + 32,$$

where x is the number of months the division has been in operation. Find the total revenue between the 12th and 36th months.

34. Milk Consumption As we saw in an earlier chapter, the average individual daily milk consumption for herds of Charolais, Angus, and Hereford calves can be described by a mathematical function. Here we write the consumption in kg/day as a function of the age of the calf in days (t) as

$$M(t) = 3.922t^{0.242}e^{-0.00357t}, \quad 7 \le t \le 182.$$

Find the total amount of milk consumed from 7 to 182 days for a calf. *Source: Animal Production.*

35. Probability The most important function in probability and statistics is the density function for the standard normal distribution, which is the familiar bell-shaped curve. The function is

$$f(x) = \frac{1}{\sqrt{2\pi}} e^{-x^2/2}.$$

(a) The area under this curve between $x = -1$ and $x = 1$ represents the probability that a normal random variable is within 1 standard deviation of the mean. Find this probability.

(b) Find the area under this curve between $x = -2$ and $x = 2$, which represents the probability that a normal random variable is within 2 standard deviations of the mean.

(c) Find the probability that a normal random variable is within 3 standard deviations of the mean.

YOUR TURN ANSWERS

1. 5.3552   **2.** 5.3477

7 CHAPTER REVIEW

SUMMARY

Earlier chapters dealt with the derivative, one of the two main ideas of calculus. This chapter deals with integration, the second main idea. There are two aspects of integration. The first is indefinite integration, or finding an antiderivative; the second is definite integration, which can be used to find the area under a curve. The Fundamental Theorem of Calculus unites these two ideas by showing that the way to find the area under a curve is to use the antiderivative. Substitution is a technique for finding antiderivatives. Numerical integration can be used to find the definite integral when finding an antiderivative is not feasible. The idea of the definite integral can also be applied to finding the area between two curves.

Antidifferentiation Formulas

Power Rule
$$\int x^n \, dx = \frac{x^{n+1}}{n+1} + C, \; n \neq -1$$

Constant Multiple Rule
$$\int k \cdot f(x) \, dx = k \int f(x) \, dx, \quad \text{for any real number } k$$

Sum or Difference Rule
$$\int [f(x) \pm g(x)] \, dx = \int f(x) \, dx \pm \int g(x) \, dx$$

Integration of x^{-1}
$$\int x^{-1} \, dx = \ln|x| + C$$

Integration of Exponential Functions
$$\int e^{kx} \, dx = \frac{e^{kx}}{k} + C, \quad k \neq 0$$
$$\int a^k \, dx = \frac{a^{kx}}{k(\ln a)} + C, \quad k \neq 0$$

Substitution Method   Choose u to be one of the following:

1. the quantity under a root or raised to a power;
2. the quantity in the denominator;
3. the exponent on e.

Definite Integrals
Definition of the Definite Integral $\int_a^b f(x) \, dx = \lim_{n \to \infty} \sum_{i=1}^n f(x_i) \Delta x$, where $\Delta x = (b - a)/n$ and x_i is any value of x in the ith interval. If $f(x)$ gives the rate of change of $F(x)$ for x in $[a, b]$, then this represents the total change in $F(x)$ as x goes from a to b.

Properties of Definite Integrals

1. $\displaystyle\int_a^b k \cdot f(x) \, dx = k \int_a^b f(x) \, dx, \quad \text{for any real number } k$

2. $\displaystyle\int_a^b [f(x) \pm g(x)] \, dx = \int_a^b f(x) \, dx \pm \int_a^b g(x) \, dx$

3. $\displaystyle\int_a^a f(x) \, dx = 0$

4. $\displaystyle\int_a^b f(x) \, dx = \int_a^c f(x) \, dx + \int_c^b f(x) \, dx, \quad \text{for any real number } c$

5. $\displaystyle\int_a^b f(x) \, dx = -\int_b^a f(x) \, dx$

Fundamental Theorem of Calculus	$\int_a^b f(x)\,dx = F(x)\big|_a^b = F(b) - F(a)$, where f is continuous on $[a, b]$ and F is any antiderivative of f
Area between Two Curves	$\int_a^b [f(x) - g(x)]\,dx$, where f and g are continuous functions and $f(x) \geq g(x)$ on $[a, b]$
Consumers' Surplus	$\int_0^{q_0} [D(q) - p_0]\,dq$, where D is the demand function and p_0 and q_0 are the equilibrium price and demand
Producers' Surplus	$\int_0^{q_0} [p_0 - S(q)]\,dq$, where S is the supply function and p_0 and q_0 are the equilibrium price and supply
Trapezoidal Rule	$\int_a^b f(x)\,dx \approx \left(\dfrac{b-a}{n}\right)\left[\dfrac{1}{2}f(x_0) + f(x_1) + \cdots + f(x_{n-1}) + \dfrac{1}{2}f(x_n)\right]$
Simpson's Rule	$\int_a^b f(x)\,dx \approx \left(\dfrac{b-a}{3n}\right)[f(x_0) + 4f(x_1) + 2f(x_2) + \cdots + 4f(x_{n-1}) + f(x_n)]$

KEY TERMS

7.1
antidifferentiation
antiderivative
integral sign
integrand
indefinite integral

7.2
integration by substitution

7.3
midpoint rule
definite integral
limits of integration
total change

7.4
Fundamental Theorem of
Calculus

7.5
consumers' surplus
producers' surplus
Lorenz curve

Gini index of inequality

7.6
numerical integration
trapezoidal rule
Simpson's rule

REVIEW EXERCISES

CONCEPT CHECK

Determine whether each of the following statements is true or false, and explain why.

1. The indefinite integral is another term for the family of all anti-derivatives of a function.

2. The indefinite integral of x^n is $x^{n+1}/(n + 1) + C$ for all real numbers n.

3. The indefinite integral $\int xf(x)\,dx$ is equal to $x\int f(x)\,dx$.

4. The velocity function is an antiderivative of the acceleration function.

5. Substitution can often be used to turn a complicated integral into a simpler one.

6. The definite integral gives the instantaneous rate of change of a function.

7. The definite integral gives an approximation to the area under a curve.

8. The definite integral of a positive function is the limit of the sum of the areas of rectangles.

9. The Fundamental Theorem of Calculus gives a relationship between the definite integral and an antiderivative of a function.

10. The definite integral of a function is always a positive quantity.

11. The area between two distinct curves is always a positive quantity.

12. The consumers' surplus and the producers' surplus equal each other.

13. In the trapezoidal rule, the number of subintervals must be even.

14. Simpson's rule usually gives a better approximation than the trapezoidal rule.

PRACTICE AND EXPLORATION

15. Explain the differences between an indefinite integral and a definite integral.

16. Explain under what circumstances substitution is useful in integration.

17. Explain why the limits of integration are changed when u is substituted for an expression in x in a definite integral.

18. Describe the type of integral for which numerical integration is useful.

In Exercises 19–40, find each indefinite integral.

19. $\int (2x + 3)\, dx$ **20.** $\int (5x - 1)\, dx$

21. $\int (x^2 - 3x + 2)\, dx$ **22.** $\int (6 - x^2)\, dx$

23. $\int 3\sqrt{t}\, dt$ **24.** $\int \frac{\sqrt{t}}{2}\, dt$

25. $\int (x^{1/2} + 3x^{-2/3})\, dx$ **26.** $\int (2x^{4/3} + x^{-1/2})\, dx$

27. $\int \frac{-4}{y^3}\, dy$ **28.** $\int \frac{5}{y^4}\, dy$ **29.** $\int -3e^{2x}\, dx$

30. $\int 5e^{-x}\, dx$ **31.** $\int xe^{3x^2}\, dx$ **32.** $\int 2xe^{x^2}\, dx$

33. $\int \frac{3u}{u^2 - 1}\, du$ **34.** $\int \frac{-u}{2 - u^2}\, du$

35. $\int \frac{x^2\, dx}{(x^3 + 5)^4}$ **36.** $\int (x^2 - 5x)^4 (2x - 5)\, dx$

37. $\int \frac{x^3}{e^{3x^4}}\, dx$ **38.** $\int e^{3x^2 + 4} x\, dx$

39. $\int \frac{(3 \ln z + 2)^4}{z}\, dz$ **40.** $\int \frac{\sqrt{5 \ln z + 3}}{z}\, dz$

41. Let $f(x) = 3x + 1$, $x_1 = -1$, $x_2 = 0$, $x_3 = 1$, $x_4 = 2$, and $x_5 = 3$. Find $\sum_{i=1}^{5} f(x_i)$.

42. Find $\int_0^4 f(x)\, dx$ for each graph of $y = f(x)$.

(a)

(b)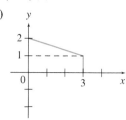

43. Approximate the area under the graph of $f(x) = 2x + 3$ and above the x-axis from $x = 0$ to $x = 4$ using four rectangles. Let the height of each rectangle be the function value on the left side.

44. Find $\int_0^4 (2x + 3)\, dx$ by using the formula for the area of a trapezoid: $A = (1/2)(B + b)h$, where B and b are the lengths of the parallel sides and h is the distance between them. Compare with Exercise 43.

45. In Exercises 31 and 32 of the section on Area and the Definite Integral, you calculated the distance that a car traveled by estimating the integral $\int_0^T v(t)\, dt$.

   (a) Let $s(t)$ represent the mileage reading on the odometer. Express the distance traveled between $t = 0$ and $t = T$ using the function $s(t)$.

   (b) Since your answer to part (a) and the original integral both represent the distance traveled by the car, the two can be set equal. Explain why the resulting equation is a statement of the Fundamental Theorem of Calculus.

46. What does the Fundamental Theorem of Calculus state?

Find each definite integral.

47. $\int_1^2 (3x^2 + 5)\, dx$ **48.** $\int_1^6 (2x^2 + x)\, dx$

49. $\int_1^5 (3x^{-1} + x^{-3})\, dx$ **50.** $\int_1^3 (2x^{-1} + x^{-2})\, dx$

51. $\int_0^1 x\sqrt{5x^2 + 4}\, dx$ **52.** $\int_0^2 x^2 (3x^3 + 1)^{1/3}\, dx$

53. $\int_0^2 3e^{-2w}\, dw$ **54.** $\int_1^5 \frac{5}{2} e^{0.4w}\, dw$

55. Use the substitution $u = 4x^2$ and the equation of a semicircle to evaluate

$$\int_0^{1/2} x\sqrt{1 - 16x^4}\, dx.$$

56. Use the substitution $u = x^2$ and the equation of a semicircle to evaluate

$$\int_0^{\sqrt{2}} 4x\sqrt{4 - x^4}\, dx.$$

In Exercises 57 and 58, use substitution to change the integral into one that can be evaluated by a formula from geometry, and then find the value of the integral.

57. $\int_1^{e^5} \frac{\sqrt{25 - (\ln x)^2}}{x}\, dx$ **58.** $\int_1^{\sqrt{7}} 2x\sqrt{36 - (x^2 - 1)^2}\, dx$

In Exercises 59–62, find the area between the x-axis and $f(x)$ over each of the given intervals.

59. $f(x) = \sqrt{4x - 3}$; $[1, 3]$

60. $f(x) = (3x + 2)^6$; $[-2, 0]$

61. $f(x) = xe^{x^2}$; $[0, 2]$

62. $f(x) = 1 + e^{-x}$; $[0, 4]$

Find the area of the region enclosed by each group of curves.

63. $f(x) = 5 - x^2$, $g(x) = x^2 - 3$

64. $f(x) = x^2 - 4x$, $g(x) = x - 6$

65. $f(x) = x^2 - 4x$, $g(x) = x + 6$, $x = -2$, $x = 4$

66. $f(t) = 5 - t^2$, $g(t) = t^2 - 3$, $t = 0$, $t = 4$

Use the trapezoidal rule with $n = 4$ to approximate the value of each integral. Then find the exact value and compare the two answers.

67. $\int_1^3 \frac{\ln x}{x}\, dx$ **68.** $\int_2^{10} \frac{x\, dx}{x - 1}$

69. $\int_0^1 e^x\sqrt{e^x + 4}\, dx$ **70.** $\int_0^2 xe^{-x^2}\, dx$

Use Simpson's rule with $n = 4$ to approximate the value of each integral. Compare your answers with the answers to Exercises 67–70.

71. $\displaystyle\int_1^3 \frac{\ln x}{x}\, dx$

72. $\displaystyle\int_2^{10} \frac{x\, dx}{x - 1}$

73. $\displaystyle\int_0^1 e^x\sqrt{e^x + 4}\, dx$

74. $\displaystyle\int_0^2 xe^{-x^2}dx$

75. Find the area of the region between the graphs of $y = \sqrt{x - 1}$ and $2y = x - 1$ from $x = 1$ to $x = 5$ in three ways.

   (a) Use antidifferentiation.

   (b) Use the trapezoidal rule with $n = 4$.

   (c) Use Simpson's rule with $n = 4$.

76. Find the area of the region between the graphs of $y = \dfrac{1}{x + 1}$ and $y = \dfrac{x + 2}{2}$ from $x = 0$ to $x = 4$ in three ways.

   (a) Use antidifferentiation.

   (b) Use the trapezoidal rule with $n = 4$.

   (c) Use Simpson's rule with $n = 4$.

77. Let $f(x) = [x(x - 1)(x + 1)(x - 2)(x + 2)]^2$.

   (a) Find $\int_{-2}^2 f(x)\, dx$ using the trapezoidal rule with $n = 4$.

   (b) Find $\int_{-2}^2 f(x)\, dx$ using Simpson's rule with $n = 4$.

   (c) Without evaluating $\int_{-2}^2 f(x)\, dx$, explain why your answers to parts (a) and (b) cannot possibly be correct.

   (d) Explain why the trapezoidal rule and Simpson's rule with $n = 4$ give incorrect answers for $\int_{-2}^2 f(x)\, dx$ with this function.

78. Given $\displaystyle\int_0^2 f(x)\, dx = 3$ and $\displaystyle\int_2^4 f(x)\, dx = 5$, calculate

$\displaystyle\int_0^2 f(2x)\, dx$. Choose one of the following. *Source: Society of Actuaries.*

   (a) 3/2   (b) 3   (c) 4

   (d) 6   (e) 8

APPLICATIONS

Business and Economics

Cost   Find the cost function for each of the marginal cost functions in Exercises 79 and 80.

79. $C'(x) = 3\sqrt{2x - 1}$; 13 units cost $270.

80. $C'(x) = \dfrac{8}{2x + 1}$; fixed cost is $18.

81. Deficit   The graph in the next column shows recent values (in trillions of dollars per year) of the U.S. deficit, which gives the rate that the U.S. debt accumulates. Use rectangles of width 2 years and height determined by the function value at the midpoint to estimate the total amount of debt accumulated from mid-2003 to mid-2013. *Source: whitehouse.gov.*

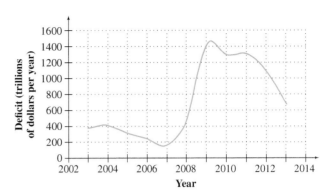

82. Utilization of Reserves   A manufacturer of electronic equipment requires a certain rare metal. He has a reserve supply of 4,000,000 units that he will not be able to replace. If the rate at which the metal is used is given by

$$f(t) = 100{,}000e^{0.03t},$$

where t is time (in years), how long will it be before he uses up the supply? (*Hint:* Find an expression for the total amount used in t years and set it equal to the known reserve supply.)

83. Sales   The rate of change of sales of a new brand of tomato soup (in thousands of dollars per month) is given by

$$S'(t) = 3\sqrt{2t + 1} + 3,$$

where t is the time (in months) that the new product has been on the market. Find the total sales after 4 months.

84. Productivity   The function defined by

$$f'(t) = -0.2048t + 3.5457$$

approximates marginal U.S. nonfarm productivity from 2000 to 2013. Productivity is measured as total output per hour compared to a measure of 100 for 2000, and t is the number of years since 2000. *Source: Bureau of Labor Statistics.*

   (a) Give the function that describes total productivity in year t.

   (b) Use your function from part (a) to find productivity at the end of 2013. Productivity actually measured 137.2. How does your value using the function compare with this?

85. Producers' and Consumers' Surplus   Suppose that the supply function for some commodity is

$$S(q) = q^2 + 5q + 100$$

and the demand function for the commodity is

$$D(q) = 350 - q^2.$$

   (a) Find the producers' surplus.

   (b) Find the consumers' surplus.

86. Net Savings   A company has installed new machinery that will produce a savings rate (in thousands of dollars per year) of

$$S'(x) = 225 - x^2,$$

where x is the number of years the machinery is to be used. The rate of additional costs (in thousands of dollars per year) to the company due to the new machinery is expected to be

$$C'(x) = x^2 + 25x + 150.$$

For how many years should the company use the new machinery? Find the net savings (in thousands of dollars) over this period.

87. Oil Production The following table shows the amount of crude oil (in billions of barrels) produced in the United States in recent years. *Source: U.S. Energy Information Administration.*

Year	Crude Oil Produced
2004	1.986
2005	1.891
2006	1.857
2007	1.853
2008	1.825
2009	1.954
2010	1.997
2011	2.063
2012	2.368
2013	2.716

In this exercise we are interested in the total amount of crude oil produced over the 9-year period from mid-2004 to mid-2013, using the data for the 10 years above.

(a) One approach is to sum up the numbers in the second column, but count only half of the first and last numbers. Give the answer to this calculation.

(b) Approximate the amount of crude oil produced over the 9-year period 2004–2013 by taking the average of the left endpoint sum and the right endpoint sum. Explain why this is equivalent to the calculation done in part (a).

(c) Explain why the answer from part (a) is the same as that obtained using the trapezoidal rule to approximate the amount of crude oil produced over the 9-year period 2004–2013.

(d) Find the equation of the least squares line for this data, letting $t = 0$ correspond to 2000. Then integrate this equation over the interval $[4, 13]$ to estimate the amount of crude oil produced over this time period. Compare with your answer to part (a).

88. Inventory At time $t = 0$, a store has 19 units of a product in inventory. The cumulative number of units sold is given by $S(t) = e^{3t} - 1$, where t is measured in weeks. The inventory will be replenished when it drops to 1 unit. The cost of carrying inventory until then is 15 per unit per week (prorated for a portion of a week). Calculate the inventory carrying cost that will be incurred before the inventory is replenished. Choose one of the following. *Source: Society of Actuaries.*

(a) 90   (b) 199   (c) 204

(d) 210   (e) 294

Life Sciences

89. Population Growth The rate of change of the population of a rare species of Australian spider for one year is given by

$$f'(t) = 100 - t\sqrt{0.4t^2 + 1},$$

where $f(t)$ is the number of spiders present at time t (in months). Find the total number of additional spiders in the first 10 months.

90. Infection Rate The rate of infection of a disease (in people per month) is given by the function

$$I'(t) = \frac{100t}{t^2 + 1},$$

where t is the time (in months) since the disease broke out. Find the total number of infected people over the first four months of the disease.

91. Insect Cannibalism In certain species of flour beetles, the larvae cannibalize the unhatched eggs. In calculating the population cannibalism rate per egg, researchers needed to evaluate the integral

$$\int_0^A c(x)\, dx,$$

where A is the length of the larval stage and $c(x)$ is the cannibalism rate per egg per larva of age x. The minimum value of A for the flour beetle *Tribolium castaneum* is 17.6 days, which is the value we will use. The function $c(x)$ starts at day 0 with a value of 0, increases linearly to the value 0.024 at day 12, and then stays constant. *Source: Journal of Animal Ecology.* Find the values of the integral using

(a) formulas from geometry;

(b) the Fundamental Theorem of Calculus.

92. Insulin in Sheep A research group studied the effect of a large injection of glucose in sheep fed a normal diet compared with sheep that were fasting. A graph of the plasma insulin levels (in pM—pico molars, or 10^{-12} of a molar) for both groups is shown below. The red graph designates the fasting sheep and the green graph the sheep fed a normal diet. The researchers compared the area under the curves for the two groups. *Source: Endocrinology.*

(a) For the fasting sheep, estimate the area under the curve using rectangles, first by using the left endpoints, then the right endpoints, and then averaging the two. Note that the width of the rectangles will vary.

(b) Repeat part (a) for the sheep fed a normal diet.

(c) How much higher is the area under the curve for the fasting sheep compared with the normal sheep?

93. Milk Production Researchers report that the average amount of milk produced (in kilograms per day) by a 4- to 5-year-old cow weighing 700 kg can be approximated by

$$y = 1.87t^{1.49}e^{-0.189\,(\ln t)^2},$$

where t is the number of days into lactation. *Source: Journal of Dairy Science.*

(a) Approximate the total amount of milk produced from $t = 1$ to $t = 321$ using the trapezoidal rule with $n = 8$.

(b) Repeat part (a) using Simpson's rule with $n = 8$.

(c) Repeat part (a) using the integration feature of a graphing calculator, and compare your answer with the answers to parts (a) and (b).

Social Sciences

94. **Automotive Accidents** The table in the next column shows the amount of property damage (in dollars) due to automobile accidents in California in recent years. In this exercise we are interested in the total amount of property damage due to automobile accidents over the 8-year period from mid-2003 to mid-2011, using the data for the 9 years. *Source: The California Highway Patrol.*

(a) One approach is to sum up the numbers in the second column, but count only half of the first and last numbers. Give the answer to this calculation.

(b) Approximate the amount of property damage over the 8-year period 2003–2011 by taking the average of the left endpoint sum and the right endpoint sum. Explain why this is equivalent to the calculation done in part (a).

Year	Property Damage ($)
2003	331,055
2004	331,208
2005	330,195
2006	325,453
2007	313,357
2008	278,986
2009	259,889
2010	252,876
2011	241,886

(c) Explain why the answer from part (a) is the same as that obtained using the trapezoidal rule to approximate the amount of property damage over the 8-year period 2003–2011.

(d) Find the equation of the least squares line for this data, letting $t = 0$ correspond to 2000. Then integrate this equation over the interval $[3, 11]$ to estimate the amount of property damage over this time period. Compare with your answer to part (a).

Physical Sciences

95. **Linear Motion** A particle is moving along a straight line with velocity $v(t) = t^2 - 2t$. Its distance from the starting point after 3 seconds is 8 cm. Find $s(t)$, the distance of the particle from the starting point after t seconds.

EXTENDED APPLICATION

ESTIMATING DEPLETION DATES FOR MINERALS

t is becoming more and more obvious that the earth contains only a finite quantity of minerals. The "easy and cheap" sources of minerals are being used up, forcing an ever more expensive search for new sources. For example, oil from the North Slope of Alaska would never have been used in the United States during the 1930s because a great deal of Texas and California oil was readily available.

We said in an earlier chapter that population tends to follow an exponential growth curve. Mineral usage also follows such a curve. Thus, if q represents the rate of consumption of a certain mineral at time t, while q_0 represents consumption when $t = 0$, then

$$q = q_0 e^{kt},$$

where k is the growth constant. For example, the world consumption of petroleum in 1970 was 16,900 million barrels. During this period energy use was growing rapidly, and by 1975 annual world consumption had risen to 21,300 million barrels. We can use these two values to make a rough estimate of the constant k, and we find that over this 5-year span the average value of k was about 0.047, representing 4.7% annual growth. If we let $t = 0$ correspond to the base year 1970, then

$$q = 16,900 e^{0.047t}.$$

is the rate of consumption at time t, assuming that all the trends of the early 1970s have continued. In 1970 a reasonable guess would have put the total amount of oil in provable reserves or likely to be discovered in the future at 1,500,000 million barrels. At the 1970–1975 rate of consumption, in how many years after 1970 would you expect the world's reserves to be depleted? We can use the integral calculus of this chapter to find out.

To begin, we need to know the total quantity of petroleum that would be used between time $t = 0$ and some future time $t = T$. Figure 34 on the next page shows a typical graph of the function $q = q_0 e^{kt}$.

Following the work we did in Section 3, divide the time interval from $t = 0$ to $t = T$ into n subintervals. Let each subinterval have width Δt. Let the rate of consumption for the ith subinterval be approximated by q_i^*. Thus, the approximate total consumption for the subinterval is given by

$$q_i^* \cdot \Delta t,$$

and the total consumption over the interval from time $t = 0$ to $t = T$ is approximated by

$$\sum_{i=1}^{n} q_i^* \cdot \Delta t.$$

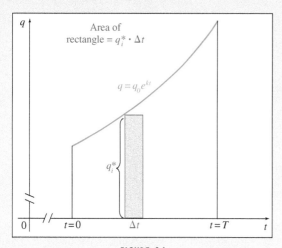

FIGURE 34

The limit of this sum as Δt approaches 0 gives the total consumption from time $t = 0$ to $t = T$. That is,

$$\text{Total consumption} = \lim_{\Delta t \to 0} \sum q_i^* \cdot \Delta t.$$

We have seen, however, that this limit is the definite integral of the function $q = q_0 e^{kt}$ from $t = 0$ to $t = T$, or

$$\text{Total consumption} = \int_0^T q_0 e^{kt}\, dt.$$

We can now evaluate this definite integral.

$$\int_0^T q_0 e^{kt}\, dt = q_0 \int_0^T e^{kt}\, dt = q_0 \left(\frac{e^{kt}}{k} \right) \Big|_0^T$$

$$= \frac{q_0}{k} e^{kt} \Big|_0^T = \frac{q_0}{k} e^{kT} - \frac{q_0}{k} e^0$$

$$= \frac{q_0}{k} e^{kT} - \frac{q_0}{k}(1)$$

$$= \frac{q_0}{k}(e^{kT} - 1) \qquad (1)$$

Now let us return to the numbers we gave for petroleum. We said that $q_0 = 16{,}900$ million barrels, where q_0 represents consumption in the base year of 1970. We have $k = 0.047$ with total petroleum reserves estimated at 1,500,000 million barrels. Thus, using Equation (1) we have

$$1{,}500{,}000 = \frac{16{,}900}{0.047}(e^{0.047T} - 1).$$

Multiply both sides of the equation by 0.047.

$$70{,}500 = 16{,}900(e^{0.047T} - 1)$$

Divide both sides of the equation by 16,900.

$$4.2 = e^{0.047T} - 1$$

Add 1 to both sides.

$$5.2 = e^{0.047T}$$

Take natural logarithms of both sides.

$$\ln 5.2 = \ln e^{0.047T}$$

$$= 0.047T$$

Finally,

$$T = \frac{\ln 5.2}{0.047} \approx 35.$$

By this result, petroleum reserves would only last 35 years after 1970, that is, until about 2005.

In fact, in the early 1970s some analysts were predicting that reserves would be exhausted before the end of the century, and this was a reasonable guess. But since 1970, more reserves have been discovered. One way to refine our model is to look at the historical data over a longer time span. The following table gives average world annual petroleum consumption in millions of barrels at 5-year intervals from 1970 to 2000. **Source: Energy Information Administration**.

Year	World Consumption (in millions of barrels)
1970	16,900
1975	21,300
1980	23,000
1985	21,900
1990	24,300
1995	25,600
2000	28,000
2005	30,700
2010	31,900

The first step in comparing these data with our exponential model is to estimate a value for the growth constant k. One simple way of doing this is to solve the equation

$$31{,}900 = 16{,}900 \cdot e^{k \cdot 40}.$$

Using natural logarithms just as we did in estimating the time to depletion for $k = 0.036$, we find that

$$k = \frac{\ln\left(\dfrac{31{,}900}{16{,}900}\right)}{40} \approx 0.016.$$

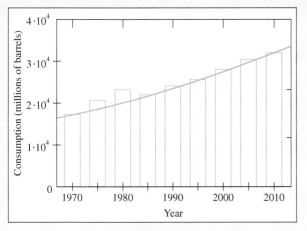

FIGURE 35

So the data from the Bureau of Transportation Statistics suggest a growth constant of about 1.6%. We can check the fit by plotting the function $16{,}900 \cdot e^{0.016t}$ along with a bar graph of the consumption data, shown in Figure 35. The fit looks reasonably good, but over this short range of 40 years, the exponential model is close to a linear model, and the growth in consumption is certainly not smooth.

The exponential model rests on the assumption of a constant growth rate. As already noted, we might expect instead that the growth rate would change as the world comes closer to exhausting its reserves. In particular, scarcity might drive up the price of oil and thus reduce consumption. We can use integration to explore an alternative model in which the factor k changes over time, so that k becomes $k(t)$, a function of time.

As an illustration, we explore a model in which the growth constant k declines toward 0 over time. We'll use 1970 as our base year, so the variable t will count years since 1970. We need a simple positive function $k(t)$ that tends toward 0 as t gets large. To get some numbers to work with, assume that the growth rate was 2% in 1970 and declined to 1% by 1995. There are many possible choices for the function $k(t)$, but a convenient one is

$$k(t) = \frac{0.5}{t + 25}.$$

Using integration to turn the instantaneous rate of consumption into the total consumption up to time T, we can write

$$\text{Total consumption} = 16{,}900 \int_0^T e^{k(t) \cdot t} \, dt$$

$$= 16{,}900 \int_0^T e^{0.5t/(t+25)} \, dt.$$

We'd like to find out when the world will use up its estimated reserves, but as just noted, the estimates have increased since the 1970s. It is estimated that the current global petroleum reserves are 3,000,000 million barrels. *Source: Geotimes.* So we need to solve

$$3{,}000{,}000 = 16{,}900 \int_0^T e^{0.5t/(t+25)} \, dt \qquad (2)$$

But this problem is much harder to solve than the corresponding problem for constant growth, because *there is no formula for evaluating this definite integral!* The function

$$g(t) = e^{0.5t/(t+25)}$$

doesn't have an antiderivative that we can write down in terms of functions that we know how to compute.

Here the numerical integration techniques discussed in Section 6 come to the rescue. We can use one of the integration rules to *approximate* the integral numerically for various values of T, and with some trial and error we can estimate how long the reserves will last. If you have a calculator or computer algebra system that does numerical integration, you can pick some T-values and evaluate the right-hand side of Equation (2). Here are the results produced by one computer algebra system:

> For $T = 120$ the integral is about 2,797,000.
>
> For $T = 130$ the integral is about 3,053,000.
>
> For $T = 140$ the integral is about 3,311,000.

So using this model we would estimate that starting in 1970 the petroleum reserves would last for about 130 years, that is, until 2100.

Our integration tools are essential in building and exploring models of resource use, but the difference in our two predictions (35 years vs. 130 years) illustrates the difficulty of making accurate predictions. A model that performs well on historical data may not take the changing dynamics of resource use into account, leading to forecasts that are either unduly gloomy or too optimistic.

EXERCISES

1. Find the number of years that the estimated petroleum reserves would last if used at the same rate as in the base year.

2. How long would the estimated petroleum reserves last if the growth constant was only 2% instead of 4.7%?

Estimate the length of time until depletion for each mineral.

3. Bauxite (the ore from which aluminum is obtained): estimated reserves in base year, 15,000,000 thousand tons; rate of consumption in base year, 63,000 thousand tons; growth constant, 6%

4. Bituminous coal: estimated world reserves, 2,000,000 million tons; rate of consumption in base year, 2200 million tons; growth constant, 4%

5. (a) Verify that the function $k(t)$ defined previously has the right values at $k = 0$ and $k = 25$.

   (b) Find a similar function that has $k(0) = 0.03$ and $k(25) = 0.02$.

6. (a) Use the function you defined in Exercise 5 (b) to write an integral for world petroleum consumption from 1970 until T years after 1970.

   (b) If you have access to a numerical integrator, compute some values of your integral and estimate the time required to exhaust the reserve of 3,000,000 million barrels.

7. A reasonable assumption is that over time scarcity might drive up the price of oil and thus reduce consumption. Comment on the fact that the rate of oil consumption actually increased in 2002, connecting current events and economic forecasts to the short-term possibility of a reduction in consumption.

8. Develop a spreadsheet that shows the time to exhaustion for various values of k.

9. Go to the website WolframAlpha.com and enter "integrate." Follow the instructions to find the time to exhaustion for various values of k. Discuss how the solution compares with the solutions provided by a graphing calculator and by Microsoft Excel.

DIRECTIONS FOR GROUP PROJECT

Suppose that you and three other students are spending a summer as interns for a local congresswoman. During your internship you realize that the information contained in your calculus class could be used to help with a new bill under consideration. The primary purpose of the bill is to require, by law, that all cars manufactured after a certain date get at least 60 miles per gallon of gasoline. Prepare a report that uses the information above to make a case for or against a bill of this nature.

8

Further Techniques and Applications of Integration

It might seem that definite integrals with infinite limits have only theoretical interest, but in fact these *improper* integrals provide answers to many practical questions. An example in Section 4 models an environmental cleanup process in which the amount of pollution entering a stream decreases by a constant fraction each year. An improper integral gives the total amount of pollutant that will ever enter the river.

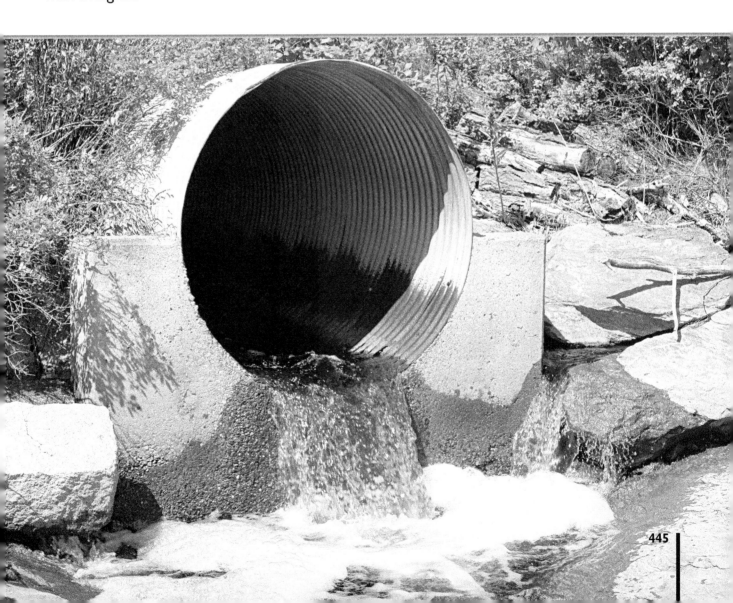

n the previous chapter we discussed indefinite and definite integrals and presented rules for finding the antiderivatives of several types of functions. We showed how numerical methods can be used for functions that cannot be integrated by the techniques presented there. In this chapter we develop additional methods of integrating functions. We also show how to evaluate an integral that has one or both limits at infinity. These new techniques allow us to consider additional applications of integration, such as volumes of solids of revolution, the average value of a function, and continuous money flow.

8.1 Integration by Parts

APPLY IT If we know the rate of growth of a patch of moss, how can we calculate the area the moss covers?

We will use integration by parts to answer this question in Exercise 42.

The technique of *integration by parts* often makes it possible to reduce a complicated integral to a simpler integral. We know that if u and v are both differentiable functions, then uv is also differentiable and, by the product rule for derivatives,

$$\frac{d(uv)}{dx} = u\frac{dv}{dx} + v\frac{du}{dx}.$$

This expression can be rewritten, using differentials, as

$$d(uv) = u\,dv + v\,du.$$

Integrating both sides of this last equation gives

$$\int d(uv) = \int u\,dv + \int v\,du,$$

or

$$uv = \int u\,dv + \int v\,du.$$

Rearranging terms gives the following formula.

Integration by Parts
If u and v are differentiable functions, then

$$\int u\,dv = uv - \int v\,du.$$

The process of finding integrals by this formula is called **integration by parts**. There are two ways to do integration by parts: the standard method and column integration. Both methods are illustrated in the following example.

EXAMPLE 1 **Integration by Parts**

Find $\int xe^{5x}\, dx$.

SOLUTION

Method 1
Standard Method

Although this integral cannot be found by using any method studied so far, it can be found with integration by parts. First write the expression $xe^{5x}\, dx$ as a product of two functions u and dv in such a way that $\int dv$ can be found. One way to do this is to choose the two functions x and e^{5x}. Both x and e^{5x} can be integrated, but $\int x\, dx$, which is $x^2/2$, is more complicated than x itself, while the derivative of x is 1, which is simpler than x. Since e^{5x} remains the same (except for the coefficient) whether it is integrated or differentiated, it is best here to choose

$$dv = e^{5x}\, dx \quad \text{and} \quad u = x.$$

Then

$$du = dx,$$

and v is found by integrating dv:

$$v = \int dv = \int e^{5x}\, dx = \frac{e^{5x}}{5}.$$

We need not introduce the constant of integration until the last step, because only one constant is needed. Now substitute into the formula for integration by parts and complete the integration.

$$\int u\, dv = uv - \int v\, du$$

$$\underbrace{\int x}_{u}\, \underbrace{e^{5x}\, dx}_{dv} = \underbrace{x}_{u}\underbrace{\left(\frac{e^{5x}}{5}\right)}_{v} - \int \underbrace{\frac{e^{5x}}{5}}_{v}\, \underbrace{dx}_{du}$$

$$= \frac{xe^{5x}}{5} - \frac{e^{5x}}{25} + C \qquad \int \frac{e^{5x}}{5}\, dx = \frac{1}{5}\int e^{5x}\, dx = \frac{e^{5x}}{25} + C$$

$$= \frac{e^{5x}}{25}(5x - 1) + C \qquad \text{Factor out } e^{5x}/25.$$

The constant C was added in the last step. As before, check the answer by taking its derivative.

FOR REVIEW

In Section 7.2 on Substitution, we pointed out that when the chain rule is used to find the derivative of the function e^{kx}, we multiply by k, so when finding the anti-derivative of e^{kx}, we divide by k. Thus $\int e^{5x}\, dx = e^{5x}/5 + C$. Keeping this technique in mind makes integration by parts simpler.

Method 2
Column Integration

A technique called **column integration**, or *tabular integration*, is equivalent to integration by parts but helps in organizing the details.* We begin by creating two columns. The first column, labeled D, contains u, the part to be differentiated in the original integral. The second column, labeled I, contains the rest of the integral: that is, the part to be integrated, but without the dx. To create the remainder of the first column, write the derivative of the function in the first row underneath it in the second row. Now write the derivative of the function in the second row underneath it in the third row. Proceed in this manner down the first column, taking derivatives until you get a 0. Form the second column in a similar manner, except take an antiderivative at each row, until the second column has the same number of rows as the first.

*This technique appeared in the 1988 movie *Stand and Deliver*.

To illustrate this process, consider our goal of finding $\int xe^{5x}\,dx$. Here $u = x$, so e^{5x} is left for the second column. Taking derivatives down the first column and antiderivatives down the second column results in the following table.

D	I
x	e^{5x}
1	$e^{5x}/5$
0	$e^{5x}/25$

Next, draw a diagonal line from each term (except the last) in the left column to the term in the row below it in the right column. Label the first such line with "+", the next with "−", and continue alternating the signs as shown.

D		I
x	+	e^{5x}
1	−	$e^{5x}/5$
0		$e^{5x}/25$

Then multiply the terms on opposite ends of each diagonal line. Finally, sum up the products just formed, adding the "+" terms and subtracting the "−" terms.

$$\int xe^{5x}\,dx = x(e^{5x}/5) - 1(e^{5x}/25) + C$$

$$= \frac{xe^{5x}}{5} - \frac{e^{5x}}{25} + C$$

$$= \frac{e^{5x}}{25}(5x - 1) + C \qquad \text{Factor out } e^{5x}/25.$$

YOUR TURN 1 Find $\int xe^{-2x}\,dx$.

Compare these steps with those of Method 1 and convince yourself that the process is the same.

TRY YOUR TURN 1

Conditions for Integration by Parts

Integration by parts can be used only if the integrand satisfies the following conditions.

1. The integrand can be written as the product of two factors, u and dv.
2. It is possible to integrate dv to get v and to differentiate u to get du.
3. The integral $\int v\,du$ can be found.

EXAMPLE 2 Integration by Parts

Find $\int \ln x\,dx$ for $x > 0$.

SOLUTION

Method 1
Standard Method

No rule has been given for integrating $\ln x$, so choose

$$dv = dx \qquad \text{and} \qquad u = \ln x.$$

Then

$$v = x \qquad \text{and} \qquad du = \frac{1}{x}\,dx,$$

and, since $uv = vu$, we have

$$\int \underbrace{\ln x}_{u} \, \underbrace{dx}_{dv} = \underbrace{x \ln x}_{v \cdot u} - \int \underbrace{x \cdot \frac{1}{x} \, dx}_{v \cdot du}$$

$$= x \ln x - \int dx$$

$$= x \ln x - x + C. \qquad \int dx = \int 1 \, dx = x + C$$

Method 2
Column Integration

Column integration works a little differently here. As in Method 1, choose $\ln x$ as the part to differentiate. The part to be integrated must be 1. (Think of $\ln x$ as $1 \cdot \ln x$.) No matter how many times $\ln x$ is differentiated, the result is never 0. In this case, stop as soon as the natural logarithm is gone.

D	I
$\ln x$	1
$1/x$	x

Draw diagonal lines with alternating $+$ and $-$ as before. On the last line, because the left column does not contain a 0, draw a horizontal line.

D		I
$\ln x$	$+$	1
$1/x$	$-$	x

The presence of a horizontal line indicates that the product is to be integrated, just as the original integral was represented by the first row of the two columns.

$$\int \ln x \, dx = (\ln x)x - \int \frac{1}{x} \cdot x \, dx$$

$$= x \ln x - \int dx$$

$$= x \ln x - x + C$$

YOUR TURN 2 Find $\int \ln 2x \, dx$.

Note that when setting up the columns, a horizontal line is drawn only when a 0 does not eventually appear in the left column.   TRY YOUR TURN 2

Sometimes integration by parts must be applied more than once, as in the next example.

EXAMPLE 3 **Integration by Parts**

Find $\int (2x^2 + 5)e^{-3x} \, dx$.

SOLUTION

Method 1
Standard Method

Choose

$$dv = e^{-3x} \, dx \qquad \text{and} \qquad u = 2x^2 + 5.$$

Then

$$v = \frac{-e^{-3x}}{3} \qquad \text{and} \qquad du = 4x \, dx.$$

Substitute these values into the formula for integration by parts.

$$\int u\,dv = uv - \int v\,du$$

$$\int (2x^2 + 5)e^{-3x}\,dx = (2x^2 + 5)\left(\frac{-e^{-3x}}{3}\right) - \int \left(\frac{-e^{-3x}}{3}\right)4x\,dx$$

$$= -(2x^2 + 5)\left(\frac{e^{-3x}}{3}\right) + \frac{4}{3}\int xe^{-3x}\,dx$$

Now apply integration by parts to the last integral, letting

$$dv = e^{-3x}\,dx \qquad \text{and} \qquad u = x,$$

so

$$v = \frac{-e^{-3x}}{3} \qquad \text{and} \qquad du = dx.$$

$$\int (2x^2 + 5)e^{-3x}\,dx = -(2x^2 + 5)\left(\frac{e^{-3x}}{3}\right) + \frac{4}{3}\int xe^{-3x}\,dx$$

$$= -(2x^2 + 5)\left(\frac{e^{-3x}}{3}\right) + \frac{4}{3}\left[x\left(\frac{-e^{-3x}}{3}\right) - \int \left(\frac{-e^{-3x}}{3}\right)dx\right]$$

$$= -(2x^2 + 5)\left(\frac{e^{-3x}}{3}\right) + \frac{4}{3}\left[-\frac{x}{3}e^{-3x} - \left(\frac{e^{-3x}}{9}\right)\right] + C$$

$$= -(2x^2 + 5)\left(\frac{e^{-3x}}{3}\right) - \frac{4}{9}xe^{-3x} - \frac{4}{27}e^{-3x} + C$$

$$= [-(2x^2 + 5)(9) - 4x(3) - 4]\frac{e^{-3x}}{27} + C \qquad \text{Factor out } e^{-3x}/27.$$

$$= (-18x^2 - 12x - 49)\frac{e^{-3x}}{27} + C \qquad \text{Simplify.}$$

Method 2
Column Integration

Choose $2x^2 + 5$ as the part to be differentiated, and put e^{-3x} in the integration column.

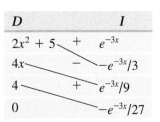

Multiplying and adding as before yields

$$\int (2x^2 + 5)e^{-3x}\,dx = (2x^2 + 5)(-e^{-3x}/3) - 4x(e^{-3x}/9) + 4(-e^{-3x}/27) + C$$

$$= (-18x^2 - 12x - 49)\frac{e^{-3x}}{27} + C. \qquad \text{Factor out } e^{-3x}/27 \text{ and simplify.}$$

YOUR TURN 3

Find $\int (3x^2 + 4)e^{2x}\,dx$.

TRY YOUR TURN 3

With the functions discussed so far in this book, choosing u and dv (or the parts to be differentiated and integrated) is relatively simple. In general, the following strategy should be used.

First see if the integration can be performed using substitution. If substitution does not work:

■ See if ln x is in the integral. If it is, set $u = \ln x$ and dv equal to the rest of the integral. (Equivalently, put ln x in the D column and the rest of the function in the I column.)

■ If ln x is not present, see if the integral contains x^k, where k is any positive integer, or any other polynomial. If it does, set $u = x^k$ (or the polynomial) and dv equal to the rest of the integral. (Equivalently, put x^k in the D column and the rest of the function in the I column.)

EXAMPLE 4   Definite Integral

Find $\displaystyle\int_1^e \frac{\ln x}{x^2}\, dx$.

SOLUTION   First find the indefinite integral using integration by parts by the standard method. (You may wish to verify this using column integration.) Whenever ln x is present, it is selected as u, so let

$$u = \ln x \qquad \text{and} \qquad dv = \frac{1}{x^2}\, dx.$$

Then

$$du = \frac{1}{x}\, dx \qquad \text{and} \qquad v = -\frac{1}{x}.$$

Substitute these values into the formula for integration by parts, and integrate the second term on the right.

$$\int u\, dv = uv - \int v\, du$$

$$\int \frac{\ln x}{x^2}\, dx = (\ln x)\frac{-1}{x} - \int\left(-\frac{1}{x}\cdot\frac{1}{x}\right) dx$$

$$= -\frac{\ln x}{x} + \int \frac{1}{x^2}\, dx$$

$$= -\frac{\ln x}{x} - \frac{1}{x} + C$$

$$= \frac{-\ln x - 1}{x} + C$$

Now find the definite integral.

$$\int_1^e \frac{\ln x}{x^2}\, dx = \left.\frac{-\ln x - 1}{x}\right|_1^e$$

$$= \left(\frac{-1-1}{e}\right) - \left(\frac{0-1}{1}\right)$$

$$= \frac{-2}{e} + 1 \approx 0.2642411177$$

TRY YOUR TURN 4

FOR REVIEW

Recall that $\int x^n\, dx = x^{n+1}/(n+1) + C$ for all $n \neq -1$, so $\int 1/x^2\, dx = \int x^{-2}\, dx = x^{-1}/(-1) + C = -1/x + C$.

YOUR TURN 4   Find

$\displaystyle\int_1^e x^2 \ln x\, dx$.

TECHNOLOGY NOTE   Definite integrals can be found with a graphing calculator using the function integral feature or by finding the area under the graph of the function between the limits. For example, using the `fnInt` feature of the TI-84 Plus C calculator to find the integral in Example 4 gives 0.2642411177. Using the area under the graph approach gives 0.26424112, the same result rounded.

Many integrals cannot be found by the methods presented so far. For example, consider the integral

$$\int \frac{1}{4 - x^2} \, dx.$$

Substitution of $u = 4 - x^2$ will not help, because $du = -2x \, dx$, and there is no x in the numerator of the integral. We could try integration by parts, using $dv = dx$ and $u = (4 - x^2)^{-1}$. Integration gives $v = x$ and differentiation gives $du = 2x \, dx/(4 - x^2)^2$, with

$$\int \frac{1}{4 - x^2} \, dx = \frac{x}{4 - x^2} - \int \frac{2x^2}{(4 - x^2)^2} \, dx.$$

The integral on the right is more complicated than the original integral, however. A second use of integration by parts on the new integral would only make matters worse. Since we cannot choose $dv = (4 - x^2)^{-1} \, dx$ because it cannot be integrated by the methods studied so far, integration by parts is not possible for this problem.

This integration can be performed using one of the many techniques of integration beyond the scope of this text.* Tables of integrals can also be used, but technology is rapidly making such tables obsolete and even reducing the importance of techniques of integration. The following example shows how the table of integrals given in the appendix of this book may be used.

EXAMPLE 5 Tables of Integrals

Find $\int \dfrac{1}{4 - x^2} \, dx$.

SOLUTION Using formula 7 in the table of integrals in the appendix, with $a = 2$, gives

$$\int \frac{1}{4 - x^2} \, dx = \frac{1}{4} \cdot \ln \left| \frac{2 + x}{2 - x} \right| + C. \qquad \text{TRY YOUR TURN 5}$$

YOUR TURN 5
Find $\displaystyle\int \frac{1}{x\sqrt{4 + x^2}} \, dx$.

TECHNOLOGY NOTE

We mentioned in the previous chapter how computer algebra systems and some calculators can perform integration. Using a TI-89, the answer to the above integral is

$$\frac{\ln\left(\dfrac{|x + 2|}{|x - 2|}\right)}{4}.$$

(The C is not included.) Verify that this is equivalent to the answer given in Example 5.

If you don't have a calculator or computer program that integrates symbolically, there is a Web site (integrals.wolfram.com), as of this writing, that finds indefinite integrals using the computer algebra system Mathematica. It includes instructions on how to enter your function. When the previous integral was entered, it returned the answer

$$\frac{1}{4}(\log(-x - 2) - \log(x - 2)).$$

Note that Mathematica does not include the C or the absolute value, and that natural logarithms are written as log. Verify that this answer is equivalent to the answer given by the TI-89 and the answer given in Example 5.

Unfortunately, there are integrals that cannot be antidifferentiated by any technique, in which case numerical integration must be used. (See the last section of the previous chapter.) In this book, for simplicity, all integrals to be antidifferentiated can be done with substitution or by parts, except for Exercises 23–28 in this section.

---

*For example, see Thomas, George B., Maurice D. Weir, and Joel Hass, *Thomas' Calculus*, 13th ed., Pearson, 2014.

8.1 EXERCISES

Use integration by parts to find the integrals in Exercises 1–10.

1. $\displaystyle\int xe^x\,dx$

2. $\displaystyle\int (x+6)e^x\,dx$

3. $\displaystyle\int (4x-12)e^{-8x}\,dx$

4. $\displaystyle\int (6x+3)e^{-2x}\,dx$

5. $\displaystyle\int x\ln x\,dx$

6. $\displaystyle\int x^3\ln x\,dx$

7. $\displaystyle\int_0^1 \dfrac{2x+1}{e^x}\,dx$

8. $\displaystyle\int_0^3 \dfrac{3-x}{3e^x}\,dx$

9. $\displaystyle\int_1^9 \ln 3x\,dx$

10. $\displaystyle\int_1^2 \ln 5x\,dx$

11. Find the area between $y=(x-2)e^x$ and the x-axis from $x=2$ to $x=4$.

12. Find the area between $y=(x+1)\ln x$ and the x-axis from $x=1$ to $x=e$.

Exercises 13–22 are mixed—some require integration by parts, while others can be integrated by using techniques discussed in the chapter on Integration.

13. $\displaystyle\int x^2 e^{2x}\,dx$

14. $\displaystyle\int \dfrac{x^2\,dx}{2x^3+1}$

15. $\displaystyle\int x^2\sqrt{x+4}\,dx$

16. $\displaystyle\int (2x-1)\ln(3x)\,dx$

17. $\displaystyle\int (8x+10)\ln(5x)\,dx$

18. $\displaystyle\int x^3 e^{x^4}\,dx$

19. $\displaystyle\int_1^2 (1-x^2)e^{2x}\,dx$

20. $\displaystyle\int_0^1 \dfrac{x^2\,dx}{2x^3+1}$

21. $\displaystyle\int_0^1 \dfrac{x^3\,dx}{\sqrt{3+x^2}}$

22. $\displaystyle\int_0^5 x\sqrt[3]{x^2+2}\,dx$

Use the table of integrals, or a computer or calculator with symbolic integration capabilities, to find each indefinite integral.

23. $\displaystyle\int \dfrac{16}{\sqrt{x^2+16}}\,dx$

24. $\displaystyle\int \dfrac{10}{x^2-25}\,dx$

25. $\displaystyle\int \dfrac{3}{x\sqrt{121-x^2}}\,dx$

26. $\displaystyle\int \dfrac{2}{3x(3x-5)}\,dx$

27. $\displaystyle\int \dfrac{-6}{x(4x+6)^2}\,dx$

28. $\displaystyle\int \sqrt{x^2+15}\,dx$

29. What rule of differentiation is related to integration by parts?

30. Explain why the two methods of solving Example 2 are equivalent.

31. Suppose that u and v are differentiable functions of x with $\displaystyle\int_0^1 v\,du = 4$ and the following functional values.

x	$u(x)$	$v(x)$
0	2	1
1	3	-4

Use this information to determine $\displaystyle\int_0^1 u\,dv$.

32. Suppose that u and v are differentiable functions of x with $\displaystyle\int_1^{20} v\,du = -1$ and the following functional values.

x	$u(x)$	$v(x)$
1	5	-2
20	15	6

Use this information to determine $\displaystyle\int_1^{20} u\,dv$.

33. Suppose we know that the functions r and s are everywhere differentiable and that $r(0) = 0$. Suppose we also know that for $0 \le x \le 2$, the area between the x-axis and the nonnegative function $h(x) = s(x)\dfrac{dr}{dx}$ is 5, and that on the same interval, the area between the x-axis and the nonnegative function $k(x) = r(x)\dfrac{ds}{dx}$ is 10. Determine $r(2)s(2)$.

34. Suppose we know that the functions u and v are everywhere differentiable and that $u(3) = 0$. Suppose we also know that for $1 \le x \le 3$, the area between the x-axis and the nonnegative function $h(x) = u(x)\dfrac{dv}{dx}$ is 15, and that on the same interval, the area between the x-axis and the nonnegative function $k(x) = v(x)\dfrac{du}{dx}$ is 20. Determine $u(1)v(1)$.

35. Use integration by parts to derive the following formula from the table of integrals.

$$\int x^n \cdot \ln |x|\ dx = x^{n+1}\left[\frac{\ln |x|}{n + 1} - \frac{1}{(n + 1)^2}\right] + C, \quad n \ne -1$$

36. Use integration by parts to derive the following formula from the table of integrals.

$$\int x^n e^{ax}\ dx = \frac{x^n e^{ax}}{a} - \frac{n}{a}\int x^{n-1} e^{ax}\ dx + C, \quad a \ne 0$$

37. (a) One way to integrate $\int x\sqrt{x + 1}\ dx$ is to use integration by parts. Do so to find the antiderivative.

    (b) Another way to evaluate the integral in part (a) is by using the substitution $u = x + 1$. Do so to find the antiderivative.

    (c) Compare the results from the two methods. If they do not look the same, explain how this can happen. Discuss the advantages and disadvantages of each method.

38. Using integration by parts,

$$\int \frac{1}{x}\ dx = \int \frac{1}{x} \cdot 1\ dx$$

$$= \frac{1}{x} \cdot x - \int \left(-\frac{1}{x^2}\right)x\ dx$$

$$= 1 + \int \frac{1}{x}\ dx.$$

Subtracting $\int \frac{1}{x}\ dx$ from both sides we conclude that $0 = 1$. What is wrong with this logic? *Source: Sam Northshield.*

APPLICATIONS

Business and Economics

39. **Rate of Change of Revenue** The rate of change of revenue (in dollars per calculator) from the sale of x calculators is

$$R'(x) = (x + 1)\ln(x + 1).$$

Find the total revenue from the sale of the first 12 calculators. (*Hint:* In this exercise, it simplifies matters to write an antiderivative of $x + 1$ as $(x + 1)^2/2$ rather than $x^2/2 + x$.)

Life Sciences

40. **Reaction to a Drug** The rate of reaction to a drug is given by

$$r'(t) = 2t^2 e^{-t},$$

where t is the number of hours since the drug was administered. Find the total reaction to the drug from $t = 1$ to $t = 6$.

41. **Growth of a Population** The rate of growth of a microbe population is given by

$$m'(t) = 27te^{3t},$$

where t is time in days. What is the total accumulated growth during the first 2 days?

42. **APPLY IT Rate of Growth** The area covered by a patch of moss is growing at a rate of

$$A'(t) = \sqrt{t}\ln t$$

cm^2 per day, for $t \ge 1$. Find the additional amount of area covered by the moss between 4 and 9 days.

43. **Thermic Effect of Food** As we saw in an earlier chapter, a person's metabolic rate tends to go up after eating a meal and then, after some time has passed, it returns to a resting metabolic rate. This phenomenon is known as the thermic effect of food, and the effect (in kJ per hour) for one individual is

$$F(t) = -10.28 + 175.9te^{-t/1.3},$$

where t is the number of hours that have elapsed since eating a meal. Find the total thermic energy of a meal for the next six hours after a meal by integrating the thermic effect function between $t = 0$ and $t = 6$. *Source: American Journal of Clinical Nutrition.*

44. **Rumen Fermentation** The rumen is the first division of the stomach of a ruminant, or cud-chewing animal. An article on the rumen microbial system reports that the fraction of the soluble material passing from the rumen without being fermented during the first hour after its ingestion could be calculated by the integral

$$\int_0^1 ke^{-kt}(1 - t)\ dt,$$

where k measures the rate that the material is fermented. *Source: Annual Review of Ecology and Systematics.*

    (a) Determine the above integral, and evaluate it for the following values of k used in the article: 1/12, 1/24, and 1/48 hour.

    (b) The fraction of intermediate material left in the rumen at 1 hour that escapes digestion by passage between 1 and 6 hours is given by

$$\int_1^6 ke^{-kt}(6 - t)/5\ dt.$$

Determine this integral, and evaluate it for the values of k given in part (a).

YOUR TURN ANSWERS

1. $-e^{-2x}(2x + 1)/4 + C$
2. $x \ln 2x - x + C$
3. $(6x^2 - 6x + 11)e^{2x}/4 + C$
4. $(2e^3 + 1)/9$
5. $(-1/2)\ln |(2 + \sqrt{4 + x^2})/x| + C$

8.2 Volume and Average Value

In this section, we discover how to find the average value of a function, and then we use this concept to answer this question in Example 4.

Volume   Figure 1 shows the regions below the graph of some function $y = f(x)$, above the x-axis, and between $x = a$ and $x = b$. We have seen how to use integrals to find the area of such a region. Suppose this region is revolved about the x-axis as shown in Figure 2. The resulting figure is called a **solid of revolution**. In many cases, the volume of a solid of revolution can be found by integration.

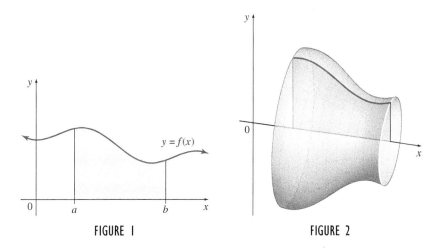

FIGURE 1   FIGURE 2

To begin, divide the interval $[a, b]$ into n subintervals of equal width Δx by the points $a = x_0, x_1, x_2, \ldots, x_i, \ldots, x_n = b$. Then think of slicing the solid into n slices of equal thickness Δx, as shown in Figure 3(a). If the slices are thin enough, each slice is very close to being a right circular cylinder, as shown in Figure 3(b). The formula for the volume

(a)   (b)

FIGURE 3

$f(x_i)$

Δx

FIGURE 4

of a right circular cylinder is $\pi r^2 h$, where r is the radius of the circular base and h is the height of the cylinder. As shown in Figure 4, the height of each slice is Δx. (The height is horizontal here, since the cylinder is on its side.) The radius of the circular base of each slice is $f(x_i)$. Thus, the volume of the slice is closely approximated by $\pi[f(x_i)]^2 \Delta x$. The volume of the solid of revolution will be approximated by the sum of the volumes of the slices:

$$V \approx \sum_{i=1}^{n} \pi[f(x_i)]^2 \Delta x.$$

By definition, the volume of the solid of revolution is the limit of this sum as the thickness of the slices approaches 0, or

$$V = \lim_{\Delta x \to 0} \sum_{i=1}^{n} \pi[f(x_i)]^2 \Delta x.$$

This limit, like the one discussed earlier for area, is a definite integral.

Volume of a Solid of Revolution

If $f(x)$ is nonnegative and R is the region between $f(x)$ and the x-axis from $x = a$ to $x = b$, the volume of the solid formed by rotating R about the x-axis is given by

$$V = \lim_{\Delta x \to 0} \sum_{i=1}^{n} \pi[f(x_i)]^2 \Delta x = \int_{a}^{b} \pi[f(x)]^2 \, dx.$$

The technique of summing disks to approximate volumes was originated by Johannes Kepler (1571–1630), a famous German astronomer who discovered three laws of planetary motion. He estimated volumes of wine casks used at his wedding by means of solids of revolution.

EXAMPLE 1 Volume

Find the volume of the solid of revolution formed by rotating about the x-axis the region bounded by $y = x + 1$, $y = 0$, $x = 1$, and $x = 4$.

SOLUTION The region and the solid are shown in Figure 5. Notice that the orientation of the x-axis is slightly different in Figure 5(b) than in Figure 5(a) to emphasize the

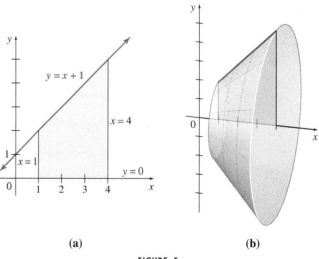

(a)   (b)

FIGURE 5

three-dimensionality of the figure. Use the formula for the volume of a solid of revolution, with $a = 1$, $b = 4$, and $f(x) = x + 1$.

$$V = \int_1^4 \pi(x + 1)^2 \, dx = \pi\left[\frac{(x + 1)^3}{3}\right]\Big|_1^4$$

$$= \frac{\pi}{3}(5^3 - 2^3)$$

$$= \frac{117\pi}{3} = 39\pi$$

TRY YOUR TURN 1

YOUR TURN 1 Find the volume of the solid of revolution formed by rotating about the x-axis the region bounded by $y = x^2 + 1$, $y = 0$, $x = -1$, and $x = 1$.

EXAMPLE 2 Volume

Find the volume of the solid of revolution formed by rotating about the x-axis the area bounded by $f(x) = 4 - x^2$ and the x-axis.

SOLUTION The region and the solid are shown in Figure 6 below. First, determine a and b by finding the x-intercepts. If $y = 0$, then $4 - x^2 = (2 + x)(2 - x) = 0$, and $x = -2$ or $x = 2$. With $a = -2$ and $b = 2$, the volume is

$$V = \int_{-2}^2 \pi(4 - x^2)^2 \, dx$$

$$= \int_{-2}^2 \pi(16 - 8x^2 + x^4) \, dx$$

$$= \pi\left(16x - \frac{8x^3}{3} + \frac{x^5}{5}\right)\Big|_{-2}^2$$

$$= \frac{512\pi}{15}.$$

TECHNOLOGY NOTE A graphing calculator with the `fnInt` feature gives the value as 107.2330292, which agrees with the approximation of $512\pi/15$ to the 7 decimal places shown.

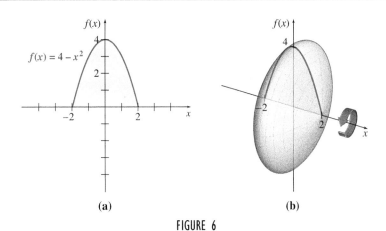

$f(x)$

$f(x) = 4 - x^2$

-2 2 x

$f(x)$

4

2 2 x

(a)   (b)

FIGURE 6

EXAMPLE 3 Volume

Find the volume of a right circular cone with height h and base radius r.

SOLUTION Figure 7(a) on the next page shows the required cone, while Figure 7(b) shows an area that could be rotated about the x-axis to get such a cone. The cone formed by the rotation is shown in Figure 7(c). Here $y = f(x)$ is the equation of the line through $(0, 0)$ and (h, r). The slope of this line is r/h, and since the y-intercept is 0, the equation of the line is

$$y = \frac{r}{h}x.$$

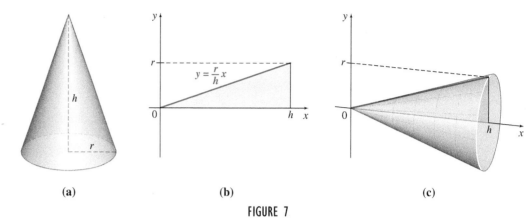

(a)

(b)

(c)

FIGURE 7

Then the volume is

$$V = \int_0^h \pi \left(\frac{r}{h}x\right)^2 dx = \pi \int_0^h \frac{r^2 x^2}{h^2} dx$$

$$= \pi \frac{r^2 x^3}{3h^2} \bigg|_0^h \qquad \text{Since } r \text{ and } h \text{ are constants}$$

$$= \frac{\pi r^2 h}{3}.$$

This is the familiar formula for the volume of a right circular cone.

Average Value of a Function

The average of the n numbers $v_1, v_2, v_3, \ldots,$ $v_i, \ldots, v_n$ is given by

$$\frac{v_1 + v_2 + v_3 + \cdots + v_n}{n} = \frac{\sum_{i=1}^n v_i}{n}.$$

For example, to compute an average temperature, we could take readings at equally spaced intervals and average the readings.

The average value of a function f on $[a, b]$ can be defined in a similar manner; divide the interval $[a, b]$ into n subintervals, each of width Δx. Then choose an x-value, x_i, in each subinterval, and find $f(x_i)$. The average function value for the n subintervals and the given choices of x_i is

$$\frac{f(x_1) + f(x_2) + \cdots + f(x_n)}{n} = \frac{\sum_{i=1}^n f(x_i)}{n}.$$

Since $(b - a)/n = \Delta x$, multiply the expression on the right side of the equation by $(b - a)/(b - a)$ and rearrange the expression to get

$$\frac{b - a}{b - a} \cdot \frac{\sum_{i=1}^n f(x_i)}{n} = \frac{1}{b - a} \sum_{i=1}^n f(x_i) \left(\frac{b - a}{n}\right) = \frac{1}{b - a} \sum_{i=1}^n f(x_i) \Delta x.$$

Now, take the limit as $n \to \infty$. If the limit exists, then

$$\lim_{n \to \infty} \frac{1}{b - a} \sum_{i=1}^n f(x_i) \Delta x = \frac{1}{b - a} \lim_{n \to \infty} \sum_{i=1}^n f(x_i) \Delta x = \frac{1}{b - a} \int_a^b f(x) \, dx.$$

The following definition summarizes this discussion.

Average Value of a Function

The **average value of a function** f on the interval $[a, b]$ is

$$\frac{1}{b-a}\int_a^b f(x)\,dx,$$

provided the indicated definite integral exists.

In Figure 8 the quantity $\bar{y}$ represents the average height of the irregular region. The average height can be thought of as the height of a rectangle with base $b - a$. For $f(x) \geq 0$, this rectangle has area $\bar{y}(b - a)$, which equals the area under the graph of $f(x)$ from $x = a$ to $x = b$, so that

$$\bar{y}(b - a) = \int_a^b f(x)\,dx.$$

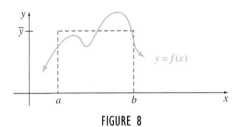

FIGURE 8

EXAMPLE 4   Average Price

A stock analyst plots the price per share of a certain common stock as a function of time and finds that it can be approximated by the function

$$S(t) = 25 - 5e^{-0.01t},$$

where t is the time (in years) since the stock was purchased. Find the average price of the stock over the first six years.

APPLY IT

SOLUTION   Use the formula for average value with $a = 0$ and $b = 6$. The average price is

$$\frac{1}{6-0}\int_0^6 (25 - 5e^{-0.01t})\,dt = \frac{1}{6}\left(25t - \frac{5}{-0.01}e^{-0.01t}\right)\Big|_0^6$$

$$= \frac{1}{6}(25t + 500e^{-0.01t})\Big|_0^6$$

$$= \frac{1}{6}(150 + 500e^{-0.06} - 500)$$

$$\approx 20.147,$$

or approximately \$20.15.

YOUR TURN 2   Find the average value of the function $f(x) = x + \sqrt{x}$ on the interval $[1, 4]$.

TRY YOUR TURN 2

8.2 WARM-UP EXERCISES

Find the following.

W1. $\displaystyle\int_0^2 (x + 5)^2 \, dx$ *(Sec. 7.4)*

W2. $\displaystyle\int_0^3 (6 - x^2)^2 \, dx$ *(Sec. 7.4)*

W3. $\displaystyle\int_{-2}^1 \sqrt{x + 3} \, dx$ *(Sec. 7.4)*

W4. $\displaystyle\int_2^7 \frac{7}{\sqrt{x + 2}} \, dx$ *(Sec. 7.4)*

W5. $\displaystyle\int_0^1 (e^x)^2 \, dx$ *(Sec. 7.4)*

W6. $\displaystyle\int_1^3 x \ln x \, dx$ *(Sec. 8.1)*

8.2 EXERCISES

Find the volume of the solid of revolution formed by rotating about the x-axis each region bounded by the given curves.

1. $f(x) = x,\quad y = 0,\quad x = 0,\quad x = 3$

2. $f(x) = 3x,\quad y = 0,\quad x = 0,\quad x = 2$

3. $f(x) = 2x + 1,\quad y = 0,\quad x = 0,\quad x = 4$

4. $f(x) = x - 4,\quad y = 0,\quad x = 4,\quad x = 10$

5. $f(x) = \dfrac{1}{3}x + 2,\quad y = 0,\quad x = 1,\quad x = 3$

6. $f(x) = \dfrac{1}{2}x + 4,\quad y = 0,\quad x = 0,\quad x = 5$

7. $f(x) = \sqrt{x},\quad y = 0,\quad x = 1,\quad x = 4$

8. $f(x) = \sqrt{x + 5},\quad y = 0,\quad x = 1,\quad x = 3$

9. $f(x) = \sqrt{2x + 1},\quad y = 0,\quad x = 1,\quad x = 4$

10. $f(x) = \sqrt{4x + 2},\quad y = 0,\quad x = 0,\quad x = 2$

11. $f(x) = e^x,\quad y = 0,\quad x = 0,\quad x = 2$

12. $f(x) = 2e^x,\quad y = 0,\quad x = -2,\quad x = 1$

13. $f(x) = \dfrac{2}{\sqrt{x}},\quad y = 0,\quad x = 1,\quad x = 3$

14. $f(x) = \dfrac{2}{\sqrt{x + 2}},\quad y = 0,\quad x = -1,\quad x = 2$

15. $f(x) = x^2,\quad y = 0,\quad x = 1,\quad x = 5$

16. $f(x) = \dfrac{x^2}{2},\quad y = 0,\quad x = 0,\quad x = 4$

17. $f(x) = 1 - x^2,\quad y = 0$

18. $f(x) = 2 - x^2,\quad y = 0$

The function defined by $y = \sqrt{r^2 - x^2}$ has as its graph a semicircle of radius r with center at $(0, 0)$ (see the figure in the next column). In Exercises 19–21, find the volume that results when each semicircle is rotated about the x-axis. (The result of Exercise 21 gives a formula for the volume of a sphere with radius r.)

19. $f(x) = \sqrt{1 - x^2}$

20. $f(x) = \sqrt{36 - x^2}$

21. $f(x) = \sqrt{r^2 - x^2}$

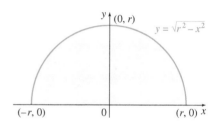

22. Find a formula for the volume of an ellipsoid. See Exercises 19–21 and the following figures.

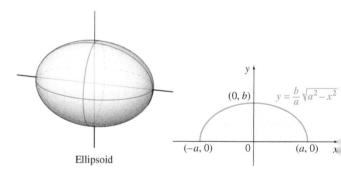

Ellipsoid

23. Use the methods of this section to find the volume of a cylinder with height h and radius r.

Find the average value of each function on the given interval.

24. $f(x) = 2 - 3x^2;\quad [1, 3]$

25. $f(x) = x^2 - 4;\quad [0, 5]$

26. $f(x) = (2x - 1)^{1/2};\quad [1, 13]$

27. $f(x) = \sqrt{x + 1};\quad [3, 8]$

28. $f(x) = e^{0.1x};\quad [0, 10]$

29. $f(x) = e^{x/7};\quad [0, 7]$

30. $f(x) = x \ln x;\quad [1, e]$

31. $f(x) = x^2 e^{2x};\quad [0, 2]$

In Exercises 32 and 33, use the integration feature on a graphing calculator to find the volume of the solid of revolution by rotating about the *x*-axis each region bounded by the given curves.

32. $f(x) = \dfrac{1}{4 + x^2}, \quad y = 0, \quad x = -2, \quad x = 2$

33. $f(x) = e^{-x^2}, \quad y = 0, \quad x = -1, \quad x = 1$

APPLICATIONS

Business and Economics

34. Average Price David Lopez plots the price per share of a stock that he owns as a function of time and finds that it can be approximated by the function

$$S(t) = t(25 - 5t) + 18,$$

where *t* is the time (in years) since the stock was purchased. Find the average price of the stock over the first five years.

35. Average Price A stock analyst plots the price per share of a certain common stock as a function of time and finds that it can be approximated by the function

$$S(t) = 37 + 6e^{-0.03t},$$

where *t* is the time (in years) since the stock was purchased. Find the average price of the stock over the first six years.

36. Average Inventory The Yasuko Okada Fragrance Company (YOFC) receives a shipment of 400 cases of specialty perfume early Monday morning of every week. YOFC sells the perfume to retail outlets in California at a rate of about 80 cases per day during each business day (Monday through Friday). What is the average daily inventory for YOFC? (*Hint:* Find a function that represents the inventory for any given business day and then integrate.)

37. Average Inventory The DeMarco Pasta Company receives 600 cases of imported San Marzano tomato sauce every 30 days. The number of cases of sauce in inventory *t* days after the shipment arrives is

$$N(t) = 600 - 20\sqrt{30t}.$$

Find the average daily inventory.

38. Public Debt In Example 5 of Section 2.3, the Federal debt held by the public as a percent of the gross domestic product (GDP) was modeled by the function

$$y = 0.00630t^3 - 0.289t^2 + 4.00t + 25.3$$

where *t* was the time (in years) since 1980. *Source: Budget of the U.S. Government*.

(a) Find the average public debt, as a percent of GDP, from 1990 to 2000.

(b) Find the average public debt, as a percent of GDP, from 2000 to 2015.

39. Wheat Production In Example 7 of Section 2.4, the U.S. corn production (in billions of bushels) was modeled by the function

$$p(t) = 1.757(1.0248)^t,$$

where *t* was the time (in years) since 1930. *Source: The New York Times Magazine*.

(a) Find the average corn production from 1930 to 1950.

(b) Find the average corn production from 2000 to 2010.

Life Sciences

40. Blood Flow The figure shows the blood flow in a small artery of the body. The flow of blood is *laminar* (in layers), with the velocity very low near the artery walls and highest in the center of the artery. In this model of blood flow, we calculate the total flow in the artery by thinking of the flow as being made up of many layers of concentric tubes sliding one on the other.

Suppose *R* is the radius of an artery and *r* is the distance from a given layer to the center. Then the velocity of blood in a given layer can be shown to equal

$$v(r) = k(R^2 - r^2),$$

where *k* is a numerical constant.

Since the area of a circle is $A = \pi r^2$, the change in the area of the cross section of one of the layers, corresponding to a small change in the radius, Δr, can be approximated by differentials. For $dr = \Delta r$, the differential of the area *A* is

$$dA = 2\pi r\, dr = 2\pi r\, \Delta r,$$

where Δr is the thickness of the layer. The total flow in the layer is defined to be the product of velocity and cross-section area, or

$$F(r) = 2\pi rk(R^2 - r^2)\Delta r.$$

(a) Set up a definite integral to find the total flow in the artery.

(b) Evaluate this definite integral.

41. Drug Reaction The intensity of the reaction to a certain drug, in appropriate units, is given by

$$R(t) = te^{-0.1t},$$

where *t* is time (in hours) after the drug is administered. Find the average intensity during the following hours.

(a) Second hour

(b) Twelfth hour

(c) Twenty-fourth hour

42. Bird Eggs The average length and width of various bird eggs are given in the following table. *Source: NCTM*.

Bird Name	Length (cm)	Width (cm)
Canada goose	8.6	5.8
Robin	1.9	1.5
Turtledove	3.1	2.3
Hummingbird	1.0	1.0
Raven	5.0	3.3

(a) Assume for simplicity that a bird's egg is roughly the shape of an ellipsoid. Use the result of Exercise 22 to estimate the volume of an egg of each bird.

   i. Canada goose

   ii. Robin

   iii. Turtledove

   iv. Hummingbird

   v. Raven

(b) In Exercise 17 of Section 1.3, we showed that the average length (in centimeters) of an egg of width w cm is given by

$$l = 1.585w - 0.487.$$

Using this result and the ideas in part (a), show that the average volume of an egg of width w centimeters is given by

$$V = \pi(1.585w^3 - 0.487w^2)/6.$$

Use this formula to calculate the average volume for the bird eggs in part (a), and compare with your results from part (a).

Social Sciences

43. Production Rate Suppose the number of items a new worker on an assembly line produces daily after t days on the job is given by

$$I(t) = 45 \ln(t + 1).$$

Find the average number of items produced daily by this employee after the following numbers of days.

   (a) 5   **(b)** 9   **(c)** 30

44. Typing Speed The function $W(t) = -3.75t^2 + 30t + 40$ describes a typist's speed (in words per minute) over a time interval $[0, 5]$.

   (a) Find $W(0)$.

   (b) Find the maximum W value and the time t when it occurs.

   (c) Find the average speed over $[0, 5]$.

Physical Sciences

45. Earth's Volume Most people assume that the earth has a spherical shape. It is actually more of an ellipsoid shape, but not an exact ellipsoid, since there are numerous mountains and valleys. Researchers have found that a *datum*, or a reference ellipsoid, that is offset from the center of the earth can be used to accurately map different regions. According to one datum, called the Geodetic Reference System 1980, this reference ellipsoid assumes an equatorial radius of 6,378,137 m and a polar radius of 6,356,752.3141 m. The ellipsoid is rotated about the polar axis. Use the result of Exercise 22 to estimate the volume of the earth. *Source: Geodesy Information System.*

YOUR TURN ANSWERS

1. $56\pi/15$ **2.** $73/18$

8.3 Continuous Money Flow

APPLY IT Given a changing rate of annual income and a certain rate of interest, how can we find the present value of the income?

We will answer this question in Example 2 using the concept of continuous money flow.

In an earlier chapter we looked at the concepts of present value and future value when a lump sum of money is deposited in an account and allowed to accumulate interest. In some situations, however, money flows into and out of an account almost continuously over a period of time. Examples include income in a store, bank receipts and payments, and highway tolls. Although the flow of money in such cases is not exactly continuous, it can be treated as though it were continuous, with useful results.

EXAMPLE 1 Total Income

The income from a soda machine (in dollars per year) is growing exponentially. When the machine was first installed, it was producing income at a rate of $1000 per year. By the end of the first year, it was producing income at a rate of $1020.20 per year. Find the total income produced by the machine during its first 3 years of operation.

SOLUTION Let t be the time (in years) since the installation of the machine. The assumption of exponential growth, coupled with the initial value of 1000, implies that the rate of change of income is of the form

$$f(t) = 1000e^{kt},$$

where k is some constant. To find k, use the value at the end of the first year.

$$f(1) = 1000e^{k(1)} = 1020.20$$
$$e^k = 1.0202 \qquad \text{Divide by 1000.}$$
$$k = \ln 1.0202 \qquad \text{Take ln of both sides.}$$
$$\approx 0.02$$

Therefore, we have

$$f(t) = 1000e^{0.02t}.$$

Since the rate of change of incomes is given, the total income can be determined by using the definite integral.

YOUR TURN 1 Find the total income over the first 2 years in Example 1 with the initial rate changed to $810 per year and the rate at the end of the first year changed to $797.94 per year.

$$\text{Total income} = \int_0^3 1000e^{0.02t}\, dt$$

$$= \frac{1000}{0.02}e^{0.02t}\Big|_0^3$$

$$= 50{,}000e^{0.02t}\Big|_0^3 = 50{,}000(e^{0.06} - 1) \approx 3091.83$$

Thus, the soda machine will produce $3091.83 total income in its first three years of operation. **TRY YOUR TURN 1**

The money in Example 1 is not received as a one-time lump sum payment of $3091.83. Instead, it comes in on a regular basis, perhaps daily, weekly, or monthly. In discussions of such problems it is usually assumed that the income is received continuously over a period of time.

Total Money Flow
Let the continuous function $f(t)$ represent the rate of flow of money per unit of time. If t is in years and $f(t)$ is in dollars per year, the area under $f(t)$ between two points in time gives the total dollar flow over the given time interval.

The function $f(t) = 2000$, shown in Figure 9, represents a uniform rate of money flow of $2000 per year. The graph of this money flow is a horizontal line; the *total money flow* over a specified time T is given by the rectangular area below the graph of $f(t)$ and above the t-axis between $t = 0$ and $t = T$. For example, the total money flow over $T = 5$ years would be $2000(5) = 10{,}000$, or $10{,}000.

The area in the uniform rate example could be found by using an area formula from geometry. For a variable function like the function in Example 1, however, a definite

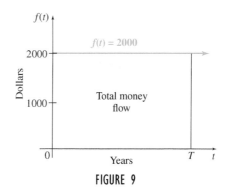

FIGURE 9

integral is needed to find the total money flow over a specific time interval. For the function $f(t) = 2000e^{0.08t}$, for example, the total money flow over a 5-year period would be given by

$$\int_0^5 2000e^{0.08t}\, dt \approx 12{,}295.62,$$

or \$12,295.62. See Figure 10.

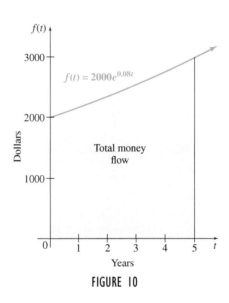

FIGURE 10

Total Money Flow
If $f(t)$ is the rate of money flow, then the **total money flow** over the time interval from $t = 0$ to $t = T$ is given by

$$\int_0^T f(t)\, dt.$$

This "total money flow" does not take into account the interest the money could earn after it is received. It is simply the total income.

Present Value of Money Flow
As mentioned earlier, an amount of money that can be deposited today at a specified interest rate to yield a given sum in the future is called the *present value* of this future sum. The future sum may be called the *future value* or *final amount*. To find the **present value of a continuous money flow** with interest compounded continuously, let $f(t)$ represent the rate of the continuous flow. In Figure 11 on the next page, the time axis from 0 to T is divided into n subintervals, each of width Δt. The amount of money that flows during any interval of time is given by the area between the t-axis and the graph of $f(t)$ over the specified time interval. The area of each subinterval is approximated by the area of a rectangle with height $f(t_i)$, where t_i is the left endpoint of the ith subinterval. The area of each rectangle is $f(t_i)\Delta t$, which (approximately) gives the amount of money flow over that subinterval.

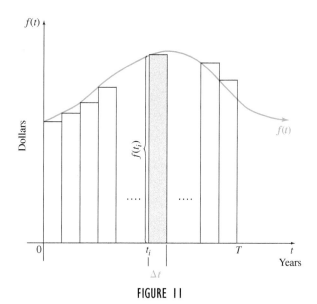

FIGURE 11

Earlier, we saw that the present value P of an amount A compounded continuously for t years at a rate of interest r is $P = Ae^{-rt}$. Letting t_i represent the time and replacing A with $f(t_i)\Delta t$, the present value of the money flow over the ith subinterval is approximately equal to

$$P_i = [f(t_i)\Delta t]e^{-rt_i}.$$

The total present value is approximately equal to the sum

$$\sum_{i=1}^{n}[f(t_i)\Delta t]e^{-rt_i}.$$

This approximation is improved as n increases; taking the limit of the sum as n increases without bound gives the present value

$$P = \lim_{n\to\infty}\sum_{i=1}^{n}[f(t_i)\Delta t]e^{-rt_i}.$$

This limit of a summation is given by the following definite integral.

Present Value of Money Flow

If $f(t)$ is the rate of continuous money flow at an interest rate r for T years, then the present value is

$$P = \int_{0}^{T} f(t)e^{-rt}\, dt.$$

To understand present value of money flow, consider an account that earns interest and has a continuous money flow. The present value of the money flow is the amount that would have to be deposited into a second account that has the same interest rate but does not have a continuous money flow, so the two accounts have the same amount of money after a specified time.

EXAMPLE 2 Present Value of Income

A company expects its rate of annual income during the next three years to be given by

$$f(t) = 75{,}000t, \quad 0 \le t \le 3.$$

What is the present value of this income over the 3-year period, assuming an annual interest rate of 8% compounded continuously?

APPLY IT

SOLUTION Use the formula for present value, with $f(t) = 75{,}000t$, $T = 3$, and $r = 0.08$.

$$P = \int_0^3 75{,}000te^{-0.08t}\, dt = 75{,}000\int_0^3 te^{-0.08t}\, dt$$

Using integration by parts, verify that

$$\int te^{-0.08t}\, dt = -12.5te^{-0.08t} - 156.25e^{-0.08t} + C.$$

Therefore,

$$75{,}000\int_0^3 te^{-0.08t}\, dt = 75{,}000\left(-12.5te^{-0.08t} - 156.25e^{-0.08t}\right)\Big|_0^3$$

$$= 75{,}000\left[-12.5(3)e^{-0.08(3)} - 156.25e^{-0.08(3)} - (0 - 156.25)\right]$$

$$\approx 75{,}000(-29.498545 - 122.910603 + 156.25)$$

$$\approx 288{,}064,$$

or about \$288,000. Notice that the actual income over the 3-year period is given by

$$\text{Total money flow} = \int_0^3 75{,}000t\, dt = \frac{75{,}000t^2}{2}\Big|_0^3 = 337{,}500,$$

YOUR TURN 2 Find the present value of an income given by $f(t) = 50{,}000t$ over the next 5 years if the interest rate is 3.5%.

or \$337,500. This means that it would take a lump-sum deposit of \$288,064 today paying a continuously compounded interest rate of 8% over a 3-year period to equal the total cash flow of \$337,500 with interest. This approach is used as a basis for determining insurance claims involving income considerations. **TRY YOUR TURN 2**

Accumulated Amount of Money Flow at Time T To find the
accumulated amount of money flow with interest at any time t, start with the formula $A = Pe^{rt}$, let $t = T$, and in place of P substitute the expression for present value of money flow. The result is the following formula.

Accumulated Amount of Money Flow at Time T
If $f(t)$ is the rate of money flow at an interest rate r at time t, the accumulated amount of money flow at time T is

$$A = e^{rT}\int_0^T f(t)e^{-rt}\, dt.$$

Here, the accumulated amount of money A represents the accumulated value or final amount of the money flow *including* interest received on the money after it comes in. (Recall, total money flow *does not* take the interest into account.)

It turns out that most money flows can be expressed as (or at least approximated by) exponential or polynomial functions. When these are multiplied by e^{-rt}, the result is a function that can be integrated. The next example illustrates uniform flow, where $f(t)$ is a constant function. (This is a special case of the polynomial function.)

| EXAMPLE 3 | Accumulated Amount of Money Flow

If money is flowing continuously at a constant rate of $2000 per year over 5 years at 6% interest compounded continuously, find the following.

(a) The total money flow over the 5-year period

SOLUTION The total money flow is given by $\int_0^T f(t)\,dt$. Here $f(t) = 2000$ and $T = 5$.

$$\int_0^5 2000\,dt = 2000t\Big|_0^5 = 2000(5) = 10{,}000$$

The total money flow over the 5-year period is $10,000.

(b) The accumulated amount of money flow, compounded continuously, at time $T = 5$

SOLUTION The accumulated amount of money flow is given by $A = e^{rT}\int_0^T f(t)e^{-rt}\,dt$. At $T = 5$ with $r = 0.06$ and $f(t) = 2000$, the amount is

$$A = e^{(0.06)5}\int_0^5 (2000)e^{-0.06t}\,dt$$

$$= (e^{0.3})(2000)\int_0^5 e^{-0.06t}\,dt$$

$$= e^{0.3}(2000)\left(\frac{1}{-0.06}\right)\left(e^{-0.06t}\Big|_0^5\right)$$

$$= \frac{2000e^{0.3}}{-0.06}(e^{-0.3} - 1)$$

$$= \frac{2000}{-0.06}(1 - e^{0.3}) \qquad\qquad (e^{0.3})(e^{-0.3}) = 1$$

$$\approx 11{,}661.96,$$

or $11,661.96. The answer to part (a), $10,000, was the amount of money flow over the 5-year period. The $11,661.96 gives that amount with interest compounded continuously over the 5-year period.

(c) The total interest earned

SOLUTION This is simply the accumulated amount of money flow minus the total amount of flow, or

$$\$11{,}661.96 - \$10{,}000.00 = \$1661.96.$$

(d) The present value of the amount with interest

SOLUTION Use $P = \int_0^T f(t)e^{-rt}\,dt$ with $f(t) = 2000$, $r = 0.06$, and $T = 5$.

$$P = \int_0^5 2000e^{-0.06t}\,dt = 2000\left(\frac{e^{-0.06t}}{-0.06}\right)\Big|_0^5$$

$$= \frac{2000}{-0.06}(e^{-0.3} - 1)$$

$$\approx 8639.39$$

The present value of the amount with interest in 5 years is $8639.39, which can be checked by substituting $11,661.96 for A in $A = Pe^{rt}$. The present value, P, could have been found by dividing the amount found in (b) by $e^{rT} = e^{0.3}$. Check that this would give the same result.

TRY YOUR TURN 3 ▬▬

If the rate of money flow is increasing or decreasing exponentially, then $f(t) = Ce^{kt}$, where C is a constant that represents the initial amount and k is the (nominal) continuous rate of change, which may be positive or negative.

FOR REVIEW

In this example we use the following two rules for exponents.

1. $a^m \cdot a^n = a^{m+n}$
2. $a^0 = 1$

YOUR TURN 3 Find the accumulated amount of money flow for the income and interest rate in Your Turn 2 in this section.

EXAMPLE 4 Accumulated Amount of Money Flow

A continuous money flow starts at a rate of $1000 per year and increases exponentially at 2% per year.

(a) Find the accumulated amount of money flow at the end of 5 years at 10% interest compounded continuously.

SOLUTION Here $C = 1000$ and $k = 0.02$, so that $f(t) = Ce^{kt} = 1000e^{0.02t}$. Using $r = 0.10$ and $T = 5$ in $A = e^{rt}\int_0^T f(t)e^{-rt}dt$ yields

$$A = e^{(0.10)5}\int_0^5 1000e^{0.02t}e^{-0.10t}\,dt$$

$$= (e^{0.5})(1000)\int_0^5 e^{-0.08t}\,dt \qquad e^{0.02t}\cdot e^{-0.10t} = e^{-0.08t}$$

$$= 1000e^{0.5}\left(\frac{e^{-0.08t}}{-0.08}\right)\Bigg|_0^5$$

$$= \frac{1000e^{0.5}}{-0.08}(e^{-0.4} - 1)$$

$$= \frac{1000}{-0.08}(e^{0.1} - e^{0.5}) \qquad e^{0.5}(e^{-0.4} - 1) = e^{0.1} - e^{0.5}$$

$$\approx 6794.38,$$

or $6794.38.

(b) Find the present value at 5% interest compounded continuously.

SOLUTION Using $f(t) = 1000e^{0.02t}$ with $r = 0.05$ and $T = 5$ in the present value expression,

$$P = \int_0^5 1000e^{0.02t}e^{-0.05t}\,dt$$

$$= 1000\int_0^5 e^{-0.03t}\,dt = 1000\left(\frac{e^{-0.03t}}{-0.03}\Bigg|_0^5\right)$$

$$= \frac{1000}{-0.03}(e^{-0.15} - 1) \approx 4643.07,$$

or $4643.07.

If the rate of change of the continuous money flow is given by the polynomial function $f(t) = a_n t^n + a_{n-1}t^{n-1} + \cdots + a_0$, the expressions for present value and accumulated amount can be integrated using integration by parts.

EXAMPLE 5 Present Value of Money Flow

The rate of change of a continuous flow of money is given by

$$f(t) = 1000t^2 + 100t.$$

Find the present value of this money flow at the end of 10 years at 10% compounded continuously.

SOLUTION Evaluate

$$P = \int_0^{10} (1000t^2 + 100t)e^{-0.10t}\,dt.$$

Using integration by parts, verify that

$$\int (1000t^2 + 100t)e^{-0.10t}\, dt =$$

$$(-10{,}000t^2 - 1000t)e^{-0.1t} - (200{,}000t + 10{,}000)e^{-0.1t} - 2{,}000{,}000e^{-0.1t} + C.$$

Thus,

$$P = (-10{,}000t^2 - 1000t)e^{-0.1t} - (200{,}000t + 10{,}000)e^{-0.1t}$$

$$\left. - 2{,}000{,}000e^{-0.1t} \right|_0^{10}$$

YOUR TURN 4 Find the present value at the end of 8 years of the continuous flow of money given by $f(t) = 200t^2 + 100t + 50$ at 5% compounded continuously.

$$= (-1{,}000{,}000 - 10{,}000)e^{-1} - (2{,}000{,}000 + 10{,}000)e^{-1}$$

$$- 2{,}000{,}000e^{-1} - (0 - 10{,}000 - 2{,}000{,}000)$$

$$\approx 163{,}245.21,$$

or \$163,245.21.

TRY YOUR TURN 4

8.3 WARM-UP EXERCISES

Evaluate.

W1. $\displaystyle \int 300e^{0.01t}\, dt$ *(Sec. 7.1)*

W2. $\displaystyle \int_0^5 100e^{0.04t}\, dt$ *(Sec. 7.4)*

W3. $\displaystyle \int 400te^{-0.04t}\, dt$ *(Sec. 8.1)*

W4. $\displaystyle \int_0^5 1000te^{-0.04t}\, dt$ *(Sec. 8.1)*

8.3 EXERCISES

Each of the functions in Exercises 1–14 represents the rate of flow of money in dollars per year. Assume a 10-year period at 8% compounded continuously and find the following: (a) the present value; (b) the accumulated amount of money flow at $t = 10$.

1. $f(t) = 1000$ **2.** $f(t) = 300$

3. $f(t) = 500$ **4.** $f(t) = 2000$

5. $f(t) = 400e^{0.03t}$ **6.** $f(t) = 800e^{0.05t}$

7. $f(t) = 5000e^{-0.01t}$ **8.** $f(t) = 1000e^{-0.02t}$

9. $f(t) = 25t$ **10.** $f(t) = 50t$

11. $f(t) = 0.01t + 100$ **12.** $f(t) = 0.05t + 500$

13. $f(t) = 1000t - 100t^2$ **14.** $f(t) = 2000t - 150t^2$

APPLICATIONS

Business and Economics

15. Accumulated Amount of Money Flow An investment is expected to yield a uniform continuous rate of money flow of \$20,000 per year for 3 years. Find the accumulated amount at an interest rate of 4% compounded continuously.

16. Present Value A real estate investment is expected to produce a uniform continuous rate of money flow of \$8000 per year for 6 years. Find the present value at the following rates, compounded continuously.

(a) 2% (b) 5% (c) 8%

17. Money Flow The rate of a continuous flow of money starts at \$5000 and decreases exponentially at 1% per year for 8 years. Find the present value and final amount at an interest rate of 8% compounded continuously.

18. Money Flow The rate of a continuous money flow starts at \$1000 and increases exponentially at 5% per year for 4 years. Find the present value and accumulated amount if interest earned is 3.5% compounded continuously.

19. Present Value A money market fund has a continuous flow of money at a rate of $f(t) = 1500 - 60t^2$, reaching 0 in 5 years. Find the present value of this flow if interest is 5% compounded continuously.

20. Accumulated Amount of Money Flow Find the amount of a continuous money flow in 3 years if the rate is given by $f(t) = 1000 - t^2$ and if interest is 5% compounded continuously.

YOUR TURN ANSWERS

1. \$1595.94 **2.** \$556,653

3. \$663,111 **4.** \$28,156.02

8.4 Improper Integrals

APPLY IT   If we know the rate at which a pollutant is dumped into a stream, how can we compute the total amount released given that the rate of dumping is decreasing over time?

In this section we will learn how to answer questions such as this one, which is answered in Example 3.

Sometimes it is useful to be able to integrate a function over an infinite period of time. For example, we might want to find the total amount of income generated by an apartment building into the indefinite future or the total amount of pollution into a bay from a source that is continuing indefinitely. In this section we define integrals with one or more infinite limits of integration that can be used to solve such problems.

The graph in Figure 12(a) shows the area bounded by the curve $f(x) = x^{-3/2}$, the x-axis, and the vertical line $x = 1$. Think of the shaded region below the curve as extending indefinitely to the right. Does this shaded region have an area?

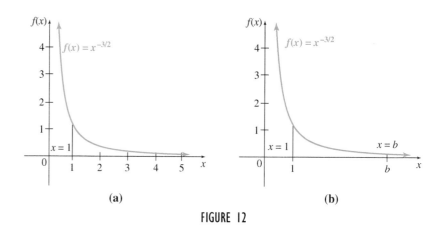

(a)   (b)

FIGURE 12

To see if the area of this region can be defined, introduce a vertical line at $x = b$, as shown in Figure 12(b). This vertical line gives a region with both upper and lower limits of integration. The area of this new region is given by the definite integral

$$\int_{1}^{b} x^{-3/2}\,dx.$$

By the Fundamental Theorem of Calculus,

$$\int_{1}^{b} x^{-3/2}\,dx = \left(-2x^{-1/2}\right)\Big|_{1}^{b}$$

$$= -2b^{-1/2} - \left(-2 \cdot 1^{-1/2}\right)$$

$$= -2b^{-1/2} + 2 = 2 - \frac{2}{b^{1/2}}.$$

FOR REVIEW

In Section 3.1 on Limits we saw that for any positive real number n,

$$\lim_{b \to \infty} \frac{1}{b^n} = 0.$$

Suppose we now let the vertical line $x = b$ in Figure 12(b) move farther to the right. That is, suppose $b \to \infty$. The expression $-2/b^{1/2}$ would then approach 0, and

$$\lim_{b \to \infty}\left(2 - \frac{2}{b^{1/2}}\right) = 2 - 0 = 2.$$

This limit is defined to be the *area* of the region shown in Figure 12(a), so that

$$\int_1^\infty x^{-3/2} \, dx = 2.$$

An integral of the form

$$\int_a^\infty f(x) \, dx, \qquad \int_{-\infty}^b f(x) \, dx, \qquad \text{or} \qquad \int_{-\infty}^\infty f(x) \, dx$$

is called an *improper integral*. These **improper integrals** are defined as follows.

Improper Integrals
If f is continuous on the indicated interval and if the indicated limits exist, then

$$\int_a^\infty f(x) \, dx = \lim_{b \to \infty} \int_a^b f(x) \, dx,$$

$$\int_{-\infty}^b f(x) \, dx = \lim_{a \to -\infty} \int_a^b f(x) \, dx,$$

$$\int_{-\infty}^\infty f(x) \, dx = \int_{-\infty}^c f(x) \, dx + \int_c^\infty f(x) \, dx,$$

for real numbers a, b, and c, where c is arbitrarily chosen.

If the expressions on the right side exist, the integrals are **convergent**; otherwise, they are **divergent**. A convergent integral has a value that is a real number. A divergent integral does not, often because the area under the curve is infinitely large.

FIGURE 13

EXAMPLE 1 **Improper Integrals**

Evaluate each integral.

(a) $\displaystyle \int_1^\infty \frac{dx}{x}$

SOLUTION A graph of this region is shown in Figure 13. By the definition of an improper integral,

$$\int_1^\infty \frac{dx}{x} = \lim_{b \to \infty} \int_1^b \frac{dx}{x}.$$

Find $\displaystyle \int_1^b \frac{dx}{x}$ by the Fundamental Theorem of Calculus.

$$\int_1^b \frac{dx}{x} = \ln |x| \Big|_1^b = \ln |b| - \ln |1| = \ln |b| - 0 = \ln |b|$$

As $b \to \infty$, $\ln |b| \to \infty$, so $\displaystyle \lim_{b \to \infty} \ln |b|$ does not exist. Since the limit does not exist, $\displaystyle \int_1^\infty \frac{dx}{x}$ is divergent.

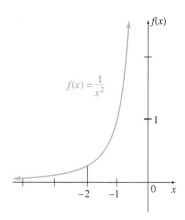

FIGURE 14

YOUR TURN 1 Find each integral.

(a) $\displaystyle\int_8^\infty \frac{1}{x^{1/3}} \, dx$ **(b)** $\displaystyle\int_8^\infty \frac{1}{x^{4/3}} \, dx$

FIGURE 15

YOUR TURN 2 Find $\displaystyle\int_0^\infty 5e^{-2x} \, dx$.

(b) $\displaystyle\int_{-\infty}^{-2} \frac{1}{x^2} \, dx$

SOLUTION A graph of this region is shown in Figure 14.

$$\int_{-\infty}^{-2} \frac{1}{x^2} \, dx = \lim_{a \to -\infty} \int_a^{-2} \frac{1}{x^2} \, dx = \lim_{a \to -\infty} \left(\frac{-1}{x} \right) \Big|_a^{-2}$$

$$= \lim_{a \to -\infty} \left(\frac{1}{2} + \frac{1}{a} \right) = \frac{1}{2}$$

Since the limit exists, this integral converges. **TRY YOUR TURN 1** ▦

It may seem puzzling that the areas under the curves $f(x) = 1/x^{3/2}$ and $f(x) = 1/x^2$ are finite, while $f(x) = 1/x$ has an infinite amount of area. At first glance the graphs of these functions appear similar. The difference is that although all three functions get small as x becomes infinitely large, $f(x) = 1/x$ does not become small enough fast enough. In the graphing calculator screen in Figure 15, notice how much faster $1/x^2$ becomes small compared with $1/x$.

| CAUTION | Since graphing calculators provide only approximations, using them to find improper integrals is tricky and requires skill and care. Although their approximations may be good in some cases, they are wrong in others, and they cannot tell us for certain that an improper integral does not exist. See Exercises 39–41. |

EXAMPLE 2 Improper Integral

Find $\int_{-\infty}^{\infty} 4e^{-3x} \, dx$.

SOLUTION In the definition of an improper integral with limits of $-\infty$ and ∞, the value of c is arbitrary, so we'll choose the simple value $c = 0$. We can then write the integral as

$$\int_{-\infty}^{\infty} 4e^{-3x} \, dx = \int_{-\infty}^{0} 4e^{-3x} \, dx + \int_{0}^{\infty} 4e^{-3x} \, dx$$

and evaluate each of the two improper integrals on the right. If they both converge, the original integral will equal their sum. To show you all the details while maintaining the suspense, we will evaluate the second integral first.

By definition,

$$\int_{0}^{\infty} 4e^{-3x} \, dx = \lim_{b \to \infty} \int_{0}^{b} 4e^{-3x} \, dx = \lim_{b \to \infty} \left(\frac{-4}{3} e^{-3x} \right) \Big|_0^b$$

$$= \lim_{b \to \infty} \left(\frac{-4}{3e^{3b}} + \frac{4}{3} \right) = 0 + \frac{4}{3} = \frac{4}{3}.$$

Similarly, the first integral is evaluated as

$$\int_{-\infty}^{0} 4e^{-3x} \, dx = \lim_{b \to -\infty} \int_{b}^{0} 4e^{-3x} \, dx = \lim_{b \to -\infty} \left(\frac{-4}{3} e^{-3x} \right) \Big|_b^0$$

$$= \lim_{b \to -\infty} \left(-\frac{4}{3} + \frac{4}{3e^{3b}} \right) = \infty.$$

Since one of the two improper integrals diverges, the original improper integral diverges.

TRY YOUR TURN 2 ▦

The following examples describe applications of improper integrals.

EXAMPLE 3   Pollution

The rate at which a pollutant is being dumped into a stream at time t is given by $P_0 e^{-kt}$, where P_0 is the rate that the pollutant is initially released into the stream. Suppose $P_0 = 1000$ and $k = 0.06$. Find the total amount of the pollutant that will be released into the stream into the indefinite future.

SOLUTION   Find

$$\int_0^\infty P_0 e^{-kt}\, dt = \int_0^\infty 1000 e^{-0.06t}\, dt.$$

This integral is similar to one of the integrals used to solve Example 2 and may be evaluated by the same method.

$$\int_0^\infty 1000 e^{-0.06t}\, dt = \lim_{b\to\infty} \int_0^b 1000 e^{-0.06t}\, dt$$

$$= \lim_{b\to\infty} \left(\frac{1000}{-0.06} e^{-0.06t} \right)\Bigg|_0^b$$

$$= \lim_{b\to\infty} \left(\frac{1000}{-0.06 e^{0.06b}} - \frac{1000}{-0.06} e^0 \right) = \frac{-1000}{-0.06} \approx 16{,}667$$

A total of approximately 16,667 units of the pollutant will be released over time.

The *capital value* of an asset is often defined as the present value of all future net earnings of the asset. In other words, suppose an asset provides a continuous money flow that is invested in an account earning a certain rate of interest. A lump sum is invested in a second account earning the same rate of interest, but with no money flow, so that as $t \to \infty$, the amounts in the two accounts approach each other. The lump sum necessary to make this happen is the capital value of the asset. If $R(t)$ gives the annual rate at which earnings are produced by an asset at time t, then the present value formula from Section 3 gives the **capital value** as

$$\int_0^\infty R(t) e^{-rt}\, dt,$$

where r is the annual rate of interest, compounded continuously.

EXAMPLE 4   Capital Value

Suppose income from a rental property is generated at the annual rate of $4000 per year. Find the capital value of this property at an interest rate of 10% compounded continuously.

SOLUTION   This is a continuous income stream with a rate of flow of $4000 per year, so $R(t) = 4000$. Also, $r = 0.10$ or 0.1. The capital value is given by

$$\int_0^\infty 4000 e^{-0.1t}\, dt = \lim_{b\to\infty} \int_0^b 4000 e^{-0.1t}\, dt$$

$$= \lim_{b\to\infty} \left(\frac{4000}{-0.1} e^{-0.1t} \right)\Bigg|_0^b$$

$$= \lim_{b\to\infty} \left(-40{,}000 e^{-0.1b} + 40{,}000 \right) = 40{,}000,$$

or $40,000.

Find each limit.

W1. $\lim\limits_{x \to \infty} \dfrac{3 - x^2}{x^4}$ (Sec. 3.1)

W2. $\lim\limits_{x \to \infty} \dfrac{4x^2 + 3x - 7}{2x^5 + 3x - 5}$ (Sec. 3.1)

W3. $\lim\limits_{x \to \infty} e^{-4x}$ (Sec. 3.1)

W4. $\lim\limits_{x \to -\infty} e^{-0.2x}$ (Sec. 3.1)

8.4 EXERCISES

Determine whether each improper integral converges or diverges, and find the value of each that converges.

1. $\displaystyle\int_3^\infty \frac{1}{x^2}\, dx$

2. $\displaystyle\int_3^\infty \frac{1}{(x + 1)^3}\, dx$

3. $\displaystyle\int_4^\infty \frac{2}{\sqrt{x}}\, dx$

4. $\displaystyle\int_{27}^\infty \frac{2}{\sqrt[3]{x}}\, dx$

5. $\displaystyle\int_{-\infty}^{-1} \frac{2}{x^3}\, dx$

6. $\displaystyle\int_{-\infty}^{-4} \frac{3}{x^4}\, dx$

7. $\displaystyle\int_1^\infty \frac{1}{x^{1.0001}}\, dx$

8. $\displaystyle\int_1^\infty \frac{1}{x^{0.999}}\, dx$

9. $\displaystyle\int_{-\infty}^{-10} x^{-2}\, dx$

10. $\displaystyle\int_{-\infty}^{-1} (x - 2)^{-3}\, dx$

11. $\displaystyle\int_{-\infty}^{-1} x^{-8/3}\, dx$

12. $\displaystyle\int_{-\infty}^{-27} x^{-5/3}\, dx$

13. $\displaystyle\int_0^\infty 8e^{-8x}\, dx$

14. $\displaystyle\int_0^\infty 50e^{-50x}\, dx$

15. $\displaystyle\int_{-\infty}^0 1000e^x\, dx$

16. $\displaystyle\int_{-\infty}^0 5e^{60x}\, dx$

17. $\displaystyle\int_{-\infty}^{-1} \ln |x|\, dx$

18. $\displaystyle\int_1^\infty \ln |x|\, dx$

19. $\displaystyle\int_0^\infty \frac{dx}{(x + 1)^2}$

20. $\displaystyle\int_0^\infty \frac{dx}{(4x + 1)^3}$

21. $\displaystyle\int_{-\infty}^{-1} \frac{2x - 1}{x^2 - x}\, dx$

22. $\displaystyle\int_1^\infty \frac{4x + 6}{x^2 + 3x}\, dx$

23. $\displaystyle\int_2^\infty \frac{1}{x \ln x}\, dx$

24. $\displaystyle\int_2^\infty \frac{1}{x(\ln x)^2}\, dx$

25. $\displaystyle\int_0^\infty xe^{4x}\, dx$

26. $\displaystyle\int_{-\infty}^0 xe^{0.2x}\, dx$ (*Hint:* Recall that $\lim\limits_{x \to -\infty} xe^x = 0$.)

27. $\displaystyle\int_{-\infty}^\infty x^3 e^{-x^4}\, dx$ (*Hint:* Recall from Exercise 68 in Section 3.1 on Limits that $\lim\limits_{x \to \infty} x^n e^{-x} = 0$.)

28. $\displaystyle\int_{-\infty}^\infty e^{-|x|}\, dx$ (*Hint:* Recall that when $x < 0$, $|x| = -x$.)

29. $\displaystyle\int_{-\infty}^\infty \frac{x}{x^2 + 1}\, dx$

30. $\displaystyle\int_{-\infty}^\infty \frac{2x + 4}{x^2 + 4x + 5}\, dx$

⬭ **Find the area between the graph of the given function and the x-axis over the given interval, if possible.**

31. $f(x) = \dfrac{1}{x - 1}$, for $(-\infty, 0]$

32. $f(x) = e^{-x}$, for $(-\infty, e]$

33. $f(x) = \dfrac{1}{(x - 1)^2}$, for $(-\infty, 0]$

34. $f(x) = \dfrac{3}{(x - 1)^3}$, for $(-\infty, 0]$

35. Find $\displaystyle\int_{-\infty}^\infty xe^{-x^2}\, dx$.

36. Find $\displaystyle\int_{-\infty}^\infty \frac{x}{(1 + x^2)^2}\, dx$.

37. Show that $\displaystyle\int_1^\infty \frac{1}{x^p}\, dx$ converges if $p > 1$ and diverges if $p \le 1$.

38. Example 1(b) leads to a paradox. On the one hand, the unbounded region in that example has an area of $1/2$, so theoretically it could be colored with ink. On the other hand, the boundary of that region is infinite, so it cannot be drawn with a finite amount of ink. This seems impossible, because coloring the region automatically colors the boundary. Explain why it is possible to color the region.

39. Consider the functions $f(x) = 1/\sqrt{1 + x^2}$ and $g(x) = 1/\sqrt{1 + x^4}$.

   (a) Use your calculator to approximate $\int_1^b f(x)\, dx$ for $b = 20, 50, 100, 1000$, and $10,000$.

   (b) Based on your answers from part (a), would you guess that $\int_1^\infty f(x)\, dx$ is convergent or divergent?

(c) Use your calculator to approximate $\int_1^b g(x)\,dx$ for $b = 20, 50, 100, 1000,$ and $10{,}000.$

(d) Based on your answers from part (c), would you guess that $\int_1^\infty g(x)\,dx$ is convergent or divergent?

(e) Show how the answer to parts (b) and (d) might be guessed by comparing the integrals with others whose convergence or divergence is known. (*Hint:* For large x, the difference between $1 + x^2$ and x^2 is relatively small.)

Note: The first integral is indeed divergent, and the second convergent, with an approximate value of 0.9270.

40. (a) Use your calculator to approximate $\int_0^b e^{-x^2}\,dx$ for $b = 1,$ 5, 10, and 20.

(b) Based on your answers to part (a), does $\int_0^\infty e^{-x^2}\,dx$ appear to be convergent or divergent? If convergent, what seems to be its approximate value?

(c) Explain why this integral should be convergent by comparing e^{-x^2} with e^{-x} for $x > 1.$

Note: The integral is convergent, with a value of $\sqrt{\pi}/2.$

41. (a) Use your calculator to approximate $\int_0^b e^{-0.00001x}\,dx$ for $b = 10, 50, 100,$ and 1000.

(b) Based on your answers to part (a), does $\int_0^\infty e^{-0.00001x}\,dx$ appear to be convergent or divergent?

(c) To what value does the integral actually converge?

APPLICATIONS

Business and Economics

Capital Value Find the capital values of the properties in Exercises 42–43.

42. A castle for which annual rent of $225,000 will be paid in perpetuity; the interest rate is 6% compounded continuously.

43. A fort on a strategic peninsula in the North Sea; the annual rent is $1,000,000, paid in perpetuity; the interest rate is 5% compounded continuously.

44. Capital Value Find the capital value of an asset that generates $7200 yearly income if the interest rate is as follows.

(a) 5% compounded continuously

(b) 10% compounded continuously

45. Capital Value An investment produces a perpetual stream of income with a flow rate of

$$R(t) = 1200e^{0.03t}.$$

Find the capital value at an interest rate of 7% compounded continuously.

46. Capital Value Suppose income from an investment starts (at time 0) at $6000 a year and increases linearly and continuously at a rate of $200 a year. Find the capital value at an interest rate of 5% compounded continuously.

47. Endowment Income Suppose income from an endowment is generated at the annual rate of $42,000 per year. Find the capital value of this endowment at an interest rate of 7.5% compounded continuously.

48. Inventory A study of the sales lost when a store has insufficient inventory to meet the demand and impatient customers are unwilling to wait for the store to reorder has found that when the demand is a decreasing exponential function, the expected sales lost per period is given by

$$E = \int_B^\infty (y - B)\frac{k^2 Be^{-ky}}{1 - e^{-kB}}\,dy,$$

where B is the base stock level and k is a positive constant. Find the value of the expected sales lost. *Source: Operations Research.*

Social Sciences

49. Drug Reaction The rate of reaction to a drug is given by

$$r'(t) = 2t^2 e^{-t},$$

where t is the number of hours since the drug was administered. Find the total reaction to the drug over all the time since it was administered, assuming this is an infinite time interval. (*Hint:* $\lim\limits_{t \to \infty} t^k e^{-t} = 0$ for all real numbers $k.$)

50. Drug Epidemic In an epidemiological model used to study the spread of drug use, a single drug user is introduced into a population of N non-users. Under certain assumptions, the number of people expected to use drugs as a result of direct influence from each drug user is given by

$$S = N\int_0^\infty \frac{a(1 - e^{-kt})}{k}e^{-bt}\,dt,$$

where $a,$ $b,$ and k are constants. Find the value of $S.$ *Source: Mathematical Biology.*

51. Present Value When harvesting a population, such as fish, the present value of the resource is given by

$$P = \int_0^\infty e^{-rt}n(t)y(t)\,dt,$$

where r is a discount factor, $n(t)$ is the net revenue at time $t,$ and $y(t)$ is the harvesting effort. Suppose $y(t) = K$ and $n(t) = at + b.$ Find the present value. *Source: Some Mathematical Questions in Biology.*

Physical Sciences

Radioactive Waste The rate at which radioactive waste is entering the atmosphere at time t is decreasing and is given by $Pe^{-kt},$ where P is the initial rate. Use the improper integral

$$\int_0^\infty Pe^{-kt}\,dt$$

with $P = 50$ to find the total amount of the waste that will enter the atmosphere for each value of $k.$

52. $k = 0.06$

53. $k = 0.04$

General Interest

54. Gamma Function An important function in many areas of mathematics, such as statistics, is the gamma function, defined as

$$\Gamma(t) = \int_0^\infty x^{t-1}e^{-x}\, dx,$$

where t is not 0 or a negative integer. Find each of the following.

(a) $\Gamma(1)$ **(b)** $\Gamma(2)$ **(c)** $\Gamma(3)$ **(d)** $\Gamma(4)$

(e) Show that

$$\Gamma(t) = (t-1)\Gamma(t-1)$$

for $t > 1$.

(f) Use the results of parts (a) through (e) to show that

$$\Gamma(n) = (n-1)!$$

for any positive integer n. This shows that the gamma function is a generalization of the factorial function.

YOUR TURN ANSWERS

1. (a) Divergent

  (b) 3/2

2. 5/2

---

8 CHAPTER REVIEW

SUMMARY

In this chapter, we introduced another technique of integration and some applications of integration. The technique is known as integration by parts, which is derived from the product rule for derivatives. We also developed definite integral formulas to calculate the volume of a solid of revolution and the average value of a function on some interval. We then used definite integrals to study continuous money flow. Finally, we learned how to evaluate improper integrals that have upper or lower limits of ∞ or $-\infty$.

Integration by Parts	If u and v are differentiable functions, then

$$\int u\, dv = uv - \int v\, du.$$

Volume of a Solid of Revolution	If $f(x)$ is nonnegative and R is the region between $f(x)$ and the x-axis from $x = a$ to $x = b$, the volume of the solid formed by rotating R about the x-axis is given by

$$V = \int_a^b \pi[f(x)]^2\, dx.$$

Average Value of a Function	The average value of a function f on the interval $[a, b]$ is

$$\frac{1}{b-a}\int_a^b f(x)\, dx,$$

provided the indicated definite integral exists.

Total Money Flow	If $f(t)$ is the rate of money flow, then the total money flow over the time interval from $t = 0$ to $t = T$ is given by

$$\int_0^T f(t)\, dt.$$

Present Value of Money Flow	If $f(t)$ is the rate of continuous money flow at an interest rate r for T years, then the present value is

$$P = \int_0^T f(t)e^{-rt}\, dt.$$

Accumulated Amount of Money Flow at Time T

If $f(t)$ is the rate of money flow at an interest rate r at time t, the accumulated amount of money flow at time T is

$$A = e^{rT} \int_0^T f(t) e^{-rt} \, dt.$$

Improper Integrals

If f is continuous on the indicated interval and if the indicated limits exist, then

$$\int_a^\infty f(x) \, dx = \lim_{b \to \infty} \int_a^b f(x) \, dx,$$

$$\int_{-\infty}^b f(x) \, dx = \lim_{a \to -\infty} \int_a^b f(x) \, dx,$$

$$\int_{-\infty}^\infty f(x) \, dx = \int_{-\infty}^c f(x) \, dx + \int_c^\infty f(x) \, dx,$$

for real numbers a, b, and c, where c is arbitrarily chosen.

Capital Value

If $R(t)$ gives the annual rate at which earnings are produced by an asset at time t, the capital value is given by

$$\int_0^\infty R(t) e^{-rt} \, dt,$$

where r is the annual rate of interest, compounded continuously.

KEY TERMS

8.1
integration by parts
column integration

8.2
solid of revolution
average value of a function

8.3
total money flow
present value of continuous
  money flow
accumulated amount of money
  flow

8.4
improper integral
convergent integral
divergent integral
capital value

REVIEW EXERCISES

CONCEPT CHECK

Determine whether each of the following statements is true or false, and explain why.

1. Integration by parts should be used to evaluate $\int_0^1 \dfrac{x^2}{x^3 + 1} \, dx$.

2. Integration by parts should be used to evaluate $\int_0^1 x e^{10x} \, dx$.

3. We would need to apply the method of integration by parts twice to determine

$$\int x^3 e^{-x^2} \, dx.$$

4. Integration by parts should be used to determine $\int \ln(4x) \, dx$.

5. The average value of the function $f(x) = 2x^2 + 3$ on $[1, 4]$ is given by

$$\frac{1}{3} \int_1^4 \pi (2x^2 + 3)^2 \, dx.$$

6. The volume of the solid formed by revolving the function $f(x) = \sqrt{x^2 + 1}$ about the x-axis on the interval $[1, 2]$ is given by

$$\int_1^2 \pi \sqrt{x^2 + 1} \, dx.$$

7. The volume of the solid formed by revolving the function $f(x) = x + 4$ about the x-axis on the interval $[-4, 5]$ is given by

$$\int_{-4}^5 \pi (x + 4)^2 \, dx.$$

8. If $f(t) = 1000e^{0.05t}$ represents the rate of flow of money for a vending machine over the first five years of income, then the total money flow for that time period is given by

$$\int_0^5 1000e^{0.05t}\, dt.$$

 9. If a company expects an annual flow of money during the next five years to be $f(t) = 1000e^{0.05t}$, the present value of this income, assuming an annual interest rate of 4.5% compounded continuously is given by

$$\int_0^5 1000e^{0.005t}\, dt.$$

10. $\displaystyle\int_{-\infty}^{\infty} xe^{-2x}\, dx = \lim_{c\to\infty} \int_{-c}^{c} xe^{-2x}\, dx$

PRACTICE AND EXPLORATIONS

11. Describe the type of integral for which integration by parts is useful.

12. Compare finding the average value of a function with finding the average of n numbers.

13. What is an improper integral? Explain why improper integrals must be treated in a special way.

Find each integral, using techniques from this or the previous chapter.

14. $\displaystyle\int x(8 - x)^{3/2}\, dx$

15. $\displaystyle\int \frac{3x}{\sqrt{x-2}}\, dx$

16. $\displaystyle\int xe^x\, dx$

17. $\displaystyle\int (3x + 6)e^{-3x}\, dx$

18. $\displaystyle\int \ln|4x + 5|\, dx$

19. $\displaystyle\int (x - 1)\ln|x|\, dx$

20. $\displaystyle\int \frac{x}{25 - 9x^2}\, dx$

21. $\displaystyle\int \frac{x}{\sqrt{16 + 8x^2}}\, dx$

22. $\displaystyle\int_1^e x^3 \ln x\, dx$

23. $\displaystyle\int_0^1 x^2 e^{x/2}\, dx$

24. Find the area between $y = (3 + x^2)e^{2x}$ and the x-axis from $x = 0$ to $x = 1$.

25. Find the area between $y = x^3(x^2 - 1)^{1/3}$ and the x-axis from $x = 1$ to $x = 3$.

26. Suppose that u and v are differentiable functions of x with

$$\int_0^1 v\, du = 2$$ and the following functional values.

x	$u(x)$	$v(x)$
0	-3	4
1	5	-1

Use this information to determine $\displaystyle\int_0^1 u\, dv.$

27. Suppose that u and v are differentiable functions of x with

$$\int_0^2 v\, du = -11$$ and the following functional values.

x	$u(x)$	$v(x)$
0	-6	-4
2	3	-2

Use this information to determine $\displaystyle\int_0^2 u\, dv.$

Find the volume of the solid of revolution formed by rotating each bounded region about the x-axis.

28. $f(x) = 3x - 1$, $y = 0$, $x = 2$

29. $f(x) = \sqrt{x - 4}$, $y = 0$, $x = 13$

30. $f(x) = e^{-x}$, $y = 0$, $x = -2$, $x = 1$

31. $f(x) = \dfrac{1}{\sqrt{x-1}}$, $y = 0$, $x = 2$, $x = 4$

32. $f(x) = 4 - x^2$, $y = 0$, $x = -1$, $x = 1$

33. $f(x) = \dfrac{x^2}{4}$, $y = 0$, $x = 4$

34. A frustum is what remains of a cone when the top is cut off by a plane parallel to the base. Suppose a right circular frustum (that is, one formed from a right circular cone) has a base with radius r, a top with radius $r/2$, and a height h. (See the figure below.) Find the volume of this frustum by rotating about the x-axis the region below the line segment from $(0, r)$ to $(h, r/2)$.

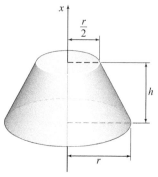

35. How is the average value of a function found?

36. Find the average value of $f(x) = \sqrt{x + 1}$ over the interval $[0, 8]$.

37. Find the average value of $f(x) = 7x^2(x^3 + 1)^6$ over the interval $[0, 2]$.

Find the value of each integral that converges.

38. $\displaystyle\int_{10}^{\infty} x^{-1}\, dx$

39. $\displaystyle\int_{-\infty}^{-5} x^{-2}\, dx$

40. $\displaystyle\int_0^{\infty} \frac{dx}{(3x + 1)^2}$

41. $\displaystyle\int_1^{\infty} 6e^{-x}\, dx$

42. $\displaystyle\int_{-\infty}^0 \frac{x}{x^2 + 3}\, dx$

43. $\displaystyle\int_4^{\infty} \ln(5x)\, dx$

Find the area between the graph of each function and the x-axis over the given interval, if possible.

44. $f(x) = \dfrac{5}{(x-2)^2}$, for $(-\infty, 1]$

45. $f(x) = 3e^{-x}$, for $[0, \infty)$

46. How is the present value of money flow found? The accumulated amount of money flow?

APPLICATIONS

Business and Economics

47. Total Revenue The rate of change of revenue from the sale of x toaster ovens is

$$R'(x) = x(x - 50)^{1/2}.$$

Find the total revenue from the sale of the 50th to the 75th ovens.

Present Value of Money Flow Each function in Exercises 48–51 represents the rate of flow of money (in dollars per year) over the given time period, compounded continuously at the given annual interest rate. Find the present value in each case.

48. $f(t) = 5000$,   8 years, 9%

49. $f(t) = 25,000$,   12 years, 10%

50. $f(t) = 150e^{0.04t}$,   5 years, 6%

51. $f(t) = 15t$,   18 months, 8%

Accumulated Amount of Money Flow at Time T Assume that each function gives the rate of flow of money in dollars per year over the given period, with continuous compounding at the given annual interest rate. Find the accumulated amount of money flow at the end of the time period.

52. $f(t) = 1000$,   5 years, 6%

53. $f(t) = 500e^{-0.04t}$,   8 years, 10%

54. $f(t) = 20t$,   6 years, 4%

55. $f(t) = 1000 + 200t$,   10 years, 9%

56. Money Flow An investment scheme is expected to produce a continuous flow of money, starting at $1000 and increasing exponentially at 5% a year for 7 years. Find the present value at an interest rate of 11% compounded continuously.

57. Money Flow The proceeds from the sale of a building will yield a uniform continuous flow of $10,000 a year for 10 years. Find the final amount at an interest rate of 10.5% compounded continuously.

58. Capital Value Find the capital value of an office building for which annual rent of $50,000 will be paid in perpetuity, if the interest rate is 9%.

Life Sciences

59. Drug Reaction The reaction rate to a new drug t hours after the drug is administered is

$$r'(t) = 0.5te^{-t}.$$

Find the total reaction over the first 5 hours.

60. Oil Leak Pollution An oil leak from an uncapped well is polluting a bay at a rate of $f(t) = 125e^{-0.025t}$ gallons per year. Use an improper integral to find the total amount of oil that will enter the bay, assuming the well is never capped.

Physical Sciences

61. Average Temperatures Suppose the temperature (degrees F) in a river at a point x meters downstream from a factory that is discharging hot water into the river is given by

$$T(x) = 160 - 0.05x^2.$$

Find the average temperature over each interval.

(a) $[0, 10]$ **(b)** $[10, 40]$ **(c)** $[0, 40]$

EXTENDED APPLICATION

ESTIMATING LEARNING CURVES IN MANUFACTURING WITH INTEGRALS

In the previous chapter you have seen how the trapezoidal rule uses sums of areas of polygons to approximate the area under a smooth curve, that is, a definite integral. In this Extended Application we look at the reverse process, using an integral to estimate a sum, in the context of estimating production costs.

As a manufacturer produces more units of a new product, the individual units generally become cheaper to produce, because with experience, production workers gain skill and speed, and managers spot opportunities for improved efficiency. This decline in unit costs is often called an *experience curve* or *learning curve*. This curve is important when a manufacturer negotiates a contract with a buyer.

Here's an example, based on an actual contract that came before the Armed Services Board of Contract Appeals. The Navy asked the ITT Defense Communications Division to bid on the manufacture of several different kinds of mobile telephone switchboards, including 280 of the model called the SB 3865. ITT figured that the cost of making a single SB 3865 was around $300,000. But they couldn't submit a bid of $300,000 × 280 or $84 million, because multiple units should have a lower unit price. So ITT used a learning curve to estimate an average unit cost of $135,300 for all 280 switchboards and submitted a bid of $135,300 × 280 or $37.9 million.

The contract gave the Navy an option to purchase 280 SB units over three years, but in fact it bought fewer. Suppose the Navy bought 140 SBs. Should it pay half of the original price of $37.9 million? No: ITT's bid was based on the efficiencies of a 280-unit run, so 140 units should be repriced to yield *more* than half the full price. A repricing clause in the contract specified that a learning curve would be used to reprice partial orders, and when the Navy ordered less than the full amount, ITT invoked this clause to reprice the switchboards. The question in dispute at the hearing was which learning curve to use.

There are two common learning curve models. The *unit learning curve* model assumes that each time the number of units doubles, say from n to $2n$, the cost of producing the last unit is some constant fraction r of the cost for the nth unit. Usually the fraction r is given as a percent. If $r = 90\%$ (typical for big pieces of hardware), then the contract would refer to a "90% learning curve." The *cumulative learning curve* model assumes that when the number of units doubles, the *average cost* of producing all $2n$ units is some constant fraction of the *average cost* of the first n units. The Navy's contract with ITT didn't specify which model was to be used—it just referred to "a 90% learning curve." The government used the unit model and ITT used the cumulative average model, and ITT calculated a fair price millions of dollars higher than the government's price!

In practice, ITT used a calculator program to make its estimate, and the government used printed tables, but both the program and the tables were derived using calculus. To see how the computation works, we'll derive the government's unit learning curve.

Each unit has a different cost, with the first unit being the most expensive and the 280th unit the least expensive. So the cost of the nth unit, call it $C(n)$, is a function of n. To find a fair price for n units we'll add up all the unit prices. That is, we will compute $C(1) + C(2) + \cdots + C(n - 1) + C(n)$. Before we can do that, we need a formula for $C(n)$ in terms of n, but all we know about $C(n)$ is that

$$C(2n) = r \cdot C(n)$$

for *every* n, with $r = 0.90$. This sort of equation is called a *functional equation:* It relates two different values of the function without giving an explicit formula for the function. In Exercise 3 you'll see how you might discover a solution to this functional equation, but here we'll just give the result:

$$C(n) = C(1) \cdot n^b, \text{ where } b = \frac{\ln r}{\ln 2} \approx -0.152.$$

Thus, to find the price for making 280 units, we need to add up all the values of $C(1) \cdot n^b$ as n ranges from 1 to 280. There's just one problem: We don't know $C(1)$! The only numbers that ITT gave were the *average* cost per unit for 280 units, namely, $135,300, and the total price of $37.9 million. But if we write out the formula for the 280-unit price in terms of $C(1)$, we can figure out $C(1)$ by dividing. Here's how it works.

The cost C is a function of an integer variable n, since the contractor can't deliver fractional units of the hardware. But the function $C(x) = C(1) \cdot x^{-0.152}$ is a perfectly good function of the real

variable x, and the sum of the first 280 values of $C(n)$ should be close to $\int_1^{280} C(x)\, dx = C(1) \cdot \int_1^{280} x^{-0.152}\, dx$. In Exercise 4 you'll see how to derive the following improved estimate:

$$C(1)\left(\int_1^{281} x^{-0.152}\, dx + \frac{1 - 281^{-0.152}}{2}\right).$$

The integrand is a power function, so you know how to evaluate the integral exactly.

$$\int_1^{281} x^{-0.152}\, dx = \frac{1}{0.848} x^{0.848} \Big|_1^{281} = \frac{1}{0.848}\left(281^{0.848} - 1^{0.848}\right)$$

Thus, the sum is approximately

$$C(1)\left[\frac{1}{0.848}\left(281^{0.848} - 1^{0.848}\right) + \frac{1 - 281^{-0.152}}{2}\right] \approx C(1) \cdot 139.75$$

Since ITT's price for 280 units was $37.9 million,

$$C(1) = \frac{\$37,900,000}{139.75} \approx \$271,000.$$

Now that we know $C(1)$, we can reprice an order of 140 units by adding up the first 140 values of $C(n)$. An estimate exactly like the one above tells us that according to the government's model, a fair price for 140 units is about $21 million. As we expected, this is more than half the 280-unit price, in fact about $2 million more. **Source: Armed Services Board of Contract Appeals.**

EXERCISES

1. According to the formula for $C(n)$, what is the unit price of the 280th unit, to the nearest thousand dollars?

2. Suppose that instead of using natural logarithms to compute b, we use logarithms with a base of 10 and define $b = (\log r)/(\log 2)$. Does this change the value of b?

3. All power functions satisfy an equation similar to our functional equation: If $f(x) = ax^b$, then $f(2x) = a(2x)^b = a2^b \cdot x^b = 2^b \cdot f(x)$. How can you choose a and b to make $C(x) = ax^b$ a solution to the functional equation $C(2n) = r \cdot C(n)$?

4. Figure 16 on the next page indicates how you could use the integral

$$\int_1^5 \frac{1}{x}\, dx + \frac{1 - \frac{1}{5}}{2} \text{ as an estimate for the sum } 1 + \frac{1}{2} + \frac{1}{3} + \frac{1}{4}$$

The graph shows the function $y = \frac{1}{x}$.

(a) Write a justification for the integral estimate. (Your argument will also justify the integral estimate for the $C(n$ sum.) Based on your explanation, does the integral expression overestimate the sum?

(b) You know how to integrate the function $1/x$. Compute the integral estimate and the actual value. What is the percentage error in the estimate?

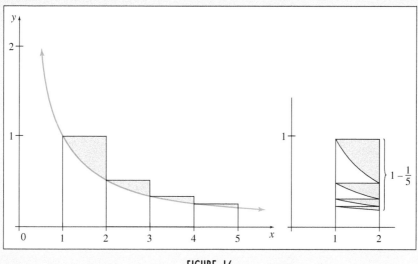

FIGURE 16

5. Go to the website WolframAlpha.com and enter: "integrate." Follow the instructions and use them along with the improved estimate to verify the fair price for 140 units of about $21 million given in the text. Then enter "sum" into Wolfram| Alpha and follow the instructions to verify the fair price by summing the first 140 values of $C(n)$. Discuss how the solution compares with the solutions provided by the integration and summation features on a graphing calculator, as well as the solution provided by adding up the values using a spreadsheet.

DIRECTIONS FOR GROUP PROJECT

Suppose that you and three other students have an internship with a manufacturing company that is submitting a bid to make several thousand units of some highly technical equipment. The one problem with the bid is that the number of units that will be purchased is only an estimate and that the actual number needed may greatly vary from the estimate. Using the information given above, prepare a presentation for an internal sales meeting that will describe the case listed above and its applications to the bid at hand. Make your presentation realistic in the sense that the product you are manufacturing should have a name, average price, and so on. Then show how integrals can be used to estimate learning curves in this situation and produce a pricing structure for the bid. Presentation software, such as Microsoft PowerPoint, should be used.

9

Multivariable Calculus

Safe diving requires an understanding of how the increased pressure below the surface affects the body's intake of nitrogen. An exercise in Section 2 of this chapter investigates a formula for nitrogen pressure as a function of two variables, depth and dive time. Partial derivatives tell us how this function behaves when one variable is held constant as the other changes. Dive tables based on the formula help divers to choose a safe time for a given depth, or a safe depth for a given time.

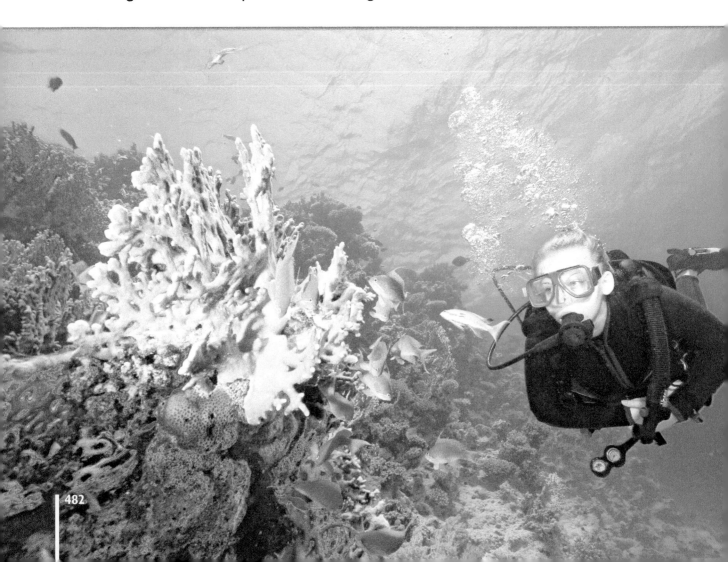

W e have thus far limited our study of calculus to functions of one variable. There are other phenomena that require more than one variable to adequately model the situation. For example, the price of an electronics device depends on how long it has been on the market, the number of competing devices, labor costs, demand, and many other factors. In this case, the price is a function of more than one variable. To analyze and better understand situations like this, we will extend the ideas of calculus, including differentiation and integration, to functions of more than one variable.

9.1 Functions of Several Variables

APPLY IT How can we determine the relationship between the amounts of labor and capital needed to produce a certain number of items?

We will study this question in Example 8 using a production function that depends on the two independent variables of labor and capital.

If a company produces x items at a cost of $10 per item, then the total cost $C(x)$ of producing the items is given by

$$C(x) = 10x.$$

The cost is a function of one independent variable, the number of items produced. If the company produces two products, with x of one product at a cost of $10 each, and y of another product at a cost of $15 each, then the total cost to the firm is a function of *two* independent variables, x and y. By generalizing $f(x)$ notation, the total cost can be written as $C(x, y)$, where

$$C(x, y) = 10x + 15y.$$

When $x = 5$ and $y = 12$ the total cost is written $C(5, 12)$, with

$$C(5, 12) = 10 \cdot 5 + 15 \cdot 12 = 230.$$

A general definition follows.

Function of Two or More Variables

The expression $z = f(x, y)$ is a **function of two variables** if a unique value of z is obtained from each ordered pair of real numbers (x, y). The variables x and y are **independent variables**, and z is the **dependent variable**. The set of all ordered pairs of real numbers (x, y) such that $f(x, y)$ exists is the **domain** of f; the set of all values of $f(x, y)$ is the **range**. Similar definitions could be given for functions of three, four, or more independent variables.

EXAMPLE 1 Evaluating Functions

Let $f(x, y) = 4x^2 + 2xy + 3/y$ and find the following.

(a) $f(-1, 3)$

SOLUTION   Replace x with -1 and y with 3.

$$f(-1, 3) = 4(-1)^2 + 2(-1)(3) + \frac{3}{3} = 4 - 6 + 1 = -1$$

(b) $f(2, 0)$

SOLUTION Replace x with 2. However, because of the quotient $3/y$, it is not possible to replace y with 0. Therefore, $f(2, 0)$ is undefined. By inspection, we see that the domain of the function is the set of all (x, y) such that $y \neq 0$.

(c) $\dfrac{f(x + h, y) - f(x, y)}{h}$

SOLUTION Calculate as follows:

$$\frac{f(x + h, y) - f(x, y)}{h} = \frac{4(x + h)^2 + 2(x + h)y + 3/y - [4x^2 + 2xy + 3/y]}{h}$$

$$= \frac{4x^2 + 8xh + 4h^2 + 2xy + 2hy + 3/y - 4x^2 - 2xy - 3/y}{h}$$

$$= \frac{8xh + 4h^2 + 2hy}{h} \qquad \text{Simplify the numerator.}$$

$$= \frac{h(8x + 4h + 2y)}{h} \qquad \text{Factor } h \text{ from the numerator.}$$

$$= 8x + 4h + 2y. \qquad\qquad \text{TRY YOUR TURN 1}$$

YOUR TURN 1 For the function in Example 1, find $f(2, 3)$.

EXAMPLE 2 Volume of a Can

Let r and h represent the radius and height of a can in cm. The volume of the can is then a function of the two variables r and h given by

$$V(r, h) = \pi r^2 h.$$

Find $V(3, 11)$.

SOLUTION Replace r with 3 and h with 11 to get

$$V(3, 11) = \pi \cdot 3^2 \cdot 11 = 99\pi \approx 311 \text{ cm}^3.$$

This says that a can with radius 3 cm and height 11 cm has a volume of approximately 311 cm^3.

EXAMPLE 3 Evaluating a Function

Let $f(x, y, z) = 4xz - 3x^2y + 2z^2$. Find $f(2, -3, 1)$.

SOLUTION Replace x with 2, y with -3, and z with 1.

$$f(2, -3, 1) = 4(2)(1) - 3(2)^2(-3) + 2(1)^2 = 8 + 36 + 2 = 46$$

TRY YOUR TURN 2

YOUR TURN 2 For the function in Example 3, find $f(1, 2, 3)$.

Graphing Functions of Two Independent Variables

Functions of one independent variable are graphed by using an x-axis and a y-axis to locate points in a plane. The plane determined by the x- and y-axes is called the *xy-plane*. A third axis is needed to graph functions of two independent variables—the z-axis, which goes through the origin in the *xy*-plane and is perpendicular to both the x-axis and the y-axis.

Figure 1 shows one possible way to draw the three axes. In Figure 1, the *yz*-plane is in the plane of the page, with the x-axis perpendicular to the plane of the page.

Just as we graphed ordered pairs earlier we can now graph **ordered triples** of the form (x, y, z). For example, to locate the point corresponding to the ordered triple $(2, -4, 3)$, start at the origin and go 2 units along the positive x-axis. Then go 4 units in a negative direction (to the left) parallel to the y-axis. Finally, go up 3 units parallel to the z-axis. The point representing $(2, -4, 3)$ is shown in Figure 1, together with several other points. The region of three-dimensional space where all coordinates are positive is called the **first octant**.

In Chapter 1 we saw that the graph of $ax + by = c$ (where a and b are not both 0) is a straight line. This result generalizes to three dimensions.

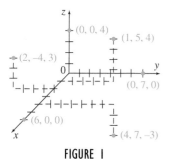

FIGURE 1

Plane

The graph of

$$ax + by + cz = d$$

is a **plane** if a, b, and c are not all 0.

EXAMPLE 4   Graphing a Plane

Graph $2x + y + z = 6$.

SOLUTION   The graph of this equation is a plane. Earlier, we graphed straight lines by finding x- and y-intercepts. A similar idea helps in graphing a plane. To find the x-intercept, which is the point where the graph crosses the x-axis, let $y = 0$ and $z = 0$.

$$2x + 0 + 0 = 6$$
$$x = 3$$

FOR REVIEW

You may want to review the methods for graphing a line found in Section 1.1.

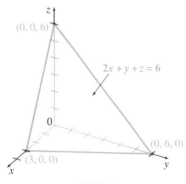

FIGURE 2

The x-intercept is the point $(3, 0, 0)$. Letting $x = 0$ and $z = 0$ gives the y-intercept $(0, 6, 0)$, while $x = 0$ and $y = 0$ lead to the z-intercept $(0, 0, 6)$. The plane through these three points includes the triangular surface shown in Figure 2. This surface is the first-octant part of the plane that is the graph of $2x + y + z = 6$. The plane does not stop at the axes but extends without bound.   **TRY YOUR TURN 3**

YOUR TURN 3 Graph $x + 2y + 3z = 12$ in the first octant.

EXAMPLE 5   Graphing a Plane

Graph $x + z = 6$.

SOLUTION   To find the x-intercept, let $y = 0$ and $z = 0$, giving $(6, 0, 0)$. If $x = 0$ and $y = 0$, we get the z-intercept $(0, 0, 6)$. Because there is no y in the equation $x + z = 6$, there can be no y-intercept. A plane that has no y-intercept is parallel to the y-axis. The first-octant portion of the graph of $x + z = 6$ is shown in Figure 3.

FIGURE 3

FIGURE 4

EXAMPLE 6 Graphing Planes

Graph each equation in the first octant.

(a) $x = 3$

SOLUTION This graph, which passes through $(3, 0, 0)$, can have no y-intercept and no z-intercept. It is, therefore, a plane parallel to the y-axis and the z-axis and, therefore, to the yz-plane. The first-octant portion of the graph is shown in Figure 4.

(b) $y = 4$

SOLUTION This graph passes through $(0, 4, 0)$ and is parallel to the xz-plane. The first-octant portion of the graph is shown in Figure 5.

(c) $z = 1$

SOLUTION The graph is a plane parallel to the xy-plane, passing through $(0, 0, 1)$. Its first-octant portion is shown in Figure 6.

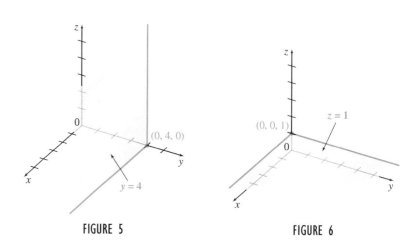

FIGURE 5   FIGURE 6

The graph of a function of one variable, $y = f(x)$, is a curve in the plane. If x_0 is in the domain of f, the point $(x_0, f(x_0))$ on the graph lies directly above or below the number x_0 on the x-axis, as shown in Figure 7.

The graph of a function of two variables, $z = f(x, y)$, is a **surface** in three-dimensional space. If (x_0, y_0) is in the domain of f, the point $(x_0, y_0, f(x_0, y_0))$ lies directly above or below the point (x_0, y_0) in the xy-plane, as shown in Figure 8.

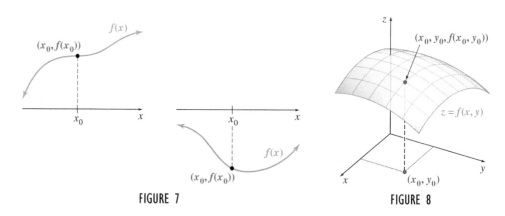

FIGURE 7   FIGURE 8

Although computer software is available for drawing the graphs of functions of two independent variables, you can often get a good picture of the graph without it by finding various **traces**—the curves that result when a surface is cut by a plane. The **xy-trace** is the intersection of the surface with the xy-plane. The **yz-trace** and **xz-trace** are defined similarly. You can also determine the intersection of the surface with planes parallel to the xy-plane. Such planes are of the form $z = k$, where k is a constant, and the curves that result when they cut the surface are called **level curves**.

EXAMPLE 7   Graphing a Function

Graph $z = x^2 + y^2$.

SOLUTION   The yz-plane is the plane in which every point has a first coordinate of 0, so its equation is $x = 0$. When $x = 0$, the equation becomes $z = y^2$, which is the equation of a parabola in the yz-plane, as shown in red in Figure 9(a). Similarly, to find the intersection of the surface with the xz-plane (whose equation is $y = 0$), let $y = 0$ in the equation. It then becomes $z = x^2$, which is the equation of a parabola in the xz-plane, as shown in blue in Figure 9(a). The xy-trace (the intersection of the surface with the plane $z = 0$) is the single point $(0, 0, 0)$ because $x^2 + y^2$ is equal to 0 only when $x = 0$ and $y = 0$.

Next, we find the level curves by intersecting the surface with the planes $z = 1, z = 2,$ $z = 3$, etc. (all of which are parallel to the xy-plane). In each case, the result is a circle:

$$x^2 + y^2 = 1, \qquad x^2 + y^2 = 2, \qquad x^2 + y^2 = 3,$$

and so on, as shown in Figure 9(b). Drawing the traces and level curves on the same set of axes suggests that the graph of $z = x^2 + y^2$ is the bowl-shaped figure, called a **paraboloid**, that is shown in Figure 9(c).

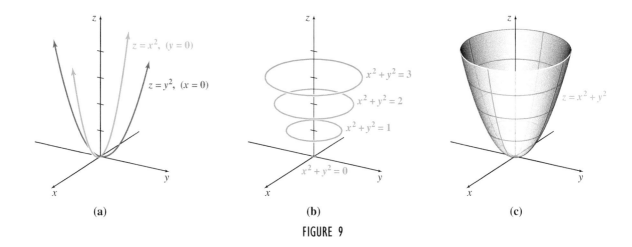

(a)   (b)   (c)

FIGURE 9

Figure 10 on the next page shows the level curves from Example 7 plotted in the xy-plane. The picture can be thought of as a topographical map that describes the surface generated by $z = x^2 + y^2$, just as the topographical map in Figure 11 describes the surface of the land in a part of New York state.

One application of level curves occurs in economics with production functions. A **production function** $z = f(x, y)$ is a function that gives the quantity z of an item produced as a function of x and y, where x is the amount of labor and y is the amount of capital (in appropriate units) needed to produce z units. If the production function has the special form $z = P(x, y) = Ax^a y^{1-a}$, where A is a constant and $0 < a < 1$, the function is called

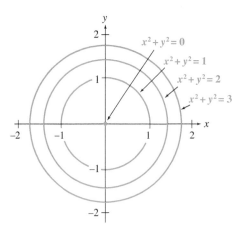

FIGURE 10   FIGURE 11

a **Cobb-Douglas production function**. This function was developed in 1928 by economist Paul H. Douglas (1892–1976), who later became a senator for the state of Illinois, and mathematician Charles W. Cobb. For production functions, level curves are used to indicate combinations of the values of x and y that produce the same value of production z.

EXAMPLE 8   Cobb-Douglas Production Function

Production of a certain item is given by the Cobb-Douglas production function

$$z = P(x, y) = x^{2/3}y^{1/3},$$

where x is the amount of labor in work-hours and y is the amount of capital, in appropriate units.

(a) What is the production when 40 work-hours and 10 units of capital are provided?

SOLUTION   Letting $x = 40$ and $y = 10$ gives the production of

$$z = P(40, 10) = 40^{2/3} \cdot 10^{1/3} \approx 25.20,$$

or about 25 units.

(b) Determine the relationship between the amounts of labor and capital required for the production of 100 units.

SOLUTION   To determine this relationship, first find the level curve for the production of 100 units by setting $z = 100$. Solving the equation for y yields

$$x^{2/3}y^{1/3} = 100$$

$$y^{1/3} = \frac{100}{x^{2/3}}$$

$$y = \frac{100^3}{x^2} \qquad \text{Cube both sides.}$$

$$y = \frac{1{,}000{,}000}{x^2}.$$

This level curve of height 100 is shown graphed in three dimensions in Figure 12(a) and on the familiar xy-plane in Figure 12(b). The points of the graph correspond to those values of labor (x) and capital (y) that lead to the production of 100 items.

(c) If 40 work-hours are available, find the amount of capital needed in the production of 100 units.

YOUR TURN 4 Find the level curve at a production of 27 items in the form $y = f(x)$ for the Cobb-Douglas production function $z = x^{1/4}y^{3/4}$.

SOLUTION Evaluating the equation from part (b) at $x = 40$ yields

$$y = \frac{1,000,000}{40^2} = 625.$$

If 40 work-hours are available, 625 units of capital will be required to produce 100 units.

TRY YOUR TURN 4

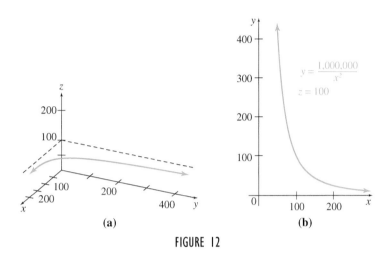

$$y = \frac{1,000,000}{x^2}$$
$$z = 100$$

(a)   **(b)**

FIGURE 12

The curve in Figure 12 is called an *isoquant,* for *iso* (equal) and *quant* (amount). In Example 8, the "amounts" all "equal" 100.

Because of the difficulty of drawing the graphs of more complicated functions, we merely list some common equations and their graphs. We encourage you to explore why these graphs look the way they do by studying their traces, level curves, and axis intercepts. These graphs were drawn by computer, a very useful method of depicting three-dimensional surfaces.

Paraboloid, $z = x^2 + y^2$

xy-trace: point
yz-trace: parabola
xz-trace: parabola

Ellipsoid, $\dfrac{x^2}{a^2} + \dfrac{y^2}{b^2} + \dfrac{z^2}{c^2} = 1$

xy-trace: ellipse
yz-trace: ellipse
xz-trace: ellipse

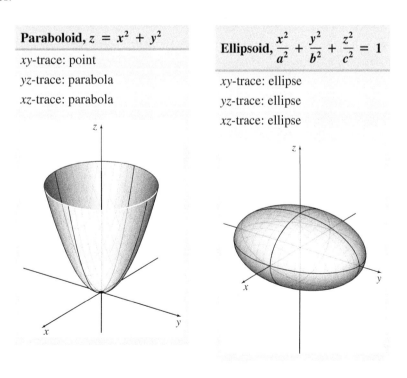

Hyperbolic Paraboloid, $z = x^2 - y^2$ (sometimes called a *saddle*)	**Hyperboloid of Two Sheets,** $-x^2 - y^2 + z^2 = 1$
xy-trace: two intersecting lines	*xy*-trace: none
yz-trace: parabola	*yz*-trace: hyperbola
xz-trace: parabola	*xz*-trace: hyperbola

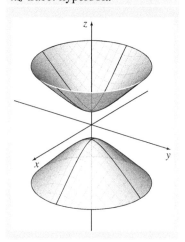

Notice that not all the graphs correspond to functions of two variables. In the ellipsoid, for example, if x and y are both 0, then z can equal c or $-c$, whereas a function can take on only one value. We can, however, interpret the graph as a **level surface** for a function of three variables. Let

$$w(x, y, z) = \frac{x^2}{a^2} + \frac{y^2}{b^2} + \frac{z^2}{c^2}.$$

Then $w = 1$ produces the level surface of the ellipsoid shown, just as $z = c$ gives level curves for the function $z = f(x, y)$.

Another way to draw the graph of a function of two variables is with a graphing calculator. Figure 13 shows the graph of $z = x^2 + y^2$ generated by a TI-89. Figure 14 shows the same graph drawn by the computer program Maple™.

FIGURE 13

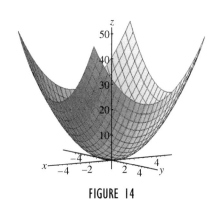

FIGURE 14

9.1 WARM-UP EXERCISES

Graph the following lines.   (*Sec. 1.1*)

W1. $2x + 3y = 6$

W2. $x = 4$

W3. $y = 2$

If $f(x) = 2x^2 + 3x - 6$, find the following.   (*Sec. 2.2*)

W4. $f(-2)$

W5. $f(3a)$

W6. $\dfrac{f(x + h) - f(x)}{h}$

9.1   EXERCISES

1. Let $f(x, y) = 2x - 3y + 5$. Find the following.

  (a) $f(2, -1)$

  (b) $f(-4, 1)$

  (c) $f(-2, -3)$

  (d) $f(0, 8)$

2. Let $g(x, y) = x^2 - 2xy + y^3$. Find the following.

  (a) $g(-2, 4)$

  (b) $g(-1, -2)$

  (c) $g(-2, 3)$

  (d) $g(5, 1)$

3. Let $h(x, y) = \sqrt{x^2 + 2y^2}$. Find the following.

  (a) $h(5, 3)$

  (b) $h(2, 4)$

  (c) $h(-1, -3)$

  (d) $h(-3, -1)$

4. Let $f(x, y) = \dfrac{\sqrt{9x + 5y}}{\log x}$. Find the following.

  (a) $f(10, 2)$

  (b) $f(100, 1)$

  (c) $f(1000, 0)$

  (d) $f\left(\dfrac{1}{10}, 5\right)$

5. Let $f(x, y) = e^x + \ln(x + y)$. Find the following.

  (a) $f(1, 0)$

  (b) $f(2, -1)$

  (c) $f(0, e)$

  (d) $f(0, e^2)$

6. Let $f(x, y) = xe^{x+y}$. Find the following.

  (a) $f(1, 0)$

  (b) $f(2, -2)$

  (c) $f(3, 2)$

  (d) $f(-1, 4)$

Graph the first-octant portion of each plane.

7. $x + y + z = 9$

8. $x + y + z = 15$

9. $2x + 3y + 4z = 12$

10. $4x + 2y + 3z = 24$

11. $x + y = 4$

12. $y + z = 5$

13. $x = 5$

14. $z = 4$

Graph the level curves in the first quadrant of the xy-plane for the following functions at heights of $z = 0$, $z = 2$, and $z = 4$.

15. $3x + 2y + z = 24$

16. $3x + y + 2z = 8$

17. $y^2 - x = -z$

18. $2y - \dfrac{x^2}{3} = z$

19. Discuss how a function of three variables in the form $w = f(x, y, z)$ might be graphed.

20. Suppose the graph of a plane $ax + by + cz = d$ has a portion in the first octant. What can be said about a, b, c, and d?

21. In the chapter on Nonlinear Functions, the vertical line test was presented, which tells whether a graph is the graph of a function. Does this test apply to functions of two variables? Explain.

22. A graph that was not shown in this section is the *hyperboloid of one sheet*, described by the equation $x^2 + y^2 - z^2 = 1$. Describe it as completely as you can.

Match each equation in Exercises 23–28 with its graph in (a)–(f) below and on the next page.

23. $z = x^2 + y^2$

24. $z^2 - y^2 - x^2 = 1$

25. $x^2 - y^2 = z$

26. $z = y^2 - x^2$

27. $\dfrac{x^2}{16} + \dfrac{y^2}{25} + \dfrac{z^2}{4} = 1$

28. $z = 5(x^2 + y^2)^{-1/2}$

  (a)

  (b)

  (c)

  (d)

(e)

(f)

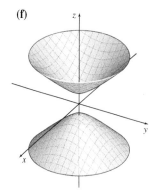

29. Let $f(x, y) = 4x^2 - 2y^2$, and find the following.

(a) $\dfrac{f(x + h, y) - f(x, y)}{h}$

(b) $\dfrac{f(x, y + h) - f(x, y)}{h}$

(c) $\lim\limits_{h \to 0} \dfrac{f(x + h, y) - f(x, y)}{h}$

(d) $\lim\limits_{h \to 0} \dfrac{f(x, y + h) - f(x, y)}{h}$

30. Let $f(x, y) = 5x^3 + 3y^2$, and find the following.

(a) $\dfrac{f(x + h, y) - f(x, y)}{h}$

(b) $\dfrac{f(x, y + h) - f(x, y)}{h}$

(c) $\lim\limits_{h \to 0} \dfrac{f(x + h, y) - f(x, y)}{h}$

(d) $\lim\limits_{h \to 0} \dfrac{f(x, y + h) - f(x, y)}{h}$

31. Let $f(x, y) = xye^{x^2+y^2}$. Use a graphing calculator or spreadsheet to find each of the following and give a geometric interpretation of the results. (*Hint:* First factor e^2 from the limit and then evaluate the quotient at smaller and smaller values of h.)

(a) $\lim\limits_{h \to 0} \dfrac{f(1 + h, 1) - f(1, 1)}{h}$

(b) $\lim\limits_{h \to 0} \dfrac{f(1, 1 + h) - f(1, 1)}{h}$

32. The following table provides values of the function $f(x, y)$. However, because of potential errors in measurement, the functional values may be slightly inaccurate. Using the statistical package included with a graphing calculator or spreadsheet and critical thinking skills, find the function $f(x, y) = a + bx + cy$ that best estimates the table where a, b, and c are integers. (*Hint:* Do a linear regression on each column with the value of y fixed and then use these four regression equations to determine the coefficient c.)

x \ y	0	1	2	3
0	4.02	7.04	9.98	13.00
1	6.01	9.06	11.98	14.96
2	7.99	10.95	14.02	17.09
3	9.99	13.01	16.01	19.02

APPLICATIONS

Business and Economics

33. Production Production of a digital camera is given by

$$P(x, y) = 100\left(\frac{3}{5}x^{-2/5} + \frac{2}{5}y^{-2/5}\right)^{-5},$$

where x is the amount of labor in work-hours and y is the amount of capital. Find the following.

(a) What is the production when 32 work-hours and 1 unit of capital are provided?

(b) Find the production when 1 work-hour and 32 units of capital are provided.

(c) If 32 work-hours and 243 units of capital are used, what is the production output?

Production Find the level curve at a production of 500 for the production functions in Exercises 34 and 35. Graph each level curve in the *xy*-plane.

34. In their original paper, Cobb and Douglas estimated the production function for the United States to be $z = 1.01x^{3/4}y^{1/4}$, where x represents the amount of labor and y the amount of capital. *Source: American Economic Review.*

35. A study of the connection between immigration and the fiscal problems associated with the aging of the baby boom generation considered a production function of the form $z = x^{0.6}y^{0.4}$, where x represents the amount of labor and y the amount of capital. *Source: Journal of Political Economy.*

36. Production For the function in Exercise 34, what is the effect on z of doubling x? Of doubling y? Of doubling both?

37. Cost If labor (x) costs \$250 per unit, materials (y) cost \$150 per unit, and capital (z) costs \$75 per unit, write a function for total cost.

Individual Retirement Accounts The multiplier function

$$M = \frac{(1 + i)^n(1 - t) + t}{[1 + (1 - t)i]^n}$$

compares the growth of an Individual Retirement Account (IRA) with the growth of the same deposit in a regular savings account. The function M depends on the three variables n, i, and t, where n represents the number of years an amount is left at interest, i represents the interest rate in both types of accounts, and t represents the income tax rate. Values of $M > 1$ indicate that the IRA grows faster than the savings account. Let $M = f(n, i, t)$ and find the following.

38. Find the multiplier when funds are left for 25 years at 5% interest and the income tax rate is 33%. Which account grows faster?

39. What is the multiplier when money is invested for 40 years at 6% interest and the income tax rate is 28%? Which account grows faster?

Life Sciences

40. Heat Loss The rate of heat loss (in watts) in harbor seal pups has been approximated by

$$H(m, T, A) = \frac{15.2m^{0.67}(T - A)}{10.23 \ln m - 10.74},$$

where m is the body mass of the pup (in kg), and T and A are the body core temperature and ambient water temperature, respectively (in °C). Find the heat loss for the following data. *Source: Functional Ecology.*

(a) Body mass = 21 kg; body core temperature = 36°C; ambient water temperature = 4°C

(b) Body mass = 29 kg; body core temperature = 38°C; ambient water temperature = 16°C

41. **Body Surface Area** The surface area of a human (in square meters) has been approximated by

$$A = 0.024265 h^{0.3964} m^{0.5378},$$

where h is the height (in cm) and m is the mass (in kg). Find A for the following data. *Source: The Journal of Pediatrics.*

(a) Height, 178 cm; mass, 72 kg

(b) Height, 140 cm; mass, 65 kg

(c) Height, 160 cm; mass, 70 kg

(d) Using your mass and height, find your own surface area.

42. **Dinosaur Running** An article titled "How Dinosaurs Ran" explains that the locomotion of different-sized animals can be compared when they have the same Froude number, defined as

$$F = \frac{v^2}{gl},$$

where v is the velocity, g is the acceleration of gravity (9.81 m per sec²), and l is the leg length (in meters). *Source: Scientific American.*

(a) One result described in the article is that different animals change from a trot to a gallop at the same Froude number, roughly 2.56. Find the velocity at which this change occurs for a ferret, with a leg length of 0.09 m, and a rhinoceros, with a leg length of 1.2 m.

(b) Ancient footprints in Texas of a sauropod, a large herbivorous dinosaur, are roughly 1 m in diameter, corresponding to a leg length of roughly 4 m. By comparing the stride divided by the leg length with that of various modern creatures, it can be determined that the Froude number for these dinosaurs is roughly 0.025. How fast were the sauropods traveling?

43. **Pollution Intolerance** According to research at the Great Swamp in New York, the percentage of fish that are intolerant to pollution can be estimated by the function

$$P(W, R, A) = 48 - 2.43W - 1.81R - 1.22A,$$

where W is the percentage of wetland, R is the percentage of residential area, and A is the percentage of agricultural area surrounding the swamp. *Source: Northeastern Naturalist.*

(a) Use this function to estimate the percentage of fish that will be intolerant to pollution if 5 percent of the land is classified as wetland, 15 percent is classified as residential, and 0 percent is classified as agricultural. (*Note*: The land can also be classified as forest land.)

(b) What is the maximum percentage of fish that will be intolerant to pollution?

(c) Develop two scenarios that will drive the percentage of fish that are intolerant to pollution to zero.

(d) Which variable has the greatest influence on P?

44. **Dengue Fever** In tropical regions, dengue fever is a significant health problem that affects nearly 100 million people each year. Using data collected from the 2002 dengue epidemic in Colima, Mexico, researchers have estimated that the incidence I (number of new cases in a given year) of dengue can be predicted by the following function.

$$I(p, a, m, n, e) = (25.54 + 0.04p - 7.92a + 2.62m + 4.46n + 0.15e)^2,$$

where p is the precipitation (mm), a is the mean temperature (°C), m is the maximum temperature (°C), n is the minimum temperature (°C), and e is the evaporation (mm). *Source: Journal of Environmental Health.*

(a) Estimate the incidence of a dengue fever outbreak for a region with 80 mm of rainfall, average temperature of 23°C, maximum temperature of 34°C, minimum temperature of 16°C, and evaporation of 50 mm.

(b) Which variable has a negative influence on the incidence of dengue? Describe this influence and what can be inferred mathematically about the biology of the fever.

45. **Deer-Vehicle Accidents** Using data collected by the U.S. Forest Service, the annual number of deer-vehicle accidents for any given county in Ohio can be estimated by the function

$$A(L, T, U, C) = 53.02 + 0.383L + 0.0015T + 0.0028U - 0.0003C,$$

where A is the estimated number of accidents, L is the road length (in kilometers), T is the total county land area (in hundred-acres (Ha)), U is the urban land area (in hundred-acres), and C is the number of hundred-acres of crop land. *Source: Ohio Journal of Science.*

(a) Use this formula to estimate the number of deer-vehicle accidents for Mahoning County, where $L = 266$ km, $T = 107{,}484$ Ha, $U = 31{,}697$ Ha, and $C = 24{,}870$ Ha. The actual value was 396.

(b) Given the magnitude and nature of the input numbers, which of the variables have the greatest potential to influence the number of deer-vehicle accidents? Explain your answer.

46. **Deer Harvest** Using data collected by the U.S. Forest Service, the annual number of deer that are harvested for any given county in Ohio can be estimated by the function

$$N(R, C) = 329.32 + 0.0377R - 0.0171C,$$

where N is the estimated number of harvested deer, R is the rural land area (in hundred-acres), and C is the number of hundred-acres of crop land. *Source: Ohio Journal of Science.*

(a) Use this formula to estimate the number of harvested deer for Tuscarawas County, where $R = 141{,}319$ Ha and $C = 37{,}960$ Ha. The actual value was 4925 deer harvested.

(b) Sketch the graph of this function in the first octant.

47. **Agriculture** Pregnant sows tethered in stalls often show high levels of repetitive behavior, such as bar biting and chain chewing, indicating chronic stress. Researchers from Great Britain have developed a function that estimates the relationship between repetitive behavior, the behavior of sows in adjacent stalls, and food allowances such that

$$\ln(T) = 5.49 - 3.00 \ln(F) + 0.18 \ln(C),$$

where T is the percent of time spent in repetitive behavior, F is the amount of food given to the sow (in kilograms per day), and C is the percent of time that neighboring sows spent bar biting and chain chewing. *Source: Applied Animal Behaviour Science*.

(a) Solve the above expression for T.

(b) Find and interpret T when $F = 2$ and $C = 40$.

48. Life Span Researchers have estimated the maximum life span (in years) for various species of mammals according to the formula

$$L(E, P) = 23E^{0.6}P^{-0.267},$$

where E is the average brain mass and P is the average body mass (both in g). Find L for the following species. Do these values seem reasonable? *Source: The Quarterly Review of Biology*.

(a) Black rat, with $E = 7.35$ g, $P = 150$ g

(b) Humans, with $E = 14{,}100$ g, $P = 68{,}700$ g

General Interest

49. Postage Rates Extra postage is charged for parcels sent by U.S. mail that are more than 84 in. in length and girth combined. (Girth is the distance around the parcel perpendicular to its length. See the figure.) Express the combined length and girth as a function of L, W, and H.

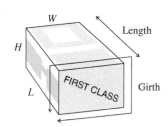

50. Required Material Refer to the figure for Exercise 49. Assume L, W, and H are in feet. Write a function in terms of L, W, and H that gives the total area of the material required to build the box.

51. Elliptical Templates The holes cut in a roof for vent pipes require elliptical templates. A formula for determining the length of the major axis of the ellipse is given by

$$L = f(H, D) = \sqrt{H^2 + D^2},$$

where D is the (outside) diameter of the pipe and H is the "rise" of the roof per D units of "run"; that is, the slope of the roof is H/D. (See the figure below.) The width of the ellipse (minor axis) equals D. Find the length and width of the ellipse required to produce a hole for a vent pipe with a diameter of 3.75 in. in roofs with the following slopes.

(a) 3/4 (b) 2/5

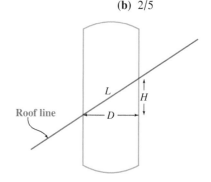

YOUR TURN ANSWERS

1. 29

2. 24

3.

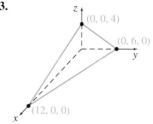

4. $y = 81/x^{1/3}$.

9.2 Partial Derivatives

APPLY IT What is the change in productivity if labor is increased by one work-hour? What if capital is increased by one unit?

We will answer this question in Example 5 using the concept of partial derivatives.

Earlier, we found that the derivative dy/dx gives the rate of change of y with respect to x. In this section, we show how derivatives are found and interpreted for multivariable functions.

A small firm makes only two products, smartphones and tablets. The profits of the firm are given by

$$P(x, y) = 40x^2 - 10xy + 5y^2 - 80,$$

where x is the number of smartphones sold and y is the number of tablets sold. How will a change in x or y affect P?

FOR REVIEW

You may want to review Chapter 4 on Calculating the Derivative for methods used to find some of the derivatives in this section.

Suppose that sales of smartphones have been steady at 10 units; only the sales of tablets vary. The management would like to find the marginal profit with respect to y, the number of tablets sold. Recall that marginal profit is given by the derivative of the profit function. Here, x is fixed at 10. Using this information, we begin by finding a new function, $f(y) = P(10, y)$. Let $x = 10$ to get

$$f(y) = P(10, y) = 40(10)^2 - 10(10)y + 5y^2 - 80$$
$$= 3920 - 100y + 5y^2.$$

The function $f(y)$ shows the profit from the sale of y tablets, assuming that x is fixed at 10 units. Find the derivative df/dy to get the marginal profit with respect to y.

$$\frac{df}{dy} = -100 + 10y$$

In this example, the derivative of the function $f(y)$ was taken with respect to y only; we assumed that x was fixed. To generalize, let $z = f(x, y)$. An intuitive definition of the *partial derivatives* of f with respect to x and y follows.

Partial Derivatives (Informal Definition)

The **partial derivative of f with respect to x** is the derivative of f obtained by treating x as a variable and y as a constant.

The **partial derivative of f with respect to y** is the derivative of f obtained by treating y as a variable and x as a constant.

The symbols $f_x(x, y)$ (no prime is used), $\partial z/\partial x$, z_x, and $\partial f/\partial x$ are used to represent the partial derivative of $z = f(x, y)$ with respect to x, with similar symbols used for the partial derivative with respect to y.

Generalizing from the definition of the derivative given earlier, partial derivatives of a function $z = f(x, y)$ are formally defined as follows.

Partial Derivatives (Formal Definition)

Let $z = f(x, y)$ be a function of two independent variables. Let all indicated limits exist. Then the partial derivative of f with respect to x is

$$f_x(x, y) = \frac{\partial f}{\partial x} = \lim_{h \to 0} \frac{f(x + h, y) - f(x, y)}{h},$$

and the partial derivative of f with respect to y is

$$f_y(x, y) = \frac{\partial f}{\partial y} = \lim_{h \to 0} \frac{f(x, y + h) - f(x, y)}{h}.$$

If the indicated limits do not exist, then the partial derivatives do not exist.

Similar definitions could be given for functions of more than two independent variables.

EXAMPLE 1 Partial Derivatives

Let $f(x, y) = 4x^2 - 9xy + 6y^3$. Find $f_x(x, y)$ and $f_y(x, y)$.

SOLUTION To find $f_x(x, y)$, treat y as a constant and x as a variable. The derivative of the first term, $4x^2$, is $8x$. In the second term, $-9xy$, the constant coefficient of x is $-9y$, so the derivative with x as the variable is $-9y$. The derivative of $6y^3$ is zero, since we are treating y as a constant. Thus,

$$f_x(x, y) = 8x - 9y.$$

YOUR TURN 1 Let
$f(x, y) = 2x^2y^3 + 6x^5y^4$. Find
$f_x(x, y)$ and $f_y(x, y)$.

Now, to find $f_y(x, y)$, treat y as a variable and x as a constant. Since x is a constant, the derivative of $4x^2$ is zero. In the second term, the coefficient of y is $-9x$ and the derivative of $-9xy$ is $-9x$. The derivative of the third term is $18y^2$. Thus,

$$f_y(x, y) = -9x + 18y^2. \qquad \text{TRY YOUR TURN 1} \blacksquare$$

The next example shows how the chain rule can be used to find partial derivatives.

EXAMPLE 2 Partial Derivatives

Let $f(x, y) = \ln |x^2 + 3y|$. Find $f_x(x, y)$ and $f_y(x, y)$.

SOLUTION Recall the formula for the derivative of a natural logarithm function. If $g(x) = \ln |x|$, then $g'(x) = 1/x$. Using this formula and the chain rule,

$$f_x(x, y) = \frac{1}{x^2 + 3y} \cdot \frac{\partial}{\partial x}(x^2 + 3y) = \frac{1}{x^2 + 3y} \cdot 2x = \frac{2x}{x^2 + 3y},$$

and

YOUR TURN 2 Let
$f(x, y) = e^{3x^2y}$. Find $f_x(x, y)$ and
$f_y(x, y)$.

$$f_y(x, y) = \frac{1}{x^2 + 3y} \cdot \frac{\partial}{\partial y}(x^2 + 3y) = \frac{1}{x^2 + 3y} \cdot 3 = \frac{3}{x^2 + 3y}.$$

$$\text{TRY YOUR TURN 2} \blacksquare$$

The notation

$$f_x(a, b) \qquad \text{or} \qquad \frac{\partial f}{\partial y}(a, b)$$

represents the value of the partial derivative when $x = a$ and $y = b$, as shown in the next example.

EXAMPLE 3 Evaluating Partial Derivatives

Let $f(x, y) = 2x^2 + 9xy^3 + 8y + 5$. Find the following.

(a) $f_x(-1, 2)$

SOLUTION First, find $f_x(x, y)$ by holding y constant.

$$f_x(x, y) = 4x + 9y^3$$

Now let $x = -1$ and $y = 2$.

$$f_x(-1, 2) = 4(-1) + 9(2)^3 = -4 + 72 = 68$$

(b) $\dfrac{\partial f}{\partial y}(-4, -3)$

SOLUTION Since $\partial f/\partial y = 27xy^2 + 8$,

$$\frac{\partial f}{\partial y}(-4, -3) = 27(-4)(-3)^2 + 8 = 27(-36) + 8 = -964.$$

(c) All values of x and y such that both $f_x(x, y) = 0$ and $f_y(x, y) = 0$. (The importance of such points will be shown in the next section.)

SOLUTION From parts (a) and (b),

$$f_x(x, y) = 4x + 9y^3 = 0 \qquad \text{and} \qquad f_y(x, y) = 27xy^2 + 8 = 0.$$

Solving the first equation for x yields $x = -9y^3/4$. Substituting this into the second equation yields

$$27\left(\frac{-9y^3}{4}\right)y^2 + 8 = 0$$

$$\frac{-243y^5}{4} + 8 = 0$$

$$\frac{-243y^5}{4} = -8$$

$$y^5 = \frac{32}{243} \qquad \text{Multiply both sides by } -4/243.$$

$$y = \frac{2}{3}. \qquad \text{Take the fifth root of both sides.}$$

Substituting $y = 2/3$ yields $x = -9y^3/4 = -9(2/3)^3/4 = -2/3$. Thus, $f_x(x, y) = 0$ and $f_y(x, y) = 0$ when $x = -2/3$ and $y = 2/3$.

(d) $f_x(x, y)$ using the formal definition of the partial derivative.

SOLUTION Calculate as follows:

$$\frac{f(x + h, y) - f(x, y)}{h} = \frac{2(x + h)^2 + 9(x + h)y^3 + 8y + 5 - (2x^2 + 9xy^3 + 8y + 5)}{h}$$

$$= \frac{2x^2 + 4xh + 2h^2 + 9xy^3 + 9hy^3 + 8y + 5 - 2x^2 - 9xy^3 - 8y - 5}{h}$$

$$= \frac{4xh + 2h^2 + 9hy^3}{h} \qquad \text{Simplify the numerator.}$$

$$= \frac{h(4x + 2h + 9y^3)}{h} \qquad \text{Factor } h \text{ from the numerator.}$$

$$= 4x + 2h + 9y^3.$$

Now take the limit as h goes to 0. Thus,

$$f_x(x, y) = \lim_{h \to 0}\frac{f(x + h, y) - f(x, y)}{h} = \lim_{h \to 0}(4x + 2h + 9y^3)$$

$$= 4x + 9y^3,$$

the same answer we found in part (a). **TRY YOUR TURN 3**

YOUR TURN 3 Let $f(x, y) = xye^{x^2+y^3}$. Find $f_x(2, 1)$ and $f_y(2, 1)$.

$z = f(x, y)$

FIGURE 15

In some cases, the difference quotient may not simplify as easily as it did in Example 3(d). In such cases, the derivative may be approximated by putting a small value for h into $[f(x + h) - f(x)]/h$. In Example 3(d), with $x = -1$ and $y = 2$, the values $h = 10^{-4}$ and 10^{-5} give approximations for $f_x(-1, 2)$ as 68.0002 and 68.00002, respectively, compared with the exact value of 68 found in Example 3(a).

The derivative of a function of one variable can be interpreted as the slope of the tangent line to the graph at that point. With some modification, the same is true of partial derivatives of functions of two variables. At a point on the graph of a function of two variables, $z = f(x, y)$, there may be many tangent lines, all of which lie in the same tangent plane, as shown in Figure 15.

In any particular direction, however, there will be only one tangent line. We use partial derivatives to find the slope of the tangent lines in the x- and y-directions as follows.

Figure 16 on the next page shows a surface $z = f(x, y)$ and a plane that is parallel to the xz-plane. The equation of the plane is $y = b$. (This corresponds to holding y fixed.) Since $y = b$ for points on the plane, any point on the curve that represents the intersection of the plane and the surface must have the form $(x, y, z) = (x, b, f(x, b))$. Thus, this curve can

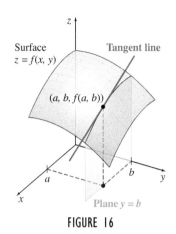

Surface
$z = f(x, y)$

Tangent line

$(a, b, f(a, b))$

a

b

y

x

Plane $y = b$

FIGURE 16

be described as $z = f(x, b)$. Since b is constant, $z = f(x, b)$ is a function of one variable. When the derivative of $z = f(x, b)$ is evaluated at $x = a$, it gives the slope of the line tangent to this curve at the point $(a, b, f(a, b))$, as shown in Figure 16. Thus, the partial derivative of f with respect to x, $f_x(a, b)$, gives the rate of change of the surface $z = f(x, y)$ in the x-direction at the point $(a, b, f(a, b))$. In the same way, the partial derivative with respect to y will give the slope of the line tangent to the surface in the y-direction at the point $(a, b, f(a, b))$.

Rate of Change

The derivative of $y = f(x)$ gives the rate of change of y with respect to x. In the same way, if $z = f(x, y)$, then $f_x(x, y)$ gives the rate of change of z with respect to x, if y is held constant.

EXAMPLE 4 **Water Temperature**

Suppose that the temperature of the water at the point on a river where a nuclear power plant discharges its hot waste water is approximated by

$$T(x, y) = 2x + 5y + xy - 40,$$

where x represents the temperature of the river water (in degrees Celsius) before it reaches the power plant and y is the number of megawatts (in hundreds) of electricity being produced by the plant.

(a) Find and interpret $T_x(9, 5)$.

SOLUTION First, find the partial derivative $T_x(x, y)$.

$$T_x(x, y) = 2 + y$$

This partial derivative gives the rate of change of T with respect to x. Replacing x with 9 and y with 5 gives

$$T_x(9, 5) = 2 + 5 = 7.$$

Just as marginal cost is the approximate cost of one more item, this result, 7, is the approximate change in temperature of the output water if input water temperature changes by 1 degree, from $x = 9$ to $x = 9 + 1 = 10$, while y remains constant at 5 (500 megawatts of electricity produced).

(b) Find and interpret $T_y(9, 5)$.

SOLUTION The partial derivative $T_y(x, y)$ is

$$T_y(x, y) = 5 + x.$$

This partial derivative gives the rate of change of T with respect to y as

$$T_y(9, 5) = 5 + 9 = 14.$$

This result, 14, is the approximate change in temperature resulting from a 1-unit increase in production of electricity from $y = 5$ to $y = 5 + 1 = 6$ (from 500 to 600 megawatts), while the input water temperature x remains constant at 9°C.

As mentioned in the previous section, if $P(x, y)$ gives the output P produced by x units of labor and y units of capital, $P(x, y)$ is a production function. The partial derivatives of this production function have practical implications. For example, $\partial P/\partial x$ gives the marginal productivity of labor. This represents the rate at which the output is changing with respect to labor for a fixed capital investment. That is, if the capital investment is held constant and labor is increased by 1 work-hour, $\partial P/\partial x$ will yield the approximate change in the production level. Likewise, $\partial P/\partial y$ gives the marginal productivity of capital, which represents the rate at which the output is changing with respect to a one-unit change in capital for a fixed labor value. So if the labor force is held constant and the capital investment is increased by 1 unit, $\partial P/\partial y$ will approximate the corresponding change in the production level.

EXAMPLE 5 Production Function

A company that manufactures computers has determined that its production function is given by

$$P(x, y) = 0.1xy^2 \ln(2x + 3y + 2),$$

where x is the size of the labor force (measured in work-hours per week) and y is the amount of capital (measured in units of $1000) invested. Find the marginal productivity of labor and capital when $x = 50$ and $y = 20$, and interpret the results.

APPLY IT

SOLUTION The marginal productivity of labor is found by taking the derivative of P with respect to x.

$$\frac{\partial P}{\partial x} = 0.1 \left[\frac{xy^2}{2x + 3y + 2} \cdot 2 + y^2 \ln(2x + 3y + 2) \right] \qquad \text{Use the product and chain rules.}$$

$$\frac{\partial P}{\partial x}(50, 20) = 0.1 \left[\frac{50(20)^2}{2(50) + 3(20) + 2} \cdot 2 + 20^2 \ln(2(50) + 3(20) + 2) \right] \approx 228$$

Thus, if the capital investment is held constant at $20,000 and labor is increased from 50 to 51 work-hours per week, production will increase by about 228 units. In the same way, the marginal productivity of capital is $\partial P / \partial y$.

$$\frac{\partial P}{\partial y} = 0.1 \left[\frac{xy^2}{2x + 3y + 2} \cdot 3 + 2xy \ln(2x + 3y + 2) \right] \qquad \text{Use the product and chain rules.}$$

$$\frac{\partial P}{\partial y}(50, 20) = 0.1 \left[\frac{50(20)^2}{2(50) + 3(20) + 2} \cdot 3 + 2(50)(20) \ln(2(50) + 3(20) + 2) \right] \approx 1055$$

If work-hours are held constant at 50 hours per week and the capital investment is increased from $20,000 to $21,000, production will increase by about 1055 units.

Second-Order Partial Derivatives

The second derivative of a function of one variable is very useful in determining relative maxima and minima. **Second-order partial derivatives** (partial derivatives of a partial derivative) are used in a similar way for functions of two or more variables. The situation is somewhat more complicated, however, with more independent variables. For example, $f(x, y) = 4x + x^2y + 2y$ has two first-order partial derivatives,

$$f_x(x, y) = 4 + 2xy \qquad \text{and} \qquad f_y(x, y) = x^2 + 2.$$

Since each of these has two partial derivatives, one with respect to y and one with respect to x, there are *four* second-order partial derivatives of function f. The notations for these four second-order partial derivatives are given below.

Second-Order Partial Derivatives

For a function $z = f(x, y)$, if the indicated partial derivative exists, then

$$\frac{\partial}{\partial x} \left(\frac{\partial z}{\partial x} \right) = \frac{\partial^2 z}{\partial x^2} = f_{xx}(x, y) = z_{xx} \qquad \frac{\partial}{\partial y} \left(\frac{\partial z}{\partial y} \right) = \frac{\partial^2 z}{\partial y^2} = f_{yy}(x, y) = z_{yy}$$

$$\frac{\partial}{\partial y} \left(\frac{\partial z}{\partial x} \right) = \frac{\partial^2 z}{\partial y \partial x} = f_{xy}(x, y) = z_{xy} \qquad \frac{\partial}{\partial x} \left(\frac{\partial z}{\partial y} \right) = \frac{\partial^2 z}{\partial x \partial y} = f_{yx}(x, y) = z_{yx}$$

NOTE For most functions found in applications and for all of the functions in this book, the second-order partial derivatives $f_{xy}(x, y)$ and $f_{yx}(x, y)$ are equal. This is always true when $f_{xy}(x, y)$ and $f_{yx}(x, y)$ are continuous. Therefore, it is not necessary to be particular about the order in which these derivatives are found.

EXAMPLE 6 Second-Order Partial Derivatives

Find all second-order partial derivatives for

$$f(x, y) = -4x^3 - 3x^2y^3 + 2y^2.$$

SOLUTION First find $f_x(x, y)$ and $f_y(x, y)$.

$$f_x(x, y) = -12x^2 - 6xy^3 \quad \text{and} \quad f_y(x, y) = -9x^2y^2 + 4y$$

To find $f_{xx}(x, y)$, take the partial derivative of $f_x(x, y)$ with respect to x.

$$f_{xx}(x, y) = -24x - 6y^3$$

Take the partial derivative of $f_y(x, y)$ with respect to y; this gives f_{yy}.

$$f_{yy}(x, y) = -18x^2y + 4$$

Find $f_{xy}(x, y)$ by starting with $f_x(x, y)$, then taking the partial derivative of $f_x(x, y)$ with respect to y.

$$f_{xy}(x, y) = -18xy^2$$

Finally, find $f_{yx}(x, y)$ by starting with $f_y(x, y)$; take its partial derivative with respect to x.

$$f_{yx}(x, y) = -18xy^2$$

EXAMPLE 7 Second-Order Partial Derivatives

Let $f(x, y) = 2e^x - 8x^3y^2$. Find all second-order partial derivatives.

SOLUTION Here $f_x(x, y) = 2e^x - 24x^2y^2$ and $f_y(x, y) = -16x^3y$. (Recall: If $g(x) = e^x$, then $g'(x) = e^x$.) Now find the second-order partial derivatives.

$$f_{xx}(x, y) = 2e^x - 48xy^2 \qquad f_{xy}(x, y) = -48x^2y$$
$$f_{yy}(x, y) = -16x^3 \qquad f_{yx}(x, y) = -48x^2y$$

TRY YOUR TURN 4 ▬

YOUR TURN 4 Let $f(x, y) = x^2e^{7y} + x^4y^5$. Find all second partial derivatives.

Partial derivatives of functions with more than two independent variables are found in a similar manner. For instance, to find $f_z(x, y, z)$ for $f(x, y, z)$, hold x and y constant and differentiate with respect to z.

EXAMPLE 8 Second-Order Partial Derivatives

Let $f(x, y, z) = 2x^2yz^2 + 3xy^2 - 4yz$. Find $f_x(x, y, z)$, $f_y(x, y, z)$, $f_{xz}(x, y, z)$, and $f_{yz}(x, y, z)$.

SOLUTION

$$f_x(x, y, z) = 4xyz^2 + 3y^2$$
$$f_y(x, y, z) = 2x^2z^2 + 6xy - 4z$$

To find $f_{xz}(x, y, z)$, differentiate $f_x(x, y, z)$ with respect to z.

$$f_{xz}(x, y, z) = 8xyz$$

Differentiate $f_y(x, y, z)$ with respect to z to get $f_{yz}(x, y, z)$.

$$f_{yz}(x, y, z) = 4x^2z - 4$$

9.2 WARM-UP EXERCISES

For each of the following functions, find the derivative.

W1. $f(x) = 2x^3 + 7x - 12$ *(Sec. 4.1)*

W2. $f(x) = \sqrt{x} + \dfrac{5}{x^2} + 6e^{2x}$ *(Sec. 4.1, 4.4)*

W3. $f(x) = xe^x$ *(Sec. 4.2, 4.4)*

W4. $f(x) = \dfrac{3x^4}{x^2 - 5}$ *(Sec. 4.2)*

W5. $f(x) = \sqrt{2x + 8}$ *(Sec. 4.3)*

W6. $f(x) = 9(e^{x^2} + 5x)^3$ *(Sec. 4.3, 4.4)*

W7. $f(x) = \ln|2x^3 + 5|$ *(Sec. 4.5)*

W8. $f(x) = (x + 1)\ln x^2$ *(Sec. 4.5)*

9.2   EXERCISES

1. Let $z = f(x, y) = 6x^2 - 4xy + 9y^2$. Find the following using the formal definition of the partial derivative.

(a) $\dfrac{\partial z}{\partial x}$ (b) $\dfrac{\partial z}{\partial y}$ (c) $\dfrac{\partial f}{\partial x}(2, 3)$ (d) $f_y(1, -2)$

2. Let $z = g(x, y) = 8x + 6x^2y + 2y^2$. Find the following using the formal definition of the partial derivative.

(a) $\dfrac{\partial g}{\partial x}$ (b) $\dfrac{\partial g}{\partial y}$ (c) $\dfrac{\partial z}{\partial y}(-3, 0)$ (d) $g_x(2, 1)$

In Exercises 3–20, find $f_x(x, y)$ and $f_y(x, y)$. Then find $f_x(2, -1)$ and $f_y(-4, 3)$. Leave the answers in terms of e in Exercises 7–10, 15–16, and 19–20.

3. $f(x, y) = -4xy + 6y^3 + 5$ **4.** $f(x, y) = 9x^2y^2 - 4y^2$

5. $f(x, y) = 5x^2y^3$ **6.** $f(x, y) = -3x^4y^3 + 10$

7. $f(x, y) = e^{x+y}$ **8.** $f(x, y) = 4e^{3x+2y}$

9. $f(x, y) = -6e^{4x-3y}$ **10.** $f(x, y) = 8e^{7x-y}$

11. $f(x, y) = \dfrac{x^2 + y^3}{x^3 - y^2}$ **12.** $f(x, y) = \dfrac{3x^2y^3}{x^2 + y^2}$

13. $f(x, y) = \ln|1 + 5x^3y^2|$ **14.** $f(x, y) = \ln|4x^4 - 2x^2y^2|$

15. $f(x, y) = xe^{x^2y}$ **16.** $f(x, y) = y^2e^{x+3y}$

17. $f(x, y) = \sqrt{x^4 + 3xy + y^4 + 10}$

18. $f(x, y) = (7x^2 + 18xy^2 + y^3)^{1/3}$

19. $f(x, y) = \dfrac{3x^2y}{e^{xy} + 2}$

20. $f(x, y) = (7e^{x+2y} + 4)(e^{x^2} + y^2 + 2)$

Find all second-order partial derivatives for the following.

21. $f(x, y) = 4x^2y^2 - 16x^2 + 4y$

22. $g(x, y) = 5x^4y^2 + 12y^3 - 9x$

23. $R(x, y) = 4x^2 - 5xy^3 + 12y^2x^2$

24. $h(x, y) = 30y + 5x^2y + 12xy^2$

25. $r(x, y) = \dfrac{6y}{x + y}$ **26.** $k(x, y) = \dfrac{-7x}{2x + 3y}$

27. $z = 9ye^x$ **28.** $z = -6xe^y$

29. $r = \ln|x + y|$ **30.** $k = \ln|5x - 7y|$

31. $z = x\ln|xy|$ **32.** $z = (y + 1)\ln|x^3y|$

For the functions defined as follows, find all values of x and y such that both $f_x(x, y) = 0$ and $f_y(x, y) = 0$.

33. $f(x, y) = 6x^2 + 6y^2 + 6xy + 36x - 5$

34. $f(x, y) = 50 + 4x - 5y + x^2 + y^2 + xy$

35. $f(x, y) = 9xy - x^3 - y^3 - 6$

36. $f(x, y) = 2200 + 27x^3 + 72xy + 8y^2$

Find $f_x(x, y, z), f_y(x, y, z), f_z(x, y, z)$, and $f_{yz}(x, y, z)$ for the following.

37. $f(x, y, z) = x^4 + 2yz^2 + z^4$

38. $f(x, y, z) = 6x^3 - x^2y^2 + y^5$

39. $f(x, y, z) = \dfrac{6x - 5y}{4z + 5}$ **40.** $f(x, y, z) = \dfrac{2x^2 + xy}{yz - 2}$

41. $f(x, y, z) = \ln|x^2 - 5xz^2 + y^4|$

42. $f(x, y, z) = \ln|8xy + 5yz - x^3|$

In Exercises 43 and 44, approximate the indicated derivative for each function by using the definition of the derivative with small values of h.

43. $f(x, y) = (x + y/2)^{x+y/2}$

(a) $f_x(1, 2)$ (b) $f_y(1, 2)$

44. $f(x, y) = (x + y^2)^{2x+y}$

(a) $f_x(2, 1)$ (b) $f_y(2, 1)$

APPLICATIONS

Business and Economics

45. Manufacturing Cost   Suppose that the manufacturing cost of a laptop is approximated by

$$M(x, y) = 45x^2 + 40y^2 - 20xy + 50,$$

where x is the cost of processors and y is the cost of labor. Find the following.

(a) $M_y(4, 2)$ (b) $M_x(3, 6)$ (c) $(\partial M/\partial x)(2, 5)$

(d) $(\partial M/\partial y)(6, 7)$

46. Revenue The revenue from the sale of x units of a sedative and y units of an antibiotic is given by

$$R(x, y) = 5x^2 + 9y^2 - 4xy.$$

Suppose 9 units of sedative and 5 units of antibiotic are sold.

(a) What is the approximate effect on revenue if 10 units of sedative and 5 units of antibiotic are sold?

(b) What is the approximate effect on revenue if the amount of antibiotic sold is increased to 6 units, while sedative sales remain constant?

47. Sales A car dealership estimates that the total weekly sales of its most popular model is a function of the car's list price, p, and the interest rate in percent, i, offered by the manufacturer. The approximate weekly sales are given by

$$f(p, i) = 99p - 0.5pi - 0.0025p^2.$$

(a) Find the weekly sales if the average list price is $19,400 and the manufacturer is offering an 8% interest rate.

(b) Find and interpret $f_p(p, i)$ and $f_i(p, i)$.

(c) What would be the effect on weekly sales if the price is $19,400 and interest rates rise from 8% to 9%?

48. Marginal Productivity Suppose the production function of a company is given by

$$P(x, y) = 250 \sqrt{x^2 + y^2},$$

where x represents units of labor and y represents units of capital. Find the following when 6 units of labor and 8 units of capital are used.

(a) The marginal productivity of labor

(b) The marginal productivity of capital

49. Marginal Productivity A manufacturer estimates that production (in hundreds of units) is a function of the amounts x and y of labor and capital used, as follows.

$$f(x, y) = \left(\frac{1}{4} x^{-1/4} + \frac{3}{4} y^{-1/4} \right)^{-4}$$

(a) Find the number of units produced when 16 units of labor and 81 units of capital are utilized.

(b) Find and interpret $f_x(16, 81)$ and $f_y(16, 81)$.

(c) What would be the approximate effect on production of increasing labor by 1 unit from 16 units of labor with 81 units of capital?

50. Marginal Productivity The production function z for the United States was once estimated as

$$z = x^{0.7} y^{0.3},$$

where x stands for the amount of labor and y the amount of capital. Find the marginal productivity of labor and of capital.

51. Marginal Productivity A similar production function for Canada is

$$z = x^{0.4} y^{0.6},$$

with x, y, and z as in Exercise 50. Find the marginal productivity of labor and of capital.

52. Marginal Productivity A manufacturer of automobile batteries estimates that his total production (in thousands of units) is given by

$$f(x, y) = 3x^{1/3} y^{2/3},$$

where x is the number of units of labor and y is the number of units of capital utilized.

(a) Find and interpret $f_x(64, 125)$ and $f_y(64, 125)$ if the current level of production uses 64 units of labor and 125 units of capital.

(b) Use your answer from part (a) to calculate the approximate effect on production of increasing labor to 65 units while holding capital at the current level.

(c) Suppose that sales have been good and management wants to increase either capital or labor by 1 unit. Which option would result in a larger increase in production?

53. Airline Competition A model of airline competition defines $f(x, y)$ as the probability that a customer will buy a ticket from Airline 1 rather than Airline 2 if the price for a ticket is x from Airline 1 and y from Airline 2. A similar function $g(x, y)$ gives the probability that a customer will buy a ticket from Airline 2 rather than Airline 1 if the price for a ticket is x from Airline 1 and y from Airline 2. For simplicity, the model assumes that $f(x, y) = g(y, x)$. *Source: The Journal of the Operational Research Society.*

(a) Explain what the last assumption means, and why it is reasonable.

(b) The researchers also assume that

$$f_x(x, y) < 0 \quad \text{and} \quad f_y(x, y) > 0.$$

Explain what this assumption means, and why it is reasonable.

(c) One version of the model assumes

$$f(x, y) = \frac{y(1 + x)e^{-x}}{x + y}.$$

Calculate $f_x(x, y)$ and $f_y(x, y)$ and verify that the assumptions of part (b) are satisfied.

54. Administrative Intensity A study defined administrative intensity as the number of administrative personnel A per production personnel X, or $I = A/X$. To seek the optimum administrative intensity for an industry, the following Cobb-Douglas production function was defined:

$$Q = CS^\sigma A^\alpha K^\beta X^\gamma,$$

where S is the number of distinct occupational specialties represented in the labor force, σ is the marginal relative productivity of specialization, K is the total productive capital, and C is a constant of proportionality. *Source: Administrative Science Quarterly.*

(a) Unlike the production functions described in this chapter, for which $\alpha + \beta + \gamma = 1$, the study assumed that increasing cumulative control losses result in decreasing returns to scale, so that

$$\alpha + \beta + \gamma = 1 - \lambda,$$

where λ is the control loss parameter. The study also defined $k = K/X$ as the capital intensity per worker. Show that the production function can then be rewritten as

$$Q = CS^\sigma I^\alpha k^\beta X^{1-\lambda}.$$

(b) The study showed that the profit P can be written as

$$P = (p - r)Q - W_0S - W_1A - F,$$

where p is the market price per unit, r is the production cost per unit, W_0 is the average wage per production worker, W_1 is the average salary per administrative worker, and F is the fixed costs. Letting $A = XI$ in the equation for P, show that $\partial P/\partial I = 0$ (which leads to the maximum profit) when the administrative intensity is given by

$$I = \left(\frac{\alpha(p - r)CS^{\sigma}k^{\beta}}{W_1X^{\lambda}}\right)^{1/(1-\alpha)}.$$

(c) Marginal productivity of administration is given by $\partial Q/\partial A$. Show that this can be written as

$$\frac{\partial Q}{\partial A} = \alpha CS^{\sigma}A^{\alpha-1}k^{\beta}X^{1-\lambda-\alpha}.$$

(d) If x and y are functions of z, and f is a function of x, y, and z, the total differential for three variables (to be discussed later in this chapter) can be converted to the chain rule for three variables:

$$\frac{dw}{dz} = f_x(x, y, z)\frac{dx}{dz} + f_y(x, y, z)\frac{dy}{dz} + f_z(x, y, z),$$

which gives the total rate of change of w with respect to z. Use this result plus the result from part (c) to show that the total rate of change of the marginal productivity of administration with respect to the number of workers can be given as

$$\frac{d}{dX}\left(\frac{\partial Q}{\partial A}\right) = \left(\frac{\partial Q}{\partial A}\right)\left[\frac{\sigma}{S}\left(\frac{dS}{dX}\right) + \frac{\beta}{k}\left(\frac{dk}{dX}\right) - \frac{\lambda}{X}\right]$$

when S and k are both functions of X.

Life Sciences

55. Calorie Expenditure The average energy expended for an animal to walk or run 1 km can be estimated by the function

$$f(m, v) = 25.92m^{0.68} + \frac{3.62m^{0.75}}{v},$$

where $f(m, v)$ is the energy used (in kcal per hour), m is the mass (in g), and v is the speed of movement (in km per hour) of the animal. *Source: Wildlife Feeding and Nutrition.*

(a) Find $f(300, 10)$.

(b) Find $f_m(300, 10)$ and interpret.

(c) If a mouse could run at the same speed that an elephant walks, which animal would expend more energy? How can partial derivatives be used to explore this question?

56. Heat Loss The rate of heat loss (in watts) in harbor seal pups has been approximated by

$$H(m, T, A) = \frac{15.2m^{0.67}(T - A)}{10.23 \ln m - 10.74},$$

where m is the body mass of the pup (in kg), and T and A are the body core temperature and ambient water temperature, respectively (in °C). Find the approximate change in heat loss under the following conditions. *Source: Functional Ecology.*

(a) The body core temperature increases from 37°C to 38°, while the ambient water temperature remains at 8°C and the body mass remains at 24 kg.

(b) The ambient water temperature increases from 10°C to 11°, while the body core temperature remains at 37°C and the body mass remains at 26 kg.

57. Body Surface Area The surface area of a human (in square meters) has been approximated by

$$A = 0.024265h^{0.3964}m^{0.5378},$$

where h is the height (in cm) and m is the mass (in kg). *Source: The Journal of Pediatrics.*

(a) Find the approximate change in surface area when the mass changes from 72 kg to 73 kg, while the height remains at 180 cm.

(b) Find the approximate change in surface area when the height changes from 160 cm to 161 cm, while the mass remains at 70 kg.

58. Blood Flow According to the Fick Principle, the quantity of blood pumped through the lungs depends on the following variables (in milliliters):

b = quantity of oxygen used by the body in one minute

a = quantity of oxygen per liter of blood that has just gone through the lungs

v = quantity of oxygen per liter of blood that is about to enter the lungs

In one minute,

Amount of oxygen used = Amount of oxygen per liter × Liters of blood pumped.

If C is the number of liters of blood pumped through the lungs in one minute, then

$$b = (a - v) \cdot C \qquad \text{or} \qquad C = \frac{b}{a - v}.$$

Source: Anaesthesia UK.

(a) Find the number of liters of blood pumped through the lungs in one minute if $a = 160$, $b = 200$, and $v = 125$.

(b) Find the approximate change in C when a changes from 160 to 161, $b = 200$, and $v = 125$.

(c) Find the approximate change in C when $a = 160$, b changes from 200 to 201, and $v = 125$.

(d) Find the approximate change in C when $a = 160$, $b = 200$, and v changes from 125 to 126.

(e) A change of 1 unit in which quantity of oxygen produces the greatest change in the liters of blood pumped?

59. Health A weight-loss counselor has prepared a program of diet and exercise for a client. If the client sticks to the program, the weight loss that can be expected (in pounds per week) is given by

$$\text{Weight loss} = f(n, c) = \frac{1}{8}n^2 - \frac{1}{5}c + \frac{1937}{8},$$

where c is the average daily calorie intake for the week and n is the number of 40-minute aerobic workouts per week.

(a) How many pounds can the client expect to lose by eating an average of 1200 cal per day and participating in four 40-minute workouts in a week?

(b) Find and interpret $\partial f/\partial n$.

(c) The client currently averages 1100 cal per day and does three 40-minute workouts each week. What would be the approximate impact on weekly weight loss of adding a fourth workout per week?

60. Health The body mass index (BMI) is a number that can be calculated for any individual as follows: Multiply a person's weight by 703 and divide by the person's height squared. That is,

$$B = \frac{703w}{h^2},$$

where w is in pounds and h is in inches. The National Heart, Lung and Blood Institute uses the body mass index to determine whether a person is "overweight" ($25 \leq B < 30$) or "obese" ($B \geq 30$). *Source: The National Institutes of Health.*

(a) Calculate the body mass index for St. Louis Rams offensive tackle Jake Long, who weighs 317 lb and is 6'7" tall.

(b) Calculate $\dfrac{\partial B}{\partial w}$ and $\dfrac{\partial B}{\partial h}$ and interpret.

(c) Using the fact that 1 in. $= 0.0254$ m, and 1 lb. $= 0.4536$ kg., transform this formula to handle metric units.

61. Body Shape Index An alternative to the body mass index (BMI) for estimating the health risks of obesity is A Body Shape Index (ABSI), given by

$$ABSI = \frac{w}{b^{2/3}h^{1/2}},$$

where w and h are the waist and height, respectively, in meters, and b is the BMI. (See the definition of BMI in the previous exercise.) The National Heart, Lung and Blood Institute uses the BMI to determine whether a person is "overweight" ($25 \leq BMI < 30$) or "obese" ($BMI \geq 30$). *Source: PLoS One.*

(a) Suppose a man is 1.85 m tall with a waist of 0.864 m and a BMI of 23.1. Find the ABSI for such a person. How does this compare with the average ABSI for a 52-year-old man of 0.0831?

(b) For the person described in part (a), find the rate that the ABSI is going up with respect to his waist.

(c) For the person described in part (a), find the rate that the ABSI is changing with respect to his height.

(d) Based on your answers to parts (b) and (c) which has a greater effect on the person's ABSI, an increase in waist or an increase in height of the same amount? Explain why this makes sense intuitively.

62. Life Span As we saw in the previous section, researchers have estimated the maximum life span (in years) for various species of mammals according to the formula

$$L(E, P) = 23E^{0.6}P^{-0.267},$$

where E is the average brain mass and P is the average body mass (both in g). Find the approximate change in life span under the following conditions. Then calculate the actual change in life span and compare. *Source: The Quarterly Review of Biology.*

(a) The brain mass stays at 7.35 g, while the body mass goes from 150 g to 151 g.

(b) The body mass stays at 68,700 g, while the brain mass goes from 14,100 g to 14,101 g.

63. Drug Reaction The reaction to x units of a drug t hours after it was administered is given by

$$R(x, t) = x^2(a - x)t^2e^{-t},$$

for $0 \leq x \leq a$ (where a is a constant). Find the following.

(a) $\dfrac{\partial R}{\partial x}$ **(b)** $\dfrac{\partial R}{\partial t}$ **(c)** $\dfrac{\partial^2 R}{\partial x^2}$ **(d)** $\dfrac{\partial^2 R}{\partial x \partial t}$

(e) Interpret your answers to parts (a) and (b).

64. Scuba Diving In 1908, J. Haldane constructed diving tables that provide a relationship between the water pressure on body tissues for various water depths and dive times. The tables were successfully used by divers to virtually eliminate decompression sickness. The pressure in atmospheres for a no-stop dive is given by the following formula:*

$$p(l, t) = 1 + \frac{l}{33}(1 - 2^{-t/5}),$$

where t is in minutes, l is in feet, and p is in atmospheres (atm). (One of the authors of this textbook, Raymond N. Greenwell, is shown here scuba diving near Key Largo, FL.) *Source: The UMAP Journal.*

(a) Find the pressure at 33 ft for a 10-minute dive.

(b) Find $p_l(33, 10)$ and $p_t(33, 10)$ and interpret. (*Hint:* $D_t(a^t) = \ln(a)a^t$.)

(c) Haldane estimated that decompression sickness for no-stop dives could be avoided if the diver's tissue pressure did not exceed 2.15 atm. Find the maximum amount of time that a diver could stay down (time includes going down and coming back up) if he or she wants to dive to a depth of 66 ft.

65. Wind Chill In 1941, explorers Paul Siple and Charles Passel discovered that the amount of heat lost when an object is exposed to cold air depends on both the temperature of the air and the velocity of the wind. They developed the *Wind Chill Index* as a way to measure the danger of frostbite while doing outdoor activities. The wind chill can be calculated as follows:

$$W(V, T) = 91.4 - \frac{(10.45 + 6.69\sqrt{V} - 0.447V)(91.4 - T)}{22}$$

where V is the wind speed in miles per hour and T is the temperature in Fahrenheit for wind speeds between 4 and 45 mph. *Source: The UMAP Journal.*

*These estimates are conservative. Please consult modern dive tables before making a dive.

(a) Find the wind chill for a wind speed of 20 mph and 10°F.

(b) If a weather report indicates that the wind chill is −25°F and the actual outdoor temperature is 5°F, use a graphing calculator to find the corresponding wind speed to the nearest mile per hour.

(c) Find $W_V(20, 10)$ and $W_T(20, 10)$ and interpret.

(d) Using the table command on a graphing calculator or a spreadsheet, develop a wind chill chart for various wind speeds and temperatures.

66. Heat Index The chart below shows the heat index, which combines the effects of temperature with humidity to give a measure of the apparent temperature, or how hot it feels to the body. For example, when the outside temperature is 90°F and the relative humidity is 40%, then the apparent temperature is approximately 93°F. Let $I = f(T, H)$ give the heat index, I, as a function of the temperature T (in degrees Fahrenheit) and the percent humidity H. Estimate the following. *Source: The Weather Channel.*

(a) $f(90, 30)$ (b) $f(90, 75)$ (c) $f(80, 75)$

Heat Index

Estimate the following by approximating the partial derivative using a value of $h = 5$ in the difference quotient.

(d) $f_T(90, 30)$ (e) $f_H(90, 30)$

(f) $f_T(90, 75)$ (g) $f_H(90, 75)$

(h) Describe in words what your answers in parts (d)–(g) mean.

67. Breath Volume The table at the bottom of this page accompanies the Voldyne® 5000 Volumetric Exerciser. The table gives the typical lung capacity (in milliliters) for women of various ages and heights. Based on the chart, it is possible to conclude that the partial derivative of the lung capacity with respect to age and with respect to height has constant values. What are those values?

Social Sciences

68. Education A developmental mathematics instructor at a large university has determined that a student's probability of success in the university's pass/fail remedial algebra course is a function of s, n, and a, where s is the student's score on the departmental placement exam, n is the number of semesters of mathematics passed in high school, and a is the student's mathematics SAT score. She estimates that p, the probability of passing the course (in percent), will be

$$p = f(s, n, a) = 0.05a + 6(sn)^{1/2}$$

for $200 \leq a \leq 800$, $0 \leq s \leq 10$, and $0 \leq n \leq 8$. Assuming that the above model has some merit, find the following.

(a) If a student scores 6 on the placement exam, has taken 4 semesters of high school math, and has an SAT score of 460, what is the probability of passing the course?

(b) Find p for a student with 5 semesters of high school mathematics, a placement score of 4, and an SAT score of 300.

(c) Find and interpret $f_n(4, 5, 480)$ and $f_a(4, 5, 480)$.

Physical Sciences

69. Gravitational Attraction The gravitational attraction F on a body a distance r from the center of Earth, where r is greater than the radius of Earth, is a function of its mass m and the distance r as follows:

$$F = \frac{mgR^2}{r^2},$$

Height (in.)		58"	60"	62"	64"	66"	68"	70"	72"	74"
	20	1900	2100	2300	2500	2700	2900	3100	3300	3500
A	25	1850	2050	2250	2450	2650	2850	3050	3250	3450
G	30	1800	2000	2200	2400	2600	2800	3000	3200	3400
E	35	1750	1950	2150	2350	2550	2750	2950	3150	3350
	40	1700	1900	2100	2300	2500	2700	2900	3100	3300
I	45	1650	1850	2050	2250	2450	2650	2850	3050	3250
N	50	1600	1800	2000	2200	2400	2600	2800	3000	3200
	55	1550	1750	1950	2150	2350	2550	2750	2950	3150
Y	60	1500	1700	1900	2100	2300	2500	2700	2900	3100
E	65	1450	1650	1850	2050	2250	2450	2650	2850	3050
A	70	1400	1600	1800	2000	2200	2400	2600	2800	3000
R	75	1350	1550	1750	1950	2150	2350	2550	2750	2950
S	80	1300	1500	1700	1900	2100	2300	2500	2700	2900

where R is the radius of Earth and g is the force of gravity—about 32 feet per second per second (ft per sec^2).

(a) Find and interpret F_m and F_r.

(b) Show that $F_m > 0$ and $F_r < 0$. Why is this reasonable?

70. Velocity In 1931, Albert Einstein developed the following formula for the sum of two velocities, x and y:

$$w(x, y) = \frac{x + y}{1 + \dfrac{xy}{c^2}},$$

where x and y are in miles per second and c represents the speed of light, 186,282 miles per second. *Source: The Mathematics Teacher.*

(a) Suppose that, relative to a stationary observer, a new super space shuttle is capable of traveling at 50,000 miles per second and that, while traveling at this speed, it launches a rocket that travels at 150,000 miles per second. How fast is the rocket traveling relative to the stationary observer?

(b) What is the instantaneous rate of change of w with respect to the speed of the space shuttle, x, when the space shuttle is traveling at 50,000 miles per second and the rocket is traveling at 150,000 miles per second?

(c) Hypothetically, if a person is driving at the speed of light, c, and she turns on the headlights, what is the velocity of the light coming from the headlights relative to a stationary observer?

71. Movement Time Fitts's law is used to estimate the amount of time it takes for a person, using his or her arm, to pick up a light object, move it, and then place it in a designated target area. Mathematically, Fitts's law for a particular individual is given by

$$T(s, w) = 105 + 265 \log_2 \left(\frac{2s}{w}\right),$$

where s is the distance (in feet) the object is moved, w is the width of the area in which the object is being placed, and T is the time (in msec). *Source: Human Factors in Engineering Design.*

(a) Calculate $T(3, 0.5)$.

(b) Find $T_s(3, 0.5)$ and $T_w(3, 0.5)$ and interpret these values. (*Hint:* $\log_2 x = \ln x/\ln 2$.)

YOUR TURN ANSWERS

1. $f_x(x, y) = 4xy^3 + 30x^4y^4$; $f_y(x, y) = 6x^2y^2 + 24x^5y^3$

2. $f_x(x, y) = 6xye^{3x^2y}$; $f_y(x, y) = 3x^2e^{3x^2y}$

3. $f_x(2, 1) = 9e^5$; $f_y(2, 1) = 8e^5$

4. $f_{xx}(x, y) = 2e^{7y} + 12x^2y^5$;
$f_{yy}(x, y) = 49x^2e^{7y} + 20x^4y^3$;
$f_{xy}(x, y) = f_{yx}(x, y) = 14xe^{7y} + 20x^3y^4$

9.3 Maxima and Minima

APPLY IT What amounts of sugar and flavoring produce the minimum cost per batch of a soft drink? What is the minimum cost?

In this section we will learn how to determine relative maxima and minima, and we will use these ideas to answer this question in Example 4.

FOR REVIEW

It may be helpful to review Section 5.2 on relative extrema at this point. The concepts presented there are basic to what will be done in this section.

One of the most important applications of calculus is finding maxima and minima of functions. Earlier, we studied this idea extensively for functions of a single independent variable; now we will see that extrema can be found for functions of two variables. In particular, an extension of the second derivative test can be defined and used to identify maxima or minima. We begin with the definitions of relative maxima and minima.

Relative Maxima and Minima

Let (a, b) be the center of a circular region contained in the xy-plane. Then, for a function $z = f(x, y)$ defined for every (x, y) in the region, $f(a, b)$ is a **relative (or local) maximum** if

$$f(a, b) \geq f(x, y)$$

for all points (x, y) in the circular region, and $f(a, b)$ is a **relative (or local) minimum** if

$$f(a, b) \leq f(x, y)$$

for all points (x, y) in the circular region.

Relative maximum at (a, b)

FIGURE 17

Relative minimum at (a, b)

FIGURE 18

As before, the word *extremum* is used for either a relative maximum or a relative minimum. Examples of a relative maximum and a relative minimum are given in Figures 17 and 18.

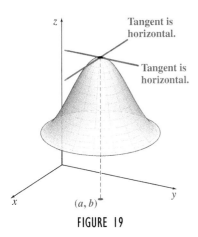

FIGURE 19

NOTE When functions of a single variable were discussed, a distinction was made between relative extrema and absolute extrema. The methods for finding absolute extrema are quite involved for functions of two variables, so we will discuss only relative extrema here. In many practical applications the relative extrema coincide with the absolute extrema. In this brief discussion of extrema for multivariable functions, we omit cases where an extremum occurs on a boundary of the domain.

As suggested by Figure 19, at a relative maximum the tangent line parallel to the xz-plane has a slope of 0, as does the tangent line parallel to the yz-plane. (Notice the similarity to functions of one variable.) That is, if the function $z = f(x, y)$ has a relative extremum at (a, b), then $f_x(a, b) = 0$ and $f_y(a, b) = 0$, as stated in the next theorem.

Location of Extrema

Let a function $z = f(x, y)$ have a relative maximum or relative minimum at the point (a, b). Let $f_x(a, b)$ and $f_y(a, b)$ both exist. Then

$$f_x(a, b) = 0 \quad \text{and} \quad f_y(a, b) = 0.$$

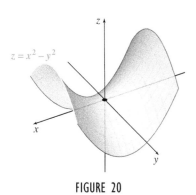

FIGURE 20

Just as with functions of one variable, the fact that the slopes of the tangent lines are 0 is no guarantee that a relative extremum has been located. For example, Figure 20 shows the graph of $z = f(x, y) = x^2 - y^2$. Both $f_x(0, 0) = 0$ and $f_y(0, 0) = 0$, and yet $(0, 0)$ leads to neither a relative maximum nor a relative minimum for the function. The point $(0, 0, 0)$ on the graph of this function is called a **saddle point**; it is a minimum when approached from one direction but a maximum when approached from another direction. A saddle point is neither a maximum nor a minimum.

The theorem on location of extrema suggests a useful strategy for finding extrema. First, locate all points (a, b) where $f_x(a, b) = 0$ and $f_y(a, b) = 0$. Then test each of these points separately, using the test given after the next example. For a function $f(x, y)$, the points (a, b) such that $f_x(a, b) = 0$ and $f_y(a, b) = 0$ are called **critical points**.

NOTE When we discussed functions of a single variable, we allowed critical points to include points from the domain where the derivative does not exist. For functions of more than one variable, to avoid complications, we will consider only cases in which the function is differentiable.

EXAMPLE 1 **Critical Points**

Find all critical points for

$$f(x, y) = 6x^2 + 6y^2 + 6xy + 36x - 5.$$

SOLUTION Find all points (a, b) such that $f_x(a, b) = 0$ and $f_y(a, b) = 0$. Here

$$f_x(x, y) = 12x + 6y + 36 \quad \text{and} \quad f_y(x, y) = 12y + 6x.$$

Set each of these two partial derivatives equal to 0.

$$12x + 6y + 36 = 0 \quad \text{and} \quad 12y + 6x = 0$$

These two equations make up a system of linear equations. We can use the substitution method to solve this system. First, rewrite $12y + 6x = 0$ as follows:

$$12y + 6x = 0$$
$$6x = -12y$$
$$x = -2y.$$

Now substitute $-2y$ for x in the other equation and solve for y.

$$12x + 6y + 36 = 0$$
$$12(-2y) + 6y + 36 = 0$$
$$-24y + 6y + 36 = 0$$
$$-18y + 36 = 0$$
$$-18y = -36$$
$$y = 2$$

From the equation $x = -2y$, we find $x = -2(2) = -4$. The solution of the system of equations is $(-4, 2)$. Since this is the only solution of the system, $(-4, 2)$ is the only critical point for the given function. By the theorem above, if the function has a relative extremum, it will occur at $(-4, 2)$. **TRY YOUR TURN 1**

The results of the next theorem can be used to decide whether $(-4, 2)$ in Example 1 leads to a relative maximum, a relative minimum, or neither.

YOUR TURN 1 Find all critical points for $f(x, y) = 4x^3 + 3xy + 4y^3$.

Test for Relative Extrema

For a function $z = f(x, y)$, let f_{xx}, f_{yy}, and f_{xy} all exist in a circular region contained in the xy-plane with center (a, b). Further, let

$$f_x(a, b) = 0 \quad \text{and} \quad f_y(a, b) = 0.$$

Define the number D, known as the **discriminant**, by

$$D = f_{xx}(a, b) \cdot f_{yy}(a, b) - [f_{xy}(a, b)]^2.$$

Then

(a) $f(a, b)$ is a relative maximum if $D > 0$ and $f_{xx}(a, b) < 0$;

(b) $f(a, b)$ is a relative minimum if $D > 0$ and $f_{xx}(a, b) > 0$;

(c) $f(a, b)$ is a saddle point (neither a maximum nor a minimum) if $D < 0$;

(d) if $D = 0$, the test gives no information.

This test is comparable to the second derivative test for extrema of functions of one independent variable. The following table summarizes the conclusions of the theorem.

	$f_{xx}(a, b) < 0$	$f_{xx}(a, b) > 0$
$D > 0$	Relative maximum	Relative minimum
$D = 0$	No information	
$D < 0$	Saddle point	

Notice that in parts (a) and (b) of the test for relative extrema, it is only necessary to test the second partial $f_{xx}(a, b)$ and not $f_{yy}(a, b)$. This is because if $D > 0$, $f_{xx}(a, b)$ and $f_{yy}(a, b)$ must have the same sign.

Part of the idea behind the test for relative extrema is that if $f_{xx}(a, b)$ and $f_{yy}(a, b)$ have opposite signs, then the surface is concave downward in one direction and concave upward in the other, so there is a saddle point. This also causes $D < 0$. If $f_{xx}(a, b)$ and $f_{yy}(a, b)$ are both positive and large enough (compared with $f_{xy}(a, b)$) so that $D > 0$, then the surface is concave upward in all directions, and the critical point leads to a relative minimum. Similarly, if $f_{xx}(a, b)$ and $f_{yy}(a, b)$ are both negative and large enough (compared with $f_{xy}(a, b)$) so that $D > 0$, then the surface is concave downward in all directions, and the critical point leads to a relative maximum.*

EXAMPLE 2 Relative Extrema

The previous example showed that the only critical point for the function

$$f(x, y) = 6x^2 + 6y^2 + 6xy + 36x - 5$$

is $(-4, 2)$. Does $(-4, 2)$ lead to a relative maximum, a relative minimum, or neither?

SOLUTION Find out by using the test for relative extrema. From Example 1,

$$f_x(-4, 2) = 0 \quad \text{and} \quad f_y(-4, 2) = 0.$$

Now find the various second partial derivatives used in finding D. From $f_x(x, y) = 12x + 6y + 36$ and $f_y(x, y) = 12y + 6x$,

$$f_{xx}(x, y) = 12, \quad f_{yy}(x, y) = 12, \quad \text{and} \quad f_{xy}(x, y) = 6.$$

(If these second-order partial derivatives had not all been constants, they would have had to be evaluated at the point $(-4, 2)$.) Now

$$D = f_{xx}(-4, 2) \cdot f_{yy}(-4, 2) - [f_{xy}(-4, 2)]^2 = 12 \cdot 12 - 6^2 = 108.$$

Since $D > 0$ and $f_{xx}(-4, 2) = 12 > 0$, part (b) of the theorem applies, showing that $f(x, y) = 6x^2 + 6y^2 + 6xy + 36x - 5$ has a relative minimum at $(-4, 2)$. This relative minimum is $f(-4, 2) = -77$. A graph of this surface drawn by the computer program Maple™ is shown in Figure 21.

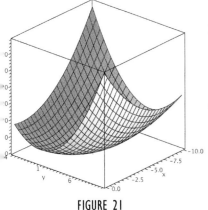

FIGURE 21

EXAMPLE 3 Relative Extrema and Saddle Point

Find all points where the function

$$f(x, y) = 9xy - x^3 - y^3 - 6$$

has any relative maxima or relative minima. Identify any saddle points.

SOLUTION First find any critical points. Here

$$f_x(x, y) = 9y - 3x^2 \quad \text{and} \quad f_y(x, y) = 9x - 3y^2.$$

Set each of these partial derivatives equal to 0.

$$\begin{array}{ll} f_x(x, y) = 0 & f_y(x, y) = 0 \\ 9y - 3x^2 = 0 & 9x - 3y^2 = 0 \\ 9y = 3x^2 & 9x = 3y^2 \\ 3y = x^2 & 3x = y^2 \end{array}$$

The substitution method can be used again to solve the system of equations

$$3y = x^2$$
$$3x = y^2.$$

*For a rigorous proof of the test for relative extrema, see *Thomas' Calculus*, 13th ed., by Thomas, George B. Jr., Maurice D. Weir, and Joel R. Hass, Pearson, 2014.

The first equation, $3y = x^2$, can be rewritten as $y = x^2/3$. Substitute this into the second equation to get

$$3x = y^2 = \left(\frac{x^2}{3}\right)^2 = \frac{x^4}{9}.$$

Solve this equation as follows.

$$27x = x^4 \qquad \text{Multiply both sides by 9.}$$
$$x^4 - 27x = 0 \qquad \text{Subtract } 27x \text{ from both sides.}$$
$$x(x^3 - 27) = 0 \qquad \text{Factor.}$$
$$x = 0 \quad \text{or} \quad x^3 - 27 = 0 \qquad \text{Set each factor equal to 0.}$$
$$x = 0 \quad \text{or} \quad x^3 = 27$$
$$x = 0 \quad \text{or} \quad x = 3 \qquad \text{Take the cube root on both sides.}$$

Use these values of x, along with the equation $3y = x^2$, rewritten as $y = x^2/3$, to find y. If $x = 0$, then $y = 0^2/3 = 0$. If $x = 3$, then $y = 3^2/3 = 3$. The critical points are $(0, 0)$ and $(3, 3)$. To identify any extrema, use the test for relative extrema. Here

$$f_{xx}(x, y) = -6x, \qquad f_{yy}(x, y) = -6y, \qquad \text{and} \qquad f_{xy}(x, y) = 9.$$

Test each of the critical points.

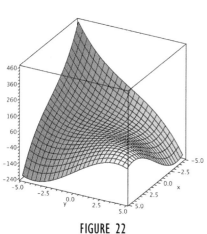

460
360
260
160
60
-40
-140
-240

FIGURE 22

YOUR TURN 2 Identify each of the critical points in YOUR TURN 1 as a relative maximum, relative minimum, or saddle point.

For $(0, 0)$:
$$f_{xx}(0, 0) = -6(0) = 0$$
$$f_{yy}(0, 0) = -6(0) = 0$$
$$f_{xy}(0, 0) = 9$$
$$D = 0 \cdot 0 - 9^2 = -81.$$
Since $D < 0$, there is a saddle point at $(0, 0)$.

For $(3, 3)$:
$$f_{xx}(3, 3) = -6(3) = -18$$
$$f_{yy}(3, 3) = -6(3) = -18$$
$$f_{xy}(3, 3) = 9$$
$$D = -18(-18) - 9^2 = 243.$$
Here $D > 0$ and $f_{xx}(3, 3) = -18 < 0$; there is a relative maximum at $(3, 3)$.

Notice that these values are in accordance with the graph generated by the computer program Maple™ shown in Figure 22. **TRY YOUR TURN 2**

EXAMPLE 4 Production Costs

A company is developing a new soft drink. The cost in dollars to produce a batch of the drink is approximated by

$$C(x, y) = 2200 + 27x^3 - 72xy + 8y^2,$$

where x is the number of kilograms of sugar per batch and y is the number of grams of flavoring per batch. Find the amounts of sugar and flavoring that result in the minimum cost per batch. What is the minimum cost?

APPLY IT

Method 1
Calculating by Hand

SOLUTION

Start with the following partial derivatives.

$$C_x(x, y) = 81x^2 - 72y \qquad \text{and} \qquad C_y(x, y) = -72x + 16y$$

Set each of these equal to 0 and solve for y.

$$81x^2 - 72y = 0 \qquad\qquad -72x + 16y = 0$$
$$-72y = -81x^2 \qquad\qquad 16y = 72x$$
$$y = \frac{9}{8}x^2 \qquad\qquad\qquad y = \frac{9}{2}x$$

Since $(9/8)x^2$ and $(9/2)x$ both equal y, they are equal to each other. Set them equal, and solve the resulting equation for x.

$$\frac{9}{8}x^2 = \frac{9}{2}x$$

$$9x^2 = 36x \qquad \text{Multiply both sides by 8.}$$

$$9x^2 - 36x = 0 \qquad \text{Subtract } 36x \text{ from both sides.}$$

$$9x(x - 4) = 0 \qquad \text{Factor.}$$

$$9x = 0 \quad \text{or} \quad x - 4 = 0 \qquad \text{Set each factor equal to 0.}$$

$$x = 0 \quad \text{or} \quad x = 4$$

Use these values of x, along with the equation $y = (9/2)x$ (or the equation $y = (9/8)x^2$), to find y. If $x = 0$, then $y = (9/2)(0) = 0$. If $x = 4$, then $y = (9/2)4 = 18$. The critical points are $(0, 0)$ and $(4, 18)$. To determine if there is a relative minimum, use the test for relative extrema. Here

$$C_{xx}(x, y) = 162x, \quad C_{yy}(x, y) = 16, \quad \text{and} \quad C_{xy}(x, y) = -72.$$

Test each of the critical points.

For $(0, 0)$:
$C_{xx}(0, 0) = 162(0) = 0$
$C_{yy}(0, 0) = 16$
$C_{xy}(0, 0) = -72$
$D = 0 \cdot 16 - (-72)^2 = -5184.$
Since $D < 0$, there is a saddle point at $(0, 0)$.

For $(4, 18)$:
$C_{xx}(4, 18) = 162(4) = 648$
$C_{yy}(4, 18) = 16$
$C_{xy}(4, 18) = -72$
$D = 648 \cdot 16 - (-72)^2 = 5184.$
Since $D > 0$ and $C_{xx}(4, 18) > 0$, the cost at $(4, 18)$ is a minimum.

To find the minimum cost, go back to the cost function and evaluate $C(4, 18)$.

$$C(x, y) = 2200 + 27x^3 - 72xy + 8y^2$$
$$C(4, 18) = 2200 + 27(4)^3 - 72(4)(18) + 8(18)^2 = 1336$$

The minimum cost for a batch of soft drink is $1336.00. A graph of this surface drawn by the computer program Maple™ is shown in Figure 23.

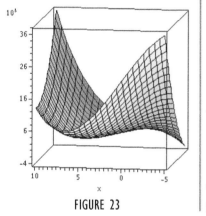

FIGURE 23

Method 2
Spreadsheets

Finding the maximum or minimum of a function of one or more variables can be done using a spreadsheet. The Solver included with Excel is located in the Tools menu and requires that cells be identified ahead of time for each variable in the problem. (On some versions of Excel, the Solver must be installed from an outside source. For details, see the *Graphing Calculator and Excel Spreadsheet Manual* available with this text.) It also requires that another cell be identified where the function, in terms of the variable cells, is placed. For example, to solve the above problem, we could identify cells A1 and B1 to represent the variables x and y, respectively. The Solver requires that we place a guess for the answer in these cells. Thus, our initial value or guess will be to place the number 5 in each of these cells. An expression for the function must be placed in another cell, with x and y replaced by A1 and B1. If we choose cell A3 to represent the function, in cell A3 we would type "`=2200+27*A1^3-72*A1*B1+8*B1^2`."

We now click on the Tools menu and choose Solver. The Solver will attempt to find a solution that either maximizes or minimizes the value of cell A3. Figure 24 on the next page illustrates the Solver box and the items placed in it.

NOTE
In figure 24, the Target Cell is written as $\$A\$3$. You can simply enter A3. The program adds the dollar signs to assure that the target cell is fixed throughout the process. This also holds for the Changing Cells A1:B1.

FIGURE 24

To obtain a solution, click on Solve. The rounded solution $x = 4$ and $y = 18$ is located in cells A1 and B1. The minimum cost $C(4, 18) = 1336$ is located in cell A3.

CAUTION One must be careful when using Solver because it will not find a maximizer or minimizer of a function if the initial guess is the exact place in which a saddle point occurs. For example, in the problem above, if our initial guess was $(0, 0)$, the Solver would have returned the value of $(0, 0)$ as the place where a minimum occurs. But $(0, 0)$ is a saddle point. Thus, it is always a good idea to run the Solver for two different initial values and compare the solutions.

9.3 WARM-UP EXERCISES

Find all relative extrema for the following functions. *(Sec. 5.2)*

W1. $f(x) = x^3 - 3x^2 - 24x + 5$

W2. $f(x) = x^4 - 8x^2 + 1$

9.3 EXERCISES

Find all points where the functions have any relative extrema. Identify any saddle points.

1. $f(x, y) = xy + y - 2x$ **2.** $f(x, y) = 3xy + 6y - 5x$

3. $f(x, y) = 3x^2 - 4xy + 2y^2 + 6x - 10$

4. $f(x, y) = x^2 + xy + y^2 - 6x - 3$

5. $f(x, y) = x^2 - xy + y^2 + 2x + 2y + 6$

6. $f(x, y) = 2x^2 + 3xy + 2y^2 - 5x + 5y$

7. $f(x, y) = x^2 + 3xy + 3y^2 - 6x + 3y$

8. $f(x, y) = 5xy - 7x^2 - y^2 + 3x - 6y - 4$

9. $f(x, y) = 4xy - 10x^2 - 4y^2 + 8x + 8y + 9$

10. $f(x, y) = 4y^2 + 2xy + 6x + 4y - 8$

11. $f(x, y) = x^2 + xy - 2x - 2y + 2$

12. $f(x, y) = x^2 + xy + y^2 - 3x - 5$

13. $f(x, y) = 3x^2 + 2y^3 - 18xy + 42$

14. $f(x, y) = 7x^3 + 3y^2 - 126xy - 63$

15. $f(x, y) = x^2 + 4y^3 - 6xy - 1$

16. $f(x, y) = 3x^2 + 7y^3 - 42xy + 5$

17. $f(x, y) = e^{x(y+1)}$ **18.** $f(x, y) = y^2 + 2e^x$

19. Describe the procedure for finding critical points of a function in two independent variables.

20. How are second-order partial derivatives used in finding extrema?

Figures (a)–(f) show the graphs of the functions defined in Exercises 21–26. Find all relative extrema for each function, and then match the equation to its graph.

21. $z = -3xy + x^3 - y^3 + \dfrac{1}{8}$ **22.** $z = \dfrac{3}{2}y - \dfrac{1}{2}y^3 - x^2y + \dfrac{1}{16}$

23. $z = y^4 - 2y^2 + x^2 - \dfrac{17}{16}$

24. $z = -2x^3 - 3y^4 + 6xy^2 + \dfrac{1}{16}$

25. $z = -x^4 + y^4 + 2x^2 - 2y^2 + \dfrac{1}{16}$

26. $z = -y^4 + 4xy - 2x^2 + \dfrac{1}{16}$

(a)

(b)

(c)

(d)

(e)

(f)

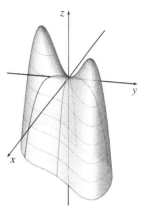

27. Show that $f(x, y) = 1 - x^4 - y^4$ has a relative maximum, even though D in the theorem is 0.

28. Show that $D = 0$ for $f(x, y) = x^3 + (x - y)^2$ and that the function has no relative extrema.

29. A friend taking calculus is puzzled. She remembers that for a function of one variable, if the first derivative is zero at a point and the second derivative is positive, then there must be a relative minimum at the point. She doesn't understand why that isn't true for a function of two variables—that is, why $f_x(x, y) = 0$ and $f_{xx}(x, y) > 0$ doesn't guarantee a relative minimum. Provide an explanation.

30. Let $f(x, y) = y^2 - 2x^2y + 4x^3 + 20x^2$. The only critical points are $(-2, 4)$, $(0, 0)$, and $(5, 25)$. Which of the following correctly describes the behavior of f at these points? *Source: Society of Actuaries.*

(a) $(-2, 4)$: local (relative) minimum
    $(0, 0)$:   local (relative) minimum
    $(5, 25)$: local (relative) maximum

(b) $(-2, 4)$: local (relative) minimum
    $(0, 0)$:   local (relative) maximum
    $(5, 25)$: local (relative) maximum

(c) $(-2, 4)$: neither a local (relative) minimum nor a local (relative) maximum
    $(0, 0)$:   local (relative) maximum
    $(5, 25)$: local (relative) minimum

(d) $(-2, 4)$: local (relative) maximum
    $(0, 0)$:   neither a local (relative) minimum nor a local (relative) maximum
    $(5, 25)$: local (relative) minimum

(e) $(-2, 4)$: neither a local (relative) minimum nor a local (relative) maximum
    $(0, 0)$:   local (relative) minimum
    $(5, 25)$: neither a local (relative) minimum nor a local (relative) maximum

31. Consider the function $f(x, y) = x^2(y + 1)^2 + k(x + 1)^2y^2$.

(a) For what values of k is the point $(x, y) = (0, 0)$ a critical point?

(b) For what values of k is the point $(x, y) = (0, 0)$ a relative minimum of the function?

32. In Exercise 9 of Section 1.3, we found the least squares line through a set of n points (x_1, y_1), (x_2, y_2), . . . , (x_n, y_n) by choosing the slope of the line m and the y-intercept b to minimize the quantity

$$S(m, b) = \sum(mx + b - y)^2,$$

where the summation symbol Σ means that we sum over all the data points. Minimize S by setting $S_m(m, b) = 0$ and $S_b(m, b) = 0$, and then rearrange the results to derive the equations from Section 1.3

$$\left(\sum x\right)b + \left(\sum x^2\right)m = \sum xy$$
$$nb + \left(\sum x\right)m = \sum y.$$

33. Suppose a function $z = f(x, y)$ satisfies the criteria for the test for relative extrema at a point (a, b), and $f_{xx}(a, b) > 0$, while $f_{yy}(a, b) < 0$. What does this tell you about $f(a, b)$? Based on the sign of $f_{xx}(a, b)$ and $f_{yy}(a, b)$, why does this seem intuitively plausible?

APPLICATIONS

Business and Economics

34. Profit Suppose that the profit (in hundreds of dollars) of a certain firm is approximated by

$$P(x, y) = 1500 + 36x - 1.5x^2 + 120y - 2y^2,$$

where x is the cost of a unit of labor and y is the cost of a unit of goods. Find values of x and y that maximize profit. Find the maximum profit.

35. Labor Costs Suppose the labor cost (in dollars) for manufacturing a precision camera can be approximated by

$$L(x, y) = \frac{3}{2}x^2 + y^2 - 2x - 2y - 2xy + 68,$$

where x is the number of hours required by a skilled craftsperson and y is the number of hours required by a semiskilled person. Find values of x and y that minimize the labor cost. Find the minimum labor cost.

36. Cost The total cost (in dollars) to produce x units of electrical tape and y units of packing tape is given by

$$C(x, y) = 2x^2 + 2y^2 - 3xy + 4x - 94y + 4200.$$

Find the number of units of each kind of tape that should be produced so that the total cost is a minimum. Find the minimum total cost.

37. Revenue The total revenue (in hundreds of dollars) from the sale of x spas and y solar heaters is approximated by

$$R(x, y) = 15 + 169x + 182y - 5x^2 - 7y^2 - 7xy.$$

Find the number of each that should be sold to produce maximum revenue. Find the maximum revenue.

38. Profit The profit (in thousands of dollars) that Aunt Mildred's Metalworks earns from producing x tons of steel and y tons of aluminum can be approximated by

$$P(x, y) = 36xy - x^3 - 8y^3.$$

Find the amounts of steel and aluminum that maximize the profit, and find the value of the maximum profit.

39. Time The time (in hours) that a branch of Amalgamated Entities needs to spend to meet the quota set by the main office can be approximated by

$$T(x, y) = x^4 + 16y^4 - 32xy + 40,$$

where x represents how many thousands of dollars the factory spends on quality control and y represents how many thousands of dollars they spend on consulting. Find the amount of money they should spend on quality control and on consulting to minimize the time spent, and find the minimum number of hours.

Social Sciences

40. Political Science The probability that a three-person jury will make a correct decision is given by

$$P(\alpha, r, s) = \alpha[3r^2(1 - r) + r^3] + (1 - \alpha)[3s^2(1 - s) + s^3],$$

where $0 < \alpha < 1$ is the probability that the person is guilty of the crime, r is the probability that a given jury member will vote "guilty" when the defendant is indeed guilty of the crime, and s is the probability that a given jury member will vote "innocent" when the defendant is indeed innocent. *Source: Frontiers of Economics.*

   (a) Calculate $P(0.9, 0.5, 0.6)$ and $P(0.1, 0.8, 0.4)$ and interpret your answers.

   (b) Using common sense and without using calculus, what value of r and s would maximize the jury's probability of making the correct verdict? Do these values depend on α in this problem? Should they? What is the maximum probability?

   (c) Verify your answer for part (b) using calculus. (*Hint:* There are two critical points. Argue that the maximum value occurs at one of these points.)

Physical Sciences

41. Computer Chips The table below, which illustrates the dramatic increase in the number of transistors in personal computers since 1985, was given in the chapter on Nonlinear Functions, Section 2.4, Exercise 54. *Source: www.intel.com.*

   (a) To fit the data to a function of the form $y = ab^t$, where t is the number of years since 1985 and y is the number of transistors (in millions), we could take natural logarithms of both sides of the equation to get $\ln y = \ln a + t \ln b$. We could then let $w = \ln y$, $r = \ln a$, and $s = \ln b$ to form $w = r + st$. Using linear regression, find values for r and s that will fit the data. Then find the function $y = ab^t$. (*Hint:* Take the natural logarithm of the values in the transistors column and then use linear regression to find values of r and s that fit the data. Once you know r and s, you can determine the values of a and b by calculating $a = e^r$ and $b = e^s$.)

Year (since 1985)	Chip	Transistors (in millions)
0	386	0.275
4	486	1.2
8	Pentium	3.1
12	Pentium II	7.5
14	Pentium III	9.5
15	Pentium 4	42
20	Pentium D	291
22	Penryn	820
24	Nehalem	1900

(b) Use the solver capability of a spreadsheet to find a function of the form $y = ab^t$ that fits the data above. (*Hint:* Using the ideas from part (a), find values for a and b that minimize the function

$$f(a, b) = [\ln(0.275) - 0\ln b - \ln a]^2$$
$$+ [\ln(1.2) - 4\ln b - \ln a]^2$$
$$+ [\ln(3.1) - 8\ln b - \ln a]^2$$
$$+ [\ln(7.5) - 12\ln b - \ln a]^2$$
$$+ [\ln(9.5) - 14\ln b - \ln a]^2$$
$$+ [\ln(42) - 15\ln b - \ln a]^2$$
$$+ [\ln(291) - 20\ln b - \ln a]^2$$
$$+ [\ln(820) - 22\ln b - \ln a]^2$$
$$+ [\ln(1900) - 24\ln b - \ln a]^2.)$$

(c) Compare your answer to this problem with the one found with a graphing calculator in the chapter on Nonlinear Functions, Section 2.4 Exercise 54.

General Interest

42. Food Frying The process of frying food changes its quality, texture, and color. According to research done at the University of Saskatchewan, the total change in color E (which is measured in the form of energy as kJ/mol) of blanched potato strips can be estimated by the function

$$E(t, T) = 436.16 - 10.57t - 5.46T - 0.02t^2 + 0.02T^2 + 0.08Tt,$$

where T is the temperature (in °C) and t is the frying time (in min). *Source: Critical Reviews in Food Science and Nutrition.*

(a) What is the value of E prior to cooking? (Assume that $T = 0$.)

(b) Use this function to estimate the total change in color of a potato strip that has been cooked for 10 minutes at 180°C.

(c) Determine the critical point of this function and determine if a maximum, minimum, or saddle point occurs at that point. Describe what may be happening at this point.

YOUR TURN ANSWERS

1. $(0, 0)$ and $(-1/4, -1/4)$

2. Saddle point at $(0, 0)$; relative maximum at $(-1/4, -1/4)$

9.4 Lagrange Multipliers

APPLY IT What dimensions for a new building will maximize the floor space at a fixed cost?

Using Lagrange multipliers, we will answer this question in Example 2.

In Section 6.2 on Applications of Extrema, it was possible to express problems involving two variables as equivalent problems requiring only a single variable. This method works well, provided that it is possible to use algebra to express the one variable in terms of the other. It is not always possible to do this, however, and most real applications require more than two variables and one or more additional restrictions, called **constraints**.

An approach that works well when there is a constraint in the problem uses an additional variable, called the **Lagrange multiplier**. For example, in the opening question, suppose a builder wants to maximize the floor space in a new building while keeping the costs fixed at $500,000. The building will be 40 ft high, with a rectangular floor plan and three stories. The costs, which depend on the dimensions of the rectangular floor plan, are given by

$$\text{Costs} = xy + 20y + 20x + 474,000,$$

where x is the width and y the length of the rectangle. Thus, the builder wishes to maximize the area $A(x, y) = xy$ and satisfy the condition

$$xy + 20y + 20x + 474,000 = 500,000.$$

In addition to maximizing area, then, the builder must keep costs at (or below) $500,000. We will see how to solve this problem in Example 2 of this section.

Problems with constraints are often solved by the method of Lagrange multipliers, named for the French mathematician Joseph Louis Lagrange (1736–1813). The method of Lagrange multipliers is used for problems of the form:

Find the relative extrema for $z = f(x, y)$,

subject to $g(x, y) = 0$.

We state the method only for functions of two independent variables, but it is valid for any number of variables.

Lagrange Multipliers

All relative extrema of the function $z = f(x, y)$, subject to a constraint $g(x, y) = 0$, will be found among those points (x, y) for which there exists a value of λ such that

$$F_x(x, y, \lambda) = 0, \qquad F_y(x, y, \lambda) = 0, \qquad F_\lambda(x, y, \lambda) = 0,$$

where

$$F(x, y, \lambda) = f(x, y) - \lambda \cdot g(x, y),$$

and all indicated partial derivatives exist.

In the theorem, the function $F(x, y, \lambda) = f(x, y) - \lambda \cdot g(x, y)$ is called the Lagrange function; λ, the Greek letter *lambda*, is the *Lagrange multiplier*.

> **CAUTION** If the constraint is not of the form $g(x, y) = 0$, it must be put in that form before using the method of Lagrange multipliers. For example, if the constraint is $x^2 + y^3 = 5$, subtract 5 from both sides to get $g(x, y) = x^2 + y^3 - 5 = 0$.

EXAMPLE 1 Lagrange Multipliers

Find the minimum value of

$$f(x, y) = 5x^2 + 6y^2 - xy,$$

subject to the constraint $x + 2y = 24$.

SOLUTION Go through the following steps.

Step 1 Rewrite the constraint in the form $g(x, y) = 0$.

In this example, the constraint $x + 2y = 24$ becomes

$$x + 2y - 24 = 0,$$

with

$$g(x, y) = x + 2y - 24.$$

Step 2 Form the Lagrange function $F(x, y, \lambda)$, the difference of the function $f(x, y)$ and the product of λ and $g(x, y)$.

Here,

$$\begin{aligned}
F(x, y, \lambda) &= f(x, y) - \lambda \cdot g(x, y) \\
&= 5x^2 + 6y^2 - xy - \lambda(x + 2y - 24) \\
&= 5x^2 + 6y^2 - xy - \lambda x - 2\lambda y + 24\lambda.
\end{aligned}$$

Step 3 Find $F_x(x, y, \lambda)$, $F_y(x, y, \lambda)$, and $F_\lambda(x, y, \lambda)$.

$$F_x(x, y, \lambda) = 10x - y - \lambda$$
$$F_y(x, y, \lambda) = 12y - x - 2\lambda$$
$$F_\lambda(x, y, \lambda) = -x - 2y + 24$$

Step 4 Form the system of equations $F_x(x, y, \lambda) = 0$, $F_y(x, y, \lambda) = 0$, and $F_\lambda(x, y, \lambda) = 0$.

$$10x - y - \lambda = 0 \qquad \text{(1)}$$
$$12y - x - 2\lambda = 0 \qquad \text{(2)}$$
$$-x - 2y + 24 = 0 \qquad \text{(3)}$$

Step 5 Solve the system of equations from Step 4 for x, y, and λ.

One way to solve this system is to begin by solving each of the first two equations for λ, then set the two results equal and simplify, as follows.

$$10x - y - \lambda = 0 \quad \text{becomes} \quad \lambda = 10x - y$$

$$12y - x - 2\lambda = 0 \quad \text{becomes} \quad \lambda = \frac{-x + 12y}{2}$$

$$10x - y = \frac{-x + 12y}{2} \qquad \text{Set the expressions for } \lambda \text{ equal.}$$

$$20x - 2y = -x + 12y \qquad \text{Multiply by 2.}$$

$$21x = 14y \qquad \text{Add } 2y \text{ and } x \text{ to both sides.}$$

$$x = \frac{14y}{21} = \frac{2y}{3}$$

Now substitute $2y/3$ for x in Equation (3).

$$-x - 2y + 24 = 0$$

$$-\frac{2y}{3} - 2y + 24 = 0 \qquad \text{Let } x = \frac{2y}{3}.$$

$$2y + 6y - 72 = 0 \qquad \text{Multiply by } -3.$$

$$8y = 72$$

$$y = \frac{72}{8} = 9$$

Since $x = 2y/3$ and $y = 9$, $x = 6$. It is not necessary to find the value of λ.

Thus, if $f(x, y) = 5x^2 + 6y^2 - xy$ has a minimum value subject to the constraint $x + 2y = 24$, it is at the point $(6, 9)$. The value of $f(6, 9)$ is 612.

We need to convince ourselves that $f(6, 9) = 612$ is indeed a minimum for the function. How can we tell that it is not a maximum? The second derivative test from the previous section does not apply to the solutions found by Lagrange multipliers. (See Exercise 21.) We could gain some insight by trying a point very close to $(6, 9)$ that also satisfies the constraint $x + 2y = 24$. For example, let $y = 9.1$, so $x = 24 - 2y = 24 - 2(9.1) = 5.8$. Then $f(5.8, 9.1) = 5(5.8)^2 + 6(9.1)^2 - (5.8)(9.1) = 612.28$, which is greater than 612. Because a nearby point has a value larger than 612, the value 612 is probably not a maximum. Another method would be to use a computer to sketch the graph of the function and see that it has a minimum but not a maximum. In practical problems, such as Example 2, it is often obvious whether a function has a minimum or a maximum. **TRY YOUR TURN 1**

YOUR TURN 1 Find the minimum value of $f(x, y) = x^2 + 2x + 9y^2 + 3y + 6xy$ subject to the constraint $2x + 3y = 12$.

NOTE In Example 1, we solved the system of equations by solving each equation with λ in it for λ. We then set these expressions for λ equal and solved for one of the original variables. This is a good general approach to use in solving these systems of equations, since we are usually not interested in the value of λ.

CAUTION Lagrange multipliers give only the relative extrema, not the absolute extrema. In many applications, the relative extrema will be the absolute extrema, but this is not guaranteed. In some cases in which the method of Lagrange multipliers finds a solution, there may not even be any absolute extrema. For example, see Exercises 18 and 19 at the end of this section.

Before looking at applications of Lagrange multipliers, let us summarize the steps involved in solving a problem by this method.

Using Lagrange Multipliers

1. Write the constraint in the form $g(x, y) = 0$.

2. Form the Lagrange function

$$F(x, y, \lambda) = f(x, y) - \lambda \cdot g(x, y).$$

3. Find $F_x(x, y, \lambda)$, $F_y(x, y, \lambda)$, and $F_\lambda(x, y, \lambda)$.

4. Form the system of equations

$$F_x(x, y, \lambda) = 0, \qquad F_y(x, y, \lambda) = 0, \qquad F_\lambda(x, y, \lambda) = 0.$$

5. Solve the system in Step 4; the relative extrema for f are among the solutions of the system.

The proof of this method is complicated and is not given here, but we can explain why the method is plausible. Consider the curve formed by points in the xy-plane that satisfy $F_\lambda(x, y, \lambda) = -g(x, y) = 0$ (or just $g(x, y) = 0$). Figure 25 shows how such a curve might look. Crossing this region are curves $f(x, y) = k$ for various values of k. Notice that at the points where the curve $f(x, y) = k$ is tangent to the curve $g(x, y) = 0$, the largest and smallest meaningful values of f occur. It can be shown that this is equivalent to $f_x(x, y) = \lambda g_x(x, y)$ and $f_y(x, y) = \lambda g_y(x, y)$ for some constant λ. In Exercise 20, you are asked to show that this is equivalent to the system of equations found in Step 4 above.

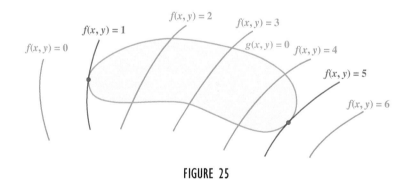

FIGURE 25

Lagrange multipliers are widely used in economics, where a frequent goal is to maximize a utility function, which measures how well consumption satisfies the consumers' desires, subject to constraints on income or time.

EXAMPLE 2   Lagrange Multipliers

Complete the solution of the problem given in the introduction to this section. Maximize the area, $A(x, y) = xy$, subject to the cost constraint

$$xy + 20y + 20x + 474{,}000 = 500{,}000.$$

APPLY IT   **SOLUTION**   Go through the five steps presented earlier.

Step 1 $g(x, y) = xy + 20y + 20x - 26{,}000 = 0$

Step 2 $F(x, y, \lambda) = xy - \lambda(xy + 20y + 20x - 26{,}000)$

Step 3 $F_x(x, y, \lambda) = y - \lambda y - 20\lambda$

$F_y(x, y, \lambda) = x - \lambda x - 20\lambda$

$F_\lambda(x, y, \lambda) = -xy - 20y - 20x + 26{,}000$

Step 4 $y - \lambda y - 20\lambda = 0$ (4)

$x - \lambda x - 20\lambda = 0$ (5)

$-xy - 20y - 20x + 26{,}000 = 0$ (6)

Step 5 Solving Equations (4) and (5) for λ gives

$$\lambda = \frac{y}{y + 20} \quad \text{and} \quad \lambda = \frac{x}{x + 20}$$

$$\frac{y}{y + 20} = \frac{x}{x + 20}$$

$$y(x + 20) = x(y + 20)$$

$$xy + 20y = xy + 20x$$

$$x = y.$$

Now substitute y for x in Equation (6) to get

$$-y^2 - 20y - 20y + 26{,}000 = 0$$

$$-y^2 - 40y + 26{,}000 = 0.$$

Use the quadratic formula to find $y \approx -182.5$ or $y \approx 142.5$. We eliminate the negative value because length cannot be negative. Since $x = y$, we know that $x \approx 142.5$.

The maximum area of $(142.5)^2 \approx 20{,}306 \text{ ft}^2$ will be achieved if the floor plan is a square with a side of 142.5 ft. You can verify that this answer is a maximum using the method at the end of Example 1.

As mentioned earlier, the method of Lagrange multipliers works for more than two independent variables. The next example shows how to find extrema for a function of three independent variables.

EXAMPLE 3   Volume of a Box

Find the dimensions of the closed rectangular box of maximum volume that can be produced from 6 ft² of material.

SOLUTION

Method 1
Lagrange Multipliers

In Chapter 6 on Applications of the Derivative, we were able to solve problems such as this by adding an extra constraint, such as requiring the bottom of the box to be square. Here we have no such constraint. Let x, y, and z represent the dimensions of the box, as shown in Figure 26 on the next page. The volume of the box is given by

$$f(x, y, z) = xyz.$$

FIGURE 26

As shown in Figure 26, the total amount of material required for the two ends of the box is $2xy$, the total needed for the sides is $2xz$, and the total needed for the top and bottom is $2yz$. Since 6 ft^2 of material is available,

$$2xy + 2xz + 2yz = 6 \qquad \text{or} \qquad xy + xz + yz = 3.$$

In summary, $f(x, y, z) = xyz$ is to be maximized subject to the constraint $xy + xz + yz = 3$. Go through the steps that were given.

Step 1 $g(x, y, z) = xy + xz + yz - 3 = 0$

Step 2 $F(x, y, z, \lambda) = xyz - \lambda(xy + xz + yz - 3)$

Step 3 $F_x(x, y, z, \lambda) = yz - \lambda y - \lambda z$

$\qquad F_y(x, y, z, \lambda) = xz - \lambda x - \lambda z$

$\qquad F_z(x, y, z, \lambda) = xy - \lambda x - \lambda y$

$\qquad F_\lambda(x, y, z, \lambda) = -xy - xz - yz + 3$

Step 4 $yz - \lambda y - \lambda z = 0$

$\qquad xz - \lambda x - \lambda z = 0$

$\qquad xy - \lambda x - \lambda y = 0$

$\qquad -xy - xz - yz + 3 = 0$

Step 5 Solve each of the first three equations for λ. You should get

$$\lambda = \frac{yz}{y + z}, \qquad \lambda = \frac{xz}{x + z}, \qquad \text{and} \qquad \lambda = \frac{xy}{x + y}.$$

Set these expressions for λ equal, and simplify as follows. Notice in the second and last steps that since none of the dimensions of the box can be 0, we can divide both sides of each equation by x or z.

$$\frac{yz}{y + z} = \frac{xz}{x + z} \qquad \text{and} \qquad \frac{xz}{x + z} = \frac{xy}{x + y}$$

$$\frac{y}{y + z} = \frac{x}{x + z} \qquad\qquad \frac{z}{x + z} = \frac{y}{x + y}$$

$$xy + yz = xy + xz \qquad\qquad zx + zy = yx + yz$$

$$yz = xz \qquad\qquad\qquad zx = yx$$

$$y = x \qquad\qquad\qquad\quad z = y$$

(Setting the first and third expressions equal gives no additional information.) Thus $x = y = z$. From the fourth equation in Step 4, with $x = y$ and $z = y$,

$$-xy - xz - yz + 3 = 0$$

$$-y^2 - y^2 - y^2 + 3 = 0$$

$$-3y^2 = -3$$

$$y^2 = 1$$

$$y = \pm 1.$$

The negative solution is not applicable, so the solution of the system of equations is $x = 1, y = 1, z = 1$. In other words, the box with maximum volume under the constraint is a cube that measures 1 ft on each side. As in the previous examples, verify that this is a maximum.

Method 2 Spreadsheets

Finding extrema of a constrained function of one or more variables can be done using a spreadsheet. As in the previous section, we identify cells to represent the variables. In this example, we will let A1, B1, and C1 represent x, y, and z, respectively. We place a guess for the answer in these cells, selecting a solution that satisfies the constraint.

Next, an expression for the function to be optimized is entered, with x, y, and z replaced by A1, B1, and C1, respectively. If we choose A3 to represent the function, we would type "=A1*B1*C1" in cell A3.

In addition, we need to input the constraint into the Excel Solver. To do this, we first input the left-hand or variable part of the constraint into a designated cell. If A5 is the designated cell, then in cell A5 we would type "=A1*B1+A1*C1+B1*C1." We now click on the Tools menu and choose Solver. To add the constraint, we select Add next to the constraint box. We enter A5 for the Cell Reference, select the equal sign, and enter 3 (the right-hand side or constant part) in the Constraint box. Selecting OK enters the constraint. Figure 27 illustrates the Solver box and the items placed in it.

YOUR TURN 2 Solve Example 3 with the box changed so that the front and the top are missing.

To obtain a solution, Set Target Cell equal to A3, select Max, and enter A1:C1 in the By Changing Cells Box. Click on Solve and the solution $x = 1$, $y = 1$, and $z = 1$ appears in cells A1, B1, and C1, respectively. The maximum volume $f(1, 1, 1) = 1$ is located in cell A3.   **TRY YOUR TURN 2**

CAUTION One must be careful when using Solver because the solution may depend on the initial value. Thus, it is always a good idea to run the Solver for two different initial values and compare the solutions.

FIGURE 27

Utility Functions

A **utility function** of two variables is a function $z = f(x, y)$ in which x and y give the quantity of items that a consumer might value, such as cereal and milk, and z is a measure of the value that the consumer places on the combination of items represented by the point (x, y). (Naturally, this definition can be extended to any number of variables, but for simplicity we will restrict this discussion to two variables.) For example, if $z = f(x, y) = x^2y^4$, where x represents the number of quarts of milk and y represents the number of pounds of cereal, then 4 quarts of milk and 3 pounds of cereal has a utility of $4^2 \cdot 3^4 = 1296$. The consumer would value this combination as much as 36 quarts of milk and 1 pound of cereal, since this combination also has a utility of $36^2 \cdot 1^4 = 1296$, but less so than 3 quarts of milk and 4 pounds of cereal, which has a utility of $3^2 \cdot 4^4 = 2304$. The set of points $c = f(x, y)$ form an **indifference curve**, since the consumer considers all points on this curve to be equally desirable.

Now suppose a quart of milk costs \$2 and a pound of cereal costs \$3, so that the combination of milk and cereal represented by the point (x, y) costs $2x + 3y$. Suppose further that the consumer has \$90 to spend. A natural question would be how much of each quantity to buy to maximize the consumer's utility. In other words, the consumer wishes to maximize $f(x, y) = x^2y^4$ subject to the constraint that $2x + 3y = 90$, or $g(x, y) = 2x + 3y - 90 = 0$. This is exactly the type of problem that Lagrange multipliers were designed to solve. Use Lagrange multipliers to verify that the function $f(x, y) = x^2y^4$ subject to the constraint $g(x, y) = 2x + 3y - 90 = 0$ has a maximum value of 36,000,000 when $x = 15$ and $y = 20$, so the consumer should purchase 15 quarts of milk and 20 pounds of cereal.

W1. Find $f_x(x, y)$ and $f_y(x, y)$ for
$f(x, y) = 4x^3y^2 - 6x^4 - 3y^5$. *(Sec. 9.2)*

W2. Find $f_x(x, y, z)$, $f_y(x, y, z)$, and $f_z(x, y, z)$ for
$f(x, y, z) = x^2y^3z^5 - 4x^2y^5 + 3xz^6 - 2y^8z^7$. *(Sec. 9.2)*

9.4 EXERCISES

Find the relative maxima or minima in Exercises 1–10.

1. Maximum of $f(x, y) = 4xy$, subject to $x + y = 16$

2. Maximum of $f(x, y) = 2xy + 4$, subject to $x + y = 20$

3. Maximum of $f(x, y) = xy^2$, subject to $x + 2y = 15$

4. Maximum of $f(x, y) = 8x^2y$, subject to $3x - y = 9$

5. Minimum of $f(x, y) = x^2 + 2y^2 - xy$, subject to $x + y = 8$

6. Minimum of $f(x, y) = 3x^2 + 4y^2 - xy - 2$, subject to $2x + y = 21$

7. Maximum of $f(x, y) = x^2 - 10y^2$, subject to $x - y = 18$

8. Maximum of $f(x, y) = 12xy - x^2 - 3y^2$, subject to $x + y = 16$

9. Maximum of $f(x, y, z) = xyz^2$, subject to $x + y + z = 6$

10. Maximum of $f(x, y, z) = xy + 2xz + 2yz$, subject to $xyz = 32$

11. Find positive numbers x and y such that $x + y = 24$ and $3xy^2$ is maximized.

12. Find positive numbers x and y such that $x + y = 48$ and $5x^2y + 10$ is maximized.

13. Find three positive numbers whose sum is 90 and whose product is a maximum.

14. Find three positive numbers whose sum is 240 and whose product is a maximum.

15. Find the maximum and minimum values of $f(x, y) = x^3 + 2xy + 4y^2$ subject to $x + 2y = 12$. Be sure to use the method at the end of Example 1 to determine whether each solution is a maximum or a minimum.

16. Explain the difference between the two methods we used in Sections 3 and 4 to solve extrema problems.

17. Why is it unnecessary to find the value of λ when using the method explained in this section?

18. Show that the function $f(x, y) = xy^2$ in Exercise 3, subject to $x + 2y = 15$, does not have an absolute minimum or maximum. (*Hint*: Solve the constraint for x and substitute into f.)

19. Show that the function $f(x, y) = 8x^2y$ in Exercise 4, subject to $3x - y = 9$, does not have an absolute minimum or maximum. (*Hint*: Solve the constraint for y and substitute into f.)

20. Show that the three equations in Step 4 of the box "Using Lagrange Multipliers" are equivalent to the three equations
$f_x(x, y) = \lambda g_x(x, y), f_y(x, y) = \lambda g_y(x, y), g(x, y) = 0.$

21. Consider the problem of minimizing $f(x, y) = x^2 + 2x + 9y^2 + 4y + 8xy$ subject to $x + y = 1$.

    (a) Find the solution using the method of Lagrange multipliers.

    (b) Experiment with points very near the point from part (a) to convince yourself that the point from part (a) actually gives a minimum. (*Hint*: See the last paragraph of Example 1.)

    (c) Solve $x + y = 1$ for y and substitute the expression for y into $f(x, y)$. Then explain why the resulting expression in x has a minimum but no maximum.

    (d) Suppose you erroneously applied the method of finding the discriminant D from the previous section to determine whether the point found in part (a) is a minimum. What does the test erroneously tell you about the point?

22. Discuss the advantages and disadvantages of the method of Lagrange multipliers compared with solving the equation $g(x, y) = 0$ for y (or x), substituting that expression into f and then minimizing or maximizing f as a function of one variable. You might want to try some examples both ways and consider what happens when there are more than two variables.

APPLICATIONS

Business and Economics
Utility Maximize each of the following utility functions, with the cost of each commodity and total amount available to spend given.

23. $f(x, y) = xy^2$, cost of a unit of x is \$1, cost of a unit of y is \$2, and \$60 is available.

24. $f(x, y) = x^2y^3$, cost of a unit of x is \$2, cost of a unit of y is \$1, and \$80 is available.

25. $f(x, y) = x^4y^2$, cost of a unit of x is \$2, cost of a unit of y is \$4, and \$60 is available.

26. $f(x, y) = x^3y^4$, cost of a unit of x is \$3, cost of a unit of y is \$3, and \$42 is available.

27. **Maximum Area for Fixed Expenditure** Because of terrain difficulties, two sides of a fence can be built for \$6 per ft, while the other two sides cost \$4 per ft. (See the sketch.) Find the field of maximum area that can be enclosed for \$1200.

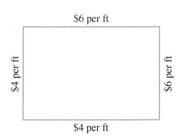

28. Maximum Area for Fixed Expenditure To enclose a yard, a fence is built against a large building, so that fencing material is used only on three sides. Material for the ends costs $15 per ft; material for the side opposite the building costs $25 per ft. Find the dimensions of the yard of maximum area that can be enclosed for $2400.

29. Cost The total cost to produce x large jewelry-making kits and y small ones is given by

$$C(x, y) = 2x^2 + 6y^2 + 4xy + 10.$$

If a total of ten kits must be made, how should production be allocated so that total cost is minimized?

30. Profit The profit from the sale of x units of radiators for automobiles and y units of radiators for generators is given by

$$P(x, y) = -x^2 - y^2 + 4x + 8y.$$

Find values of x and y that lead to a maximum profit if the firm must produce a total of 6 units of radiators.

31. Production A manufacturing firm estimates that its total production of automobile batteries in thousands of units is

$$f(x, y) = 3x^{1/3}y^{2/3},$$

where x is the number of units of labor and y is the number of units of capital utilized. Labor costs are $80 per unit, and capital costs are $150 per unit. How many units each of labor and capital will maximize production, if the firm can spend $40,000 for these costs?

32. Production For another product, the manufacturing firm in Exercise 31 estimates that production is a function of labor x and capital y as follows:

$$f(x, y) = 12x^{3/4}y^{1/4}.$$

If $25,200 is available for labor and capital, and if the firm's costs are $100 and $180 per unit, respectively, how many units of labor and capital will give maximum production?

33. Area A farmer has 500 m of fencing. Find the dimensions of the rectangular field of maximum area that can be enclosed by this amount of fencing.

34. Area Find the area of the largest rectangular field that can be enclosed with 600 m of fencing. Assume that no fencing is needed along one side of the field.

35. Surface Area A cylindrical can is to be made that will hold 250π in^3 of candy. Find the dimensions of the can with minimum surface area.

36. Surface Area An ordinary 12-oz beer or soda pop can holds about 25 in^3. Find the dimensions of a can with minimum surface area. Measure a can and see how close its dimensions are to the results you found.

37. Volume A rectangular box with no top is to be built from 500 m^2 of material. Find the dimensions of such a box that will enclose the maximum volume.

38. Surface Area A 1-lb soda cracker box has a volume of 185 in^3. The end of the box is square. Find the dimensions of such a box that has minimum surface area.

39. Cost A rectangular closed box is to be built at minimum cost to hold 125 m^3. Since the cost will depend on the surface area, find the dimensions that will minimize the surface area of the box.

40. Cost Find the dimensions that will minimize the surface area (and hence the cost) of a rectangular fish aquarium, open on top, with a volume of 32 ft^3.

41. Container Construction RD Enterprises is designing an open top box to transport 400 yd^3 of material from one place to another in several trips. Two of the sides and bottom of the box can be made of a free material that the company has on hand, but only 8 yd^2 of the material is available. Because of the hauling regulations associated with transporting this material, the two ends of the box must be made from a heavyweight material that costs $1000 per yd^2. Each trip costs $10.

(a) Let x, y, and z represent the length, width, and height (in yards) of the box, respectively. If we want to use all of the free material, show that the total cost in dollars is given by the function

$$f(x, y, z) = \frac{4000}{xyz} + 2000yz$$

subject to the constraint $2xz + xy = 8$.

(b) Use the Solver feature on a spreadsheet to find the dimensions of the box that minimize the transportation cost, subject to the constraint.

Social Sciences

42. Political Science The probability that the majority of a three-person jury will convict a guilty person is given by the formula:

$$P(r, s, t) = rs(1 - t) + (1 - r)st + r(1 - s)t + rst$$

subject to the constraint that

$$r + s + t = \alpha,$$

where r, s, and t represent each of the three jury members' probability of reaching a guilty verdict and α is some fixed constant that is generally less than or equal to the number of jurors. *Source: Mathematical Social Sciences.*

(a) Form the Lagrange function.

(b) Find the values of r, s, and t that maximize the probability of convicting a guilty person when $\alpha = 0.75$.

(c) Find the values of r, s, and t that maximize the probability of convicting a guilty person when $\alpha = 3$.

YOUR TURN ANSWERS

1. $f(12, -4) = 12$

2. The box should be 2 ft wide and 1 ft high and long.

9.5 Total Differentials and Approximations

APPLY IT How do errors in measuring the length and radius of a blood vessel affect the calculation of its volume?

In Example 3 in this section, we will see how to answer this question using a total differential.

In the second section of this chapter we used partial derivatives to find the marginal productivity of labor and of capital for a production function. The marginal productivity approximates the change of production for a 1-unit change in labor or capital. To estimate the change in productivity for a small change in both labor and capital, we can extend the concept of differential, introduced in an earlier chapter for functions of one variable, to the concept of *total differential*.

Total Differential for Two Variables

Let $z = f(x, y)$ be a function of x and y. Let dx and dy be real numbers. Then the **total differential** of z is

$$dz = f_x(x, y) \cdot dx + f_y(x, y) \cdot dy.$$

(Sometimes dz is written df.)

Recall that the differential for a function of one variable $y = f(x)$ is used to approximate the function by its tangent line. This works because a differentiable function appears very much like a line when viewed closely. Similarly, the differential for a function of two variables $z = f(x, y)$ is used to approximate a function by its tangent plane. A differentiable function of two variables looks like a plane when viewed closely, which is why the earth looks flat when you are standing on it.

FOR REVIEW

In Chapter 6 on Applications of the Derivative, we introduced the differential. Recall that the differential of a function defined by $y = f(x)$ is

$$dy = f'(x) \cdot dx,$$

where dx, the differential of x, is any real number (usually small). We saw that the differential dy is often a good approximation of Δy, where $\Delta y = f(x + \Delta x) - f(x)$ and $\Delta x = dx$.

YOUR TURN 1 For the function $f(x, y) = 3x^2y^4 + 6\sqrt{x^2 - 7y^2}$, find (a) dz, and (b) the value of dz when $x = 4$, $y = 1$, $dx = 0.02$, and $dy = -0.03$.

EXAMPLE 1 Total Differentials

Consider the function $z = f(x, y) = 9x^3 - 8x^2y + 4y^3$.

(a) Find dz.

SOLUTION First find $f_x(x, y)$ and $f_y(x, y)$.

$$f_x(x, y) = 27x^2 - 16xy \quad \text{and} \quad f_y(x, y) = -8x^2 + 12y^2$$

By the definition,

$$dz = (27x^2 - 16xy) \, dx + (-8x^2 + 12y^2) \, dy.$$

(b) Evaluate dz when $x = 1$, $y = 3$, $dx = 0.01$, and $dy = -0.02$.

SOLUTION Substituting these values into the result from part (a) gives

$$dz = [27(1)^2 - 16(1)(3)](0.01) + [-8(1)^2 + 12(3)^2](-0.02)$$
$$= (-21)(0.01) + (100)(-0.02)$$
$$= -2.21.$$

This result indicates that an increase of 0.01 in x and a decrease of 0.02 in y, when $x = 1$ and $y = 3$, will produce an approximate *decrease* of 2.21 in $f(x, y)$.

TRY YOUR TURN 1

Approximations

Recall that with a function of one variable, $y = f(x)$, the differential dy approximates the change in y, Δy, corresponding to a change in x, Δx or dx. The approximation for a function of two variables is similar.

Approximations

For small values of dx and dy,

$$dz \approx \Delta z,$$

where $\Delta z = f(x + dx, y + dy) - f(x, y)$.

EXAMPLE 2 Approximations

Approximate $\sqrt{2.98^2 + 4.01^2}$.

SOLUTION Notice that $2.98 \approx 3$ and $4.01 \approx 4$, and we know that $\sqrt{3^2 + 4^2} = \sqrt{25} = 5$. We, therefore, let $f(x, y) = \sqrt{x^2 + y^2}$, $x = 3$, $dx = -0.02$, $y = 4$, and $dy = 0.01$. We then use dz to approximate $\Delta z = \sqrt{2.98^2 + 4.01^2} - \sqrt{3^2 + 4^2}$.

$$dz = f_x(x, y) \cdot dx + f_y(x, y) \cdot dy$$

$$= \left(\frac{1}{2\sqrt{x^2 + y^2}} \cdot 2x \right) dx + \left(\frac{1}{2\sqrt{x^2 + y^2}} \cdot 2y \right) dy$$

$$= \left(\frac{x}{\sqrt{x^2 + y^2}} \right) dx + \left(\frac{y}{\sqrt{x^2 + y^2}} \right) dy$$

$$= \frac{3}{5}(-0.02) + \frac{4}{5}(0.01) \qquad {\scriptstyle x = 3, dx = -0.02, y = 4, dy = 0.01}$$

$$= -0.004$$

YOUR TURN 2
Approximate $\sqrt{5.03^2 + 11.99^2}$.

Thus, $\sqrt{2.98^2 + 4.01^2} \approx 5 + (-0.004) = 4.996$. A calculator gives $\sqrt{2.98^2 + 4.01^2} \approx 4.996048$. The error is approximately 0.000048.　　　　TRY YOUR TURN 2

For small values of dx and dy, the values of Δz and dz are approximately equal. Since $\Delta z = f(x + dx, y + dy) - f(x, y)$,

$$f(x + dx, y + dy) = f(x, y) + \Delta z$$

or

$$f(x + dx, y + dy) \approx f(x, y) + dz.$$

Replacing dz with the expression for the total differential gives the following result.

Approximations by Differentials

For a function f having all indicated partial derivatives, and for small values of dx and dy,

$$f(x + dx, y + dy) \approx f(x, y) + dz,$$

or

$$f(x + dx, y + dy) \approx f(x, y) + f_x(x, y) \cdot dx + f_y(x, y) \cdot dy.$$

The idea of a total differential can be extended to include functions of three or more independent variables.

Total Differential for Three Variables

If $w = f(x, y, z)$, then the total differential dw is

$$dw = f_x(x, y, z)\, dx + f_y(x, y, z)\, dy + f_z(x, y, z)\, dz,$$

provided all indicated partial derivatives exist.

EXAMPLE 3 Blood Vessels

A short length of blood vessel is in the shape of a right circular cylinder (see Figure 28).

42 mm ± 0.9 mm

2.5 mm ± 0.2 mm

FIGURE 28

(a) The length of the vessel is measured as 42 mm, and the radius is measured as 2.5 mm. Suppose the maximum error in the measurement of the length is 0.9 mm, with an error of no more than 0.2 mm in the measurement of the radius. Find the maximum possible error in calculating the volume of the blood vessel.

APPLY IT

SOLUTION The volume of a right circular cylinder is given by $V = \pi r^2 h$. To approximate the error in the volume, find the total differential, dV.

$$dV = (2\pi rh) \cdot dr + (\pi r^2) \cdot dh$$

Here, $r = 2.5, h = 42, dr = 0.2$, and $dh = 0.9$. Substitution gives

$$dV = [(2\pi)(2.5)(42)](0.2) + [\pi(2.5)^2](0.9) \approx 149.6.$$

The maximum possible error in calculating the volume is approximately 149.6 mm^3.

(b) Suppose that the errors in measuring the radius and length of the vessel are at most 1% and 3%, respectively. Estimate the maximum percent error in calculating the volume.

SOLUTION To find the percent error, calculate dV/V.

$$\frac{dV}{V} = \frac{(2\pi rh)dr + (\pi r^2)dh}{\pi r^2 h} = 2\frac{dr}{r} + \frac{dh}{h}$$

YOUR TURN 3 In Example 3, estimate the maximum percent error in calculating the volume if the errors in measuring the radius and length of the vessel are at most 4% and 2%, respectively.

Because $dr/r = 0.01$ and $dh/h = 0.03$,

$$\frac{dV}{V} = 2(0.01) + 0.03 = 0.05.$$

The maximum percent error in calculating the volume is approximately 5%.

TRY YOUR TURN 3

EXAMPLE 4 Volume of a Can of Beer

The formula for the volume of a cylinder given in Example 3 also applies to cans of beer, for which $r \approx 1.5$ in. and $h \approx 5$ in. How sensitive is the volume to changes in the radius compared with changes in the height?

SOLUTION Using the formula for dV from the previous example with $r = 1.5$ and $h = 5$ gives

$$dV = (2\pi)(1.5)(5)dr + \pi(1.5)^2 dh = \pi(15dr + 2.25dh).$$

The factor of 15 in front of dr in this equation, compared with the factor of 2.25 in front of dh, shows that a small change in the radius has almost 7 times the effect on the volume as a small change in the height. One author argues that this is the reason that beer cans are so tall and thin. The brewers can reduce the radius by a tiny amount and compensate by making the can taller. The resulting can appears larger in volume than the shorter, wider can. (Others have argued that a shorter, wider can does not fit as easily in the hand.) *Source: The College Mathematics Journal.*

Find dy for the given values of x and Δx.   (*Sec. 6.6*)

W1. $y = 2x^3 + 3x$; $x = 2, \Delta x = 0.1$

W2. $y = \ln(x^2 + 1)$; $x = 2, \Delta x = -0.02$

Use the differential to approximate each quantity.   (*Sec. 6.6*)

W3. $y = \sqrt{24}$

W4. $y = e^{0.01}$

9.5 EXERCISES

Evaluate dz using the given information.

1. $z = 2x^2 + 4xy + y^2$; $x = 5, y = -1, dx = 0.03, dy = -0.02$

2. $z = 5x^3 + 2xy^2 - 4y$; $x = 1, y = 3, dx = 0.01, dy = 0.02$

3. $z = \dfrac{y^2 + 3x}{y^2 - x}$; $x = 4, y = -4, dx = 0.01, dy = 0.03$

4. $z = \ln(x^2 + y^2)$; $x = 2, y = 3, dx = 0.02, dy = -0.03$

Evaluate dw using the given information.

5. $w = \dfrac{5x^2 + y^2}{z + 1}$; $x = -2, y = 1, z = 1, dx = 0.02,$
$dy = -0.03, dz = 0.02$

6. $w = x \ln(yz) - y \ln\dfrac{x}{z}$; $x = 2, y = 1, z = 4,$
$dx = 0.03, dy = 0.02, dz = -0.01$

Use the total differential to approximate each quantity. Then use a calculator to approximate the quantity, and give the absolute value of the difference in the two results to 4 decimal places.

7. $\sqrt{8.05^2 + 5.97^2}$

8. $\sqrt{4.96^2 + 12.06^2}$

9. $(1.92^2 + 2.1^2)^{1/3}$

10. $(2.93^2 - 0.94^2)^{1/3}$

11. $1.03e^{0.04}$

12. $0.98e^{-0.04}$

13. $0.99 \ln 0.98$

14. $2.03 \ln 1.02$

APPLICATIONS

Business and Economics

15. Manufacturing   Approximate the volume of aluminum needed for a beverage can of radius 2.5 cm and height 14 cm. Assume the walls of the can are 0.08 cm thick.

16. Manufacturing   Approximate the volume of material needed to make a water tumbler of diameter 3 cm and height 9 cm. Assume the walls of the tumbler are 0.2 cm thick.

17. Volume of a Coating   An industrial coating 0.1 in. thick is applied to all sides of a box of dimensions 10 in. by 9 in. by 18 in. Estimate the volume of the coating used.

18. Manufacturing Cost   The manufacturing cost of a smartphone is approximated by
$$M(x, y) = 45x^2 + 40y^2 - 20xy + 50,$$
where x is the cost of the parts and y is the cost of labor. Right now, the company spends $8 on parts and $14 on labor. Use differentials to approximate the change in cost if the company spends $8.25 on parts and $13.75 on labor.

19. Production   The production function for one country is
$$z = x^{0.65}y^{0.35},$$
where x stands for units of labor and y for units of capital. At present, 50 units of labor and 29 units of capital are available. Use differentials to estimate the change in production if the number of units of labor is increased to 52 and capital is decreased to 27 units.

20. Production   The production function for another country is
$$z = x^{0.8}y^{0.2},$$
where x stands for units of labor and y for units of capital. At present, 20 units of labor and 18 units of capital are being provided. Use differentials to estimate the change in production if an additional unit of labor is provided and if capital is decreased to 16 units.

Life Sciences

21. Bone Preservative Volume   A piece of bone in the shape of a right circular cylinder is 7 cm long and has a radius of 1.4 cm. It is coated with a layer of preservative 0.09 cm thick. Estimate the volume of preservative used.

22. Blood Vessel Volume   A portion of a blood vessel is measured as having length 7.9 cm and radius 0.8 cm. If each measurement could be off by as much as 0.15 cm, estimate the maximum possible error in calculating the volume of the vessel.

23. Blood Volume   In Exercise 58 of Section 2 in this chapter, we found that the number of liters of blood pumped through the lungs in one minute is given by
$$C = \frac{b}{a - v}.$$
Suppose $a = 160, b = 200$, and $v = 125$. Estimate the change in C if a becomes 145, b becomes 190, and v changes to 130.

24. Heat Loss   In Exercise 56 of Section 2 of this chapter, we found that the rate of heat loss (in watts) in harbor seal pups could be approximated by
$$H(m, T, A) = \frac{15.2m^{0.67}(T - A)}{10.23 \ln m - 10.74},$$
where m is the body mass of the pup (in kg), and T and A are the body core temperature and ambient water temperature, respectively (in °C). Suppose m is 25 kg, T is 36.0°, and A is 12.0°C. Approximate the change in H if m changes to 26 kg, T to 36.5°, and A to 10.0°C.

25. Life Span   As we saw in Exercise 62 of Section 9.2, researchers have estimated the maximum life span (in years) for various species of mammals according to the formula
$$L(E, P) = 23E^{0.6}P^{-0.267},$$
where E is the average brain mass and P is the average body mass (both in g). Consider humans, with $E = 14,100$ g and

$P = 68,700$ g. Find the approximate change in life span if the brain mass increases to 14,300 g and the body mass decreases to 68,400 g. Then calculate the actual change in life span and compare. *Source: The Quarterly Review of Biology.*

26. **Horn Volume** The volume of the horns from bighorn sheep was estimated by researchers using the equation

$$V = \frac{h\pi}{3}(r_1^2 + r_1 r_2 + r_2^2),$$

where h is the length of a horn segment (in centimeters) and r_1 and r_2 are the radii of the two ends of the horn segment (in centimeters). *Source: Conservation Biology.*

(a) Determine the volume of a segment of horn that is 40 cm long with radii of 5 cm and 3 cm, respectively.

(b) Use the total differential to estimate the volume of the segment of horn if the horn segment from part (a) was actually 42 cm long with radii of 5.1 cm and 2.9 cm, respectively. Compare this with the actual volume.

27. **Eastern Hemlock** Ring shake, which is the separation of the wood between growth rings, is a serious problem in hemlock trees. Researchers have developed the following function that estimates the probability P that a given hemlock tree has ring shake.

$$P(A, B, D) = \frac{1}{1 + e^{3.68 - 0.016A - 0.77B - 0.12D}},$$

where A is the age of the tree (yr), B is 1 if bird pecking is present and 0 otherwise, and D is the diameter (in.) of the tree at breast height. *Source: Forest Products Journal.*

(a) Estimate the probability that a 150-year-old tree, with bird pecking present and a breast height diameter of 20 in., will have ring shake.

(b) Estimate the probability that a 150-year-old tree, with no presence of bird pecking and a breast height diameter of 20 in., will have ring shake.

(c) Develop a statement about what can be said about the influence that the three variables have on the probability of ring shake.

(d) Using the total differential, estimate the probability if the actual age of the tree was 160 years and the diameter at breast height was 25 in. Assume that no bird pecking was present. Compare your answer to the actual value. (*Hint:* Assume that $B = 0$ and exclude that variable from your calculations.)

(e) Comment on the practicality of using differentials in part (d).

28. **Dialysis** A model that estimates the concentration of urea in the body for a particular dialysis patient, following a dialysis session, is given by

$$C(t, g) = 0.6(0.96)^{(210t/1500) - 1}$$
$$+ \frac{gt}{126t - 900}[1 - (0.96)^{(210t/1500) - 1}],$$

where t represents the number of minutes of the dialysis session and g represents the rate at which the body generates urea in mg per minute. *Source: Clinical Dialysis.*

(a) Find $C(180, 8)$.

(b) Using the total differential, estimate the urea concentration if the dialysis session of part (a) was cut short by 10 minutes and the urea generation rate was 9 mg per minute. Compare this with the actual concentration. (*Hint:* First, replace the variable g with the number 8, thus reducing the function to one variable. Then use your graphing calculator to calculate the partial derivative $C_t(180, 8)$. A similar procedure can be done for $C_g(180, 8)$.)

Physical Sciences

29. **Swimming** The amount of time in seconds it takes for a swimmer to hear a single, hand-held, starting signal is given by the formula

$$t(x, y, p, C) = \frac{\sqrt{x^2 + (y - p)^2}}{331.45 + 0.6C},$$

where (x, y) is the location of the starter (in meters), $(0, p)$ is the location of the swimmer (in meters), and C is the air temperature (in degrees Celsius). Assume that the starter is located at the point $(x, y) = (5, -2)$. See the diagram. *Source: COMAP.*

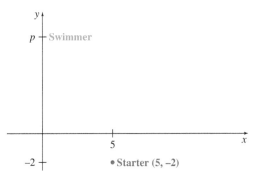

(a) Calculate $t(5, -2, 20, 20)$ and $t(5, -2, 10, 20)$. Could the difference in time change the outcome of a race?

(b) Calculate the total differential for t if the starter remains stationary, the swimmer moves from 20 m to 20.5 m away from the starter in the y direction, and the temperature decreases from 20°C to 15°C. Interpret your answer.

General Interest

30. **Estimating Volume** The height of a cone is measured as 9.3 cm and the radius as 3.2 cm. Each measurement could be off by as much as 0.1 cm. Estimate the maximum possible error in calculating the volume of the cone.

31. **Estimating Area** The height of a triangle is measured as 37.5 cm, with the base measured as 15.8 cm. The measurement of the height can be off by as much as 0.8 cm and that of the base by no more than 1.1 cm. Estimate the maximum possible error in calculating the area of the triangle.

32. **Estimating Volume** Suppose there is a maximum error of $a\%$ in measuring the radius of a cone and a maximum error of $b\%$ in measuring the height. Estimate the maximum percent error in calculating the volume of the cone, and compare this value with the maximum percent error in calculating the volume of a cylinder.

33. **Estimating Volume** Suppose that in measuring the length, width, and height of a box, there is a maximum 1% error in each measurement. Estimate the maximum error in calculating the volume of the box.

34. Hose A hose has a radius of approximately 0.5 in. and a length of approximately 20 ft. By what factor does a change in the radius affect the volume compared with a change in the length?

35. Ice Cream Cone An ice cream cone has a radius of approximately 1 in. and a height of approximately 4 in. By what factor does a change in the radius affect the volume compared with a change in the height?

9.6 Double Integrals

APPLY IT How can we find the volume of a bottle with curved sides?

We will answer this question in Example 6 using a double integral, the key idea in this section.

In an earlier chapter, we saw how integrals of functions with one variable may be used to find area. In this section, this idea is extended and used to find volume. We found partial derivatives of functions of two or more variables at the beginning of this chapter by holding constant all variables except one. A similar process is used in this section to find antiderivatives of functions of two or more variables. For example, in

> **FOR REVIEW**
>
> You may wish to review the key ideas of indefinite and definite integrals from Chapter 7 on Integration before continuing with this section. See the review problems at the end of that chapter.

$$\int (5x^3y^4 - 6x^2y + 2)\, dy$$

the notation dy indicates integration with respect to y, so we treat y as the variable and x as a constant. Using the rules for antiderivatives gives

$$\int (5x^3y^4 - 6x^2y + 2)\, dy = x^3y^5 - 3x^2y^2 + 2y + C(x).$$

The constant C used earlier must be replaced with $C(x)$ to show that the "constant of integration" here can be any function involving only the variable x. Just as before, check this work by taking the derivative (actually the partial derivative) of the answer:

$$\frac{\partial}{\partial y}[x^3y^5 - 3x^2y^2 + 2y + C(x)] = 5x^3y^4 - 6x^2y + 2 + 0,$$

which shows that the antiderivative is correct.

We can use this antiderivative to evaluate a definite integral.

EXAMPLE 1 Definite Integral

Evaluate $\displaystyle\int_1^2 (5x^3y^4 - 6x^2y + 2)\, dy$.

SOLUTION

$$\int_1^2 (5x^3y^4 - 6x^2y + 2)\, dy = \left[x^3y^5 - 3x^2y^2 + 2y + C(x)\right]\Big|_1^2 \qquad \text{Use the indefinite integral previously found.}$$

$$= [x^3 2^5 - 3x^2 2^2 + 2 \cdot 2 + C(x)] \qquad \text{Use the Fundamental Theorem of Calculus.}$$
$$\quad - [x^3 1^5 - 3x^2 1^2 + 2 \cdot 1 + C(x)]$$

$$= 32x^3 - 12x^2 + 4 + C(x)$$
$$\quad - [x^3 - 3x^2 + 2 + C(x)]$$

$$= 31x^3 - 9x^2 + 2 \qquad \text{Simplify.}$$

YOUR TURN 1 Evaluate
$$\int_1^3 (6x^2y^2 + 4xy + 8x^3 + 10y^4 + 3)\, dy.$$

In the second step, we substituted $y = 2$ and $y = 1$ and subtracted, according to the Fundamental Theorem of Calculus. Notice that $C(x)$ does not appear in the final answer, just as the constant does not appear in a regular definite integral. Therefore, from now on we will not include $C(x)$ when we find the antiderivative for a definite integral with respect to y.

TRY YOUR TURN 1 ▬

By integrating the result from Example 1 with respect to x, we can evaluate a double integral.

EXAMPLE 2 Definite Integral

Evaluate $\displaystyle\int_0^3 \left[\int_1^2 (5x^3y^4 - 6x^2y + 2)\, dy \right] dx.$

SOLUTION

$$\int_0^3 \left[\int_1^2 (5x^3y^4 - 6x^2y + 2)\, dy \right] dx = \int_0^3 [31x^3 - 9x^2 + 2]\, dx \qquad \text{Use the result from Example 1.}$$

$$= \frac{31}{4}x^4 - 3x^3 + 2x \Big|_0^3 \qquad \text{Use the Fundamental Theorem of Calculus.}$$

$$= \left(\frac{31}{4} \cdot 3^4 - 3 \cdot 3^3 + 2 \cdot 3 \right) - \left(\frac{31}{4} \cdot 0^4 - 3 \cdot 0^3 + 2 \cdot 0 \right)$$

$$= \frac{2211}{4} \qquad\qquad ▬$$

We can integrate the inner integral with respect to y and the outer integral with respect to x, as in Example 2, or in the reverse order. The next example shows the same integral done both ways.

EXAMPLE 3 Definite Integrals

Evaluate each integral.

(a) $\displaystyle\int_1^2 \left[\int_3^5 (6xy^2 + 12x^2y + 4y)\, dx \right] dy$

SOLUTION

$$\int_1^2 \left[\int_3^5 (6xy^2 + 12x^2y + 4y)\, dx \right] dy = \int_1^2 \left[(3x^2y^2 + 4x^3y + 4xy) \Big|_3^5 \right] dy \qquad \text{Integrate with respect to } x.$$

$$= \int_1^2 [(3 \cdot 5^2 \cdot y^2 + 4 \cdot 5^3 \cdot y + 4 \cdot 5 \cdot y) \qquad \text{Substitute } x = 5 \text{ and } x = 3, \text{ and subtract.}$$

$$- (3 \cdot 3^2 \cdot y^2 + 4 \cdot 3^3 \cdot y + 4 \cdot 3 \cdot y)]\, dy$$

$$= \int_1^2 [(75y^2 + 500y + 20y)$$

$$- (27y^2 + 108y + 12y)]\, dy$$

$$= \int_1^2 (48y^2 + 400y)\, dy$$

$$= (16y^3 + 200y^2) \Big|_1^2 \qquad \text{Integrate with respect to } y.$$

$$= (16 \cdot 2^3 + 200 \cdot 2^2) - (16 \cdot 1^3 + 200 \cdot 1^2)$$

$$= 128 + 800 - (16 + 200)$$

$$= 712.$$

(b) $\int_3^5 \left[\int_1^2 (6xy^2 + 12x^2y + 4y)\, dy \right] dx$

SOLUTION (This is the same integrand with the same limits of integration as in part (a), but the order of integration is reversed.)

$$\int_3^5 \left[\int_1^2 (6xy^2 + 12x^2y + 4y)dy \right] dx = \int_3^5 \left[(2xy^3 + 6x^2y^2 + 2y^2) \Big|_1^2 \right] dx \qquad \text{Integrate with respect to } y.$$

$$= \int_3^5 [(2x \cdot 2^3 + 6x^2 \cdot 2^2 + 2 \cdot 2^2) \qquad \begin{array}{l}\text{Substitute}\\ y = 2 \text{ and}\\ y = 1, \text{ and}\\ \text{subtract.}\end{array}$$

$$\qquad - (2x \cdot 1^3 + 6x^2 \cdot 1^2 + 2 \cdot 1^2)] dx$$

$$= \int_3^5 [(16x + 24x^2 + 8)$$

$$\qquad - (2x + 6x^2 + 2)]\, dx$$

$$= \int_3^5 (14x + 18x^2 + 6)\, dx$$

$$= (7x^2 + 6x^3 + 6x) \Big|_3^5 \qquad \text{Integrate with respect to } x.$$

$$= (7 \cdot 5^2 + 6 \cdot 5^3 + 6 \cdot 5) - (7 \cdot 3^2 + 6 \cdot 3^3 + 6 \cdot 3)$$

$$= 175 + 750 + 30 - (63 + 162 + 18) = 712$$

TRY YOUR TURN 2

YOUR TURN 2 Evaluate
$$\int_0^2 \left[\int_1^3 (6x^2y^2 + 4xy + 8x^3 + 10y^4 + 3)\, dy \right] dx, \text{ and then integrate}$$
with the order of integration changed.

NOTE In the second step of Example 3 (a), it might help you avoid confusion as to whether to put the limits of 3 and 5 into x or y by writing the integral as

$$\int_1^2 \left[(3x^2y^2 + 4x^3y + 4xy) \Big|_{x=3}^{x=5} \right] dy.$$

The brackets we have used for the inner integral in Example 3 are not essential because the order of integration is indicated by the order of $dx\, dy$ or $dy\, dx$. For example, if the integral is written as

$$\int_1^2 \int_3^5 (6xy^2 + 12x^2y + 4y)\, dx\, dy,$$

we first integrate with respect to x, letting x vary from 3 to 5, and then with respect to y, letting y vary from 1 to 2, as in Example 3(a).

The answers in the two parts of Example 3 are equal. It can be proved that for a large class of functions, including most functions that occur in applications, the following equation holds true.

Fubini's Theorem

$$\int_a^b \int_c^d f(x, y)\, dy\, dx = \int_c^d \int_a^b f(x, y)\, dx\, dy$$

Either of these integrals is called an **iterated integral**, since it is evaluated by integrating twice, first using one variable and then using the other. The fact that the iterated integrals above are equal makes it possible to define a *double integral*. First, the set of points (x, y), with $a \le x \le b$ and $c \le y \le d$, defines a rectangular region R in the plane, as shown in Figure 29. Then, the *double integral over R* is defined as follows.

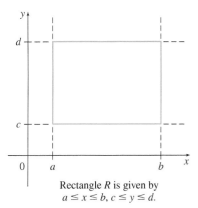

Rectangle R is given by
$a \le x \le b, c \le y \le d$.

FIGURE 29

Double Integral

The **double integral** of $f(x, y)$ over a rectangular region R is written

$$\iint_R f(x,y)\, dy\, dx \qquad \text{or} \qquad \iint_R f(x,y)\, dx\, dy,$$

and equals either

$$\int_a^b \int_c^d f(x,y)\, dy\, dx \qquad \text{or} \qquad \int_c^d \int_a^b f(x,y)\, dx\, dy.$$

Extending earlier definitions, $f(x, y)$ is the **integrand** and R is the **region of integration**.

EXAMPLE 4 Double Integrals

Find $\displaystyle \iint_R \frac{3\sqrt{x}\cdot y}{y^2 + 1}\, dx\, dy$ over the rectangular region R defined by $0 \le x \le 4$ and $0 \le y \le 2$.

SOLUTION Integrate first with respect to x; then integrate the result with respect to y.

$$\iint_R \frac{3\sqrt{x}\cdot y}{y^2 + 1}\, dx\, dy = \int_0^2 \int_0^4 \frac{3\sqrt{x}\cdot y}{y^2 + 1}\, dx\, dy$$

$$= \int_0^2 \frac{2x^{3/2}\cdot y}{y^2 + 1}\Big|_0^4\, dy \qquad \text{Use the power rule with } x^{1/2}.$$

$$= \int_0^2 \left(\frac{2(4)^{3/2}\cdot y}{y^2 + 1} - \frac{2(0)^{3/2}\cdot y}{y^2 + 1} \right) dy$$

$$= 8 \int_0^2 \frac{2y}{y^2 + 1}\, dy \qquad \text{Factor out } 4^{3/2} = 8.$$

$$= 8 \int_1^5 \frac{du}{u} \qquad \text{Let } u = y^2 + 1. \text{ Change limits of integration.}$$

$$= 8 \ln u \Big|_1^5$$

$$= 8 \ln 5 - 8 \ln 1 = 8 \ln 5$$

As a check, integrate with respect to y first. The answer should be the same.

TRY YOUR TURN 3

YOUR TURN 3 Find
$$\iint_R \frac{1}{\sqrt{x + y + 3}}\, dx\, dy$$
over the rectangular region R defined by $0 \le x \le 5$ and $1 \le y \le 6$.

Volume As shown earlier, the definite integral $\int_a^b f(x)\, dx$ can be used to find the area under a curve. In a similar manner, double integrals are used to find the *volume under a surface*. Figure 30 on the next page shows that portion of a surface $f(x, y)$ directly over a rectangle R in the xy-plane. Just as areas were approximated by a large number of small rectangles, volume could be approximated by adding the volumes of a large number of properly drawn small boxes. The height of a typical box would be $f(x, y)$ with the length and width given by dx and dy. The formula for the volume of a box would then suggest the following result.

Volume

Let $z = f(x, y)$ be a function that is never negative on the rectangular region R defined by $a \le x \le b, c \le y \le d$. The volume of the solid under the graph of f and over the region R is

$$\iint_R f(x,y)\, dx\, dy.$$

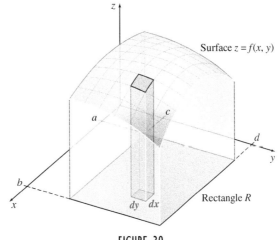

Surface $z = f(x, y)$

Rectangle R

FIGURE 30

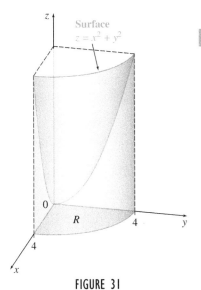

Surface
$z = x^2 + y^2$

FIGURE 31

EXAMPLE 5   Volume

Find the volume under the surface $z = x^2 + y^2$ shown in Figure 31.

SOLUTION   By the equation just given, the volume is

$$\iint_R f(x, y) \, dx \, dy,$$

where $f(x, y) = x^2 + y^2$ and R is the region $0 \le x \le 4, 0 \le y \le 4$. By definition,

$$
\begin{aligned}
\iint_R f(x, y) \, dx \, dy &= \int_0^4 \int_0^4 (x^2 + y^2) \, dx \, dy \\
&= \int_0^4 \left(\frac{1}{3} x^3 + xy^2 \right) \Big|_0^4 \, dy \qquad \text{Integrate with respect to } x. \\
&= \int_0^4 \left(\frac{64}{3} + 4y^2 - 0 \right) dy \\
&= \left(\frac{64}{3} y + \frac{4}{3} y^3 \right) \Big|_0^4 \qquad \text{Integrate with respect to } y. \\
&= \frac{64}{3} \cdot 4 + \frac{4}{3} \cdot 4^3 - 0 = \frac{512}{3}.
\end{aligned}
$$

TRY YOUR TURN 4

YOUR TURN 4   Find the volume under the surface $z = 4 - x^3 - y^3$ over the rectangular region $0 \le x \le 1, 0 \le y \le 1$.

4 cm   3 cm

FIGURE 32

EXAMPLE 6   Perfume Bottle

A product design consultant for a cosmetics company has been asked to design a bottle for the company's newest perfume. The thickness of the glass is to vary so that the outside of the bottle has straight sides and the inside has curved sides, with flat ends shaped like parabolas on the 4-cm sides, as shown in Figure 32. Before presenting the design to management, the consultant needs to make a reasonably accurate estimate of the amount each bottle will hold. If the base of the bottle is to be 4 cm by 3 cm, and if a cross section of its interior is to be a parabola of the form $z = -y^2 + 4y$, what is its internal volume?

APPLY IT

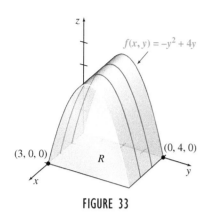

FIGURE 33

SOLUTION The interior of the bottle can be graphed in three-dimensional space, as shown in Figure 33, where $z = 0$ corresponds to the base of the bottle. Its volume is simply the volume above the region R in the xy-plane and below the graph of $f(x, y) = -y^2 + 4y$. This volume is given by the double integral

$$\int_0^3 \int_0^4 (-y^2 + 4y) \, dy \, dx = \int_0^3 \left(\frac{-y^3}{3} + \frac{4y^2}{2} \right) \Big|_0^4 dx \qquad \text{Integrate with respect to } y.$$

$$= \int_0^3 \left(\frac{-64}{3} + 32 - 0 \right) dx$$

$$= \frac{32}{3} x \Big|_0^3 \qquad \text{Integrate with respect to } x.$$

$$= 32 - 0 = 32.$$

The bottle holds 32 cm³.

Double Integrals Over Other Regions

In this section, we found double integrals over rectangular regions by evaluating iterated integrals with constant limits of integration. We can also evaluate iterated integrals with *variable* limits of integration. (Notice in the following examples that the variable limits always go on the *inner* integral sign.)

The use of variable limits of integration permits evaluation of double integrals over the types of regions shown in Figure 34. Double integrals over more complicated regions are discussed in more advanced books. Integration over regions such as those in Figure 34 is done with the results of the following theorem.

Double Integrals over Variable Regions

Let $z = f(x, y)$ be a function of two variables. If R is the region (in Figure 34(a)) defined by $a \le x \le b$ and $g(x) \le y \le h(x)$, then

$$\iint_R f(x, y) \, dy \, dx = \int_a^b \int_{g(x)}^{h(x)} f(x, y) \, dy \, dx.$$

If R is the region (in Figure 34(b)) defined by $g(y) \le x \le h(y)$ and $c \le y \le d$, then

$$\iint_R f(x, y) \, dx \, dy = \int_c^d \int_{g(y)}^{h(y)} f(x, y) \, dx \, dy.$$

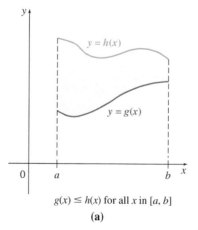

$g(x) \le h(x)$ for all x in $[a, b]$

(a)

$g(y) \le h(y)$ for all y in $[c, d]$

(b)

FIGURE 34

EXAMPLE 7   Double Integrals

Evaluate $\displaystyle\int_1^2\int_y^{y^2} xy \, dx \, dy$.

SOLUTION   The region of integration is shown in Figure 35. Integrate first with respect to x, then with respect to y.

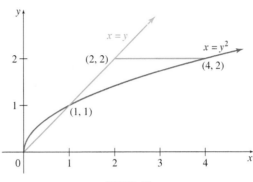

FIGURE 35

$$\int_1^2\int_y^{y^2} xy \, dx \, dy = \int_1^2\left(\int_y^{y^2} xy \, dx\right) dy = \int_1^2\left(\frac{1}{2}x^2y\right)\Big|_y^{y^2} dy$$

Replace x first with y^2 and then with y, and subtract.

$$\int_1^2\int_y^{y^2} xy \, dx \, dy = \int_1^2\left[\frac{1}{2}(y^2)^2y - \frac{1}{2}(y)^2y\right] dy$$

$$= \int_1^2\left(\frac{1}{2}y^5 - \frac{1}{2}y^3\right) dy = \left(\frac{1}{12}y^6 - \frac{1}{8}y^4\right)\Big|_1^2$$

$$= \left(\frac{1}{12}\cdot 2^6 - \frac{1}{8}\cdot 2^4\right) - \left(\frac{1}{12}\cdot 1^6 - \frac{1}{8}\cdot 1^4\right)$$

$$= \frac{64}{12} - \frac{16}{8} - \frac{1}{12} + \frac{1}{8} = \frac{27}{8}$$

YOUR TURN 5   Find $\displaystyle\iint_R (x^3 + 4y) \, dy \, dx$ over the region bounded by $y = 4x$ and $y = x^3$ for $0 \le x \le 2$.

TRY YOUR TURN 5

EXAMPLE 8   Double Integrals

Let R be the shaded region in Figure 36, and evaluate

$$\iint_R (x + 2y) \, dy \, dx.$$

SOLUTION   Region R is bounded by $h(x) = 2x$ and $g(x) = x^2$, with $0 \le x \le 2$. By the first result in the previous theorem,

$$\iint_R (x + 2y) \, dy \, dx = \int_0^2\int_{x^2}^{2x} (x + 2y) \, dy \, dx$$

$$= \int_0^2 (xy + y^2)\Big|_{x^2}^{2x} dx \qquad \text{Integrate with respect to } y.$$

$$= \int_0^2 [x(2x) + (2x)^2 - [x\cdot x^2 + (x^2)^2]] \, dx$$

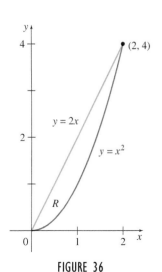

FIGURE 36

$$= \int_0^2 [2x^2 + 4x^2 - (x^3 + x^4)] \, dx$$

$$= \int_0^2 (6x^2 - x^3 - x^4) \, dx$$

$$= \left(2x^3 - \frac{1}{4}x^4 - \frac{1}{5}x^5 \right) \Big|_0^2 \qquad \text{Integrate with respect to } x.$$

$$= 2 \cdot 2^3 - \frac{1}{4} \cdot 2^4 - \frac{1}{5} \cdot 2^5 - 0$$

$$= 16 - 4 - \frac{32}{5} = \frac{28}{5}.$$

Interchanging Limits of Integration

Sometimes it is easier to integrate first with respect to x and then y, while with other integrals the reverse process is easier. The limits of integration can be reversed whenever the region R is like the region in Figure 36, which has the property that it can be viewed as either type of region shown in Figure 34. In practice, this means that all boundaries can be written in terms of y as a function of x, or by solving for x as a function of y.

For instance, in Example 8, the same result would be found if we evaluated the double integral first with respect to x and then with respect to y. In that case, we would need to define the equations of the boundaries in terms of y rather than x, so R would be defined by $y/2 \le x \le \sqrt{y}, 0 \le y \le 4$. The resulting integral is

$$\int_0^4 \int_{y/2}^{\sqrt{y}} (x + 2y) \, dx \, dy = \int_0^4 \left(\frac{x^2}{2} + 2xy \right) \Big|_{y/2}^{\sqrt{y}} dy \qquad \text{Integrate with respect to } x.$$

$$= \int_0^4 \left[\left(\frac{y}{2} + 2y\sqrt{y} \right) - \left(\frac{y^2}{8} + 2 \left(\frac{y}{2} \right) y \right) \right] dy$$

$$= \int_0^4 \left(\frac{y}{2} + 2y^{3/2} - \frac{9}{8}y^2 \right) dy$$

$$= \left(\frac{y^2}{4} + \frac{4}{5}y^{5/2} - \frac{3}{8}y^3 \right) \Big|_0^4 \qquad \text{Integrate with respect to } y.$$

$$= 4 + \frac{4}{5} \cdot 4^{5/2} - 24 - 0$$

$$= \frac{28}{5}.$$

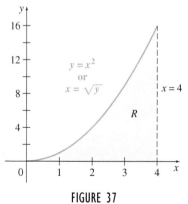

FIGURE 37

EXAMPLE 9 Interchanging Limits of Integration

Evaluate

$$\int_0^{16} \int_{\sqrt{y}}^4 \sqrt{x^3 + 4} \, dx \, dy.$$

SOLUTION Notice that it is impossible to first integrate this function with respect to x. Thus, we attempt to interchange the limits of integration.

For this integral, region R is given by $\sqrt{y} \le x \le 4, 0 \le y \le 16$. A graph of R is shown in Figure 37.

The same region R can be written in an alternate way. As Figure 37 shows, one boundary of R is $x = \sqrt{y}$. Solving for y gives $y = x^2$. Also, Figure 37 shows that $0 \le x \le 4$. Since R can be written as $0 \le y \le x^2, 0 \le x \le 4$, the double integral above can be written

$$\int_0^4 \int_0^{x^2} \sqrt{x^3 + 4}\, dy\, dx = \int_0^4 y\sqrt{x^3 + 4}\, \Big|_0^{x^2} dx \qquad \text{Integrate with respect to } y.$$

$$= \int_0^4 x^2\sqrt{x^3 + 4}\, dx$$

$$= \frac{1}{3} \int_0^4 3x^2\sqrt{x^3 + 4}\, dx$$

$$= \frac{1}{3} \int_4^{68} u^{1/2}\, du \qquad \text{Let } u = x^3 + 4. \text{ Change limits of integration.}$$

$$= \frac{2}{9} u^{3/2} \Big|_4^{68}$$

$$= \frac{2}{9}[68^{3/2} - 4^{3/2}]$$

$$\approx 122.83.$$

CAUTION Fubini's Theorem cannot be used to interchange the order of integration when the limits contain variables, as in Example 9. Notice in Example 9 that after the order of integration was changed, the new limits were completely different. It would be a serious error to rewrite the integral in Example 9 as

$$\int_{\sqrt{y}}^4 \int_0^{16} \sqrt{x^3 + 4}\, dy\, dx.$$

9.6 WARM-UP EXERCISES

Evaluate each definite integral.   *(Sec. 7.4)*

W1. $\displaystyle\int_0^2 (3x^3 - 6x^2)\, dx$

W2. $\displaystyle\int_{-2}^3 \sqrt{x + 6}\, dx$

W3. $\displaystyle\int_2^4 \frac{3}{x}\, dx$

W4. $\displaystyle\int_0^1 e^{6x}\, dx$

W5. $\displaystyle\int_1^2 x^2 e^{x^3}\, dx$

W6. $\displaystyle\int_0^4 x\sqrt{x^2 + 9}\, dx$

9.6   EXERCISES

Evaluate each integral.

1. $\displaystyle\int_0^5 (x^4 y + y)\, dx$

2. $\displaystyle\int_1^2 (xy^3 - x)\, dy$

3. $\displaystyle\int_4^5 x\sqrt{x^2 + 3y}\, dy$

4. $\displaystyle\int_3^6 x\sqrt{x^2 + 3y}\, dx$

5. $\displaystyle\int_4^9 \frac{3 + 5y}{\sqrt{x}}\, dx$

6. $\displaystyle\int_2^7 \frac{3 + 5y}{\sqrt{x}}\, dy$

7. $\displaystyle\int_2^6 e^{2x + 3y}\, dx$

8. $\displaystyle\int_{-1}^1 e^{2x + 3y}\, dy$

9. $\displaystyle\int_0^3 y e^{4x + y^2}\, dy$

10. $\displaystyle\int_1^5 y e^{4x + y^2}\, dx$

Evaluate each iterated integral. (Many of these use results from Exercises 1–10.)

11. $\int_1^2 \int_0^5 (x^4 y + y)\, dx\, dy$

12. $\int_0^3 \int_1^2 (xy^3 - x)\, dy\, dx$

13. $\int_0^1 \int_3^6 x\sqrt{x^2 + 3y}\, dx\, dy$

14. $\int_0^3 \int_4^5 x\sqrt{x^2 + 3y}\, dy\, dx$

15. $\int_1^2 \int_4^9 \frac{3 + 5y}{\sqrt{x}}\, dx\, dy$

16. $\int_{16}^{25} \int_2^7 \frac{3 + 5y}{\sqrt{x}}\, dy\, dx$

17. $\int_1^3 \int_1^3 \frac{1}{xy}\, dy\, dx$

18. $\int_1^5 \int_2^4 \frac{1}{y}\, dx\, dy$

19. $\int_2^4 \int_3^5 \left(\frac{x}{y} + \frac{y}{3}\right) dx\, dy$

20. $\int_3^4 \int_1^2 \left(\frac{6x}{5} + \frac{y}{x}\right) dx\, dy$

Find each double integral over the rectangular region R with the given boundaries.

21. $\iint_R (3x^2 + 4y)\, dx\, dy; \quad 0 \le x \le 3, 1 \le y \le 4$

22. $\iint_R (x^2 + 4y^3)\, dy\, dx; \quad 1 \le x \le 2, 0 \le y \le 3$

23. $\iint_R \sqrt{x + y}\, dy\, dx; \quad 1 \le x \le 3, 0 \le y \le 1$

24. $\iint_R x^2 \sqrt{x^3 + 2y}\, dx\, dy; \quad 0 \le x \le 2, 0 \le y \le 3$

25. $\iint_R \frac{3}{(x + y)^2}\, dy\, dx; \quad 2 \le x \le 4, 1 \le y \le 6$

26. $\iint_R \frac{y}{\sqrt{2x + 5y^2}}\, dx\, dy; \quad 0 \le x \le 2, 1 \le y \le 3$

27. $\iint_R ye^{x+y^2}\, dx\, dy; \quad 2 \le x \le 3, 0 \le y \le 2$

28. $\iint_R x^2 e^{x^3 + 2y}\, dx\, dy; \quad 1 \le x \le 2, 1 \le y \le 3$

Find the volume under the given surface $z = f(x, y)$ and above the rectangle with the given boundaries.

29. $z = 8x + 4y + 10; \; -1 \le x \le 1, 0 \le y \le 3$

30. $z = 3x + 10y + 20; \quad 0 \le x \le 3, -2 \le y \le 1$

31. $z = x^2; \; 0 \le x \le 2, 0 \le y \le 5$

32. $z = \sqrt{y}; \; 0 \le x \le 4, 0 \le y \le 9$

33. $z = x\sqrt{x^2 + y}; \; 0 \le x \le 1, 0 \le y \le 1$

34. $z = yx\sqrt{x^2 + y^2}; \; 0 \le x \le 4, 0 \le y \le 1$

35. $z = \frac{xy}{(x^2 + y^2)^2}; \quad 1 \le x \le 2, 1 \le y \le 4$

36. $z = e^{x+y}; \; 0 \le x \le 1, 0 \le y \le 1$

Although it is often true that a double integral can be evaluated by using either dx or dy first, sometimes one choice over the other makes the work easier. Evaluate the double integrals in Exercises 37 and 38 in the easiest way possible.

37. $\iint_R xe^{xy}\, dx\, dy; \quad 0 \le x \le 2, 0 \le y \le 1$

38. $\iint_R 2x^3 e^{x^2 y}\, dx\, dy; \quad 0 \le x \le 1, 0 \le y \le 1$

Evaluate each double integral.

39. $\int_2^4 \int_2^{x^2} (x^2 + y^2)\, dy\, dx$

40. $\int_0^2 \int_0^{3y} (x^2 + y)\, dx\, dy$

41. $\int_0^4 \int_0^x \sqrt{xy}\, dy\, dx$

42. $\int_1^4 \int_0^x \sqrt{x + y}\, dy\, dx$

43. $\int_2^6 \int_{2y}^{4y} \frac{1}{x}\, dx\, dy$

44. $\int_1^4 \int_x^{x^2} \frac{1}{y}\, dy\, dx$

45. $\int_0^4 \int_1^{e^x} \frac{x}{y}\, dy\, dx$

46. $\int_0^1 \int_{2x}^{4x} e^{x+y}\, dy\, dx$

Use the region R with the indicated boundaries to evaluate each double integral.

47. $\iint_R (5x + 8y)\, dy\, dx; \quad 1 \le x \le 3, 0 \le y \le x - 1$

48. $\iint_R (2x + 6y)\, dy\, dx; \quad 2 \le x \le 4, 2 \le y \le 3x$

49. $\iint_R (4 - 4x^2)\, dy\, dx; \quad 0 \le x \le 1, 0 \le y \le 2 - 2x$

50. $\iint_R \frac{1}{x}\, dy\, dx; \quad 1 \le x \le 2, 0 \le y \le x - 1$

51. $\iint_R e^{x/y^2}\, dx\, dy; \quad 1 \le y \le 2, 0 \le x \le y^2$

52. $\iint_R (x^2 - y)\, dy\, dx; \quad -1 \le x \le 1, -x^2 \le y \le x^2$

53. $\iint_R x^3 y\, dy\, dx; \quad R \text{ bounded by } y = x^2, y = 2x$

54. $\iint_R x^2 y^2\, dx\, dy; \quad R \text{ bounded by } y = x, y = 2x, x = 1$

55. $\iint_R \frac{1}{y}\, dy\, dx; \quad R \text{ bounded by } y = x, y = \frac{1}{x}, x = 2$

56. $\iint_R e^{2y/x}\, dy\, dx; \quad R \text{ bounded by } y = x^2, y = 0, x = 2$

Evaluate each double integral. If the function seems too difficult to integrate, try interchanging the limits of integration, as in Exercises 37 and 38.

57. $\int_0^{\ln 2} \int_{e^y}^2 \frac{1}{\ln x}\, dx\, dy$

58. $\int_0^2 \int_{y/2}^1 e^{x^2}\, dx\, dy$

59. Recall from the Volume and Average Value section in the previous chapter that volume could be found with a single integral. In this section volume is found using a double integral. Explain when volume can be found with a single integral and when a double integral is needed.

60. Give an example of a region that cannot be expressed by either of the forms shown in Figure 34. (One example is the disk with a hole in the middle between the graphs of $x^2 + y^2 = 1$ and $x^2 + y^2 = 2$ in Figure 10.)

The idea of the average value of a function, discussed earlier for functions of the form $y = f(x)$, can be extended to functions of more than one independent variable. For a function $z = f(x, y)$, the average value of f over a region R is defined as

$$\frac{1}{A} \iint\limits_{R} f(x, y) \, dx \, dy,$$

where A is the area of the region R. Find the average value for each function over the regions R having the given boundaries.

61. $f(x, y) = 6xy + 2x; \quad 2 \leq x \leq 5, 1 \leq y \leq 3$

62. $f(x, y) = x^2 + y^2; \quad 0 \leq x \leq 2, 0 \leq y \leq 3$

63. $f(x, y) = e^{-5y + 3x}; \quad 0 \leq x \leq 2, 0 \leq y \leq 2$

64. $f(x, y) = e^{2x + y}; \quad 1 \leq x \leq 2, 2 \leq y \leq 3$

APPLICATIONS

Business and Economics

65. Packaging   The manufacturer of a fruit juice drink has decided to try innovative packaging in order to revitalize sagging sales. The fruit juice drink is to be packaged in containers in the shape of tetrahedra in which three edges are perpendicular, as shown in the figure below. Two of the perpendicular edges will be 3 in. long, and the third edge will be 6 in. long. Find the volume of the container. (*Hint:* The equation of the plane shown in the figure is $z = f(x, y) = 6 - 2x - 2y$.)

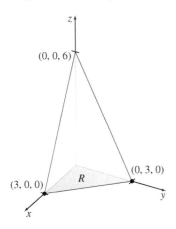

66. Average Cost   A company's total cost for operating its two warehouses is

$$C(x, y) = \frac{1}{9}x^2 + 2x + y^2 + 5y + 100$$

dollars, where x represents the number of units stored at the first warehouse and y represents the number of units stored at the

second. Find the average cost to store a unit if the first warehouse has between 40 and 80 units, and the second has between 30 and 70 units. (*Hint:* Refer to Exercises 61–64.)

67. Average Production   A production function is given by

$$P(x, y) = 500x^{0.2}y^{0.8},$$

where x is the number of units of labor and y is the number of units of capital. Find the average production level if x varies from 10 to 50 and y from 20 to 40. (*Hint:* Refer to Exercises 61–64.)

68. Average Profit   The profit (in dollars) from selling x units of one product and y units of a second product is

$$P = -(x - 100)^2 - (y - 50)^2 + 2000.$$

The weekly sales for the first product vary from 100 units to 150 units, and the weekly sales for the second product vary from 40 units to 80 units. Estimate average weekly profit for these two products. (*Hint:* Refer to Exercises 61–64.)

69. Average Revenue   A company sells two products. The demand functions of the products are given by

$$q_1 = 300 - 2p_1 \quad \text{and} \quad q_2 = 500 - 1.2p_2,$$

where q_1 units of the first product are demanded at price p_1 and q_2 units of the second product are demanded at price p_2. The total revenue will be given by

$$R = q_1p_1 + q_2p_2.$$

Find the average revenue if the price p_1 varies from \$25 to \$50 and the price p_2 varies from \$50 to \$75. (*Hint:* Refer to Exercises 61–64.)

70. Time   In Exercise 39 of Section 9.3, we saw that the time (in hours) that a branch of Amalgamated Entities needs to spend to meet the quota set by the main office can be approximated by

$$T(x, y) = x^4 + 16y^4 - 32xy + 40,$$

where x represents how many thousands of dollars the factory spends on quality control and y represents how many thousands of dollars they spend on consulting. Find the average time if the amount spent on quality control varies from \$0 to \$4000 and the amount spent on consulting varies from \$0 to \$2000. (*Hint:* Refer to Exercises 61–64.)

71. Profit   In Exercise 38 of Section 9.3, we saw that the profit (in thousands of dollars) that Aunt Mildred's Metalworks earns from producing x tons of steel and y tons of aluminum can be approximated by

$$P(x, y) = 36xy - x^3 - 8y^3.$$

Find the average profit if the amount of steel produced varies from 0 to 8 tons, and the amount of aluminum produced varies from 0 to 4 tons. (*Hint:* Refer to Exercises 61–64.)

9

CHAPTER REVIEW

SUMMARY

In this chapter, we extended our study of calculus to include functions of several variables. We saw that it is possible to produce three-dimensional graphs of functions of two variables and that the process is greatly enhanced using level curves. Level curves are formed by determining the values of x and y that produce a particular functional value. We also saw the graphs of several surfaces, including the

- paraboloid, whose equation is $z = x^2 + y^2$,
- ellipsoid, whose general equation is $\dfrac{x^2}{a^2} + \dfrac{y^2}{b^2} + \dfrac{z^2}{c^2} = 1$,
- hyperbolic paraboloid, whose equation is $z = x^2 - y^2$, and
- hyperboloid of two sheets, whose equation is $-x^2 - y^2 + z^2 = 1$.

Level curves are also important in economics and are used to indicate combinations of the values of x and y that produce the same value of production z. This procedure was used to analyze the Cobb-Douglas production function, which has the general form

$$z = P(x, y) = Ax^a y^{1-a}, \text{ where } A \text{ is constant and } 0 < a < 1.$$

Partial derivatives are the extension of the concept of differentiation with respect to one of the variables while the other variables are held constant. Partial derivatives were used to identify extrema of a function of several variables. In particular, we identified all points where the partial with respect to x and the partial with respect to y are both zero, which we called critical points. We then classified each critical point as a relative maximum, a relative minimum, or a saddle point. Recall that a saddle point is a minimum when approached from one direction but a maximum when approached from another direction. We introduced the method of Lagrange multipliers to determine extrema in problems with constraints. Differentials, introduced earlier for functions of one variable, were generalized to define the total differential. We saw that total differentials can be used to approximate the value of a function using its tangent plane. We concluded the chapter by introducing double integrals, which are simply two iterated integrals, one for each variable. Double integrals were then used to find volume.

Function of Two Variables	The expression $z = f(x, y)$ is a function of two variables if a unique value of z is obtained from each ordered pair of real numbers (x, y). The variables x and y are independent variables, and z is the dependent variable. The set of all ordered pairs of real numbers (x, y) such that $f(x, y)$ exists is the domain of f; the set of all values of $f(x, y)$ is the range.
Plane	The graph of $ax + by + cz = d$ is a plane if a, b, and c are not all 0.
Partial Derivatives (Informal Definition)	The partial derivative of f with respect to x is the derivative of f obtained by treating x as a variable and y as a constant. The partial derivative of f with respect to y is the derivative of f obtained by treating y as a variable and x as a constant.
Partial Derivatives (Formal Definition)	Let $z = f(x, y)$ be a function of two independent variables. Let all indicated limits exist. Then the partial derivative of f with respect to x is $$f_x(x, y) = \frac{\partial f}{\partial x} = \lim_{h \to 0} \frac{f(x + h, y) - f(x, y)}{h},$$ and the partial derivative of f with respect to y is $$f_y(x, y) = \frac{\partial f}{\partial y} = \lim_{h \to 0} \frac{f(x, y + h) - f(x, y)}{h}.$$ If the indicated limits do not exist, then the partial derivatives do not exist.
Second-Order Partial Derivatives	For a function $z = f(x, y)$, if the partial derivative exists, then $$\frac{\partial}{\partial x}\left(\frac{\partial z}{\partial x}\right) = \frac{\partial^2 z}{\partial x^2} = f_{xx}(x, y) = z_{xx} \qquad \frac{\partial}{\partial y}\left(\frac{\partial z}{\partial y}\right) = \frac{\partial^2 z}{\partial y^2} = f_{yy}(x, y) = z_{yy}$$ $$\frac{\partial}{\partial y}\left(\frac{\partial z}{\partial x}\right) = \frac{\partial^2 z}{\partial y \partial x} = f_{xy}(x, y) = z_{xy} \qquad \frac{\partial}{\partial x}\left(\frac{\partial z}{\partial y}\right) = \frac{\partial^2 z}{\partial x \partial y} = f_{yx}(x, y) = z_{yx}$$

Relative Extrema   Let (a, b) be the center of a circular region contained in the xy-plane. Then, for a function $z = f(x, y)$ defined for every (x, y) in the region, $f(a, b)$ is a relative maximum if

$$f(a, b) \geq f(x, y)$$

for all points (x, y) in the circular region, and $f(a, b)$ is a relative minimum if

$$f(a, b) \leq f(x, y)$$

for all points (x, y) in the circular region.

Location of Extrema   Let a function $z = f(x, y)$ have a relative maximum or relative minimum at the point (a, b). Let $f_x(a, b)$ and $f_y(a, b)$ both exist. Then

$$f_x(a, b) = 0 \text{ and } f_y(a, b) = 0.$$

Test for Relative Extrema   For a function $z = f(x, y)$, let f_{xx}, f_{yy}, and f_{xy} all exist in a circular region contained in the xy-plane with center (a, b). Further, let

$$f_x(a, b) = 0 \text{ and } f_y(a, b) = 0.$$

Define D, known as the discriminant, by

$$D = f_{xx}(a, b) \cdot f_{yy}(a, b) - [f_{xy}(a, b)]^2.$$

Then

(a) $f(a, b)$ is a relative maximum if $D > 0$ and $f_{xx}(a, b) < 0$;
(b) $f(a, b)$ is a relative minimum if $D > 0$ and $f_{xx}(a, b) > 0$;
(c) $f(a, b)$ is a saddle point (neither a maximum nor a minimum) if $D < 0$;
(d) if $D = 0$, the test gives no information.

Lagrange Multipliers   All relative extrema of the function $z = f(x, y)$, subject to the constraint $g(x, y) = 0$, will be found among those points (x, y) for which there exists a value of λ such that

$$F_x(x, y, \lambda) = 0, F_y(x, y, \lambda) = 0, \text{ and } F_\lambda(x, y, \lambda) = 0,$$

where

$$F(x, y, \lambda) = f(x, y) - \lambda \cdot g(x, y),$$

and all indicated partial derivatives exist.

Using Lagrange Multipliers   1. Write the constraint in the form $g(x, y) = 0$.

2. Form the Lagrange function

$$F(x, y, \lambda) = f(x, y) - \lambda \cdot g(x, y).$$

3. Find $F_x(x, y, \lambda)$, $F_y(x, y, \lambda)$, and $F_\lambda(x, y, \lambda)$.
4. Form the system of equations

$$F_x(x, y, \lambda) = 0, F_y(x, y, \lambda) = 0, \text{ and } F_\lambda(x, y, \lambda) = 0.$$

5. Solve the system in Step 4; the relative extrema for f are among the solutions of the system.

Total Differential for Two Variables   Let $z = f(x, y)$ be a function of x and y. Let dx and dy be real numbers. Then the total differential of z is

$$dz = f_x(x, y) \cdot dx + f_y(x, y) \cdot dy.$$

(Sometimes dz is written df.)

Approximations   For small values of dx and dy,

$$dz \approx \Delta z$$

where $\Delta z = f(x + dx, y + dy) - f(x, y).$

Approximations by Differentials
For a function f having all indicated partial derivatives, and for small values of dx and dy,

$$f(x + dx, y + dy) \approx f(x, y) + dz,$$

or

$$f(x + dx, y + dy) \approx f(x, y) + f_x(x, y) \cdot dx + f_y(x, y) \cdot dy.$$

Total Differential for Three Variables
If $w = f(x, y, z)$, then the total differential dw is

$$dw = f_x(x, y, z) \cdot dx + f_y(x, y, z) \cdot dy + f_z(x, y, z) \cdot dz,$$

provided all indicated partial derivatives exist.

Double Integral
The double integral of $f(x, y)$ over a rectangular region R defined by $a \leq x \leq b, c \leq y \leq d$ is written

$$\iint_R f(x, y) \, dy \, dx \quad \text{or} \quad \iint_R f(x, y) \, dx \, dy,$$

and equals either

$$\int_a^b \int_c^d f(x, y) \, dy \, dx \quad \text{or} \quad \int_c^d \int_a^b f(x, y) \, dx \, dy.$$

Volume
Let $z = f(x, y)$ be a function that is never negative on the rectangular region R defined by $a \leq x \leq b, c \leq y \leq d$. The volume of the solid under the graph of f and over the region R is

$$\iint_R f(x, y) \, dy \, dx.$$

Double Integrals over Variable Regions
Let $z = f(x, y)$ be a function of two variables. If R is the region defined by $a \leq x \leq b$ and $g(x) \leq y \leq h(x)$, then

$$\iint_R f(x, y) \, dy \, dx = \int_a^b \int_{g(x)}^{h(x)} f(x, y) \, dy \, dx.$$

If R is the region defined by $g(y) \leq x \leq h(y)$ and $c \leq y \leq d$, then

$$\iint_R f(x, y) \, dx \, dy = \int_c^d \int_{g(y)}^{h(y)} f(x, y) \, dx \, dy.$$

KEY TERMS

9.1
function of two variables
independent variable
dependent variable
domain
range
ordered triple
first octant
plane
surface
trace
level curves

paraboloid
production function
Cobb-Douglas production
   function
level surface
ellipsoid
hyperbolic paraboloid
hyperboloid of two sheets

9.2
partial derivative
second-order partial
   derivative

9.3
relative maximum
relative minimum
saddle point
critical point
discriminant

9.4
constraints
Lagrange multiplier
utility function
indifference curve

9.5
total differential

9.6
Fubini's Theorem
iterated integral
double integral
integrand
region of integration

REVIEW EXERCISES

CONCEPT CHECK

Determine whether each of the following statements is true or false, and explain why.

1. The graph of $6x - 2y + 7z = 14$ is a plane.

2. The graph of $2x + 4y = 10$ is a plane that is parallel to the z-axis.

3. A level curve for a paraboloid could be a single point.

4. If the partial derivatives with respect to x and y at some point are both 0, the tangent plane to the function at that point is horizontal.

5. If $f(x, y) = 3x^2 + 2xy + y^2$, then $f(x + h, y) = 3(x + h)^2 + 2xy + h + y^2$.

6. For a function $z = f(x, y)$, suppose that the point (a, b) has been identified such that $f_x(a, b) = f_y(a, b) = 0$. We can conclude that a relative maximum or a relative minimum must exist at (a, b).

7. A saddle point can be a relative maximum or a relative minimum.

8. A function of two variables may have both a relative maximum and an absolute maximum at the same point.

9. The method of Lagrange multipliers tells us whether a point identified by the method is a maximum or minimum.

10. $\displaystyle\int_2^4 \int_1^5 (3x + 4y)\, dy\, dx = \int_2^4 \int_1^5 (3x + 4y)\, dx\, dy$

11. $\displaystyle\int_0^1 \int_{-2}^2 xe^y\, dy\, dx = \int_{-2}^2 \int_0^1 xe^y\, dx\, dy$

12. $\displaystyle\int_0^4 \int_1^x (x + xy^2)\, dy\, dx = \int_1^x \int_0^4 (x + xy^2)\, dx\, dy$

PRACTICE AND EXPLORATIONS

13. Describe in words how to take a partial derivative.

14. Describe what a partial derivative means geometrically.

15. Describe what a total differential is and how it is useful.

16. Suppose you are walking through the region of New York state shown in the topographical map in Figure 11 in the first section of this chapter. Assume you are heading north, toward the top of the map, over the western side of the mountain at the left, but not directly over the peak. Explain why you reach your highest point when you are going in the same direction as a contour line. Explain how this relates to Lagrange multipliers. (*Hint:* See Figure 25.)

Find $f(-1, 2)$ and $f(6, -3)$ for the following.

17. $f(x, y) = -4x^2 + 6xy - 3$

18. $f(x, y) = 2x^2y^2 - 7x + 4y$

19. $f(x, y) = \dfrac{x - 2y}{x + 5y}$

20. $f(x, y) = \dfrac{\sqrt{x^2 + y^2}}{x - y}$

Graph the first-octant portion of each plane.

21. $x + y + z = 4$

22. $x + 2y + 6z = 6$

23. $5x + 2y = 10$

24. $4x + 3z = 12$

25. $x = 3$

26. $y = 4$

27. Let $z = f(x, y) = 3x^3 + 4x^2y - 2y^2$. Find the following.

(a) $\dfrac{\partial z}{\partial x}$ (b) $\dfrac{\partial z}{\partial y}(-1, 4)$ (c) $f_{xy}(2, -1)$

28. Let $z = f(x, y) = \dfrac{x + y^2}{x - y^2}$. Find the following.

(a) $\dfrac{\partial z}{\partial y}$ (b) $\dfrac{\partial z}{\partial x}(0, 2)$ (c) $f_{xx}(-1, 0)$

Find $f_x(x, y)$ and $f_y(x, y)$.

29. $f(x, y) = 6x^2y^3 - 4y$

30. $f(x, y) = 5x^4y^3 - 6x^5y$

31. $f(x, y) = \sqrt{4x^2 + y^2}$

32. $f(x, y) = \dfrac{2x + 5y^2}{3x^2 + y^2}$

33. $f(x, y) = x^3e^{3y}$

34. $f(x, y) = (y - 2)^2e^{x+2y}$

35. $f(x, y) = \ln|2x^2 + y^2|$

36. $f(x, y) = \ln|2 - x^2y^3|$

Find $f_{xx}(x, y)$ and $f_{xy}(x, y)$.

37. $f(x, y) = 5x^3y - 6xy^2$

38. $f(x, y) = -3x^2y^3 + x^3y$

39. $f(x, y) = \dfrac{3x}{2x - y}$

40. $f(x, y) = \dfrac{3x + y}{x - 1}$

41. $f(x, y) = 4x^2e^{2y}$

42. $f(x, y) = ye^{x^2}$

43. $f(x, y) = \ln|2 - x^2y|$

44. $f(x, y) = \ln|1 + 3xy^2|$

Find all points where the functions defined below have any relative extrema. Find any saddle points.

45. $z = 2x^2 - 3y^2 + 12y$

46. $z = x^2 + y^2 + 9x - 8y + 1$

47. $f(x, y) = x^2 + 3xy - 7x + 5y^2 - 16y$

48. $z = x^3 - 8y^2 + 6xy + 4$

49. $z = \dfrac{1}{2}x^2 + \dfrac{1}{2}y^2 + 2xy - 5x - 7y + 10$

50. $f(x, y) = 2x^2 + 4xy + 4y^2 - 3x + 5y - 15$

51. $z = x^3 + y^2 + 2xy - 4x - 3y - 2$

52. $f(x, y) = 7x^2 + y^2 - 3x + 6y - 5xy$

53. Describe the different types of points that might occur when $f_x(x, y) = f_y(x, y) = 0$.

Use Lagrange multipliers to find the extrema of the functions defined in Exercises 54 and 55.

54. $f(x, y) = x^2y; \quad x + y = 4$

55. $f(x, y) = x^2 + y^2; \quad x = y - 6$

56. Find positive numbers x and y, whose sum is 80, such that x^2y is maximized.

57. Find positive numbers x and y, whose sum is 75, such that xy^2 is maximized.

58. Notice in the previous two exercises that we specified that x and y must be positive numbers. Does a maximum exist without this requirement? Explain why or why not.

Evaluate dz using the given information.

59. $z = 6x^2 - 7y^2 + 4xy$; $\quad x = 3, y = -1, dx = 0.03, dy = 0.01$

60. $z = \dfrac{x + 5y}{x - 2y}$; $\quad x = 1, y = -2, dx = -0.04, dy = 0.02$

Use the total differential to approximate each quantity. Then use a calculator to approximate the quantity, and give the absolute value of the difference in the two results to 4 decimal places.

61. $\sqrt{5.1^2 + 12.05^2}$

62. $\sqrt{4.06}\, e^{0.04}$

Evaluate the following.

63. $\displaystyle\int_1^4 \dfrac{4y - 3}{\sqrt{x}}\, dx$

64. $\displaystyle\int_1^5 e^{3x+5y}\, dx$

65. $\displaystyle\int_0^5 \dfrac{6x}{\sqrt{4x^2 + 2y^2}}\, dx$

66. $\displaystyle\int_1^3 \dfrac{y^2}{\sqrt{7x + 11y^3}}\, dy$

Evaluate each iterated integral.

67. $\displaystyle\int_0^2 \int_0^4 (x^2y^2 + 5x)\, dx\, dy$

68. $\displaystyle\int_0^3 \int_0^5 (2x + 6y + y^2)\, dy\, dx$

69. $\displaystyle\int_3^4 \int_2^5 \sqrt{6x + 3y}\, dx\, dy$

70. $\displaystyle\int_1^2 \int_3^5 e^{2x-7y}\, dx\, dy$

71. $\displaystyle\int_2^4 \int_2^4 \dfrac{1}{y}\, dx\, dy$

72. $\displaystyle\int_1^2 \int_1^2 \dfrac{1}{x}\, dx\, dy$

Find each double integral over the region R with boundaries as indicated.

73. $\displaystyle\iint_R (x^2 + 2y^2)\, dx\, dy$; $\quad 0 \le x \le 5, 0 \le y \le 2$

74. $\displaystyle\iint_R \sqrt{2x + y}\, dx\, dy$; $\quad 1 \le x \le 3, 2 \le y \le 5$

75. $\displaystyle\iint_R \sqrt{y + x}\, dx\, dy$; $\quad 0 \le x \le 7, 1 \le y \le 9$

76. $\displaystyle\iint_R ye^{y^2+x}\, dx\, dy$; $\quad 0 \le x \le 1, 0 \le y \le 1$

Find the volume under the given surface $z = f(x, y)$ and above the given rectangle.

77. $z = x + 8y + 4$; $\quad 0 \le x \le 3, 1 \le y \le 2$

78. $z = x^2 + y^2$; $\quad 3 \le x \le 5, 2 \le y \le 4$

Evaluate each double integral. If the function seems too difficult to integrate, try interchanging the limits of integration.

79. $\displaystyle\int_0^1 \int_0^{2x} xy\, dy\, dx$

80. $\displaystyle\int_1^2 \int_2^{2x^2} y\, dy\, dx$

81. $\displaystyle\int_0^1 \int_{x^2}^{x} x^3y\, dy\, dx$

82. $\displaystyle\int_0^1 \int_y^{\sqrt{y}} x\, dx\, dy$

83. $\displaystyle\int_0^2 \int_{x/2}^1 \dfrac{1}{y^2 + 1}\, dy\, dx$

84. $\displaystyle\int_0^8 \int_{x/2}^4 \sqrt{y^2 + 4}\, dy\, dx$

Use the region R, with boundaries as indicated, to evaluate the given double integral.

85. $\displaystyle\iint_R (2x + 3y)\, dx\, dy$; $\quad 0 \le y \le 1, y \le x \le 2 - y$

86. $\displaystyle\iint_R (2 - x^2 - y^2)\, dy\, dx$; $\quad 0 \le x \le 1, x^2 \le y \le x$

APPLICATIONS

Business and Economics

87. Charge for Auto Painting The charge (in dollars) for painting a sports car is given by

$$C(x, y) = 4x^2 + 5y^2 - 4xy + \sqrt{x},$$

where x is the number of hours of labor needed and y is the number of gallons of paint and sealant used. Find the following.

(a) The charge for 10 hours and 5 gal of paint and sealant

(b) The charge for 15 hours and 10 gal of paint and sealant

(c) The charge for 20 hours and 20 gal of paint and sealant

88. Manufacturing Costs The manufacturing cost (in dollars) for a certain computer is given by

$$c(x, y) = 2x + y^2 + 4xy + 25,$$

where x is the memory capacity of the computer in gigabytes (GB) and y is the number of hours of labor required. For 640 GB and 6 hours of labor, find the following.

(a) The approximate change in cost for an additional 1 GB of memory

(b) The approximate change in cost for an additional hour of labor

89. Productivity The production function z for one country is

$$z = x^{0.7}y^{0.3},$$

where x represents the amount of labor and y the amount of capital. Find the marginal productivity of the following.

(a) Labor $\qquad\qquad$ **(b)** Capital

90. Cost The cost (in dollars) to manufacture x solar cells and y solar collectors is

$$c(x, y) = x^2 + 5y^2 + 4xy - 70x - 164y + 1800.$$

(a) Find values of x and y that produce minimum total cost.

(b) Find the minimum total cost.

Utility   Maximize each of the following utility functions, with the cost of each commodity and total amount available to spend given.

91. $f(x, y) = xy^3$, cost of a unit of x is \$2, cost of a unit of y is \$4, and \$80 is available.

92. $f(x, y) = x^5y^2$, cost of a unit of x is \$10, cost of a unit of y is \$6, and \$42 is available.

93. Cost   The cost (in dollars) to produce x satellite receiving dishes and y transmitters is given by

$$C(x, y) = 100 \ln(x^2 + y) + e^{xy/20}.$$

Production schedules now call for 15 receiving dishes and 9 transmitters. Use differentials to approximate the change in costs if 1 more dish and 1 fewer transmitter are made.

94. Production Materials   Approximate the volume of material needed to manufacture a cone of radius 2 cm, height 8 cm, and wall thickness 0.21 cm.

95. Production Materials   A sphere of radius 2 ft is to receive an insulating coating 1 in. thick. Approximate the volume of the coating needed.

96. Production Error   The height of a sample cone from a production line is measured as 11.4 cm, while the radius is measured as 2.9 cm. Each of these measurements could be off by 0.2 cm. Approximate the maximum possible error in the volume of the cone.

97. Profit   The total profit from 1 acre of a certain crop depends on the amount spent on fertilizer, x, and on hybrid seed, y, according to the model

$$P(x, y) = 0.01(-x^2 + 3xy + 160x - 5y^2 + 200y + 2600).$$

The budget for fertilizer and seed is limited to \$280.

(a) Use the budget constraint to express one variable in terms of the other. Then substitute into the profit function to get a function with one independent variable. Use the method shown in Chapter 6 on Applications of the Derivative to find the amounts spent on fertilizer and seed that will maximize profit. What is the maximum profit per acre? (*Hint:* Throughout this exercise you may ignore the coefficient of 0.01 until you need to find the maximum profit.)

(b) Find the amounts spent on fertilizer and seed that will maximize profit using the first method shown in this chapter. (*Hint:* You will not need to use the budget constraint.)

(c) Use the Lagrange multiplier method to solve the original problem.

(d) Look for the relationships among these methods.

98. Average Production   A production function is given by

$$P(x, y) = 400x^{0.3}y^{0.7},$$

where x is the number of units of labor and y is the number of units of capital. Find the average production level if x varies from 40 to 50 units of labor and y varies from 20 to 40 units of capital. (See Exercises 61–64 in Section 9.6.)

99. Average Cost   The cost (in dollars) for producing x units of one product and y units of a second product is

$$C(x, y) = 0.03x^2 + 6y + 2xy + 10.$$

The weekly production of the first product varies from 100 units to 150 units, and the weekly production of the second product varies from 50 to 100 units. Estimate the average weekly cost for these two products. (See Exercises 61–64 in Section 9.6.)

Life Sciences

100. Blood Vessel Volume   A length of blood vessel is measured as 2.7 cm, with the radius measured as 0.7 cm. If each of these measurements could be off by 0.1 cm, estimate the maximum possible error in the volume of the vessel.

101. Total Body Water   Accurate prediction of total body water is critical in determining adequate dialysis doses for patients with renal disease. For African American males, total body water can be estimated by the function

$$T(A, M, S) = -18.37 - 0.09A + 0.34M + 0.25S,$$

where T is the total body water (in liters), A is age (in years), M is mass (in kilograms), and S is height (in centimeters). *Source: Kidney International.*

(a) Find $T(65, 85, 180)$.

(b) Find and interpret $T_A(A, M, S)$, $T_M(A, M, S)$, and $T_S(A, M, S)$.

102. Brown Trout   Researchers from New Zealand have determined that the length of a brown trout depends on both its mass and age and that the length can be estimated by

$$L(m, t) = (0.00082t + 0.0955)e^{(\ln m + 10.49)/2.842},$$

where $L(m, t)$ is the length of the trout (in centimeters), m is the mass of the trout (in grams), and t is the age of the trout (in years). *Source: Transactions of the American Fisheries Society.*

(a) Find $L(450, 4)$.

(b) Find $L_m(450, 7)$ and $L_t(450, 7)$ and interpret.

103. Survival Curves   The following figure shows survival curves (percent surviving as a function of age) for people in the United States in 1900 and 2000. Let $f(x, y)$ give the proportion surviving at age x in year y. Use the graph to estimate the following. Interpret each answer in words. *Source: National Vital Statistics Report.*

(a) $f(60, 1900)$ **(b)** $f(70, 2000)$

(c) $f_x(60, 1900)$ **(d)** $f_x(70, 2000)$

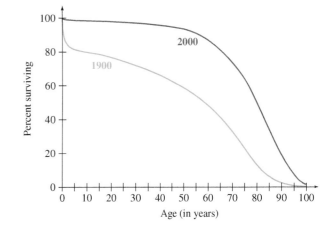

General Interest

104. Area The bottom of a planter is to be made in the shape of an isosceles triangle, with the two equal sides 3 ft long and the third side 2 ft long. The area of an isosceles triangle with two equal sides of length a and third side of length b is

$$f(a, b) = \frac{1}{4} b \sqrt{4a^2 - b^2}.$$

(a) Find the area of the bottom of the planter.

(b) The manufacturer is considering changing the shape so that the third side is 2.5 ft long. What would be the approximate effect on the area?

105. Surface Area A closed box with square ends must have a volume of 125 in³. Use Lagrange multipliers to find the dimensions of such a box that has minimum surface area.

106. Area Use Lagrange multipliers to find the maximum rectangular area that can be enclosed with 400 ft of fencing, if no fencing is needed along one side.

EXTENDED APPLICATION

USING MULTIVARIABLE FITTING TO CREATE A RESPONSE SURFACE DESIGN

Suppose you are designing a flavored drink with orange and banana flavors. You want to find the ideal concentrations of orange and banana flavoring agents, but since the concentrations could range from 0% to 100%, you can't try every possibility. A common design technique in the food industry is to make up several test drinks using different combinations of flavorings and have them rated for taste appeal by a panel of tasters. Such ratings are called *hedonic responses* and are often recorded on a 10-point scale from 0 (worst) to 9 (best). One combination will most likely get the highest average score, but since you have only tried a few of the infinite number of flavor combinations, the winning combination on the taste test might be far from the mix that would be the most popular in the market. How can you use the information from your test to locate the best point on the *flavor plane*?

One approach to this problem uses *response surfaces*, three-dimensional surfaces that approximate the data points from your flavor test.* For your test, you might choose mixtures that are spread out over the flavor plane. For example, you could combine low, medium, and high orange with low, medium, and high banana to get 9 different flavors. If you had 15 tasters and used intensities of 20, 50, and 80 for each fruit, the test data might look like those in the table.

Average Hedonic Scores ($n = 15$)			
		Banana Intensity (0 to 100)	
		20	50	80
Orange	*20*	3.2	4.9	2.8
Intensity	*50*	6.0	7.2	5.1
(0 to 100)	*80*	4.5	5.5	4.8

For example, the table shows that the drink with orange intensity 20 and banana intensity 80 got an average flavor rating of 2.8 from the test panel (they didn't like it).

Your test results are points in space, where you can think of the x-axis as the orange axis, the y-axis as banana, and the z-axis as taste score. A three-dimensional bar chart is a common way of displaying data of this kind. Figure 38 below is a bar chart of the flavor test results.

Looking at the bar chart, we can guess that the best flavor mix will be somewhere near the middle. We'd like to "drape" a smooth surface over the bars and see where that surface has a maximum. But as with any sample, our tasters are not perfectly representative of the whole population: Our test results give the general shape of the true population response, but each bar includes an error that results from our small sample size. The solution is to fit a *smooth* surface to the data points.

A simple type of function for modeling such data sets is a quadratic function. You've seen many quadratic functions of two variables in the examples and exercises for this chapter, and you know

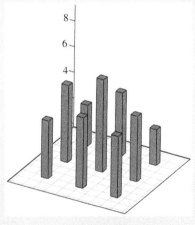

FIGURE 38

*For a brief introduction to response surfaces, see Devore, Jay L., Nicholas R. Farnum and Jimmy A. Doi, *Applied Statistics for Engineers and Scientists*, 3rd edition, Duxbury, 2013.

that they can have maxima, minima, and saddle points. We don't know in advance which quadratic shape will give us the best fit, so we'll use the most general quadratic,

$$G(x, y) = Ax^2 + By^2 + Cxy + Dx + Ey + F.$$

Our job is to find the six coefficients, A through F, that give the best fit to our nine data points. As with the least squares line formula you used in Section 1.3, there are formulas for these six coefficients. Most statistical software packages will generate them directly from your data set, and here is the best-fitting quadratic found by one such program:

$$G(x, y) = -0.00202x^2 - 0.00163y^2 + 0.000194xy$$
$$+ 0.21380x + 0.14768y - 2.36204.$$

In this case the response surface shows how the dependent variable, taste rating, *responds* to the two independent variables, orange and banana intensity. Figures 39 and 40 are two views of the surface together with the data: a surface superimposed on the bar chart, and the same surface with the data shown as points in space.

In research papers, response surface models are often reported using level curves. A contour map for the surface we have found looks like Figure 41 at the top of the next column, with orange increasing from left to right and banana from bottom to top.

FIGURE 39

FIGURE 40

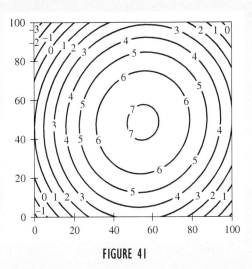

FIGURE 41

It's quite easy to estimate the location of the maximum by marking a point in the middle of the central ellipse and finding its coordinates on the two axes (try it!). You can also use the techniques you learned in the section on Maxima and Minima, computing partial derivatives of G with respect to x and y and solving the resulting linear system. The numbers are awkward, but with some help from a calculator you'll find that the maximum occurs at approximately $(55.3, 48.6)$. So the quadratic model predicts that the most popular drink will have an orange concentration of 55.3 and a banana concentration of 48.6. The model also predicts the public's flavor rating for this drink: We would expect it to be $G(55.3, 48.6)$, which turns out to be about 7.14. When food technologists design a new food, this kind of modeling is often a first step. The next step might be to make a new set of test drinks with concentrations clustered around the point $(55.3, 48.6)$, and use further tests to explore this region of the "flavor plane" in greater detail.

Response surfaces are also helpful for constrained optimization. In the section on Lagrange Multipliers, you saw how Lagrange multipliers could solve problems of the form:

> Find the relative extrema for $z = f(x, y)$,
> subject to the constraint $g(x, y) = 0$.

Sometimes the constraints have a different form: You may have *several* dependent variables that respond to the same inputs, and the design goal is to keep each variable *within a given range*. Here's an example based on the data in U.S. Patent No. 4,276,316, which is titled *Process for Treating Nuts*. The patent granted to researcher Shri C. Sharma and assigned to CPC International, Inc. covers a method for preparing nuts for blanching (that is, having their skins removed). The patent summary reads in part:

> *The nuts are heated with a gas at a temperature of 125° to 175°C for 30 to 180 seconds and then immediately cooled to below 35°C within 5 minutes prior to blanching. This provides improved blanching, sorting and other steps in a process for producing products ranging from nuts per se to peanut butters or spreads.* **Source: United States Patent and Trademark Office.**

In support of the effectiveness of the method, the patent offers data that describe the effects of nine different combinations of air temperature and treatment time on three variables of interest for blanched peanuts: blanching efficiency, roasted peanut flavor, and overall flavor. Efficiency is given in percent, and the two hedonic variables were rated by tasters on a scale of 0 to 9.

The time variable has been converted into a natural logarithm because treatment time effects typically scale with the log of the time. The problem is now to pick a temperature and time range that give the optimum combination of efficiency, roasted flavor, and overall flavor. Each of these three dependent variables responds to the inputs in a different way, and the patent documentation includes quadratic response surfaces for each variable. The lighter shading in Figures 42–44 indicates higher values, which are more desirable. Temperature is plotted across the bottom, and the log of treatment time increases from bottom to top.

Sometimes process designers faced with this kind of problem will combine the dependent variables into a single function by taking a weighted average of their values, and then use a single response surface to optimize this function. Here we look at a different scenario. Suppose we set the following process goals: We want blanching efficiency of at least 93%, a roasted flavor rating

The "pointed" contours are 5.5

FIGURE 44

of at least 5.2, and an overall flavor rating of at least 5.5. Is there a combination of time and temperature that meets these criteria? If so, what is it?

The first step is to identify the "successful" area on each response surface, which we can do by shading the corresponding region in the contour plot, shown in Figures 45–47 on the next page.

Now the strategy is clear: We want to stack the three plots on top of each other and see if the shaded regions overlap. Figure 48 is the result.

So we can see that there are two regions on the temperature–time plane that will work. For example, the upper area of overlap suggests a processing temperature of 140°C to 150°C, with a processing time between 90 and 150 seconds (remember that the numbers on the vertical axis are *natural logarithms* of the time in seconds).

Response surfaces are a standard tool in designing everything from food to machine parts, and we have touched on only a small part of the theory here. Frequently a process depends on more than two independent variables. For example, a soft-drink formula might include three flavorings, an acidifying agent, and a sweetener. The response "surface" now lives in six dimensions and we can no longer draw nice pictures, but the same multivariable mathematics that generated our quadratic response surfaces will lead us to the optimal combination of variables.

The innermost contour is 93%

FIGURE 42 — Blanching efficiency

The innermost contours are 5.2

FIGURE 43 — Roasted flavor

Air Temperature, °C	Treatment Time, Seconds	Log of Treatment Time	Blanching Efficiency	Roasted Peanut Flavor	Overall Flavor
138	45	3.807	93.18	4.94	5.51
160	120	4.787	94.99	5.24	5.37
149	75	4.317	98.43	5.27	5.10
138	120	4.787	96.42	5.05	5.71
160	45	3.807	96.48	5.17	5.62
127	75	4.317	93.56	4.64	5.04
149	180	5.193	94.99	5.24	5.37
149	30	3.401	87.30	5.43	5.44
171	45	3.807	94.40	4.37	5.18

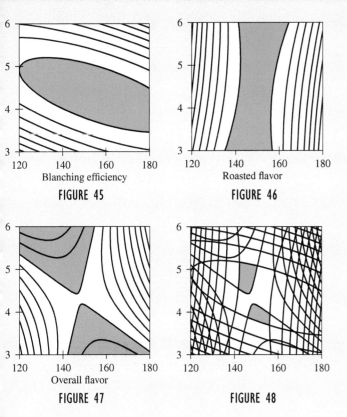

6
5
4
3
120   140   160   180
Blanching efficiency

FIGURE 45

6
5
4
3
120   140   160   180
Roasted flavor

FIGURE 46

6
5
4
3
120   140   160   180
Overall flavor

FIGURE 47

6
5
4
3
120   140   160   180

FIGURE 48

EXERCISES

1. The general quadratic function of two variables has six terms. How many terms are in the general cubic function of two variables?

2. Use the contour plot of orange-banana flavor to estimate the "flavor coordinates" of the best-tasting drink.

3. Find the maximum on the flavor response surface by finding the critical point of the function $G(x, y)$.

4. Without shading or numbers on the contours, how would you know that the point you found in Exercises 2 and 3 represents the best flavor rather than the worst flavor? (*Hint:* Compute the discriminant D as described in the section on Maxima and Minima.)

5. Our best drink has a predicted flavor rating of 7.14, but one of our test drinks got a *higher* rating, 7.2. What's going on?

6. Blanching efficiency has a maximum near the center of the temperature–time plane. What is going on near the center of the plane for the roasted flavor and overall flavor response surfaces? Within the domain plotted, where does overall flavor reach a maximum?

7. In the overall flavor contour plot, if we move one contour toward higher flavor from the "pointed" 5.5 contours, we find curved contours that represent an overall flavor rating of about 5.6. If instead of requiring an overall flavor rating of 5.5 we decided to require a rating of 5.6, what would happen to our process design?

8. Use the last figure to describe the other region in the temperature–time plane that delivers a successful process for preparing nuts for blanching.

9. At the website WolframAlpha.com, you can enter "maximize" followed by an expression $f(x, y)$, such as $4 - x^2 - y^2$, and you will be told the maximum value of the expression, as well as the point where the maximum occurs. You will also be shown a graph of the surface $z = f(x, y)$ and a contour plot. Try this for the function $G(x, y)$ given in this Extended Application, and compare with your answers to Exercises 2 and 3.

DIRECTIONS FOR GROUP PROJECT

Perform an experiment that is similar to the flavored drink example from the text on some other product. For example, you could perform an experiment in which you develop hedonic responses for various levels of salt and butter on popcorn. Using technology, to the extent that it is available to you, carry out the analysis of your experiment to determine an optimal mixture of each ingredient.

Appendix A

Solutions to Prerequisite Skills Diagnostic Test *(with references to Ch. R)*

For more practice on the material in questions 1–4, see *Beginning and Intermediate Algebra* (6th ed.) by Margaret L. Lial, John Hornsby, and Terry McGinnis, Pearson, 2016.

1. $10/50 = 0.20 = 20\%$

2.

$$\frac{13}{7} - \frac{2}{5} = \frac{13}{7} \cdot \frac{5}{5} - \frac{2}{5} \cdot \frac{7}{7} \qquad \text{Get a common denominator.}$$

$$= \frac{65}{35} - \frac{14}{35}$$

$$= \frac{51}{35}$$

3. The total number of apples and oranges is $x + y$, so $x + y = 75$.

4. The sentence can be rephrased as "The number of students is at least four times the number of professors," or $s \geq 4p$.

5.

$$7k + 8 = -4(3 - k)$$

$$7k + 8 = -12 + 4k \qquad \text{Distributive property.}$$

$$7k - 4k + 8 - 8 = -12 - 8 + 4k - 4k \qquad \text{Subtract 8 and } 4k \text{ from both sides.}$$

$$3k = -20 \qquad \text{Simplify.}$$

$$k = -20/3 \qquad \text{Divide both sides by 3.}$$

For more practice, see Sec. R.4.

6.

$$\frac{5}{8}x + \frac{1}{16}x = \frac{11}{16} + x$$

$$16\left(\frac{5}{8}x + \frac{1}{16}x\right) = 16\left(\frac{11}{16} + x\right) \qquad \text{Multiply by common denominator.}$$

$$16 \cdot \frac{5}{8}x + 16 \cdot \frac{1}{16}x = 16 \cdot \frac{11}{16} + 16x \qquad \text{Distributive property.}$$

$$10x + x = 11 + 16x \qquad \text{Simplify.}$$

$$11x = 11 + 16x \qquad \text{Simplify.}$$

$$-5x = 11 \qquad \text{Subtract } 16x \text{ from both sides.}$$

$$x = \frac{-11}{5} \qquad \text{Divide both sides by } -5.$$

For more practice, see Sec. R.4.

7. The interval $-2 < x \leq 5$ is written as $(-2, 5]$. For more practice, see Sec. R.5.

8. The interval $(-\infty, -3]$ is written as $x \leq -3$. For more practice, see Sec. R.5.

9.

$$5(y - 2) + 1 \leq 7y + 8$$

$$5y - 9 \leq 7y + 8 \qquad \text{Distribute and simplify.}$$

$$5y - 7y - 9 + 9 \leq 7y - 7y + 8 + 9 \qquad \text{Subtract } 7y \text{ from both sides and add 9.}$$

$$-2y \leq 17 \qquad \text{Simplify.}$$

$$y \geq -17/2 \qquad \text{Divide both sides by 3.}$$

10.

$$\frac{2}{3}(5p - 3) > \frac{3}{4}(2p + 1)$$

$$\frac{10p}{3} - 2 > \frac{3p}{2} + \frac{3}{4} \qquad \text{Distribute and simplify.}$$

$$12\left(\frac{10p}{3} - 2\right) > 12\left(\frac{3p}{2} + \frac{3}{4}\right) \qquad \text{Multiply by common denominator.}$$

$$40p - 24 > 18p + 9 \qquad \text{Distribute and simplify.}$$

$$22p > 33 \qquad \text{Subtract } 18p \text{ from both sides and add 24.}$$

$$p > \frac{33}{22} \qquad \text{Divide both sides by 22.}$$

$$p > \frac{3}{2} \qquad \text{Simplify.}$$

For more practice, see Sec. R.5.

11. $(5y^2 - 6y - 4) - 2(3y^2 - 5y + 1) = 5y^2 - 6y - 4 - 6y^2 + 10y - 2 = -y^2 + 4y - 6$. For more practice, see Sec. R.1.

12. $(x^2 - 2x + 3)(x + 1) = (x^2 - 2x + 3)x + (x^2 - 2x + 3) = x^3 - 2x^2 + 3x + x^2 - 2x + 3 = x^3 - x^2 + x + 3$. For more practice, see Sec. R.1.

13. $(a - 2b)^2 = a^2 - 4ab + 4b^2$. For more practice, see Sec. R.1.

14. $3pq + 6p^2q + 9pq^2 = 3pq(1 + 2p + 3q)$. For more practice, see Sec. R.2.

15. $3x^2 - x - 10 = (3x + 5)(x - 2)$. For more practice, see Sec. R.2.

16.
$$\frac{a^2 - 6a}{a^2 - 4} \cdot \frac{a - 2}{a} = \frac{a(a - 6)}{(a - 2)(a + 2)} \cdot \frac{a - 2}{a} \qquad \text{Factor.}$$

$$= \frac{a - 6}{a + 2} \qquad \text{Simplify.}$$

For more practice, see Sec. R.3.

17.
$$\frac{x + 3}{x^2 - 1} + \frac{2}{x^2 + x} = \frac{x + 3}{(x - 1)(x + 1)} + \frac{2}{x(x + 1)} \qquad \text{Factor.}$$

$$= \frac{x + 3}{(x - 1)(x + 1)} \cdot \frac{x}{x} + \frac{2}{x(x + 1)} \cdot \frac{x - 1}{x - 1} \qquad \text{Get a common denominator.}$$

$$= \frac{x^2 + 3x}{x(x - 1)(x + 1)} + \frac{2x - 2}{x(x + 1)(x - 1)} \qquad \text{Multiply out.}$$

$$= \frac{x^2 + 5x - 2}{x(x - 1)(x + 1)} \qquad \text{Add fractions.}$$

For more practice, see Sec. R.3.

18.
$$3x^2 + 4x = 1$$

$$3x^2 + 4x - 1 = 0 \qquad \text{Subtract 1 from both sides.}$$

$$x = \frac{-4 \pm \sqrt{4^2 - 4 \cdot 3(-1)}}{2 \cdot 3} \qquad \text{Use the quadratic formula.}$$

$$= \frac{-4 \pm \sqrt{28}}{6} \qquad \text{Simplify.}$$

$$= \frac{-2 \pm \sqrt{7}}{2} \qquad \text{Simplify.}$$

For more practice, see Sec. R.4.

19. First solve the corresponding equation

$$\frac{8x}{z + 3} = 2$$

$$8z = 2(z + 3) \qquad \text{Multiply both sides by } (z + 3).$$

$$8z = 2z + 6 \qquad \text{Distribute.}$$

$$6z = 6 \qquad \text{Subtract } 2z \text{ from both sides.}$$

$$z = 1 \qquad \text{Divide both sides by 6.}$$

The fraction may also change from being less than 2 to being greater than 2 when the denominator equals 0, namely, at $z = -3$. Testing each of the intervals determined by the numbers -3 and 0 shows that the fraction on the left side of the inequality is less than or equal to 2 on $(-3, 1]$. We do not include $x = -3$ in the solution because that would make the denominator 0. For more practice, see Sec. R.5.

20.
$$\frac{4^{-1}(x^2y^3)^2}{x^{-2}y^5} = \frac{x^2(x^4y^6)}{4y^5} \qquad \text{Simplify.}$$

$$= \frac{x^6y}{4} \qquad \text{Simplify.}$$

For more practice, see Sec. R.6.

21.
$$\frac{4^{1/4}(p^{2/3}q^{-1/3})^{-1}}{4^{-1/4}p^{4/3}q^{4/3}} = \frac{4^{1/2}p^{-2/3}q^{1/3}}{p^{4/3}q^{4/3}} \qquad \text{Simplify.}$$

$$= \frac{2}{p^2 q} \qquad \text{Simplify.}$$

For more practice, see Sec. R.6.

22.
$$k^{-1} - m^{-1} = \frac{1}{k} \cdot \frac{m}{m} - \frac{1}{m} \cdot \frac{k}{k} \qquad \text{Get a common denominator.}$$

$$= \frac{m - k}{km} \qquad \text{Simplify.}$$

For more practice, see Sec. R.6.

23. $(x^2 + 1)^{-1/2}(x + 2) + 3(x^2 + 1)^{1/2} = (x^2 + 1)^{-1/2}[x + 2 + 3(x^2 + 1)] = (x^2 + 1)^{-1/2}(3x^2 + x + 5)$. For more practice, see Sec. R.6.

24. $\sqrt[3]{64b^6} = 4b^2$. For more practice, see Sec. R.7.

25. $\dfrac{2}{4 - \sqrt{10}} \cdot \dfrac{4 + \sqrt{10}}{4 + \sqrt{10}} = \dfrac{2(4 + \sqrt{10})}{16 - 10} = \dfrac{4 + \sqrt{10}}{3}$. For more practice, see Sec. R.7.

26. $\sqrt{y^2 - 10y + 25} = \sqrt{(y - 5)^2} = |y - 5|$. For more practice, see Sec. R.7.

Appendix B
Table I—Formulas from Geometry

PYTHAGOREAN THEOREM

For a right triangle with legs of lengths a and b and hypotenuse of length c, $a^2 + b^2 = c^2$.

CIRCLE

Area: $A = \pi r^2$

Circumference: $C = 2\pi r$

RECTANGLE

Area: $A = lw$

Perimeter: $P = 2l + 2w$

TRIANGLE

Area: $A = \dfrac{1}{2}bh$

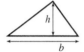

SPHERE

Volume: $V = \dfrac{4}{3}\pi r^3$

Surface area: $A = 4\pi r^2$

CONE

Volume: $V = \dfrac{1}{3}\pi r^2 h$

RECTANGULAR BOX

Volume: $V = lwh$

Surface area: $A = 2lh + 2wh + 2lw$

CIRCULAR CYLINDER

Volume: $V = \pi r^2 h$

Surface area: $A = 2\pi r^2 + 2\pi rh$

TRIANGULAR PRISM

Volume: $V = \dfrac{1}{2}bhl$

GENERAL INFORMATION ON SURFACE AREA

To find the surface area of a figure, break down the total surface area into the individual components and add up the areas of the components. For example, a rectangular box has six sides, each of which is a rectangle. A circular cylinder has two ends, each of which is a circle, plus the side, which forms a rectangle when opened up.

Table 2—Integrals

(*C* is an arbitrary constant.)

---

1. $\displaystyle \int x^n\, dx = \frac{x^{n+1}}{n+1} + C \quad (\text{if } n \ne -1)$

2. $\displaystyle \int e^{kx}\, dx = \frac{e^{kx}}{k} + C$

3. $\displaystyle \int \frac{a}{x}\, dx = a \ln |x| + C$

4. $\displaystyle \int \ln |ax|\, dx = x(\ln |ax| - 1) + C$

5. $\displaystyle \int \frac{1}{\sqrt{x^2 + a^2}}\, dx = \ln \left| x + \sqrt{x^2 + a^2} \right| + C$

6. $\displaystyle \int \frac{1}{\sqrt{x^2 - a^2}}\, dx = \ln \left| x + \sqrt{x^2 - a^2} \right| + C$

7. $\displaystyle \int \frac{1}{a^2 - x^2}\, dx = \frac{1}{2a} \cdot \ln \left| \frac{a+x}{a-x} \right| + C \quad (a \ne 0)$

8. $\displaystyle \int \frac{1}{x^2 - a^2}\, dx = \frac{1}{2a} \cdot \ln \left| \frac{x-a}{x+a} \right| + C \quad (a \ne 0)$

9. $\displaystyle \int \frac{1}{x\sqrt{a^2 - x^2}}\, dx = -\frac{1}{a} \cdot \ln \left| \frac{a + \sqrt{a^2 - x^2}}{x} \right| + C \quad (a \ne 0)$

10. $\displaystyle \int \frac{1}{x\sqrt{a^2 + x^2}}\, dx = -\frac{1}{a} \cdot \ln \left| \frac{a + \sqrt{a^2 + x^2}}{x} \right| + C \quad (a \ne 0)$

11. $\displaystyle \int \frac{x}{ax+b}\, dx = \frac{x}{a} - \frac{b}{a^2} \cdot \ln |ax + b| + C \quad (a \ne 0)$

12. $\displaystyle \int \frac{x}{(ax+b)^2}\, dx = \frac{b}{a^2(ax+b)} + \frac{1}{a^2} \cdot \ln |ax + b| + C \quad (a \ne 0)$

13. $\displaystyle \int \frac{1}{x(ax+b)}\, dx = \frac{1}{b} \cdot \ln \left| \frac{x}{ax+b} \right| + C \quad (b \ne 0)$

14. $\displaystyle \int \frac{1}{x(ax+b)^2}\, dx = \frac{1}{b(ax+b)} + \frac{1}{b^2} \cdot \ln \left| \frac{x}{ax+b} \right| + C \quad (b \ne 0)$

15. $\displaystyle \int \sqrt{x^2 + a^2}\, dx = \frac{x}{2} \sqrt{x^2 + a^2} + \frac{a^2}{2} \cdot \ln \left| x + \sqrt{x^2 + a^2} \right| + C$

16. $\displaystyle \int x^n \cdot \ln |x|\, dx = x^{n+1} \left[\frac{\ln |x|}{n+1} - \frac{1}{(n+1)^2} \right] + C \quad (n \ne -1)$

17. $\displaystyle \int x^n e^{ax}\, dx = \frac{x^n e^{ax}}{a} - \frac{n}{a} \cdot \int x^{n-1} e^{ax}\, dx + C \quad (a \ne 0)$

Answers to Selected Exercises

Answers to selected writing exercises are provided.

Answers to Prerequisite Skills Diagnostic Test

1. 20% **2.** 51/35 **3.** $x + y = 75$ **4.** $s \geq 4p$ **5.** $-20/3$ (Sec. R.4) **6.** $-11/5$ (Sec. R.4) **7.** $(-2, 5]$ (Sec. R.5)
8. $x \leq -3$ (Sec. R.5) **9.** $y \geq -17/2$ (Sec. R.5) **10.** $p > 3/2$ (Sec. R.5)

Chapter R Algebra Reference

Exercises R.1 (page R-5)

For exercises . . .	1–6	7,8,15–22	9–14	23–26
Refer to example . . .	2	3	4	5

1. $-x^2 + x + 9$ **2.** $-6y^2 + 3y + 10$ **3.** $-16q^2 + 4q + 6$ **4.** $9r^2 - 4r + 19$
5. $-0.327x^2 - 2.805x - 1.458$ **6.** $0.8r^2 + 3.6r - 1.5$ **7.** $-18m^3 - 27m^2 + 9m$ **8.** $-12x^4 + 30x^2 + 36x$
9. $9t^2 + 9ty - 10y^2$ **10.** $18k^2 - 7kq - q^2$ **11.** $4 - 9x^2$ **12.** $36m^2 - 25$ **13.** $(6/25)y^2 + (11/40)yz + (1/16)z^2$
14. $(15/16)r^2 - (7/12)rs - (2/9)s^2$ **15.** $27p^3 - 1$ **16.** $15p^3 + 13p^2 - 10p - 8$ **17.** $8m^3 + 1$
18. $12k^4 + 21k^3 - 5k^2 + 3k + 2$ **19.** $3x^2 + xy + 2xz - 2y^2 - 3yz - z^2$ **20.** $2r^2 + 2rs - 5rt - 4s^2 + 8st - 3t^2$
21. $x^3 + 6x^2 + 11x + 6$ **22.** $x^3 - 2x^2 - 5x + 6$ **23.** $x^2 + 4x + 4$ **24.** $4a^2 - 16ab + 16b^2$
25. $x^3 - 6x^2y + 12xy^2 - 8y^3$ **26.** $27x^3 + 27x^2y + 9xy^2 + y^3$

Exercises R.2 (page R-8)

For exercises . . .	1–4	5–15	16–20	21–32
Refer to example . . .	1	2,3	3, 2nd CAUTION	4

1. $7a^2(a + 2)$ **2.** $3y(y^2 + 8y + 3)$ **3.** $13p^2q(p^2q - 3p + 2q)$
4. $10m^2(6m^2 - 12mn + 5n^2)$ **5.** $(m + 2)(m - 7)$ **6.** $(x + 5)(x - 1)$ **7.** $(z + 4)(z + 5)$ **8.** $(b - 7)(b - 1)$
9. $(a - 5b)(a - b)$ **10.** $(s - 5t)(s + 7t)$ **11.** $(y - 7z)(y + 3z)$ **12.** $(3x + 7)(x - 1)$ **13.** $(3a + 7)(a + 1)$
14. $(5y + 2)(3y - 1)$ **15.** $(7m + 2n)(3m + n)$ **16.** $6(a - 10)(a + 2)$ **17.** $3m(m + 3)(m + 1)$ **18.** $2(2a + 3)(a + 1)$
19. $2a^2(4a - b)(3a + 2b)$ **20.** $12x^2(x - y)(2x + 5y)$ **21.** $(x + 8)(x - 8)$ **22.** $(3m + 5)(3m - 5)$
23. $10(x + 4)(x - 4)$ **24.** Prime **25.** $(z + 7y)^2$ **26.** $(s - 5t)^2$ **27.** $(3p - 4)^2$ **28.** $(a - 6)(a^2 + 6a + 36)$
29. $(3r - 4s)(9r^2 + 12rs + 16s^2)$ **30.** $3(m + 5)(m^2 - 5m + 25)$ **31.** $(x - y)(x + y)(x^2 + y^2)$
32. $(2a - 3b)(2a + 3b)(4a^2 + 9b^2)$

Exercises R.3 (page R-11)

For exercises . . .	1–12	13–38
Refer to example . . .	1	2

1. $v/7$ **2.** $5p/2$ **3.** $8/9$ **4.** $2/(t + 2)$ **5.** $x - 2$ **6.** $4(y + 2)$ **7.** $(m - 2)/(m + 3)$
8. $(r + 2)/(r + 4)$ **9.** $3(x - 1)/(x - 2)$ **10.** $(z - 3)/(z + 2)$ **11.** $(m^2 + 4)/4$ **12.** $(2y + 1)/(y + 1)$ **13.** $3k/5$
14. $25p^2/9$ **15.** $9/(5c)$ **16.** 2 **17.** $1/4$ **18.** $3/10$ **19.** $2(a + 4)/(a - 3)$ **20.** $2/(r + 2)$ **21.** $(k - 2)/(k + 3)$
22. $(m + 6)/(m + 3)$ **23.** $(m - 3)/(2m - 3)$ **24.** $2(2n - 1)/(3n - 5)$ **25.** 1 **26.** $(6 + p)/(2p)$
27. $(12 - 15y)/(10y)$ **28.** $137/(30m)$ **29.** $(3m - 2)/[m(m - 1)]$ **30.** $(r - 6)/[r(2r + 3)]$ **31.** $14/[3(a - 1)]$
32. $23/[20(k - 2)]$ **33.** $(7x + 1)/[(x - 2)(x + 3)(x + 1)]$ **34.** $(y^2 + 1)/[(y + 3)(y + 1)(y - 1)]$
35. $k(k - 13)/[(2k - 1)(k + 2)(k - 3)]$ **36.** $m(3m - 19)/[(3m - 2)(m + 3)(m - 4)]$ **37.** $(4a + 1)/[a(a + 2)]$
38. $(5x^2 + 4x - 4)/[x(x - 1)(x + 1)]$

Exercises R.4 (page R-16)

For exercises . . .	1–8	9–26	27–37
Refer to example . . .	2	3–5	6,7

1. -12 **2.** $3/4$ **3.** 12 **4.** $-3/8$ **5.** $-7/8$ **6.** $-6/11$ **7.** 4 **8.** $-10/19$ **9.** $-3, -2$
10. $-1, 3$ **11.** 7 **12.** $-2, 5/2$ **13.** $-1/4, 2/3$ **14.** 2, 5 **15.** $-3, 3$ **16.** $-4, 1/2$ **17.** 0, 4 **18.** $(5 + \sqrt{13})/6 \approx 1.434$,
$(5 - \sqrt{13})/6 \approx 0.232$ **19.** $(2 + \sqrt{10})/2 \approx 2.581, (2 - \sqrt{10})/2 \approx -0.581$ **20.** $(-1 + \sqrt{5})/2 \approx 0.618$,
$(-1 - \sqrt{5})/2 \approx -1.618$ **21.** $5 + \sqrt{5} \approx 7.236, 5 - \sqrt{5} \approx 2.764$ **22.** $(4 + \sqrt{6})/5 \approx 1.290, (4 - \sqrt{6})/5 \approx 0.310$
23. 1, 5/2 **24.** No real number solutions **25.** $(-1 + \sqrt{73})/6 \approx 1.257, (-1 - \sqrt{73})/6 \approx -1.591$ **26.** $-1, 0$ **27.** 3
28. 12 **29.** $-59/6$ **30.** 6 **31.** 3 **32.** $-5/2$ **33.** $2/3$ **34.** 1 **35.** 2 **36.** No solution **37.** No solution

Exercises R.5 (page R-21)

For exercises . . .	1–14		15–26	27–38	39–42	43–54
Refer to example . . .	Figure 1, Example 2	2	3	4	5–7	

1. $(-\infty, 4)$ ⟵ ○ (0, 4)

2. $[-3, \infty)$ (−3, 0) **3.** $[1, 2)$ (0, 1, 2) **4.** $[-2, 3]$ (−2, 0, 3)

5. $(-\infty, -9)$ (−9, 0) **6.** $[6, \infty)$ (0, 6) **7.** $-7 \leq x \leq -3$ **8.** $4 \leq x < 10$

9. $x \leq -1$ **10.** $x > 3$ **11.** $-2 \leq x < 6$ **12.** $0 < x < 8$ **13.** $x \leq -4$ or $x \geq 4$ **14.** $x < 0$ or $x \geq 3$

15. $(-\infty, 2]$ (0, 2) **16.** $(-\infty, 1)$ (0, 1) **17.** $(3, \infty)$ (0, 3)

18. $(-\infty, 1]$ (0, 1) **19.** $(1/5, \infty)$ $(0, \frac{1}{5}, 1)$ **20.** $(1/3, \infty)$ $(0, \frac{1}{3}, 1)$

21. $(-4, 6)$ -4 0 6 **22.** $[7/3, 4]$ 0 $\frac{7}{3}$ 4 **23.** $[-5, 3)$ -5 0 3

24. $[-1, 2]$ -1 0 2 **25.** $[-17/7, \infty)$ $-\frac{17}{7}$ 0 **26.** $(-\infty, 50/9]$ 0 $\frac{50}{9}$

27. $(-5, 3)$ -5 0 3 **28.** $(-\infty, -6] \cup [1, \infty)$ -6 $0\ 1$ **29.** $(1, 2)$ 0 1 2

30. $(-\infty, -4) \cup (1/2, \infty)$ -4 $0\,\frac{1}{2}$ **31.** $(-\infty, -4) \cup (4, \infty)$ -4 0 4

32. $[-3/2, 5]$ $-\frac{3}{2}$ 0 5 **33.** $(-\infty, -1] \cup [5, \infty)$ $-1\ 0$ 5

34. $[-1/2, 2/5]$ $-\frac{1}{2}$ 0 $\frac{2}{5}$ **35.** $(-\infty, -1) \cup (1/3, \infty)$ -1 $0\,\frac{1}{3}$

36. $(-\infty, -2) \cup (5/3, \infty)$ -2 0 $\frac{5}{3}$ **37.** $(-\infty, -3] \cup [3, \infty)$ -3 0 3

38. $(-\infty, 0) \cup (16, \infty)$ 0 16 **39.** $[-2, 0] \cup [2, \infty)$ -2 0 2

40. $(-\infty, -4] \cup [-3, 0]$ $-4\ -3$ 0 **41.** $(-\infty, 0) \cup (1, 6)$ 0 1 6

42. $(-1, 0) \cup (4, \infty)$ -1 0 4 **43.** $(-5, 3]$ **44.** $(-\infty, -1) \cup (1, \infty)$ **45.** $(-\infty, -2)$

46. $(-2, 3/2)$ **47.** $[-8, 5)$ **48.** $(-\infty, -3/2) \cup [-13/9, \infty)$ **49.** $[2, 3)$ **50.** $(-\infty, -1)$ **51.** $(-2, 0] \cup (3, \infty)$
52. $(-4, -2) \cup (0, 2)$ **53.** $(1, 3/2]$ **54.** $(-\infty, -2) \cup (-2, 2) \cup [4, \infty)$

Exercises R.6 (page R-25)

For exercises . . .	1–8	9–26	27–36	37–50	51–56
Refer to example . . .	1	2	3,4	5	6

1. $1/64$ **2.** $1/81$ **3.** 1 **4.** 1 **5.** $-1/9$ **6.** $1/9$ **7.** 36 **8.** $27/64$
9. $1/64$ **10.** 8^5 **11.** $1/10^8$ **12.** 7 **13.** x^2 **14.** 1 **15.** $8k^3$ **16.** $1/(3z^7)$ **17.** $x^5/(3y^3)$ **18.** $m^3/5^4$ **19.** a^3b^6
20. $49/(c^6d^4)$ **21.** $(a + b)/(ab)$ **22.** $(1 - ab^2)/b^2$ **23.** $2(m - n)/[mn(m + n^2)]$ **24.** $(3n^2 + 4m)/(mn^2)$
25. $xy/(y - x)$ **26.** $y^4/(xy - 1)^2$ **27.** 11 **28.** 3 **29.** 4 **30.** -25 **31.** $1/2$ **32.** $4/3$ **33.** $1/16$ **34.** $1/5$ **35.** $4/3$
36. $1000/1331$ **37.** 9 **38.** 3 **39.** 64 **40.** 1 **41.** x^4/y^4 **42.** b/a^3 **43.** r **44.** $12^3/y^8$ **45.** $3k^{3/2}/8$ **46.** $1/(2p^2)$
47. $a^{2/3}b^2$ **48.** $y^2/(x^{1/6}z^{5/4})$ **49.** $h^{1/3}t^{1/5}/k^{2/5}$ **50.** m^3p/n **51.** $3x(x^2 + 3x)^2(x^2 - 5)$ **52.** $6x(x^3 + 7)(-2x^3 - 5x + 7)$
53. $5x(x^2 - 1)^{-1/2}(x^2 + 1)$ **54.** $3(6x + 2)^{-1/2}(27x + 5)$ **55.** $(2x + 5)(x^2 - 4)^{-1/2}(4x^2 + 5x - 8)$
56. $(4x^2 + 1)(2x - 1)^{-1/2}(36x^2 - 16x + 1)$

Exercises R.7 (page R-29)

For exercises . . .	1–22	23–26	27–40	41–44
Refer to example . . .	1,2	3	4	5

1. 5 **2.** 6 **3.** -5 **4.** $5\sqrt{2}$ **5.** $20\sqrt{5}$ **6.** $4y^2\sqrt{2y}$ **7.** 9 **8.** 8
9. $7\sqrt{2}$ **10.** $9\sqrt{3}$ **11.** $9\sqrt{7}$ **12.** $-2\sqrt{7}$ **13.** $5\sqrt[3]{2}$ **14.** $3\sqrt[3]{5}$ **15.** $xyz^2\sqrt{2x}$ **16.** $4r^3s^4t^6\sqrt{10rs}$ **17.** $4xy^2z^3\sqrt[3]{2y^2}$
18. $x^2yz^2\sqrt[4]{y^3z^3}$ **19.** $ab\sqrt{ab}(b - 2a^2 + b^3)$ **20.** $p^2\sqrt{pq}(pq - q^4 + p^2)$ **21.** $\sqrt[6]{a^5}$ **22.** $b^2\sqrt[4]{b}$ **23.** $|4 - x|$
24. $|3y + 5|$ **25.** Cannot be simplified   **26.** Cannot be simplified   **27.** $5\sqrt{7}/7$ **28.** $\sqrt{10}/2$ **29.** $-\sqrt{3}/2$ **30.** $\sqrt{2}$
31. $-3(1 + \sqrt{2})$ **32.** $-5(2 + \sqrt{6})/2$ **33.** $3(2 - \sqrt{2})$ **34.** $(5 - \sqrt{10})/3$ **35.** $(\sqrt{r} + \sqrt{3})/(r - 3)$
36. $5(\sqrt{m} + \sqrt{5})/(m - 5)$ **37.** $\sqrt{y} + \sqrt{5}$ **38.** $(z + \sqrt{5z} - \sqrt{z} - \sqrt{5})/(z - 5)$ **39.** $-2x - 2\sqrt{x(x + 1)} - 1$
40. $[p^2 + p + 2\sqrt{p(p^2 - 1)} - 1]/(-p^2 + p + 1)$ **41.** $-1/[2(1 - \sqrt{2})]$ **42.** $1/(3 + \sqrt{3})$
43. $-1/[2x - 2\sqrt{x(x + 1)} + 1]$ **44.** $2/[p + \sqrt{p(p - 2)}]$

Chapter 1 Linear Functions

Exercises 1.1 (page 13)

For exercises . . .	1–4	5–8	13,29,30	14,31–34	15–17	18,27	19–24	25,26	28	45–60	61–75
Refer to example . . .	1	3	8	9	4	7	5	2	6	11,12	10,13

W1. -3 **W2.** $y = -2x - 13$
W3. $y = (2/5)x + 19/30$
W4. $y = (2/3)x - 7/3$
1. $3/5$ **3.** Not defined   **5.** 1 **7.** $5/9$ **9.** Not defined   **11.** 0 **13.** 2 **15.** $y = -2x + 5$ **17.** $y = -7$
19. $y = -(1/3)x + 10/3$ **21.** $y = 6x - 7/2$ **23.** $x = -8$ **25.** $x + 2y = -6$ **27.** $x = -6$
29. $3x + 2y = 0$ **31.** $x - y = 7$ **33.** $5x - y = -4$ **35.** No   **39.** (a)   **41.** -4

45. **47.** **49.** **51.**

53. **55.** **57.** **59.**

61. (a) 12,000; $y = 12,000x + 3000$ **(b)** 8 years 1 month   **63. (a)** Yes

(b) $y = 1078.6t + 16,072$; the slope indicates that the annual cost of tuition and fees at private four-year colleges is increasing by about \$1079 per year.   **(c)** The year 2025 is too far in the future to rely on this equation to predict costs; too many other factors may influence these costs by then.   **65. (a)** $y = 4.469t + 86.593$ **(b)** 176.0, which is slightly more than the actual CPI   **(c)** It is increasing at a rate of approximately 4.5 per year.   **67. (a)** $u = 0.85(220 - x) = 187 - 0.85x, l = 0.7(220 - x) = 154 - 0.7x$ **(b)** 140 to 170 beats per minute   **(c)** 126 to 153 beats per minute   **(d)** The women are 16 and 52. Their pulse is 143 beats per minute.   **69.** About 86 yr   **71. (a)** $y = 12,620.06t - 381,816$ **(b)** About 1,069,491   **73. (a)** There appears to be a linear relationship.   **(b)** $y = 76.9x$ **(c)** About 780 megaparsecs (about 1.5×10^{22} mi)   **(d)** About 12.4 billion yr   **75. (a)** $y_n = -1.83t + 50.98$ **(b)** $y_o = 2.67t + 6.98$ **(c)** The number of respondents who got their news from newspapers is decreasing by about 1.83% per year, while the number of respondents who got their news online is increasing by about 2.67% per year.

Exercises 1.2 (page 24)

W1. 60   **W2.**

For exercises . . .	1–10	19–22	23–26,37,38,45,46	27–32	33–36,39–44	47–49
Refer to example . . .	1	4	5,6	2,3	7	8

1. -3 **3.** 22   **5.** 0   **7.** -4 **9.** $7 - 5t$ **11.** True   **13.** True   **19.** If $R(x)$ is the cost of renting a snowboard for x hours, then $R(x) = 2.25x + 10$.   **21.** If $C(x)$ is the cost of parking a car for x half-hours, then $C(x) = 0.75x + 2$.   **23.** $C(x) = 30x + 100$ **25.** $C(x) = 75x + 550$ **27. (a)** \$16   **(b)** \$11   **(c)** \$6   **(d)** 640 watches   **(e)** 480 watches   **(f)** 320 watches   **(g)** 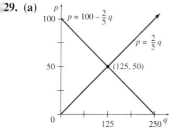   **(h)** 0 watches   **(i)** About 1333 watches   **(j)** About 2667 watches   **(k)**   **(l)** 800 watches, \$6

29. (a)   **(b)** 125 tubs, \$50   **31.** $D(q) = 6.9 - 0.4q$ **33. (a)** 2 units   **(b)** \$980   **(c)** 52 units   **35. (a)** $C(x) = 3.50x + 90$ **(b)** 17 shirts   **(c)** 108 shirts   **37. (a)** $C(x) = 0.097x + 1.32$ **(b)** \$1.32   **(c)** \$98.32   **(d)** \$98.417 (or \$98.42)   **(e)** 9.7¢   **(f)** 9.7¢, the cost of producing one additional cup of coffee would be 9.7¢.   **39.** Break-even quantity is 45 units; don't produce; $P(x) = 20x - 900$ **41.** Break-even quantity is -50 units; impossible to make a profit when $C(x) > R(x)$ for all positive x; $P(x) = -10x - 500$ (always a loss)   **43.** 5   **45. (a)** About 23 acorns per square meter   **(b)** 34; the number of deer tick larvae per 400 squares meters in the spring will increase by 34 for each additional acorn per square meter in the preceding fall.   **47. (a)** 14.4°C   **(b)** -28.9°C   **(c)** 122°C   **49.** $-40°$

Exercises 1.3 (page 34)

For exercises . . .	3(b),4,7(a), 10(d),11(d), 12(e),13(e),14(e), 16(b),17(d), 18(b),19(d), 20(c),21(b), 23(d),24(c),26(a), 27(d),28(d),29(d)	3(c),10(a),11(a), 12(a),13(a),14(a), 16(c),17(b),18(c), 19(a),20(a),21(c), 23(a),24(b),25(a), 26(b),27(a)(b), 28(a)(b),29(c)	3(d),10(b), 11(b),12(c), 13(c),14(c), 19(b),20(b), 21(d),23(b)(c), 25(b)(c),26(c)	5(a)(b),6(a) (b),8(a), 15(b)(d), 22(a)(b)	5(c), 6(c), 16(d)	10(c),11(c), 12(d),13(d), 14(d),15(f), 19(c),27(c), 28(c)
Refer to example . . .	4	1	2	1,4	5	3

3. (a)   **(b)** 0.993   **(c)** $Y = 0.555x - 0.5$ **(d)** 5.6

5. (a) $Y = 0.9783x + 0.0652, 0.9783$ **(b)** $Y = 1.5, 0$ **(c)** The point $(9,9)$ is an outlier that has a strong effect on the least squares line and the correlation coefficient.

7. (a) 0.7746   **(b)**

11. (a) $Y = -0.1897x + 8.469$ **(b)** About 4675   **(c)** 2024
(d) -0.9847; strong negative correlation
13. (a) $Y = -4.26x + 111.0$ **(b)** By about 4.26 percent per year
(c) 47.1%   **(d)** 2017   **(e)** -0.9973; strong negative correlation

15. (a)   Yes   **(b)** $Y = 703.91x - 45,220$; 0.9964; strong positive correlation   **(c)** By about $704 per year   **(d)** $Y = 1404.76x - 94,521$; 0.9943; strong positive correlation **(e)** By about $1405 per year   **(f)** 2071; 2021

17. (a)   Yes   **(b)** $Y = 1.585x - 0.487$ **(c)** No; it gives negative values for small widths.   **(d)** 0.999

19. (a) $Y = 0.212x - 0.309$ **(b)** 15.2 chirps per second   **(c)** 86.4° F   **(d)** 0.835
21. (a)   Yes   **(b)** 0.9996; yes   **(c)** $Y = 0.4565x + 8.612$ **(d)** About $25,959
23. (a) $Y = -0.0067x + 14.75$ **(b)** 12   **(c)** 11   **(d)** -0.13 **(e)** There is no linear relationship.   **25. (a)** $Y = 14.9x + 2820$ **(b)** 5060, compared to actual 5000; 6990, compared to actual 7000; 9080, compared to actual 9000   **(c)** 6250 BTUs; 6500 BTU air conditioner

27. (a) $Y = -0.1271x + 113.61$ **(b)** $Y = -0.3458x + 146.65$ **(c)** $x \approx 151$; the women will catch up to the men in the year 2051.   **(d)** $r_{men} = -0.9762$; $r_{women} = -0.9300$; both sets of data points closely fit a line with negative slope.

(e)

29. (a) 4.298 miles per hour **(b)**   Yes **(c)** $Y = 4.317x + 3.419$
(d) 0.9971; yes
(e) 4.317 miles per hour

Chapter 1 Review Exercises (page 42)

For exercises . . .	1–6,15–47,57(a)(b),63(a)(b)	7–10,13, 48–56,59,62	11,12,14,57(c)(f),58, 60,61,63(c)(e),64,65
Refer to section . . .	1	2	3

1. False **2.** False **3.** True **4.** False **5.** True **6.** False **7.** True **8.** False **9.** False **10.** False **11.** False
12. True **15.** 1 **17.** $-2/11$ **19.** $-4/3$ **21.** 0 **23.** 5 **25.** $y = (2/3)x - 13/3$ **27.** $y = -x - 3$ **29.** $y = -10$
31. $2x - y = 10$ **33.** $x = -1$ **35.** $y = -5$

37.   **39.**   **41.**   **43.**

45. (a) $y = 26.2t + 100$ **(b)** Imports from China are increasing by about \$26.2 billion per year. **(c)** \$493 billion **(d)** 2020
47. (a) $y = -922t + 62,081$ **(b)** The median income for all U.S. households is decreasing by about \$922 per year. **(c)** \$48,251
(d) 2024 **49. (a)** $p = S(q) = 0.5q + 10$ **(b)** $p = D(q) = -0.5q + 72.50$ **(c)** \$41.25, 62.5 dietary supplement pills
51. $C(x) = 180x + 2000$ **53.** $C(x) = 46x + 120$ **55. (a)** 40 pounds **(b)** \$280 **57. (a)** $y = 476t + 25,198$
(b) $y = 357.5t + 26,620$ **(c)** $y = 386t + 25,975$ **(d)**   **(f)** 0.9356

59. (a) $b(t) = -0.833t + 64.5; p(t) = -0.433t + 47.8; c(t) = 0.2t + 54.2$ **(b)** Beef is decreasing by about 0.833 lb/year; pork
is decreasing by about 0.433 lb/yr; chicken is increasing by about 0.2 lb/yr. **(c)** 2010 **(d)** Beef: 52.0 lb; pork: 41.3 lb; chicken: 57.2 lb
61. (a) $Y = 0.9724x + 31.43$ **(b)** About 216 **(c)** 0.93 **63. (a)** $y = 260t + 6400$ **(b)** $y = 210.6t + 6992.6$
(c) $Y = 247.1t + 6532.0$ **(e)** 0.9515 **65. (a)** 0.4529 **(b)** 0.3955 **(c)** -0.4768

Chapter 2 Nonlinear Functions

Exercises 2.1 (page 58)

For exercises . . .	1–8	9–16	17–32	33–40	41–56,78,79	57–62	63–70	71,72,77	73–76	80,81
Refer to example . . .	2,3	4(b)	4(a),(c)–(e)	3(d)	5	6	7	1	8	9

W1. $3/2, -3/2$ **W2.** $1/2, -5$ **W3.** $[-4, 4]$ **W4.** $[-3, 7]$ **1.** Not a function **3.** Function **5.** Function **7.** Not a function
9. $(-2, -1), (-1, 1), (0, 3), (1, 5), (2, 7), (3, 9)$;
range: $\{-1, 1, 3, 5, 7, 9\}$
11. $(-2, 3/2), (-1, 2), (0, 5/2),$
$(1, 3), (2, 7/2), (3, 4)$;
range: $\{3/2, 2, 5/2, 3, 7/2, 4\}$
13. $(-2, 0), (-1, -1), (0, 0), (1, 3),$
$(2, 8), (3, 15)$; range: $\{-1, 0, 3, 8, 15\}$

15. $(-2, 4), (-1, 1), (0, 0),$
$(1, 1), (2, 4), (3, 9);$
range: $\{0, 1, 4, 9\}$

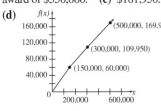

17. $(-\infty, \infty)$ **19.** $(-\infty, \infty)$ **21.** $[-2, 2]$ **23.** $[3, \infty)$ **25.** $(-\infty, -1) \cup (-1, 1) \cup (1, \infty)$
27. $(-\infty, -4) \cup (4, \infty)$ **29.** $(-\infty, -1] \cup [5, \infty)$ **31.** $(-\infty, -1) \cup (1/3, \infty)$ **33.** Domain: $[-5, 4)$;
range: $[-2, 6]$ **35.** Domain: $(-\infty, \infty)$; range: $(-\infty, 12]$ **37.** Domain: $[-2, 4]$; range: $[0, 4]$
(a) 0 **(b)** 4 **(c)** 3 **(d)** $-1.5, 1.5, 2.5$ **39.** Domain: $(-2, 4]$; range: $[-3, 2]$ **(a)** -3 **(b)** -2
(c) -1 **(d)** 2.5 **41. (a)** 33 **(b)** 15/4 **(c)** $3a^2 - 4a + 1$ **(d)** $12/m^2 - 8/m + 1$ or
$(12 - 8m + m^2)/m^2$ **(e)** $0, 4/3$ **43. (a)** 7 **(b)** 0 **(c)** $(2a + 1)/(a - 4)$ if $a \neq 4, 7$ if $a = 4$
(d) $(4 + m)/(2 - 4m)$ if $m \neq 1/2, 7$ if $m = 1/2$ **(e)** -5 **45.** $6t^2 + 12t + 4$
47. $r^2 + 2rh + h^2 - 2r - 2h + 5$ **49.** $9/q^2 - 6/q + 5$ or $(9 - 6q + 5q^2)/q^2$
51. (a) $2x + 2h + 1$ **(b)** $2h$ **(c)** 2 **53. (a)** $2x^2 + 4xh + 2h^2 - 4x - 4h - 5$
(b) $4xh + 2h^2 - 4h$ **(c)** $4x + 2h - 4$ **55. (a)** $1/(x + h)$ **(b)** $-h/[x(x + h)]$

(c) $-1/[x(x + h)]$ **57.** Function **59.** Not a Function **61.** Function **63.** Odd **65.** Even **67.** Even **69.** Odd **71. (a)** No
(b) Year, t **(c)** Price of silver, $S(t)$ **(d)** $[2000, 2013]$ **(e)** $[4, 35]$ **(f)** $15 **(g)** 2011 **73. (a)** $36 **(b)** $36 **(c)** $64 **(d)** $120
(e) $120 **(f)** $148 **(g)** $148 **(i)** x, the number of full and partial days **(j)** S, the cost of renting a saw **75. (a)** $93,300. Attorneys
can receive a maximum of $93,300 on a jury award of $250,000. **(b)** $124,950. Attorneys can receive a maximum of $124,950 on a jury
award of $350,000. **(c)** $181,950. Attorneys can receive a maximum of $181,950 on a jury award of $550,000.

(d)

77. (a) About 140 m **(b)** About 250 m **79. (a) (i)** 3.6 kcal/km **(ii)** 61 kcal/km
(b) $x = g(z) = 1000z$ **(c)** $y = 4.4z^{0.88}$ **81. (a)** $P(w) = 1000/w + 2w$ **(b)** $(0, \infty)$
(c)

Exercises 2.2 (page 71)

For exercises ...	3–8	9–12	13–24	25–38	39–46	49–53,57,58, 60–63,66–72	54–56	59,64,65
Refer to example ...	1–3	4	4–6	10	Before Example 10	8	7	9

W1. $(2x - 9)^2$ **W2.** $(x + 5/8)^2$ **W3.** $2, -5/3$ **W4.** $(-5 \pm \sqrt{57})/4$ **3.** D **5.** A **7.** C **9.** $y = 3(x + 3/2)^2 - 7/4,$
$(-3/2, -7/4)$ **11.** $y = -2(x - 2)^2 - 1, (2, -1)$ **13.** Vertex is $(-5/2, -1/4)$; axis is $x = -5/2$; x-intercepts are -3 and -2;
y-intercept is 6. **15.** Vertex is $(-3, 2)$; axis is $x = -3$; x-intercepts are -4 and -2;
y-intercept is -16.

17. Vertex is $(-2, -16)$; axis is $x = -2$;
x-intercepts are $-2 \pm 2\sqrt{2} \approx 0.83$
and -4.83; y-intercept is -8.

19. Vertex is $(1, 3)$; axis is $x = 1$; no
x-intercepts; y-intercept is 5.

21. Vertex is $(4, 11)$; axis is $x = 4$;
x-intercepts are $4 \pm \sqrt{22}/2 \approx 6.35$ and
1.65; y-intercept is -21.

23. Vertex is $(4, -5)$; axis is $x = 4$; x-intercepts are $4 \pm \sqrt{15} \approx 7.87$ and 0.13; y-intercept is $1/3$.

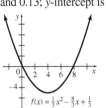

25. D   **27.** C   **29.** E   **31.**

33.

35.

37.

39.

41.

43.

45.

47. (a) r **(b)** $-r$ **(c)** $-r$ **49. (a)**

(b) 1 batch
(c) \$16,000
(d) \$4000

51. (a)

(b) 2.5 batches   **(c)** \$31,250   **(d)** \$5000   **53.** Maximum revenue is \$9225; 35 seats are unsold.
55. (a) $800 + 25x$ **(b)** $80 - x$ **(c)** $R(x) = 64{,}000 + 1200x - 25x^2$ **(d)** 24   **(e)** \$78,400
57. (a) $R(x) = x(500 - x) = 500x - x^2$ **(b)**

(c) \$250   **(d)** \$62,500

59. (a) Quadratic

(b) $f(t) = -2.04(t - 15)^2 + 1037$
(c) $f(t) = -2.69t^2 + 72.4t + 548$

(d)

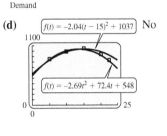

No

61. (a) 87 yr   **(b)** 98 yr   **63. (a)** 28.5 weeks   **(b)** 0.81   **(c)** 0 weeks or 57 weeks of gestation; no
65. (a)

(b) Quadratic   **(c)** $f(t) = 0.00207t^2 - 0.226t + 26.8$

(d) $f(t) = 0.00248(t - 60)^2 + 20.3$ **(e)**

67. (a) 16 ft   **(b)** 2 sec   **69.** 95 ft by 190 ft
71. $y = -(1/15)x^2$; $10\sqrt{3}$ m ≈ 17.32 m

Exercises 2.3 (page 83)

For exercises . . .	3–6	7–15,21–26	16–20,27–42,57,58	46,47,50–52	53,61,62	54–56,59
Refer to example . . .	1	4	7	8	5	2,3

W1. The graph of $f(x)$ shifted 2 units to the left and 3 units down.   **W2.** The graph of $f(x)$ reflected across the x-axis and shifted 3 units to the right.   **W3.** The graph of $f(x)$ reflected across the y-axis and shifted 2 units up.   **W4.** The graph of $f(x)$ reflected across the y-axis and shifted 2 units to the right.

3.

5.

7. D   **9.** E   **11.** I   **13.** G   **15.** A   **17.** D   **19.** E   **21.** 4, 6, etc. (true degree $= 4$); $+$ **23.** 5, 7, etc. (true degree $= 5$); $+$ **25.** 7, 9, etc. (true degree $= 7$); $-$

27. Horizontal asymptote: $y = 0$; vertical asymptote: $x = -2$; no x-intercept; y-intercept $= -2$

29. Horizontal asymptote: $y = 0$; vertical asymptote: $x = -3/2$; no x-intercept; y-intercept $= 2/3$

31. Horizontal asymptote: $y = 2$; vertical asymptote: $x = 3$; x-intercept $= 0$; y-intercept $= 0$

33. Horizontal asymptote: $y = 1$; vertical asymptote: $x = 4$; x-intercept $= -1$; y-intercept $= -1/4$

35. Horizontal asymptote: $y = -1/2$; vertical asymptote: $x = -5$; x-intercept $= 3/2$; y-intercept $= 3/20$

37. Horizontal asymptote: $y = -1/3$; vertical asymptote: $x = -2$; x-intercept $= -4$; y-intercept $= -2/3$

39. No asymptotes; hole at $x = -4$; x-intercept $= -3$; y-intercept $= 3$

41. One possible answer is $y = 2x/(x - 1)$.   **43.** (a) 0   (b) 2, -3 (d) $(x + 1)(x - 1)(x + 2)$ (e) $3(x + 1)(x - 1)(x + 2)$ (f) $(x - a)$ **45.** (a) Two; one at $x = -1.4$ and one at $x = 1.4$ (b) Three; one at $x = -1.414$, one at $x = 1.414$, and one at $x = 1.442$

47. (a) $440; $419; $383; $326; $284; $251   (b) Vertical asymptote at $x = -475$; horizontal asymptote at $y = 0$ (c) $y = 463.2$ (d)

49. $f_1(x) = x(100 - x)/25$, $f_2(x) = x(100 - x)/10$, $f(x) = x^2(100 - x)^2/250$

51. (a) $6700; $15,600; $26,800; $60,300; $127,300; $328,300; $663,300   (b) No   (c)

53. (a)

(b) $y = -0.03132t^2 + 12.46t + 383.1$

(c) $y = -0.03132t^2 + 12.46t + 383.1$

(d) $y = -0.01681t^3 + 1.548t^2 - 24.10t + 499.1$

(e) $y = -0.01681t^3 + 1.548t^2 - 24.10t + 499.1$

55. (a) $A(x) = 0.003631x^3 - 0.03746x^2 + 0.1012x + 0.009$
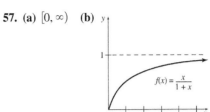

(b) Close to 2 hours
(c) About 1.1 to 2.7 hours

57. (a) $[0, \infty)$ **(b)**

$f(x) = \dfrac{x}{1+x}$

(c)

$f(x) = \dfrac{x}{1+x^2}$

(d) The population of the next generation, $f(x)$, gets smaller when the current generation, x, is larger.

59. (a) 220 g; 602.5 g; 1220 g   **(b)** $c < 19.68$
(c) $m(c) = \dfrac{c^3}{100} - \dfrac{1500}{c}$ **(d)** 41.9 cm

61. (a)

(b) $y = 531.27t^2 - 20,425t + 712,450$

(c) $y = -49.713t^3 + 4785.5t^2 - 123,025t + 1,299,100$

Exercises 2.4 (page 96)

W1. 2^{-6x} **W2.** 3^{35x-15} **W3.** 5^{3x+1} **W4.** x^{a-b}

For exercises . . .	3–11,29–32	13–28	37–46	48,49,51,52	47,50,53–56
Refer to example . . .	1,2	3	4,5	6	7

1. $2, 4, 8, 16, 32, \ldots, 1024; 1.125899907 \times 10^{15}$ **3.** E   **5.** C   **7.** F   **9.** A   **11.** C   **13.** 5   **15.** -4 **17.** -3 **19.** $21/4$
21. $-12/5$ **23.** $2, -2$ **29.**   **25.** $0, -1$ **27.** $0, 1/2$

29.

31.

37. (a) $2166.53   **(b)** $2189.94   **(c)** $2201.90
(d) $2209.97   **(e)** $2214.03   **39.** She should choose the 5.9% investment, which would yield $23.74 additional interest.   **41. (a)** $10.94
(b) $11.27   **(c)** $11.62   **43.** 5.22%
45. (a) $209,162   **(b)** $56.29

47. (a)

(b) $f(t) = 92(1.0657)^t$
(c) 6.57%

(d) 2021

49. (a) The function gives a population of about 3660 million, which is close to the actual population.   **(b)** 6022 million   **(c)** 7725 million

Exponentially

51. (a) 41.93 million **(b)** 12.48 million **(c)** 2.1%, 2.3%; Asian **(d)** 37.99 million
(e) Hispanic: 2039; Asian: 2036; Black: 2079

53. (a) $P = 1013e^{-1.34 \times 10^{-4}x}$;
$P = -0.0748x + 1013$;
$P = 1/(2.79 \times 10^{-7}x + 9.87 \times 10^{-4})$
(b) $P = 1013e^{-1.34 \times 10^{-4}x}$ is the best fit.
(c) 829 millibars, 232 millibars

(d) $P = 1038(0.99998661)^x$. This is slightly different from the function found in part (b), which can be rewritten as $P = 1013(0.99998660)^x$.

55. (a) $C = 21{,}269.4t + 3052.6$;
$C = 1772.45t^2 + 22{,}549.55$;
$C = 19{,}369.64(1.2557)^t$

(b) Exponential

(c) $C = 19{,}259.86(1.2585)^t$
(d) 258,285; 277,782; 297,676; 304,013

Exercises 2.5 (page 109)

For exercises . . .	1–12	13–24	27–36	37–40 90,92, 94,96	41–56	57–64	65–68	69–70	75–77,79,80, 89,90(d)	75(d),76(c), 78	83–85
Refer to example . . .	1	2	3	4	5	6	7	first CAUTION	8	9	10

W1. $3, -2$ **W2.** $20/3$ **W3.** $-2/3$ **W4.** 4 **1.** $\log_5 125 = 3$ **3.** $\log_3 81 = 4$ **5.** $\log_3(1/9) = -2$ **7.** $2^5 = 32$
9. $e^{-1} = 1/e$ **11.** $10^5 = 100{,}000$ **13.** 2 **15.** 3 **17.** -4 **19.** $-2/3$ **21.** 1 **23.** $5/3$ **25.** $\log_3 4$ **27.** $\log_5 3 + \log_5 k$
29. $1 + \log_3 p - \log_3 5 - \log_3 k$ **31.** $\ln 3 + (1/2)\ln 5 - (1/3)\ln 6$ **33.** $5a$ **35.** $2c + 3a + 1$ **37.** 2.113 **39.** -0.281
41. $x = 1/6$ **43.** $z = 4/3$ **45.** $r = 25$ **47.** $x = 1$ **49.** No solution **51.** $x = 3$ **53.** $x = 5$ **55.** $x = 1/\sqrt{3e} \approx 0.3502$
57. $x = (\ln 6)/(\ln 2) \approx 2.5850$ **59.** $k = 1 + \ln 6 \approx 2.7918$ **61.** $x = (\ln 3)/(\ln (5/3)) \approx 2.1507$
63. $x = (\ln 1.25)/(\ln 1.2) \approx 1.224$ **65.** $e^{(\ln 10)(x+1)}$ **67.** 20.09^x **69.** $x < 5$ **75. (a)** 23.4 yr **(b)** 11.9 yr **(c)** 9.0 yr
(d) 23.3 yr; 12 yr; 9 yr **77.** 5.19% **79.** 2021 **81. (b)** 2015: 0.01207, 2020: 0.01207, 2025: 0.01228 **83. (a)** About 0.693
(b) $\ln 2$ **(c)** Yes **85. (a)** About 1.099 **(b)** About 1.386 **87.** About every 7 hr, $T = (3\ln 5)/\ln 2$ **89. (a)** 2039 **(b)** 2036
91. $s/n = 2^{C/B} - 1$ **93.** No; 1/10 **95. (a)** 1000 times greater **(b)** 1,000,000 times greater

Exercises 2.6 (page 118)

W1. $\ln (5)/2 \approx 0.8047$
W2. $\ln (3) - 2 \approx -0.9014$
W3. $\ln (4.5)/3 \approx 0.5014$

For exercises . . .	6,28(b), 29(b)	7–10, 15–17,21(c)	11–14, 18–20	19(d), 20(d)	23,24	25–28, 30,31,42	21(d),(e), 22,29,41	32–40
Refer to example . . .	3	4	6	7	8	1	5	2

7. 4.06% **9.** 8.33% **11.** $6209.93 **13.** $6283.17 **15.** 9.20% **17.** 6.17% **19. (a)** $257,107.67 **(b)** $49,892.33
(c) $68,189.54 **(d)** 14.28% **21. (a)** The 8% investment compounded quarterly. **(b)** $759.26 **(c)** 8.24% and 8.06%
(d) 3.71 yr **(e)** 3.75 yr **23. (a)** 200 **(b)** About 1/2 year **(c)** No **(d)** Yes; 1000 **25. (a)** $P(t) = 0.003285e^{0.007232t}$ **(b)** 3300
(c) No; it is too small. Exponential growth does not accurately describe population growth for the world over a long period of time.
27. (a) $y = 25{,}000e^{0.047t}$ **(b)** $y = 25{,}000(1.048)^t$ **(c)** About 18.6 hours **29. (a)** 17.9 days **(b)** January 17
31. (a) 2, 5, 24, and 125 **(b)** 0.061 **(c)** 0.24 **(d)** No, the values of k are different. **(e)** Between 3 and 4 **33.** About 13 years
35. (a) 0.0193 gram **(b)** 69 years **37. (a)** $y = 500e^{-0.0863t}$ **(b)** $y = 500(386/500)^{t/3}$ **(c)** 8.0 days **39. (a)** 19.5 watts
(b) 173 days **41. (a)** 67% **(b)** 37% **(c)** 23 days **(d)** 46 days **43.** 18.02° **45.** 1 hour

Chapter 2 Review Exercises (page 124)

1. True **2.** False **3.** True
4. True **5.** False **6.** False
7. True **8.** False **9.** False
10. False **11.** False **12.** False
13. False **14.** True **15.** False
16. False **17.** True **18.** True

For exercises . . .	1,2,6,20,21, 35–42,87,110, 115,122	3,4,31–34, 105,109	5,9,43–46, 51–54,83, 88–93,101,108, 111,112,114, 116–118,120	7,10–17,22, 29,30,47–50, 55–82,84, 94–95,121	8,23–28,86	18,96–100, 102–104
Refer to section . . .	3	2	4	5	1	6

23. $(-3, 14), (-2, 5), (-1, 0), (0, -1), (1, 2), (2, 9), (3, 20)$;
range: $\{-1, 0, 2, 5, 9, 14, 20\}$

25. (a) 17 **(b)** 4 **(c)** $5k^2 - 3$ **(d)** $-9m^2 + 12m + 1$
(e) $5x^2 + 10xh + 5h^2 - 3$ **(f)** $-x^2 - 2xh - h^2 + 4x + 4h + 1$
(g) $10x + 5h$ **(h)** $-2x - h + 4$ **27.** $(-\infty, 0) \cup (0, \infty)$
29. $(-7, \infty)$ **31.**

33.

35.

37.

39.

41.

43.

45.

47.

49.

51. -5 **53.** -6 **55.** $\log_3 243 = 5$ **57.** $\ln 2.22554 \approx 0.8$ **59.** $2^5 = 32$ **61.** $e^{4.41763} \approx 82.9$ **63.** 4 **65.** 3/2
67. $\log_5(21k^4)$ **69.** $\log_3(y^4/x^2)$ **71.** $p \approx 1.581$ **73.** $m \approx -1.807$ **75.** $x \approx -3.305$ **77.** $m \approx 2.156$ **79.** $k = 2$
81. $p = 3/4$ **83. (a)** $(-\infty, \infty)$ **(b)** $(0, \infty)$ **(c)** 1 **(d)** $y = 0$ **(e)** Greater than 1 **(f)** Between 0 and 1
87. (a) \$28,000 **(b)** \$7000 **(c)** \$63,000 **(d)**

(e) No **89.** \$921.95 **91.** \$15,510.79
93. \$17,901.90 **95.** 70 quarters or 17.5 years;
111 quarters or 27.75 years **97.** 6.17% **99.** \$1494.52
101. \$17,339.86 **103.** About 9.59%
105. (a) $n = 1500 - 10p$ **(b)** $R = p(1500 - 10p)$
(c) $50 \leq p \leq 150$ **(d)** $R = (1500n - n^2)/10$
(e) $0 \leq n \leq 1000$ **(f)** \$75 **(g)** 750 **(h)** \$56,250

(i)

(j) The revenue starts at \$50,000 when the price is \$50, rises to a maximum of \$56,250 when the price is \$75, and falls to 0 when the price is \$150. **107. (a)**

(b) $2x + 5$
(c) $A(x) = x + 4 + 7/x$
(d) $1 - 7/[x(x + 1)]$
109. The third day; 104.2°F

111. (a)   **(b)** $y = 2.384t^2 - 42.55t + 269.2$, $y = -0.4931t^3 + 11.26t^2 - 82.00t + 292.9$, $y = 213.8(0.9149)^t$ **(c)** $\boxed{y = -0.4931t^3 + 11.26t^2 - 82.00t + 292.9}$ **(d)** $372, -788, 36$

113. 187.9 cm; 345 kg   **115. (a)** 10.915 billion; this is about 4.006 billion more than the estimate of 6.909 billion.
(b) 26.56 billion; 96.32 billion   **117.** 0.25; 0.69 minute   **119. (a)** $S = 21.35 + 104.6 \ln A$ **(b)** $S = 85.49 A^{0.3040}$
(c)   **(d)** 742.2, 694.7   **121. (a)** 0 yr   **(b)** 1.85×10^9 yr   **(c)** As r increases, t increases, but at a slower and slower rate. As r decreases, t decreases at a faster and faster rate.

Chapter 3 The Derivative

Exercises 3.1 (page 147)
W1. $(2x + 3)(4x + 5)$
W2. $(3x - 4)(4x + 3)$
W3. $(3x + 7)/(x + 2)$
W4. $(2x - 5)/(x - 3)$

For exercises . . .	1,7(b), 8(b),9(a), 84–86	2,7(a), 10(a), 53–56	3,5(a), 6(a), 17–20	4	5(a),6(b), 8(a),9(b), 10(b),15,16	11, 12	21–30	31–38, 41,42	39, 40	43–52, 77–83, 87–95	59–65, 67, 73–76
Refer to example . . .	5	3	2	4	1	11	6	8	9	12	10

1. c   **3.** b   **5. (a)** 3   **(b)** 1   **7. (a)** 0   **(b)** Does not exist   **9. (a) (i)** -1;   **(ii)** $-1/2$ **(iii)** Does not exist;   **(iv)** Does not exist   **(b) (i)** $-1/2$ **(ii)** $-1/2$ **(iii)** $-1/2$ **(iv)** $-1/2$ **11.** 3   **15.** 4   **17.** 10   **19.** Does not exist   **21.** -18 **23.** 1/3
25. 3   **27.** 512   **29.** 2/3   **31.** 6   **33.** 3/2   **35.** -5 **37.** $-1/9$ **39.** 1/10   **41.** $2x$ **43.** 3/7   **45.** 3/2   **47.** 0
49. ∞ (does not exist)   **51.** $-\infty$ (does not exist)   **53.** 1   **55. (a)** 2   **(b)** Does not exist   **57.** $-5, 7$ **59.** 6   **61.** 1.5
63. (a) Does not exist   **(b)** $x = -2$ **(c)** If $x = a$ is a vertical asymptote for the graph of $f(x)$, then $\lim_{x \to a} f(x)$, does not exist.
67. (a) 0   **(b)** $y = 0$ **69. (a)** $-\infty$ (does not exist)   **(b)** $x = 0$ **73.** 5   **75.** 0.3333 or 1/3   **77. (a)** 1.5   **79. (a)** -2
81. (a) 8   **85. (a)** 7.25 cents   **(b)** 7.25 cents   **(c)** 7.5 cents   **(d)** Does not exist   **(e)** 7.5 cents   **87.** $6; the average cost approaches $6 as the number of DVDs becomes very large.   **89.** 63 items; the number of items a new employee produces gets closer and closer to 63 as the number of days of training increases.   **91.** $R/(i - g)$ **93. (a)** 36.2 cm; the depth of the sediment layer deposited below the bottom of the lake in 1970 is 36.2 cm.   **(b)** 155 cm; the depth of the sediment approaches 155 cm going back in time.   **95. (a)** 0.572   **(b)** 0.526   **(c)** 0.503   **(d)** 0.5; the numbers in (a), (b), and (c) give the probability that the legislator will vote yes on the second, fourth, and eighth votes. In (d), as the number of roll calls increases, the probability of a yes vote approaches 0.5 but is never less than 0.5.

Exercises 3.2 (page 158)
W1. 3   **W2.** 10   **W3.** 4   **W4.** Does not exist   **W5.** 25

For exercises . . .	1–6,34,35	7–18,31–33	19–28	36–41
Refer to example . . .	1	2	3	4

1. $a = -1$:   **(a)** $f(-1)$ does not exist.   **(b)** 1/2   **(c)** 1/2   **(d)** 1/2
(e) $f(-1)$ does not exist.   **3.** $a = 1$:   **(a)** 2   **(b)** -2 **(c)** -2 **(d)** -2 **(e)** $f(1)$ does not equal the limit.   **5.** $a = -5$:
(a) $f(-5)$ does not exist   **(b)** ∞ (does not exist)   **(c)** $-\infty$ (does not exist)   **(d)** Limit does not exist.   **(e)** $f(-5)$ does not exist and the limit does not exist; $a = 0$:   **(a)** $f(0)$ does not exist.   **(b)** 0   **(c)** 0   **(d)** 0   **(e)** $f(0)$ does not exist.
7. $a = 0$, limit does not exist; $a = 2$, limit does not exist.   **9.** $a = 2$, limit is 4   **11.** Nowhere   **13.** $a = -2$, limit does not exist.   **15.** $a < 1$, limit does not exist.   **17.** $a = 0, -\infty$ (limit does not exist); $a = 1, \infty$ (limit does not exist).
19. (a)   **(b)** 2   **(c)** 1, 5   **21. (a)**   **(b)** -1 **(c)** 11, 3   **23. (a)**

(b) None   **25.** 2/3   **27.** 4   **31. (a)** Discontinuous at $x = 1.2$ **33. (a)**   **35. (a)** $500   **(b)** $1500   **(c)** $1000

(d) Does not exist   **(e)** Discontinuous at $x = 10$; a change in shifts   **(f)** 15   **37. (a)** $100   **(b)** $150   **(c)** $125
(d) At $x = 100$ **39. (a)** $1.40   **(b)** $1.61   **(c)** Does not exist   **(d)** $1.40   **(e)** $2.66   **(f)** $2.66   **(g)** $2.66
(h) $2.66   **(i)** $1, 2, 3, \ldots, 12$ **41. (a)** About 687 g   **(b)** No   **(c)**

Exercises 3.3 (page 170)

W1. $2h^2 + 23h + 69$ **W2.** $2h + 11$ **W3.** $2/(4 + h)$
W4. $-2/[5(5 + h)]$ **1.** 6   **3.** -15 **5.** $1/3$ **7.** 0.4323
9. 17   **11.** 18   **13.** 5   **15.** 2   **17.** 2   **19.** 6.773   **21.** 1.121

For exercises ...	1–8	9–18,33, 36,38,44	19–22, 28,29,37	25–27	30–32, 34,35, 39–41	42,43
Refer to example ...	1	3	6	4 and 5	2	7

25. (a) $700 per item   **(b)** $500 per item   **(c)** $300 per item   **(d)** $1100 per item   **27. (a)** -25 boxes per dollar
(b) -20 boxes per dollar   **(c)** -30 boxes per dollar   **(d)** Demand is decreasing. Yes, a higher price usually reduces demand.
29. (a) $32.37 per year   **(b)** $37.60 per year   **(c)** $34.86 per year   **31. (a)** 19.5 cents per gallon per month
(b) -4.7 cents per gallon per month   **(c)** -0.3 cents per gallon per month   **33. (a)** Increased about 302,000 people per year
(b) Increase about 356,000 people per year   **35. (a)** -400 per year   **(b)** -150 per year   **(c)** -1682 per year
37. (a)

(b) 81.51 kilojoules per hour per hour   **(c)** 18.81 kilojoules per hour per hour   **(d)** 1.3 hours
39. (a) $-15,540$ immigrants per year   **(b)** 15,560 immigrants per year   **(c)** 10 immigrants
per year   **(d)** They are equal; no   **(e)** About 1,090,000 immigrants   **41. (a)** $4°$ per 1000 ft
(b) $1.75°$ per 1000 ft   **(c)** $-4/3°$ per 1000 ft   **(d)** $0°$ per 1000 ft   **(e)** 3000 ft; 1000 ft; if
7000 ft is changed to 10,000 ft, the lowest temperature would be at 10,000 ft.   **(f)** 9000 ft
43. (a) 50 mph   **(b)** 44 mph

Exercises 3.4 (page 188)

W1. $6x + 3h - 2$ **W2.** $-3/[(x + h - 2)(x - 2)]$
W3. $y = -(5/6)x + 7/3$ **W4.** $y = -(3/8)x + 17/4$
1. (a) 0   **(b)** 1   **(c)** -1 **(d)** Does not exist   **(e)** m

For exercises ...	5–10, 54,55, 58–61	11–14	15, 16	17, 18	19, 20	21–26	27–34, 42–45, 53	35–38	49–52, 56–57
Refer to example ...	3	4	6	7	5	1 and 9	2	10	8

3. At $x = -2$ **5.** 2   **7.** $1/4$ **9.** 0   **11.** $3; 3; 3; 3$ **13.** $-8x + 9; 25; 9; -15$ **15.** $-12/x^2; -3$; does not exist; $-4/3$
17. $1/(2\sqrt{x})$; does not exist; does not exist; $1/(2\sqrt{3})$ **19.** $6x^2; 24; 0; 54$ **21. (a)** $y = 10x - 15$ **(b)** $y = 8x - 9$
23. (a) $y = -(1/2)x + 7/2$ **(b)** $y = -(5/4)x + 5$ **25. (a)** $y = (4/7)x + 48/7$ **(b)** $y = (2/3)x + 6$ **27.** $-5; -117; 35$
29. $7.389; 8,886,112; 0.0498$ **31.** $1/2; 1/128; 2/9$ **33.** $1/(2\sqrt{2}); 1/8$; does not exist   **35.** 0   **37.** $-3; -1; 0; 2; 3; 5$
39. (a) $(a, 0)$ and (b, c) **(b)** $(0, b)$ **(c)** $x = 0$ and $x = b$ **41. (a)** Distance   **(b)** Velocity   **43.** 56.66   **45.** -0.0158
49. (a) $-4p - 4$ **(b)** -44; demand is decreasing at a rate of about 44 items for each increase in price of $1.   **51. (a)** $16 per table
(b) $15.996 (or $16)   **(c)** $15.998 (or $16)   **(d)** The marginal revenue gives a good approximation of the actual revenue from the
sale of the 1001st table.   **53.** Answers are in trillion of dollars.   **(a)** $2.54; 2.03; 1.02$ **(b)** $0.061; -0.019; -0.079; -0.120; -0.133$
55. 1000; the population is increasing at a rate of 1000 shellfish per time unit. 570; the population is increasing more slowly at 570
shellfish per time unit. 200; the population is increasing at a much slower rate of 200 shellfish per time unit.   **57. (a)** 1690 m per sec
(b) 4.84 days per m per sec; an increase in velocity from 1700 m per sec to 1701 m per sec indicates an approximate increase in the
age of the cheese of 4.84 days.   **59. (a)** About 270; the temperature was increasing at a rate of about $270°$ per hour at 9:00 A.M.
(b) About ±150; the temperature was decreasing at a rate of about $150°$ per hour at 11:30 A.M.   **(c)** About 0; the temperature
staying constant at 12:30 P.M.   **(d)** About 11:15 A.M.   **61. (a)** 0; about 0.5 mph/oz

Exercises 3.5 (page 196)

W1. 2   **W2.** -5 **3.** $f:Y_2; f':Y_1$ **5.** $f:Y_1; f':Y_2$

For exercises ...	2–6	7–24
Refer to example ...	4	1,2, and 3

7.

9.

11.

13.

15.

17.

19.

21. 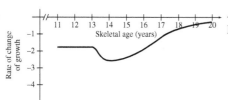 About 9 cm; about 2.6 cm less per year

23.

Chapter 3 Review Exercises (page 201)

1. True **2.** True **3.** True **4.** False **5.** True **6.** False
7. False **8.** True **9.** True **10.** True **11.** False **12.** False
17. (a) 4 **(b)** 4 **(c)** 4 **(d)** 4 **19. (a)** ∞ **(b)** −∞
(c) Does not exist **(d)** Does not exist **21.** ∞ **23.** 19/9 **25.** 8
27. −13 **29.** 1/6 **31.** 2/5 **33.** 3/8 **35.** Discontinuous at x_2 and x_4 **37.** 0, does not exist, does not exist; −1/3,
does not exist, does not exist **39.** −5, does not exist, does not exist **41.** Continuous everywhere **43. (a)**

For exercises . . .	1–3, 17–34, 45–46, 61,66	4–6,15, 35–44, 63,74	7–9, 47–50, 63–65	10–12,16, 51–58,62, 68–69,73	59–60,67, 70–72,74
Refer to section . . .	1	2	3	4	5

(b) 1 **(c)** 0, 2 **45.** 2 **47.** 126; 18 **49.** 9/77; 18/49 **51. (a)** $y = 13x − 17$ **(b)** $y = 7x − 5$
53. (a) $y = −x + 9$ **(b)** $y = −3x + 15$ **55.** $8x + 3$ **57.** 1.332 **59.**
61. (e) **63. (a)** $150 **(b)** $187.50 **(c)** $189
(d)

(e) Discontinuous at $x = $125
(f) $1.50 **(g)** $1.50 **(h)** $1.35
(i) 1.5; when 100 lb are purchased, an
additional pound will cost about $1.50 more. **(j)** 1.35; when 140 lb are purchased, an additional
pound will cost about $1.35 more. **65. (b)** $x = 7.5$ **(c)** The marginal cost equals the average
cost at the point where the average cost is smallest.

67.

about 8.2; about −0.5

69. (a)

(b) $[0.8, 5.2]$ **(c)** 3 weeks; 500 cases
(d) $V'(t) = −2t + 6$ **(e)** 0 **(f)** +; −

71. (a)   **(b)**

73. (a) Nowhere **(b)** 50, 130, 230, 770 **(c)**

Chapter 4 Calculating the Derivative

Exercises 4.1 (page 221)

For exercises . . .	1–22,27–30,55	47,56	51	53	52,54	57,58,60–62, 64,68–77	59,63, 65,67	31–43
Refer to example . . .	1,2,3,5	6	8	7	9	10	4	1 (in 4th section of previous chapter)

W1. $y = -2x + 11$

W2. $y = (2/5)x + 26/5$

1. $dy/dx = 36x^2 - 16x + 7$

3. $dy/dx = 12x^3 - 18x^2 + (1/4)x$ **5.** $f'(x) = 21x^{2.5} - 5x^{-0.5}$ or $21x^{2.5} - 5/x^{0.5}$

7. $dy/dx = 4x^{-1/2} + (9/2)x^{-1/4}$ or $4/x^{1/2} + 9/(2x^{1/4})$ **9.** $dy/dx = -30x^{-4} - 20x^{-5} - 8$ or $-30/x^4 - 20/x^5 - 8$

11. $f'(t) = -7t^{-2} + 15t^{-4}$ or $-7/t^2 + 15/t^4$ **13.** $dy/dx = -24x^{-5} + 21x^{-4} - 3x^{-2}$ or $-24/x^5 + 21/x^4 - 3/x^2$

15. $p'(x) = 5x^{-3/2} - 12x^{-5/7}$ or $5/x^{3/7} - 12/x^{5/7}$ **17.** $dy/dx = (-3/2)x^{-5/4}$ or $3/(2x^{5/4})$ **19.** $f'(x) - 2x - 5x^{-2}$ or $2x - 5/x^2$

21. $g'(x) = 256x^3 - 192x^2 + 32x$ **23.** (b) **27.** $-(9/2)x^{-3/2} - 3x^{-5/2}$ or $-9/(2x^{3/2}) - 3/x^{5/2}$ **29.** $-25/3$

31. -28; $y = -28x + 34$ **33.** $25/6$ **35.** $(4/9, 20/9)$ **37.** $-5, 2$ **39.** $(4 \pm \sqrt{37})/3$ **41.** $(-1/2, -19/2)$ **43.** $(-2, -24)$

45. 7 **51.** (a) 30 (b) 0 (c) -10 **53.** $990 **57.** (a) $18¢; 36¢$ (b) $0.732¢$ per year; $1.04¢$ per year

(c) $C(t) = -0.0001331t^3 + 0.02458t^2 - 0.5841t + 3.101$; $0.876¢$ per year; $0.901¢$ per year **59.** (a) 0.4824 (b) 2.216

61. (a) 1232.62 cm³ (b) 948.08 cm³/yr **63.** $5.00t^{0.86}$ **65.** (a) 3 minutes, 58.1 seconds (b) 0.118 sec/m; at 100 meters,

the fastest possible time increases by 0.118 second for each additional meter. (c) Yes **67.** (a) 27.5 (b) 23 pounds

(c) $-175,750/h^3$ (d) -0.64; for a 125-lb female with a height of 65 in., the BMI decreases by 0.64 for each additional inch of height.

(f) $B_m = w_m/h_m^2$ **69.** (a) $v(t) = 36t - 13$ (b) $-13; 167; 347$ **71.** (a) $v(t) = -9t^2 + 8t - 10$ (b) $-10; -195; -830$

73. 0 ft/sec; -32 ft/sec (b) 2 seconds (c) 64 ft **75.** 340 cycles/sec, -680 cycles/sec/m **77.** (a) 35, 36 (b) When $x = 5$,

$dy_1/dx = 4.13$ and $dy_2/dx \approx 4.32$. These values are fairly close and represent the rate of change of four years for a dog for one year

of a human, for a dog that is actually 5 years old. (c) $y = 4x + 16$

Exercises 4.2 (page 230)

For exercises . . .	1–10,29,34,39, 41,42,44,46–49	11–26,31–33,35, 40,45,47,50–55	27,28,30	41,42
Refer to example . . .	1,2	3	4	5

W1. $f'(x) = 12x^3 + 12x^2$ **W2.** $f'(x) = -6/x^4 + 3/\sqrt{x}$

W3. $f'(x) = 6x^{-1/3} - 2x^{-3/2}$ **1.** $dy/dx = 18x^2 - 6x + 4$

3. $dy/dx = 8x - 20$ **5.** $k'(t) = 4t^3 - 4t$

7. $dy/dx = (3/2)x^{1/2} + (1/2)x^{-1/2} + 2$ or $3x^{1/2}/2 + 1/(2x^{1/2}) + 2$ **9.** $p'(y) = -8y^{-5} + 15y^{-6} + 30y^{-7}$

11. $f'(x) = 57/(3x + 10)^2$ **13.** $dy/dt = -17/(4 + t)^2$ **15.** $dy/dx = (x^2 - 2x - 1)/(x - 1)^2$

17. $f'(t) = 2t/(t^2 + 3)^2$ **19.** $g'(x) = (4x^2 + 2x - 12)/(x^2 + 3)^2$ **21.** $p'(t) = [-\sqrt{t}/2 - 1/(2\sqrt{t})]/(t - 1)^2$ or

$(-t - 1)/[2\sqrt{t}(t - 1)^2]$ **23.** $dy/dx = (5/2)x^{-1/2} - 3x^{-3/2}$ or $(5x - 6)/(2x\sqrt{x})$ **25.** $h'(z) = (-z^{4.4} + 11z^{1.2})/(z^{3.2} + 5)^2$

27. $f'(x) = (60x^3 + 57x^2 - 24x + 13)/(5x + 4)^2$ **29.** 77 **31.** In the first step, the numerator should be

$(x^2 - 1)2 - (2x + 5)(2x)$. **33.** $y = -2x + 9$ **35.** (a) $f'(x) = (7x^3 - 4)/x^{5/3}$ (b) $f'(x) = 7x^{4/3} - 4x^{-5/3}$

39. $0, -1.307$, and 1.307 **41.** (a) $2cz + bc$ (b) $3cz^2 + 2z(d + bc) + bd$ **43.** (a) $22.86 per unit (b) $12.92 per unit

(c) $(3x + 2)/(x^2 + 4x)$ per unit (d) $\bar{C}'(x) = (-3x^2 - 4x - 8)/(x^2 + 4x)^2$ **45.** (a) $M'(d) = 2000d/(3d^2 + 10)^2$ (b) 8.3;

the new employee can assemble about 8.3 additional bicycles per day after 2 days of training. 1.4; the new employee can assemble about

1.4 additional bicycles per day after 5 days of training. **49.** Increasing at a rate of $0.03552 per gallon per month **51.** (a) $AK/(A + x)^2$

(b) $K/(4A)$ **53.** (a) 8.57 min (b) 16.36 min (c) 6.12 min²/kcal; 2.48 min²/kcal **55.** (a) 0.1173 (b) 2.625

Exercises 4.3 (page 240)

For exercises . . .	1–6	7–14,53,59	15–20	21–28,54,57, 62,63,66,67	29–34,64	35–40,56	58,65	55	60,61	45–50
Refer to example . . .	1	2	3	5,6	7	8	9	10	4	1 (in 3rd section of previous chapter)

W1. $f'(x) = (2x - x^4)/(x^3 + 1)^2$ **W2.** $f'(x) = (3 - 21x^4)/[\sqrt{x}(x^4 + 1)^2]$ **W3.** $f'(x) = (1 + 4x^{-1/3})/[3(x^{1/3} + 1)^2]$

1. 1767 **3.** 131 **5.** $320k^2 + 224k + 39$ **7.** $(6x + 55)/8; (3x + 164)/4$ **9.** $1/x^2; 1/x^2$ **11.** $\sqrt{8x^2 - 4}; 8x + 10$

13. $\sqrt{(x - 1)/x}; -1/\sqrt{x + 1}$ **15.** If $f(x) = x^{3/5}$ and $g(x) = 5 - x^2$, then $y = f[g(x)]$. **17.** If $f(x) = -\sqrt{x}$ and

$g(x) = 13 + 7x$, then $y = f[g(x)]$. **19.** If $f(x) = x^{1/3} - 2x^{2/3} + 7$ and $g(x) = x^2 + 5x$, then $y = f[g(x)]$.

21. $dy/dx = 4(8x^4 - 5x^2 + 1)^3(32x^3 - 10x)$ **23.** $k'(x) = 288x(12x^2 + 5)^{-7}$ **25.** $s'(t) = (1215/2)t^2(3t^3 - 8)^{1/2}$

27. $g'(t) = -63t^2/(2\sqrt{7t^3 - 1})$ **29.** $m'(t) = -6(5t^4 - 1)^3(85t^4 - 1)$ **31.** $dy/dx = 3x^2(3x^4 + 1)^3(19x^4 + 64x + 1)$

33. $q'(y) = 2y(y^2 + 1)^{1/4}(9y^2 + 4)$ **35.** $dy/dx = 60x^2/(2x^3 + 1)^3$ **37.** $r'(t) = 2(5t - 6)^3(15t^2 + 18t + 40)/(3t^2 + 4)^2$

39. $dy/dx = (-18x^2 + 2x + 1)/(2x - 1)^6$ **43.** (a) -2 (b) $-24/7$ **45.** $y = (3/5)x + 16/5$ **47.** $y = x$ **49.** 1, 3

53. $D(c) = (-c^2 + 10c + 12,475)/25$ **55.** (a) $101.22 (b) $111.86 (c) $117.59 **57.** (a) $-$10,500 (b) $-$4570.64

59. $P[f(a)] = 18a^2 + 24a + 9$ **61.** (a) $A[r(t)] = A(t) = 4\pi t^2$; this function gives the area of the pollution in terms of the time

since the pollutants were first emitted. (b) 32π; at 12 P.M. the area of pollution is changing at the rate of 32π mi² per hour

63. (a) -0.5 (b) $-1/54 \approx -0.02$ (c) $-1/128 \approx -0.008$ (d) Always decreasing; the derivative is negative for all $t \geq 0$.
65. (a) 34 minutes (b) $-(108/17)\pi$ mm^3 per minute, $-(72/17)\pi$ mm^2 per minute **67.** (a) $12x^{11}$ (b) x^{12}; $12x^{11}$

Exercises 4.4 (page 247)

W1. $f'(x) = 3x^5/\sqrt{x^6 + 5}$
W2. $f'(x) = 10(4x^2 + 3x + 2)^9(8x + 3)$
W3. $f'(x) = 10x(x^2 - 1)^{3/2}/(x^2 + 1)^{1/2}$

For exercises . . .	1–12,25–30,33,34, 40,41,45,46,50–54, 56,59–65	13–16,23,24, 39,42,44,49	17–22,31,32,51	43,47,48,55,57	58
Refer to example . . .	1	2	3	5	4

1. $dy/dx = 4e^{4x}$ **3.** $dy/dx = -24e^{3x}$ **5.** $dy/dx = -32e^{2x+1}$ **7.** $dy/dx = 2xe^{x^2}$ **9.** $dy/dx = 12xe^{2x^2}$
11. $dy/dx = 16xe^{2x^2-4}$ **13.** $dy/dx = xe^x + e^x = e^x(x + 1)$ **15.** $dy/dx = 2(x + 3)(2x + 7)e^{4x}$
17. $dy/dx = (2xe^x - x^2e^x)/e^{2x} = x(2 - x)/e^x$ **19.** $dy/dx = [x(e^x - e^{-x}) - (e^x + e^{-x})]/x^2$
21. $dp/dt = 8000e^{-0.2t}/(9 + 4e^{-0.2t})^2$ **23.** $f'(z) = 4(2z + e^{-z})(1 - ze^{-z})$ **25.** $dy/dx = 3(\ln 7)7^{3x+1}$
27. $dy/dx = 6x(\ln 4)4^{x^2+2}$ **29.** $ds/dt = (\ln 3)3^{\sqrt{t}}/\sqrt{t}$ **31.** $dy/dt = [(1 - t)e^{3t} - 4e^{2t} + (1 + t)e^t]/(e^{2t} + 1)^2$
33. $f'(x) = (9x + 4)e^{x\sqrt{3x+2}}/[2\sqrt{3x + 2}]$ **39.** (a) \$3.81 (b) \$0.20 (c) $C'(x)$ approaches zero.
41. (a)

(b) \$48,843, 6.694 (c) \$55,464, 2.722 **43.** (a) $G(t) = 250/(1 + 124e^{-0.45t})$ (b) 17.8 million; 7.4 million/yr (c) 105.2 million; 27.4 million/yr (d) 246.2 million; 1.7 million/yr (e) It increases for a while and then gradually decreases to 0. **45.** (a) 51,600,000 (b) 1,070,000 people/year **47.** (a) $G(t) = 31.4/(1 + 12.5e^{-0.393t})$ (b) 3.33 g; 1.17 g/day (c) 11.4 g; 2.85 g/day (d) 25.2 g; 1.95 g/day (e) It increases for a while and then gradually decreases to 0. **49.** (a) 3.857 cm^3 (b) 0.973 cm (c) 18 years (d) 1100 cm^3 (e) 0.282; at 240 months old, the tumor is increasing in volume at the instantaneous rate of 0.282 cm^3/month. **51.** (a) 0.180 (b) 2024 (c) The marginal increase in the proportion per year in 2014 is approximately 0.038. **53.** (a) 509.7 kg, 498.4 kg (b) 1239 days, 1095 days (c) 0.22 kg/day, 0.22 kg/day (d) The growth patterns of the two functions are very similar. (e) The graphs of the rates of change of the two functions are also very similar.

55. (a) $G(t) = 1/(1 + 270e^{-3.5t})$ (b) 0.109, 0.341 per century (c) 0.802, 0.555 per century (d) 0.993, 0.0256 per century (e) It increases for a while and then gradually decreases to 0. **57.** (a) $G(t) = 6.8/(1 + 3.242e^{-0.2992t})$ (b) 3.435 million students; 0.509 million students/yr (c) 5.247 million students; 0.359 million students/yr (d) 6.242 million students; 0.153 million students/yr (e) The rate increases at first and then decreases toward 0. **59.** (a) $(V/R)e^{-t/RC}$ (b) 1.35×10^{-7} amps
61. (a) $29.24°$, 0.3361 (b) $35.18°$, 0.6918 (c) $44.46°$, 1.205 **63.** (a) 625.14 ft (b) 0; yes (c) -1.476 ft per ft from center
65. (a) 218.4 seconds (b) The record is decreasing by 0.025 second per year at the end of 2014. (c) 218 seconds. If the estimate is correct, then this is the least amount of time that it will ever take for a human to run a mile.

Exercises 4.5 (page 255)

W1. $f'(x) = 2xe^{x^2}$ **W2.** $f'(x) = 2x(2x + 1)e^{4x}$
W3. $f'(x) = e^{2x}(2 - e^{3x})/(e^{3x} + 1)^2$

For exercises . . .	1,2,59,64,66	3,4,7–10,31,32,60,67	5,6,11–30,33–44, 57,58,61,62,65	68
Refer to example . . .	1	2	3	4

1. $dy/dx = 1/x$ **3.** $dy/dx = -3/(8 - 3x)$ or $3/(3x - 8)$ **5.** $dy/dx = (8x - 9)/(4x^2 - 9x)$ **7.** $dy/dx = 1/[2(x + 5)]$
9. $dy/dx = 3(2x^2 + 5)/[x(x^2 + 5)]$ **11.** $dy/dx = -15x/(3x + 2) - 5\ln(3x + 2)$ **13.** $ds/dt = t + 2t \ln|t|$
15. $dy/dx = [2x - 4(x + 3)\ln(x + 3)]/[x^3(x + 3)]$ **17.** $dy/dx = (4x + 7 - 4x \ln x)/[x(4x + 7)^2]$
19. $dy/dx = (6x \ln x - 3x)/(\ln x)^2$ **21.** $dy/dx = 4(\ln|x + 1|)^3/(x + 1)$ **23.** $dy/dx = 1/(x \ln x)$ **25.** $dy/dx = e^{x^2}/x + 2xe^{x^2} \ln x$
27. $dy/dx = (xe^x \ln x - e^x)/[x(\ln x)^2]$ **29.** $g'(z) = 3(e^{2z} + \ln z)^2(2e^{2z} + 1/z)$ **31.** $dy/dx = 1/(x \ln 10)$
33. $dy/dx = -1/[(\ln 10)(1 - x)]$ or $1/[(\ln 10)(x - 1)]$ **35.** $dy/dx = 5/[(2 \ln 5)(5x + 2)]$
37. $dy/dx = 3(x + 1)/[(\ln 3)(x^2 + 2x)]$ **39.** $dw/dp = (\ln 2)2^p/[(\ln 8)(2^p - 1)]$
41. $f'(x) = e^{\sqrt{x}}\{1/[2\sqrt{x}(\sqrt{x} + 5)] + \ln(\sqrt{x} + 5)/[2\sqrt{x}]\}$
43. $f'(t) = [(t^2 + 1)\ln(t^2 + 1) - t^2 + 2t + 1]/\{(t^2 + 1)[\ln(t^2 + 1) + 1]^2\}$
55. $h(x) = (x^2 + 1)^{5x}[10x^2/(x^2 + 1) + 5\ln(x^2 + 1)]$ **57.** (a) $dR/dq = 100 + 50(\ln q - 1)/(\ln q)^2$
(b) \$112.48 (c) To decide whether it is reasonable to sell additional items **59.** (a) $-\$0.19396$ (b) $-\$0.06099$
61. (a) $N(t) = 1000e^{9.8901e^{-e^{-0.25t}} - 0.216t}$ (b) 1,307,416 bacteria per hour; the number of bacteria is increasing at a rate of 1,307,416 per hour, 20 hours after the experiment began.

(c)

(d)

(e) 9.8901; $1000e^{9.8901} \approx$ 19,734,033   **63. (b) (i)** 3.343 imagoes per mated female per day   **(ii)** 1.466 imagoes per mated female per day   **(c) (i)** −0.172 imagoes per mated female per day per flies per bottle   **(ii)** −0.0511 imagoes per mated female per day per flies per bottle   **65.** 26.9 ants/day; 13.1 ants/day

67. (a) 1.567×10^{11} kWh   **(b)** 63.4 months   **(c)** 4.14×10^{-26} **(d)** dM/dE decreases and approaches zero.

Chapter 4 Review Exercises (page 259)

For exercises . . .	1,2,4,11–16, 53,54,75,81–83	3,17–20,55,56,67,68 76,77,90	5,21–30,51,52,57, 58,69–72,78,93	6–7,31–36,43–46,59,60, 73,79,84–89,91,92	8–10,37–42,47–50, 61,62,74,80
Refer to section . . .	1	2	3	4	5

1. False   **2.** True   **3.** False   **4.** True   **5.** False   **6.** False   **7.** False   **8.** True   **9.** True   **10.** False
11. $dy/dx = 15x^2 - 14x - 9$ **13.** $dy/dx = 24x^{5/3}$ **15.** $f'(x) = -12x^{-5} + 3x^{-1/2}$ or $-12/x^5 + 3/x^{1/2}$
17. $k'(x) = 21/(4x + 7)^2$ **19.** $dy/dx = (x^2 - 2x)/(x - 1)^2$ **21.** $f'(x) = 24x(3x^2 - 2)^3$ **23.** $dy/dt = 7t^6/(2t^7 - 5)^{1/2}$
25. $dy/dx = 3(2x + 1)^2(8x + 1)$ **27.** $r'(t) = (-15t^2 + 52t - 7)/(3t + 1)^4$ **29.** $p'(t) = t(t^2 + 1)^{3/2}(7t^2 + 2)$
31. $dy/dx = -12e^{2x}$ **33.** $dy/dx = -6x^2 e^{-2x^3}$ **35.** $dy/dx = 10xe^{2x} + 5e^{2x} = 5e^{2x}(2x + 1)$ **37.** $dy/dx = 2x/(2 + x^2)$
39. $dy/dx = (x - 3 - x \ln |3x|)/[x(x - 3)^2]$ **41.** $dy/dx = [e^x(x + 1)(x^2 - 1) \ln(x^2 - 1) - 2x^2 e^x]/[(x^2 - 1)[\ln(x^2 - 1)]^2]$
43. $ds/dt = 2(t^2 + e^t)(2t + e^t)$ **45.** $dy/dx = -6x(\ln 10) \cdot 10^{-x^2}$ **47.** $g'(z) = (3z^2 + 1)/[(\ln 2)(z^3 + z + 1)]$
49. $f'(x) = (x + 1)e^{3x}/(xe^x + 1) + 2e^{2x} \ln(xe^x + 1)$ **51. (a)** −3/2   **(b)** −24/11   **53.** −2; $y = -2x - 4$
55. −3/4; $y = -(3/4)x - 9/4$ **57.** 3/4; $y = (3/4)x + 7/4$ **59.** 1; $y = x + 1$ **61.** 1; $y = x - 1$
63. No points if $k > 0$; exactly one point if $k = 0$ or if $k < -1/2$; exactly two points if $-1/2 \leq k < 0$.   **65.** 5%; 5.06%
67. $abn^2 x^{n-1}/(n - x)^{n+1}$ **69.** $\overline{C}'(x) = (-x - 2)/[2x^2(x + 1)^{1/2}]$ **71.** $\overline{C}'(x) = (x^2 + 3)^2(5x^2 - 3)/x^2$
73. $\overline{C}'(x) = [e^{-x}(x + 1) - 10]/x^2$ **75. (a)** 22; sales will increase by $22 million when $1000 more is spent on research.
(b) 19.5; sales will increase by $19.5 million when $1000 more is spent on research.   **(c)** 18; sales will increase by
$18 million when $1000 more is spent on research.   **(d)** As more is spent on research, the increase in sales is decreasing.
77. (a) −2.201; costs will decrease by $2201 for the next $100 spent on training.   **(b)** −0.564; costs will decrease
by $564 for the next $100 spent on training.   **(c)** Decreasing   **79.** $218.65. The balance increases by roughly
$218.65 for every 1% increase in the interest rate when the rate is 5%.   **81. (a)** 4.64 billion/yr; the volume of mail
is increasing by about 4,640,000,000 pieces per year.   **(b)** −2.59 billion/yr; the volume of mail is decreasing by
about 2,590,000,000 pieces per year.   **83. (a)** $y = 1.799 \times 10^{-5}t^3 + 3.177 \times 10^{-4}t^2 - 0.06866t + 2.504$,
$y = -1.112 \times 10^{-6}t^4 + 2.920 \times 10^{-4}t^3 - 0.02238t^2 + 0.6483t - 4.278$ **(b)** $0.59/yr, $0.46/yr
85. (a) $G(t) = 30,000/(1 + 14e^{-0.15t})$ **(b)** 4483; 572   **87. (a)** 3493.76 grams   **(b)** 3583 grams   **(c)** 84 days   **(d)** 1.76 g/day
(e) $M(t) = 3583e^{-e^{-0.020(t-66)}}$ Growth is initially rapid, then tapers off.   **(f)**

Day	Weight	Rate
50	904	24.90
100	2159	21.87
150	2974	11.08
200	3346	4.59
250	3494	1.76
300	3550	0.66

89. (a) 13,980 megawatts/yr   **(b)** 44,133 megawatts/yr   **(c)** 139,325 megawatts/yr   **91.** 0.239; the production of corn is increasing
at a rate of 0.239 billion bushels a year in 2000.   **93. (a)** −0.4677 fatalities per 1000 licensed drivers per 100 million miles per year; at
the age of 20, each extra year results in a decrease of 0.4677 fatalities per 1000 licensed drivers per 100 million miles.
(b) 0.003672 fatalities per 1000 licensed drivers per 100 million miles per year; at the age of 60, each extra year results in an
increase of 0.003672 fatalities per 1000 licensed drivers per 100 million miles.

Chapter 5 Graphs and the Derivative

Exercises 5.1 (page 275)
W1. −3/2, 5/4   **W2.** 0, −1/3, 15/4   **W3.** −3, 3
W4. 0, 2, −2   **W5.** $f'(x) = -x(x^3 + 12x - 10)/(x^3 + 5)^2$
W6. $f'(x) = -x/\sqrt{3 - x^2}$ **W7.** $f'(x) = x(15x^3 + 2)e^{5x^3}$
W8. $f'(x) = 3x^{1/2}/(x^{3/2} + 5)$ **1. (a)** $(1, \infty)$ **(b)** $(-\infty, 1)$
3. (a) $(-\infty, -2)$ **(b)** $(-2, \infty)$ **5. (a)** $(-\infty, -4), (-2, \infty)$ **(b)** $(-4, -2)$ **7. (a)** $(-7, -4), (-2, \infty)$ **(b)** $(-\infty, -7), (-4, -2)$

For exercises . . .	1–8, 43,44 50,60, 61	13–22,25, 29–34,38, 39,42,49, 51,54–59	21–24	26–28, 35,36	45,52,53	46–48
Refer to example . . .	1	2	4	3	6	5

9. (a) $(-\infty, -1), (3, \infty)$ **(b)** $(-1, 3)$ **11. (a)** $(-\infty, -8), (-6, -2.5), (-1.5, \infty)$ **(b)** $(-8, -6), (-2.5, -1.5)$ **13. (a)** 17/12
(b) $(-\infty, 17/12)$ **(c)** $(17/12, \infty)$ **15. (a)** $-3, 4$ **(b)** $(-\infty, -3), (4, \infty)$ **(c)** $(-3, 4)$ **17. (a)** $-3/2, 4$ **(b)** $(-\infty, -3/2), (4, \infty)$
(c) $(-3/2, 4)$ **19. (a)** $-2, -1, 0$ **(b)** $(-2, -1), (0, \infty)$ **(c)** $(-\infty, -2), (-1, 0)$ **21. (a)** None **(b)** None **(c)** $(-\infty, \infty)$
23. (a) None **(b)** None **(c)** $(-\infty, -1), (-1, \infty)$ **25. (a)** 0 **(b)** $(0, \infty)$ **(c)** $(-\infty, 0)$ **27. (a)** 0 **(b)** $(0, \infty)$ **(c)** $(-\infty, 0)$
29. (a) 7 **(b)** $(7, \infty)$ **(c)** $(3, 7)$ **31. (a)** 1/3 **(b)** $(-\infty, 1/3)$ **(c)** $(1/3, \infty)$ **33. (a)** 0, 2/ln 2 **(b)** $(0, 2/\ln 2)$
(c) $(-\infty, 0), (2/\ln 2, \infty)$ **35. (a)** 0, 2/5 **(b)** $(0, 2/5)$ **(c)** $(-\infty, 0), (2/5, \infty)$ **39.** Vertex: $(-b/(2a), (4ac - b^2)/(4a))$;
increasing on $(-\infty, -b/(2a))$, decreasing on $(-b/(2a), \infty)$ **41.** On $(0, \infty)$; nowhere; nowhere **43. (a)** About $(567, \infty)$
(b) About $(0, 567)$ **45. (a)** Nowhere **(b)** $(0, \infty)$ **47.** $(0, 2200)$ **49. (a)** Increasing on $(0, 7.4)$ or from 2000 to
about the middle of 2007 **(b)** Decreasing on $(7.4, 50)$ or from about the middle of 2007 to 2050 **51. (a)** $(0, 1.85)$
(b) $(1.85, 5)$ **53. (a)** $(0, 1)$ **(b)** $(1, \infty)$ **55. (a)** $F'(t) = 175.9e^{-t/1.3}(1 - 0.769t)$ **(b)** $(0, 1.3); (1.3, \infty)$
57. (a) $dR/dN = -N_0/\left[2N^{3/2}\sqrt{2\pi D(N_0 - N)}\right]$ **(b)** Decreases **(c)** It approaches $-\infty$; it approaches $-\infty$. **59.** $(-\infty, 0); (0, \infty)$
61. (a) $(2500, 5750)$ **(b)** $(5750, 6000)$ **(c)** $(2800, 4800)$ **(d)** $(2500, 2800)$ and $(4800, 6000)$

Exercises 5.2 (page 287)

W1. Increasing on $(-\infty, -2)$ and $(3, \infty)$; decreasing
on $(-2, 3)$ **W2.** Increasing on $(-3/2, 0)$ and $(4, \infty)$;
decreasing on $(-\infty, -3/2)$ and $(0, 4)$

For exercises . . .	1–8, 37–39, 52	13–20, 35,36, 53	21–24	25–34, 54–56, 58	43–46, 48–51	47,57,59
Refer to example . . .	2	3(a)	3(b)	3(c)	4	1

1. Relative minimum of -4 at 1 **3.** Relative maximum of 3 at -2 **5.** Relative maximum of 3 at -4; relative minimum of 1 at -2
7. Relative maximum of 3 at -4; relative minimum of -2 at -7 and -2 **9.** Relative maximum at -1; relative minimum at 3
11. Relative maxima at -8 and -2.5; relative minima at -6 and -1.5 **13.** Relative minimum of 8 at 5 **15.** Relative maximum of
-8 at -3; relative minimum of -12 at -1 **17.** Relative maximum of 827/96 at $-1/4$; relative minimum of $-377/6$ at -5
19. Relative maximum of -4 at 0; relative minimum of -85 at 3 and -3 **21.** Relative maximum of 3 at $-8/3$ **23.** Relative
maximum of 1 at -1; relative minimum of 0 at 0 **25.** No relative extrema **27.** Relative maximum of 0 at 1; relative minimum of
8 at 5 **29.** Relative maximum of -2.46 at -2; relative minimum of -3 at 0 **31.** No relative extrema **33.** Relative minimum of
$e \ln 2$ at $1/\ln 2$ **35.** $(3, 13)$ **37.** Relative maximum of 6.211 at 0.085; relative minimum of -57.607 at 2.161
39. Relative minimum at $x = 5$

41. Relative maximum at -1; relative minimum at 3 **43. (a)** 13
(b) \$44 **(c)** \$258 **45. (a)** 100 **(b)** \$14.72 **(c)** \$635.76
47. Relative maximum of 18,400 megawatts at midnight $(t = 0)$;
relative minimum of 16,700 megawatts at 3:22 A.M.; relative
maximum of 25,200 megawatts at 5:29 P.M.; relative minimum of
18,100 megawatts at midnight $(t = 24)$. **49.** $q = 10; p \approx \$73.58$
51. 120 units **53.** 5:04 P.M.; 6:56 A.M. **55.** 4.96 years; 458.22 kg
57. 10 **59. (a)** 28 ft **(b)** 2.57 sec

Exercises 5.3 (page 301)

W1. Relative maximum of 196 at $x = -4$; relative
minimum of -304 at $x = 6$ **W2.** Relative maximum
of 10 at $x = 0$; relative minimum of -38 at $x = -2$
and of -2038 at $x = 8$ **1.** $f''(x) = 30x - 14; -14; 46$

For exercises . . .	1–16	17–26, 76,77	29–50,56, 67–70,79, 80,82,87–91	59–66, 83,84,95	72–75 81,85,86	93,94,96
Refer to example . . .	2	3	5	6	7	4

3. $f''(x) = 48x^2 - 18x - 4; -4; 152$ **5.** $f''(x) = 6; 6; 6$ **7.** $f''(x) = 2/(1 + x)^3; 2; 2/27$
9. $f''(x) = 4/(x^2 + 4)^{3/2}; 1/2; 1/(4\sqrt{2})$ **11.** $f''(x) = -6x^{-5/4}$ or $-6/x^{5/4}$; $f''(0)$ does not exist; $-3/2^{1/4}$
13. $f''(x) = 20x^2e^{-x^2} - 10e^{-x^2}; -10; 70e^{-4} \approx 1.282$ **15.** $f''(x) = (-3 + 2 \ln x)/(4x^3)$; does not exist; -0.050
17. $f'''(x) = 168x + 36; f^{(4)}(x) = 168$ **19.** $f'''(x) = 300x^2 - 72x + 12; f^{(4)}(x) = 600x - 72$ **21.** $f'''(x) = 18(x + 2)^{-4}$ or
$18/(x + 2)^4; f^{(4)}(x) = -72(x + 2)^{-5}$ or $-72/(x + 2)^5$ **23.** $f'''(x) = -36(x - 2)^{-4}$ or $-36/(x - 2)^4; f^{(4)}(x) = 144(x - 2)^{-5}$ or
$144/(x - 2)^5$ **25. (a)** $n!$ **(b)** 0 **27.** $f''(x) = e^x; f'''(x) = e^x; f^{(n)}(x) = e^x$ **29.** Concave upward on $(2, \infty)$; concave downward
on $(-\infty, 2)$; inflection point at $(2, 3)$ **31.** Concave upward on $(-\infty, -1)$ and $(8, \infty)$; concave downward on $(-1, 8)$; inflection points
at $(-1, 7)$ and $(8, 6)$ **33.** Concave upward on $(2, \infty)$; concave downward on $(-\infty, 2)$; no inflection points **35.** Always concave
upward; no inflection points **37.** Concave upward on $(-\infty, 3/2)$; concave downward on $(3/2, \infty)$; inflection point at $(3/2, 525/2)$
39. Concave upward on $(5, \infty)$; concave downward on $(-\infty, 5)$; no inflection points **41.** Concave upward on $(-10/3, \infty)$; concave
downward on $(-\infty, -10/3)$; inflection point at $(-10/3, -250/27)$ **43.** Never concave upward; always concave downward; no
inflection points **45.** Concave upward on $(-\infty, 0)$ and $(1, \infty)$; concave downward on $(0, 1)$; inflection points at $(0, 0)$ and $(1, -3)$
47. Concave upward on $(-1, 1)$; concave downward on $(-\infty, -1)$ and $(1, \infty)$; inflection points at $(-1, \ln 2)$ and $(1, \ln 2)$
49. Concave upward on $(-\infty, -e^{-3/2})$ and $(e^{-3/2}, \infty)$; concave downward on $(-e^{-3/2}, 0)$ and $(0, e^{-3/2})$; inflection points at
$(-e^{-3/2}, -3e^{-3}/(2 \ln 10))$ and $(e^{-3/2}, -3e^{-3}/(2 \ln 10))$ **51.** Concave upward on $(-\infty, 0)$ and $(4, \infty)$; concave downward
on $(0, 4)$; inflection points at 0 and 4 **53.** Concave upward on $(-7, 3)$ and $(12, \infty)$; concave downward on $(-\infty, -7)$ and $(3, 12)$;
inflection points at $-7, 3,$ and 12 **55.** Choose $f(x) = x^k$ where $1 < k < 2$. For example, $f(x) = x^{4/3}$ has a relative minimum at
$x = 0$, and $f(x) = x^{5/3}$ has an inflection point at $x = 0$. **57. (a)** Close to 0 **(b)** Close to 1 **59.** Relative maximum at -5

61. Relative maximum at 0; relative minimum at 2/3   **63.** Relative minimum at −3   **65.** Relative maximum at −4/7; relative minimum at 0   **67. (a)** Minimum at about −0.4 and 4.0; maximum at about 2.4   **(b)** Increasing on about (−0.4, 2.4) and about (4.0, ∞); decreasing on about (−∞, −0.4) and (2.4, 4.0)   **(c)** About 0.7 and 3.3   **(d)** Concave upward on about (−∞, 0.7) and (3.3, ∞); concave downward on about (0.7, 3.3)   **69. (a)** Maximum at 1; minimum at −1   **(b)** Increasing on (−1, 1); decreasing on (−∞, −1) and (1, ∞)   **(c)** About −1.7, 0, and about 1.7   **(d)** Concave upward on about (−1.7, 0) and (1.7, ∞); concave downward on about (−∞, −1.7) and (0, 1.7)   **73.** (22, 6517.9)   **75.** (2.06, 20.7)   **79. (a)** $R'(x) = C(xke^{-kx} + 1 - e^{-kx})$, all x in $[0, 1]$ **(b)** $R''(x) = Cke^{-kx}(2 - kx), 0 \le x < 2/k$ **81. (a)** Initial population   **(b)** Inflection point   **(c)** Maximum carrying capacity   **83. (a)** After 2 hours   **(b)** 3/4%   **85.** (6.427, 15.7)   **87.** Inflection point at $t = (\ln c)/k \approx 2.96$ years; this signifies the time when the rate of growth begins to slow down, since L changes from concave up to concave down at this inflection point.
89. Always concave down   **91.** $1/\sqrt{6\pi D}$, 110   **93. (a)** −96 ft/sec   **(b)** −160 ft/sec   **(c)** −256 ft/sec   **(d)** −32 ft/sec²
95. $v(t) = 256 - 32t; a(t) = -32$; 1024 ft; 16 seconds after being shot

Exercises 5.4 (page 312)
W1. $(-1, 0), (2, -27)$ **W2.** $(3, -140.7), (4, -160.7)$

For exercises . . .	3–10, 29–30	11,12	13–20	21–28
Refer to example . . .	1	2	3	4

1. 0   **3.**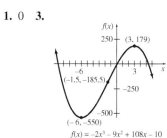

$f(x) = -2x^3 - 9x^2 + 108x - 10$

5.

$f(x) = -3x^3 + 6x^2 - 4x - 1$

7.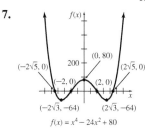

$f(x) = x^4 - 24x^2 + 80$

9.

$f(x) = x^4 - 4x^3$

11.

$f(x) = 2x + \frac{10}{x}$

13.

$f(x) = \frac{-x + 4}{x + 2}$

15.

$f(x) = \frac{1}{x^2 + 4x + 3}$

17.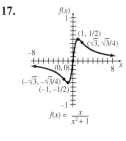

$f(x) = \frac{x}{x^2 + 1}$

19.

$f(x) = \frac{1}{x^2 - 9}$

21.

$f(x) = x \ln |x|$

23.

$f(x) = \frac{\ln x}{x}$

25.

$f(x) = xe^{-x}$

27.

$f(x) = (x - 1)e^{-x}$

29.

$f(x) = x^{2/3} - x^{5/3}$

31. 3, 7, 9, 11, 15   **33.** 17, 19, 23, 25, 27

In Exercises 35–39, other answers are possible.   **35.**   **37.**   **39.**

Chapter 5 Review Exercises (page 315)

For exercises . . .	1,2,12,13, 17–24,70	3,4,14,15, 25–32,63,64, 74,75(a),76(b),(c)	5–9,16,33–38, 61,62,65–68,73, 75(b),76(a)	10,11, 39–60,63(e), 69–71
Refer to section . . .	1	2	3	4

1. True   **2.** False   **3.** False   **4.** True   **5.** False
6. True   **7.** False   **8.** False   **9.** False
10. False   **11.** True   **12.** False
17. Increasing on $(-9/2, \infty)$; decreasing on $(-\infty, -9/2)$ **19.** Increasing on $(-5/3, 3)$; decreasing on $(-\infty, -5/3)$ and
$(3, \infty)$ **21.** Never decreasing; increasing on $(-\infty, 3)$ and $(3, \infty)$ **23.** Decreasing on $(-\infty, -1)$ and $(0, 1)$; increasing on
$(-1, 0)$ and $(1, \infty)$ **25.** Relative maximum of -4 at 2   **27.** Relative minimum of -7 at 2   **29.** Relative maximum of 101
at -3; relative minimum of -24 at 2   **31.** Relative maximum at $(-0.618, 0.206)$; relative minimum at $(1.618, 13.203)$
33. $f''(x) = 36x^2 - 10$; 26; 314   **35.** $f''(x) = 180(3x - 6)^{-3}$ or $180/(3x - 6)^3$; $-20/3$; $-4/75$
37. $f''(t) = (t^2 + 1)^{-3/2}$ or $1/(t^2 + 1)^{3/2}$; $1/2^{3/2} \approx 0.354$; $1/10^{3/2} \approx 0.032$

39.
$f(x) = -2x^3 - \frac{1}{2}x^2 + x - 3$

41.
$f(x) = x^4 - \frac{4}{3}x^3 - 4x^2 + 1$

43.
$f(x) = \frac{x-1}{2x+1}$

45.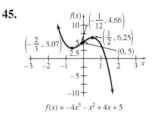
$f(x) = -4x^3 - x^2 + 4x + 5$

47.
$f(x) = x^4 + 2x^2$

49.
$f(x) = \frac{x^2 + 4}{x}$

51.
$f(x) = \frac{2x}{3 - x}$

53.
$f(x) = xe^{2x}$

55.
$f(x) = \ln(x^2 + 4)$

57.
$f(x) = 4x^{1/3} + x^{4/3}$

In Exercise 59, other answers are possible.   **59.**

61. (a) Both are negative.   **63. (a)** $P(q) = -q^3 + 7q^2 + 49q$ **(b)** 7 brushes   **(c)** \$229   **(d)** \$343   **(e)** $q = 7/3$; between
2 and 3 brushes.   **65.** $f(t)$ is increasing and concave downward. $f'(t)$ is positive and decreasing. $f''(t)$ is negative.   **67. (a)** The
first derivative has many critical numbers.   **(b)** The curve is always decreasing except at frequent inflection points.

69.

71.

73. 7.6108 yr; the age at which the rate of learning to pass the test
begins to slow down   **75. (a)** At 1965, 1973, 1976, 1983, 1986,
and 1988   **(b)** Concave upward; this means that the stockpile was
increasing at an increasingly rapid rate.   **77. (b)** $[0, 44], [56, 65],$
$[75, 84], [97, 100]$.   **(c)** $(50, 1.5), (60, 3.0), (70, 3.5), (80, 4.0),$
$(90, 4.5)$

Chapter 6 Applications of the Derivative

Exercises 6.1 (page 327)

For exercises . . .	1–6,31–38	7,8,11–30,39	41–46,52–62	47–51
Refer to example . . .	2	1	3	On Graphical Optimization

W1. $-4/3, 5/2$ **W2.** $0, 5/3$
1. Absolute maximum at x_3; no absolute minimum
3. No absolute extrema   **5.** Absolute minimum at x_1; no absolute maximum   **7.** Absolute maximum at x_1; absolute minimum at x_2
11. Absolute maximum of 12 at $x = 5$; absolute minimum of -8 at $x = 0$ and $x = 3$ **13.** Absolute maximum of 19.67 at $x = -4$;

absolute minimum of -1.17 at $x = 1$ **15.** Absolute maximum of 1 at $x = 0$; absolute minimum of -80 at $x = -3$ and $x = 3$ **17.** Absolute maximum of $1/3$ at $x = 0$; absolute minimum of $-1/3$ at $x = 3$ **19.** Absolute maximum of 0.21 at $x = 1 + \sqrt{2} \approx 2.4$; absolute minimum of 0 at $x = 1$ **21.** Absolute maximum of 1.710 at $x = 3$; absolute minimum of -1.587 at $x = 0$ **23.** Absolute maximum of 7 at $x = 1$; absolute minimum of 0 at $x = 0$ **25.** Absolute maximum of 4.910 at $x = 4$; absolute minimum of -1.545 at $x = 2$ **27.** Absolute maximum of 19.09 at $x = -1$; absolute minimum of 0.6995 at $x = (\ln 3)/3$ **29.** Absolute maximum of 1.356 at $x = 0.6085$; absolute minimum of 0.5 at $x = -1$ **31.** Absolute minimum of 7 at $x = 2$; no absolute maximum   **33.** Absolute maximum of 137 at $x = 3$; no absolute minimum   **35.** Absolute maximum of 0.1 at $x = 4$; absolute minimum of -0.5 at $x = -2$ **37.** Absolute maximum of 0.1226 at $x = e^{1/3}$; no absolute minimum   **39. (a)** Absolute minimum of -5 at $x = -1$; absolute maximum of 0 at $x = 0$ **(b)** Absolute maximum of about -0.76 at $x = 2$; absolute minimum of -1 at $x = 1$ **41. (a)** Relative maxima of 8496 in 2001, 7556 in 2004, 6985 in 2006, and 6700 in 2008; relative minima of 7127 in 2000, 7465 in 2003, 6748 in 2005, 5933 in 2007, and 5014 in 2011   **(b)** Absolute maximum of 8496 in 2001 and absolute minimum of 5014 in 2011   **43.** The maximum profit is \$700,000 when 1,000,000 tires are sold.   **45. (a)** 112   **(b)** 162   **47.** About 11.5 units   **49.** 100 units   **53.** 6 mo; 6%   **55.** About 7.2 mm   **57.** Maximum of 25 mpg at 45 mph; minimum of 16.1 mpg at 65 mph   **59.** The piece formed into a circle should have length $12\pi/(4 + \pi)$ ft, or about 5.28 ft.

Exercises 6.2 (page 336)

For exercises . . .	1–8,15–18, 27,35,36, 39,48	9–14, 19–24, 28,51	25,26, 29–32, 47	33,34, 43,44, 49,50	37,38, 40–42, 45,46
Refer to example . . .	1	3	4	2	5

W1. $0, -4/5, 7/2$ **W2.** $\left(-1 \pm \sqrt{13}\right)/2$
1. (a) $y = 180 - x$ **(b)** $P = x(180 - x)$ **(c)** $[0, 180]$
(d) $dP/dx = 180 - 2x$; $x = 90$ **(e)** $P(0) = 0$; $P(180) = 0$;
$P(90) = 8100$ **(f)** 8100; 90 and 90   **3. (a)** $y = 90 - x$ **(b)** $P = x^2(90 - x)$ **(c)** $[0, 90]$ **(d)** $dP/dx = 180x - 3x^2$;
$x = 0, x = 60$ **(e)** $P(0) = 0, P(60) = 108,000, P(90) = 0$ **(f)** 108,000; 60 and 30   **5.** $\overline{C}(x) = x^2/2 + 2x - 3 + 35/x$;
$x = 2.722$ **7. (a)** $R(x) = 160,000x - 100x^2$ **(b)** 800   **(c)** \$640,000   **9. (a)** $1400 - 2x$ **(b)** $A(x) = 1400x - 2x^2$
(c) 350 m   **(d)** 245,000 m^2 **11.** 405,000 m^2 **13.** \$960   **15. (a)** 125 passengers   **(b)** \$156,250   **17.** In 10 days; \$960
19. 4 in. by 4 in. by 2 in.   **21.** 2/3 ft (or 8 in.)   **23.** 20 cm by 20 cm by 40 cm; \$7200   **27.** 12.5 ft per side; \$28,906.25
29. Radius $= 5.206$ cm, height $= 11.75$ cm   **31.** Radius $= 5.242$ cm; height $= 11.58$ cm   **33.** 1 mile from point A
35. (a) 15 days   **(b)** 16.875%   **37.** 12.98 thousand   **39. (a)** 12 days   **(b)** 50 per mL   **(c)** 1 day   **(d)** 81.365 per mL
41. 49.37   **43.** Point P is $3\sqrt{7}/7 \approx 1.134$ mi from Point A.   **45. (a)** Replace a with e^r and b with r/P.
(b) Shepherd: $f'(S) = a[1 + (1 - c)(S/b)^c]/[1 + (S/b)^c]^2$; Ricker: $f'(S) = ae^{-bS}(1 - bS)$; Beverton–Holt:
$f'(S) = a/[1 + (S/b)]^2$ **(c)** Shepherd: a; Ricker: a; Beverton–Holt: a; the constant a represents the slope of the graph of
$f(S)$ at $S = 0$.   **(d)** 194,000 tons   **(e)** 256,000 tons   **47.** 2.18 μm, 4.36 μm   **49.** $\left(56 - 2\sqrt{21}\right)/7 \approx 6.7$ mi
51. 36 in. by 18 in. by 18 in.

Exercises 6.3 (page 347)

For exercises . . .	1,2,9–12, 15–18	3,13,14	4,27,32	5,7,8,25,26, 28–30	19–24,31
Refer to example . . .	1	2	3	4	5

W1. $f'(x) = -k/x^2$ **W2.** $f'(x) = -2a/x^3$ **3.** (c)
5. It is negative.   **9.** 10,000   **11.** 10   **13.** 4899
19. (a) $E = p/(200 - p)$ **(b)** 25   **21. (a)** $E = 2p^2/(7500 - p^2)$ **(b)** 25,000   **23. (a)** $E = 5/q$ **(b)** 5   **25. (a)** $E = 0.5$;
inelastic; total revenue increases as price increases.   **(b)** $E = 8$; elastic; total revenue decreases as price increases.   **27.** 0.06; the
demand is inelastic.   **29.** 2.826; the demand is elastic.   **31. (a)** 0.071   **(b)** Inelastic   **(c)** \$1255

Exercises 6.4 (page 353)

For exercises . . .	1–16, 38–42	17–37	43–53
Refer to example . . .	1,2	3	4

W1. $f'(x) = \ln(x^2 + 1) + 2x^2/(x^2 + 1)$ **W2.** $f'(x) = e^{x^3}(3x^3 - 2)/x^3$
1. $dy/dx = -6x/(5y)$ **3.** $dy/dx = (8x - 5y)/(5x - 3y)$ **5.** $dy/dx = 15x^2/(6y + 4)$
7. $dy/dx = -3x(2 + y)^2/2$ **9.** $dy/dx = \sqrt{y}/\left[\sqrt{x}(5\sqrt{y} - 2)\right]$ **11.** $dy/dx = (4x^3y^3 + 6x^{1/2})/(9y^{1/2} - 3x^4y^2)$
13. $dy/dx = (5 - 2xye^{x^2y})/(x^2e^{x^2y} - 4)$ **15.** $dy/dx = y(2xy^3 - 1)/(1 - 3x^2y^3)$ **17.** $y = (3/4)x + 25/4$
19. $y = x + 2$ **21.** $y = x/64 + 7/4$ **23.** $y = (11/12)x - 5/6$ **25.** $y = -(37/11)x + 59/11$ **27.** $y = (5/2)x - 1/2$
29. $y = 1$ **31.** $y = -2x + 7$ **33.** $y = -x + 2$ **35.** $y = (8/9)x + 10/9$ **37. (a)** $y = -(3/4)x + 25/2$; $y = (3/4)x - 25/2$
(b)

39. (a) $du/dv = -2u^{1/2}/(2v + 1)^{1/2}$ **(b)** $dv/du = -(2v + 1)^{1/2}/(2u^{1/2})$
(c) They are reciprocals.   **41.** $dy/dx = -x/y$; there is no function $y = f(x)$ that satisfies
$x^2 + y^2 + 1 = 0$.   **43. (a)** \$0.94; the approximate increase in cost of an additional unit
(b) \$0; the approximate change in revenue for a unit increase in sales   **45. (a)** 0.44; inelastic
(b) 0.44   **47.** $R'(w) = -29.0716w^{-1.43}$ **49.** $1/\left(3\sqrt{3}\right)$
53. $ds/dt = (4s - 6t^2 + 5)/(3s^2 - 4t)$

Exercises 6.5 (page 360)

For exercises . . .	1–8, 16–22	9–14	15	23,24,29, 31,32	25	26–28, 30
Refer to example . . .	1	5	6	3	2	4

W1. $dy/dx = -(3x^2y + y^4)/(x^3 + 4xy^3)$
W2. $dy/dx = (3y^3 - 2x)/(2y - 9xy^2)$
1. -64 **3.** $-9/7$ **5.** $1/5$ **7.** $-3/2$ **9.** \$384 per month   **11.** (a) Revenue is increasing at a rate of \$180 per day.
(b) Cost is increasing at a rate of \$50 per day.   (c) Profit is increasing at a rate of \$130 per day.   **13.** Demand is decreasing at a rate
of approximately 98 units per unit time.   **15.** 0.067 mm per min   **17.** About 1.9849 g per day   **19.** (a) $105.15 \text{ m}^{-0.25} dm/dt$
(b) About 52.89 kcal per day^2 **21.** 25.6 crimes per month   **23.** 24/5 ft/min   **25.** 16π ft^2/min   **27.** 2/27 cm/min
29. 62.5 ft per min   **31.** $\sqrt{2} \approx 1.41$ ft per sec

Exercises 6.6 (page 367)

For exercises . . .	1–8	9–16	17–20	21–35	36–41
Refer to example . . .	1	2	3	4	5

W1. $f'(x) = 2x^3/\sqrt{x^4 + 2}$ **W2.** $f'(x) = 2e^{2x}/(e^{2x} + 1)$
1. 1.9   **3.** 0.1   **5.** 0.060   **7.** -0.023 **9.** 12.0417; 12.0416; 0.0001   **11.** 0.995; 0.9950; 0   **13.** 1.01; 1.0101; 0.0001
15. 0.05; 0.0488; 0.0012   **17.** (a) -4.4 thousand lb   (b) -52.2 thousand lb   **19.** \$60   **21.** About 9600 in^3
23. (a) 0.007435   (b) -0.005105 **25.** (a) 0.347 million   (b) -0.022 million   **27.** 1568π mm^3 **29.** 80π mm^2
31. (a) About 9.3 kg   (b) About 9.5 kg   **33.** -7.2π cm^3 **35.** 0.472 cm^3 **37.** 0.00125 cm   **39.** ± 1.273 in^3 **41.** ± 0.116 in^3

Chapter 6 Review Exercises (page 371)

For exercises . . .	1–3, 11–18	4,5, 49–54	6,19–30, 43,44,56	7,8, 31–38, 55,59–61	9,10, 39–42, 62,63	45–48, 57,58, 64–68
Refer to section . . .	1	3	4	5	6	2

1. False   **2.** True   **3.** False   **4.** True   **5.** True
6. True   **7.** True   **8.** True   **9.** True   **10.** True
11. Absolute maximum of 33 at 4; absolute minimum of 1 at 0
and 6   **13.** Absolute maximum of 39 at -3; absolute minimum of $-319/27$ at 5/3   **17.** (a) Maximum $= 0.37$; minimum $= 0$
(b) Maximum $= 0.35$; minimum $= 0.13$ **21.** $dy/dx = (2x - 9x^2y^4)/(8y + 12x^3y^3)$ **23.** $dy/dx = 6\sqrt{y - 1}/[x^{1/3}(1 - \sqrt{y - 1})]$
25. $dy/dx = -(30 + 50x)/3$ **27.** $dy/dx = (2xy^4 + 2y^3 - y)/(x - 6x^2y^3 - 6xy^2)$ **29.** $y = (-16/23)x + 94/23$ **33.** 272
35. -2 **37.** $-8e^3$ **41.** 0.00204   **43.** (a) $(2, -5)$ and $(2, 4)$ (b) $(2, -5)$ is a relative minimum; $(2, 4)$ is a relative maximum.
(c) No   **45.** (a) 600 boxes   (b) \$720   **47.** 3 in.   **49.** 1789   **51.** 80   **53.** 0.47; inelastic   **55.** 56π ft^2 per min
57. (a)
(b) About the 15th day   **59.** 8/3 ft per min   **61.** $21/16 = 1.3125$ ft per min   **63.** ± 0.736 in^2
65. $1.25 + 2 \ln 1.5$ **67.** 10 ft; 18.67 sec

Chapter 7 Integration

Exercises 7.1 (page 385)

For exercises . . .	5–24, 27,28, 35,36, 41,42, 60,63, 66	25,26, 29–32, 61	33,34, 37–40,	43,44	45–52, 57–59, 65	53–56	67–74
Refer to example . . .	4,5	7,8	6	12	9	10	11

W1. $f'(x) = 20x^3 + 3/\sqrt{x}$
W2. $f'(x) = 21e^{3x} - 20x^{3/2}$
1. They differ only by a constant.   **5.** $6k + C$
7. $z^2 + 3z + C$ **9.** $2t^3 - 4t^2 + 7t + C$
11. $z^4 + z^3 + z^2 - 6z + C$ **13.** $10z^{3/2}/3 + \sqrt{2}z + C$ **15.** $5x^4/4 - 20x^2 + C$ **17.** $8v^{3/2}/3 - 6v^{5/2}/5 + C$
19. $4u^{5/2} - 4u^{7/2} + C$ **21.** $-7/z + C$ **23.** $-\pi^3/(2y^2) - 2\sqrt{\pi y} + C$ **25.** $6t^{-1.5} - 2 \ln |t| + C$ **27.** $-1/(3x) + C$
29. $-15e^{-0.2x} + C$ **31.** $-3 \ln |x| - 10e^{-0.4x} + e^{0.1}x + C$ **33.** $(1/4)\ln |t| + t^3/6 + C$ **35.** $e^{2u}/2 + 2u^2 + C$
37. $x^3/3 + x^2 + x + C$ **39.** $6x^{7/6}/7 + 3x^{2/3}/2 + C$ **41.** $10^x/(\ln 10) + C$ **43.** $f(x) = 3x^{5/3}/5$ **45.** $C(x) = 2x^2 - 5x + 8$
47. $C(x) = 3e^{0.01x} + 5$ **49.** $C(x) = 3x^{5/3}/5 + 2x + 114/5$ **51.** $C(x) = 5x^2/2 - \ln |x| - 153.50$
53. $p = 175 - 0.01x - 0.01x^2$ **55.** $p = 500 - 0.1\sqrt{x}$ **57.** (a) $p(t) = 21.57t^2 - 143.5t + 595.85$
(b) Approximately 2,376,000   **59.** (a) $P(x) = 25x^4/2 + 10x^3 - 40$ (b) \$240   **61.** $a \ln x - bx + C$
63. (a) $N(t) = 155.3e^{0.3219t} + 144.7$ (b) 7537   **65.** (a) $B(t) = 0.01588t^3 - 0.4574t^2 + 13.20t + 792.3$ (b) About 2,294,000
67. $v(t) = 5t^3/3 + 4t + 6$ **69.** $s(t) = -16t^2 + 6400$; 20 sec   **71.** $s(t) = 2t^{5/2} + 3e^{-t} + 1$ **73.** 160 ft/sec, 12 ft

Exercises 7.2 (page 394)

For exercises . . .	3,4,29, 33,34	5–8, 21,22	9,10,27, 28,30	11–16, 31,35, 36	17–20, 32	23–26	39–46
Refer to example . . .	1	3	2	5	4	6	7

W1. $2x^5 + 4x^{3/2} + C$ **W2.** $5 \ln |x| - 2/x^2 + C$
W3. $e^{6x}/6 + C$ **3.** $2(2x + 3)^5/5 + C$
5. $-(2m + 1)^{-2}/2 + C$ **7.** $-(x^2 + 2x - 4)^{-3}/3 + C$
9. $(4z^2 - 5)^{3/2}/12 + C$ **11.** $e^{2x^3}/2 + C$ **13.** $e^{2t-t^2}/2 + C$
15. $-e^{1/z} + C$ **17.** $[\ln(t^2 + 2)]/2 + C$ **19.** $[\ln(x^4 + 4x^2 + 7)]/4 + C$ **21.** $-1/[2(x^2 + x)^2] + C$
23. $(p + 1)^7/7 - (p + 1)^6/6 + C$ **25.** $2(u - 1)^{3/2}/3 + 2(u - 1)^{1/2} + C$ **27.** $(x^2 + 12x)^{3/2}/3 + C$

29. $(1 + 3\ln x)^3/9 + C$ **31.** $(1/2)\ln(e^{2x} + 5) + C$ **33.** $(\ln 10)(\log x)^2/2 + C$ **35.** $8^{3x^2+1}/(6\ln 8) + C$
39. (a) $R(x) = 6(x^2 + 27{,}000)^{1/3} - 180$ **(b)** 150 players **41. (a)** $C(x) = 6\ln(5x^2 + e) + 4$ **(b)** Yes
43. (a) $f(t) = 4.0674 \times 10^{-4}[(t - 1970)^{2.3}/2.3 + 1970(t - 1970)^{1.3}/1.3] + 61.298$ **(b)** About 162,000
45. $P(t) = 3250 - 1250e^{-t^2/5}$ **(b)** 3043

Exercises 7.3 (page 403)

For exercises ...	5–14, 17–22, 26,27	15,16,24–25, 28–32,35–37	33,34,38–40
Refer to example ...	1,2	3	4

3. (a) 88 **(b)** $\int_0^8 (2x + 5)\,dx$ **5. (a)** 21 **(b)** 23 **(c)** 22 **(d)** 22
7. (a) 10 **(b)** 10 **(c)** 10 **(d)** 11 **9. (a)** 8.22 **(b)** 15.48 **(c)** 11.85
(d) 10.96 **11. (a)** 6.70 **(b)** 3.15 **(c)** 4.93 **(d)** 4.17 **13. (a)** 4 **(b)** 4 **15. (a)** 4 **(b)** 5 **17.** 4π **19.** 24
21. (b) 0.385 **(c)** 0.33835 **(d)** 0.334334 **(e)** 0.333333 **25.** Left: 4428 trillion BTUs; right: 9003 trillion BTUs;
average: 6715.5 trillion BTUs **27.** Left: $235; right: $185; average: $210 **29. (a)** Left: about 582,000 cases; right: about
580,000 cases; average: about 581,000 cases **(b)** Left: about 146,000 cases; right: about 144,000 cases; average: about 145,000 cases
31. About 1300 ft; yes **33.** 4161 ft, 4607 ft, 4384 ft **35. (a)** About 680 BTU/ft² **(b)** About 320 BTU/ft² **37. (a)** 9 ft
(b) 2 sec **(c)** 4.6 ft **(d)** Between 3 and 3.5 sec **39.** 22.5 and 18 ft

Exercises 7.4 (page 415)

For exercises ...	1–6, 9–12, 15–20, 23,24	7,8,13,14, 21,22, 27–30,54	31–44,47	55–61, 63–72
Refer to example ...	1,2,3	4	5,6	7

W1. $2(x^2 + 4)^{3/2} + C$ **W2.** $2\ln(x^2 + 2) + C$ **W3.** $5e^{x^3} + C$
1. -18 **3.** $-3/2$ **5.** $28/3$ **7.** 13 **9.** $-16/3$ **11.** 76 **13.** 4/5
15. 108/25 **17.** $20e^{0.3} - 20e^{0.2} + 3\ln 2 - 3\ln 3 \approx 1.353$
19. $e^8/4 - e^4/4 - 1/6 \approx 731.4$ **21.** 91/3 **23.** $447/7 \approx 63.86$
25. $(\ln 2)^2/2 \approx 0.2402$ **27.** 49 **29.** $1/8 - 1/[2(3 + e^2)] \approx 0.07687$ **31.** 10 **33.** 76 **35.** 41/2
37. $e^2 - 3 + 1/e \approx 4.757$ **39.** $e - 2 + 1/e \approx 1.086$ **41.** 23/3 **43.** $e^2 - 2e + 1 \approx 2.952$
45. $\int_a^c f(x)\,dx = \int_a^b f(x)\,dx + \int_b^c f(x)\,dx$ **47.** -8 **51.** -12 **53. (a)** $x^5/5 - 1/5$ **(c)** $f'(1) \approx 2.746$, and $g(1) = e \approx 2.718$
55. (a) $(9000/8)(17^{4/3} - 2^{4/3}) \approx \$46{,}341$ **(b)** $(9000/8)(26^{4/3} - 17^{4/3}) \approx \$37{,}477$ **(c)** It is slowly increasing without bound.
57. No **59. (a)** 0.8778 ft **(b)** 0.6972 ft **61. (a)** 18.12 **(b)** 8.847 **63. (b)** $\int_0^{60} n(x)\,dx$ **(c)** $2(51^{3/2} - 26^{3/2})/15 \approx 30.89$ million
65. (a) $Q(R) = \pi kR^4/2$ **(b)** $0.04k$ mm per min **67. (b)** About 505,000 kJ/W$^{0.67}$ **69. (a)** About 286 million; the total
population aged 0 to 90 **(b)** About 64 million **71. (a)** $c'(t) = 1.2e^{0.04t}$ **(b)** $\int_0^{10} 1.2e^{0.04t}\,dt$ **(c)** $30e^{0.4} - 30 \approx 14.75$ billion
(d) About 12.8 yr **(e)** About 14.4 yr

Exercises 7.5 (page 425)

For exercises ...	1,2	3–6, 11–18,20, 23,24	7–10, 19,21,22, 25,26,41,42	27–30, 39,40	31–37
Refer to example ...	1	3	2	4	5

W1. $1839/28 \approx 65.68$ **W2.** $(e^6 - 1)/2 \approx 201.2$
1. 21 **3.** 20 **5.** 23/3 **7.** 366.2 **9.** 4/3
11. $2\ln 2 - \ln 6 + 3/2 \approx 1.095$ **13.** $6\ln(3/2) - 6 + 2e^{-1} + 2e \approx 2.605$
15. $(e^{-2} + e^4)/2 - 2 \approx 25.37$ **17.** 1/2 **19.** 1/20 **21.** $3(2^{4/3})/2 - 3(2^{7/3})/7 \approx 1.620$
23. $(e^9 + e^6 + 1)/3 \approx 2836$ **25.** $-1.9241, -0.4164, 0.6650$ **27. (a)** 8 yr **(b)** About $148
(c) About $771 **29. (a)** 39 days **(b)** $3369.18 **(c)** $484.02 **(d)** $2885.16 **31.** $12,931.66 **33.** $54
35. (a)

(b) $(15, 375)$ **(c)** $4500 **(d)** $3375 **37. (a)** 12 **(b)** $5616, $1116
(c) $1872, $1503 **(d)** $387 **39. (a)** About 71.25 gal **(b)** About 25 hr
(c) About 105 gal **(d)** About 47.91 hr **41. (a)** 0.019; the lower 10% of the
income producers earn 1.9% of the total income of the population. **(b)** 0.184;
the lower 40% of the income producers earn 18.4% of the total income of the population.
(c) **(d)** 0.15 **(e)** Income is distributed less equally in 2012
than in 1968.

Exercises 7.6 (page 433)

For exercises . . .	1a–12a,15, 23a–24a,29	1b–12b,17,19,20, 23b–24b,29,30, 33–35	21,22,25–28, 31,32
Refer to example . . .	1	2	3

1. (a) 12.25 **(b)** 12 **(c)** 12 **3. (a)** 3.35 **(b)** 3.3
(c) $3 \ln 3 \approx 3.296$ **5. (a)** 11.34 **(b)** 10.5 **(c)** 10.5
7. (a) 0.9436 **(b)** 0.8374 **(c)** $4/5 = 0.8$
9. (a) 1.236 **(b)** 1.265 **(c)** $2 - 2e^{-1} \approx 1.264$ **11. (a)** 5.991 **(b)** 6.167 **(c)** 6.283; Simpson's rule **13. (b)** is true.
15. (a) 0.2 **(b)** 0.220703, 0.205200, 0.201302, 0.200325, 0.020703, 0.005200, 0.001302, 0.000325 **(c)** $p = 2$ **17. (a)** 0.2
(b) 0.2005208, 0.2000326, 0.2000020, 0.2000001, 0.0005208, 0.0000326, 0.0000020, 0.0000001 **(c)** $p = 4$ **19.** $M = 0.7355$;
$S = 0.8048$ **21. (a)** 6715.5 trillion BTUs **(b)** 6607 trillion BTUs **23. (a)** 1.831 **(b)** 1.758 **25.** About 30 mcg(h)/ml;
this represents the total amount of drug available to the patient for each ml of blood. **27.** About 9 mcg(h)/ml; this represents the
total effective amount of the drug available to the patient for each ml of blood. **29. (a)** $y = b_0(t/7)^{b_1}e^{-b_2 t/7}$ **(b)** About 1212 kg;
about 1231 kg **(c)** About 1224 kg; about 1250 kg **31. (a)** 71.5 **(b)** 69.0 **33.** 3979 **35. (a)** 0.6827 **(b)** 0.9545 **(c)** 0.9973

Chapter 7 Review Exercises (page 437)

For exercises . . .	1–4, 19–30, 79,80, 84	5,16,17, 31–40, 87,94	6–8,15, 41–43, 81,92, 96	9,10, 45–62, 82,83, 89–91, 95	11,12, 63–66, 85,86	13,14, 18,44, 67–77, 93
Refer to section . . .	1	2	3	4	5	6

1. True **2.** False **3.** False **4.** True **5.** True
6. False **7.** False **8.** True **9.** True **10.** False
11. True **12.** False **13.** False **14.** True **19.** $x^2 + 3x + C$
21. $x^3/3 - 3x^2/2 + 2x + C$ **23.** $2t^{3/2} + C$
25. $2x^{3/2}/3 + 9x^{1/3} + C$ **27.** $2y^{-2} + C$ **29.** $-3e^{2x}/2 + C$
31. $e^{3x^2}/6 + C$ **33.** $(3 \ln |u^2 - 1|)/2 + C$ **35.** $-(x^3 + 5)^{-3}/9 + C$ **37.** $-e^{-3x^4}/12 + C$ **39.** $(3 \ln z + 2)^5/15 + C$
41. 20 **43.** 24 **45. (a)** $s(T) - s(0)$ **(b)** $\int_0^T v(t)\,dt = s(T) - s(0)$ is equivalent to the Fundamental Theorem with $a = 0$ and
$b = T$ because $s(t)$ is an antiderivative of $v(t)$. **47.** 12 **49.** $3 \ln 5 + 12/25 \approx 5.308$ **51.** 19/15 **53.** $3(1 - e^{-4})/2 \approx 1.473$
55. $\pi/32$ **57.** $25\pi/4$ **59.** 13/3 **61.** $(e^4 - 1)/2 \approx 26.80$ **63.** 64/3 **65.** 149/3 **67.** 0.5833; 0.6035 **69.** 4.187; 4.155
71. 0.6011 **73.** 4.156 **75. (a)** 4/3 **(b)** 1.146 **(c)** 1.252 **77. (a)** 0 **(b)** 0 **79.** $C(x) = (2x - 1)^{3/2} + 145$
81. $7000 trillion **83.** $38,000 **85. (a)** $916.67 **(b)** $666.67 **87. (a)** 18.159 billion barrels **(b)** 18.159 billion barrels
(d) $y = 0.06970t + 1.459$; 18.46 billion barrels **89.** 782 **91. (a)** 0.2784 **(b)** 0.2784 **93. (a)** About 8208 kg
(b) About 8430 kg **(c)** About 8558 kg **95.** $s(t) = t^3/3 - t^2 + 8$.

Chapter 8 Further Techniques and Applications of Integration

Exercises 8.1 (page 453)

For exercises . . .	1–4,36,40,41,43,44	5,6,35,39,42	7–10,20–22	13–19	23–28
Refer to example . . .	1,3	2	4	1–3	5

W1. $3/x$ **W2.** $x^6/2 + 1/x^2 + C$
W3. $3x^{5/3}/5 + 2x^{1/2} + C$ **W4.** $e^{2x}/2 - 5/e^x + C$
W5. $e^{x^2}/5 + C$ **W6.** $38 + e^3 \approx 58.09$ **1.** $xe^x - e^x + C$ **3.** $(-x/2 + 23/16)\,e^{-8x} + C$
5. $(x^2 \ln x)/2 - x^2/4 + C$ **7.** $-5e^{-1} + 3 \approx 1.161$ **9.** $26 \ln 3 - 8 \approx 20.56$ **11.** $e^4 + e^2 \approx 61.99$
13. $x^2e^{2x}/2 - xe^{2x}/2 + e^{2x}/4 + C$ **15.** $(2/7)(x + 4)^{7/2} - (16/5)(x + 4)^{5/2} + (32/3)(x + 4)^{3/2} + C$ or
$(2/3)x^2(x + 4)^{3/2} - (8/15)x(x + 4)^{5/2} + (16/105)(x + 4)^{7/2} + C$ **17.** $(4x^2 + 10x)\ln 5x - 2x^2 - 10x + C$
19. $(-e^2/4)(3e^2 + 1) \approx -42.80$ **21.** $2\sqrt{3} - 10/3 \approx 0.1308$ **23.** $16 \ln |x + \sqrt{x^2 + 16}| + C$
25. $-(3/11) \ln \left|(11 + \sqrt{121 - x^2})/x\right| + C$ **27.** $-1/(4x + 6) - (1/6) \ln |x/(4x + 6)| + C$ **31.** -18 **33.** 15
37. (a) $(2/3)x(x + 1)^{3/2} - (4/15)(x + 1)^{5/2} + C$ **(b)** $(2/5)(x + 1)^{5/2} - (2/3)(x + 1)^{3/2} + C$
39. $(169/2)\ln 13 - 42 \approx \$174.74$ **41.** $15e^6 + 3 \approx 6054$ **43.** About 219 kJ

Exercises 8.2 (page 460)

For exercises . . .	1–23,32,33,42,45	24–31,34–39,41,43,44	40
Refer to example . . .	1–3	4	Derivation of volume formula

W1. 218/3 **W2.** 243/5 **W3.** 14/3
W4. 14 **W5.** $(e^2 - 1)/2 \approx 3.1945$
W6. $(9 \ln 3)/2 - 2 \approx 2.9438$ **1.** 9π **3.** $364\pi/3$ **5.** $386\pi/27$ **7.** $15\pi/2$ **9.** 18π **11.** $\pi(e^4 - 1)/2 \approx 84.19$
13. $4\pi \ln 3 \approx 13.81$ **15.** $3124\pi/5$ **17.** $16\pi/15$ **19.** $4\pi/3$ **21.** $4\pi r^3/3$ **23.** $\pi r^2 h$ **25.** $13/3 \approx 4.333$
27. $38/15 \approx 2.533$ **29.** $e - 1 \approx 1.718$ **31.** $(5e^4 - 1)/8 \approx 34.00$ **33.** 3.758 **35.** $42.49 **37.** 200 cases
39. (a) 2.27 billion **(b)** 11.06 billion **41. (a)** $110e^{-0.1} - 120e^{-0.2} \approx 1.284$ **(b)** $210e^{-1.1} - 220e^{-1.2} \approx 3.640$
(c) $330e^{-2.3} - 340e^{-2.4} \approx 2.241$ **43. (a)** $9(6 \ln 6 - 5) \approx 51.76$ **(b)** $5(10 \ln 10 - 9) \approx 70.13$
(c) $3(31 \ln 31 - 30)/2 \approx 114.7$ **45.** $1.083 \times 10^{21}\,\text{m}^3$

Exercises 8.3 (page 469)

W1. $30,000\, e^{0.01t} + C$ **W2.** 553.51

W3. $-10,000\, te^{-0.04t} - 250,000\, e^{-0.04t} + C$ **W4.** 10,951.94

For exercises . . .	1(a)–8(a),16	1(b)–8(b),15,17,18	9–14,19,20
Refer to example . . .	2	3,4	5

1. (a) \$6883.39 (b) \$15,319.26 **3.** (a) \$3441.69 (b) \$7659.63 **5.** (a) \$3147.75 (b) \$7005.46

7. (a) \$32,968.35 (b) \$73,372.42 **9.** (a) \$746.91 (b) \$1662.27 **11.** (a) \$688.64 (b) \$1532.59

13. (a) \$11,351.78 (b) \$25,263.84 **15.** \$63,748.43 **17.** \$28,513.76, \$54,075.81 **19.** \$4560.94

Exercises 8.4 (page 474)

W1. 0 **W2.** 0 **W3.** 0 **W4.** ∞ **1.** 1/3 **3.** Divergent

For exercises . . .	1–26,31–34,37	27–30,35,36	42–48,51	49,50,52–54
Refer to example . . .	1	2	4	3

5. 1 **7.** 10,000 **9.** 1/10 **11.** 3/5 **13.** 1 **15.** 1000

17. Divergent **19.** 1 **21.** Divergent **23.** Divergent **25.** Divergent **27.** 0 **29.** Divergent **31.** Divergent **33.** 1

35. 0 **39.** (a) 2.808, 3.724, 4.417, 6.720, 9.022 (b) Divergent (c) 0.8770, 0.9070, 0.9170, 0.9260, 0.9269 (d) Convergent

41. (a) 9.9995, 49.9875, 99.9500, 995.0166 (b) Divergent (c) 100,000 **43.** \$20,000,000 **45.** \$30,000 **47.** \$30,000

49. 4 **51.** $K(a + br)/r^2$ **53.** 1250

Chapter 8 Review Exercises (page 477)

1. False **2.** True **3.** False **4.** True **5.** False

For exercises . . .	1–4,14–27	5–7,28–37,61	8,9,46–57,59	10,38–45,58,60
Refer to section . . .	1	2	3	4

6. False **7.** True **8.** True **9.** True **10.** False

15. $6x(x - 2)^{1/2} - 4(x - 2)^{3/2} + C$ **17.** $-(x + 2)e^{-3x} - (1/3)e^{-3x} + C$ **19.** $(x^2/2 - x) \ln |x| - x^2/4 + x + C$

21. $(1/8)\sqrt{16 + 8x^2} + C$ **23.** $10e^{1/2} - 16 \approx 0.4872$ **25.** $234/7 \approx 33.43$ **27.** -19 **29.** $81\pi/2 \approx 127.2$

31. $\pi \ln 3 \approx 3.451$ **33.** $64\pi/5 \approx 40.21$ **37.** 2,391,484/3 **39.** 1/5 **41.** $6/e \approx 2.207$ **43.** Divergent **45.** 3

47. $16,250/3 \approx \$5416.67$ **49.** \$174,701.45 **51.** \$15.58 **53.** \$5354.97 **55.** \$30,035.17 **57.** \$176,919.15

59. 0.4798 **61.** (a) 158.3° (b) 125° (c) 133.3°

Chapter 9 Multivariable Calculus

Exercises 9.1 (page 491)

For exercises . . .	1–6,29,30, 38–48,51	7–14	15–18	23–28	33–36
Refer to example . . .	1–3	4–6	7,8	material after Example 8	8

W1.   **W2.** **W3.**

W4. -4 **W5.** $18a^2 + 9a - 6$ **W6.** $4x + 2h + 3$ **1.** (a) 12 (b) -6 (c) 10 (d) -19 **3.** (a) $\sqrt{43}$ (b) 6 (c) $\sqrt{19}$

(d) $\sqrt{11}$ **5.** (a) e (b) e^2 (c) 2 (d) 3

7.   **9.**   **11.**   **13.**

15. **17.**

23. (c) **25.** (e) **27.** (b) **29.** (a) $8x + 4h$

(b) $-4y - 2h$ (c) $8x$ (d) $-4y$ **31.** (a) $3e^2$;

slope of tangent line in the direction of x at $(1, 1)$

(b) $3e^2$; slope of tangent line in the direction of y at $(1, 1)$

33. (a) 1987 (rounded) (b) 595 (rounded)

(c) 359,768(rounded)

35. $y = 500^{5/2}/x^{3/2} \approx 5,590,170/x^{3/2}$

37. $C(x, y, z) = 250x + 150y + 75z$ **39.** 1.416; the IRA account grows faster.
41. (a) 1.89 m^2 **(b)** 1.62 m^2 **(c)** 1.78 m^2 **43. (a)** 8.7% **(b)** 48% **(c)** Multiple solutions: $W = 19.75, R = 0, A = 0$ or $W = 10, R = 10, A = 4.59$ **(d)** Wetland percentage **45. (a)** 397 accidents **47. (a)** $T = 242.257\, C^{0.18}/F^3$ **(b)** 58.82; a tethered sow spends nearly 59% of the time doing repetitive behavior when she is fed 2 kg of food a day and neighboring sows spend 40% of the time doing repetitive behavior.
49. $f(L, W, H) = L + 2H + 2W$ **51. (a)** 4.69 in. by 3.75 in. **(b)** 4.04 in. by 3.75 in.

Exercises 9.2 (page 501)

For exercises ...	1,2,33–36,43,44	3–20,53	21–32	37–42	45–47,55–71	48–52,54
Refer to example ...	3	1–3	6,7	8	4	5

W1. $f'(x) = 6x^2 + 7$ **W2.** $f'(x) = 1/(2\sqrt{x}) - 10/x^3 + 12e^{2x}$ **W3.** $f'(x) = e^x + xe^x$ **W4.** $f'(x) = (6x^5 - 60x^3)/(x^2 - 5)^2$
W5. $f'(x) = 1/\sqrt{2x + 8}$ **W6.** $f'(x) = 27(e^{x^2} + 5x)^2(2xe^{x^2} + 5)$ **W7.** $f'(x) = 6x^2/(2x^3 + 5)$ **W8.** $f'(x) = \ln x^2 + 2(x + 1)/x$
1. (a) $12x - 4y$ **(b)** $-4x + 18y$ **(c)** 12 **(d)** -40 **3.** $f_x(x, y) = -4y$; $f_y(x, y) = -4x + 18y^2$; 4; 178 **5.** $f_x(x, y) = 10xy^3$;
$f_y(x, y) = 15x^2y^2$; -20; 2160 **7.** $f_x(x, y) = e^{x+y}$; $f_y(x, y) = e^{x+y}$; e^1 or e; e^{-1} or $1/e$ **9.** $f_x(x, y) = -24e^{4x-3y}$; $f_y(x, y) = 18e^{4x-3y}$;
$-24e^{11}$; $18e^{-25}$ **11.** $f_x(x, y) = (-x^4 - 2xy^2 - 3x^2y^3)/(x^3 - y^2)^2$; $f_y(x, y) = (3x^3y^2 - y^4 + 2x^2y)/(x^3 - y^2)^2$; $-8/49$; $-1713/5329$
13. $f_x(x, y) = 15x^2y^2/(1 + 5x^3y^2)$; $f_y(x, y) = 10x^3y/(1 + 5x^3y^2)$; 60/41; 1920/2879 **15.** $f_x(x, y) = e^{xy}(2x^2y + 1)$; $f_y(x, y) = x^3e^{xy}$;
$-7e^{-4}$; $-64e^{48}$ **17.** $f_x(x, y) = (1/2)(4x^3 + 3y)/(x^4 + 3xy + y^4 + 10)^{1/2}$; $f_y(x, y) = (1/2)(3x + 4y^3)/(x^4 + 3xy + y^4 + 10)^{1/2}$;
$29/(2\sqrt{21})$; $48/\sqrt{311}$ **19.** $f_x(x, y) = [6xy(e^{xy} + 2) - 3x^2y^2e^{xy}]/(e^{xy} + 2)^2$; $f_y(x, y) = [3x^2(e^{xy} + 2) - 3x^3ye^{xy}]/(e^{xy} + 2)^2$;
$-24(e^{-2} + 1)/(e^{-2} + 2)^2$; $(624e^{-12} + 96)/(e^{-12} + 2)^2$ **21.** $f_{xx}(x, y) = 8y^2 - 32$; $f_{yy}(x, y) = 8x^2$; $f_{xy}(x, y) = f_{yx}(x, y) = 16xy$
23. $R_{xx}(x, y) = 8 + 24y^2$; $R_{yy}(x, y) = -30xy + 24x^2$; $R_{xy}(x, y) = R_{yx}(x, y) = -15y^2 + 48xy$ **25.** $r_{xx}(x, y) = 12y/(x + y)^3$;
$r_{yy}(x, y) = -12x/(x + y)^3$; $r_{xy}(x, y) = r_{yx}(x, y) = (6y - 6x)/(x + y)^3$ **27.** $z_{xx} = 9ye^x$; $z_{yy} = 0$; $z_{xy} = z_{yx} = 9e^x$
29. $r_{xx} = -1/(x + y)^2$; $r_{yy} = -1/(x + y)^2$; $r_{xy} = r_{yx} = -1/(x + y)^2$ **31.** $z_{xx} = 1/x$; $z_{yy} = -x/y^2$; $z_{xy} = z_{yx} = 1/y$
33. $x = -4, y = 2$ **35.** $x = 0, y = 0$; or $x = 3, y = 3$ **37.** $f_x(x, y, z) = 4x^3$; $f_y(x, y, z) = 2z^2$; $f_z(x, y, z) = 4yz + 4z^3$;
$f_{yz}(x, y, z) = 4z$ **39.** $f_x(x, y, z) = 6/(4z + 5)$; $f_y(x, y, z) = -5/(4z + 5)$; $f_z(x, y, z) = -4(6x - 5y)/(4z + 5)^2$;
$f_{yz}(x, y, z) = 20/(4z + 5)^2$ **41.** $f_x(x, y, z) = (2x - 5z^2)/(x^2 - 5xz^2 + y^4)$; $f_y(x, y, z) = 4y^3/(x^2 - 5xz^2 + y^4)$;
$f_z(x, y, z) = -10xz/(x^2 - 5xz^2 + y^4)$; $f_{yz}(x, y, z) = 40xy^3z/(x^2 - 5xz^2 + y^4)^2$ **43. (a)** 6.773 **(b)** 3.386 **45. (a)** 80
(b) 150 **(c)** 80 **(d)** 440 **47. (a)** $902,100 **(b)** $f_p(p, i) = 99 - 0.5i - 0.005p$; $f_i(p, i) = -0.5p$; the rate at which weekly sales are changing per unit of change in price when the interest rate remains constant $(f_p(p, i))$ or per unit change in interest rate when the price remains constant $(f_i(p, i))$ **(c)** A weekly sales decrease of $9700 **49. (a)** 50.57 hundred units **(b)** $f_x(16, 81) = 1.053$ hundred units and is the rate at which production is changing when labor changes by 1 unit (from 16 to 17) and capital remains constant; $f_y(16, 81) = 0.4162$ hundred units and is the rate at which production is changing when capital changes by 1 unit (from 81 to 82) and labor remains constant. **(c)** Production would increase by approximately 105 units. **51.** $0.4x^{-0.6}y^{0.6}$; $0.6x^{0.4}y^{-0.4}$
53. (c) $f_x(x, y) = y(-x^2 - x - xy - 1)e^{-x}/(x + y)^2$, $f_y(x, y) = x(1 + x)e^{-x}/(x + y)^2$ **55. (a)** 1279 kcal per hr
(b) 2.906 kcal per hr per g; the instantaneous rate of change of energy usage for a 300-kg animal traveling at 10 km per hr is about 2.9 kcal per hr per g. **57. (a)** 0.0142 m^2 **(b)** 0.00442 m^2 **59. (a)** 4.125 lb **(b)** $\partial f/\partial n = n/4$; the rate of change of weight loss per unit change in workouts **(c)** An additional loss of 3/4 lb **61. (a)** 0.0783 **(b)** 0.0906 per m **(c)** -0.0212 per m
(d) waist **63. (a)** $(2ax - 3x^2)t^2e^{-t}$ **(b)** $x^2(a - x)(2t - t^2)e^{-t}$ **(c)** $(2a - 6x)t^2e^{-t}$ **(d)** $(2ax - 3x^2)(2t - t^2)e^{-t}$
(e) $\partial R/\partial x$ gives the rate of change of the reaction per unit of change in the amount of drug administered. $\partial R/\partial t$ gives the rate of change of the reaction for a 1-hour change in the time after the drug is administered. **65. (a)** $-24.9°$F **(b)** 15 mph
(c) $W_V(20, 10) = -1.114$; while holding the temperature fixed at 10°F, the wind chill decreases approximately 1.1°F when the wind velocity increases by 1 mph; $W_T(20, 10) = 1.429$; while holding the wind velocity fixed at 20 mph, the wind chill increases approximately 1.429°F if the actual temperature increases from 10°F to 11°F.
(d) Sample table

T\V	5	10	15	20
30	27	16	9	4
20	16	3	-5	-11
10	6	-9	-18	-25
0	-5	-21	-32	-39

67. -10 ml per year, 100 ml per in. **69. (a)** $F_m = gR^2/r^2$; the rate of change in force per unit change in mass while the distance is held constant; $F_r = -2mgR^2/r^3$; the rate of change in force per unit change in distance while the mass is held constant **71. (a)** 1055 **(b)** $T_s(3, 0.5) = 127.4$ msec per ft. If the distance to move an object increases from 3 ft to 4 ft, while keeping w fixed at 0.5, the approximate increase in movement time is 127.4 msec. $T_w(3, 0.5) = -764.6$ msec per ft. If the width of the target area increases by 1 ft, while keeping s fixed at 3 ft, the approximate decrease in movement time is 764.6 msec.

Exercises 9.3 (page 512)

For exercises . . .	1–18,21–28	34–40,42
Refer to example . . .	1–3	4

W1. Relative maximum at $(-2, 33)$; relative minimum at $(4, -75)$
W2. Relative maximum at $(0, 1)$; relative minima at $(-2, -15)$ and $(2, -15)$
1. Saddle point at $(-1, 2)$ **3.** Relative minimum at $(-3, -3)$ **5.** Relative minimum at $(-2, -2)$ **7.** Relative minimum at $(15, -8)$
9. Relative maximum at $(2/3, 4/3)$ **11.** Saddle point at $(2, -2)$ **13.** Saddle point at $(0, 0)$; relative minimum at $(27, 9)$
15. Saddle point at $(0, 0)$; relative minimum at $(9/2, 3/2)$ **17.** Saddle point at $(0, -1)$ **21.** Relative maximum of 9/8 at $(-1, 1)$;
saddle point at $(0, 0)$; (a)   **23.** Relative minima of $-33/16$ at $(0, 1)$ and at $(0, -1)$; saddle point at $(0, 0)$; (b)   **25.** Relative maxima
of $17/16$ at $(1, 0)$ and $(-1, 0)$; relative minima of $-15/16$ at $(0, 1)$ and $(0, -1)$; saddle points at $(0, 0)$, $(-1, 1)$, $(1, -1)$, $(1, 1)$, and
$(-1, -1)$; (e)   **31.** (a) all values of k (b) $k \geq 0$ **35.** Minimum cost of \$59 when $x = 4$, $y = 5$ **37.** Sell 12 spas and 7 solar
heaters for a maximum revenue of \$166,600.   **39.** \$2000 on quality control and \$1000 on consulting, for a minimum time of 8 hours
41. (a) $r = -1.722$, $s = 0.3652$, $y = 0.1787(1.441)^r$ (b) Same as (a)   (c) Same as (a)

Exercises 9.4 (page 522)

For exercises . . .	1–8,11,12,15,21	9,10,13,14,37–42	23–26	27–36
Refer to example . . .	1	3	material after Example 3	2

W1. $f_x(x, y) = 12x^2y^2 - 24x^3$; $f_y(x, y) = 8x^3y - 15y^4$
W2. $f_x(x, y, z) = 2xy^3z^5 - 8xy^5 + 3z^6$; $f_y(x, y, z) = 3x^2y^2z^5 - 20x^2y^4 - 16y^7z^7$; $f_z(x, y, z) = 5x^2y^3z^4 + 18xz^5 - 14y^8z^6$
1. $f(8, 8) = 256$ **3.** $f(5, 5) = 125$ **5.** $f(5, 3) = 28$ **7.** $f(20, 2) = 360$ **9.** $f(3/2, 3/2, 3) = 81/4 = 20.25$ **11.** $x = 8$, $y = 16$
13. 30, 30, 30   **15.** Minimum value of 128 at $(2, 5)$, maximum value of 160 at $(-2, 7)$ **21.** (a) Minimum value of -5
at $(3, -2)$ (d) $(3, -2)$ is a saddle point.   **23.** Purchase 20 units of x and 20 units of y for a maximum utility of 8000.
25. Purchase 20 units of x and 5 units of y for a maximum utility of 4,000,000.   **27.** 60 feet by 60 feet   **29.** 10 large kits and no
small kits   **31.** 167 units of labor and 178 units of capital   **33.** 125 m by 125 m   **35.** Radius = 5 in.; height = 10 in.
37. 12.91 m by 12.91 m by 6.455 m   **39.** 5 m by 5 m by 5 m   **41.** (b) 4 yd by 1 yd by 1/2 yd

Exercises 9.5 (page 527)

For exercises . . .	1–6,18–20,23–29	7–14	15–17,21,22,30–33	34,35
Refer to example . . .	1	2	3	4

W1. 2.7   **W2.** -0.016 **W3.** 4.9   **W4.** 1.01   **1.** 0.12
3. 0.0311   **5.** -0.335 **7.** 10.022; 10.0221; 0.0001
9. 2.0067; 2.0080; 0.0013   **11.** 1.07; 1.0720; 0.0020   **13.** -0.02; -0.0200; 0   **15.** 20.73 cm³   **17.** 86.4 in³   **19.** 0.07694 unit
21. 6.65 cm³   **23.** 2.98 liters   **25.** Increase of 3.51 yr; increase of 3.51 yr   **27.** (a) 87%   (b) 75%   (d) 89%; 87%
29. (a) 0.06569 sec; 0.0379 sec; in a close race, this could certainly affect the outcome.   (b) 0.001993 sec; this is the approximate change in time when the temperature decreases 5 degrees and the swimmer stands 0.5 m farther away from the starter in the y
direction while the starter's position is held fixed.   **31.** 26.945 cm²   **33.** 3%   **35.** 8

Exercises 9.6 (page 537)

For exercises . . .	1–10	11–20	21–28,61–64,66–71	29–38,65	39–46	47–56	57,58
Refer to example . . .	1	2,3	4	5,6	7	8	9

W1. -4 **W2.** 38/3   **W3.** $\ln 8 \approx 2.0794$ **W4.** $(e^6 - 1)/6 \approx 67.0715$ **W5.** $(e^8 - e)/3 \approx 992.7466$ **W6.** 98/3
1. $630y$ **3.** $(2x/9)[(x^2 + 15)^{3/2} - (x^2 + 12)^{3/2}]$ **5.** $6 + 10y$ **7.** $(1/2)(e^{12+3y} - e^{4+3y})$ **9.** $(1/2)(e^{4x+9} - e^{4x})$
11. 945   **13.** $(2/45)(39^{5/2} - 12^{5/2} - 7533)$ **15.** 21   **17.** $(\ln 3)^2$ **19.** $8 \ln 2 + 4$ **21.** 171   **23.** $(4/15)(33 - 2^{5/2} - 3^{5/2})$
25. $-3 \ln(3/4)$ or $3 \ln(4/3)$ **27.** $(1/2)(e^7 - e^6 - e^3 + e^2)$ **29.** 96   **31.** 40/3   **33.** $(2/15)(2^{5/2} - 2)$ **35.** $(1/4) \ln(17/8)$
37. $e^2 - 3$ **39.** 97,632/105   **41.** 128/9   **43.** $\ln 16$ or $4 \ln 2$ **45.** 64/3   **47.** 34   **49.** 10/3   **51.** $7(e - 1)/3$ **53.** 16/3
55. $4 \ln 2 - 2$ **57.** 1   **61.** 49   **63.** $(e^6 + e^{-10} - e^{-4} - 1)/60$ **65.** 9 in³   **67.** 14,753 units   **69.** \$34,833   **71.** \$32,000

Chapter 9 Review Exercises (page 543)

For exercises . . .	1–5,17–26,87, 89,101–103	6,13,14,27–44, 105,106	7,8,45–53,90, 91,92	9,54–57, 98,99,102	10–12,67–86,	15,59–66,88, 93–96,100
Refer to section . . .	1	2	3	4	6	5

1. True   **2.** True   **3.** True   **4.** True   **5.** False   **6.** False   **7.** False   **8.** True   **9.** False   **10.** False   **11.** True   **12.** False
17. -19; -255 **19.** $-5/9$; $-4/3$ **21.**   **23.**   **25.**

$x + y + z = 4$

$5x + 2y = 10$

$x = 3$

27. (a) $9x^2 + 8xy$ (b) -12 (c) 16   **29.** $f_x(x, y) = 12xy^3$; $f_y(x, y) = 18x^2y^2 - 4$ **31.** $f_x(x, y) = 4x/(4x^2 + y^2)^{1/2}$;
$f_y(x, y) = y/(4x^2 + y^2)^{1/2}$ **33.** $f_x(x, y) = 3x^2e^{3y}$; $f_y(x, y) = 3x^3e^{3y}$ **35.** $f_x(x, y) = 4x/(2x^2 + y^2)$; $f_y(x, y) = 2y/(2x^2 + y^2)$

37. $f_{xx}(x, y) = 30xy$; $f_{xy}(x, y) = 15x^2 - 12y$ **39.** $f_{xx}(x, y) = 12y/(2x - y)^3$; $f_{xy}(x, y) = (-6x - 3y)/(2x - y)^3$
41. $f_{xx}(x, y) = 8e^{2y}$; $f_{xy}(x, y) = 16xe^{2y}$ **43.** $f_{xx}(x, y) = (-2x^2y^2 - 4y)/(2 - x^2y)^2$; $f_{xy}(x, y) = -4x/(2 - x^2y)^2$
45. Saddle point at $(0, 2)$ **47.** Relative minimum at $(2, 1)$ **49.** Saddle point at $(3, 1)$ **51.** Saddle point at $(-1/3, 11/6)$; relative minimum at $(1, 1/2)$ **55.** Minimum of 18 at $(-3, 3)$ **57.** $x = 25, y = 50$ **59.** 1.22 **61.** 13.0846; 13.0848; 0.0002
63. $8y - 6$ **65.** $(3/2)[(100 + 2y^2)^{1/2} - (2y^2)^{1/2}]$ **67.** 1232/9 **69.** $(2/135)[(42)^{5/2} - (24)^{5/2} - (39)^{5/2} + (21)^{5/2}]$
71. $2 \ln 2$ or $\ln 4$ **73.** 110 **75.** $(4/15)(782 - 8^{5/2})$ **77.** 105/2 **79.** 1/2 **81.** 1/48 **83.** $\ln 2$ **85.** 3
87. (a) $\$(325 + \sqrt{10}) \approx \$328.16$ **(b)** $\$(800 + \sqrt{15}) \approx \$803.87$ **(c)** $\$(2000 + \sqrt{20}) \approx \$2004.47$
89. (a) $0.7y^{0.3}/x^{0.3}$ **(b)** $0.3x^{0.7}/y^{0.7}$ **91.** Purchase 10 units of x and 15 units of y for a maximum utility of 33,750.
93. Decrease by \$243.82 **95.** 4.19 ft^3 **97. (a)** \$200 spent on fertilizer and \$80 spent on seed will produce a maximum profit of \$266 per acre. **(b)** Same as (a). **(c)** Same as (a) **99.** \$19,685 **101. (a)** 49.68 liters **(b)** -0.09, the approximate change in total body water if age is increased by 1 yr and mass and height are held constant is -0.09 liter; 0.34, the approximate change in total body water if mass is increased by 1 kg and age and height are held constant is 0.34 liter; 0.25, the approximate change in total body water if height is increased by 1 cm and age and mass are held constant is 0.25 liter. **103. (a)** 50; in 1900, 50% of those born 60 years earlier are still alive. **(b)** 75; in 2000, 75% of those born 70 years earlier are still alive. **(c)** -1.25; in 1900, the percent of those born 60 years earlier who are still alive was dropping at a rate of 1.25 percent per additional year of life. **(d)** -2; in 2000, the percent of those born 70 years earlier who are still alive was dropping at a rate of 2 percent per additional year of life. **105.** 5 in. by 5 in. by 5 in.

Credits

Text Credits

ALL CHAPTERS: TI-84 Plus C screenshots from Texas Instruments. Courtesy of Texas Instruments.

CHAPTER 1: **34** Exercise 4: From Course 120 Examination, Applied Statistical Methods of the Education and Examination Committee of the Society of Actuaries (November 1989). Copyright © Society of Actuaries. Used by permission of Society of Actuaries. **37** Exercise 19: From *The Songs of Insects* by George Washington Pierce. Published by Harvard University Press. Copyright © 1948. **43** Exercise 57: From *NADA Data State of the Industry Report 2013*. Published by National Automobile Dealers Association. Copyright © 2013.

CHAPTER 2: **61** Exercise 77: Peter Tyack, Woods Hole Oceanographic Institution. Copyright © by Woods Hole Oceanographic Institution. Used by permission of Woods Hole Oceanographic Institution. **97** Exercise 46: From Course 140 Examination, Mathematics of Compound Interest of the Education and Examination Committee of the Society of Actuaries (November 1989). Copyright © 1989 by Society of Actuaries. Used by permission of Society of Actuaries.

CHAPTER 3: **160** Exercise 33: From Course 1 Examination, Mathematics of Compound Interest of the Education and Examination Committee of the Society of Actuaries (May 2003). Copyright © 2003 by Society of Actuaries. Used by permission of Society of Actuaries. **203** Exercise 61: From Course 1 Examination, Mathematics of Compound Interest of the Education and Examination Committee of the Society of Actuaries (May 2003). Copyright © 2003 by Society of Actuaries. Used by permission of Society of Actuaries. **204** Exercise 70: Peter Tyack, Woods Hole Oceanographic Institution. Copyright © by Woods Hole Oceanographic Institution. Used by permission of Woods Hole Oceanographic Institution.

CHAPTER 4: **248** Exercise 44: From Course 1 Examination, Education and Examination Committee of the Society of Actuaries (May 2003). Copyright © 2003 by Society of Actuaries. Used by permission of Society of Actuaries.

CHAPTER 5: **267** Figure 1: From *SLEEP* by Sergio Garbarino, Lino Nobili, Manolo Beelke, Fabrizio De Carli, and Franco Ferrillo. Copyright © by Associated Professional Sleep Societies. Used by permission of Associated Professional Sleep Societies.

CHAPTER 6: **328** Exercise 40: From Course 1 Examination, Education and Examination Committee of the Society of Actuaries (May 2003). Copyright © 2003 by Society of Actuaries. Used by permission of Society of Actuaries. **348** Exercise 3: From Uniform CPA Examination Questions and Unofficial Answers. Copyright © 1991 by The American Institute of Certified Public Accountants. Used by permission of The American Institute of Certified Public Accountants.

CHAPTER 7: **393** Example 8: From *Chartmasters' Rock 100: An Authoritative Ranking of the Most Popular Songs for Each Year, 1954 Through 1991* by Jim Quirin and Barry Cohen. Copyright ©1992 by Chartmasters. Used by permission of Chartmasters. **439** Exercise 78: From Course 1 Examination, Education and Examination Committee of the Society of Actuaries (May 2003). Copyright © 2003 by Society of Actuaries. Used by permission of Society of Actuaries. **440** Exercise 88: From Course 1 Examination, Education and Examination Committee of the Society of Actuaries (May 2003). Copyright © 2003 by Society of Actuaries. Used by permission of Society of Actuaries.

CHAPTER 9: **513** Exercise 30: From Course 1 Examination, Mathematics of Compound Interest of the Education and Examination Committee of the Society of Actuaries (May 2003). Copyright © 2003 by Society of Actuaries. Used by permission of Society of Actuaries.

Photo Credits

COVER: Hans Huber/Getty Images

CHAPTER R: R-1 Zuchero/Fotolia

CHAPTER 1: **1** Syda Productions/Fotolia **46** Forster Forest/Shutterstock

CHAPTER 2: **48** Dominik Michalek/Shutterstock **111** Cuson/Shutterstock **129** (top) Andrea Danti/Shutterstock; (bottom) John Good/National Park Service

CHAPTER 3: **133** Bikeriderlondon/Shutterstock **169** Courtesy of Raymond N. Greenwell **184** Courtesy of Raymond N. Greenwell **191** Tony Brindley/Shutterstock **206** BananaStock/Getty Images

CHAPTER 4: **209** Warren Goldswain/123RF **225** Spirit of America/Shutterstock **232** Pack-Shot/Shutterstock **257** AardLumens/Fotolia

CHAPTER 5: **266** Pavel Losevsky/Fotolia **319** Stockbyte/Thinkstock

CHAPTER 6: **321** Michael Ireland/Fotolia **374** Auremar/Shutterstock

CHAPTER 7: **375** Majeczka/Shutterstock **384** Rudy Balasko/Shutterstock **407** (top) Courtesy of Raymond N. Greenwell; (bottom) Gary Kemper/AP Images

CHAPTER 8: **445** BW Folson/Shutterstock

CHAPTER 9: **482** Rich Carey/Shutterstock **504** Courtesy of Raymond N. Greenwell

Chapter opener photos are repeated at a smaller size on the Contents and Preface pp. v–ix.

Index of Applications

Index

Note: A complete Index of Applications appears on pp. I-1 to I-4.

Absolute extrema
  explanation of, 322
  graphical optimization of, 326–327
  method for finding, 323–327, 369
Absolute maximum, 322, 323–325
Absolute minimum, 322, 323–325, 335
Absolute value, R-27
Absolute value function, 186–187
Absolute value sign, 382
Acceleration, 293–294, 384–385
Accumulated amount of money flow, 466–468, 477
Addition
  in order of operations, R-2
  of polynomials, R-2–R-3
  of rational expressions, R-8
Addition property
  of equality, R-11
  of inequality, R-17
Algebra, importance of, R-2
Algebra review
  equations, R-11–R-16
  exponents, R-21–R-25
  factoring, R-5–R-8
  inequalities, R-16–R-21
  polynomials, R-2–R-5
  radicals, R-26–R-29
  rational expressions, R-8–R-11
Annual yield, 115
Antiderivatives, 375–385
  evaluation of, 412
  explanation of, 376
  Fundamental Theorem of Calculus and, 408–414
  methods for finding, 376–385
Antidifferentiation, 376
Antidifferentiation formulas, 436
Approximations
  by differentials, 524–526, 541–542
  for functions of two variables, 524–525
  linear, 363–367, 370
  midpoint rule and, 397
  of area, 396–402
  of definite integral, 400
  Simpson's rule and, 430–433
  trapezoidal rule and, 397, 427–430
Archimedes, 396, 398n
Area
  approximation of, 396–402
  between two curves, 418–425, 437
  definite integrals and, 396–402, 412–414
  finding of, 414
  minimizing, 334–335
  under curves, 399
Area formula, 57–58
Associative properties, R-2
Asymptotes
  explanation of, 81, 122
  method for finding, 143–144, 305
  oblique, 308

Average cost
  cumulative learning model and, 480
  explanation of, 83, 229–230
  marginal, 229
  minimum, 230
Average rate of change
  explanation of, 161–162, 200
  formula for, 162
Average value, 458–459, 476
Axes, 2
Axis of symmetry, 63

Bernoulli, Jakob, 378
Binomial theorem, 211
Binomials, R-4
Brahmagupta, 398n
Break-even analysis, 22–23
Break-even point, 22
Break-even quantity, 22, 41
Briggs, Henry, 102

Calculator exercises. *See* Graphing calculator exercises
Calculus
  differential, 376
  Fundamental Theorem of, 408–414, 437
  historical background of, 135n
  integral, 376
  multivariable, 482–549
Capital value, 473, 477
Carbon dating, 114–115
Cartesian coordinate system, 2
Cauchy, Augustin-Louis, 135n
Caution notes, R-5, R-6, R-7, R-9, R-13, R-16, R-17, R-19, R-20, R-23, R-27, 4, 28, 30, 53, 55, 101, 104, 105, 116, 135–136, 143, 165, 182, 186, 194, 228–229, 237, 244, 271, 273, 281, 285, 287, 291, 297, 299, 322, 331, 357, 380, 382, 393, 412, 431, 472, 512, 516, 518, 521, 537
Celsius, Anders, 23n
Chain rule, 233–239
  alternative form of, 237, 259
  composition of functions, 233–235
  explanation of, 236, 259
  use of, 235–239
Change
  average rate of, 161–163, 200
  instantaneous rate of, 163–165
  rate of. *See* Rates of change
  total, 402–403
Change in x, 3
Change in y, 3
Change-of-base theorem
  for exponentials, 106–107, 123
  for logarithms, 103–104, 123
Closed interval, R-18, 155
Cobb, Charles W., 488
Cobb-Douglas production function, 488–490

Coefficients, R-2
  explanation of, 76
  leading, 76
Cohen, Bill, 393
Column integration, 447–451
Common denominators, R-10
Common logarithms, 103
Commutative properties, R-2
Completing the square, 64
Composite function, 234–235
Compound amount, 91
Compound interest
  chain rule and, 239
  continuous, 92–93, 123
  effective rate for, 115–116
  explanation of, 90–91
Concave downward, 294–295
Concave upward, 294–295
Concavity
  of graphs, 294–297
  test for, 296, 305–306, 314
Constant
  decay, 113
  derivatives of, 211–212
  growth, 113
  integration, 377, 379–380
Constant functions, 52, 258
Constant multiple rule, 379–380, 436
Constant rule, 211
Constant times a function, 213–214, 259
Constraints, 515
Consumers' surplus, 423–425, 437
Continuity, 152–158
  at $x = c$, 153
  explanation of, 153, 200
  from right/left, 155
  Intermediate Value Theorem and, 158
  on an open interval, 155
  on closed interval, 155
Continuous compounding
  effective rate for, 116, 123
  explanation of, 92–93, 123
Continuous functions, 153, 155–156
Continuous money flow, 462–469
Convergent integrals, 471
Correlation, 31–33, 41
Correlation coefficient, 31–33, 41
Cost analysis, 20–21, 158
Cost-benefit models, 82–83
Cost function, 21–22, 402
Critical numbers
  explanation of, 269
  in domain of function, 285
  method for finding, 269–270
  relative extrema at, 284–285
Critical point theorem, 325–326, 335
Critical points, 269, 507–508
Cube root, R-23
Cubes
  difference of two, R-7
  sum of two, R-7

Sources

Chapter 1

Section 1.1

1. Example 10 from *Morbidity and Mortality Weekly Report*, Centers for Disease Control and Prevention, Vol. 63, No. 02, Jan. 17, 2014, pp. 29–34.
2. Example 13 from *Trends in College Pricing 2013*, College Board, p. 15.
3. Exercise 63 from *Trends in College Pricing 2013*, College Board, p. 15.
4. Exercise 64 from *The World Almanac and Book of Facts 2014*, p. 367.
5. Exercise 65 from *The World Almanac and Book of Facts 2014*, p. 47.
6. Exercise 66 from Alcabes, P., A. Munoz, D. Vlahov, and G. Friedland, "Incubation Period of Human Immunodeficiency Virus," *Epidemiologic Review*, Vol. 15, No. 2, The Johns Hopkins University School of Hygiene and Public Health, 1993, pp. 303–318.
7. Exercise 67 from Hockey, Robert V., *Physical Fitness: The Pathway to Healthful Living*, Times Mirror/Mosby College Publishing, 1989, pp. 85–87.
8. Exercise 68 from *Science*, Vol. 253, No. 5017, July 19, 1991, pp. 306–308.
9. Exercise 69 from *Science*, Vol. 254, No. 5034, Nov. 15, 1991, pp. 936–938, and www.cdc.gov/nchs/data.
10. Exercise 70 from *Levels and Trends in Child Mortality Report 2013*, World Health Organization, p. 1.
11. Exercise 71 from *2012 Yearbook of Immigration Statistics*, Office of Immigration Statistics, Table 1.
12. Exercise 72 from *Estimated Median Age at First Marriage, by Sex: 1890 to Present*, U.S. Census Bureau, Current Population Survey, March and Annual Social and Economic Supplements, 2011 and earlier.
13. Exercise 73 from Acker, A. and C. Jaschek, *Astronomical Methods and Calculations*, John Wiley & Sons, 1986; Karttunen, H. (ed.), *Fundamental Astronomy*, Springer-Verlag, 1994.
14. Exercise 74 from *Science News*, June 23, 1990, p. 391.
15. Exercise 75 from *The State of the News Media 2013*, The Pew Research Center, Sept. 27, 2012.

Section 1.2

1. Page 18 from www.agmrc.org/media/cms/oceanspray_4BB99D38246C8.pdf.
2. Exercise 45 from *Science*, Vol. 281, July 17, 1998, pp. 350–351.
3. Exercise 46 from *Science*, Vol. 290, Nov. 17, 2000, p. 1291.
4. Exercise 48 from *Science News*, Sept. 26, 1992, p. 195; *Science News*, Nov. 7, 1992, p. 399.

5. Exercise 50 from www.calstate.edu/budget/fybudget/2009–2010/supportbook2/challenges-off-campus-costs.shtml.

Section 1.3

1. Page 27 from U.S. Department of Health and Human Services, National Center for Health Statistics, found in *New York Times 2010 Almanac*, p. 394, and www.cdc.gov/nchs/faststats/acc-inj.htm.
2. Example 5 from *Survey of Local Government Finances—School Systems, Table 8*; National Center for Education Statistics, U.S. Department of Education, *National Assessment of Educational Progress (NAEP), 1998–2011 Reading Assessments*, U.S. Census Bureau.
3. Exercise 4 from "November 1989 Course 120 Examination Applied Statistical Methods" of the *Education and Examination Committee of The Society of Actuaries*. Reprinted by permission of The Society of Actuaries.
4. Exercise 10 from Bureau of Economic Analysis, *Survey of Current Business Online*, Vol. 93, No. 10, Oct., 2013, p. 1.
5. Exercise 11 from *Federal Deposit Insurance Corporation Number of Institutions, Branches and Total Offices FDIC-Insured Commercial Banks, Balances at Year End, 1934–2012*, Federal Deposit Insurance Corporation, Table CB01.
6. Exercise 12 from *Current Population Survey*, "Table 4: Households with a Computer and Internet Use: 1984 to 2012," U.S. Census Bureau, Jan., 2014.
7. Exercise 13 from *Wireless Substitution: Early Release of Estimates from the National Health Interview Survey, July–December 2012*, Table 1.
8. Exercise 14 from the Federal Reserve, *Current Credit—G.19*, Feb. 2014, Centers for Disease Control and Prevention, www.federalreserve.gov/release/g19/current/.
9. Exercise 15 from *Table A-3. Mean Earnings of Workers 18 Years and Over, by Educational Attainment, Race, Hispanic Origin, and Sex: 1975 to 2012*, U.S. Census Bureau.
10. Exercise 16 from Expedia.com.
11. Exercise 17 from www.nctm.org/wlme/wlme6/five.htm.
12. Exercise 18 from Stanford, Craig B., "Chimpanzee Hunting Behavior and Human Evolution," *American Scientist*, Vol. 83, May–June 1995, pp. 256–261; and Goetz, Albert, "Using Open-Ended Problems for Assessment," *The Mathematics Teacher*, Vol. 99, No. 1, August 2005, pp. 12–17.
13. Exercise 19 from Pierce, George W., *The Songs of Insects*, Cambridge, Mass., Harvard University Press, Copyright © 1948 by the President and Fellows of Harvard College.
14. Exercise 20 from *Digest of Education Statistics 2012*, National Center for Education Statistics, Table 64.

15. Exercise 21 from *Historical Poverty Tables*, U.S. Census Bureau.
16. Exercise 22 from Lee, Grace, Paul Velleman, and Howard Wainer, "Giving the Finger to Dating Services," *Chance*, Vol. 21, No. 3, 2008, pp. 59–61.
17. Exercise 24 from data provided by Gary Rockswold, Mankato State University, Minnesota.
18. Exercise 26 from Carter, Virgil and Robert E. Machol, *Operations Research*, Vol. 19, 1971, pp. 541–545.
19. Exercise 27 from Whipp, Brian J. and Susan Ward, "Will Women Soon Outrun Men?" *Nature*, Vol. 355, Jan. 2, 1992, p. 25. The data are from Peter Matthews, *Track and Field Athletics: The Records*, Guinness, 1986, pp. 11, 44; from Robert W. Schultz and Yuanlong Liu, in *Statistics in Sports*, edited by Bennett, Jay and Jim Arnold, 1998, p. 189; and from *The World Almanac and Book of Facts 2012*, p. 974.
20. Exercise 28 from Rendell, M. (ed.), *The Olympics; Athens to Athens 1896–2004*, Weidenfeld and Nicolson, 2003, pp. 338–340; and *The World Almanac and Book of Facts 2014*, pp. 864, 868.
21. Exercise 29 from www.run100s.com/HR/.

Review Exercises

1. Exercises 45 and 46 from *Trade in Goods with China*, U.S. Census Bureau, Foreign Trade.
2. Exercise 47 from *Table H-5. Race and Hispanic Origin of Householder—Households by Median and Mean Income: 1967 to 2012*, U.S. Census Bureau, Current Population Survey, Annual Social and Economic Supplements.
3. Exercise 57 from *NADA Data 2013*, National Automobile Dealers Association, p. 9.
4. Exercise 58 from *OASDI Recipients and Monthly Payments, 1940–2012*, Social Security Administration.
5. Exercise 59 from U.S. Department of Agriculture, Economic Research Service, "Food Consumption, Prices and Expenditures, Annual," Feb. 1, 2014.
6. Exercise 60 from Food and Agriculture Organization of the United Nations, FAOSTAT, and UNDESA (2011), "2010 Revision of World Population Prospects."
7. Exercise 62 from U.S. Census Bureau, Current Population Survey, March, and Annual Social and Economic Supplements, 2013 and earlier.
8. Exercise 63 from U.S. Census Bureau, Current Population Survey, Annual Social and Economic Supplements.
9. Exercise 64 from U.S. Census Bureau, Population Division, Dec. 2013; The World Almanac and Book of Facts 2014, p. 38.

10. Exercise 65 from Moore, Thomas L., "Paradoxes in Film Rating," *Journal of Statistics Education*, Vol. 14, 2006, www.amstat.org/publications/jse/v14n1/datasets.moore.html.

Extended Application

1. Page 47 from *Health, United States, 2012*, National Center for Health Statistics, U.S. Department of Health and Human Services, Table 18, p. 76.

Chapter 2

Section 2.1

1. Page 49 from kitco.com/charts/historicalgold.html.
2. Example 1 uses data from Yahoo!Finance: www.yahoo.com.
3. Exercise 71 from kitco.com/charts/historicalsilver.html.
4. Exercise 72 from U.S. Energy Information Administration, *International Energy Outlook 2011*, DOE/EIA-0480 (2011), September 2011.
5. Exercise 75 from Gawande, Atul, "The Malpractice Mess," *The New Yorker*, Nov. 14, 2005, p. 65.
6. Exercise 76 from www.tax.ny.gov.
7. Exercise 77 from Peter Tyack, © Woods Hole Oceanographic Institution.
8. Exercise 78 from Robbins, Charles T., *Wildlife Feeding and Nutrition*, 2nd ed., Academic Press, 1993, p. 125.
9. Exercise 79 from Robbins, Charles T., *Wildlife Feeding and Nutrition*, 2nd ed., Academic Press, 1993, p. 142.

Section 2.2

1. Example 9 from Bureau of Labor Statistics, *Monthly Labor Review*, January 2012.
2. Exercise 59 from U.S. Department of Energy, *Monthly Energy Review*, Aug. 2013, and *Annual Coal Report*, 2012.
3. Exercise 60 from Harris, Edward F., Joseph D. Hicks, and Betsy D. Barcroft, "Tissue Contributions to Sex and Race: Differences in Tooth Crown Size of Deciduous Molars," *American Journal of Physical Anthropology*, Vol. 115, 2001, pp. 223–237.
4. Exercise 61 from Ralph DeMarr, University of New Mexico.
5. Exercise 62 from seer.cancer.gov/faststats/selections.php?#output.
6. Exercise 63 from Abuhamad, A. Z., et al., "Doppler Flow Velocimetry of the Splenic Artery in the Human Fetus: Is It a Marker of Chronic Hypoxia?" *American Journal of Obstetrics and Gynecology*, Vol. 172, No. 3, March 1995, pp. 820–825.
7. Exercise 64 from U.S. Census Bureau, Current Population Survey, Annual Social and Economic Supplement, June 2011, and *The New York Times 2010 Almanac*, p. 294.
8. Exercise 65 from U.S. Census Bureau and *The World Almanac and Book of Facts 2013*, p. 201.
9. Exercise 66 from Ralph DeMarr, University of New Mexico.
10. Exercise 68 from *National Traffic Safety Institute Student Workbook*, 1993, p. 7.

Section 2.3

1. Example 5 from the Budget of the U.S. Government, *Fiscal Year 2014 Historical Tables*, pp. 143–144.
2. Exercises 44 and 45 from Donley, Edward and Elizabeth Ann George, "Hidden Behavior in Graphs," *Mathematics Teacher*, Vol. 86, No. 6, Sept. 1993.
3. Exercise 50 from Dana Lee Ling, College of Micronesia-FSM.
4. Exercise 53 from U.S. Department of Energy, *Annual Coal Report, 2012*.
5. Exercise 55 from Garriott, James C. (ed.), *Medical Aspects of Alcohol Determination in Biological Specimens*, PSG Publishing Company, 1988, p. 57.
6. Exercise 56 from Association of American Medical Colleges, "MCAT Scores and GPAs for Applicants and Matriculants to U.S. Medical Schools."
7. Exercise 57 from Smith, J. Maynard, *Models in Ecology*, Oxford: Cambridge University Press, 1974.
8. Exercise 58 from Edelstein-Keshet, Leah, *Mathematical Models in Biology*, Random House, 1988.
9. Exercise 59 from Dobbing, John and Jean Sands, "Head Circumference, Biparietal Diameter and Brain Growth in Fetal and Postnatal Life," *Early Human Development*, Vol. 2, No. 1, April 1978, pp. 81–87.
10. Exercise 60 from Bausch & Lomb. The original chart gave all data to 2 decimal places.
11. Exercise 61 from Head Start, Early Childhood Learning and Knowledge Center, "Head Start Program Facts Fiscal Year 2012," p.10.
12. Exercise 62 from Gary Rockswold, Mankato State University, Mankato, Minnesota.

Section 2.4

1. Example 7 from Pollan, Michael, "The (Agri) Cultural Contradictions of Obesity," *The New York Times Magazine*, Oct. 12, 2003, p. 41; USDA-National Agriculture Statistics Service, 2006; and *The World Almanac and Book of Facts 2013*, p. 127.
2. Exercise 1 from Thomas, Jamie, "Exponential Functions," *The AMATYC Review*, Vol. 18, No. 2, Spring 1997.
3. Exercise 45 from Bureau of Labor Statistics, Aug. 19, 2014.
4. Exercise 46 from Problem 5 from "November 1989 Course 140 Examination, Mathematics of Compound Interest" of the Education and Examination Committee of The Society of Actuaries. Reprinted by permission of The Society of Actuaries.
5. Exercise 47 from U.S. Department of Commerce, Bureau of Economic Analysis, "Current-Dollar and 'Real' Gross Domestic Product," bea.gov/national/index.htm#gdp.
6. Exercise 49 from esa.un.org/unpp/index.asp.
7. Exercise 50 from *The Complexities of Physician Supply and Demand: Projections Through 2025*, Association of American Medical Colleges, Nov. 2008, p. 20.
8. Exercise 51 from U.S. Census Bureau, U.S. Interim Projections by Age, Sex, Race, and Hispanic Origin, www.census.gov/ipc/www/usinterimproj/.
9. Exercise 53 from Miller, A. and J. Thompson, *Elements of Meteorology*, Charles Merrill, 1975.
10. Exercise 54 from www.intel.com/technology/mooreslaw/.
11. Exercise 55 from www.wwindea.org/.
12. Exercise 56 from Marland, G., T. A. Boden, and R. J. Andres, "Global CO_2 Emissions from Fossil-Fuel Burning, Cement Manufacture, and Gas Flaring: 1751–2010," *2013. Global, Regional, and National CO_2 Emissions*, Oak Ridge National Laboratory, U.S. Department of Energy, Oak Ridge, TN.

Section 2.5

1. Example 10 from Ludwig, John and James Reynolds, *Statistical Ecology: A Primer on Methods and Computing*, New York: Wiley, 1988, p. 92.
2. Exercise 71 from Lucky Larry #16 by Joan Page, *The AMATYC Review*, Vol. 16, No. 1, Fall 1994, p. 67.
3. Exercise 82 from *Science*, Vol. 284, June 18, 1999, p. 1937.
4. Exercise 86 from Huxley, J. S., *Problems of Relative Growth*, Dover, 1968.
5. Exercise 87 from Horelick, Brindell and Sinan Koont, "Applications of Calculus to Medicine: Prescribing Safe and Effective Dosage," *UMAP Module 202*, 1977.
6. Exercise 88 from McNab, Brian K., "Complications Inherent in Scaling the Basal Rate of Metabolism in Mammals," *The Quarterly Review of Biology*, Vol. 63, No.1, Mar. 1988, pp. 25–54.
7. Exercise 89 from U.S. Census Bureau, U.S. Interim Projections by Age, Sex, Race, and Hispanic Origin, www.census.gov/ipc/www/usinterimproj/.
8. Exercise 91 from *Scientific American*, Oct. 1999, p. 103.
9. Exercise 92 from *The New York Times*, June 6, 1999, p. 41.
10. Exercise 93 from www.npr.org/templates/rundowns/rundown.php?prgId=2&prgDate=5-7-2002.
11. Exercise 94(c) from www.west.net/rperry/PueblaTlaxcala/puebla.html.
12. Exercise 94(d) from www.history.com/this-day-in-history/earthquake-shakes-mexico-city.
13. Exercise 94(g) from *The New York Times*, Jan. 13, 1995.
14. Exercise 96 from Tymoczko, Dmitri, "The Geometry of Music Chords," *Science*, Vol. 313, July 7, 2006, p. 72–74.

Section 2.6

1. Exercise 25 from www.census.gov/ipc/www/idb/worldpopinfo.php.
2. Exercise 26 from Brody, Jane, "Sly Parasite Menaces Pets and Their Owners," *The New York Times*, Dec. 21, 1999, p. F7.
3. Exercise 30 from Speroff, Theodore, et al., "A Risk-Benefit Analysis of Elective Bilateral Oophorectomy: Effect of Changes in Compliance with Estrogen Therapy on Outcome," *American Journal of Obstetrics and Gynecology*, Vol. 164, Jan. 1991, pp. 165–174.
4. Exercise 31 from downsyndrome.about.com/od/diagnosingdownsyndrome/a/Matagechart.htm.
5. Exercise 40 from *Science*, Vol. 277, July 25, 1997, p. 483.

Review Exercises

1. Exercise 108 from Bureau of Labor Statistics, "Table 24. Historical Consumer Price Index for All Urban Consumers (CPI-U) U.S. City Average, All Items," www.bls.gov/cpi/cpid1504.pdf.
2. Exercise 110 from *Family Practice*, May 17, 1993, p. 55.
3. Exercise 111 from www.cdc.gov/hiv/topics/surveillance/resources/reports/2007report/table25.htm.
4. Exercise 112 from Rusconi, Franca, et al., "Reference Values for Respiratory Rate in the First 3 Years of Life," *Pediatrics*, Vol. 94, No. 3, Sept. 1994, pp. 350–355.
5. Exercise 113 from Cattet, Marc R. L., et al., "Predicting Body Mass in Polar Bears: Is Morphometry Useful?" *Journal of Wildlife Management*, Vol. 61. No. 4, 1997, pp. 1083–1090.
6. Exercise 115 from Von Foerster, Heinz, Patricia M. Mora, and Lawrence W. Amiot, "Doomsday: Friday, 13 November, A.D. 2026," *Science*, Vol. 132, Nov. 4, 1960, pp. 1291–1295.
7. Exercise 115a from esa.un.org/unpp/index.asp.
8. Exercise 119 from Johnson, Michael P. and Daniel S. Simberloff, "Environmental Determinants of Island Species Numbers in the British Isles," *Journal of Biogeography*, Vol. 1, 1974, pp. 149–154.
9. Exercise 122 from Ronan, C., *The Natural History of the Universe*, Macmillan, 1991.

Extended Application

1. Page 130 from Bornstein, Marc H. and Helen G. Bornstein, "The Pace of Life," *Nature*, Vol. 259, Feb. 19, 1976, pp. 557–559.
2. Page 131 from Johnson, Michael P., and Daniel S. Simberloff, "Environmental Determinants of Island Species Numbers in the British Isles," *Journal of Biogeography*, Vol. 1, 1974, pp. 149–154.
3. Page 131 from Juan Camilo Bohorquez, "Common ecology quantifies human insurgency," *Nature*, Vol. 462, Dec. 17, 2009, pp. 911–914.
4. Exercise 1 from Gwartney, James D., Richard L. Stroup, and Russell S. Sobel, *Economics: Private and Public Choice*, 9th ed., The Dryden Press, 2000, p. 59.
5. Exercise 2 from White, Craig R., et al., "Phylogenetically informed analysis of the allometry of Mammalian Basal metabolic rate supports neither geometric nor quarter-power scaling," *Evolution*, Vol. 63, Oct. 2009, pp. 2658–2667.

Chapter 3

Section 3.1

1. Exercise 85 from www.boe.ca.gov/sutax/taxrateshist.htm.
2. Exercise 86 from about.usps.com/who-we-are/postal-history/domestic-letter-rates-since-1863.pdf.
3. Exercise 88 from American Airlines, www.aa.com.
4. Exercises 90 and 91 contributed by Robert D. Campbell of the Frank G. Zarb School of Business at Hofstra University.
5. Exercise 92 from Kulesa, P., G. Cruywagen et al. "On a Model Mechanism for the Spatial Patterning of Teeth Primordia in the Alligator," *Journal of Theoretical Biology*, Vol. 180, 1996, pp. 287–296.
6. Exercise 93 from Nord, Gail and John Nord, "Sediment in Lake Coeur d'Alene, Idaho," *The Mathematics Teacher*, Vol. 91, No. 4, April 1998, pp. 292–295.
7. Exercise 95 from Bishir, John W. and Donald W. Drewes, *Mathematics in the Behavioral and Social Sciences*, New York: Harcourt Brace Jovanovich, 1970, p. 538.

Section 3.2

1. Page 152 from U.S. Department of Labor.
2. Exercise 33 from Problem 26 from May 2003 Course 1 Examination of the *Education and Examination Committee of the Society of Actuaries*. Reprinted by permission of the Society of Actuaries.
3. Exercise 39 from U.S. Postal Service.
4. Exercise 41 from Xin, H., I. Berry, T. Barton, and G. Tabler, "Feed and Water Consumption, Growth, and Mortality of Male Broiler," *Poultry Science*, Vol. 73, No. 5, May 1994, pp. 610–616.

Section 3.3

1. Example 1 from U.S. Census Bureau.
2. Example 2 from www2.ed.gov/about/overview/budget/history/index.html.
3. Exercise 30 from Bureau of Labor Statistics.
4. Exercise 31 from www.eia.gov/dnav/pet/pet_pri_gnd_dcus_nus_m.htm.
5. Exercise 32 from Lew, Jacob J., et al., "2013 Annual Report of the Boards of Trustees of the Federal Hospital Insurance and Federal Supplementary Medical Insurance Trust Funds," Table V.B1, May 31, 2013, p. 191.

6. Exercise 33 from U.S. Census Bureau.
7. Exercise 34 from Carl Haub, Population Reference Bureau, 2000.
8. Exercise 35 from www.avert.org/usa-hiv-aids-statistics.htm.
9. Exercise 36 from Harris, E. F., J. D. Hicks, and B. D. Barcroft, "Tissue Contributions to Sex and Race: Differences in Tooth Crown Size of Deciduous Molars," *American Journal of Physical Anthropology*, Vol. 115, 2001, pp. 223–237.
10. Exercise 37 from Reed, G. and J. Hill, "Measuring the Thermic Effect of Food," *American Journal of Clinical Nutrition*, Vol. 63, 1996, pp. 164–169.
11. Exercise 38 from Jorgenson, J., M. Festa-Bianchet, M. Lucherini, and W. Wishart, "Effects of Body Size, Population Density, and Maternal Characteristics on Age at First Reproduction of Bighorn Ewes," *Canadian Journal of Zoology*, Vol. 71, No. 12, Dec. 1993, pp. 2509–2517.
12. Exercise 39 from www.dhs.gov/yearbook-immigration-statistics-2013-lawful-permanent-residents.
13. Exercise 40 from www.monitoringthefuture.org/data/13data/13drtbl1.pdf.

Section 3.4

1. Example 3 from downsyndrome.about.com.
2. Example 10 from Gannon, Megan, "Nova Star Explosion Is Visible to the Naked Eye: Where to Look," Aug. 21, 2013, www.space.com/22453-nova-delphinus-star-explosion-naked-eye.html.
3. Exercise 52 from Michael W. Ecker in "Controlling the Discrepancy in Marginal Analysis Calculations," *The College Mathematics Journal*, Vol. 37, No. 4, Sept. 2006, pp. 299–300.
4. Exercise 53 from 2009 OASDI Trustees Report, Social Security Administration, http://www.ssa.gov/OACT/TR/2009/LD_figIVB1.html.
5. Exercise 54 from Brighton, Caroline H., "Mechanical Power Output of Cockatiel Flight in Relation to Flight Speed" *Biolog-e: The Undergraduate Bioscience Research Journal*, 2007, www.fbs.leeds.ac.uk/students/ejournal/Biolog-e/uploads/Caroline%20Brighton_synopsis.pdf.
6. Exercise 56 from Kissileff, H. R. and J. L. Guss, "Microstructure of Eating Behavior in Humans," *Appetite*, Vol. 36, No. 1, Feb. 2001, pp. 70–78.
7. Exercise 57 from Benedito, J., J. Carcel, M. Gisbert, and A. Mulet, "Quality Control of Cheese Maturation and Defects Using Ultrasonics," *Journal of Food Science*, Vol. 66, No. 1, 2001, pp. 100–104.
8. Exercise 59 from heatkit.com/html/bakeov03.htm.
9. Exercises 60 and 61 from Bahill, A. T. and W. J. Karnavas, "Determining Ideal Baseball Bat Weights Using Muscle Force-Velocity Relationships," Biological Cybernetics, Vol. 62, 1989, pp. 89–97.

Section 3.5

1. Exercise 18 from Centers for Disease Control, www.cdc.gov/nchs/about/major/nhanes/growthcharts/charts.htm.
2. Exercise 20 from Brighton, Caroline H., "Mechanical Power Output of Cockatiel Flight in Relation to Flight Speed" *Biolog-e: The Undergraduate Bioscience Research Journal*, 2007, www.fbs.leeds.ac.uk/students/ejournal/Biolog-e/uploads/Caroline%20Brighton_synopsis.pdf.
3. Exercise 21 from Hensinger, Robert, *Standards in Pediatric Orthopedics: Tables, Charts, and Graphs Illustrating Growth*, New York: Raven Press, 1986, p. 192.
4. Exercise 22 from Sinclair, David, *Human Growth After Birth*, New York: Oxford University Press, 1985.
5. Exercise 23 from waterdata.usgs.gov/usa/nwis/uv?04282500.
6. Exercise 24 from Winston, Patrick Henry, "Skills, Big Ideas, and Getting Grades," *MIT Faculty Newsletter*, Vol. XX, No. 4, March/April 2008.

Review Exercises

1. Exercise 61 from Problem 3 from May 2003 Course 1 Examination of the *Education and Examination Committee of the Society of Actuaries*. Reprinted by permission of the Society of Actuaries.
2. Exercise 66 from Murray, Alan, "Winners? Losers? Estimates Show How Impact of Tax Proposal Varies," *Wall Street Journal*, May 9, 1986, p. 29.
3. Exercise 67 from data.bls.gov/timeseries/LNS14000000.
4. Exercise 68 from 2011 *Alzheimer's Disease Facts and Figures*, Alzheimer's Association, 2011, p. 17.
5. Exercise 70 from Peter Tyack, © Woods Hole Oceanographic Institution.
6. Exercise 71 from Centers for Disease Control, www.cdc.gov/nchs/about/major/nhanes/growthcharts/charts.htm.
7. Exercise 72 from Hensinger, Robert, *Standards in Pediatric Orthopedics: Tables, Charts, and Graphs Illustrating Growth*, New York: Raven Press, 1986, p. 193.

Chapter 4

Section 4.1

1. Example 4 from Finke, M., "Energy Requirements of Adult Female Beagles," *Journal of Nutrition*, Vol. 124, 1994, pp. 2604s–2608s.
2. Example 10 from Table 2. Projections of the Population by Selected Age Groups and Sex for the United States: 2010 to 2050 (NP2008-T2), Population Division, U.S. *Census Bureau*, August 14, 2008.
3. Exercise 55 from Pingle, Mark, "Introducing Dynamic Analysis Using Malthus's Principle of Population," *The Journal of Economic Education*, Vol. 34, 2003, pp. 3–20.

4. Exercise 57 from "Rates for Domestic Letters, 1863–2014," about.usps.com/who-we-are/postal-history/domestic-letter-rates-since-1863.htm.
5. Exercise 58 from "Introduction: U.S. Currency and Coin Outstanding and in Circulation," Financial Management Service, U.S. Department of the Treasury, June 2014. www.fms.treas.gov/bulletin/index.html.
6. Exercise 59 from Walker, A., *Observation and Inference: An Introduction to the Methods of Epidemiology*, Epidemology Resources, Inc., 1991.
7. Exercise 61 from Fitzsimmons, N., S. Buskirk, and M. Smith. "Population History, Genetic Variability, and Horn Growth in Bighorn Sheep," *Conservation Biology*, Vol. 9, No. 2, April 1995, pp. 314–323.
8. Exercise 62 from Dobbing, John and Jean Sands, "Head Circumference, Biparietal Diameter and Brain Growth in Fetal and Postnatal Life," *Early Human Development*, Vol. 2, No. 1, April 1978, pp. 81–87.
9. Exercise 63 from Okubo, Akira, "Fantastic Voyage into the Deep: Marine Biofluid Mechanics," in *Mathematical Topics in Population Biology Morphogenesis and Neurosciences*, edited by E. Teramoto and M. Yamaguti, Springer-Verlag, 1987, pp. 32–47.
10. Exercise 64 from Tan J., N. Silverman, J. Hoffman, M. Villegas, and K. Schmidt, "Cardiac Dimensions Determined by Cross-Sectional Echocardiography in the Normal Human Fetus From 18 Weeks to Term," *American Journal of Cardiology*, Vol. 70, No. 18, Dec. 1, 1992, pp. 1459–1467.
11. Exercise 65 from Kennelly, A., "An Approximate Law of Fatigue in Speeds of Racing Animals," *Proceedings of the American Academy of Arts and Sciences*, Vol. 42, 1906, pp. 275–331.
12. Exercise 66 from Tuchinsky, Philip, "The Human Cough," *UMAP Module 211*, Lexington, MA, COMAP, Inc., 1979, pp. 1–9.
13. Exercise 67 from win.niddk.nih.gov/publications/tools.htm.
14. Exercise 74 from Yechieli, Yoseph, Ittai Gavrieli, Brian Berkowitz, and Daniel Ronen, "Will the Dead Sea Die?" *Geology*, Vol. 26, No. 8, Aug. 1998, pp. 755–758. These researchers have predicted that the Dead Sea will not die but reach an equilibrium level.
15. Exercise 75 from the Acoustical Society of America, Orange County Regional Chapter, www.ocasa.org/MayanPyramid.htm.
16. Exercise 76 from Ewing, Maureen, Wayne J. Camara, and Roger E. Millsap, *The Relationship Between PSAT/NMSQT® Scores and AP® Examination Grades: A Follow-Up Study*, The College Board, 2006.
17. Exercise 77 from Vennebush, Patrick, "Media Clips: A Dog's Human Age," *The Mathematics Teacher*, Vol. 92, 1999, pp. 710–712.

Section 4.2

1. Exercise 41 and 42 from Cupillari, Antonella, "Another Look at the Rules of Differentiation,"

PRIMUS: Problems, Resources, and Issues in Mathematics Undergraduate Studies, Vol. 14, 2004, pp. 193–200.
2. Exercise 52 from Kellar, Brian and Heather Thompson, "Whelk-come to Mathematics," *The Mathematics Teacher*, Vol. 92, No. 6, September 1999, pp. 475–481.
3. Exercise 53 from Sanders, Mark and Ernest McCormick, *Human Factors in Engineering and Design*, 7th ed., New York: McGraw-Hill, 1993, pp. 243–246.
4. Exercise 55 from Mannering, F. and W. Kilareski, *Principles of Highway Engineering and Traffic Control*, New York: Wiley, 1990.
5. Exercise 56 from www.baseball-reference.com/bullpen/Pythagorean_Theorem_of_Baseball.

Section 4.3

1. Exercise 65 from Guffey, Roger, "The Life Expectancy of a Jawbreaker: The Application of the Composition of Functions," *The Mathematics Teacher*, Vol. 92, No. 2, Feb. 1999, pp. 125–127.
2. Exercises 66 and 67 from Chrisomalis, Stephen, "Numerical Prefixes," 2007, phrontistery.info/numbers.html.

Section 4.4

1. Exercise 41 from Dant, Rajiv P. and Paul D. Berger, "Modelling Cooperative Advertising Decisions in Franchising," *The Journal of the Operational Research Society*," Vol. 47, 1996, pp. 1120–1136.
2. Exercise 43 from World Bank, World Development Indicators. www.google.com/publicdata?ds=wb-wdi&met=it_net_user&idim=country:USA&dl=en&hl=en&q=number+of+internet+users.
3. Exercise 44 from Problem 11 from May 2003 Course 1 Examination of the *Education and Examination Committee of the Society of Actuaries*. Reprinted by permission of the Society of Actuaries.
4. Exercise 45 from U.S. Census Bureau, U.S. Interim Projections by Age, Sex, Race, and Hispanic Origin, www.census.gov/ipc/www/usinterimproj/.
5. Exercise 46 from esa.un.org/unpp/index.asp.
6. Exercise 47 from Ricklefs, Robert E., "A Graphical Method of Fitting Equations to Growth Curves," *Ecology*, Vol. 48, Nov. 1967, pp. 978–983.
7. Exercise 48 from *International Recovery Plan for the Whooping Crane (Grus Americana)*, third revision, U.S. Fish and Wildlife Service, March 2007.
8. Exercise 49 from Spratt, John, et al., "Decelerating Growth and Human Breast Cancer," *Cancer*, Vol. 71, No. 6, 1993, pp. 2013–2019.
9. Exercise 50 from U.S. Vital Statistics, 1995.
10. Exercise 51 from Friedman, Simon H. and Jens O. Karlsson, "A Novel Paradigm," *Nature*, Vol. 385, No. 6616, Feb. 6, 1997, p. 480.
11. Exercise 52 from Prestrud, Pal and Kjell Nilssen, "Growth, Size, and Sexual Dimorphism in Arctic

Foxes," *Journal of Mammalogy*, Vol. 76, No. 2, May 1995, pp. 522–530.

12. Exercise 53 from DeNise, R. and J. Brinks, "Genetic and Environmental Aspects of the Growth Curve Parameters in Beef Cows," *Journal of Animal Science*, Vol. 61, No. 6, 1985, pp. 1431–1440.

13. Exercise 54 from LaRosa, John C., et al., "The Cholesterol Facts: A Joint Statement by the American Heart Association and the National Heart, Lung, and Blood Institute," *Circulation*, Vol. 81, No. 5, May 1990, p. 1722.

14. Exercise 55 from Cisne, John L., "How Science Survived: Medieval Manuscripts 'Demography' and Classic Texts' Extinction," *Science*, Vol. 307, Feb. 25, 2005, pp. 1305–1307.

15. Exercise 57 from Allen, I. E. and J. Seaman, "Staying the Course—Online Education in the United States, 2008," The Sloan Consortium, Nov. 2008.

16. Exercise 59 from Kevin Friedrich, Sharon, PA.

17. Exercise 60 from Schoen, Carl, "A New Empirical Model of the Temperature-Humidity Index," *Journal of Applied Meteorology*, American Meteorological Society, Vol. 44, Sept. 2005, pp. 1413–1420.

18. Exercise 61 from Government of Canada, climate.weather.gc.ca/climate_normals/normals_documentation_e.html.

19. Exercise 63 from Osserman, Robert, "Mathematics of the Gateway Arch, *Notices of the AMS*, Vol. 57, No. 2, Feb. 2010, pp. 220–229.

20. Exercise 64 from Adams, Mike, "The Dead Grandmother/Exam Syndrome," *Annals of Improbably Research*, Nov.—Dec. 1999, posted online at www.neatorama.com/2011/04/05/the-dead-grandmotherexam-syndrome/.

21. Exercise 65 from Bennett, Jay, "Statistical Modeling in Track and Field," *Statistics in Sports*, Arnold, 1998, p. 179.

Section 4.5

1. Example 4 from *Kelley Blue Book*, www.kbb.com/new-cars/best-resale-value-awards/best-resale-top-10-cars/.

2. Exercise 61 from Zanoni, B., C. Garzaroli, S. Anselmi, and G. Rondinini, "Modeling the Growth of *Enterococcus faecium* in Bologna Sausage," *Applied and Environmental Microbiology*, Vol. 59, No. 10, Oct. 1993, pp. 3411–3417.

3. Exercise 62 from Sharkey, I., et al., "Body Surface Area Estimation in Children Using Weight Alone: Application in Pediatric Oncology," *British Journal of Cancer*, Vol. 85, No. 1, 2001, pp. 23–28.

4. Exercise 63 from Pearl, R. and S. Parker, *Proc. Natl. Acad. Sci.*, Vol. 8, 1922, p. 212, quoted in *Elements of Mathematical Biology* by Alfred J. Lotka, Dover Publications, 1956, pp. 308–311.

5. Exercise 64 from Miller, Michelle N. and John A. Byers, "Energetic Cost of Locomotor Play in Pronghorn Fawns," *Animal Behavior*, Vol. 41, 1991, pp. 1007–1013.

6. Exercise 66 from "Health, United States, 2009," U.S. Census.

7. Exercise 67 from Bradley, Christopher, "Media Clips," *The Mathematics Teacher*, Vol. 93, No. 4, April 2000, pp. 300–303.

8. Exercise 68 from Bender, Edward, *An Introduction to Mathematical Modeling*, New York: Wiley, 1978, p. 213.

Review Exercises

1. Exercise 63 from "Japanese University Entrance Examination Problems in Mathematics," edited by Ling-Erl Eileen T. Wu, published by the Mathematical Association of America, copyright 1993, pp. 18–19.

2. Exercise 64 from Maurer, Stephen B., "Hat Derivatives," *The College Mathematics Journal*, Vol. 33, No. 2, Jan. 2002, pp. 32–37.

3. Exercises 67 and 68 from Cupillari, Antonella, "Another Look at the Rules of Differentiation," *PRIUMS: Problems, Resources, and Issues in Mathematics Undergraduate Studies*, Vol. 14, 2004, pp. 193–200.

4. Exercise 81 from U.S. Postal Service, *Postal Facts 2014*.

5. Exercise 82 from *The World Almanac and Book of Facts 2014*, p. 109.

6. Exercise 83 from stats.bls.gov/data/inflation_calculator.htm.

7. Exercise 86 from Marshall, William H. and Tina Wylie Echeverria, "Characteristics of the Monkeyface Prickleback," *California Fish and Game*, Vol. 78, No. 2, Spring 1992.

8. Exercise 87 from Pestrud, Pal and Kjell Nilssen, "Growth, Size, and Sexual Dimorphism in Arctic Foxes," *Journal of Mammalogy*, Vol. 76, No. 2, May 1995, pp. 522–530.

9. Exercise 88 from U.S Census Bureau. 2004, *US Interim Projections by Age, Sex, Race and Hispanic Origin.*

10. Exercise 89 from www.wwindea.org.

11. Exercise 92 from Lo Bello, Anthony and Maurice Weir, "Glottochronology: An Application of Calculus to Linguistics," *The UMAP Journal*, Vol. 3, No. 1, Spring 1982, pp. 85–99.

12. Exercise 93 from www-nrd.nhtsa.dot.gov/pdf/nrd-30/NCSA/RNotes/1998/AgeSex96.pdf.

Chapter 5

Section 5.1

1. Page 267 from Garbarino, S., L. Nobili, M. Beelke, F. Phy, and F. Ferrillo, "The Contribution Role of Sleepiness in Highway Vehicle Accidents," *Sleep*, Vol. 24, No. 2, 2001, pp. 203–206. © 2001 American Academy of Sleep Medicine. Reproduced with permission of the American Academy of Sleep Medicine via Copyright Clearance Center.

2. Exercise 42 from Berry, Andrew, "Root Average Equals Turning Point Average," *The Mathematics Teacher*, Vol. 99, No. 9, May 2006, p. 595.

3. Exercise 49 from 2009 OASDI Trustees Report, Social Security Administration, www.ssa.gov/OACT/TR/2009/LD_figIVB1.html.

4. Exercise 50 from Employment Status of the Civilian Noninstitutional Population, 1943 to Date, data.bls.gov/cps/cpsaat01.htm.

5. Exercise 51 from Garriott, James C., ed., *Medicolegal Aspects of Alcohol Determination in Biological Specimens*, PSG Publishing Company, 1988, p. 57.

6. Exercise 54 from Stefanadis, C., J. Dernellis, et al., "Assessment of Aortic Line of Elasticity Using Polynomial Regression Analysis," *Circulation*, Vol. 101, No. 15, April 18, 2000, pp. 1819–1825.

7. Exercise 55 from Reed, George and James Hill, "Measuring the Thermic Effect of Food," *American Journal of Clinical Nutrition*, Vol. 63, 1996, pp. 164–169.

8. Exercise 56 from Perotto, D., R. Cue, and A. Lee, "Comparison of Nonlinear Functions of Describing the Growth Curve of Three Genotypes of Dairy Cattle," *Canadian Journal of Animal Science*, Vol. 73, Dec. 1992, pp. 773–782.

9. Exercise 57 from Slatkin, Montgomery and D. John Anderson, "A Model of Competition for Space," *Ecology*, Vol. 65, 1984, pp. 1840–1845.

10. Exercise 60 from Kristensen, Hans M., "Nuclear Weapons Status and Options Under a START Follow-On Agreement," Federation of American Scientists Presentation to Arm Control Association Briefing, April 27, 2009.

11. Exercise 61 from www.fle-online.com/Dyno.asp.

Section 5.2

1. Exercise 40 from Dubinsky, Ed. "Is Calculus Obsolete?" *The Mathematics Teacher*, Vol. 88, No. 2, Feb. 1995, pp. 146–148.

2. Exercise 47 from currentenergy.ucdavis.edu/details.php?region=newyork.

3. Exercise 52 from Employment Status of the Civilian Noninstitutional Population, 1943 to Date, data.bls.gov/cps/cpsaat01.htm.

4. Exercise 54 from Mezzadra, C., R. Paciaroni, S. Vulich, E. Villarreal, and L. Melucci, "Estimation of Milk Consumption Curve Parameters for Different Genetic Groups of Bovine Calves," *Animal Production*, Vol. 49, 1989, pp. 83–87.

5. Exercise 55 from Schwartz, C. and Kris Hundertmark, "Reproductive Characteristics of Alaskan Moose," *Journal of Wildlife Management*, Vol. 57, No. 3, July 1993, pp. 454–468.

6. Exercise 56 from Reed, George and James Hill, "Measuring the Thermic Effects of Food," *American Journal of Clinical Nutrition*, Vol. 63, 1996, pp. 164–169.

7. Exercise 57 from Eagly, A. H. and K. Telaak, "Width of the Latitude of Acceptance as a Determinant of Attitude Change," *Journal of Personality and Social Psychology*, Vol. 23, 1972, pp. 388–397.

Section 5.3

1. Exercise 77 from Fudenberg, Drew and Jean Tirole, "Learning by Doing and Market Performance," *Bell Journal of Economics*, Vol. 14, 1983, pp. 522–530.
2. Exercise 78 from 2009 OASDI Trustees Report, Social Security Administration, www.ssa.gov/OACT/TR/2009/LD_figIVB1.html.
3. Exercise 79 from Ou, J., M. Parlar, and M. Sharafali, "A Differentiated Service Scheme to Optimize Website Revenues," *The Journal of the Operational Research Society*, Vol. 57, 2006, pp. 1323–1340.
4. Exercise 80 from Won, Eugene J.S., "A Theoretical Investigation of the Effects of Similarity on Brand Choice Using the Elimination-by-Tree Model," *Marketing Science*, Vol. 26, 2007, pp. 868–875.
5. Exercise 82 from *The New York Times*, Aug. 29, 1993, p. E2.
6. Exercise 85 from Ricklefs, Robert E., "A Graphical Method of Fitting Equations to Growth Curves," *Ecology*, Vol. 48, Nov. 1967, pp. 978–983.
7. Exercise 86 from *International Recovery Plan for the Whooping Crane (Grus Americana)*, third revision, U.S. Fish and Wildlife Service, March 2007.
8. Exercise 87 from Weymouth, F. W., H. C. McMillin, and Willis H. Rich, "Latitude and Relative Growth in the Razor Clam," *Journal of Experimental Biology*, Vol. 8, 1931, pp. 228–249.
9. Exercise 88 from Speer, John F. et al., "A Stochastic Numerical Model of Breast Cancer Growth That Simulates Clinical Data," *Cancer Research*, Vol. 44, Sept. 1984, pp. 4124–4130.
10. Exercise 89 from Song, A. and S. Eckhoff, "Optimum Popping Moisture Content for Popcorn Kernels of Different Sizes," *Cereal Chemistry*, Vol. 71, No. 5, 1994, pp. 458–460.
11. Exercise 90 from Kulesa, P., et al., "On a Model Mechanism for the Spatial Performing of Teeth Primordia in the Alligator," *Journal of Theoretical Biology*, Vol. 180, 1996, pp. 287–296.
12. Exercise 91 from Slatkin, Montgomery and D. John Anderson, "A Model of Competition for Space," *Ecology*, Vol. 65, 1984, pp. 1840–1845.
13. Exercise 92 from *The New York Times*, Dec. 17, 1995, p. 49.
14. Exercise 96 from Larry Taylor of North Dakota State University.

Review Exercises

1. Exercise 64 from www.eia.gov/dnav/pet/pet_pri_gnd_dcus_nus_m.htm.
2. Exercise 65 from *The New York Times*, Jan. 18, 2010, p. A11.
3. Exercise 66 from Rossi, Hugo, "Mathematics is an Edifice, Not a Toolbox," *Notices of the American Mathematical Society*, Vol. 43, Oct. 1996, p. 1108.

4. Exercise 67 from Meltzer, David E., "Age Dependence of Olympic Weightlifting," *Medicine and Science in Sports and Exercise*, Vol. 26, No. 8, Aug. 1994, p. 1053.
5. Exercise 68 from West, Geoffrey B., James H. Brown, and Brian J. Enquist, "A General Model for the Origin of Allometric Scaling Laws in Biology," *Science*, Vol. 276, April 4, 1997, pp. 122–126.
6. Exercise 69 from Davie, A. and D. Evans, "Blood Lactate Responses to Submaximal Field Exercise Tests in Thoroughbred Horses," *The Veterinary Journal*, Vol. 159, 2000, pp. 252–258.
7. Exercise 70 from Murray, J. D., *Mathematical Biology*, Springer–Verlag, 1989, p. 163.
8. Exercise 71 from Pearl, R. and S. Parker, *Proc. Natl. Acad. Sci.*, Vol. 8, 1922, p. 212, quoted in *Elements of Mathematical Biology* by Alfred J. Lotka, Dover Publications, 1956, pp. 308–311.
9. Exercise 72 from Hurley, Peter J., "Red Cell Plasma Volumes in Normal Adults," *Journal of Nuclear Medicine*, Vol. 16, 1975, pp. 46–52, and Pearson, T. C., et al. "Interpretation of Measured Red Cell Mass and Plasma Volume in Adults," *British Journal of Haematology*, Vol. 89, 1995, pp. 748–756.
10. Exercise 73 from Courtis, S. A., "Maturation Units for the Measurement of Growth," *School and Society*, Vol. 30, 1929, pp. 683–690.
11. Exercise 74 from *Science*, Vol. 299, March 28, 2003, p. 1991.
12. Exercise 75 from Kristensen, Hans M., "Nuclear Weapons Status and Options Under a START Follow-on Agreement," Federation of American Scientists Presentation to Arms Control Association Briefing, April 27, 2009.
13. Exercise 77 from Winston, Patrick Henry, "Skills, Big Ideas, and Getting Grades Out of the Way," *MIT Faculty Newsletter*, Vol. XX, March/April 2008.

Chapter 6

Section 6.1

1. Example 3 from Federal Reserve. www.federalreserve.gov/releases/h10/hist/dat00_ca.htm.
2. Exercise 40 from Problem 19 from the May 2003 Course 1 Examination of the *Education and Examination Committee of the Society of Actuaries*. Reprinted by permission of the Society of Actuaries.
3. Exercises 41 and 42 from Bank Crime Statistics, Federal Bureau of Investigation. www.fbi.gov/stats-services/publications.
4. Exercise 52 from Pondy, Louis R., "Effects of Size, Complexity, and Ownership on Administrative Intensity," *Administrative Science Quarterly*, Vol. 14, 1969, pp. 47–60.
5. Exercise 55 from Harris, Edward F., Joseph D. Hicks, and Betsy D. Barcroft, "Tissue Contributions to Sex and Race: Differences

in Tooth Crown Size of Deciduous Molars," *American Journal of Physical Anthropology*, Vol. 115, 2001, pp. 223–237.
6. Exercise 56 from Rowan, N., C. Johnstone, R. McLean, J. Anderson, and J. Clarke, "Prediction of Toxigenic Fungal Growth in Buildings by Using a Novel Modelling System," *Applied and Environmental Microbiology*, Vol. 65, No. 11, Nov. 1999, pp. 4814–4821.

Section 6.2

1. Page 331 from Biddle, Wayne, "Skeleton Alleged in the Stealth Bomber's Closet," *Science*, Vol. 244, May 12, 1989.
2. Page 335 from Cullen, Michael R., *Mathematics for the Biosciences*. Copyright © 1983 PWS Publishers. Reprinted by permission.
3. Exercise 40 from Ricker, W. E., "Stock and Recruitment," *Journal of the Fisheries Research Board of Canada*, Vol. 11, 1957, pp. 559–623.
4. Exercise 45 from Cook, R. M., A. Sinclair, and G. Stefánsson, "Potential Collapse of North Sea Cod Stocks," *Nature*, Vol. 385, Feb. 6, 1997, pp. 521–522.
5. Exercise 47 from El-Mofty, S. K. and D. W. Lug, "Prevalence of High-Risk Human Papillomavirus DNA in Nonkeratinizing (Cylindrical Cell) Carcinoma of the Sinonasal Tract: A Distinct Clinicopathologic and Molecular Disease Entity," *American Journal of Surgical Pathology*, Vol. 29, No. 10, 2005, pp. 1367–1372.
6. Exercise 48 from *A Concrete Approach to Mathematical Modelling* by Michael Mesterton-Gibbons, Wiley-Interscience, 1995, pp. 93–96.
7. Exercise 51 from faq.usps.com/adaptivedesktop/faq.jsp?ef=USPSFAQ.

Section 6.3

1. Page 345 from Gwartney, James D., Richard L. Stroup, Russell S. Sobel, *Economics: Private and Public Choice*, 9th ed., The Dryden Press, 2000, p. 510.
2. Example 3 from Wales, Terrence J., "Distilled Spirits and Interstate Consumption Efforts," *The American Economic Review*, Vol. 57, No. 4, 1968, pp. 853–863.
3. Example 4 from Hogarty, T. F. and K. G. Elzinga, "The Demand for Beer," *The Review of Economics and Statistics*, Vol. 54, No. 2, 1972, pp. 195–198.
4. Exercise 3 from the Uniform CPA Examination Questions and Unofficial Answers, Copyright © 1991 by the American Institute of Certified Public Accountants, Inc. Reprinted (or adapted) with permission.
5. Exercise 5 from www.investopedia.com/terms/g/giffen-good.asp.
6. Exercise 27 from Cooper, John, C., "Price Elasticity of Demand for Crude Oil: Estimates for 23 Countries," *OPEC Review*, March 2003, pp. 1–8.

7. Exercise 28 from Kako, T., Gemma, M., and Ito, S., "Implications of the Minimum Access Rice Import on Supply and Demand Balance of Rice in Japan," *Agricultural Economics*, Vol. 16, 1997, pp. 193–204.

8. Exercise 29 from Atwood, Jeff, "Coding Horror," August 5, 2009. www.codinghorror.com.

9. Exercise 30 from support.sas.com/rnd/app/ examples/ets/simpelast/.

10. Exercise 31 from Battersby, B. and E. Oczkowski, "An Econometric Analysis of the Demand for Domestic Air Travel in Australia," *International Journal of Transport Economics*, Vol. XXVIII, No. 2, June 2001, pp. 193–204.

11. Exercise 32 from my-forest.com/ economics/index.html.

12. Exercise 33 from Gordon, Warren B., "The Calculus of Elasticity," *The AMATYC Review*, Vol. 26, No. 2, Spring 2006, pp. 53–55.

Section 6.4

1. Exercise 42 from Packel, Edward W. and David S. Yuen, "Projectile Motion with Resistance and the Lambert W Function," *The College Mathematics Journal*, Vol. 35, No. 5, Nov. 2004, pp. 337–350.

2. Exercises 45 and 46 from Tanyeri-Aber, A. and Rosson, C., "Demand for Dairy Products in Mexico," *Agricultural Economics*, Vol. 16, 1997, pp. 6–76.

3. Exercise 47 from Gagliardi, L. and F. Rusconi, "Respiratory Rate and Body Mass in the First Three Years of Life," *Archives of Disease in Children*, Vol. 76, 1997, pp. 151–154.

4. Exercise 48 from Lotka, Alfred J., *Elements of Mathematical Biology*, Dover Publications, 1956, p. 313.

5. Exercise 49 from Murray, J. D., *Mathematical Biology*, New York: Springer-Verlag, 1989, pp. 156–158.

6. Exercise 50 from Jiang, Lin, Oscar M. E. Schofield, and Paul G. Falkowski, "Adaptive Evolution of Phytoplankton Cell Size," *The American Naturalist*, Vol. 166, 2005, pp. 496–505.

7. Exercise 51 from Abrams, Peter A., "The Effects of Adaptive Behavior on the Type-2 Functional Response," *Ecology*, Vol. 71, 1990, pp. 877–885.

Section 6.5

1. Exercise 17 from Wanderley, S., M. Costa-Neves, and R. Rega, "Relative Growth of the Brain in Human Fetuses: First Gestational Trimester," *Archives d'Anatomie, d'Histologie et d'Embryologie*, Vol. 73, 1990, pp. 43–46.

2. Exercise 18 from Robbins, C., *Wildlife Feeding and Nutrition*, New York: Academic Press, 1983, p. 119.

3. Exercise 19 from Robbins, C., *Wildlife Feeding and Nutrition*, New York: Academic Press, 1983, p. 133.

4. Exercise 20 from Robbins, C., *Wildlife Feeding and Nutrition*, New York: Academic Press, 1983, p. 114.

Section 6.6

1. Exercise 23 from Garriott, James C. (ed.), *Medicolegal Aspects of Alcohol Determination in Biological Specimens*, PSG Publishing Company, 1988, p. 57.

2. Exercise 30 from Landon, D., C. Waite, R. Peterson, and L. Mech, "Evaluation of Age Determination Techniques for Gray Wolves," *Journal of Wildlife Management*, Vol. 62, No. 2, 1998, pp. 674–682.

3. Exercise 31 from Van Lunen, T. and D. Cole, "Growth and Body Composition of Highly Selected Boars and Gilts," *Animal Science*, Vol. 67, 1998, pp. 107–116.

Review Exercises

1. Exercise 53 from Tanyeri-Aber, A. and Rosson, C., "Demand for Dairy Products in Mexico," *Agricultural Economics*, Vol. 16, 1997, pp. 67–76.

2. Exercise 57 from Matsumoto, B., K. Nonaka, and M. Nakata, "A Genetic Study of Dentin Growth in the Mandibular Second and Third Molars of Male Mice," *Journal of Craniofacial Genetics and Developmental Biology*, Vol. 16, No. 3, July–Sept. 1996, pp. 137–147.

3. Exercise 58 from Voros, E., C. Robert, and A. Robert, "Age-Related Changes of the Human Skin Surface Microrelief," *Gerontology*, Vol. 36, 1990, pp. 276–285.

4. Exercise 65 from Redheffer, Ray, *Differential Equations: Theory and Applications*, Jones and Bartlett Publishers, 1991, pp. 107–108.

Extended Application

1. Based on "A Total Cost Model for a Training Program" by P. L. Goyal and S. K. Goyal, Faculty of Commerce and Administration, Concordia University.

Chapter 7

Section 7.1

1. Exercise 57 from State Intellectual Property Office of the P.R.C., english.sipo.gov.cn/laws/ annualreports/2013/.

2. Exercise 61 from Dennis, Brian and Robert F. Costantino, "Analysis of Steady-State Populations with the Gamma Abundance Model: Application to *Tribolium*," *Ecology*, Vol. 69, No. 4, Aug. 1988, pp. 1200–1213.

3. Exercise 65 and 66 from the National Center for Education Statistics, Digest of Education Statistics:2012, nces.ed.gov/programs/digest/d12/.

Section 7.2

1. Example 8 from Quirin, Jim and Barry Cohen, *Chartmasters' Rock 100*, 5th ed. Copyright 1992 by Chartmasters. Reprinted by permission.

2. Exercise 43 from *National Transportation Statistics 2014*, Bureau of Transportation Statistics.

3. Exercise 44 from *Hospital Statistics*, American Hospital Association.

Section 7.3

1. Exercise 23 from MacLaskey, Michael, *All About Lawns*, ed. by Alice Mace, Ortho Information Services, © 1980, p.108.

2. Exercise 25 from U.S. Energy Information Administration, Monthly Energy Review July 2014, Table 10.1, "Renewable Energy Production and Consumption by Source." www.eia.gov/totalenergy/data/monthly/.

3. Exercise 29 from Department of Environment, Food, and Rural Affairs, Foot and Mouth Disease, footandmouth.csl.gov.uk.

4. Exercise 30 from California Highway Patrol, Table 1A; "Fatal Collisions by Month 2003–2012." www.chp.ca.gov/switrs/.

5. Exercises 31 and 32 from www.roadandtrack .com.

6. Exercises 33 and 34 from www.caranddriver .com/reviews.

7. Exercises 35 and 36 from Sustainable by Design, www.susdesign.com/windowheatgain.

8. Exercise 37 based on an example given by Stephen Monk of the University of Washington.

9. Exercise 40 from Wildbur, Peter, *Information Graphics*, Van Nostrand Reinhold, 1989, pp. 126–127.

Section 7.4

1. Exercise 62 from Gompertz, Benjamin, "On the Nature of the Function Expressive of the Law of Human Mortality," *Philosophical Transactions of the Royal Society of London*, 1825.

2. Exercise 66 from Jorgenson, Jon T., et al., "Effects of Population Density on Horn Development in Bighorn Rams," *Journal of Wildlife Management*, Vol. 62, No. 3, 1998, pp. 1011–1020.

3. Exercise 67 from Finke, M., "Energy Requirements of Adult Female Beagles," *Journal of Nutrition*, Vol. 124, 1994, pp. 2604s–2608s.

4. Exercise 68 from Nord, Gail and John Nord, "Sediment in Lake Coeur d'Alene, Idaho," *The Mathematics Teacher*, Vol. 91, No. 4, April 1998, pp. 292–295.

5. Exercise 69 was originally contributed by Ralph DeMarr, University of New Mexico. It was updated with information from the U.S. Census Bureau. www.census.gov/popest/ national/asrh/2013.

6. Exercise 70 was originally contributed by Ralph DeMarr, University of New Mexico. It was updated with information from the U.S. Census Bureau, Current Population Reports, P60-245, September 2013. www.census.gov/ prod/2013pubs/p60-245.pdf.

Section 7.5

1. Exercise 38 from www.census.gov/hhes/ www/income/data/historical/household/.

2. Exercise 41 from www.census.gov/ prod/2013pubs/p60-245.pdf.

Section 7.6

1. Exercise 21 from U.S. Energy Information Administration, Monthly Energy Review July 2014, Table 10.1, "Renewable Energy Production and Consumption by Source." www.eia.gov/totalenergy/data/monthly/.
2. Exercises 25–28 from Chodos, D. J. and A. R. DeSantos, *Basics of Bioavailability*, Upjohn Company, 1978.
3. Exercise 29 from Mezzadra, C., R. Paciaroni, S. Vulich, E. Villarreal, and L. Melucci, "Estimation of Milk Consumption Curve Parameters for Different Genetic Groups of Bovine Calves," *Animal Production*, Vol. 49, 1989, pp. 83–87.
4. Exercise 30 from Department of Environment, Food, and Rural Affairs, Foot and Mouth Disease, footandmouth.csl.gov.uk.
5. Exercise 34 from Mezzadra, C., R. Paciaroni, S. Vulich, E. Villarreal, and L. Melucci, "Estimation of Milk Consumption Curve Parameters for Different Genetic Groups of Bovine Calves, *Animal Production*, Vol. 49, 1989, pp. 83–87.

Review Exercises

1. Exercise 78 from Problem 7 from May 2003 Course 1 Examination of the *Education and Examination Committee of the Society of Actuaries*. Reprinted by permission of the Society of Actuaries.
2. Exercise 81 from www.whitehouse.gov/omb/budget/historicals.
3. Exercise 84 from data.bls.gov/pdq/SurveyOutputServlet.
4. Exercise 87 from U.S. Energy Information Administration, Monthly Energy Review July 2014, www.eia.gov/totalenergy/data/monthly/.
5. Exercise 88 from Problem 38 from May 2003 Course 1 Examination of the *Education and Examination Committee of the Society of Actuaries*. Reprinted by permission of the Society of Actuaries.
6. Exercise 91 from Hastings, Alan and Robert F. Costantino, "Oscillations in Population Numbers: Age-Dependent Cannibalism," *Journal of Animal Ecology*, Vol. 60, No. 2, June 1991, pp. 471–482.
7. Exercise 92 from Oliver, M. H., et al., "Material Undernutrition During the Periconceptual Period Increases Plasma Taurine Levels and Insulin Response to Glucose but not Arginine in the Late Gestation Fetal Sheep," *Endocrinology*, Vol 14, No. 10, 2001, pp. 4576–4579.
8. Exercise 93 from Freeze, Brian S. and Timothy J. Richards, "Lactation Curve Estimation for Use in Economic Optimization Models in the Dairy Industry," *Journal of Dairy Science*, Vol. 75, 1992, pp. 2984–2989.
9. Exercise 94 from Table IF "Property-Damage-Only Collisions by Month 1999–2008. The California Highway Patrol." www.chp.ca.gov/switrs/.

Extended Application

1. Page 442 from www.eia.gov/cfapps/ipdbproject/IEDIndex3.cfm?tid=5&pid=5&aid=2.
2. Page 443 from www.geotimes.org/nov02/feature_oil.html.

Chapter 8

Section 8.1

1. Exercise 38 from Sam Northshield, Plattsburgh State University.
2. Exercise 43 from Reed, George and James Hill, "Measuring the Thermic Effect of Food," *American Journal of Clinical Nutrition*, Vol. 63, 1996, pp. 164–169.
3. Exercise 44 from Hungate, R. E., "The Rumen Microbial Ecosystem," *Annual Review of Ecology and Systematics*, Vol. 6, 1975, pp. 39–66.

Section 8.2

1. Exercise 38 from the Budget of the U.S. Government, *Fiscal Year 2014 Historical Tables*, pp. 143–144.
2. Exercise 39 from Pollan, Michael, "the (Agri) Cultural Contradictions of Obesity," *The New York Times Magazine*, Oct. 12, 2003, p. 41; USDA–National Agriculture Statistics Service, 2006; and *The World Almanac and Book of Facts 2013*, p. 127.
3. Exercise 42 from www.nctm.org/wlme/wlme6/five.htm.
4. Exercise 45 from www.nctm.org/eresources/view_article.asp?article_id=6219&page=5.

Section 8.4

1. Exercise 48 from Gaver, D. P. Jr., "On Base-Stock Level Inventory Control," *Operations Research*, Vol. 7, 1959, pp. 689–703.
2. Exercise 50 from Murray, J. D., *Mathematical Biology*, Springer-Verlag, 1989, p. 642, 648.
3. Exercise 51 from Ludwig, Donald, "An Unusual Free Boundary Problem from the Theory of Optimal Harvesting," in *Lectures on Mathematics in the Life Sciences, Vol. 12: Some Mathematical Questions in Biology*, American Mathematical Society, 1979, pp. 173–209.

Extended Application

1. For a report on this case, see ASBCA No. 44791, 1994. For an introduction to learning curves, see Heizer, Jay and Barry Render, *Operations Management*, Prentice-Hall, 2001, or Argote, Linda and Dennis Epple, "Learning Curves in Manufacturing," *Science*, Feb. 23, 1990.

Chapter 9

Section 9.1

1. Exercise 34 from Cobb, Charles W. and Paul H. Douglas, "A Theory of Production," *American Economic Review*, Vol. 18, No. 1, Supplement, March 1928, pp. 139–165.

2. Exercise 35 from Storesletten, Kjetil, "Sustaining Fiscal Policy Thorough Immigration," *Journal of Political Economy*, Vol. 108, No. 2, April 2000, pp. 300–323.
3. Exercise 40 from Harding, K. C., et al., "Mass-Dependent Energetics and Survival in Harbour Seal Pups," *Functional Ecology*, Vol. 19, No.1, Feb., 2005, pp. 129–135.
4. Exercise 41 from Haycock G. B., G. J. Schwartz, and D. H. Wisotsky, "Geometric Method for Measuring Body Surface Area: A Height Weight Formula Validated in Infants, Children and Adults," *The Journal of Pediatrics*, Vol. 93, No.1, 1978, pp. 62–66.
5. Exercise 42 from Alexander, R. McNeill, "How Dinosaurs Ran," *Scientific American*, Vol. 264, April 1991, p. 4.
6. Exercise 43 from Van Holt, T., D. Murphy, and L. Chapman, "Local and Landscape Predictors of Fish-assemblage Characteristics in the Great Swamp, New York," *Northeastern Naturalist*, Vol. 12, No. 3, 2006, pp. 353–374.
7. Exercise 44 from Chowell, F. and F. Sanchez, "Climate-based Descriptive Models of Dengue Fever: The 2002 Epidemic in Colima, Mexico," *Journal of Environmental Health*, Vol. 68, No. 10, June 2006, pp. 40–44.
8. Exercises 45 and 46 from Iverson, Aaron and Louis Iverson, "Spatial and Temporal Trends of Deer Harvest and Deer-Vehicle Accidents in Ohio," *Ohio Journal of Science*, 99, 1999, pp. 84–94.
9. Exercise 47 from Appleby, M., A. Lawrence, and A. Illius, "Influence of Neighbours on Stereotypic Behaviour of Tethered Sows," *Applied Animal Behaviour Science*, Vol. 24, 1989, pp. 137–146.
10. Exercise 48 from Hofman, Michel A., "Energy Metabolism, Brain Size and Longevity in Mammals." *The Quarterly Review of Biology*, Vol. 58, 1983, pp. 495–512.

Section 9.2

1. Exercise 53 from Currie, C. S. M., R. C. H Cheng, and H. K. Smith, "Dynamic Pricing of Airline Tickets with Competition," *The Journal of the Operational Research Society*, Vol. 59, No. 8, 2008, pp. 1026–1037.
2. Exercise 54 from Pondy, Louis R., "Effects of Size, Complexity, and Ownership on Administrative Intensity," *Administrative Science Quarterly*, Vol. 14, 1969, pp. 47–60.
3. Exercise 55 from Robbins, C., *Wildlife Feeding and Nutrition*, New York: Academic Press, 1983, p. 114.
4. Exercise 56 from Harding, K. C., et al., "Mass-Dependent Energetics and Survival in Harbour Seal Pups," *Functional Ecology*, Vol. 19, No. 1, Feb., 2005, pp. 129–135.
5. Exercise 57 from Haycock G. B., G. J. Schwartz, and D. H. Wisotsky, "Geometric Method for Measuring Body Surface Area: A Height Weight Formula Validated in Infants, Children and Adults," *The Journal of Pediatrics*, Vol. 93, No.1, 1978, pp. 62–66.

6. Exercise 58 from www.anaesthesiauk.com/ article.aspx?articleid=251.

7. Exercise 60 from www.win.niddk.nih.gov/ publications/tools.htm.

8. Exercise 61 from *PLoS One*, www .plosone.org/article/info%3Adoi/10.1371/ journal.pone.0039504.

9. Exercise 62 from Hofman, Michel A., "Energy Metabolism, Brain Size and Longevity in Mammals," *The Quarterly Review of Biology*, Vol. 58, 1983, pp. 495–512.

10. Exercise 64 from Westbrook, David, "The Mathematics of Scuba Diving," *The UMAP Journal*, Vol. 18, No. 2, 1997, pp. 2–19.

11. Exercise 65 from Bosch, William and L. Cobb, "Windchill," *The UMAP Journal*, Vol. 13, No. 3, 1990, pp. 481–489.

12. Exercise 66 from www.weather.com (May 9, 2000).

13. Exercise 70 from Fiore, Greg, "An Out-of-Math Experience: Einstein, Relativity, and the Developmental Mathematics Student," *The Mathematics Teacher*, Vol. 93, No. 3, 2000, pp. 194–199.

14. Exercise 71 from Sanders, Mark and Ernest McCormick, *Human Factors in Engineering Design*, 7th ed., New York: McGraw-Hill, 1993, pp. 290–291.

Section 9.3

1. Exercise 30 from Problem 35 from May 2003 Course 1 Examination of the *Education and Examination Committee of the Society of Actuaries*. Reprinted by permission of the Society of Actuaries.

2. Exercise 40 from Grofman, Bernard, "A Preliminary Model of Jury Decision Making as a Function of Jury Size, Effective Jury Decision Rule, and Mean Juror Judgmental Competence," *Frontiers of Economics*, Vol. 3, 1980, pp. 98–110.

3. Exercise 41 from www.intel.com/ technology/mooreslaw/.

4. Exercise 42 from Hindra, F. and Oon-Doo Baik, "Kinetics of Quality Changes During Food Frying," *Critical Reviews in Food Science and Nutrition*, Vol. 46, 2006, pp. 239–258.

Section 9.4

1. Exercise 42 from Owen, Guillermo, et al., "Proving a Distribution-Free Generalization of the Condorcet Jury Theorem," *Mathematical Social Sciences*, Vol. 17, 1989, pp. 1–16.

Section 9.5

1. Example 4 from Colley, Susan Jane, "Calculus in the Brewery," *The College Mathematics Journal*, Vol. 25, No. 3, May 1994, p. 227.

2. Exercise 25 from Hofman, Michel A., "Energy Metabolism, Brain Size and Longevity in Mammals," *The Quarterly Review of Biology*, Vol. 58, 1983, pp. 495–512.

3. Exercise 26 from Fitzsimmons, N., S. Buskirk, and M. Smith, "Population History, Genetic Variability, and Horn Growth in Bighorn Sheep," *Conservation Biology*, Vol. 9, No. 2, April 1995, pp. 314–323.

4. Exercise 27 from Brown, J. P. and P. E. Sendak, "Association of Ring Shake in Eastern Hemlock with Tree Attributes,"

Forest Products Journal, Vol. 56, No. 10, October 2006, pp. 31–36.

5. Exercise 28 from Gotch, Frank, "Clinical Dialysis: Kinetic Modeling in Hemodialysis," *Clinical Dialysis*, 3rd ed., Norwalk: Appleton & Lange, 1995, pp. 156–186.

6. Exercise 29 from Walker, Anita, "Mathematics Makes a Splash: Evaluating Hand Timing Systems," *The HiMAP Pull-Out Section*, Spring 1992, COMAP.

Review Exercises

1. Exercise 101 from Chumlea, W., S. Guo, C. Zellar, et al., "Total Body Water Reference Values and Prediction Equations for Adults," *Kidney International*, Vol. 59, 2001, pp. 2250–2258.

2. Exercise 102 from Hayes, J., J. Stark, and K. Shearer, "Development and Test of a Whole-Lifetime Foraging and Bio-energetics Growth Model for Drift-Feeding Brown Trout," *Transactions of the American Fisheries Society*, Vol. 129, 2000, pp. 315–332.

3. Exercise 103 from *National Vital Statistics Reports*, Vol. 51, No. 3, December 19, 2002.

Extended Application

1. Page 547 from www.uspto.gov/patents. You can locate patents by number or carry out a text search of the full patent database.

KEY DEFINITIONS, THEOREMS, AND FORMULAS

3.1 Rules for Limits

Let a, A, and B be real numbers, and let f and g be functions such that

$$\lim_{x \to a} f(x) = A \quad \text{and} \quad \lim_{x \to a} g(x) = B.$$

1. If k is a constant, then $\lim_{x \to a} k = k$ and $\lim_{x \to a} [k \cdot f(x)] = k \cdot \lim_{x \to a} f(x) = k \cdot A$.

2. $\lim_{x \to a} [f(x) \pm g(x)] = \lim_{x \to a} f(x) \pm \lim_{x \to a} g(x) = A \pm B$

   (The limit of a sum or difference is the sum or difference of the limits.)

3. $\lim_{x \to a} [f(x) \cdot g(x)] = \left[\lim_{x \to a} f(x)\right] \cdot \left[\lim_{x \to a} g(x)\right] = A \cdot B$

   (The limit of a product is the product of the limits.)

4. $\lim_{x \to a} \dfrac{f(x)}{g(x)} = \dfrac{\lim_{x \to a} f(x)}{\lim_{x \to a} g(x)} = \dfrac{A}{B}$ if $B \neq 0$

   (The limit of a quotient is the quotient of the limits, provided the limit of the denominator is not zero.)

5. If $p(x)$ is a polynomial, then $\lim_{x \to a} p(x) = p(a)$.

6. For any real number k, $\lim_{x \to a} [f(x)]^k = \left[\lim_{x \to a} f(x)\right]^k = A^k$, provided this limit exists.

7. $\lim_{x \to a} f(x) = \lim_{x \to a} g(x)$ if $f(x) = g(x)$ for all $x \neq a$.

8. For any real number $b > 0$, $\lim_{x \to a} b^{f(x)} = b^{\left[\lim_{x \to a} f(x)\right]} = b^A$.

9. For any real number b such that $0 < b < 1$ or $1 < b$,

   $$\lim_{x \to a} [\log_b f(x)] = \log_b [\lim_{x \to a} f(x)] = \log_b A \text{ if } A > 0.$$

3.1 Limits at Infinity

For any positive real number n,

$$\lim_{x \to \infty} \frac{1}{x^n} = 0 \quad \text{and} \quad \lim_{x \to -\infty} \frac{1}{x^n} = 0.$$

3.3 Instantaneous Rate of Change

The instantaneous rate of change for a function f when $x = a$ is

$$\lim_{h \to 0} \frac{f(a + h) - f(a)}{h}, \quad \text{provided this limit exists.}$$

3.4 Derivative

The derivative of the function f at x, written $f'(x)$, is defined as

$$f'(x) = \lim_{h \to 0} \frac{f(x + h) - f(x)}{h}, \quad \text{provided this limit exists.}$$

Rules for Derivatives

The following rules for derivatives are valid when all the indicated derivatives exist.

4.1

Constant Rule If $f(x) = k$, where k is any real number, then $f'(x) = \dfrac{d[k]}{dx} = 0$.

4.1

Power Rule If $f(x) = x^n$ for any real number n, then $f'(x) = \dfrac{d[x^n]}{dx} = nx^{n-1}$.

4.1

Constant Times a Function Let k be a real number. Then the derivative of $f(x) = k \cdot g(x)$ is

$$f'(x) = \frac{d}{dx}[k \cdot g(x)] = k \cdot g'(x).$$

4.1

Sum or Difference Rule If $f(x) = u(x) \pm v(x)$, then

$$f'(x) = \frac{d}{dx}[u(x) \pm v(x)] = u'(x) \pm v'(x).$$

4.2   *Product Rule*   If $f(x) = u(x) \cdot v(x)$, then

$$f'(x) = \frac{d}{dx}[u(x) \cdot v(x)] = u(x) \cdot v'(x) + v(x) \cdot u'(x).$$

4.2   *Quotient Rule*   If $f(x) = u(x)/v(x)$, and $v(x) \neq 0$, then

$$f'(x) = \frac{d}{dx}\left[\frac{u(x)}{v(x)}\right] = \frac{v(x) \cdot u'(x) - u(x) \cdot v'(x)}{[v(x)]^2}.$$

4.3   *Chain Rule*   If y is a function of u, say $y = f(u)$, and if u is a function of x, say $u = g(x)$, then $y = f(u) = f[g(x)]$, and

$$\frac{dy}{dx} = \frac{dy}{du} \cdot \frac{du}{dx}.$$

4.3   *Chain Rule (Alternative Form)*   If $y = f[g(x)]$, then $dy/dx = d/dx[f[g(x)]] = f'[g(x)] \cdot g'(x)$.

4.4   *Exponential Function*

$$\frac{d}{dx}[a^{g(x)}] = (\ln a)a^{g(x)}g'(x)$$

$$\frac{d}{dx}[e^{g(x)}] = e^{g(x)}g'(x)$$

4.5   *Logarithmic Function*

$$\frac{d}{dx}[\log_a|g(x)|] = \frac{1}{\ln a} \cdot \frac{g'(x)}{g(x)}$$

$$\frac{d}{dx}[\ln |g(x)|] = \frac{g'(x)}{g(x)}$$

5.2 First Derivative Test   Let c be a critical number for a function f. Suppose that f is continuous on (a, b) and differentiable on (a, b) except possibly at c, and that c is the only critical number for f in (a, b).

1. $f(c)$ is a relative maximum of f if the derivative $f'(x)$ is positive in the interval (a, c) and negative in the interval (c, b).

2. $f(c)$ is a relative minimum of f if the derivative $f'(x)$ is negative in the interval (a, c) and positive in the interval (c, b).

5.3 Second Derivative Test   Let f'' exist on some open interval containing c (except possibly at c itself), and let $f'(c) = 0$.

1. If $f''(c) > 0$, then $f(c)$ is a relative minimum.

2. If $f''(c) < 0$, then $f(c)$ is a relative maximum.

3. If $f''(c) = 0$ or $f''(c)$ does not exist, then the test gives no information about extrema, so use the first derivative test.

7.2 Substitution

Each of the following forms can be integrated using the substitution $u = f(x)$.

Form of the Integral   **Result**

1. $\displaystyle\int [f(x)]^n f'(x)\, dx, \quad n \neq -1$

$\displaystyle\int u^n\, du = \frac{u^{n+1}}{n+1} + C = \frac{[f(x)]^{n+1}}{n+1} + C$

2. $\displaystyle\int \frac{f'(x)}{f(x)}\, dx$

$\displaystyle\int \frac{1}{u}\, du = \ln |u| + C = \ln |f(x)| + C$

3. $\displaystyle\int e^{f(x)} f'(x)\, dx$

$\displaystyle\int e^u\, du = e^u + C = e^{f(x)} + C$

7.4 Fundamental Theorem of Calculus

Let f be continuous on the interval $[a, b]$, and let F be *any* antiderivative of f. Then

$$\int_a^b f(x)\, dx = F(b) - F(a) = F(x)\Big|_a^b.$$

7.6 Trapezoidal Rule

Let f be a continuous function on $[a, b]$ and let $[a, b]$ be divided into n equal subintervals by the points $a = x_0, x_1, x_2, \ldots, x_n = b$. Then, by the trapezoidal rule,

$$\int_a^b f(x)\, dx \approx \left(\frac{b-a}{n}\right)\left[\frac{1}{2}f(x_0) + f(x_1) + \cdots + f(x_{n-1}) + \frac{1}{2}f(x_n)\right].$$

7.6 Simpson's Rule

Let f be a continuous function on $[a, b]$ and let $[a, b]$ be divided into an even number n of equal subintervals by the points $a = x_0, x_1, x_2, \ldots, x_n = b$. Then, by Simpson's rule,

$$\int_a^b f(x)\, dx \approx \left(\frac{b-a}{3n}\right)[f(x_0) + 4f(x_1) + 2f(x_2) + 4f(x_3) + \cdots + 2f(x_{n-2}) + 4f(x_{n-1}) + f(x_n)].$$

8.1 Integration by Parts

If u and v are differentiable functions, then

$$\int u\, dv = uv - \int v\, du.$$

8.4 Improper Integrals

If f is continuous on the indicated interval and if the indicated limits exist, then

$$\int_a^\infty f(x)\, dx = \lim_{b \to \infty} \int_a^b f(x)\, dx,$$

$$\int_{-\infty}^b f(x)\, dx = \lim_{a \to -\infty} \int_a^b f(x)\, dx,$$

$$\int_{-\infty}^\infty f(x)\, dx = \int_{-\infty}^c f(x)\, dx + \int_c^\infty f(x)\, dx,$$

for real numbers a, b, and c, where c is arbitrarily chosen.

9.3 Test for Relative Extrema

For a function $z = f(x, y)$, let f_{xx}, f_{yy}, and f_{xy} all exist in a circular region contained in the xy-plane with center (a, b). Further, let

$$f_x(a, b) = 0 \qquad \text{and} \qquad f_y(a, b) = 0.$$

Define the number D, known as the discriminant, by

$$D = f_{xx}(a, b) \cdot f_{yy}(a, b) - [f_{xy}(a, b)]^2.$$

Then

(a) $f(a, b)$ is a relative maximum if $D > 0$ and $f_{xx}(a, b) < 0$;

(b) $f(a, b)$ is a relative minimum if $D > 0$ and $f_{xx}(a, b) > 0$;

(c) $f(a, b)$ is a saddle point (neither a maximum nor a minimum) if $D < 0$;

(d) if $D = 0$, the test gives no information.

9.6 Double Integral

The double integral of $f(x, y)$ over a rectangular region R is written

$$\iint_R f(x, y)\, dy\, dx \qquad \text{or} \qquad \iint_R f(x, y)\, dx\, dy,$$

and equals either

$$\int_a^b \int_c^d f(x, y)\, dy\, dx \qquad \text{or} \qquad \int_c^d \int_a^b f(x, y)\, dx\, dy.$$

10.2 Solving a Linear First-Order Differential Equation

1. Put the equation in the linear form $dy/dx + P(x)y = Q(x)$.
2. Find the integrating factor $I(x) = e^{\int P(x)\, dx}$.
3. Multiply each term of the equation from Step 1 by $I(x)$.
4. Replace the sum of terms on the left with $D_x[I(x)y]$.
5. Integrate both sides of the equation.
6. Solve for y.

10.3 Euler's Method

Let $y = f(x)$ be the solution of the differential equation

$$dy/dx = g(x, y), \quad \text{with} \quad y(x_0) = y_0,$$

for $x_0 \le x \le x_n$. Let $x_{i+1} = x_i + h$, where $h = (x_n - x_0)/n$ and

$$y_{i+1} = y_i + g(x_i, y_i)h,$$

for $0 \le i \le n - 1$. Then

$$f(x_{i+1}) \approx y_{i+1}.$$

12.1 Geometric Sequence

If a geometric sequence has first term a and common ratio r, then the sum of the first n terms, S_n, is given by

$$S_n = \frac{a(r^n - 1)}{r - 1}, \quad \text{where } r \ne 1.$$

13.2 Basic Trigonometric Derivatives

$$D_x(\sin x) = \cos x \qquad\qquad D_x(\cos x) = -\sin x$$
$$D_x(\tan x) = \sec^2 x \qquad\qquad D_x(\cot x) = -\csc^2 x$$
$$D_x(\sec x) = \sec x \tan x \qquad\qquad D_x(\csc x) = -\csc x \cot x$$

13.3 Basic Trigonometric Integrals

$$\int \sin x\, dx = -\cos x + C \qquad\qquad \int \cos x\, dx = \sin x + C$$

$$\int \sec^2 x\, dx = \tan x + C \qquad\qquad \int \csc^2 x\, dx = -\cot x + C$$

$$\int \sec x \tan x\, dx = \sec x + C \qquad\qquad \int \csc x \cot x\, dx = -\csc x + C$$

$$\int \tan x\, dx = -\ln|\cos x| + C \qquad\qquad \int \cot x\, dx = -\ln|\sin x| + C$$